Human Development
A Life-Span Approach

Life is a great bundle of little things.

Oliver Wendell Holmes,
Professor at the Breakfast Table

Human Development

A Life-Span Approach

THIRD EDITION

Karen L. Freiberg, R.N., Ph.D.
The University of Maryland
Baltimore County
Department of Psychology

Jones and Bartlett Publishers, Inc.
Boston *Monterey*

Editorial offices: Jones and Bartlett Publishers, Inc., 23720 Spectacular Bid, Monterey, CA 93940.

Sales and customer service offices: Jones and Bartlett Publishers, Inc., 20 Park Plaza, Boston, MA 02116.

Printed in the United States of America

10 9 8 7 6 5 4 3 2 1

Library of Congress Cataloging-in-Publication Data

Freiberg, Karen, L., 1944–
Human development.

Includes bibliographies and index.
1. Developmental psychology. 2. Human growth.
I. Title. [DNLM: 1. Growth. 2. Human Development.
BF 713 F862h]
BF713.F74 1987 155 86-21392

ISBN 0-86720-385-4

Production Services: Del Mar Associates
Production Editor: Jackie Estrada
Illustrations: Paul Slick
Interior and Cover Design: Paul Slick
Cover: Henry Moore, *Family Group* (1946), courtesy of the Hirshhorn Museum and Sculpture Garden, Smithsonian Institution, gift of Joseph H. Hirshhorn, 1966
Typesetting: Allservice Phototypesetting Co.

The selection and dosage of drugs presented in this book are in accord with standards accepted at the time of publication. The authors and publisher have made every effort to provide accurate information. However, research, clinical practice, and government regulations often change the accepted standard in this field. Before administering any drug, the reader is advised to check the manufacturer's product information sheet for the most up-to-date recommendations on dosage, precautions, and contraindications. This is especially important in the case of drugs that are new or seldom used.

To Michael, Kenneth, and Signe Lise,
and my extended family

Photo Credits

Jane Zich, 4
Department of Health and Human Services, 12
Leslie Davis, 18, 121, 133, 137, 146, 151, 171, 205, 213, 229, 234, 273, 282, 314, 379, 442, 464, 495, 501, 507, 555, 576, 589, 592, 597
Raymond H. Starr, Jr., 99
Scribe, 21, 53, 132, 147, 169, 200, 220, 233, 235, 271, 288, 304, 325, 339, 352, 360, 364, 370, 374, 419, 474, 485, 497, 503, 527, 553, 603, 636
J. Michael Freiberg, 139, 179, 252, 435, 460, 491, 524, 550, 569
Robert Dwyer, 149, 187
WJZ-TV, Baltimore, 296
Joan C. Netherwood, 354, 596
Wide World Photos, 400
Laurence Chiu, 410
Walter Woingast, 416
Y'Lonn Jackson, 430
Barbara Davis, 433
Elizabeth Hamlin/Stock Boston, 472
Kim Barnes, 489
Stephen Dunn, 504
NIH Gerontology Research Center, 483, 522, 572, 574, 586
The Oneonta Star, 536
Paul Sakuma/Wide World Photos, 599

Preface

Human Development: A Life-Span Approach, 3rd edition, presents the interactions of physical growth, health, cognitive maturation, and family and social networks in the development of persons of all ages. This text is useful to all students, faculty members, professionals, and lay readers who seek an integration of the multiple influences on human development. It draws on research from many fields: biology, genetics, medicine, psychology, sociology, anthropology, education, philosophy, even political science.

Two introductory chapters examine determinants of development, research methodologies, and theoretical viewpoints. Five chapters cover the time span from conception to young adulthood and carefully emphasize the significance of biological maturation on cognitive and psychosocial growth and change. Extensive references document the limiting, expanding, and determining effects of the environment, past and present, on unique individuals. Four additional chapters bring together new information and research on changes during the years of adulthood. A section in each chapter details the primary health concerns that characterize and profoundly effect each phase of life. The concluding chapter of the text looks at death and bereavement throughout the life span, introducing the reader to a fairly new and rapidly expanding area of research. The chapter, and book, ends with a valuable, thought-provoking discussion of perspectives on living that brings the reader to a full-cycle view of each person's unique participation in life.

Changes in this text, which in its first and second editions won the *American Journal of Nursing*'s Books of the Year award (1979 and 1984), include the addition of over 500 new 1980s references and twelve intriguing case studies to stimulate reflections on commonly held assumptions about life. Over 300 classic references have been retained, as have been the detailed tables of contents and updated chapter summaries. The graphics, which include functional photographs, line drawings, charts, tables, and boxed topics, have been greatly revised and expanded. Key concepts are listed at the beginning of each chapter and are set in bold type when they first occur in the text. Definitions of these terms are also handily provided at the bottom of the page. Questions for review at the end of each chapter assist the reader to examine feelings about important topics. Extensive references enable the reader to further explore all topics of interest. A glossary at the end of the book is an additional aid to understanding terminology.

Thanks go to all of the users and reviewers of the earlier editions of this text who provided the justification for a third edition. Thanks must also be extended to my family and friends who were patient during the lengthy revision process. Special kudos go to Madelon Kellough for her faithful typing assistance, Jackie Estrada for her careful editing, and Nancy Sjöberg and Del Mar Associates for turning my manuscript into a beautiful finished product.

Contents

7 Adolescence 337

8 Young Adulthood 397

9 The 30s and 40s 457

10 The 50s and 60s 519

Human Development

A Life-Span Approach

The Study of Human Development

1

Chapter Outline

Specific Determinants of Development
- *Heredity and Environment*
- *The Family*
- *The Community*

Historical Study of Human Behavior

Life-Span Study

Research Methodology
- *Cross-sectional and Longitudinal Studies*
- *Experimental Studies*
- *Observational Studies*
- *Correlational Studies*
- *Interviews and Surveys*
- *Ethical Principles*
- *Statistical Analysis*

Key Concepts

analysis of variance
applied research
basic research
case study
chi-square test
clinical investigation
cohort
confidence intervals
control group
correlational research
correlation coefficient
cross-sectional studies
deductive reasoning
dependent variable
descriptive psychology
descriptive statistics
empiricism
experiment
experimental psychology
field study
genotype
independent variable
inductive reasoning
inferential statistics
interview
longitudinal studies
mean
median
mode
mores
objectivity
observation
ontogeny
participant observation
phenotype
phylogeny
probability
random sample
range
rationalism
regression analysis
replication
retrospective study
scientific method
standard deviation
statistical significance
subjectivity
t-test

Case Study

Martha, a recently graduated nurse, had become very attached to Alonzo, a small boy hospitalized for complications of Christmas disease, a form of hemophilia (failure of the blood to form clots; bleeders' disease).

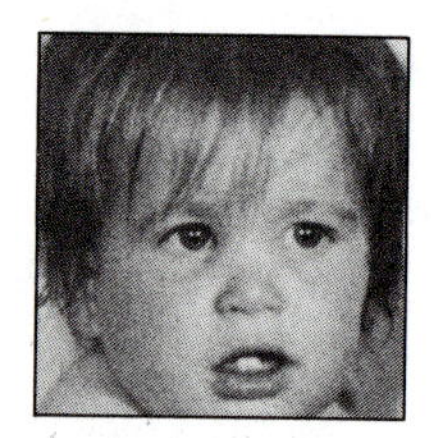

Although she had a bad cold, she continued to care for him. When Alonzo got an upper respiratory tract infection, which quickly worsened to bronchitis, Martha felt responsible.

Alonzo's bronchitis was being treated with oral antibiotics. Martha was asked to give an intramuscular injection of the antibiotic instead. Because he had hemophilia, any injection would result in prolonged bleeding at the injection site. Martha realized that she could probably control the bleeding by applying pressure with a Gelfoam sponge soaked in thrombin. However, she wondered why she should subject Alonzo to this extra trauma and risk. Martha read and reread the doctor's order. She agonized over what to do—obey it or question it. Only three days earlier, she had questioned another doctor's order and that doctor had become annoyed with her. Finally, Martha decided to risk another physician's ire rather than to traumatize little Alonzo. When she found the physician, she was informed that her reaction was part of a research study on obedience to authority and that in fact she would have been prevented from giving the injection had she opted to obey the order.

Martha was irate. When she continued to fume several hours later, the research investigator counseled her about the need for research, the need for some deception, and the importance of her contribution to the data being collected. Martha continued to vent her rage. She had been especially stressed by the need to decide between hurting Alonzo and questioning an order because of her guilt over giving Alonzo his infection and because of her recent experience with questioning a doctor's order. When the research investigator realized how upset Martha was, she made an appointment for Martha to see a psychologist and agreed to pay for Martha's therapy until she recovered from her emotional upheaval. The researcher explained that ethical considerations of research necessitate the correction or removal of any undesirable consequences after traumatic participation. What do you think?

"Larry King, popular radio and television broadcaster, was arrested today on charges of stealing five thousand dollars from financier Louis Wolfson . . ." Larry King (1982) gave this quotation at the outset of his autobiography. He then explained the financial fiasco that resulted in his being booked for grand larceny. (The charges were eventually dismissed.) What led to his notoriety as well as his celebrity? Through the book Larry King progressively outlined the many determinants of his development: his father's death, his family's relocation, his living on welfare, his mother's permissiveness, his school experiences, his friends, his Jewish identity, his maturation, his work, his social acceptance, his play, his marriages.

As King's life story illustrates, human development is never static; it is always in a state of flux. Actions in one area necessitate reactions in other areas. But there is continuity in human development as well. Individuals tend to form patterns of responses. And groups of persons, whether grouped by sex, age, ethnic, or cultural characteristics, tend to follow role models from within the same group and to behave in similar ways. Finally, genetic factors assure that there is a great deal of sameness in the rates and principles underlying both **ontogeny**, the unfolding life history of the individual human organism, and **phylogeny**, the unfolding life history of the species.

Although one goal of human development is to achieve a degree of independence—freedom from influence, control, or determination by others—all people have some degree of dependence, or reliance on others as well: "No man is an island, entire of itself; / every man is a piece of the continent, / a part of the main." We are all interdependent in the drama of human development.

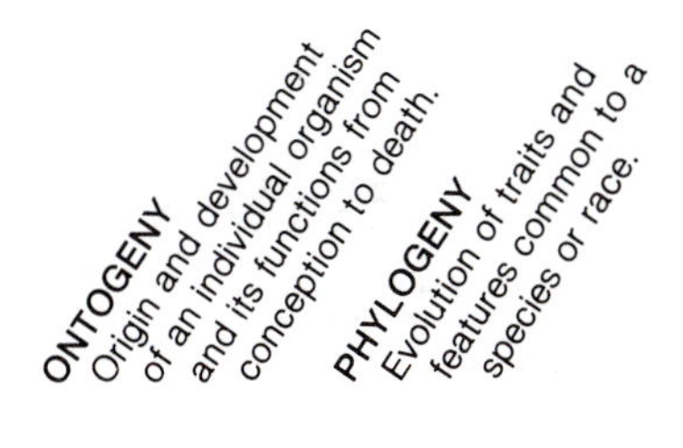

Specific Determinants of Development

In the past century, researchers have fragmented the study of human development into so many parts that a truly integrated account of all the systems known to influence development would require large tomes. In specialized courses students can apply themselves to a critical investigation of human development from any number of specialized points of view, including genetics, embryology, anatomy, physiology, neurology, behavioral endocrinology, anthropology, sociology, economics, civics, history, philosophy, and psychology. Even within specialized disciplines there are further subcategories of interest. The field of psychology has over forty subdivisions at present, including physiological and comparative psychology, educational psychology, personality and social psychology, and clinical neuropsychology (see Figure 1–1). These specialized areas of study are necessary for a fuller understanding of the processes making humans human.

Contributions from research in various specialized subfields make it increasingly evident that human development is influenced by a great variety of factors. These factors range from the family and culture to climate and nutrition (see Figure 1–2). Among the major determinants of development to be examined here are heredity and environment, the family, and the community.

PSYCHOLOGY THROUGH THE AGES

AT AGE 20: As an undergraduate I majored in psychology.

AT AGE 25: As a graduate student I focused on neuropsychology.

AT AGE 30: I did a postdoc in psychoneuroimmunology.

AT AGE 35: I branched off into psychoneuroimmunoendocrinology.

AT AGE 40: I've been examining the psychoneuroimmunoendocrinological aspects of cardiovascular disease.

Figure 1–1
Some scientists are more concerned with specific areas of development than with development in general.

HEREDITY AND ENVIRONMENT

Figure 1–2 can be divided into forces that have primarily a hereditary base, such as sex, biological maturation, race, and predisposition or resistance to certain illnesses, and forces that are more environmentally based, such as nutrition, education, social acceptance, and culture. However, we cannot say that any one aspect of human development is determined exclusively by either heredity or environment. There are always ways in which the two interact.

Predisposition/resistance to illness Genes Friends Nutrition Climate
Education Family Relocations Nurturance Culture IQ
Biological maturation Emotions Discipline Religion Social acceptance
Travel Play Work Racial identity Finances Sex Fashions
Persona

Figure 1–2
A multitude of influences shape personality.

Can we say "A boy is a boy is a boy" the way Gertrude Stein said "A rose is a rose is a rose"? The answer is a resounding *no* for several reasons. Inheritance of one X chromosome and one Y chromosome confers the **genotype**, or genetic make-up, of a male. But if a genetic male has a rare disease called androgen-insensitivity syndrome, he will have a **phenotype**, or external appearance, of a female. In this case, because his body cells are insensitive to androgens (masculin-

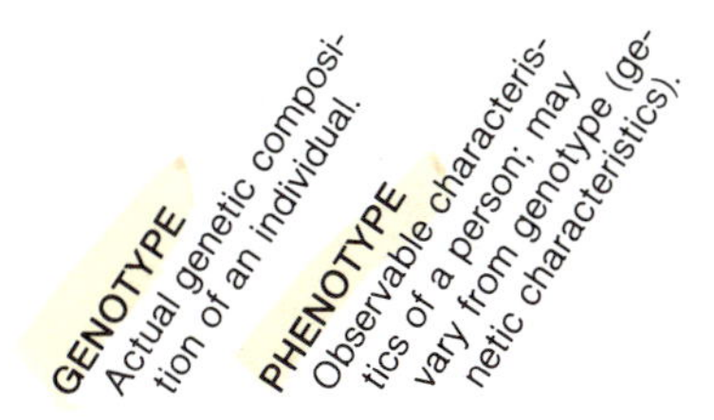

izing hormones), he will have undescended testes and a penis small enough to be mistaken for a clitoris, and at puberty his breasts will enlarge under the influence of the estrogens (feminizing hormones) that are present in all humans, whether male or female. In spite of the female phenotype, the individual with androgen-insensitivity syndrome is a chromosomal male. Gender may be defined by gonads as well as by chromosomes. If a child has testes, he is a male. Is he then female if he has been castrated? No, because he does not have ovaries, the female gonads. Is a boy a female if his body produces mostly estrogens and few androgens, or if his androgens are converted by his body into estrogens? Again, the answer is no. Sex hormone balance differs in each unique human and in each individual from day to day. A male's hormones may make him more feminine than another male, but he is still male. Heredity confers the chromosomes, the gonads, and even the hormones, but not all males receive equally.

Although gender (or sex) has a strong hereditary basis, it is affected by the environment in many ways. The mother's prenatal status, for example, may induce changes in her developing embryo/fetus. Diabetic mothers may be given supplemental estrogen during pregnancy, which has the potential of feminizing male offspring. Similarly, some pregnant women are given artificial progestins (another sex hormone that sometimes has androgenic properties) in order to prevent spontaneous abortions. Progestin-affected female fetuses are often masculinized. These prenatal environmental events pale, however, in comparison to all of the postnatal environmental influences on sexual identity. Boys reared to be extremely masculine according to their own culture's norms for masculinity may be considered feminine according to another culture's norms. Boys can also be reared to be very feminine or to be androgynous (having positive aspects of both male and female behaviors). Boys may wish they were girls and consciously adopt feminine behaviors despite their rearing; similarly, girls may consciously adopt masculine behaviors. Sex hormone secretions may also be affected by many environmental events, including stress, diet, drugs, accidents, and illnesses. If as simple a "hereditary" dimension as being male or female can be so obfuscated by environmental events, imagine how much more complicated are dimensions that are not as clearly determined by genes!

Now let's look at the other side of the coin and ask whether a personality characteristic with a strong underlying environmental base, such as career choice, is free of heredity influence. Again the answer is a hearty *no.* Career choice can be affected by a number of genetic and biological factors: one's physical size and strength, one's health, one's sex, one's intelligence, one's state of biological maturation, or possibly by effects of hormones (high prenatal estrogen exposure has been associated with lower assertiveness, while high postpubertal testosterone levels have been associated with more physically aggressive behaviors; Tieger, 1980).

In short, we cannot label any human characteristic as determined exclusively by either heredity or environment but must consider the interactive effects of both.

THE FAMILY

The family is, at various times, both held sacred and considered a scourge. For many people it is regarded more often as one than the other. It is difficult to cast the contemporary North American family into one mold, for, although the functions that families perform may be similar, their social and emotional climates can be vastly dissimilar.

A primary function of families in relation to their children is physical care. Human children require years of feeding, clothing, sheltering, and protecting before they can take care of themselves without adult help. A second important function of families is socialization of children. Social values and appropriate social behaviors need to be taught. The family more than any other institution has the responsibility for instilling religious and moral training and for providing discipline. A third important function of the family is to provide a sense of belonging for all its members. A family connotes ancestors, traditions, intimacy, family rituals, even private passwords or jokes. The tie to one's family is felt and enduring, even if family members feud or change frequently or if the home is crisis prone. A fourth important function of families is the provision of an affection bond. Parents should be a source of love to each other, to their offspring, and to all those who reside in the family household. Provision of a tender, loving, caring, supportive feeling is one of the most important ways in which a family can optimize the development of its members.

THE COMMUNITY

Sociology and anthropology have shown us that community customs are important determinants of development. Although one culture may be appreciably influenced by a given set of variables (for example, sex roles and social status), it does not follow that all cultures will be affected by the same standards. Nor does it follow that every individual within a culture will be moved to behave like every other person due to the prevailing **mores** (folkways that are considered conducive to the welfare of society and so develop the force of law).

Although the family may start the work of instilling society's values in its children, the community is the testing ground on which all of these values and beliefs are tried. Various community settings tend to increase or decrease certain behaviors (see Figure 1–3). All aspects of the socialization process are affected by community interactions. As an illustration, consider two aspects of psychosocial development described by the psychoanalyst Erik Erikson. He believed that the first, most central social lesson learned by human infants is trust in caregivers. Community members, in the form of baby-sitters, daycare center staff, neighbors, friends, relatives, religious group members and interested others, either enhance an infant's trust in adults as providers or instill feelings of mistrust. The second, central social lesson, which is learned in toddlerhood, is autonomy. The child begins to view himself or herself as a separate person, able to stand alone, with a sense of self-control, pride, and inner goodness. Community members can either enhance the child's sense of autonomy or foster in the child a sense of shame and doubt. Community

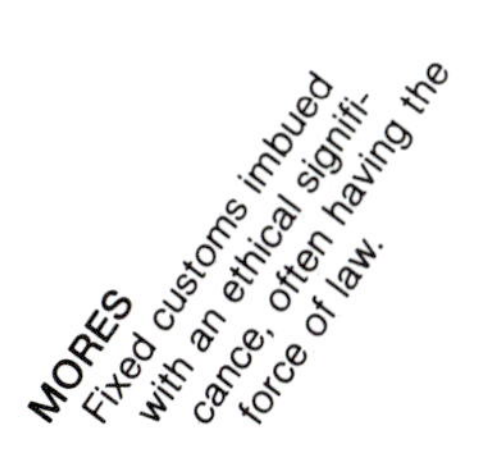

Open country

City apartment

Figure 1–3
Family and community settings vary considerably even within a small radius of land space.

members who are critical, overprotective, exploitive, or abusive can make a child feel foolish, mistrustful, and dependent again.

Social networking is essential to the well-being of the human organism. Children and adults alike, when deprived of social contact for prolonged periods, will suffer breakdowns in their psychosocial health. Law enforcers, for example, have long been aware that solitary confinement is the most severe form of punishment. Community members supplement or sometimes replace family members as sources of social contact; they provide companionship, reassurance, and emotional security. They cooperate in tasks, share ideas, and provide feedback to one another on patterns of acceptable and unacceptable behavior.

Community members share in the tasks of social control by making and altering rules, adhering to them, and providing law enforcement institutions in their community. They share the task of assuring economic welfare by providing jobs for one another. They give one another mutual support when tasks or problems arise that are too great for any one or two individuals to handle alone (for example, emergency disaster relief).

Throughout the following chapters you will find references to family influences and community influences. Remember, human existence is complex. No research methodology, no theory, and no textbook can ever fully describe all its complexities. Consider yourself. Can anyone else ever fully understand you? Do you completely understand yourself?

Historical Study of Human Behavior

In the fourth century B.C., Greek philosophers argued long about the nature and origins of human behavior. They were concerned especially with determinism versus free will, the thymos (emotion), the soma (the body), the psyche (the soul), whether the body and the psyche were monistic (one unified whole) or dualistic (two independent parts), whether mind (idealism) or matter (materialism) was the basis for both, whether things should be known by their functions (actions) or by their structures (form), and whether one could best know about all these things by using reasoning processes

Suburbia

City slum

(**rationalism**) or by conducting planned observations and experiments (**empiricism**).

Aristotle favored rationalism. His influence weighed heavily on later Hellenistic, Roman, and European philosophers. He also favored reasoning about the activities of the organism—psyche is as psyche does! He held that knowledge was gained through **deductive reasoning**—thinking about experiences from generalities to particulars, or from the universal to the individual.

The Renaissance philosophers, particularly René Descartes, disagreed. Descartes preferred to discuss the mind, not the psyche, and proposed that the powers of the mind, volition (free will), and understanding could control passions. A dualist, he felt that matter (as opposed to the mind) could and should be studied mathematically and mechanically. Whereas Renaissance empiricist scientists such as da Vinci, Copernicus, Galileo, Newton, and Bacon made great discoveries about the nature of matter, Renaissance philosophers such as Hobbes, Locke, Berkeley, Kant, and especially Hermann von Helmholtz searched for the locale of psychological functions in the mind (brain) and reasoned from the particulars they found toward generalities or from the individual to the universal, (**inductive reasoning**).

Darwin added new fuel to the fires of debate about the nature and origin of human behavior with his theories of evolution and "survival of the fittest." His baby biography, "A Biographical Sketch of an Infant" (1877), emphasized the potency of heredity over environment. It not only argued for genetic determination of many human behaviors but also made baby biographies a subject of scientific inquiry and initiated the study of child development. His cousin, Sir

Frances Galton, supported his theses with a book called *Hereditary Genius* (1892). Galton compared 977 eminent men and contrasted them to some women (whom he believed to be not only physically weaker but also intellectually inferior) and some mentally retarded persons. He pointed out that ability seemed to depend on descent. On the basis of chance, 1 out of the 977 men should have had an eminent relation. Instead, 322 of them had relatives as famous as themselves. In the Galton-Darwin family, for example, members of the clan served in Parliament, were landed gentry, and were physicians, clergymen, or military officers. Galton's attempts to prove the preponderance of heredity over environment failed to consider external factors such as social class, education, nutrition, social support networks, patronage, and luck common to all the eminent men he considered. Galton, in an unsuccessful attempt to separate heredity from environment, also made the first known scientific study of twins and applied statistical measures to his work, developing one of the most important measurements, the correlation coefficient (see p. 26).

Psychology emerged as an independent study separate from philosophy scarcely more than 100 years ago. It built upon the foundation provided by earlier great thinkers. From the laboratories of Wilhelm Wundt in Germany came an insistence that human behavior be studied through rigorous experimentation using the scientific method rather than through reasoning processes. The **scientific method** is to state a problem, form a hypothesis, experiment, observe, and draw conclusions. All statements are considered mere supposition by the true scientist until they can be proved true or false by experimental testing. Furthermore, tests must be explicit enough so that other scientists can repeat (replicate) them. Wundt insisted that a dependable knowledge of psychology could be established only in this fashion. No opinions or long-held beliefs could be substituted for facts. Wundt wanted nothing to be taken for granted.

Experimental psychology, especially Wundt's method of structuralism (isolating the elements of mental content) with subjective introspection (self-analysis), was soon augmented by behaviorism, created by Watson in the United States (after the work of the Russian physiologist Pavlov). This approach is described in more detail in Chapter 2.

Descriptive psychology, which attempts to describe the unique complexity of individuals rather than discover general laws about psychological processes, developed alongside experimental psychology, but often in opposition to it. Descriptive psychologists criti-

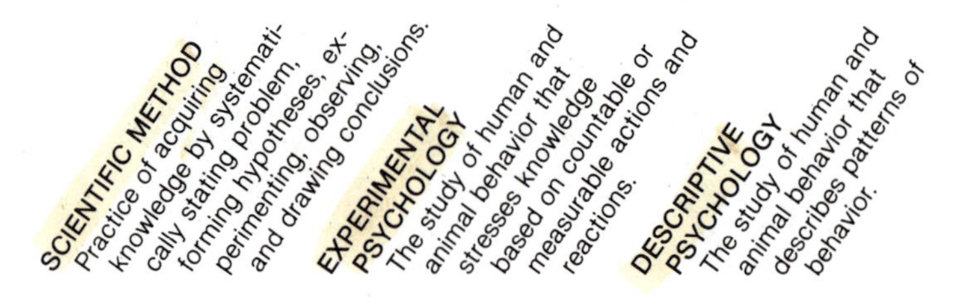

cized rigorous experimentation for losing sight of the idiosyncrasies of human behavior. Branches of psychology that used descriptive research techniques rather than the pure scientific method of experimentation included psychoanalysis and clinical psychology, following work done by Freud in Austria; cognitive psychology, following work done by Piaget in Switzerland and Luria and Vygotsky in Russia; and humanist psychology, following work done by existential philosophers and exemplified by the theories of Maslow and Rogers. (These approaches will also be described in more detail in Chapter 2.) One criticism of descriptive psychology is that although it does attend to the uniqueness of human beings it fails to explain or predict typical human actions and reactions.

Urie Bronfenbrenner (1979) described an ecological perspective for understanding human development. He suggested that psychological processes should be viewed as properties of environmental systems. Each person actively helps shape these systems. The larger, enduring system, or *macrosystem,* involves ideologies of people and various cultural beliefs. It affects the *exosystems* (systems in which family members do not function directly, such as politics, mass media, quality of the environment, and economics) and *microsystems* (systems in which family members play direct roles, such as home, peer groups, school, neighborhood, health services, and religious meetings). Relationships between systems such as home and religion or health services and school are called *mesosystems.* Many interrelationships between microsystems are beneficial to human development. Systems should be analyzed with an eye to the whole macrosystem with its interdependent exosystems, mesosystems, and microsystems in order to get a clearer view of the ecology of human development. In addition, characteristics of the people, environment, and ideologies of past systems must be considered to understand their effects on present and future systems.

Life-Span Study

Life-span study is a relatively new discipline. In fact, one hundred years ago, children were looked on as miniature adults, and many were employed in mines and factories for long hours each day (Aries, 1962). In portraits of children by such artists as Gainsborough and Renoir, affluent children were dressed in adult clothing.

Infancy and childhood emerged as important fields of separate study in the 1920s, especially through the work of Jean Piaget in Switzerland and two Americans, Arnold Gesell at Yale and John Watson at Johns Hopkins. But their views were contradictory; Piaget and Gesell stressed heredity and maturation, whereas Watson emphasized learning and conditioning as foundations for new child behaviors.

Adolescence as a subject of scientific study became conspicuous in the 1920s, especially as G. Stanley Hall's *Adolescence* (1904) became widely read. He described adolescence as a period of *Sturm und Drang* (storm and stress). Later, anthropologists, especially Margaret Mead in *Coming of Age in Samoa* (1928) and *Growing Up in New Guinea* (1953), suggested that storm and stress are not a universal phenomenon. Rather, there are cultural variations in the period called adolescence. In some societies there is very little upheaval but rather a smooth transition from childhood to adulthood often marked by some form of initiation rite.

G. Stanley Hall pioneered work in the psychology of old age as well as adolescence. At the age of 78, he added *Senescence: The Last Half of Life* (1922) to his list of publications. Gerontology (the study of old age) became an independent discipline in the 1940s with the creation of a Gerontology Research Center in Baltimore, headed by Nathan Shock, under the auspices of the National Institutes of Health (see Figure 1–4). Now called the National Institute of Aging, this facility has been studying the aging process intellectually, biomedically, socially, and emotionally with an internationally known longitudinal program (following the same subjects over a period of time) for over 30 years. As the elderly population has increased, research on aging is being correspondingly augmented.

Figure 1–4
Nathan W. Shock, Ph.D, Scientist Emeritus, National Institutes of Health, outside the modern Gerontology Research Center, National Institute on Aging, Baltimore, Maryland. Dr. Shock, as the former center director, helped design and conceive the center building opened in 1968. Prior to retirement in 1977, he directed aging research at the Baltimore unit for over 35 years.

A study of the total life span is a monumental task for any researcher to contemplate. However, some ongoing longitudinal studies are providing information about change over large segments of the life span in such areas as physical, intellectual, and personality functioning. Future studies of human development will, it is hoped, expand our knowledge of the total repertoire of everyday behaviors from conception to death in the broadest ecological settings.

Research Methodology

Research is a controlled, objective, systematic, and patient study of some phenomena in some field that is carried out to learn more about a subject. If the research is undertaken for the primary satisfaction of knowing and understanding the subject, it is called **basic** or **pure research**. If it is entered into for the purpose of knowing a subject well enough to make changes, such as creating more benefits for some recipient group, it is called **applied research**. The findings of basic research often have applications, however, and applied research may make important contributions to the basic knowledge of a field.

In the field of human development, questions asked by basic or applied researchers reflect many concerns. Some common inquiries might be stated in the following ways. Try substituting a question that interests you (such as jealousy, competition, or altruism) in the blank spaces:

How does _______ get its start?
How often does _______ occur?
Why does _______ occur?
What is the result of _______ occurring?
What is a normal pattern for _______?
What constitutes "abnormal" for _______?
Can "abnormal" _______ be prevented?
Can "abnormal" _______ be reversed to normal?
What is the most efficient way to achieve normal _______?
Could occurrence of _______ with A, B, and C persons predict the occurrence of _______ with X, Y, and Z persons? Under what conditions?

It is important that you understand some of the basic methods of research. In textbooks, journals, newspapers, television reports, or

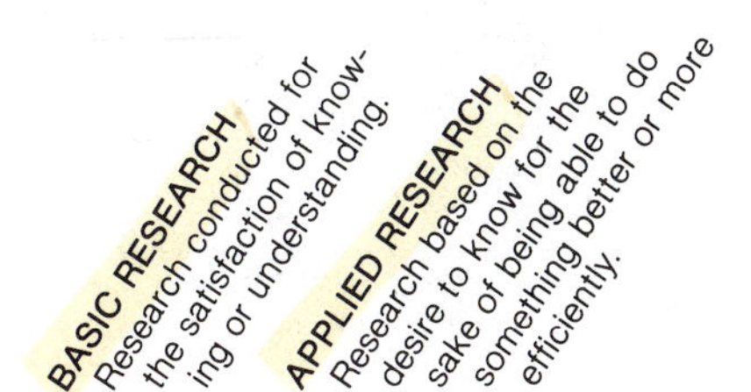

even word-of-mouth communications, people will report research findings to you. Some of the reports will be quite reliable and worthy of your trust. Others will be contrived or based on faulty logic. You, as a consumer of research, should be able to question how conclusions were reached. You should also be able to make an intelligent estimate of how reliable the results are, based on the research methodologies used. Very often the same question can be researched in different ways and the resulting answers will be in conflict. Which study should you believe? The following brief introduction to research methodology cannot make you an expert judge of research quality. However, it can make you aware of problems inherent to all research methodologies used to study human development. If you emerge from a careful reading of the following descriptions feeling skeptical about the 100 percent accuracy of any human-relations research result, this chapter will have accomplished its first goal. If you begin to question how research results are obtained in the future, a second important goal will have been met.

CROSS-SECTIONAL AND LONGITUDINAL STUDIES

Cross-sectional studies, which measure different subjects at the same time, are used more frequently than **longitudinal studies**, which measure the same subjects over a period of time, because they can be completed more quickly (see Figure 1–5). However, cross-sectional studies tend to ignore the individual and his or her unique growth and development. This drawback can be especially hazardous if the subjects under study are going through an especially rapid period of growth and change, such as infancy or adolescence. For example, when infants ranging in age from birth through three months are grouped together for study, their individual differences, which are vast, are ignored.

Figure 1–5
Schematic representation of longitudinal and cross-sectional studies.

	Year of birth			Longitudinal															
	'83	Jim at 2	Kay at 2	Ann at 2	Al at 2	Eve at 2	Bob at 2	Joe at 2	Ina at 2	Ed at 2	Sue at 2	Lee at 2	Ted at 2	Mae at 2	Ken at 2	Liz at 2	Deb at 2	Don at 2	Pat at 2
	'84	Jim at 3	Kay at 3	Ann at 3	Al at 3	Eve at 3	Bob at 3	Joe at 3	Ina at 3	Ed at 3	Sue at 3	Lee at 3	Ted at 3	Mae at 3	Ken at 3	Liz at 3	Deb at 3	Don at 3	Pat at 3
	'85	Jim at 4	Kay at 4	Ann at 4	Al at 4	Eve at 4	Bob at 4	Joe at 4	Ina at 4	Ed at 4	Sue at 4	Lee at 4	Ted at 4	Mae at 4	Ken at 4	Liz at 4	Deb at 4	Don at 4	Pat at 4
	'86	Jim at 5	Kay at 5	Ann at 5	Al at 5	Eve at 5	Bob at 5	Joe at 5	Ina at 5	Ed at 5	Sue at 5	Lee at 5	Ted at 5	Mae at 5	Ken at 5	Liz at 5	Deb at 5	Don at 5	Pat at 5
Cross-sectional	'87	Jim at 6	Kay at 6	Ann at 6	Al at 6	Eve at 6	Bob at 6	Joe at 6	Ina at 6	Ed at 6	Sue at 6	Lee at 6	Ted at 6	Mae at 6	Ken at 6	Liz at 6	Deb at 6	Don at 6	Pat at 6
	'88	Jim at 7	Kay at 7	Ann at 7	Al at 7	Eve at 7	Bob at 7	Joe at 7	Ina at 7	Ed at 7	Sue at 7	Lee at 7	Ted at 7	Mae at 7	Ken at 7	Liz at 7	Deb at 7	Don at 7	Pat at 7
	'89	Jim at 8	Kay at 8	Ann at 8	Al at 8	Eve at 8	Bob at 8	Joe at 8	Ina at 8	Ed at 8	Sue at 8	Lee at 8	Ted at 8	Mae at 8	Ken at 8	Liz at 8	Deb at 8	Don at 8	Pat at 8

Another hazard of cross-sectional research involves the dangers of comparing groups of individuals of different ages. Findings may be determined by social and generational differences between the age groups as much as by the variable under study. Consider, for example, the impact on people of World War II in the 1940s, the relative peace and prosperity of the 1950s, and the discontent, riots, protests, and Vietnam War of the 1960s. These situations could not help but influence the social and emotional development of people who reached adulthood during those decades. Would a research finding such as "A belief in war as a means of solving problems increases with age" make you skeptical if you knew the persons studied were all either forty-five, fifty-five, or sixty-five years old? Wouldn't you question whether the positive evaluation of violence was really a result of what each subject had experienced in his or her youth?

When cross-sectional research is used to study groups of individuals who are similar in age and background, the criticism of differential experiences is less meaningful. Comparisons of associates that are carefully designed to eliminate problems of intragroup or social and cultural differences can be fruitful as well as convenient ways of learning a great deal about human behavior.

Longitudinal research is desirable for ascertaining the stability or instability of individual characteristics over time. Many concerns of human development research (such as emotional response patterns, physical functions, or intelligence) are better studied longitudinally than by means of a one-time investigation. The Baltimore Longitudinal Study of Aging, for example, has been studying 650 male and 600 female volunteers since 1958. Each volunteer has his or her own biological profile, so that individual changes can be measured as each one ages. Longitudinal research, like cross-sectional research, has many problems. One obvious difficulty is that longitudinal research takes so much time. Subjects in longitudinal research get "lost" (quit, move away, die), and the expenses incurred by the researchers over time are enormous. It is also possible that subjects in a longitudinal study will become "investigation wise" (anticipate what the researchers want to see), will lose their motivation to give their best efforts to the research, or will remember how they responded in the last session and imitate that response even though it no longer characterizes their behavior or beliefs.

Recently, several methods of combining cross-sectional and longitudinal research have emerged. Appelbaum and McCall (1983) have presented three general approaches: cohort-sequential, time-sequential, and cross-sequential. A **cohort** is a person born about the

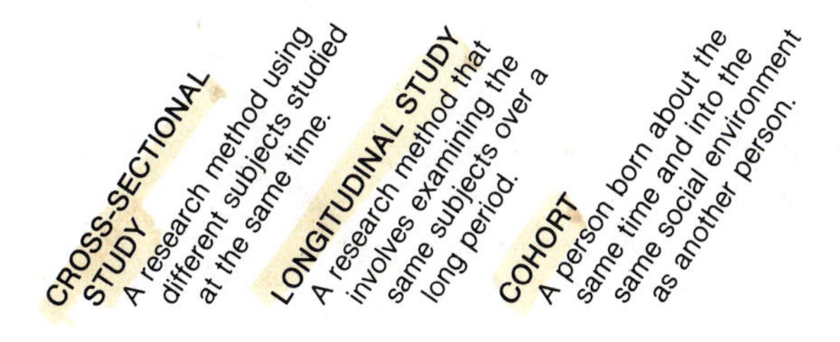

same time and into the same society as the individual under study. A person born ten or more years earlier or later than you might experience a kind of "generation gap" from you and therefore would not be your cohort. *Cohort-sequential* research looks at cohorts longitudinally with sequential times of measurement and is replicated using cohorts born in different years, usually until the oldest subjects in the last study reach the age of the oldest subjects in the first study (see Figure 1–6A). A *time-sequential* study is done cross-sectionally and is replicated at different times of measurement, usually until the oldest subjects in the last study reach the age of the oldest subjects in the first study (see Figure 1–6B). Finally, in a *cross-sequential* study several cross-sections of cohorts are studied longitudinally with the same times of measurement without regard to age (see Figure 1–6C). These methods combine the desirability of analyzing the unique individual differences of subjects with the relative ease of looking at a greater number of subjects in a shorter time. Again, however, there is a risk that subjects may become investigation wise or lose their motivation over time.

Figure 1–6
Schematic representation of three combinations of longitudinal and cross-sectional studies: cohort-sequential, time-sequential, and cross-sequential.

Year of measurement

Year of birth	'84	'85	'86	'87	'88	'89	'90	'91	'92	'93	'94	'95	'96	'97	'98	'99	'00	'01
'83	Cal at 1	Cal at 2	Cal at 3	Cal at 4	Cal at 5	Cal at 6	Cal at 7	Cal at 8	Cal at 9	Cal at 10	Cal at 11	Cal at 12	Cal at 13	Cal at 14	Cal at 15	Cal at 16	Cal at 17	Cal at 18
'84		Nan at 1	Nan at 2	Nan at 3	Nan at 4	Nan at 5	Nan at 6	Nan at 7	Nan at 8	Nan at 9	Nan at 10	Nan at 11	Nan at 12	Nan at 13	Nan at 14	Nan at 15	Nan at 16	Nan at 17
'85			Ray at 1	Ray at 2	Ray at 3	Ray at 4	Ray at 5	Ray at 6	Ray at 7	Ray at 8	Ray at 9	Ray at 10	Ray at 11	Ray at 12	Ray at 13	Ray at 14	Ray at 15	Ray at 16
'86				Zoe at 1	Zoe at 2	Zoe at 3	Zoe at 4	Zoe at 5	Zoe at 6	Zoe at 7	Zoe at 8	Zoe at 9	Zoe at 10	Zoe at 11	Zoe at 12	Zoe at 13	Zoe at 14	Zoe at 15
'87					Hal at 1	Hal at 2	Hal at 3	Hal at 4	Hal at 5	Hal at 6	Hal at 7	Hal at 8	Hal at 9	Hal at 10	Hal at 11	Hal at 12	Hal at 13	Hal at 14
'88						Flo at 1	Flo at 2	Flo at 3	Flo at 4	Flo at 5	Flo at 6	Flo at 7	Flo at 8	Flo at 9	Flo at 10	Flo at 11	Flo at 12	Flo at 13
'89							Vi at 1	Vi at 2	Vi at 3	Vi at 4	Vi at 5	Vi at 6	Vi at 7	Vi at 8	Vi at 9	Vi at 10	Vi at 11	Vi at 12

A. Cohort-sequential B. Time-sequential C. Cross-sequential

EXPERIMENT A study in which the investigator manipulates one or more variables to determine the effect on another variable.

INDEPENDENT VARIABLE The variable that is controlled by the experimenter to determine its effect on the dependent variable.

DEPENDENT VARIABLE In an experiment, the behavior that is measured and is expected to change with manipulation of an independent variable.

EXPERIMENTAL STUDIES

The "science" of psychology, advocated by Wundt and thousands of psychologists over the last century, insists on the rigor and precision of the scientific method and the **experiment**. Such explicit, quantitative, data-analytic experimental techniques are the realm of experimental psychologists. Although not all experimental studies are conducted in a laboratory, the lab is the location of choice for them. It is easier to control a whole range of irrelevant stimuli in a carefully prepared laboratory environment. When an experimental study is conducted, a single behavior (or set of behaviors) is allowed to vary; all others are held constant (unchanging). In this way it is possible to know quite precisely what, if any, effect a manipulation introduced by the researcher will have on the subject or subjects. The behaviors being manipulated by the experimenter are called the **independent variables**, or antecedent (coming before) variables. The changes that occur in the subject as a result of the introduction of the independent variables are known as the **dependent variables**, or consequent variables.

In the ideal experiment, in order to demonstrate that there is an antecedent-consequent relationship between the variables, a **control group** is used. The control group is not subjected to the manipulation of the independent variable under study as the experimental group is. Both groups are observed carefully. If the dependent variables occur only in the experimental group, then it can be assumed that the manipulations of the independent variables were effective in bringing about a consequence that ordinarily would not occur.

The results of experimental research can be checked over and over again in **replication** (reproduction of the original) studies because the experimental conditions are specified so carefully. This lends a degree of reliability to the research results, especially if replications are carried out by different, unbiased researchers.

It is always necessary to have enough subjects in both the experimental and the control groups to ensure that the phenomenon under study is something that characterizes all humans or all people of a special group (women, sixth graders, Navajo Indians). It is difficult to have confidence in research when an experimenter has selected only a few special persons as experimental subjects. Instead, there should be a **random sample** selected from the total population of the group under investigation. This selection process gives every member of the population an equal chance of being included in the

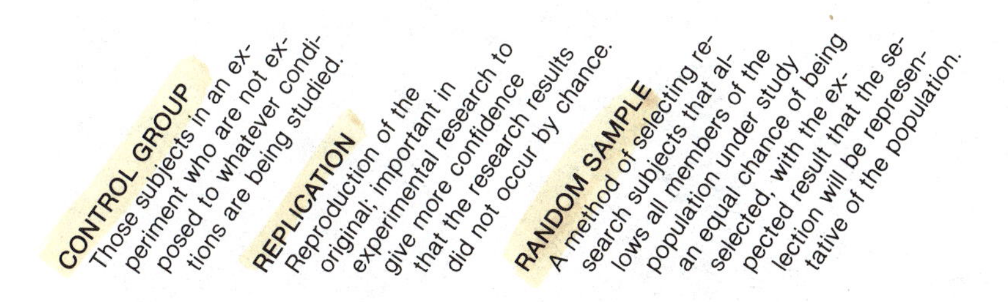

sample and also makes every possible combination of subjects selected equally likely. Most statistics textbooks have tables of random numbers that aid in the selection of random samples. First, every member of the population under investigation is assigned a number. Then the table is entered at a random point. As the experimenter moves from the starting point down the column of numbers, each member of the population whose number comes up is included in the sample until a sufficiently large number of subjects has been selected. Control subjects should be matched as closely as possible to the experimental subjects in age, sex, race, culture, education, and environment to further ensure that it is the independent variable of the experiment that brings about a result, not any chance external factors that make the experimental subjects different from the control subjects.

One criticism frequently voiced about experimental research is that, because it is so highly manipulated and controlled, it does not give a very good picture of what actually occurs in the real world (see Figure 1–7). It is true that there is a degree of artificiality about experimental research, but this does not always negate its value. Consider the following experiment. Myers (1982) was interested in helping parents learn about their babies and gain confidence in, and satisfaction from, their parenting abilities. She recruited forty-two middle-class married couples just after the birth of their first child. All mothers had had normal, uncomplicated vaginal deliveries of full-term, healthy, normal-birth-weight infants. The married couples included blacks, whites, and Hispanics. All had had a minimum of high school education and were over 20 years of age. The couples

Figure 1–7 *Laboratory experimental research provides greater certainty that subjects do not differ in treatment conditions than does naturalistic research. Naturalistic research, however, gives a better picture of what actually occurs in the real world. Both methods have advantages, depending on the nature of the research question.*

were randomly assigned to one of three groups (two treatment variations and a control group) by the throw of a die. Both treatment (experimental) groups of parents were taught to administer the Brazelton Neonatal Behavioral Assessment Scale and observe their own infant's behavior (this scale assesses twenty-seven behavioral items and twenty reflex items to see how a neonate responds to the environment). In one treatment group, the father of the infant was trained to administer the Brazelton test. In the other treatment group, the mother was trained to administer the test. In the control group, neither parent was given training in test administration but simply became acquainted with the experimenter for about the same time that test training took. Six hours after the training (or acquaintance session) and again four weeks after hospital discharge, assistants blind to any knowledge of which couples belonged to which group recorded how the couples interacted with their infant. They watched for instances of looking, talking to, smiling, touching, or positioning the infant. They also gave the couples questionnaires about knowledge of infant behavior, confidence in their parenting abilities, satisfaction with their infant, father's caregiving involvement, and perception of their infant compared with an average infant. The findings of the study were that the treatment couples had more knowledge about infants and more confidence in, and satisfaction from, their parenting abilities than the control couples, and the fathers who had learned to administer the Brazelton test were more involved with their infant's care than were control fathers.

This research suggests a way to help parents increase their knowledge, confidence, satisfaction, and interactions with their infants. The criticism of artificiality is not important in this case. The artificial situation may in itself be useful to help increase knowledge of, and interest in, infants. However, the criticism of artificiality does demonstrate other limitations of experimental research. This study has a limited range of applicability (to married middle-class couples giving vaginal birth to normal full-term first-born infants in settings where Brazelton test training can be conducted). Similar limitations are true for many experimental studies. This study also cannot be generalized to other populations: to nonmarried parents, lower-class parents, less well educated parents, younger parents, nonvaginally delivered infants, or infants who were not full-term, normal birth weight, or first-born. What is true for one group of infants or group of parents is not necessarily true for others.

This study told us *what* happened. It did not try to explain all the reasons *why* it happened. The study illustrates still another problem with experimental research. The kinds of stimuli that can be used as independent variables and the kinds of responses that can be used as dependent variables in experimental situations are confined within narrow bounds. An experiment will rarely employ more than two or three independent variables, and these are usually of a fairly simple nature. This kind of research cannot contribute much to our knowledge of such things as thought processes, emotional feeling tones, or unconscious inhibitors of behavior. It is a rigorous, precise, replicable, fairly objective way of doing research on a limited number of topics in human development.

OBSERVATIONAL STUDIES

Observational studies, also called naturalistic or descriptive studies, try to investigate behaviors as they occur in the real world. An **observation** offers the potential for studying many actions and events that would be impossible to reproduce in the laboratory. From observational or naturalistic studies comes descriptive psychology, which attempts to capture the uniqueness and idiosyncratic make-up of each individual. In the late 1960s and early 1970s descriptive psychology experienced a flood of new research with the increased popularity of the humanistic theories (to be described in Chapter 2), emphasizing maximal development of each human being's unique potentialities.

Electronic recording devices (audiovisual equipment, videotapes, sound spectograms, sound movies) make it possible to record behavior as it occurs and then to analyze segments of the behavior carefully and repeatedly at a later date. Newer techniques are continually being developed that make the analysis of naturally occurring behaviors even easier (videodisks, computers). However, observational studies still have many methodological problems. The behaviors to be isolated for analysis must be carefully identified and defined. All persons trained to extract these elements must understand the definitions fully. They must reach a high level of agreement (called interrater reliability) among themselves about what they are seeing. Further, they should not have any preconceived notions about what they ought to be seeing, as preconceptions can distort their perception of what they see. This state or quality of observing phenomena without bias, in a detached, impersonal way, is called **objectivity**. The opposite is **subjectivity**: a state or quality of imposing personal prejudices, thoughts, and feelings into one's work. It is hard to overcome subjectivity in an observational study of human behavior. Experimental studies that deal with counting responses or gathering measures of quantity avoid subjectivity better than observational studies that rely more heavily on measures of the quality of responses.

Condry has provided good illustrations of how subjectivity contaminates research results. Condry and Condry (1976) asked over 200 male and female college students to rate the emotional responses of the same baby to presentations of a teddy bear, a jack-in-the-box, a doll, and a buzzer (the sequence was filmed on videotape). Half of the students were told that the baby was male. The other half were told that the baby was female. There was a significant

OBSERVATION An act of noting and recording facts and events as they occur in the real world; also the data so noted and recorded.

OBJECTIVITY State or quality of observing phenomena without bias, in a detached, impersonal way.

SUBJECTIVITY State or quality of imposing personal prejudices, thoughts, and feelings into one's work.

FIELD STUDY Observational study in which only specified activities and events are cataloged.

PARTICIPANT OBSERVATION A method of research in which the researcher attempts to become immersed in the way of life to be studied to the point of temporarily becoming part of the social context.

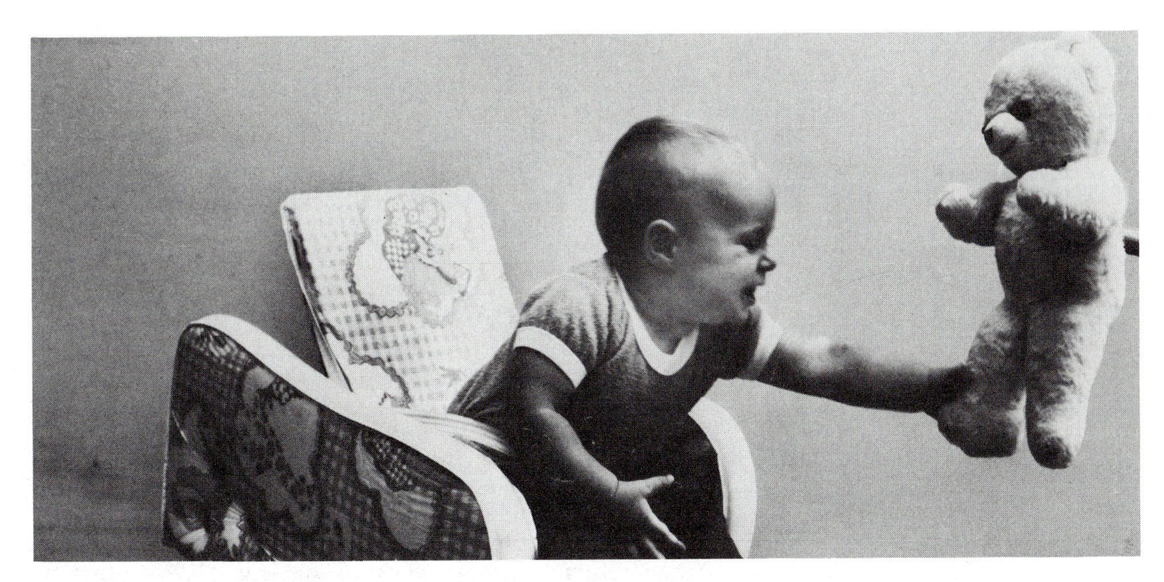

Figure 1–8
Subjective judgments enter into the way one "sees" behavior of boys versus girls, one's own child versus another's child, and even a child seen in the morning versus one seen in the afternoon.

difference between the ratings of the baby's emotional responses depending on whether the students believed the baby to be a boy or a girl. Condry and Ross (1985) asked 175 male and female college students to rate the videotaped behavior sequence of two preschool children playing outdoors on a winter day. Snowsuits disguised their sex. The students who were told that both children were boys rated the play as significantly less aggressive than did students told that they were watching two girls, or a boy and a girl. The boy-boy condition was perceived as positive play. The other conditions were perceived to be negative interacting. The data supported the hypothesis that observers may see differences in a child's behavior as a function of the sex-type label alone (see Figure 1–8). Such errors can often happen in research when investigators have personal prejudices about any kind of qualitative responses.

In **field studies**, there are limits on the activities and events that are catalogued in any instances of observing naturally occurring behaviors. The researcher must first have a good idea of the variables she or he is looking for. This requirement imposes a restriction on how much information can be gleaned from the data after they are gathered. It also may prevent looking at a random sample of a population if one or more variables under study are person specific (for example, male breast cancer). Comparing certain behaviors observed in different groups (such as men with breast cancer and men without) may result in invalid conclusions. How can the researcher be sure that any group differences observed are due to the cancer and not some extraneous variable? Although results of field studies are generally more suggestive than conclusive, they contribute a great deal more information to our knowledge of human development than we could ascertain with experimental studies alone.

Participant observation is a special form of field study in which the researcher attempts to immerse herself or himself in the way of life to be studied to the point that she or he becomes temporarily a part of the social context. Anthropologists and sociologists are most apt to use this technique to study human behavior. Most people tend to change what they do and say (away from what is representative of their normal views and modes of behavior) when they know they are being observed. The participant observer, by becoming a part of the entity being studied, tries to reduce to a minimum the impact of her or his presence on others. The researcher must take care to assure that her or his own behaviors do not significantly affect the actions and reactions of others. The great disadvantage of the participant observer approach is that it involves a considerable investment of time to become a part of a "real life" situation and to become accepted by people more or less on their own terms before the uncontaminated data can be collected.

Clinical investigations are usually studies done in facilities where people come to receive advice or treatment for problems (medical, social, educational, and so on). Consequently, they often deal with maladjustments of some kind. They may, however, also be used to study healthy people. Clinical investigations may involve in-depth laboratory tests, physical examinations, radiological scans, interviews, questionnaires, or surveys, as well as observations by physicians, nurses, social workers, psychologists, educational specialists, nutritionists, researchers, or other professional specialists, depending on the nature of the study. Sharing of information by an interdisciplinary team allows each participant to have a better picture of the individual under study.

The result of intensive investigation of the symptoms, past history, and/or current influences on the subject is most apt to be a **case study**. The focus is on the individual and his or her uniqueness. Research using case studies may try to find commonalities among persons with particular problems or experiences. There are dangers inherent in this kind of research. The investigator may try to force different experiences into the same classification for the sake of doing statistical analyses. Or the researcher may affect the results of the analyses by interpreting the data according to some preconceived notions. Nevertheless, case studies can provide us with a wealth of new information. An alert, receptive member of a clinical team who is involved in studying subjects day after day may be able to see patterns developing over time. Such an investigator may be able to draw on stored bits of information in recorded case studies and synthesize them into a unified interpretation of some phenomenon. The collections of data can be used first to evoke hypotheses and later to test them.

Some case studies are done by lay persons in the form of autobiographies, biographies, diaries, collections of letters, record books, baby books, and the like. These kinds of data, though very subjective, can be useful to researchers in the formulation of hypotheses about factors that affect human lives. Many insights have been discovered, or rediscovered, in such sources. Some case studies are designed by researchers to study normal developmental processes. Piaget, for example, whose theory will be explained in the next chapter, used this method to study cognitive development.

CORRELATIONAL STUDIES

Correlational studies do not attempt to prove that one phenomenon or set of phenomena causes another but rather that there exists some relationship between the occurrences of behaviors. Try to guess the relationship between the following variables:

- Income level (high/low) and purchase of prescription drugs brand-name/generic substitute).
- Convincingness of a lie (high/low) and characteristics of the person to whom you are lying (attractive, opposite-sexed person/unattractive, same-sexed person).
- Occupational status (college student/working woman) and the incidence of bulimic binge eating (high/low).

These kinds of relationships, measuring phenomena that already exist, are questioned in **correlational research**. No manipulation of behavior is necessary as it is in experimental research, nor is it necessary to have any time-consuming observations of how behavior unfolds. It is possible to look at the interrelationships of several hundred variables at once, seeing which ones co-relate positively (in the same direction), which ones co-relate negatively (in the opposite direction), and which ones are unrelated. Statistical techniques allow researchers to measure the strength of co-relationships in numerical terms (discussed later in this chapter).

The correlations found for the preceding examples are:

- Low-income persons buy more brand-name drugs (Lambert et al., 1980).
- You will lie more convincingly to an unattractive person of the same sex (DePaulo, Stone, and Lassiter, 1985).
- College students have a higher incidence of bulimia than working women of the same age (Hart and Ollendick, 1985).

Although correlation does not necessarily imply causation, it may suggest it, which is a danger inherent in this research strategy. The finding that poor people buy more brand-name drugs might suggest to a person who did not know prices that brand-name drugs are cheaper than generic substitutes (they are not). If the lower price tag did not cause the selection of the brand-name product, what might have? Willingness to rely on impressions of superiority gained from advertising of the brand-name product? Fear of the unknown (generic substitute)? Some other factor? Some combination of factors? Correlational studies often lead researchers to pursue more information about the causes of relationships through experimental or observational studies. In many cases knowledge of the correlation is the best a researcher can do. Further research may be unethical. For example, researchers cannot ask one group of bulimic women to quit college in order to see whether this stops their binge eating. Nor could researchers follow bulimic students longitudinally to see whether the eating disorder disappears once they graduate and become working women. Once the syndrome is diagnosed, an effort must be made to cure it at once, not just observe it. It is feasible that other factors are responsible for bulimia—factors that are unrelated to the college experience but that may attract the female to a student life rather than to full-time employment. Certainly not all female college students experience bulimic episodes, nor do all working women escape the syndrome.

Correlational studies can take a multitude of variables into account and look at the positive and negative directions and the strengths of these multiple variables. What they cannot do is control variables, manipulate events, or assign causality with much confidence.

INTERVIEWS AND SURVEYS

Interviews are used to collect data about people's behaviors in the past (called **retrospective studies**) or about their current attitudes or behaviors. They are usually done in conjunction with observational or correlational studies. Although a great deal of information can be obtained in this way, there are a number of factors that can make responses somewhat less than accurate. First, there is the obvious fact that most people like to cast themselves in a favorable light. Second, many persons, in an attempt to please the interviewer, will give the answer they feel is desired. Third, the way that a question is worded or asked can often influence what response will be given. Fourth, people are forgetful. Even if memories are excellent, no one perceives every aspect of every experience. Many times people remember only what they want to remember. The phenomena that are most meaningful are most apt to be remembered. Consequently, memories are subjective as well as occasionally unclear about many events. Finally, everyone experiences changes of mind. Truthful an-

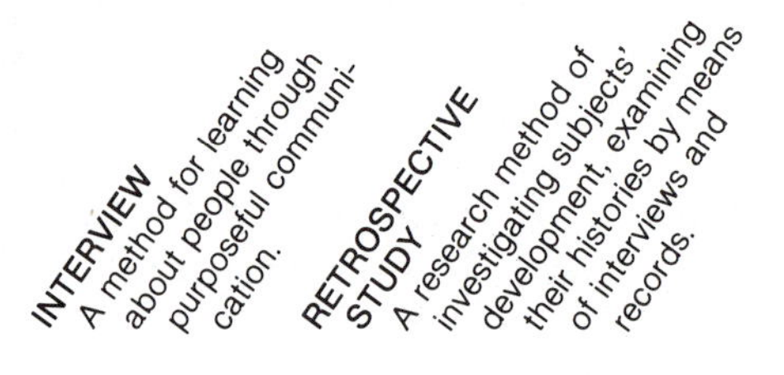

swers that are given one day may no longer reflect how one feels after a day, a week, or a month.

Surveys, questionnaires, and paper-and-pencil tests are susceptible to many of the weaknesses of interviews. Answers may reflect lack of frankness, forgetfulness, selective memory, confusion about or misunderstanding of wording, or vacillating attitudes. Subjects filling in the blanks may be troubled by nervousness, which affects their ability to remember or reason, by boredom, which can lead to carelessness and decreased efforts to concentrate on the questions, or by lack of motivation to participate, which can lead to carelessness or to intentional falsification of answers. Nevertheless, when respondents strive to be honest and are motivated to add to a body of research data, these methods can amass a great deal of information that cannot be gathered in other ways.

ETHICAL PRINCIPLES

Regardless of the research methodology adopted, any study using human participants must observe caution to protect the welfare of the subjects. The researcher must also strive to preserve and protect their dignity, worth, and fundamental human rights. The American Psychological Association (1981) has proposed the following ethical considerations for research with human participants:

- Prior to participation, the investigator clarifies the obligations and responsibilities of each subject and explains all aspects of the research about which the participants inquire.
- If concealment or deception is necessary, the investigator ensures that the participants are provided with sufficient explanation as soon as possible.
- The investigator respects the individual's freedom to decline to participate in or to withdraw from the research at any time.
- The investigator protects the participant from physical and mental discomfort, harm, and danger unless the research has great potential benefit and fully informed and voluntary consent is obtained from each participant. Procedures for contacting the investigator within a reasonable time following participation if stress, harm, or related concerns arise must be given.
- The investigator is responsible for the ethical treatment of subjects by collaborators, assistants, students, and employees.
- After the data are collected, the investigator provides the participant with information about the nature of the study and attempts to remove any misconceptions that may have arisen.
- If research results in undesirable consequences, the investigator has the responsibility to remove or correct these consequences.
- Confidential information obtained about a research participant during an investigation is protected.

STATISTICAL ANALYSIS

Methods of statistical analysis have been developed, and are continuing to develop, to evaluate data obtained in research. Methods of statistical analysis help to shape the format of research and, in so doing, are part and parcel of research methods. Although statistics is a branch of applied mathematics in its own right, with roots in probability theory, its value is stressed here as a tool of human development research. The following brief overview will help you understand some of the statistical terms in research reports.

Descriptive statistics describe and summarize or help a reader understand the characteristics of data. Observations are translated into quantitative data or numerical measures. *Means, medians,* and *modes* are descriptive statistics used frequently to represent a set of measures as a single number. They give information about the central tendencies or averages of a set of data. The **mean** is the arithmetic average. The **median** is the middle number in any ordered series. The **mode** is the numerical term that occurs most frequently in the data. For instance, if you have scores of 62, 62, 63, 84, and 94, the mean value will be 73, the median 63, and the mode 62. In addition to the average score, it is usually necessary to learn the distribution or spread of a set of scores. Descriptive statistics that assess the spread of a set of data are the range and the standard deviation. The highest and lowest values of the data are indicated by the **range**. The degree of spread from the center of the data is described by the **standard deviation**. The standard deviation is calculated for each set of data by determining the distance of each score from the mean, squaring it, summing the squares, dividing by one less than the total number of scores, and finding the square root of the resulting figure. For normal distributions approximately 68 percent of all scores will be expected to fall within one standard deviation above and below the mean. Approximately 95 percent will fall within two standard deviations and 99 percent within three standard deviations above and below the mean (see Figure 1–9).

A frequently used term in statistics is the **correlation coefficient**. It is a numerical value that indicates the degree and direction of the relationship of two variables to one another. When two variables change in the same direction (such as height and weight), they are said to be positively correlated. When two variables change in the opposite direction (such as high blood pressure and life expectancy), they are said to be negatively correlated. Perfect positive

DESCRIPTIVE STATISTICS Numerical data assembled, classified, and tabulated to describe and summarize the characteristics of the data.

MEAN An arithmetic average.

MEDIAN The measure of central tendency that has half of the cases above it and half below it; the 50th percentile.

MODE The numerical term that occurs most frequently in a set of data.

RANGE In statistics, the highest and lowest values of the data.

STANDARD DEVIATION A statistical technique for expressing the extent of variation of a group of scores from the mean.

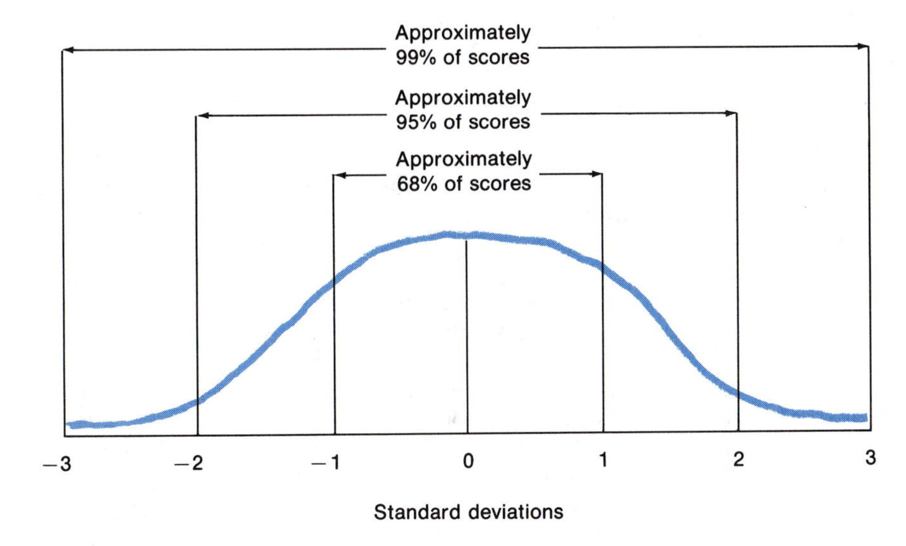

Figure 1–9
The bell-shaped normal frequency distribution.

and negative correlations between two variables are indicated by a correlation coefficient of +1.0 and −1.0 respectively. Most correlation coefficients range between +1.0 and −1.0 and are therefore fractions (for example, +.25, −.55, +.90) that indicate some low, moderate, or high degree of positive or negative association between two variables.

Inferential statistics go beyond description. The characteristics of a population are inferred from the characteristics of a *sample* (a part intended to show the kind and quality of the whole) of that population. For example, from a knowledge of the birth weights of a sample of infants, one can infer the birth weights of the larger population of infants from which the sample was drawn. Based on sample values, statistical methods can be employed to estimate population values with varying degrees of confidence.

There are many kinds of statistical tests used on different sets of data to answer different types of questions. For example, **t-tests** are often used in evaluating differences in mean outcomes of two treatments. **Analyses of variance** are used to evaluate differences in mean outcomes among three or more treatments. **Chi-square tests** are used to evaluate how variables are distributed in a population and to evaluate the independence or dependence among various qualitative

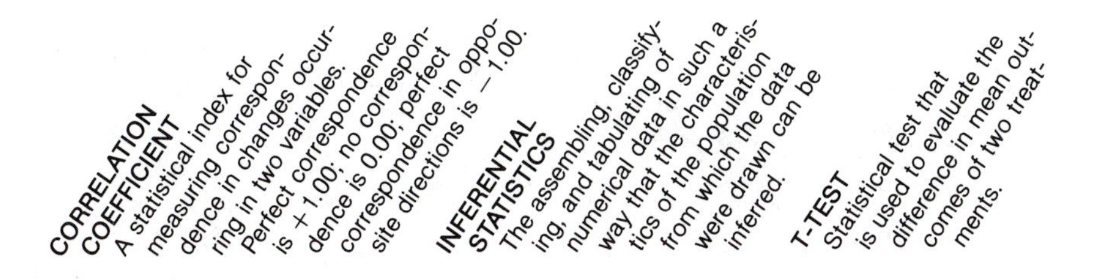

features within the elements of a population. **Regression analysis** is used in testing for significance of relationships among two or more variables. You may learn about these and other statistical tests in a statistics course.

It is important for you to understand what is meant by confidence and significance, whether or not you ever take a statistics course. **Confidence intervals** are mathematical expressions that define a range into which a certain proportion of the values of a variable (or feature of a population) will fall. They are derived from a knowledge of the characteristics of a sample. For instance, we could mathematically define a 95 percent confidence interval that would include the birth weights of 95 percent of infants in a population, based on our knowledge of the mean and standard deviation of the birth weights of a random sample of infants drawn from that population. Weights falling outside of this 95 percent confidence interval would be considered distinctly unusual. They could be expected to occur by chance with a **probability** (expressed as p) of no more than 5 percent (expressed as $p \leq .05$). The highest and lowest values of the confidence interval are called the confidence limits. Researchers in a field somewhat arbitrarily decide what the stringency of their confidence limits will be. Influenced by experience, researchers may say that a value of a variable is unusual and has **statistical significance** if its chance of occurring is less than 5 percent or less than 1 percent. Generally, in studies of human development, confidence limits are found in which the probability of a variable having a value greater than the upper limit or less than the lower limit of the confidence interval is less than 5 percent ($p < .05$) or less than 1 percent ($p < .01$).

Summary

In presenting a view of human development over the life span, this text focuses on many determinants of development. The process of growth and change should be examined as the interaction of the individual in his or her physical, cognitive, and psychosocial spheres. The impact of the primary setting of the family is vital, as are the roles of the larger systems of the community and the culture. It is important to integrate and interrelate all systems, inherited and acquired, with one another in order to understand individual development.

Human behavior was primarily the field of inquiry of philosophers until the nineteenth century. Now it is the concern of several

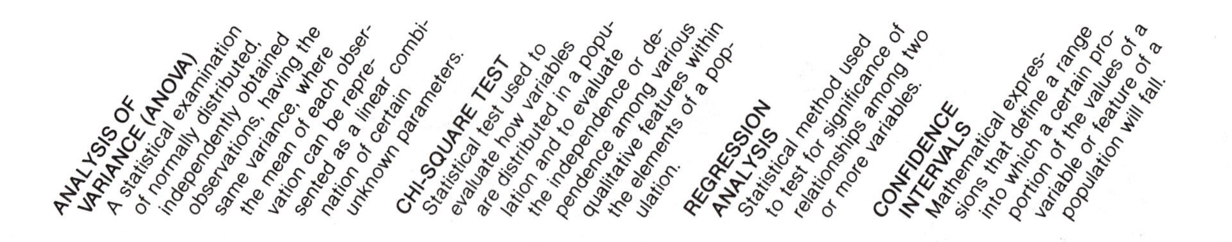

of the biological sciences as well as the social sciences. Specialized studies frequently concentrate on only one aspect of human life such as age (embryology, infancy, gerontology), sex, or culture in order to learn as much as possible about a specific area. Interdisciplinary sharing helps round out our picture of how development proceeds.

Research can be cross-sectional (studying many different persons at the same time) or longitudinal (studying the same persons over a period of time). These methods can be combined in cohort-sequential, time-sequential, and cross-sequential studies.

Three major kinds of research—experimental, observational, and correlational—have been applied to the study of human development. Experimental research is contrived. It can provide control of variables, is objective, and is replicable; however, it has a limited application to the study of areas such as thought processes and feelings. Observational research is a look at "real life." It can contribute a great deal of information about many different subjects; however, it must rely on some subjective judgments and may not be easily replicated. Correlational studies show positive or negative correspondence between two or more variables but do not necessarily imply causation. They can usually be conducted with a minimum of interference in people's lives.

Ethical considerations must always be observed to protect the participants in research studies of human development.

The data obtained from any type of research are usually analyzed with statistics. Results are expressed numerically in terms of their mean, median, mode, and standard deviation. Relations between variables are expressed in terms of correlation coefficients or by chi-square tests. Inferential statistics extend analysis beyond simple description to predict characteristics of the population from those of a sample.

Questions for Review

1. Write down ten factors that you feel have had a great force in shaping your destiny. Order them from most important to least important to you. Do they have more family or more community emphases?
2. Some parents feel it is important to limit their child's exposure to the community and its influences as much as possible,

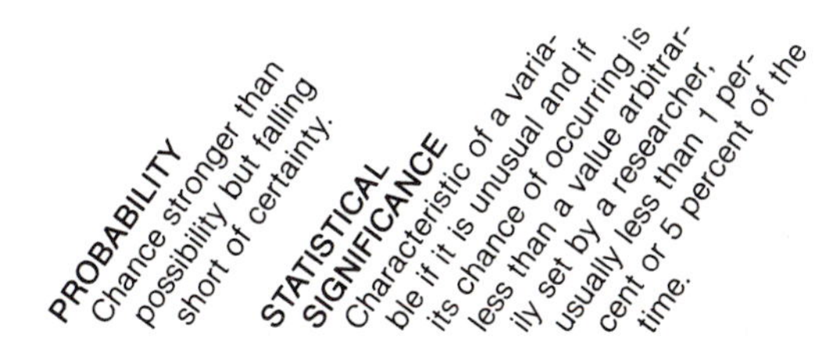

especially when the child is quite young. Do you agree or disagree? Why?

3. Sir Frances Galton felt women were intellectually inferior to men. What environmental factors may have led him to this conclusion?
4. Describe an experimental, observational, or correlational study you would be interested in doing. Discuss some of the problems you might have in conducting this piece of research.
5. List the pros and cons of using the experimental method in doing research in human development.
6. Why are experiments often considered more scientific than observational or correlational forms of research?
7. Some feminists say that very little of the research already done in human development is useful for statements about women since most of it was done by men using male subjects, who have had different experiences than women. Do you agree or disagree with this criticism? Why?

References

American Psychological Association. Ethical principles of psychologists. *American Psychologist,* 1981, *36,* 633–638.

Applebaum, M. I., and McCall, R. B. Design and analysis in developmental psychology. In W. Kessen (Ed.), *History, theory and methods.* Vol. 1 of P. H. Mussen (Ed.), *Handbook of child psychology,* 4th ed. New York: Wiley, 1983.

Aries, P. *Centuries of childhood.* New York: Knopf, 1962.

Bronfenbrenner, U. *The ecology of human development.* Cambridge, Mass.: Harvard University Press, 1979.

Condry, J. C., and Condry, S. Sex differences: A study of the eye of the beholder. *Child Development,* 1976, *47*(3), 812–819.

Condry, J. C., and Ross, D. F. Sex and aggression: The influence of gender label on the perception of aggression in children. *Child Development,* 1985, *56,* 225–233.

Darwin, C. *On the origin of species by means of natural selection.* London: John Murray, 1859.

Darwin, C. A biographical sketch of an infant. *Mind,* 1877, *7,* 285–294.

DePaulo, B. M.; Stone, J. I.; and Lassiter, G. D. Telling ingratiating lies—effects of target sex and target attractiveness on verbal and nonverbal deceptive success. *Journal of Personality and Social Psychology,* 1985, *48*(5), 1191–1203.

Freeman, M. History, narrative, and life-span developmental knowledge. *Human Development,* 1984, *27,* 1–19.

Galton, F. *Hereditary genius.* London: Macmillan Ltd., 1892.

Hall, G. S. *Adolescence* (2 vols.). New York: Appleton, 1904.

Hall, G. S. *Senescence: The last half of life.* New York: Appleton, 1922.

Hart, K. J., and Ollendick, T. H. Prevalence of bulimia in working and university women. *American Journal of Psychiatry,* 1985, *142*(7), 851–854.

King, L. *Larry King by Larry King.* New York: Simon & Schuster, 1982.

Lambert, Z.; Doering, P.; Goldstein, E.; and McCormick, W. Predisposition toward generic drug acceptance. *Journal of Consumer Research,* 1980, *7*(1), 14–23.

Mead, M. *Coming of age in Samoa.* New York: Morrow, 1928.

Mead, M. *Growing up in New Guinea.* New York: New American Library, 1953.

Myers, B. J. Early intervention using Brazelton training with middle-class mothers and fathers of newborns. *Child Development,* 1982, *53,* 462–471.

Tieger, T. On the biological basis of sex differences in aggression. *Child Development,* 1980, *51,* 943–963.

♥

Theories of Human Development

2

Chapter Outline

Psychodynamic Theories
- *Freud's Psychosexual Theory*
- *Erikson's Psychosocial Theory*
- *Contributions of Other Psychodynamic Theorists*

Behaviorist Theories
- *Watson's Classical Conditioning*
- *Skinner's Operant Conditioning*

Social-Learning Theories
- *Bandura's Learning by Modeling*
- *Sears's Developmental Learning*

Cognitive-Developmental Theories
- *Piaget's Genetic Epistemology*
- *Contributions of Other Cognitive-Developmental Theorists*

Humanistic Theories
- *Maslow's Hierarchy of Needs*
- *Rogers's Phenomenological Approach*
- *Contributions of Other Humanistic Theorists*

Comparisons of Theories and Their Influence

Key Concepts

accommodation
adaptation
archetypes
assimilation
behaviorism
behavior modification
classical conditioning
client-centered therapy
cognition
collective unconscious
defense mechanisms
disequilibrium
ego
equilibrium
Eros
extinction
extravert
fixation
free association
genetic epistemology
humanistic psychology
id
identification
inferiority
inner speech
interpersonal relationships
introvert
libido
modeling
moral realism
moral relativism
neuroses
nuclear conflict
operant conditioning
organization
peak experiences
phenomenological approach
psychodynamic
pychoses
rational-emotive theory
rudimentary behaviors
schema
secondary behavioral systems
secondary motivational systems
self-actualization
social-learning theory
stimulus-response learning
subconscious
superego
superiority
Thanatos
traits (common, cardinal, central, secondary)
unconscious

♥

Case Study

Genevieve's older brother and sister knew how to ride a bike, even if the family didn't own one. A neighbor boy taught them on his bike. They had a lot of fun taking turns riding it down the long hill in front of their homes. One day Genevieve begged to be taught. The three older children took her to the top of the hill, put her on the bike, told her to hold on, and shoved her on her way. The speed, and her frozen posture due to sheer fright, kept the bicycle from tipping over. She sailed down the hill and finally began slowing somewhat as she approached the business district of town. Genevieve wanted to stop, but she didn't know how. She turned the handlebars of the bike to the right, went up a small embankment, and hit the front fender of a car in a used car lot. She fell off the bike, unhurt. She picked it up; it was not damaged. The fender of the car she hit was also unscathed. She walked the bike home, gave it back to the three older children, assured them she'd had fun but wanted to do something different now, and went into her house.

Two days later a woman who lived across the street from the used car lot saw Genevieve's older sister riding the bicycle and called for her to come over to her porch. She scolded the girl for the "hit and run" accident two days earlier and insisted that she go to the owner of the car lot and report her behavior now. Genevieve's sister denied any knowledge of such an accident. She immediately began to suspect Genevieve and went home and asked her about it. Genevieve also denied any part in, or knowledge of, such an accident.

A psychodynamic theorist, following in the Freudian tradition, would probably explain Genevieve's denial as a defense mechanism against her anxiety at being discovered. Freud might even suggest that she had felt some libidinous pleasure in the ride preceding the accident, which increased her anxiety about being discovered. A behaviorist would probably explain Genevieve's denial as a learned response. She had learned that admission of guilt would bring punishment; denial, unless her guilt could be proved, would not. A social-learning theorist would probably explain Genevieve's denial as a modeled response. She had probably observed other persons avoid punishment by denying behaviors. A cognitive-developmental theorist would probably be interested to learn that Genevieve was only five years old. At this age, it is difficult for children to reverse their thought processes and reconstruct events as they actually happened in the past. Genevieve's lie might thus be seen more as a cognitive limitation than a deliberate falsehood—she could have confused truth with fantasy about what happened or did not happen. A humanist would probably explain Genevieve's denial as meeting her greater need for safety. The fact that she denied the experience would also suggest that the hit-and-run behavior was incongruent with her self-concept and consequently couldn't be incorporated into her sense of phenomenological reality. Many other theorists would be able to give alternate explanations for Genevieve's behavior. What do you think?

Theories are not fact; they are merely speculations or hypotheses that individuals make for the purpose of explaining and predicting phenomena of interest to them. Theories include definitions and propositions that specify the interrelationships of variables in a systematic fashion. You have theories of your own. You may not have formalized them or put them in writing, but you have them. Some of your theories are quite firmly entrenched in your thought processes; others may change each time you discuss a topic, meet someone new, or read a book. Some of your theories are exclusively yours; others you may have borrowed or adopted from friends, from professors, or from reading. Because of the status of theories as "not well understood" or "not proved," many are frequently altered or exchanged for others.

Human development is a complicated and difficult area in which to construct research. The theories presented here are respected because of the *empirical* (founded on experiment or observation) evidence that supports their propositions. This is not to say that they are correct or incorrect; rather, they have received the attention and support of some experts and some of the public at large. Every time a person sets his or her theories down in print for others to read, that person leaves himself or herself open to agreement, praise, approval, and criticism. Carefully researched, well-organized, systematic theories with definitions of special terminology are more apt to gain support than poorly constructed or undocumented theories. Feel free to agree or disagree with part or all of any of the theories that follow.

Psychodynamic Theories

Psychodynamic psychology refers to the study of the conscious, the subconscious, and especially the unconscious elements in human behavior. There are many theories that fall under the umbrella term psychodynamic, each with its own slant but all sharing certain similarities. Psychodynamic theorists study what happened in childhood to learn about the adult. They look for instinctual drives to explain human actions and reactions, and they study how these basic drives have been gratified (successfully or unsuccessfully). All believe that people use defense mechanisms to avoid pain or anxiety and to modify the pressures of various drives. The underlying assumption

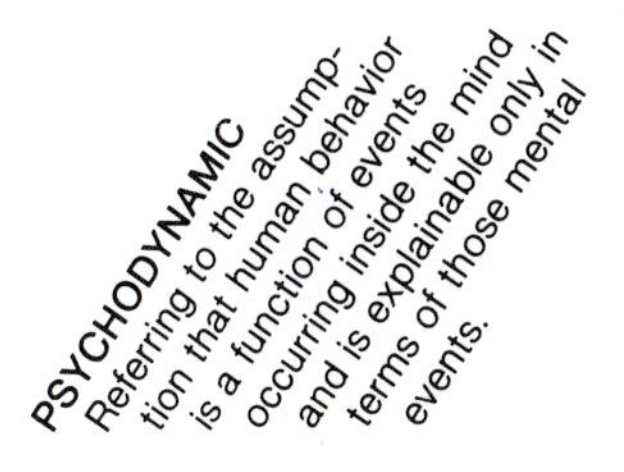

of all the psychodynamic theorists is that conscious, subconscious, and unconscious mental processes regulate how energy is distributed or deployed and how drives are satisfied or modified.

FREUD'S PSYCHOSEXUAL THEORY

The father of all psychodynamic theories was Sigmund Freud. He is considered one of the most influential contributors to the fields of psychology and psychiatry. Many of his assumptions have become so firmly established in the language of our times (for instance, libido, id, unconscious) that we regard them as facts rather than suppositions.

Libido Freud believed that sexual desire is the primary motivator of behavior. He described **libido** as a life force, a source of energy, especially dominated by "demands of sexual desire . . . for pleasure" (Freud, 1920/1966, p. 142). Sexual desire, he felt, is present in some form in most physical pleasures. He believed most adulthood **neuroses** (partial disorganization of the personality, characterized by anxieties, compulsions, obsessions, tics, phobias) and **psychoses** (drastic distortions of a person's perception of reality) can be traced back to sexually oriented conflicts in childhood. His psychosexual theory has recently been under fire because of his views of sex differences and male supremacy. Many of the sexual conflicts he believed to be developmental were probably once true of children raised during his lifetime (1856–1939) in Europe. However, they may not be true of all children today.

Freud devised the technique of **free association** of ideas in a "relaxed" setting to counteract his patients' tendencies to organize and censor their thoughts before speaking them aloud. The image of a psychiatrist sitting back in a chair while the patient reclines on a couch originated in his Vienna office. Early in his career, Freud also hypnotized patients to help them review their earlier life experiences. He often used dream analysis to help them gain more insight into their needs, desires, fears, and anxieties.

Id, Ego, Superego Freud postulated that personality has three components: the id, the ego, and the superego. The **id**, according to Freud, is the hedonistic (pleasure-seeking) part of the personality. Instincts, sexual impulses, the hunger drive, and aggressive urges are all housed there. Freud felt that the psychic energy, or libido, of the id is composed of two types of drives (or instincts). One, **Eros**, is

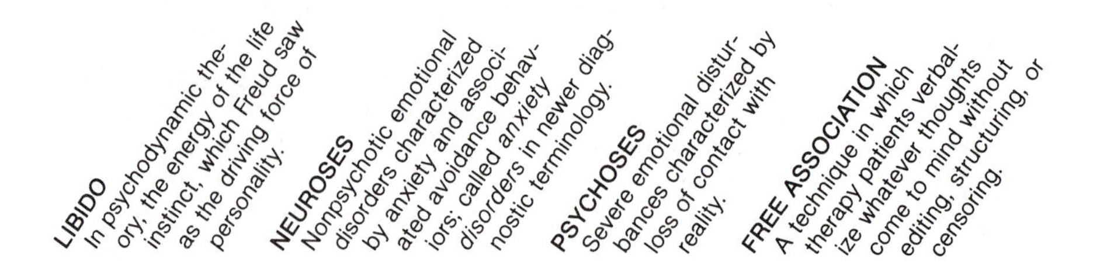

Figure 2–1
Freud believed that behavior is determined by urges of the id and the superego and by unconscious and subconscious thoughts as well as by conscious decisions.

a life force, aimed at survival and self-propagation. It motivates one to seek food, shelter, creature comforts, and sexual satisfaction. The other, **Thanatos**, is a death force, aimed against the world and the self. The energy of Thanatos builds up until it must be discharged either against others (aggressive drive) or against the self in some self-destructive manner. While a death force may seem the antithesis of hedonism, it is not. Many people will attest to the pleasure they derive from putting their lives in danger (drag racing, sky diving, and so on). Freud believed that id impulses, with their untamed, animal-like drives, are in constant need of control. The **superego**, he believed, is the controlling, moral arm of the personality. According to Freud, it serves as a watchdog, ensuring that our dominant id impulses do not motivate us to engage in socially unacceptable behaviors (see Figure 2–1). It is derived in accordance with the reli-

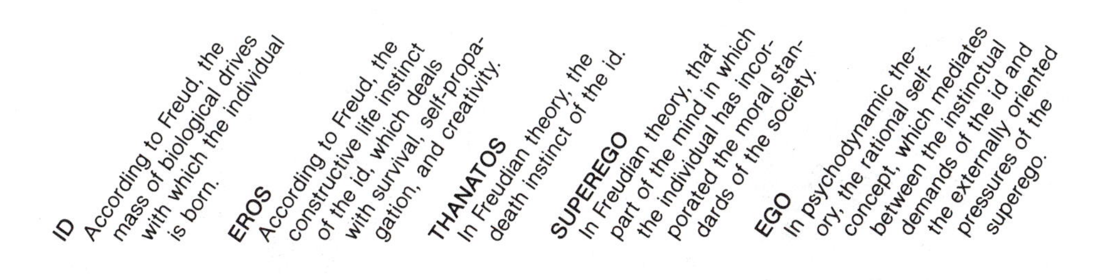

gious teachings, moral standards, and ethics and mores of our parents and culture. It is composed of the conscience, which is a prohibitive influence on our behavior, and the ego-ideal, which motivates us to do good deeds. The superego supposedly develops slowly over the course of one's childhood and adolescence. The id, in contrast, is the aspect of the personality that houses and directs psychic energy from birth. Between the id and the superego is the ego. Freud saw the **ego** as the realistic aspect of the personality. Ego is the self: the problem-solving, perceiving, remembering, judging, speaking mediator of all conscious, subconscious, and unconscious drives. Freud believed that personality is a dynamic and changing phenomenon, influenced by the development of each person's id, ego, and superego. At times a person is more id controlled and at other times he or she is more influenced by the superego. However, Freud felt that these personality components, and the relative way they balance and counterbalance each other, become established by the end of adolescence. By this point, one has a typical response pattern to frustrations, conflicts, and threats. Freud felt that major changes in adult personality can be accomplished only with great difficulty.

The Unconscious and the Subconscious Freud believed that all human behavior is rooted in mental processes at one of the levels of consciousness: the conscious mind, the unconscious, and the subconscious. He felt that ideas buried in the **unconscious** mind (those not readily available to perception) can influence us in ways that are beyond our ability to control. Some unconscious memories are simply forgotten. The ego has repressed others and prevents them from gaining access to the conscious (knowing) mind. Unconscious ideas, however, have the power to affect behaviors. At the **subconscious** level people may be only vaguely aware of the roots of their ideas. For example, you may have feelings about a person or a place without being able to get a firm grasp on what those feelings are or where they came from. In therapy some of the unconscious or subconscious thoughts may be revealed to the therapist through free association, dream analysis, slips of the tongue ("Freudian slips"), flashes of insight (the "aha" experience), old memories, or hypnosis.

As Freud sought to understand how childhood conflicts create problems in adulthood, he proposed a sequence of developmental stages (see Table 2–1). Each stage, he believed, is based on instincts related to body zones that bring sexual pleasure. If a child passes

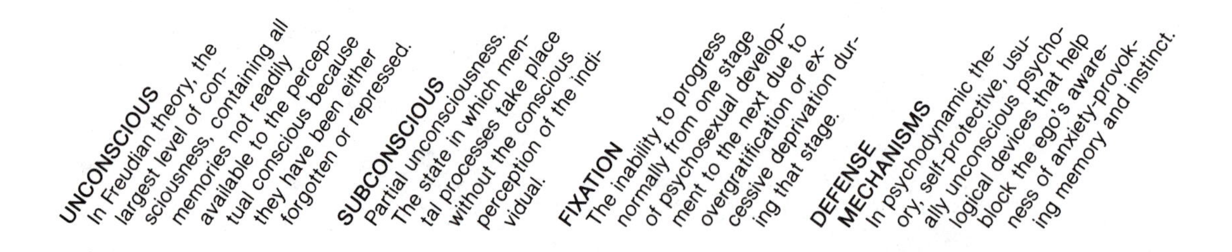

Table 2–1 Freudian Stages of Psychosexual Development

Stage	Sex drives related to:	Possible results of trauma:
Oral	Feeding and weaning	Features of oral fixations: nail biting, cigarette smoking, gum chewing
Anal	Toilet training	Features of anal fixations: parsimony, orderliness, punctiliousness, obstinacy, pedantry, possessiveness
Phallic	Genital manipulation	Oedipal complex: boy overly attached to mother, fears father may castrate him (castration complex); or Electra complex: girl overly attached to father, wishes to be a man, has "penis envy"
Latent	(Sex drives repressed)	Prolonged or exaggerated Oedipal or Electra complex
Genital	Puberty	Failure to develop mature, socially acceptable modes of attaining sexual gratification

through each stage without trauma (and few can, in Freud's opinion), he or she will become a "well-adjusted" adult. Trauma in any given stage will result in some **fixation** (arrest of psychosexual development) or complex of the personality based on that stage. The origin of neurotic symptoms in adults can often be traced back to too much or too little gratification during one or more of these developmental stages of childhood.

Freud's childhood stages, derived entirely from his experience with adult patients, are among his most criticized premises. The personality fixations and complexes he described may occur for many other reasons than those he suggested. Conversely, they may *not* occur even in unfavorable childhood circumstances.

Some of the **defense mechanisms** that Freud proposed as guards against sexual anxieties have received wide acceptance. Many of these hypothesized behaviors are now accepted as learned reactions to anxiety that an individual habitually adopts when frustrated. Some of the most common defense mechanisms are described in Box 2–1. All of these defense mechanisms are characteristic ways in which the ego avoids pain, modifies expression of strong emotions, and relieves excessive anxiety. Overuse of any of the defense mechanisms, however, can be self-defeating. They can interfere with one's coping and adapting abilities and restrain one in disabled neurotic postures.

Freud based his theory primarily on what he saw and heard from neurotic personalities who sought his psychiatric treatment. He did not use personality tests, experiment, or collect data for quantitative analysis. However, he did check his theory continually against the evidence as he saw it. This resulted in many revisions of his concepts and eventually in a theory that has endured.

BOX 2–1

Defense Mechanisms

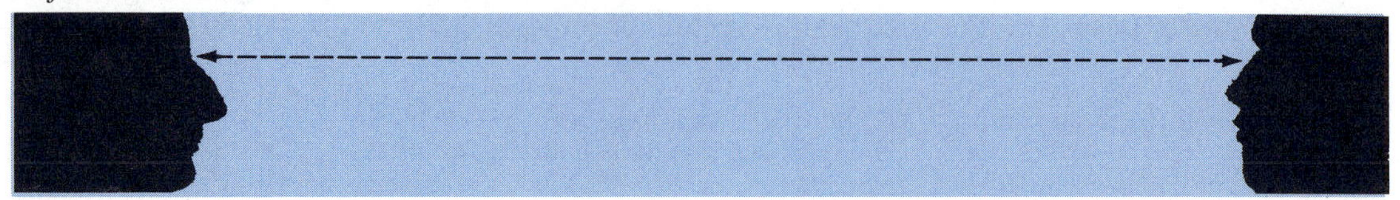

Compensation: Effort to win respect or prestige in one activity as a substitute for inability to achieve in some other realm or endeavor.

Denial: A refusal to believe or accept something as it is but rather to perceive it as one wishes it would be.

Displacement: Transfer of emotion away from the person or situation that incurred the strong feeling and toward an inappropriate person or object.

Fantasy: Imagination of what one could have said or will say, could have done or will do; daydreaming.

Intellectualization: The separation of events and/or ideas or concepts from the emotions that impinge on them.

Introjection: Taking into one's own personality the characteristics of another.

Projection: Attributing one's own motives, emotions, or characteristics to someone else.

Rationalization: Concealment of motive for behavior by assigning some socially acceptable reasons for the action.

Reaction formation: Distortion of a drive to its opposite—for example, acting kindly toward a person whom one dislikes.

Regression: Reactivation of behaviors more appropriate to an earlier stage of development.

Repression: Burying something in the subconscious or unconscious levels of thought.

Sublimation: Discharge of libido in socially acceptable activity rather than using it to obtain sexual gratification.

ERIKSON'S PSYCHOSOCIAL THEORY

Erik Erikson believes that social and cultural factors influence the manner in which an individual resolves the various conflicts brought about by biological maturation. Thus, his theory is called *psychosocial* rather than psychosexual. Erikson, who met Freud and studied psychoanalysis under Freud's daughter, Anna, agrees with Freud's postulation that maturation brings about conflicts revolving around oral, anal, phallic, latent, and genital stages of psychosexual development. However, he believes that psychosocial development continues after the resolution of the sexual conflicts. He has proposed eight conflicts of life: the first five parallel Freud's psychosexual stages, the last three occur during adulthood (see Figure 2–2). Erikson was the first theorist to write about development across the entire life span.

The eight conflicts proposed by Erikson are referred to as **nuclear conflicts**. By nuclear, he means central: the issues about which behaviors are focused; those of crucial importance to the personality at that stage of life. In infancy, for example, the conflict revolves around the biological need for food and the maturation of the sucking reflex associated with feeding. Erikson, like Freud, believes that oral gratification brings sexual (libidinous) gratification as well. Erikson calls his first nuclear conflict *trust versus mistrust,* however, reflecting his belief that social forces are more powerful mediating forces than sexual forces in bringing about the gratification of needs.

NUCLEAR CONFLICT Eriksonian term applied to major conflicts characterizing each of his eight stages.

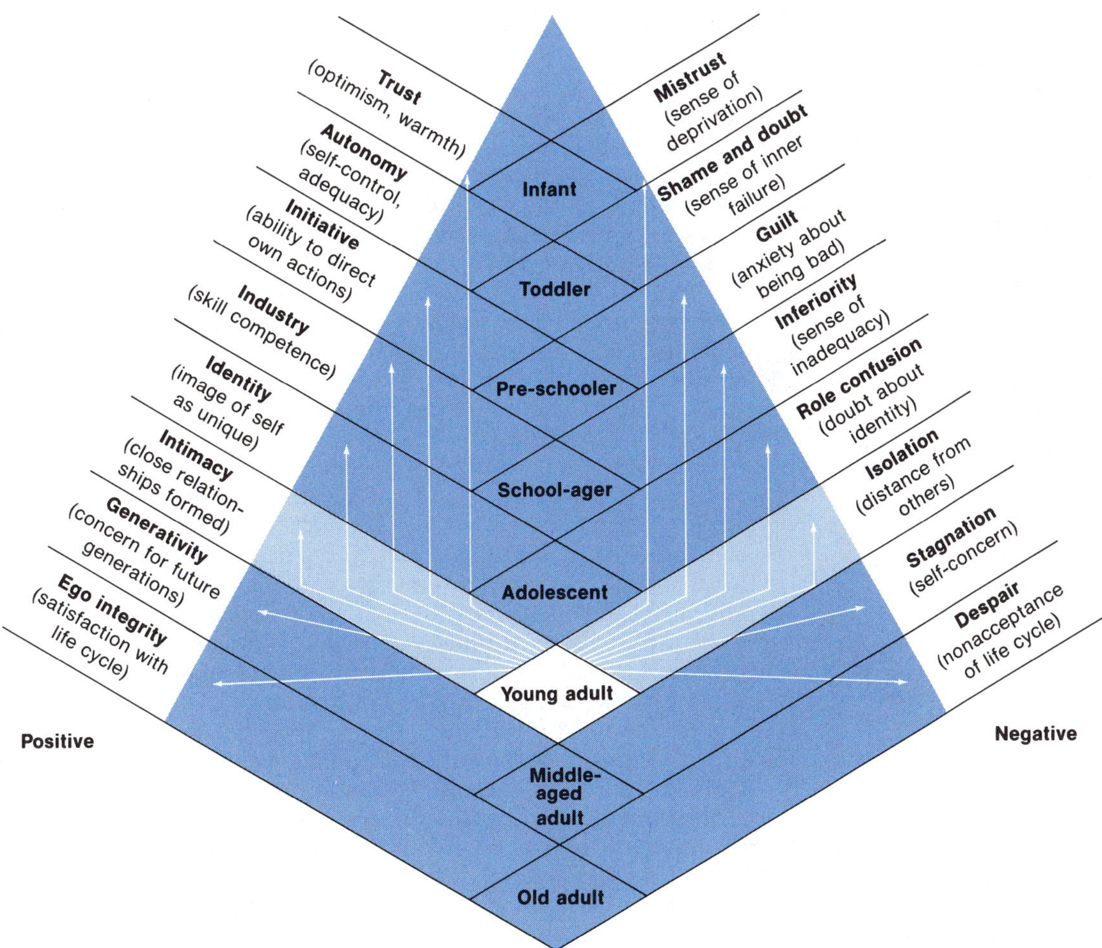

Figure 2–2
Erikson's theory of the eight ages of man is cumulative. At each age of life, one nuclear conflict is at the forefront. The major conflict facing the young adult, for example, is achieving intimacy and avoiding isolation. Attempts to resolve other conflicts continue (see arrows) but to a lesser degree.
SOURCE: Reprinted and adapted from Childhood and Society, *2nd Edition, by Erik H. Erikson, by permission of W. W. Norton & Company, Inc. Copyright 1950, © 1963 by W. W. Norton & Company, Inc.*

If an infant is fed lovingly and is supported in other biological needs, that infant will develop a sense of trust in his or her world. Experiences of being left hungry will result in a sense of mistrust. Each of Erikson's eight nuclear conflicts will be discussed in detail in Chapters 4 through 10.

It is possible—in fact, probable, in Erikson's view—that at any stage of life, the solving of a conflict will be somewhere between a perfect positive and a total negative. Obviously, it is desirable to have a more positive than negative resolution of each conflict. The way in which each conflict is resolved will affect adjustments to, and development during, all subsequent conflicts. Also, during subsequent conflicts the individual will still spend some time re-resolving earlier conflicts. Erikson (1968) wrote that every adult carries residues of childhood conflicts in the recesses of his or her personality, and that these conflicts are constantly being re-resolved. It is never too late in life to change a negative resolution of an old conflict into a more positive resolution. The handling of all conflicts, both the nuclear and the less central ones, shapes the developing person's *ego identity*. Both the concept of ego and the concept of identity are important in Erikson's theory.

Freud felt that only the id component of personality is present at birth. Erikson disagrees. He suggests that newborn babies also have an immature ego. This ego is reshaped by the self and the social world during each progressive nuclear conflict. The ego, as it matures, is a unifying force between the perceived identity and all societal forces.

Identity versus role confusion is the conflict that Erikson proposed as the central focus during the adolescent years. A quest for identity occurs all through life, however, even if it is of crucial importance to the personality during pubescence. Identity involves an acceptance of oneself in one's social world. It is more easily obtained when one feels continuity between one's original ego during the earliest nuclear conflicts and one's transformed ego with the resolution of each successive nuclear conflict. Remember, at each stage the person not only confronts and resolves new biological needs and social forces but also reexamines and redefines old nuclear conflicts. Identity builds on these transformations of past to present. Erikson, especially after having studied American Indians being assimilated into the Anglo culture and American servicemen readjusting to peace after World War II, stresses the impact of cultural identity on one's personal quest for identity. One's identity is more easily maintained when one looks toward a future that is similar to one's past. If a person has to become oriented to a culture unlike the one experienced in the past, more work is needed to re-resolve all earlier nuclear conflicts as well as deal with the current one.

Erikson's theory is considered an *epigenetic* theory. Epigenesis refers to that which is built upon (epi) the original (genesis). Past experiences affect present behavior. Each new nuclear conflict is resolved by examining (consciously or unconsciously) the results of behaviors or experiences of the past. One's behavior thus becomes much more differentiated the more one has experienced socially and culturally in the process of maturing biologically. More positive resolutions of each past conflict contribute to an easier evolution of a new, well-adjusted ego identity. More negative resolutions contribute to more neuroses in the quest for identity. However, even negative experiences contribute to the learning, changing, maturing process. Erikson holds out hope that, by working at all conflicts to some degree throughout life, all humans can achieve a sense of inner unity and a healthy personality.

CONTRIBUTIONS OF OTHER PSYCHODYNAMIC THEORISTS

Carl Jung (1916/1968), a student of Freud, introduced the concept of inherited **archetypes** to psychodynamic theory. Archetypes are unconscious figures typically seen in fantasies, dreams, or other situations when the imagination is active. These archetypes (Great Mother, Wise Old Man, Hero, Divine Child) predispose a person to act in the way one's ancestors acted. Jung (1931) theorized that a deeper level of consciousness exists than Freud's unconscious. He called it the **collective unconscious**. He proposed that human beings have inherited the wisdom and experience of previous centuries, even from our evolutionary prehuman existence, through this collective unconscious. Religious or mystic experiences, deep trances, dreams, or hallucinations, Jung felt, might occasionally shed some light on these collective memories.

Jung disagreed with Freud that libido is predominantly a sexual force. He defined libido as a general life energy behind all human action that is often expressed as the urge to excel. Jung's most famous concepts are those of extraversion and introversion (see Figure 2–3). The **extravert** directs his or her interest to phenomena outside the self rather than to his or her own experiences and feelings. An **introvert** directs his or her interest toward the self rather than to outside phenomena.

Extraversion
Adaptable, sociable
Attentive to sensations
Governed by objective reality
Tendency to hysteria

More "masculine"

Introversion
Obsessive-compulsive tendencies
Self-centered, rigid
Governed by subjective feelings,
Attentive to intuition

More "feminine"

Figure 2–3
Jung's opposite orientations toward life.

Jung stressed that a person with a predominantly introverted orientation would be unconsciously extraverted and have extraverted tendencies operating in his or her make-up. "Normal" personalities are better able to bring together the complementary polarities that divide them than are the more "neurotic," "obsessive-compulsive" introverts or "hysterical" extraverts.

Jung agreed with Freud about the importance of the unconscious and past experiences in determining current behaviors. However, he felt that humans' aims, aspirations, and orientation toward the future also guide actions. Although Jung agreed with Freud's developmental stages as far as they went, he believed that develop-

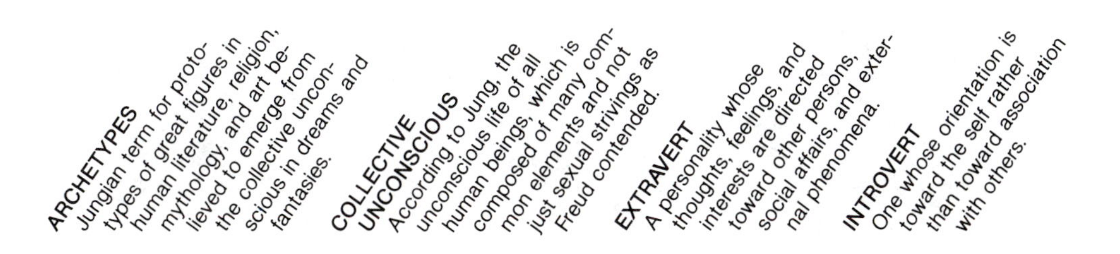
ARCHETYPES Jungian term for prototypes of great figures in human literature, religion, mythology, and art believed to emerge from the collective unconscious in dreams and fantasies.

COLLECTIVE UNCONSCIOUS According to Jung, the unconscious life of all human beings, which is composed of many common elements and not just sexual strivings as Freud contended.

EXTRAVERT A personality whose thoughts, feelings, and interests are directed toward other persons, social affairs, and external phenomena.

INTROVERT One whose orientation is toward the self rather than toward association with others.

ment continues throughout life rather than becoming relatively fixed at young adulthood.

Alfred Adler, another associate and contemporary of Sigmund Freud, developed his own rival psychodynamic theory called *individual psychology.* Adler (1929/1971) believed that people are motivated to move from a position of **inferiority** (feeling of inadequacy) toward security, power, and **superiority** (feeling of high value and worth). He saw the drive for superiority and power to be a more potent motivator of behavior than sexual libido. He wrote extensively about inferiority complexes and the mechanisms of compensation and overcompensation that some people use to hide their feelings of inferiority. In his later writings Adler also stressed the importance of social relationships to the search for meaning in life. The well-adjusted individual overcomes his or her self-absorption and places the interests of others ahead of selfish personal strivings.

The psychodynamic theory of Freud's daughter, Anna (1946), put more emphasis on ego development in relation to the id and superego. She also placed a far greater stress on the period of puberty and adolescence as a shaper of adult psychosexual behaviors. Her concerns centered on what the individual does to adjust socially, to put defense mechanisms into practice, to solve problems, and to control impulses.

Karen Horney (1937) saw basic anxiety rather than sexual desire as the primary motivator of behavior. She felt that this anxiety begins with the infant's birth into a frightening world. Unless adequate defenses are developed against this anxiety, one of three neurotic trends may develop: moving toward people, moving away from people, or moving against people. Horney believed that normal people have some or all three personality characteristics affecting their behavior, depending on the situations in which they find themselves. Neurotic people, on the other hand, function in only one of the stereotypic modes.

Harry Stack Sullivan emphasized **interpersonal relationships** in society as a motivator of behavior. Fear of social disapproval, he felt, or poor interpersonal relationships create more problems than do lack of satisfaction of sexual urges. Most important are interfamilial relationships, especially in childhood. If a child is repulsed when he or she tries to win love, approval, or acceptance, the result is not only a low self-concept but also a malevolent transformation of personality, where the child finds fault with and tears down others as well as himself or herself. Personality can change, Sullivan wrote (1953), when patterns of social interactions change.

INFERIORITY Term used to describe the feelings of inadequacy that characterize some people.

SUPERIORITY Adlerian term used to describe the exaggerated feeling of high value and worth that characterizes some people.

INTERPERSONAL RELATIONSHIPS Close interactions with other persons in one's social network.

CLASSICAL CONDITIONING A basic learning process in which a previously neutral stimulus (conditioned stimulus) is paired with one that elicits a known response (unconditioned stimulus) until the neutral stimulus comes to elicit a similar response (conditioned response).

BEHAVIORISM School of psychology based on the study of observable behavior and the patterns of stimulus and response that govern it.

Behaviorist Theories

Psychodynamic theories are based on the premise that the roots of a person's behavior lie in that individual's earlier experiences. They suggest that changing behavior takes a considerable amount of time and requires an understanding of memories acquired over time. In contrast, behaviorist theories are minimally concerned with the past and its influence on a person's behavior. They believe that new behaviors can be substituted for old behaviors after a relatively short period of time. New behaviors can be taught to, or conditioned in, the individual.

WATSON'S CLASSICAL CONDITIONING

At the turn of the century, a Russian physiologist, Ivan Pavlov, introduced the idea of changing behavior through a process known as **classical conditioning**. He trained a dog to salivate at the sound of a bell by coupling the bell with every presentation of food over an interval of time. After a while the dog associated the sound of the bell with the arrival of food and, whenever he heard the bell, began salivating (a natural reflex) in anticipation. The food was called the *unconditioned stimulus* (US), the salivation that it initiated the *unconditioned response* (UR). The sound of the bell (or any other arbitrary stimulus) was labeled the *conditioned stimulus* (CS). After the CS was paired with the US a number of times, learning took place such that the CS brought about salivation all by itself. Salivation to the bell was called the *conditioned response* (CR) (see Figure 2–4).

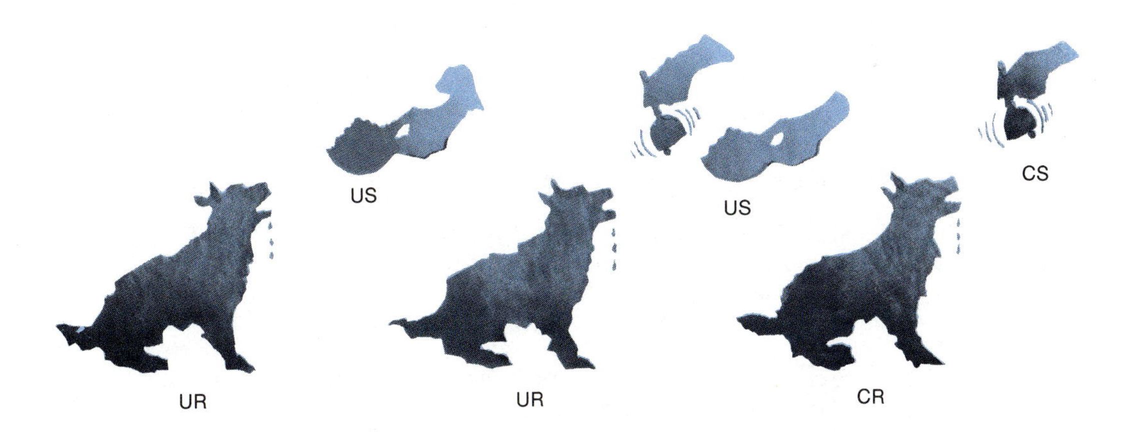

Figure 2–4
Pavlov's classical conditioning paradigm.

An American psychologist, John Watson, introduced **behaviorism** (the objective, experimental study of human behavior) to the field of psychology in the 1920s. He wanted more of a separation between psychology and the realms of philosophy and religion. Behaviorists, using classical conditioning techniques, could observe, measure, and unambiguously state what was learned in technical, scientific language. Behaviorists did not try to speculate about the hidden causes of behavior for which no proof could be supplied.

Instead, the new breed of behavioral psychologists produced evidence that anxieties and phobias could be both created and cured by behavioral means. Watson and Raynor (1920) classically conditioned fear in an infant by pairing the presentation of a white furry rat (the CS) with the noise of a sharp blow on a steel bar (the US). Their subject was an 11-month-old infant who reached for everything brought near him. The noise of the steel bar being struck frightened little Albert (UR). Since the noise was always paired with the presentation of the rat, Albert soon learned to fear the presentation of the rat (CR). He generalized this learned fear to other white, furry things as well (a rabbit, cotton, wool, and so on).*

Mary Cover Jones (1924), a student of Watson's, later published a paper in which she detailed her success in helping a small boy overcome his fear of furry things. She conditioned him to tolerate a rabbit by coupling his lunches with the presentation of the animal. After a while the boy was able to eat calmly with the rabbit on his lap.

Watson firmly believed that all human behaviors are learned by association (pairing of CS with US to bring about a CR) processes. Watson (1924) boasted that he could take any healthy infant and train him to become any kind of person: beggarman, thief, doctor, lawyer, merchant, chief; regardless of his talents, penchants, tendencies, or race of his ancestors. He advised parents to stop their mawkish, sentimental treatment of children, warning that mother love is a dangerous instrument. He advised parents never to hug and kiss children or let them sit in their laps (1928). Affection, even a pat on the head, he felt, should only be given to "condition" socially useful behaviors. Watson's advice had some impact on American parenting. It was modified to a great extent, however, by competing advice from Freud and his followers in the psychodynamic movement who stressed the developing child's need for a sense of security and a reduction of frustrations, pains, and anxieties, as she or he moves through biologically predetermined stages.

SKINNER'S OPERANT CONDITIONING

B. F. Skinner popularized the concept of **stimulus-response learning** (S-R learning) and introduced the concept of operant conditioning (defined on p. 47) procedures to behaviorism. His carefully controlled experiments and numerous publications of orderly data, his theoretical explanations of behaviorism, and his arguments for the uses of operant conditioning to modify and/or control human behavior made him a recognized leader of behaviorists from the 1940s to the 1980s. He published several influential and controversial books, including two best-sellers in which he argued that operant conditioning should be made an integral part of the social order: *Walden Two* (1948) and *Beyond Freedom and Dignity* (1971).

The idea of stimulus-response learning is that certain stimulus variables can be used to bring about overt, observable responses.

---

*The ethics of this experiment has been a matter of much controversy. Such research would probably not be approved today.

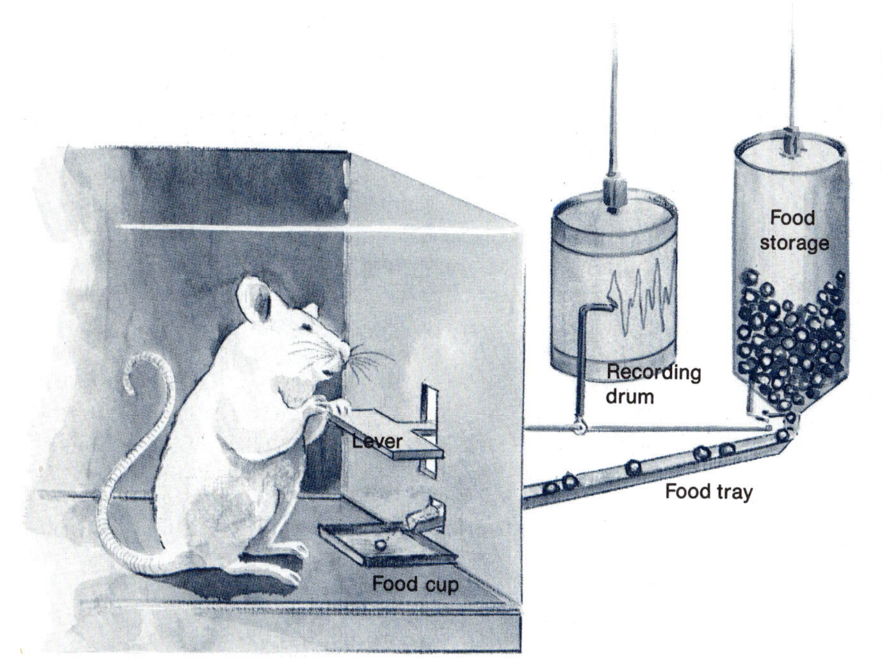

Figure 2–5
Schematic representation of a "Skinner box," used for conditioning animals.

In Pavlov's classical conditioning paradigm, a stimulus that brings about an involuntary response (for example, salivation when food appears) is paired with a second stimulus, which is finally substituted for the original stimulus. This form of conditioning is not always feasible, however, because many of the responses desired for modifying behaviors are voluntary rather than involuntary. Skinner's operant conditioning approach provides a feasible method of altering voluntary responses.

In **operant conditioning** a voluntary response is associated with a stimulus by being rewarded (*reinforced*) when it occurs. Operant conditioning works because animals and humans tend to repeat actions that bring them pleasurable consequences. Skinner advocated the use of positive reinforcement or negative reinforcement (withholding rewards) to bring overt responses under the control of stimuli. Some other behaviorists introduced the use of some form of punishment to help terminate undesirable voluntary responses.

Skinner designed and built special electronic cages now referred to as "Skinner boxes" for use in many of his orderly experiments of conditioned responses in animals (see Figure 2–5). Into

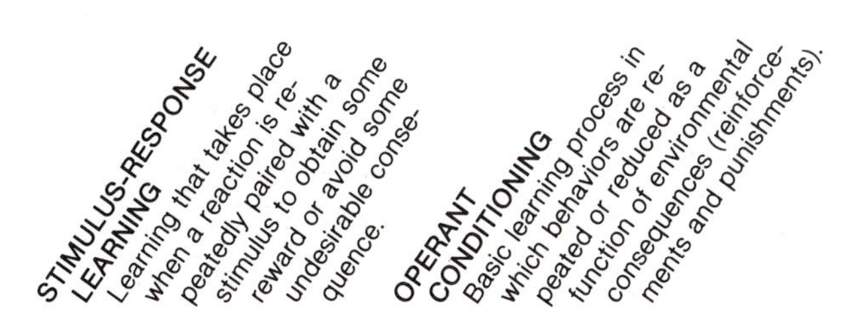

this apparatus he would place a rat or a pigeon or a monkey. When the animal pressed a lever (pigeons needed to peck a disk), the specially constructed box would deliver a small quantity of food to the animal and record the response. At first responses were shaped by rewarding every response. Eventually, reinforcers were given on a fixed or variable schedule of reinforcement. Intermittent reinforcement, Skinner discovered, makes a conditioned response stronger and more resistant to fading away than giving a reward for every response (continuous reinforcement).

Behavioral psychologists who apply Skinnerian principles to bring about operant conditioning in humans or to teach more complex tasks to animals may have to spend more time shaping desirable responses. At first, a response that approximates the behavior that the psychologist wants to develop is reinforced. In successive steps the rewards are made contingent on a response that more closely resembles, and finally becomes, the desired behavior. Box 2–2 presents a problem and its solution as an example of operant conditioning in action. Procedures of operant conditioning are also known as **behavior modification**.

In laboratory experiments using operant conditioning, behavioral scientists have been able to demonstrate that their conditioning procedures, not any extraneous factors, are responsible for modifying responses. They can **extinguish** responses (cause them to die out) by deconditioning and bring them back by reconditioning. Although intermittent reinforcements strengthen responses in the conditioning process, they do not guarantee that the conditioning will be permanent. In order for behaviors to last in the course of everyday living, they must continue to evoke positive rewards for the person performing them.

Programs of behavior modification have been very useful in studying child development in experimental situations (especially nursery schools) in the past. Behavior modification programs have also been used successfully to alter adult behaviors: overeating, daydreaming, smoking, alcoholism, poor study habits, and aggression. However, the same procedures that are so beneficial in helping humans overcome undesirable habits or in training animals can also be put to destructive uses, as has happened with brainwashing. In the Korean War, for example, many American soldiers were made to confess to crimes that they never committed or to accept Communist ideology by operant conditioning. They were deconditioned from their own values with punishments: various forms of physical and

BOX 2–2
Shaping Behavior

The problem: Johnny won't approach his peers in nursery school but follows the teacher or her assistants around, demanding their attention.

Step 1. Obtain an operant level. Observe and count the number of approaches Johnny makes to any peer in a certain time interval. Also count the number of advances he makes to the teacher and her assistants.

Step 2. Shape the behavior desired. To begin, each time Johnny plays alone without following a teacher or demanding adult attention, reinforce him with attention. Ignore him when he makes advances toward the teachers. As soon as he can spend short periods of time alone, make reinforcement contingent on a higher level of behavior (such as watching his peers play). When this is achieved, make the reward of attention contingent on a still higher level of behavior (such as interacting with peers) until Johnny reaches the desired level of behavior.

Step 3. Now make the reinforcement schedule variable rather than continuous. Never reward undesirable behavior, but use an intermittent reinforcement schedule for the conditioned behavior (in this case interacting with peers).

mental torture. They were then reinforced for espousing the desired beliefs: that they had committed crimes, that Americans were evil imperialists. Behavior modification has also been cast in a negative light in futuristic novels such as George Orwell's *1984* and Aldous Huxley's *Brave New World.* Humans are portrayed as becoming so conditioned or standardized that they lose their free will and individuality. Although conditioning can conceivably be put to evil uses, it is also a technique with potential for producing many desirable behaviors in humans and animals. Skinner, in *Beyond Freedom and Dignity* (1971), advocated its use to eliminate problems such as aggression, wastefulness, and injustice in our society.

Social-Learning Theories

Social-learning theories evolved out of behaviorism. They differ from behaviorist theories in that they do not hold that reinforcement is always necessary for learning. In operant conditioning procedures reinforcements must be carefully controlled. No rewards must be given for undesirable acts, and the desired behaviors must be attended to continuously at first. Some semblance of a response must also occur naturally before it can be shaped into a more desirable response with reinforcement. **Social-learning theory** accounts for learning that takes place when responses, or some approximation of them, do not occur naturally. Neal Miller and John Dollard (1941) suggested imitation as an explanation of how novel behaviors are acquired. This idea was demonstrated and expanded by many psychologists in the following two decades.

BANDURA'S LEARNING BY MODELING

The writings of Bandura and Walters (1963), Bandura and Rosenthal (1966), and Bandura (1969) exemplify the belief in learning by imitation, or **modeling**. Many of the experiments of Bandura and his colleagues involved observing the behaviors of children after they had watched models demonstrating various actions or feelings. Models were varied, from real life, to filmed, to cartoon or animal models. They were nurturant, hostile, or neutral at different times and behaved under conditions of positive, neutral, or negative reinforcement. Children were found to imitate the behavior of models more frequently if the models were perceived to be powerful or if some form of positive reinforcement was given to the models for the behaviors or emotions they demonstrated. Some modeled behaviors elicited *similar* behaviors rather than exact imitations. Much important learning was observed to take place vicariously (experienced by one person through his or her imagined participation in another person's experience). Further, some behaviors could be inhibited or extinguished if children observed that they brought about undesirable consequences for another person (see Figure 2–6).

There are many things that children and adults may perceive as positive reinforcers for given behaviors: material rewards (money, food, gifts), praise, nurturance, inclusion in a group, attention, even criticism if it is seen as a form of attention. Bandura identified two other kinds of reinforcement that may influence social learning: *vicarious reinforcement* and *self-reinforcement.* In vicarious reinforcement the person imitating some modeled behavior obtains pleasure from the act not because he or she is rewarded but because the model or some other person was rewarded. The reinforcement of the model takes the place of one's own reward. In self-reinforcement the learner actually rewards himself or herself for behaviors or emotions deemed meritorious. Self-administered reinforcers are considered very important in the learning of complex social behaviors.

Bandura's theory moves beyond the suggestion that imitation accounts for the acquisition of novel behaviors to an explanation based on observational learning, or "no-trial" learning, in which behaviors are not necessarily imitated to be learned. Children (and adults) are exposed to a vast number of social behaviors that they learn without overt modeling. They watch behaviors on television, in movies, and in real life of parents, peers, teachers, neighbors, and the like. They listen to interactions. They read about interactions.

These exposures result in observational learning whenever conditions are right.

Bandura (1977) has suggested four intervening conditions between a modeled (observed) event and its acquisition (or learning): attention, retention, motor reproduction, and motivation. Learning may not occur if there is inadequate attending, inadequate retention, inadequate motor reproduction, or insufficient motivation to learn. Bandura has identified many subcomponents of these four component processes underlying observational learning. Adequate attending, for example, requires the observer to be aroused, to be able to take in information through the senses, to be able to perceive it, and

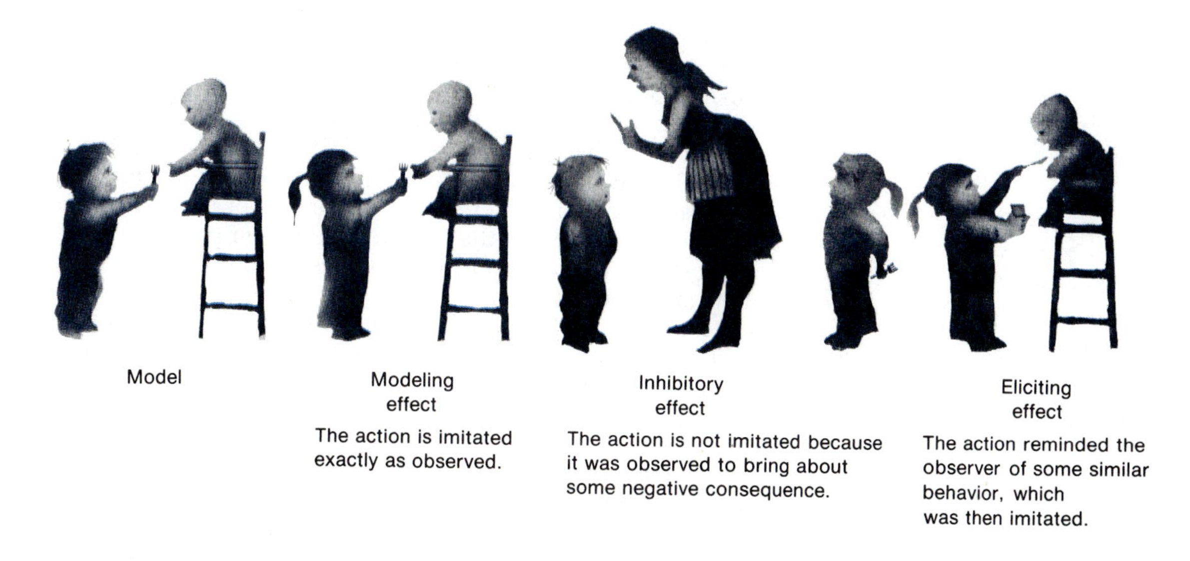

Figure 2–6
Effects of modeling on social learning.

to have some past experiences of reinforcement for similar attention. The model, in order to maintain the observer's attention, should engage in behaviors that are not too complex but that are distinctive, should create a positive affect in the observer, and should be perceived as effective. Adequate retention requires some cognitive processes. The observer must be able to organize what he or she is observing, retain it in some symbolic way (visual image, verbal), and rehearse it in some way (verbal, visual, or motor). To have adequate motor reproduction of the modeled behavior, the observer must have the physical abilities to perform the action, some self-

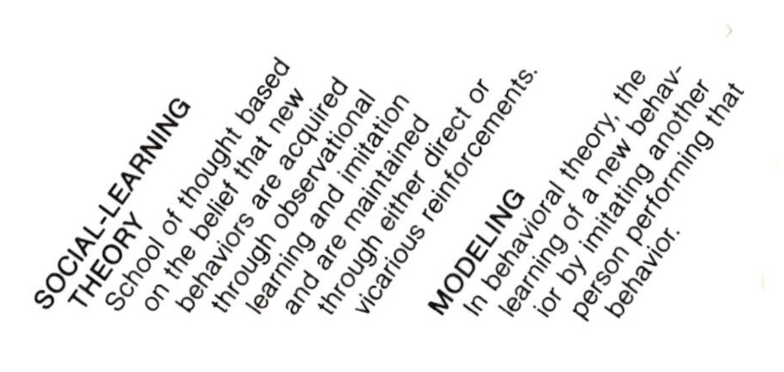
SOCIAL-LEARNING THEORY School of thought based on the belief that new behaviors are acquired through observational learning and imitation and are maintained through either direct or vicarious reinforcements.

MODELING In behavioral theory, the learning of a new behavior by imitating another person performing that behavior.

monitoring of the quality of the reproduction, and some feedback as to its accuracy. Finally, to be sufficiently motivated to learn through observation, the person must have some anticipation of a desirable reinforcement for an overt performance of the behavior. The expected reinforcement can be a vicarious reinforcement, a self-reinforcement, or an external reinforcement.

Bandura does not view the person as a mere point between some environmental stimulus and some conditioned response. He suggests a process of reciprocal determinism in learning: The person, the behavior, and the environment all interact with each other. Each affects the others in any actions or reactions. The person, for example, shapes the environment and the behavior; the behavior shapes the person and the environment; the environment shapes the person and the behavior.

SEARS'S DEVELOPMENTAL LEARNING

Robert Sears applied social-learning principles to the socialization of children. Sears felt that parents are the earliest and most frequent models for children's imitations. Consequently, their behaviors and their child-rearing practices determine the nature of the child's development. Children do what they see and hear their parents do, not just what their parents tell them to do. Physical punishment is, in Sears's view, an ineffective way to change children's behaviors. In addition to leaving a child in conflict about the misbehavior and the parental retaliation, it serves as a model of parental aggression that will be imitated by the child.

Sears and his colleagues (1965) proposed three phases of social learning from infancy through childhood. In the earliest form of social learning, which he called **rudimentary behaviors**, the infant is motivated to interact with others to avoid pain and reduce the displeasures brought about by physiological processes (hunger, elimination, temperature regulation). When these needs are met, the infant learns which behaviors (crying, kicking, grasping, smiling) produce the most rapid reduction of pain and which ones are least effective in obtaining gratification. Each infant behavior becomes associated with a parental response that affects the quality of the infant's next repetition of that behavior and with the quality of the parents' next response to it. In other words, Sears suggested that behavior is both the cause and the effect of later behavior. Intervening between a behavior and a response are variables associated with the parents' attitudes about child-rearing, parenting practices, and the home environment. Individual differences emerge in each parent-infant pair in both the quality of their interactive responses and in the value of the reinforcers they use with each other.

Sears stressed the importance of looking at *dyads* (units of two) in order to understand the process of developmental learning. In his second phase of social learning, which he called **secondary behavioral systems**, these dyadic interactions lead to expectations of behaviors on the part of both participants (usually mother and child). Discrete behaviors become linked together into a behavioral system in which one behavior acquires the power to elicit the other behaviors that are linked in the same chain. For example, a hunger call not only brings mother and food but also a whole chain of interactive behaviors that the particular dyad engages in while eating. Gradually a mother will give more positive reinforcement (affection, praise) for the child's efforts to imitate the family's behavioral patterns. The child relies on the family for these reinforcements and imitates increasingly more complex social behaviors (self-feeding, dressing, toileting).

Sears (1957) used social learning by imitation to explain how identification occurs. The child, dependent on the mother, father, or primary caregiver, comes to enjoy her or his affectionate nurture. The child imitates this powerful and rewarding model. The act of imitation itself then acquires a secondary rewarding (reinforcing) value. Imitating the gestures and actions of the loved one will bring pleasure to the child in the loved one's absence. This imitation then becomes habitual and is known as **identification**.

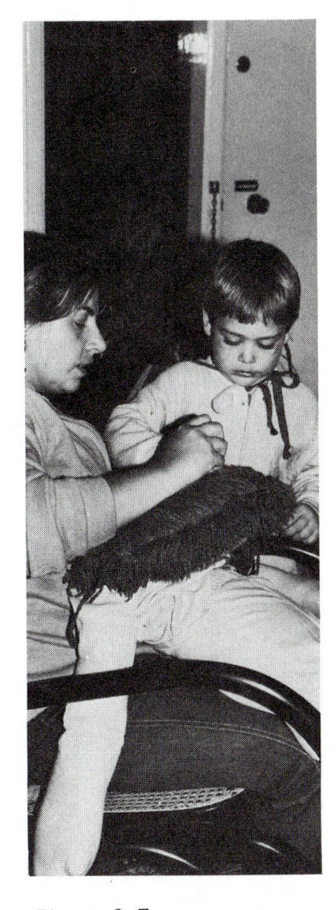

Figure 2–7
Identification grows out of imitation of a powerful and rewarding model.

Like the psychodynamic theorists, Sears and his colleagues (1965) found anxiety and psychic tension to be motivators of behavior. Learning, however, is very important. Any activity that reduces the child's level of psychic tension is satisfying. Behaviors that precede or accompany a tension-reducing act acquire secondary rewarding (reinforcing) value and are thus imitated and learned. Much accidental learning takes place this way.

Sears stressed that the expectancies of the primary caregivers should be in keeping with the maturational level of the child. The learning process is enhanced when the child strongly identifies with the primary caregivers and is rewarded for compliance with requested behaviors (see Figure 2–7). The strength of identification varies positively with:

1. The amount of affectionate nurture given the child.
2. The severity of the demands placed on the child.
3. The extent to which the primary caregiver uses withdrawal of love as a disciplinary technique.
4. The amount of absence of the persons with whom the child identifies.

Sears's third phase of developmental learning, that of acquiring **secondary motivational systems**, begins when the child moves beyond reliance on the family for reinforcements of behavior. Persons in schools, peer groups, and networks such as religious, sports, mu-

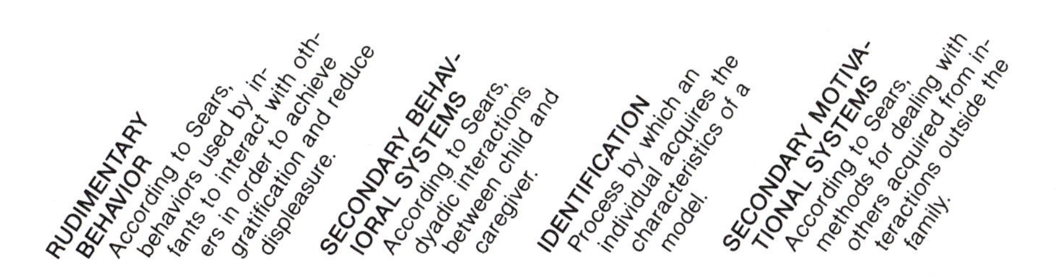

sic, or service groups become potential reinforcers of behavior. The child is motivated to learn and imitate many of the behaviors of group members to obtain gratifications. By this time the child can also use self-reinforcement as much as desired and becomes quite self-reliant.

Social-learning theorists are continually modifying their theories to account for the acquisition of more and more behavioral phenomena. Because they limit their research to carefully controlled experiments, observe ethical considerations, and report their findings completely, they are highly respected. However, social-learning theories tend to ignore what they cannot explain in behavioral terms or demonstrate in experimental situations. Because past experiences, especially the highly charged emotional experiences that are the concern of the psychodynamic theorists, cannot generally be investigated in experimental situations and because any recounting of them is so subjective, they are intentionally left unstudied.

Cognitive-Developmental Theories

Cognitive-developmental theories deal with perception and thinking but concentrate on the process of *how* an individual comes to perceive, think about, and understand his or her environment. They emphasize biological changes and maturation and only peripherally touch on emotional aspects of behavior. The most systematic explorations of **cognition** (the process of knowing and perceiving) were carried out by Jean Piaget between 1920 and 1980.

PIAGET'S GENETIC EPISTEMOLOGY

Piaget conducted most of his research on cognition in Geneva, Switzerland, using children and adolescents as subjects. He began his career as a biologist, earning a Ph.D. for a study of mollusks. He then found employment in Paris, standardizing intelligence tests for humans. He became fascinated with the confusing way children reason to arrive at answers to questions. His biological background led him to question how it is that humans come to adapt to their world and to reason with mature, objective, and abstract logic. Thus began his investigations of cognitive development and **genetic epistemology** (the study of the origin, nature, method, and limits of knowledge). Piaget was more concerned with how we come to know what we know than with our actual range of stored information. How does the infant who "knows" the world through reflex actions (root-

ing, sucking, grasping, looking) come to eventually understand mature hypothetical concepts? Piaget's theory explains the qualitative differences in methods of thinking used by children and adolescents rather than enumerating the quantity of right answers that children can provide at different ages.

Piaget used his own three children as subjects for many of his early descriptive studies. He was an astute observer of changes in their behavior that indicated a more advanced stage of intellectual development. In addition to observing, he invented many little problems or games to test their reasoning abilities. For example, when his daughter Jacqueline was five months and five days old, Piaget (1962) reported:

> I then tried the experiment of alternately separating and bringing together my hands as I stood in front of her. She watched me attentively and reproduced the movement three times. She stopped when I stopped and began again when I did, never looking at her hands, but keeping her eyes fixed on mine.
>
> At 0;5(6) (five months, six days) and 0;5(7) (five months, seven days) she failed to react, perhaps because I was at the side and not in front of her. At 0;5(8), however, when I resumed my movement in front of her, she imitated me fourteen times in just under two minutes. I myself only did it about forty times. After I stopped, she only did it three times in five minutes. It was thus a clear case of imitation. (pp. 15–16)

Organization and Adaptation Piaget, like the social-learning theorists, found that imitation is one of the important skills that children have to help them learn about and adapt to the adult world. He saw **organization** and **adaptation** as two basic functions of all organisms. In their adaptation to the world, humans practice the interrelated processes of *assimilation* and *accommodation*. **Assimilation** refers to an individual's taking in (perceiving) some new pieces of information from the environment and using the current cognitive framework to deal with the information. **Accommodation** refers to modifications of the current cognitive framework that the individual must make due to the impact of some newly assimilated pieces of information. Previous concepts may be radically altered in the process of accommodation. Children are continually revising and refining their thought processes through assimilation and accommodation, thus becoming cognitively more mature.

Equilibrium/Disequilibrium Piaget introduced the concepts of **equilibrium** and **disequilibrium** to help explain cognitive development. Equilibrium refers to a state of relative balance between assimilation and accommodation of environmental stimuli. In equilibrium one does not have to disturb one's thought processes too much to assimilate the information at hand. However, when a relative state of equilibrium exists, the little inconsistencies in one's knowledge are more readily apparent. When inconsistencies cannot be accounted for by preexisting concepts, and when the mind must work to form new concepts into which the information can fit, a state of disequilibrium is present. The mind continually fluctuates between relative equilibrium and disequilibrium as learning takes place. This portion of Piaget's theory parallels the concept of *stimulus-response (S-R) learning* of behavioral theory. The information that upsets a state of equilibrium is comparable to a stimulus, for which a response (new learning accommodation) is necessary to restore the balance.

Development by Stages Piaget postulated that intellectual development occurs by stages (see Table 2–2). At the end of each hypothetical stage, children attain a feeling of near equilibrium in their assimilation and accommodation of environmental events. The remaining inconsistencies then serve to usher in a new phase of higher learning. (Each of these stages is explained in more detail in

Table 2–2 Piaget's Four Stages of Cognitive Development

Stage	Substages	Characteristics
Sensorimotor period	Reflexes Primary circular reactions Secondary circular reactions Coordination of secondary schemas Tertiary circular reactions Invention of new means through mental combinations	The apparatus of the senses and of the musculature become increasingly operative.
Preoperational period	Preconceptual phase Intuitive phase	The child has the emerging ability to think mentally.
Concrete operations period		The child learns to reason about what he or she sees and does in the here-and-now world
Formal operations period		The individual has the ability to see logical relationships among diverse properties and to reason in the abstract.

the cognitive development sections of Chapters 4, 5, 6, and 7.) Piaget found that progression through these stages is gradual and orderly. In his opinion, the development of intelligence is biologically determined. Every child goes through all of the stages at his or her own pace.

Piaget's goal was to study epistemology—how we come to know things—not education. However, he preferred educational programs that allow children free exploration of materials rather than providing them with too many structured learning experiences. Through the manipulation of materials and the experience of experimenting with them, children assimilate and accommodate the aspects of the stimuli that are novel to them yet familiar enough to be fitted into preexisting schemas. A **schema**, in Piaget's theory, is a basic cognitive unit (such as an activity or a thought process) that the child or adult is capable of performing. Many concepts that adults try to teach children in formal lessons either may lack novelty and thus be repetitious and boring or may be too novel to fit into any of the categories of knowledge that the children have so far acquired.

Piaget's Clinical Method Piaget arrived at many of his conclusions about cognitive development by asking different questions of each child, rather than by uniformly testing a large number of children on the same concepts, and by observing children in natural settings, as he did his own. Questions can be used to explore a child's reasoning abilities in depth without suggesting that any answers are right or wrong. Questions can be kept at a level that each child understands, and any spontaneous, interesting answers can be pursued. In short, this clinical method allows the examiner to explore each child's unique reasoning processes with very few constraints. Many of Piaget's students at the research institute in Geneva have used this approach. One of his students (Opper, 1966) stated that it takes at least a year of daily practice before an examiner can achieve any proficiency at this method of clinical questioning.

Piaget is recognized as "the great child psychologist of the twentieth century" (Cohen, 1983). His descriptions of the development of reasoning abilities during infancy, childhood, and adolescence are rich and voluminous (he authored over 30 books and more than 400 papers). However, he asked no standardized questions, collected no systematic observational data, and did not submit his research to any kind of statistical analysis. Further, many of his beliefs have been empirically questioned. Infants, for example, seem

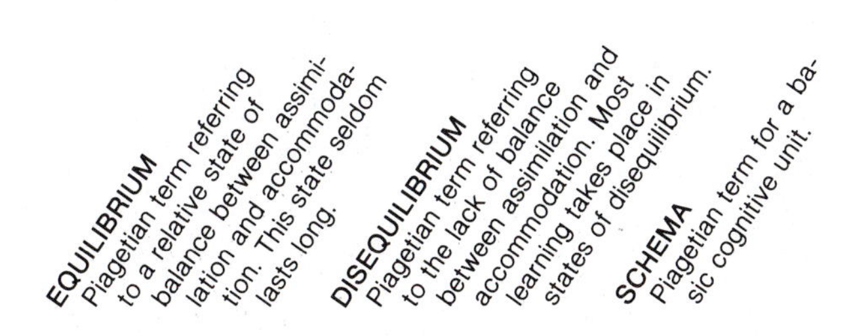

to be more intelligent than he allowed, and adolescents seem to be less logical. Piaget also neglected to consider social influences on development. He disregarded Freudian concepts such as the unconscious and psychosexual conflicts. He looked only at children's rational utterances, not at their irrational behaviors. He collected data predominantly on white subjects from the area around Geneva. He did not indicate whether he found sex differences, cultural differences, or differences related to factors such as the adequacy of nutrition. In spite of drawbacks, replications of his studies have supported much of his theory of cognitive development. The techniques of conceptualizing that some children acquire early in life remain relatively unchanged as each child assimilates and accommodates new information and attains cognitive maturity. Although Piagetian-stagelike acquisition of concepts characterizes some children, some cognitive domains, everyday cognition, and initial reactions to input, there is a decidedly nonstagelike character in other cognitive domains, in formal test situations, and in intermediary reactions to input (Flavell, 1982). Some children also have their own individual styles of problem solving that do not conform to the Piagetian stage model.

CONTRIBUTIONS OF OTHER COGNITIVE-DEVELOPMENTAL THEORISTS

Jerome Bruner (1966) suggested that cognitive development can be viewed as an acquisition of increasingly more complex modes of representation of external objects, events, and experiences. He proposed three representational systems: the enactive, the iconic, and the symbolic. These modes are acquired in the fixed order given but do not necessarily supersede each other. All modes remain active and involved in information processing throughout life. At birth, the infant has an innate ability to exercise certain reflexes (grasping, head turning, looking, listening, sucking) that enable him or her to process some information from the environment. These behaviors and other sensory and motor activities that mature in the first two years (smelling, tasting, touching, crawling, walking, pushing, pulling, manipulating objects) allow for an *enactive mode* (through activity) of representing information. Soon the toddler acquires the ability to form mental images (icons) of objects, events, and experiences to store in his or her memory. Bruner calls this the *iconic mode* of representation. Finally, language is learned and a symbol (word) can be used for processing and storing information—the *symbolic mode.* As the child acquires each of these modes of representation, his or her internal structures for processing information become more sophisticated, with more use of rules and strategies.

Bruner's theory, in addition to describing internal cognitive changes, stresses the external environmental incentives for change. Parents and teachers need to challenge children to exercise their representational strategies through exploration and discovery learning. They need to encourage children to use language to classify objects and events into related mental categories and to be ever more efficient in symbol use. They should encourage children to attend to more than one dimension of objects, events, or problems and to generate alternative solutions to problems. Children's cognitive development is enhanced by increased language and symbol usage (as with computer programming languages, for example). However, children must adjust their behavior to fit cultural conventions (not all adults are conversant in computer languages). Cultures that encourage sophisticated symbol use by children also encourage greater cognitive development.

Two Russian psychologists, A. R. Luria and L. S. Vygotsky, working on cognitive studies concurrently with Piaget in the 1920s and 1930s, preceded Bruner in their emphasis on the role of language in cognition. Vygotsky (1934/1962) felt that without language thinking would be impossible. At first an infant's behaviors are controlled by the directions and speech of the adult caregivers. As the child acquires language, he or she begins to control some actions with self-directions given overtly (publicly). Later the child controls his or her own behavior with covert (private) language. Thinking and reasoning then develop, based on internalized language, or **inner speech**. Luria (1976) tested areas of cognition such as perception, generalization, deduction, reasoning, imagination, and ability to analyze one's own inner life. He found that people in more primitive social structures had thought processes that were concrete, based on practical activities, and centered on situational, person-oriented, or object-oriented experiences. More technological systems produced people whose cognitions were more abstract, more objective, and more theoretical and who had more self-awareness and social awareness. Luria theorized that cognitive development must be viewed in terms of the dynamic environment to which each individual is exposed. Specifically, the use of a language that allows one to abstract, codify, and generalize signs and objects is linked with more complex cognitive processes (Luria, 1982).

Lawrence Kohlberg extended Piaget's theory of how children learn the moral standards of their social order. Both proposed that children move from moral egocentricity toward social consciousness as they mature. However, whereas Piaget wrote of two basic moral orientations (that of **moral realism** in which rules are seen as sacred and unalterable and that of **moral relativism** in which rules are seen as modifiable to bring the greatest good to the greatest number), Kohlberg (1969) described six stages of moral development (see Table 2–3). In a twenty-year longitudinal follow-up study, Colby, Kohlberg, and others (1983) documented the basic assumptions of Kohlberg's theory and the stage sequence he proposed. Acquisition of the highest levels, Kohlberg wrote, depends on previous cogni-

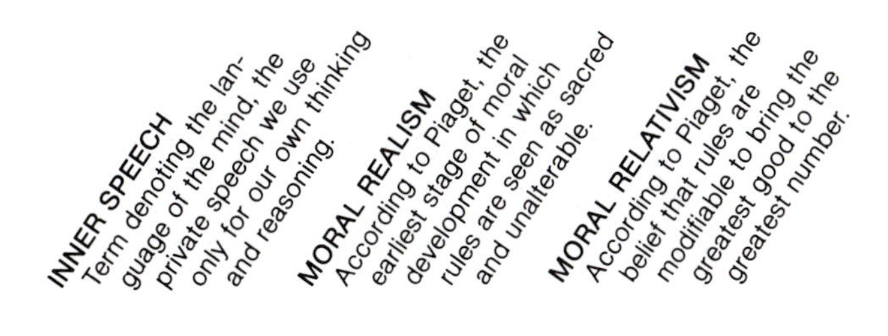

Table 2–3 Brief Summation of Kohlberg's Stages of Moral Development

Level	Stage	Characteristics
Premoral	Stage 1	Punishment and obedience orientation
	Stage 2	Acts that are satisfying to self and occasionally satisfying to others defined as right
Conventional morality	Stage 3	Morality of maintaining good relations and approval of others
	Stage 4	Orientation to showing respect for authority and maintaining social order for social order's sake
Postconventional morality	Stage 5	Morality of accepting democratically contracted laws
	Stage 6	Morality of individual principles of conscience—doing what seems right regardless of reactions of others

tion, experience, and acquisition of each of the lower levels. Kohlberg also found that moral judgments are positively correlated with age, socioeconomic status, IQ, and education. Each of Kohlberg's stages will be discussed in more detail in Chapters 6 and 7.

Jane Loevinger (1976) has proposed that both moral and ego development can continue into adulthood. She suggests a six-stage adult continuum, which can be used to characterize different personality types. Up through the adolescent years the ego is impulse ridden. The individual's concerns center on acceptance, approval, the reactions of others, blame, guilt, and punishment—the *impulse stage.* In a subsequent developmental stage, the *opportunistic* or *self-protective stage,* the person is concerned with self-control out of self-interest. Rules are obeyed in order to avoid getting into trouble or to avoid becoming dominated by some moral superior. Continued development leads to a *conformist stage,* where the individual follows rules to avoid any sense of shame or disapproval from the whole social order. Loevinger believes that some adults continue to turn inward, toward more self-criticism. In this *conscientious stage* the person uses self-standards to measure the rightness or wrongness of behaviors. Ideals, however, may change when the social context

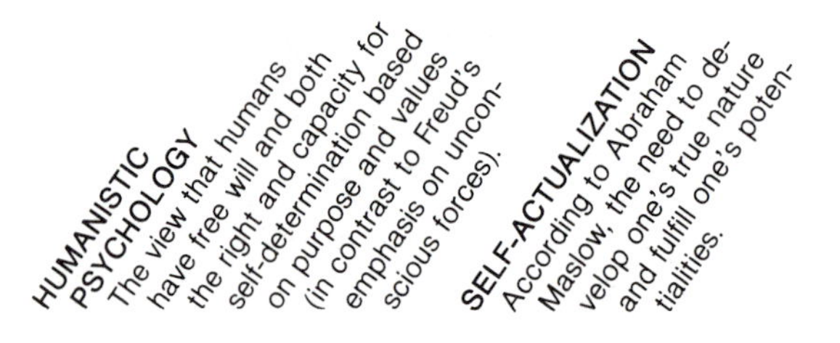

changes. The next stage, the *autonomous period,* occurs when the adult can comprehend that inconsistencies or contradictions in traits or behaviors occur and may not necessarily be judged good or bad. During this stage it is easier for the adult to accept himself or herself and others even when behaviors are not up to some preconceived standards. A final stage, Loevinger suggests, may be achieved by about 1 percent of the adult population. At this *integrated stage* the person not only accepts the differences between self and others but also enjoys these variances.

Humanistic Theories

There is no one master builder of *humanistic theory,* as Freud was of psychodynamic theory and Piaget of the study of cognition. It grew in part out of existential philosophies such as those of Sören Kierkegaard, Jean-Paul Sartre, and Martin Heidegger. Humanistic theories are also often called *phenomenological* theories, referring to the fact that subjective reality—what a person perceives his or her world to be—is more important than external reality. Let us define what is meant by **humanistic psychology** and then briefly discuss some of the contributions of select humanistic theorists.

The humanistic viewpoint is that people are born basically good and strive throughout their lives to become all they are capable of being. The ways in which people define their experiences within their family and their society may either hinder or enhance their potential for growth. All humans are different because of their varying experiences and the unique meanings that these experiences have for them. Consequently, humanists believe that theory builders, philosophers, educators, parents, and the like should not try to fit human beings into roles, stages, or categories and predict how they should or will behave.

A central tenet of humanistic theory is that people cannot love others unless they first love themselves. Child-rearing, according to the humanists, should involve fewer rules, fewer roles, and fewer expectations of what constitutes a "good father," a "good mother," or "good children." Each member of the family unit needs freedom to grow and become what he or she can be. Guilt feelings, hostilities, punishments, and the like are counterproductive to the development of healthy personalities. Conversely, when family members feel loved and accepted for what they are, their defenses will disappear and they will feel freer to pursue joy, love, rational behavior, and self-actualization.

MASLOW'S HIERARCHY OF NEEDS

Abraham Maslow exemplified the positive approach of the humanists. In Maslow's view, basic needs are organized into a hierarchy of relative potency. At the top of the hierarchy is the need for **self-actualization**, a state of being open, autonomous, spontaneous, accepting, loving, creative, and democratic—in short, being a happy, fulfilled human being. Unfortunately, self-actualized people are rarely found, because this state is not manifest in people who have lower-order needs unfilled. The most potent (most basic) needs are

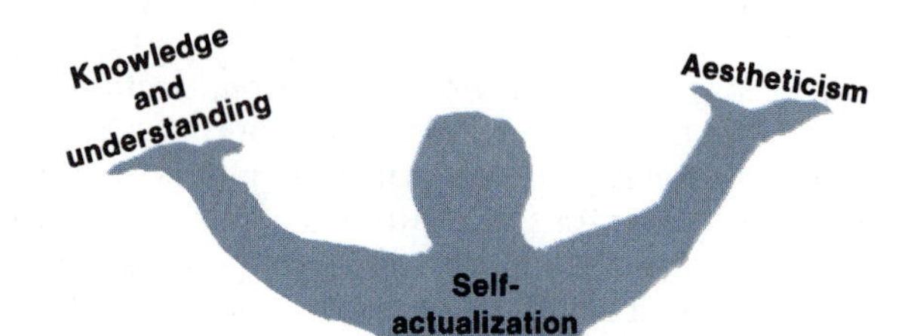

Figure 2–8
Maslow's needs hierarchy, from lower-order requirements to higher-order desires.

the physiological requirements for food, water, and shelter (see Figure 2–8). Maslow (1954) saw the need for safety as secondary to physiological necessities. While most North Americans have their physiological needs fulfilled (with the exception of street people and those living in abject poverty), many persons do not feel safe. The high crime rate contributes to this insecurity, but so does the fierce competition in schools, colleges, and places of employment. Many people feel in jeopardy of losing their positions in situations where only the "fittest" survive.

If and when safety needs are satisfied, a person seeks out kinship and belonging and a sense that he or she is loved, wanted, and needed. Maslow saw many adults in contemporary society struggling to fill the intermediate-level requirement for love and belonging that is prerequisite for the higher-order sense of self-esteem. Maslow also saw many people struggling with the need for a greater sense of self-esteem, self-confidence, prestige, and recognition from others. One may, Maslow felt, also have resurgences of lower needs, which predominate until they are satisfied. One does not have to accomplish great deeds or wield power to have self-esteem. One simply needs to feel satisfied with one's own role and skills and feel that others also find his or her accomplishments adequate. Given this sense of competence, a person can finally be motivated to achieve the higher-order need of self-actualization.

Self-actualized people are focused on problems outside themselves. They are characterized by passions for justice, honesty, equality, and personal freedoms. Their friendships tend to cut across racial, ethnic, age, sex, educational, political, and social class barriers. Maslow saw self-actualized people not as normal people with something extra but rather as normal people with nothing taken away. He stressed the potential for everyone to become self-actualized given attainment of their prepotent needs for food and water, safety, love and belongingness, and self-esteem. Maslow saw two other higher-order needs that motivate behavior in some people: the need to know and understand, and the need for art and beauty (aesthetics) in one's life.

Maslow (1965) included in his theory the concept of **peak experiences**, which he defined as moments of intense happiness or ecstasy with a mystical quality. During a peak experience one has a momentary sense of unity with the whole world. The grass seems greener, the sky bluer; the secrets of living seem within reach, and one feels more noble, free, and at peace. Maslow felt this phenomenon of peaking is within the reach of everyone, not only highly creative or successful persons. Some people may not recognize the brief peak experience when it occurs.

Maslow stressed healthful behaviors and attainment of one's full potential. He encouraged people to help one another move toward the highest level of growth through mutual acceptance:

> If we would be helpers, counselors, teachers, guiders, or psychotherapists, what we must do is to accept the person and help him learn what kind of person he is already. What is his style, what are his aptitudes, what is he good at, not good for, what can we build upon, what are his good raw materials, his good potentialities? (Maslow, 1968, p. 693)

The holistic health movement today is an application of Maslow's needs theory. Rather than focusing on the treatment of ill health, holistic medicine anticipates people's lower physiological needs and higher psychosocial needs. It concentrates on preventing illness (both physical and mental) by helping people assume responsibilities for safeguarding their health by adequately meeting all of their needs.

ROGERS'S PHENOMENOLOGICAL APPROACH

Another important contribution to the humanist movement has been made by Carl Rogers. He believes that we all help shape our own personalities and that in order to change we must acknowledge to ourselves what we are. Thus, self-acceptance marks the beginning of change. He developed the method of **client-centered therapy** in

PEAK EXPERIENCE Maslow's term for a momentary state of ecstasy and a sense of unity with the whole world.

CLIENT-CENTERED THERAPY A nonjudgmental, nondirective, humanistic psychotherapy developed by Carl Rogers.

which the therapist refrains from making value judgments or giving advice. Instead the client is helped to discover for himself or herself what behaviors to change and how to change them.

Central to Rogers's personality theory is the **phenomenological approach**, the belief that only the person involved can know his or her own phenomenological world. (*Phenomenological* refers to what is apparent to the senses and is perceived or experienced as real, rather than the external reality of the thing itself.) What is important for understanding a person's behavior is knowledge of how he or she perceives and experiences the phenomenological world that constitutes his or her "reality." Rogers (1951) believes that as experiences occur in one's life they are handled in one of three ways:

1. They are perceived and organized into a congruent (corresponding, agreeing) relationship with the self.
2. They are ignored because they are perceived to have no relationship with the self.
3. They are denied perception and organization (symbolization) or distorted in their symbolization because they are inconsistent with the self.

When one's phenomenologic field of experiences is predominantly congruent with one's self-concept, there is a relative freedom from strain and anxiety and one is said to be well-adjusted (see Figure 2–9). However, when many experiences are distorted in their symbolization or are denied awareness, there is psychological tension and one is said to be maladjusted.

Rogers (1961) found that the best vantage point for understanding a person's perceptions and organizations, distortions, or denials of experience is the individual's own internal frame of reference. As a therapist, he attempts to facilitate the individual's abilities

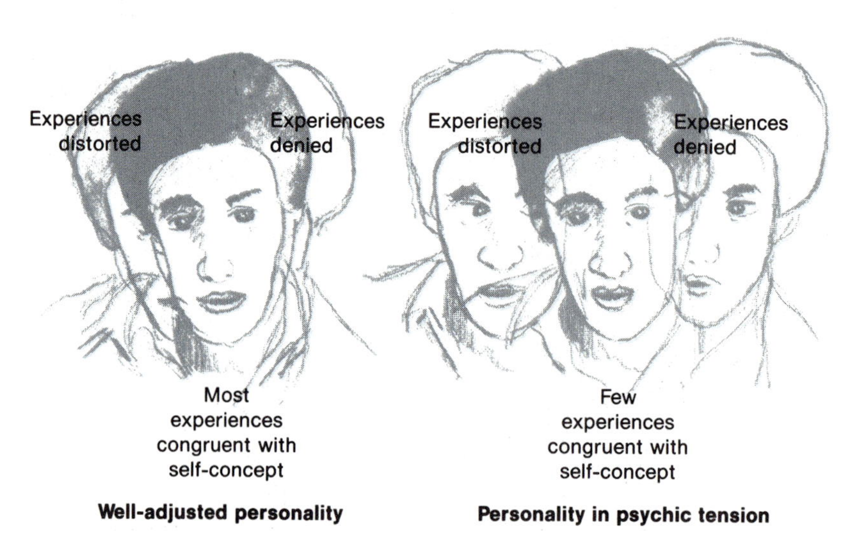

Figure 2–9 *Schematic representation of a person's varied reactions to phenomenological fields of experiences.*

to discover incongruencies, relate hurts, pinpoint problems, and decide on action by using *active listening*. He tries to see the world of the other person as that person sees it. He never offers a negative evaluation of others but maintains *unconditional positive regard* for them. His theory proposes that, if people can stop imposing their negative opinions, their directions, and their value systems on the individual, that person can, through organization of personal experiences, learn to become his or her best self.

CONTRIBUTIONS OF OTHER HUMANISTIC THEORISTS

The humanistic bent of Albert Ellis led him to develop an approach called **rational-emotive theory**. Ellis believes that neurotic behaviors are often the result of irrational thought processes (Ellis and Harper, 1961). Many people, he feels, talk nonsense to themselves. They convince themselves that they are afraid, unattractive, or in some way inferior. Ellis suggests that it is not the event (A) that causes an emotional reaction (C) but rather the phenomenological world (B) known only to the person experiencing it (what one thinks, how one interprets a happening, what one feels) that is the real cause of behavior. This A-B-C theory is schematically represented in Figure 2–10.

Ellis (1974) lists four human life goals: to survive, to be happy, to get along with members of a social group, and to relate intimately with a few select members of this group. These four goals regulate an enormous part of one's life. Humans tend to make assumptions

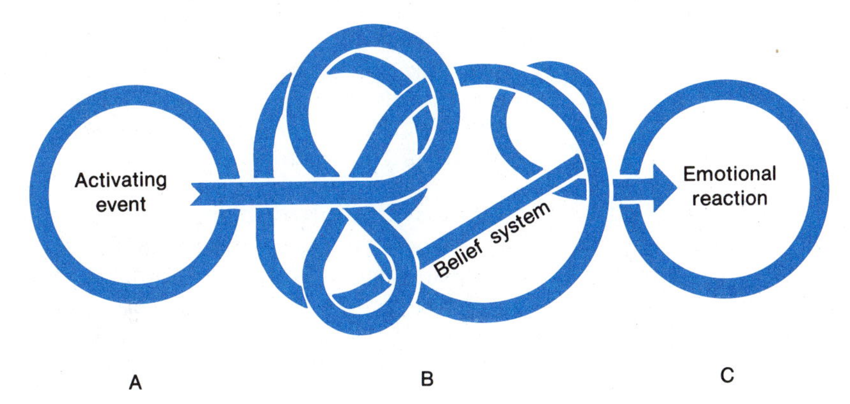

Figure 2–10
Schematic representation of Ellis's A-B-C theory of personality. The event (A) does not cause the emotional reaction (C). Rather, one's beliefs (B) about A cause the reaction (C).

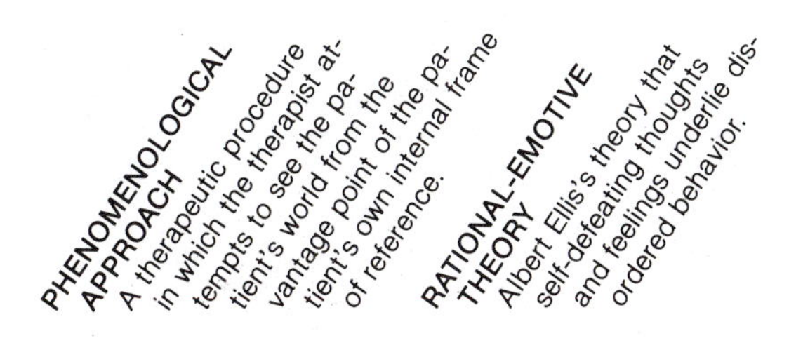

about how to achieve these goals and to evaluate meanings attached to these goals. Some of these valuations are rational (based on sound judgment); others may be irrational (based on faulty logic). When a person's belief systems are based on many irrational assumptions, their emotional reactions to events are disturbed. Neurotic, maladaptive emotional responses, Ellis believes, can be modified or eliminated by changing belief systems.

Gordon Allport (1937) built a theory of personality based on **traits**. Traits are consistent modes of behavior. There are *common traits* that characterize all members of a cultural group to varying degrees, such as the Navajo's endurance or the Hopi's accommodation. There are *cardinal traits,* such as Martin Luther King's quest for freedom and equality, that are all-pervasive and dominant (a ruling passion) in the person's life. There are *central traits,* such as those mentioned in letters of recommendation (honesty, industriousness), which are consistently seen in an individual. Finally, there are *secondary traits,* characteristic ways of behaving that are of minor significance and usually known only to close associates (such as fear of speaking in public). No two persons ever have precisely the same trait. Each trait is distinct and individualized within its possessor. Each personality is also comprised of a cluster of traits, unique to that individual alone.

Allport's theory of personality stresses the concept of a *functional autonomy of motives* as well as traits. Simply stated, a motive that is originally based on one tension or set of tensions (for example, the need to do well in school to earn allowance from parents) can become autonomous (self-governing). One can eventually be motivated to engage in a behavior simply because carrying it through brings a sense of self-satisfaction. Personality, Allport believes, continually develops and changes in response to hereditary, temperamental, social, and psychological factors.

Comparisons of Theories and Their Influence

There are many other theorists who have made important contributions to the study of human development. Interested students may especially want to read about the Gestalt theories of Wertheimer, Köhler, and Koffka; the dynamic field theory of Lewin; the perceptual development theory of the Gibsons; the ethological theory of Lorenz, Tinbergen, and Bowlby; the maturational theory of Gesell; the organismic theory of Werner; the sociobiological theory of Wilson; or the ecological theory of Bronfenbrenner. Portions of some of these theories are presented elsewhere in this textbook. Most theories do not so much contradict each other as speak to separate aspects of the developmental process. Let us review and compare the theories we have covered, and their areas of special influence.

TRAITS Distinctive and enduring characteristics of a person or his or her behavior.

Psychodynamic theories are oriented toward explaining the reasons for development of abnormal behaviors in psychosexual and psychosocial aspects of living. They look to the past to explain present behaviors. They have been influential not only in the development of therapies to ameliorate abnormal behaviors of children and adults but also in the development of advice for parents on aspects of child-rearing that will promote healthy personality development.

Behaviorist theories are oriented toward understanding the development of responsive behaviors to various stimuli. They look at the pattern of reinforcers in a person's environment to explain specific behaviors. They have been influential in helping us understand not only how specific behaviors are acquired but also how specific behaviors can be extinguished and how new behaviors can be gradually shaped and learned through reinforcers. Behavior theory can be applied to the learning of thousands of specific behaviors, which can be either classically or operantly conditioned. This theory is not as useful in explaining such aspects of learning as cognitive processes, creativity, or universal behaviors characteristic of the whole species. It intentionally ignores concepts such as the unconscious, the ego, libido, and defense mechanisms.

Social-learning theories are oriented toward understanding the development of behaviors for which reinforcers may not obviously be present. They focus on observational learning experiences within the social world and identification processes. They have been influential in helping us understand the role of modeling in behavior acquisition. They bridge the gap between psychodynamic personality theories and cognitive-developmental theories, incorporating concepts of both psychic tensions and cognitive processing. They have given us a great deal of information on more positive ways to rear children using concepts of modeling, reinforcements, and identification strength.

Cognitive-developmental theories are oriented toward explaining how humans acquire and process information and become knowledgeable about the world around them. They look for changes in cognitive processes from infancy, through childhood and adolescence, and into adulthood. They have been influential not only in helping us learn what to expect qualitatively from children, adolescents, and adults at various cognitive levels but also in helping us to understand how to challenge and stimulate learning within human capabilities.

Humanistic theories are oriented toward explaining the maintenance of healthy personalities. They focus on the here-and-now aspects of behavioral functioning, on personal experiencing, and on the meanings that the individual assigns to feelings, events, and subjective phenomena. By studying healthy personalities, humanists have been influential in showing us the potential that all people have for enhancing their emotional lives.

Summary

Theories are not facts. They are "best guesses" put forth to explain a phenomenon or set of phenomena. There are many theories of human development. Most are systematic statements of the principles involved in and the apparent relationships that exist for understanding human behavior. This chapter has not covered all theories of human development but has tried to present a flavor of some of the popular theories, including the psychodynamic, behaviorist, social-learning, cognitive-developmental, and humanistic.

Psychodynamic theories usually focus on the impact and influences of past experiences. Freud described unconscious motivators of behavior, the id, ego, and superego, defense mechanisms, and stages of psychosexual development. Erikson proposed eight nuclear conflicts of life, the first three paralleling Freud's first three stages of psychosexual development, the latter five stressing interactions within society. Jung emphasized lifelong personality development, a collective unconscious, and the concepts of introversion and extraversion. Adler emphasized felt inferiority and the mechanism of compensation. Anna Freud stressed ego development, and Karen Horney stressed basic anxiety. Sullivan focused on the impact of interpersonal relationships.

Unlike psychodynamic theorists, behaviorists pay little attention to past experiences. They effect behavior changes after relatively short periods of behavior therapy using conditioning. Watson popularized the use of classical conditioning to effect behavioral changes in human subjects. Skinner popularized operant conditioning and stimulus-response (S-R) learning. He used positive reinforcement (reward) to change behavior. Some behaviorists also use punishment.

Social-learning theorists explain how social behaviors are acquired in childhood in the absence of direct reinforcement. Bandura proposed modeling as a fundamental means for acquiring new behaviors. Sears added identification to explain the learning of new social actions.

Cognitive-developmental theories are more concerned with cognition than social behaviors. Piaget stressed the need for maturation and experience for cognitive growth. He described the learning processes of assimilation and accommodation and postulated four stages of cognitive development. Bruner suggested three representational systems to explain increased cognitive awareness and more sophisticated information processing: the enactive mode, the iconic mode, and the symbolic mode. Luria and Vygotsky emphasized the role of language in cognitive development. Kohlberg extended Piaget's theory to explain how children acquire moral behaviors. Loevinger emphasized the continuous cognitive changes that take place in moral and ego development and internalization of moral judgments in adulthood.

Humanists study healthy personalities to see how people constantly strive to be all they can be. Humanistic theories stress people's goodness and need for freedom in order to become open, spontaneous, and independent. The ideal, as described by Maslow, is to be *self-actualized.* Rogers stressed the phenomenological sense of self and each person's abilities to change behaviors in a supportive, nonjudgmental climate. Ellis wrote that people should learn to be more rational in their belief systems in order to become fully functioning. Allport contributed the idea of individualized clusters of traits in personalities and the functional autonomy of motives.

Questions for Review

1. Imagine you would like to see a therapist for some problem you are having. Would you choose a psychodynamic therapist, a behaviorist, or a humanist? Explain.
2. Compare Freud's psychodynamic theory with behaviorism. Are there any shared beliefs? In what areas are the viewpoints diametrically opposed?
3. Do you agree or disagree with the humanistic view that child-rearing should involve fewer rules, fewer roles, and fewer expectations of what constitutes a good father, a good mother, or good child, thus allowing for the self-actualization of each person involved? Discuss.
4. Contrast psychodynamic theory with humanistic theory. What beliefs are shared? Did psychoanalysis fuel the humanistic movement? Why or why not?
5. Develop an eclectic (composed of material from various sources) theory of your own. Choose ten ideas from this chapter that best characterize your beliefs about human development. Upon which theoretical orientation have you leaned most heavily?

References

Adler, A. *The practice and theory of individual psychology.* New York: Humanities Press, 1971. (Originally published, 1929.)

Allport, G. *Personality: A psychological interpretation.* New York: Holt, Rinehart & Winston, 1937.

Bandura, A. *Principles of behavior modification.* New York: Holt, Rinehart & Winston, 1969.

Bandura, A. *Social learning theory.* Englewood Cliffs, N.J.: Prentice-Hall, 1977.

Bandura, A., and Rosenthal, T. Vicarious classical conditioning as a function of arousal level. *Journal of Personality and Social Psychology,* 1966, *3,* 54–62.

Bandura, A., and Walters, R. *Social learning and personality development.* New York: Holt, Rinehart & Winston, 1963.

Bruner, J. S. *Toward a theory of instruction.* Cambridge, Mass.: Harvard University Press, 1966.

Cohen, D. *Piaget: Critique and assessment.* New York: St. Martin's Press, 1983.

Colby, A.; Kohlberg, L.; Gibbs, J.; and Lieberman, M. A longitudinal study of moral judgment. *Monographs of the Society for Research in Child Development,* 1983, *48* (1–2, Serial No. 200).

Ellis, A. Rational-emotive theory. In A. Burton (Ed.), *Operational theories of personality.* New York: Brunner/Mazel, 1974.

Ellis, A., and Harper, R. *A guide to rational living.* Englewood Cliffs, N.J.: Prentice-Hall, 1961.

Erikson, E. H. *Childhood and society,* 2nd ed. New York: Norton, 1963.

Erikson, E. H. *Identity, youth and crisis.* New York: Norton, 1968.

Flavell, J. H. On cognitive development. *Child Development,* 1982, *53,* 1–10.

Freud, A. *The ego and the mechanism of defense.* (C. Baines, trans.) New York: International Universities Press, 1946.

Freud, S. *Introductory lectures on psychoanalysis.* (J. Strachey, Ed. and trans.) New York: Norton, 1966. (Originally published, 1920.)

Horney, K. *The neurotic personality of our time.* New York: Norton, 1937.

Jones, M. C. A laboratory study of fear: The case of Peter. *Journal of Genetic Psychology,* 1924, *31,* 308–315.

Jung, C. *Psychology of the unconscious.* New York: Dodd, Mead, 1931.

Jung, C. *Analytical psychology.* New York: Pantheon, 1968 (originally published 1916).

Kohlberg, L. Stage and sequence: The cognitive-developmental approach to socialization. In D. A. Goslin (Ed.), *Handbook of socialization: Theory and research.* Chicago: Rand McNally, 1969.

Loevinger, J. *Ego development: Conceptions and theories.* San Francisco: Jossey-Bass, 1976.

Luria, A. R. *Cognitive development: Its cultural and social foundations.* Cambridge, Mass.: Harvard University Press, 1976.

Luria, A. R. *Language and cognition.* New York: Wiley, 1982.

Maslow, A. *Motivation and personality.* New York: Harper Brothers, 1954.

Maslow, A. Lessons from the peak experience. In R. Farson (Ed.), *Science and human affairs.* Palo Alto, Calif.: Science and Behavior Books, 1965.

Maslow, A. Some educational implications of the humanistic psychologies. *Harvard Educational Review,* 1968, *38,* 685–696.

Miller, N., and Dollard, J. *Social learning and imitation.* New Haven, Conn.: Yale University Press, 1941.

Opper, S. Personal communication, November 12, 1966.

Piaget, J. *Play, dreams and imitation in childhood.* (C. Gattegno and F. Hodgson, trans.) New York: Norton, 1962.

Rogers, C. *Client-centered therapy.* Boston: Houghton Mifflin, 1951.

Rogers, C. *On becoming a person.* Boston: Houghton Mifflin, 1961.

Sears, R. Identification as a form of behavioral development. In D. B. Harris (Ed.), *The concept of development.* Minneapolis: University of Minnesota Press, 1957, 149–161.

Sears, R.; Rau, L.; and Alpert, R. *Identification and child-rearing.* Stanford, Calif.: Stanford University Press, 1965.

Skinner, B. F. *Walden two.* New York: Macmillan, 1948.

Skinner, B. F. *Beyond freedom and dignity.* New York: Knopf, 1971.

Sullivan, H. S. *The interpersonal theory of psychiatry.* (H. Perry and M. Gawel, Eds.) New York: Norton, 1953.

Vygotsky, L. *Thought and language.* (E. Hanfmann and G. Vakar, trans.) Cambridge, Mass.: MIT Press, 1962. (Originally published, 1934.)

Watson, J. B. *Behaviorism.* New York: Norton, 1924.

Watson, J. B. *Psychological care of infant and child.* New York: Norton, 1928.

Watson, J. B., and Raynor, R. Conditioned emotional reactions. *Journal of Experimental Psychology,* 1920, *3,* 1–4.

♥

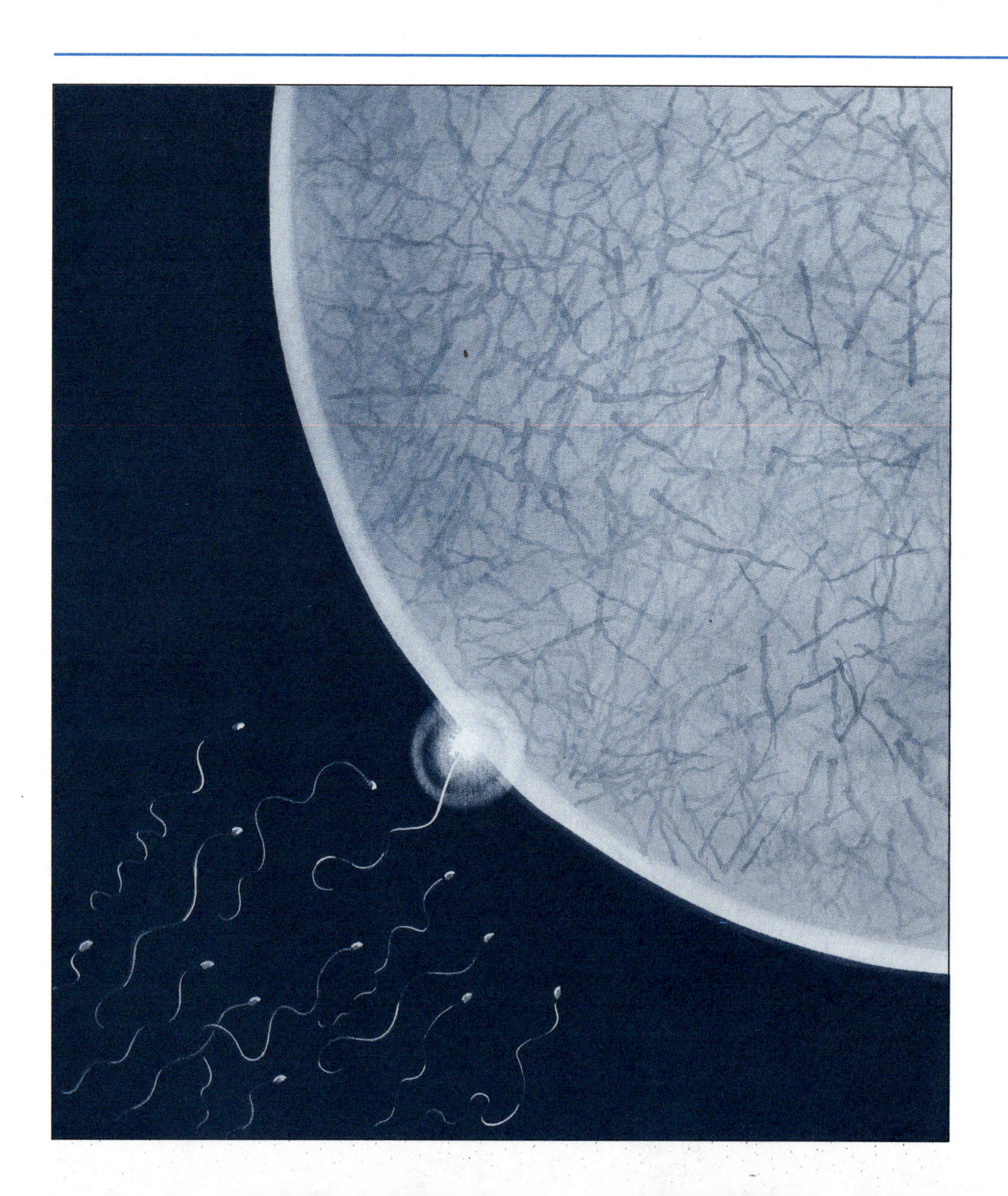

Prenatal Development

3

Chapter Outline

Prenatal Physiological Processes
- *Precursors of a New Human Being*
- *Genetic Inheritance*
- *Stages of Prenatal Development*

Possible Prenatal Defects
- *Chromosomal Abnormalities*
- *Genetic Abnormalities*
- *Environmental Hazards*

Birth
- *Stages of Labor*
- *Presentations*
- *Aids to Delivery*
- *Possible Birth Difficulties*

Key Concepts

ABO incompatibility
alleles
amniocentesis
androsperm
anoxia
antibodies
antigens
autosomes
blastocyst
Braxton-Hicks contractions
Caesarean section
centromere
chromosomal aberrations
chromosomes
corpus luteum
critical period
crossing over
cystic fibrosis
deoxyribonucleic acid (DNA)
dilation and effacement
diploid
dominant-recessive relationship
Down syndrome
embryonic period
endometrium
episiotomy
estrogen
fetal period
flagellum
forceps
galactosemia
gene abnormalities
genes
genotype
Graafian follicle
gynosperm
haploid
heterozygous
homozygous
implantation
Klinefelter syndrome
labor
Lamaze method of delivery
LeBoyer method of delivery
meiosis
Mendelian law
mitosis
molding
morula
mosaicism
natural family planning
neonate
ovulation
phenotype
phenylketonuria (PKU)
placenta
placental barrier
precipitous delivery
progesterone
prolonged labor
quickening
Rh factor
sex chromosomes
sex-linked diseases
sickle-cell anemia
spina bifida
spermatogenesis
stages of labor
Tay-Sachs disease
teratogenic drugs
Turner syndrome
uterus
villi
zygote

Case Study

Alethea's first baby had been born two weeks after the expected due date. Pregnant with her second child, she showed no signs of labor on the due date either. Consequently, when her husband, Frank, said he'd like to attend an out-of-state meeting for three days, Alethea said, "Go ahead, the baby won't come for another week or two." On the evening after Frank left, however, regular uterine contractions (labor) began. Alethea and Frank had attended Lamaze classes and Frank planned to coach her through prepared childbirth. Consequently, Alethea called him to come home. Frank called back two hours later to say he could not get a plane flight, and the train would take 16 hours. Alethea's contractions stopped. They began again the following day when Frank was home. Their second child arrived safely and in good health after a normal five-hour labor and delivery with both parents participating.

One obstetrician believed Alethea's first labor had been false labor—Braxton-Hicks contractions. Another obstetrician believed that her psychological desire to share the birth experience with her husband had been strong enough to postpone the physiological contractions causing dilation and effacement of the uterus until Frank's arrival (mind over body, psyche over soma). What do you think?

Think back to the birth of a baby who meant something special to you—a sibling, a niece or nephew, a friend's child, or your own child. You probably remember the date of the birth as well as such memorable details as how you got the news, the sex, the weight, and the coloring. You may celebrate the date of birth every year, for birthdays are considered anniversaries of the first day of life. Yet for nine months prior to birth, life exists as the fertilized egg develops from zygote to embryo to fetus.

The goal of this chapter is to describe *prenatal* (occurring before birth) transformations that change the fertilized egg, which is about the size of a tiny dot, into a fully developed **neonate** (newborn infant). By the time of birth the neonate has made giant strides toward becoming unique. Except in the case of identical twins, the neonate is a creature different from all others, even before he or she takes a first breath of air. We will describe male and female reproductive anatomy and physiology to trace the course of the ovum (egg) and sperm that unite to make a new human being. We will explore genetics and cell biology to see what it is about genes and chromosomes that make all humans both similar and yet unique in structure and functions. We will also discuss a number of environmental influences, both internal and external, that may possibly influence prenatal development.

Prenatal Physiological Processes

Can you recall any of your own early naive concepts about "where babies come from?" Did you ever wonder if pregnancy resulted from kissing? From overeating? From swallowing a watermelon seed? History suggests that many early people believed that some spirit entered a woman at the moment when she first felt her baby move (Hartland, 1909). They believed that impregnating spirits came from the wind, the trees, the sun, the moon, food, and water. During the Renaissance, people believed that new babies were contained in male sperm (Needham, 1959). They needed only to be planted in a woman to grow. Babies were often considered the property of men. Although a father might have his infant destroyed if the child displeased him, the care of surviving babies was left to women. Even in the seventeenth century Leeuwenhoek, in developing and experi-

NEONATE
Newborn infant up to two to four weeks.

menting with the first microscope, believed he saw miniature human beings in sperm cells. Indeed, our knowledge of the principles of genetic inheritance after the ovum-sperm union is relatively new.

PRECURSORS OF A NEW HUMAN BEING

Human development begins with the fertilization of an ovum by a sperm. The ovum and sperm have to mature separately before they merge.

The Ovum The *ovum* (egg) originates from one of the two female *ovaries* (reproductive glands) located on either side of the **uterus** (muscular organ in which the embryo/fetus is developed and protected) (see Figure 3–1). About every thirty days during a woman's fertile years (from puberty to menopause or approximately ages 12 to 47), a mature ovum is set free to be fertilized in the oviduct if sperm are present. This event is called **ovulation**. Usually only one ovum is ovulated, but sometimes two or more are released, presenting the possibility for multiple, fraternal births. The two ovaries alternate months in which they release ova (plural of ovum). The ovaries interrupt their production of ova during pregnancy and when a woman takes birth control pills, or during times of starvation, serious illness, or stress. The maturation and release of ova are brought about by hormones that are extremely sensitive and responsive to the physical condition and emotions of the woman. If one ovary

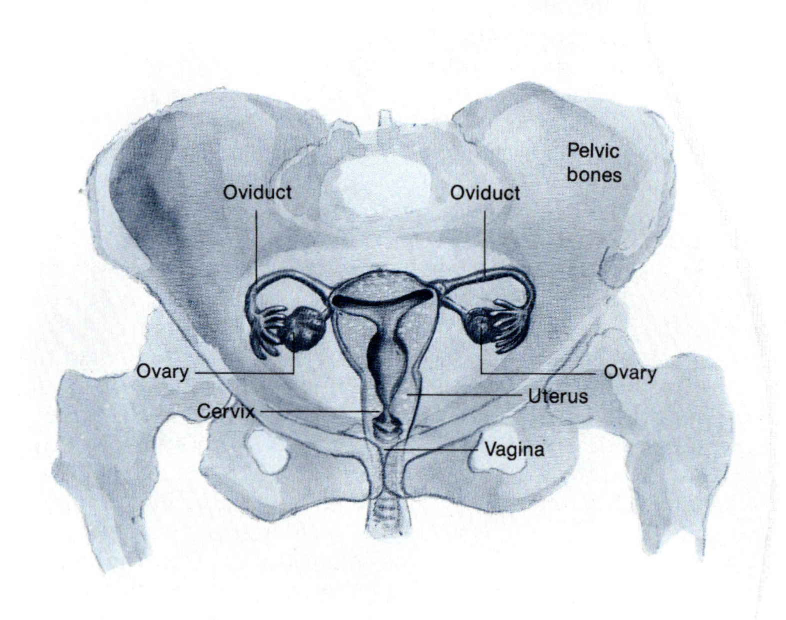

Figure 3–1
Female reproductive organs (gonads).

becomes diseased or removed, those hormones will stimulate the other ovary to take over the responsibility of releasing one ovum every thirty days on schedule. If another part of the body becomes diseased, or if the woman is under great physical or emotional stress, or is malnourished, the hormones may cause the ovaries to delay the process of maturation and release of ova temporarily, as they do during pregnancy.

The hormones that trigger ovulation are under the direct control of the hypothalamus and come from the pituitary gland in the brain. The hypothalamus regulates the production of these pituitary hormones: follicle-stimulating hormone (FSH) and luteinizing hormone (LH). Each female has approximately 700,000 rudimentary ova in her ovaries at the time of birth (Sadler, 1985), and each immature ovum is contained in a primordial (primitive) follicle. Every month from puberty through menopause from 5 to 12 of the primordial follicles begin to grow under the influence of FSH, but usually only one reaches full maturity. During the average woman's life, a maximum of about 500 follicles will fully ripen and release their ova (one per month for about 35 fertile years, subtracting months of pregnancy, birth-control pill use, and/or illnesses affecting ovarian functions). All the follicles that begin to develop each month produce another hormone, **estrogen**. This ovarian hormone triggers the release of LH by the pituitary gland. The luteinizing hormone then triggers the release of the ovum from the fully ripened follicle (called a **Graafian follicle**) and from the surface of the ovary (ovulation).

After ovulation the several partially developed follicles degenerate. Under the influence of LH, the Graafian follicle develops into a structure called the **corpus luteum** (see Figure 3–2). The corpus luteum secretes estrogen and another hormone, **progesterone**, which help the uterus prepare to accept a fertilized ovum. If the ovum is fertilized, the corpus luteum increases its production of estrogen and progesterone during pregnancy. If the ovum remains unfertilized, the corpus luteum begins to degenerate. Then, in about fourteen days, the uterus sheds, through menstruation, the lining it had prepared for the possibility of pregnancy.

As the ovum escapes from the Graafian follicle and from the surface of the ovary, it passes briefly through the abdominal cavity

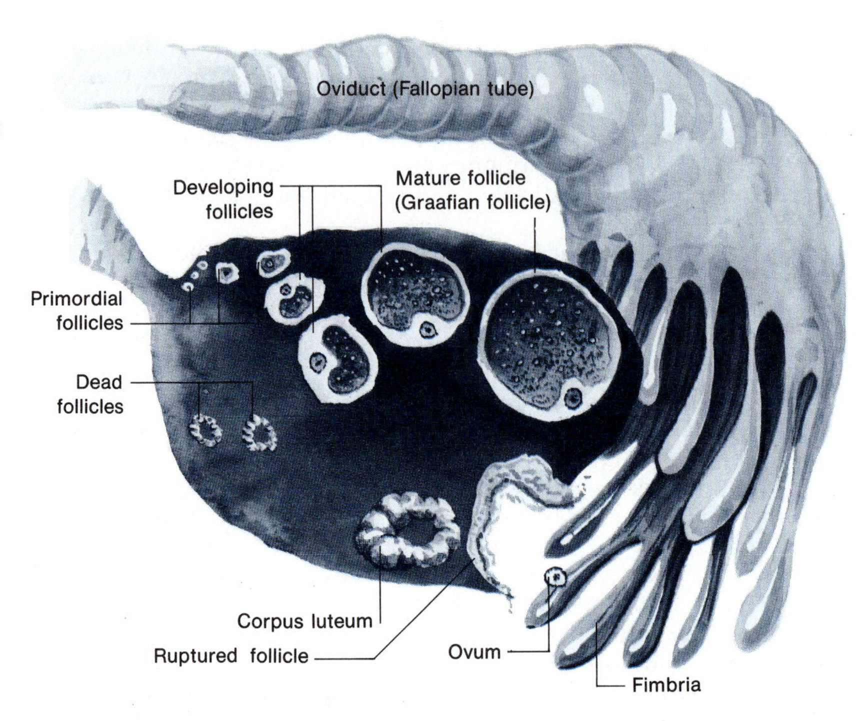

Figure 3–2 *Schematic summary of the maturation and discharge of an ovum from the ovary (ovulation).*

before it is picked up by the fimbriated (fringelike) ends of the *oviduct* (also called the Fallopian tube or uterine tube). The ovum moves slowly down the four-inch oviduct and reaches the uterus after four to seven days. Unless it is fertilized by a sperm within twelve to twenty-four hours after escaping from the ovary, it dies (Sadler, 1985). Knowledge of a woman's fertile period can be used to practice contraception and/or plan for pregnancy (see Box 3–1).

The Sperm Whereas women are born with all the ova they will ever have, men continually produce *sperm* (male sex cells) at the rate of several hundred million every few days. (Most men continually produce sperm from puberty until death.) The sperm are formed in the two *testes* (male sex glands) that lie in the scrotum suspended below the abdomen (see Figure 3–3). This location outside the abdomen is necessary for fertility because **spermatogenesis** (formation of sperm) requires a temperature lower than that of the body. The scrotum hangs loosely when warm and pulls closer to the body when cool to maintain a relatively stable temperature.

From birth, males have in their testes dormant spermatogonia or primordial germ cells that are the early precursors of sperm. Beginning at puberty these primordial germ cells grow and multiply. The differentiation and maturation process from spermatogonia to mature sperm takes two to three weeks. A sperm diminishes in size during this maturation transformation. The cell nucleus condenses and is concentrated into a "head" region, and most of the cell cytoplasm (the gluelike, semifluid matter outside the nucleus) is cast

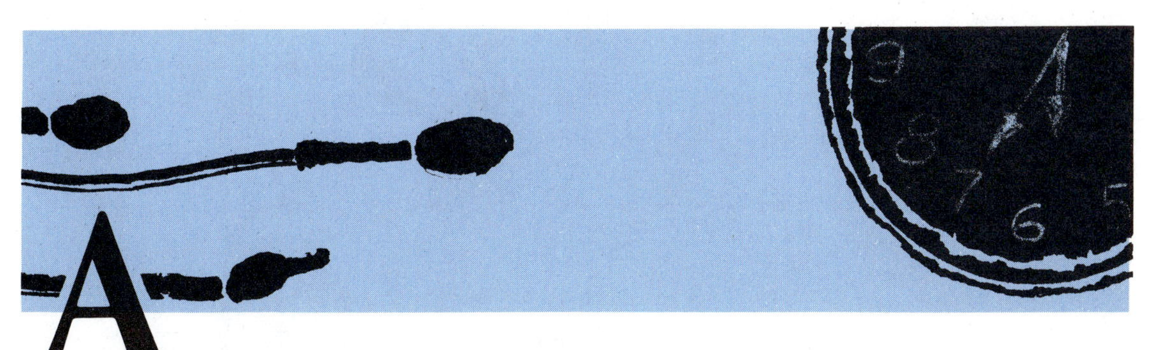

A knowledge of how to recognize one's most fertile period (around the time of ovulation) has led to several methods of natural contraception or **natural family planning**. A woman can measure her basal body temperature (BBT method), calculate her expected ovulation by the calendar, check her cervical mucous secretions (cervical mucous method), or combine these observations (symptothermic method).

In order to use the BBT method, a woman must take a daily early-morning reading of her body temperature. While follicles are ripening, prior to ovulation, BBT is slightly low, below 98.6°F. At ovulation, BBT rises about one degree and then remains elevated until shortly before the next menstrual period. A woman's BBT may also be elevated in the morning due to infection, inadequate sleep, or consumption of alcoholic beverages the night before. Thus, she must take these factors, as well as ovulation, into account when noting a rise in BBT.

The calendar method is based on the individual woman's normal menstrual cycle, which can vary from twenty-four to thirty-four days. Ovulation occurs approximately fourteen days before each menstrual period. By charting the beginning of each menses and then counting back two weeks, a woman can estimate when she ovulated retrospectively. The number of days required between the beginning of menstrual bleeding and a new ovulation counting forward is highly variable from woman to woman depending on the length of time her follicles need to mature. Ovulation can be delayed if a woman is dieting or malnourished, has an infection, has an endocrine disturbance, or has undergone unusual physical or emotional stress.

Ovulation can also be estimated as occurring when cervical mucous secretions become slippery and clear, like raw egg white. While a woman's fertile period lasts only for the life of the ovum (twelve to twenty-four hours), a man's sperm may survive in the oviduct awaiting ovulation for twenty-four to thirty-six hours. Consequently, a pregnancy may result if intercourse occurs two to three days prior to ovulation. Women trained to watch for changes in their cervical mucous secretions are better able to predict when they are two or three days prior to ovulation than are women who rely only on basal body temperature changes and the calendar.

The three methods combined (symptothermic) are more reliable than any one method alone. The natural family planning method is free and does not alter a woman's natural biologic rhythms. It has been demonstrated to be 80–90 percent effective (Tatum, 1984).

SPERMATOGENESIS The formation of mature sperm.

NATURAL FAMILY PLANNING Contraception that utilizes knowledge of ovulation for planning abstinence from intercourse.

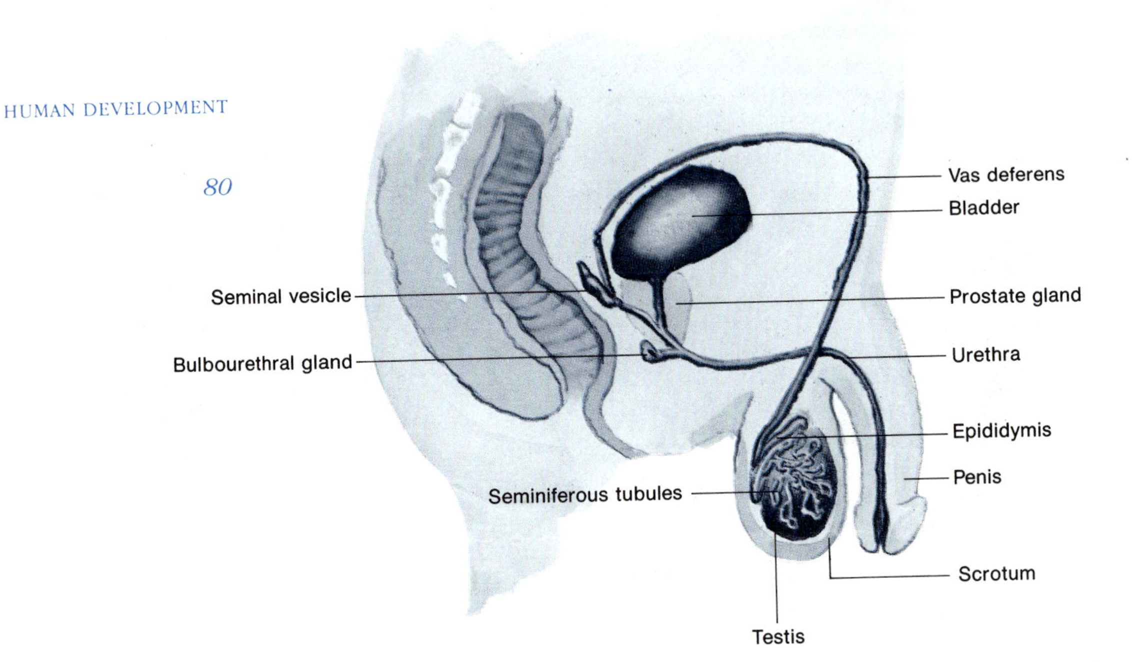

Figure 3–3
Male reproductive organs (gonads), where spermatozoa are produced, stored, lubricated, and released.

off. A long, whiplike tail, called a **flagellum** (plural = flagella), forms. This structure permits the sperm great mobility.

Fully developed sperm move from the seminiferous tubules in the testes to the *epididymis* (excretory duct at the rear, upper surface of each testis) (see Figure 3–3), where they may remain for several months. They are lubricated and acquire greater motility while in the epididymis. When ejaculation occurs, a few hundred million sperm travel from the epididymis through the long *vas deferens* (duct that carries sperm) to the penis. En route they are further lubricated by fluids from the *seminal vesicles,* the *prostate gland,* and the *bulbourethral* glands. These fluids both nourish the sperm and provide them with further means of transport. Finally, the sperm are expelled from the penis. During sexual intercourse the sperm are deposited in the vagina near the small opening to the uterus known as the cervix (see Figure 3–1). The cervix is protected by mucous secretions through which the sperm must pass. Then they must cross the uterus, enter the oviducts, and travel to the ends nearest the ovaries in order to fertilize an ovum, if present, while the ovum is still alive. While several million sperm may be deposited near the cervix, only a few hundred may reach the portion of the oviducts near the ovaries. Sperm travel fastest right after ejaculation and slow down considerably after two hours. Some sperm can be expected to reach the ovarian ends of the oviducts after about one hour, and others arrive later. The whiplike movements of the flagella help the sperm "swim." Travel is also assisted by contractile movements of the walls of the uterus and oviducts. Sperm can remain alive in the oviducts for about twenty-four to thirty-six hours waiting for an ovum (Sadler, 1985).

Sperm undergo still another transformation within the female genital tract. They lose a coating on their heads, which allows special enzymes to escape. These enzymes help break down the protec-

FLAGELLUM
Long hairlike process attached to sperm, protozoon, or bacterium that whips and thus creates movement.

tive coating on the ovum so that a sperm can penetrate it. Usually only one sperm fertilizes an ovum. On rare occasions, two sperm have penetrated the egg, resulting in an abnormal cell that dies during its development (Sadler, 1985). This occurrence is infrequent because once the head of a sperm enters the cytoplasm of the ovum a change occurs on the surface of the ovum that renders it impermeable to all other sperm.

Reduction Division One important difference between these sex cells (ova and sperm, also called germ cells or gametes) and other body cells is still to be discussed. This difference involves the way they divide and the contents of their *nuclei* (cell centers) following division. Mature germ cells have only half the normal complement of chromosomes in their nuclei. **Chromosomes** are rod-shaped structures that contain **genes**, which contain the "blueprints" for the structure and functioning of our bodies. Both chromosomes and genes will be described further in the following pages.

All the cells in the human body except ova and sperm contain forty-six chromosomes that can be arranged into twenty-three pairs. They are called **diploid** (double) cells. Mature ova and sperm contain only twenty-three unpaired chromosomes and are called **haploid** (half) cells. When fertilization occurs, the comparable chromosomes of the ovum and sperm "find each other" and can be arranged into twenty-three new pairs of chromosomes. Thus the fertilized ovum becomes a diploid cell. All the cells of the embryo and later the fetus and finally the mature human body will be formed from divisions of this fertilized ovum and will contain the same chromosomal and genetic materials as this original "parent" cell.

Body cells divide continually. In the prenatal period, cell division is extremely rapid to allow for growth of the **zygote** (fertilized ovum) into a neonate. The process by which the chromosomes in the nuclei of somatic cells divide during cell division is known as *mitosis.*

DNA and Mitosis Chromosomes are composed of compounds called *nucleotides,* which form long chains known as nucleic acids, and also of associated nucleoproteins. Each nucleotide is, in turn, composed of a phosphate group, a sugar group (*deoxyribose*), and an organic base. Strands of these phosphate-sugar-base units (nucleotides) are called **deoxyribonucleic acid (DNA)**.

In each chromosome there are two strands of nucleotides that

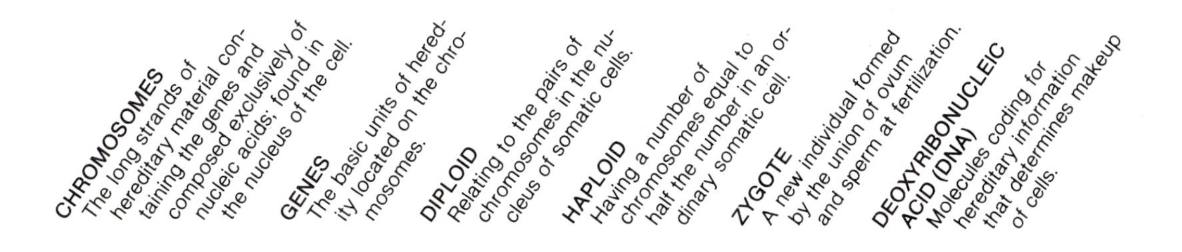

Phosphate   Sugar (deoxyribose)   One nucleotide (base + phosphate + sugar)

C — Cytosine base   T — Thymine base   A — Adenine base   G — Guanine base

Figure 3–4
Schematic representation of the deoxyribonucleic acid (DNA) molecule, showing its double helix arrangement.

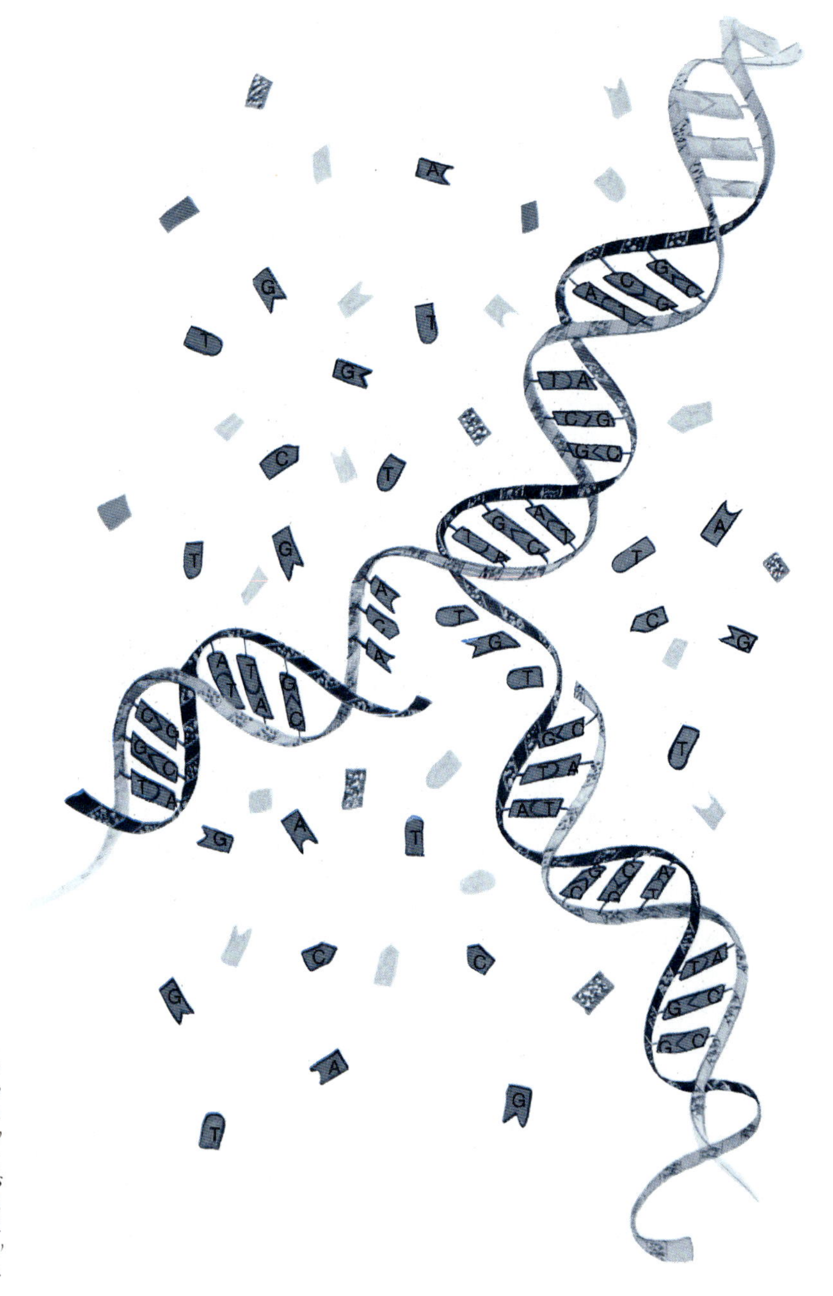

Figure 3–5
During cell division the helical strands of DNA uncoil and find new complementary bases, phosphates, and sugars in the cell material, forming new identical *strands into new* identical *chromosomes.*

wrap around each other in a double helix pattern (see Figure 3–4). Each base (of a nucleotide) in one strand faces and is bound to a complementary base on the second strand, so that the base sequence of one DNA strand determines the base sequence of its complementary DNA strand. The four bases in DNA are called *adenine, thymine, guanine,* and *cytosine.* Although these four bases can appear in various arrangements in one strand of DNA, the order of the bases is fixed on the complementary second strand of DNA. This relationship is fixed because adenine always pairs with thymine and guanine always pairs with cytosine. This base pairing is crucial to the exact replication of DNA molecules required for replication of chromosomes before cell division. The phosphate and sugar groups of each nucleotide on one strand are so attached that they contribute to the stability of the helical coils of DNA. The bases are at the center of the coils, whereas the phosphate and sugar groups are on the outside, serving as the "backbone" of the double-entwining helical structure.

In the process of replication of each chromosome, the complementary DNA strands uncoil (see Figure 3–5). New, free-floating nucleotides with complementary bases become attached to each strand. (These new nucleotides form from precursor materials within the cell.) This process results in a replication, nucleotide by nucleotide, of the two strands of complementary DNA. Each DNA strand thus serves as a pattern for its complementary strand. By this process, the new set of chromosomes carries exactly the same nucleotides in a double-helix arrangement as the original. After the process of DNA replication, the number of chromosomes per cell has doubled. The cell must then undergo a division process (called **mitosis**) to return the number of chromosomes back to the original level.

Mitosis occurs in a series of stages, termed *prophase, metaphase, anaphase,* and *telophase* (see Figure 3–6). Every chromosome separates from its replica so that a newly formed cell can have genetic material that is perfectly equivalent to that of the original cell.

Meiosis In preparation for possible later fertilization, germ cells (ova and sperm) have to undergo a *reduction* in the number of chromosomes carried in their nuclei so that, after fusion of an ovum and a sperm, the resulting product has the normal complement of forty-six chromosomes. This specialized reduction division, **meiosis**, results in only twenty-three unpaired chromosomes in each cell.

Just before the first meiotic division begins, the primitive germ cells replicate their DNA. Thus, at the start of meiosis the immature

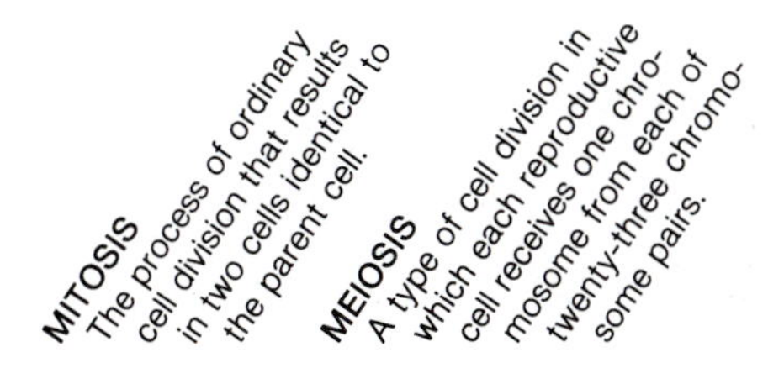

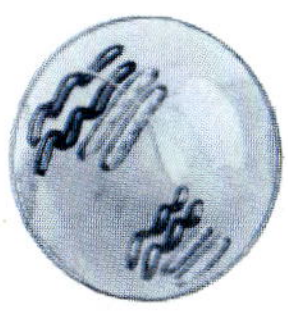

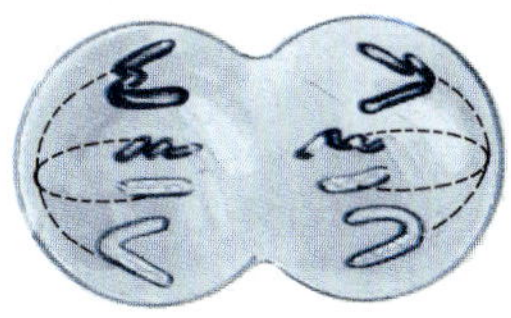
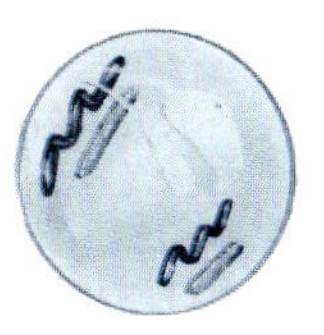
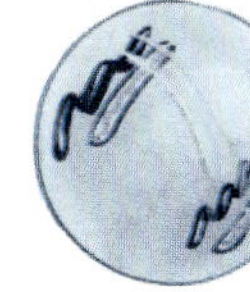

Figure 3–6
Schematic representation of cell division by mitosis. Only two pairs of chromosomes are shown per cell, rather than the full human complement of twenty-three pairs.

ova and sperm contain double the normal amount of DNA. Each of the forty-six chromosomes has a double structure, the original chromosome and its replica, bound at a structure common to both called the **centromere** (see Figure 3–7). In the first meiotic division, during prophase there is a pairing of the homologous (alike) chromosome pairs. During this characteristic lining up of the chromosomes of the germ cells, there occurs an interchange of chromosome segments, a process called the **crossing over**. This process provides an opportunity for exchange of genetic content among homologous chromosomes. During the metaphase of the first meiotic division (see Figure 3–7), there is separation of the homologous double-structured chromosome pairs. The first meiotic division continues through anaphase and telophase. After the completion of the first division, each offspring cell contains one member of each chromosome pair and has twenty-three double-structured chromosomes.

At this point the second meiotic division begins. The twenty-three double-structured chromosomes divide. Each newly formed offspring cell is referred to as a haploid (half) cell.

The purposes of the two stages of meiotic or *reduction division* are to produce germ cells with half the number of chromosomes and half the amount of DNA of somatic cells and to allow for an exchange of small amounts of genetic material between chromosomes of homologous pairs through crossing over.

In the female only one of the cells resulting from meiotic division reaches maturity. The other cells, called polar bodies, degenerate. In the male all four haploid cells become mature sperm.

In Figure 3–7 the process of reduction division is schematically illustrated for only two pairs of homologous chromosomes per cell. The same phenomena occur with the remaining twenty-one chromo-

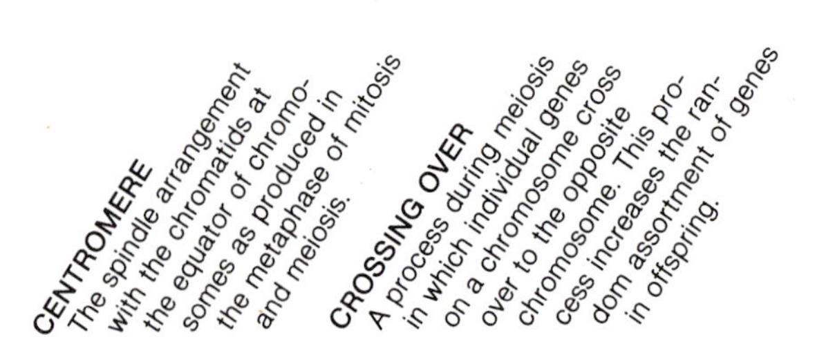

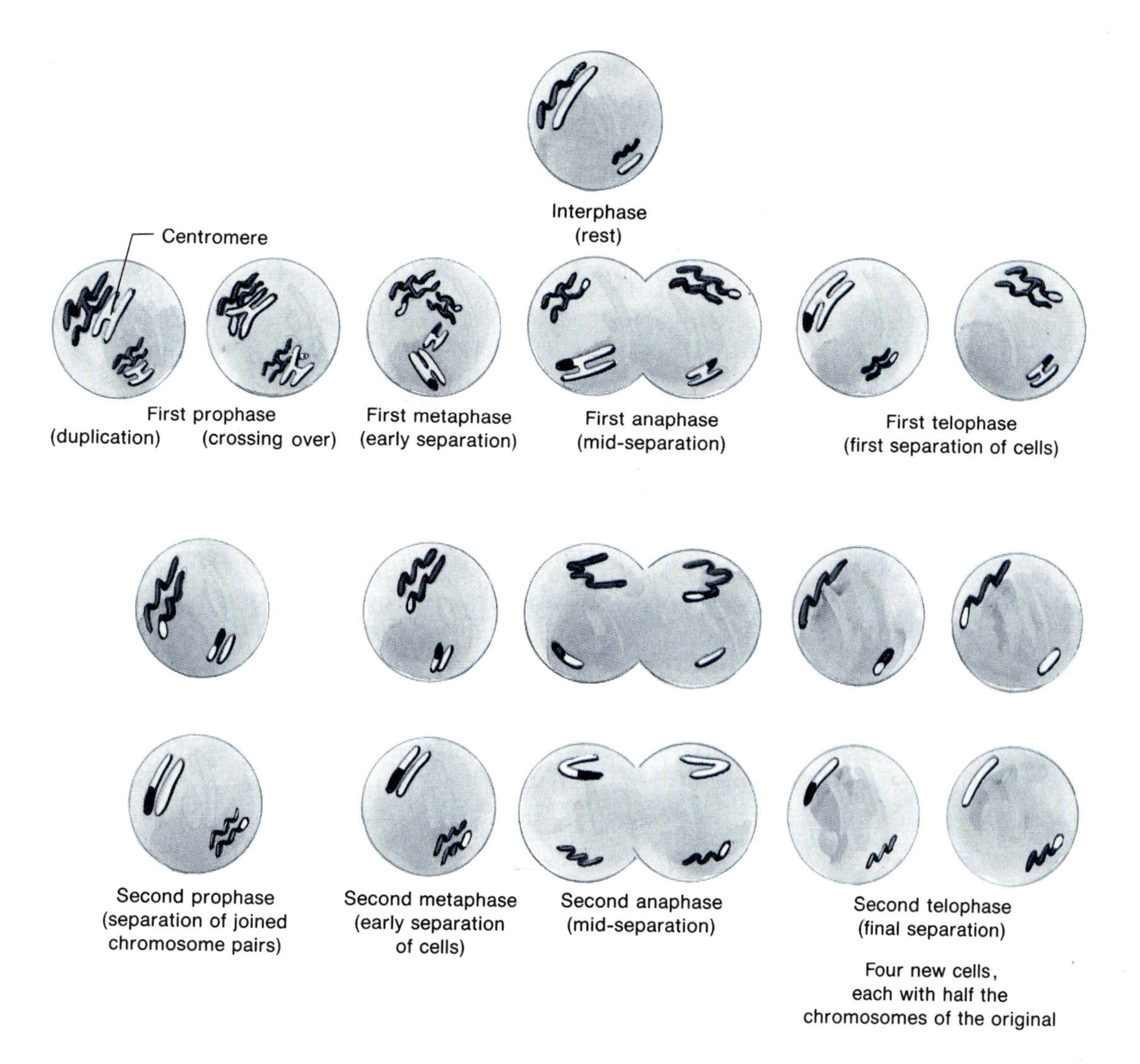

some pairs. Both crossing over and the completely random distribution of the twenty-three singlets among the four offspring cells lead to enormous numbers of different combinations of genetic material in mature germ cells. Only identical twins, who develop from one fertilized ovum, can claim to be exactly equal down to the last gene.

Figure 3–7
Schematic representation of cell division by meiosis. Crossing over may occur between some segments of the intertwined chromosome pairs during the first long and complex prophase. The final result is four haploid germ cells. In the female, only one matures to form an ovum. In the male, all four mature into sperm.

GENETIC INHERITANCE

Genes are discrete portions of the chromosomes or, specifically, of the DNA portions of the chromosomes. They carry information for building all the proteins of the body. The numbers of genes on chromosomes can only be guessed. It is possible for chromosomes to contain a large number of genes in a condensed space because of the helical coding. McAuliffe and McAuliffe (1983) report estimates of 100,000 genes in each human cell. Gene mapping (the process of finding gene locations on numbered chromosomes) is now being done by geneticists and is currently proceeding at an unprecedented pace with perhaps 200 additional gene locations being designated per year.

Genes carry messages for protein functions that determine the many activities of cells. Some genes even direct proteins that, in turn, control the functions of other genes. In this chapter only the role of genes in transmitting "inherited" characteristics will be covered. In humans, in all but the mature germ cells, there are twenty-two pairs of **autosomes** (all chromosomes other than the sex chromosomes) and two **sex chromosomes**, two X chromosomes in females and one X and one Y sex chromosome in males. Identical regions of each pair of chromosomes contain genetic information for the same function. These identical regions are called **alleles**. If the genes in the allelic portions of a chromosome pair contain different information as to how a trait should appear (for example, one contains information coding for blue eyes, the other for brown eyes), the paired genes are **heterozygous** (different). If both genes code for the same information, they are **homozygous** (alike) for the trait.

Mendelian Laws of Heredity The laws of heredity were initially worked out by an Austrian monk named Gregor Mendel in the late 1800s, long before anyone knew about genes. The principles of homozygosity and heterozygosity are still referred to as **Mendelian law** even though knowledge of the essential elements of inheritance has vastly improved since Mendel's time. He worked out his laws with sweet pea strains that he grew in his monastery garden. The following illustration of Mendelian law uses eye color as an example.

A person with brown eyes may be homozygous for brown eyes. If so, that person is both genotypically brown eyed (**genotype** refers to one's genetic make-up for a trait) and phenotypically brown eyed (**phenotype** refers to the *visible* expression of inherited traits).

A person with brown eyes may be heterozygous for brown eyes rather than homozygous. While the *phenotype* (visible expression) of the genes may be a brown color, one region on one of the paired chromosomes may carry a message for blue eye color. Paired heterozygous genes can work in a **dominant-recessive relationship**. The dominant gene overshadows the effect of the recessive gene in the phenotype. However, the recessive gene remains in the genotype and can be passed on, unaltered and uncancelled, to offspring. Genetic information determining brown eyes is dominant over such information for blue eyes.

Suppose you have brown eyes phenotypically but are heterozy-

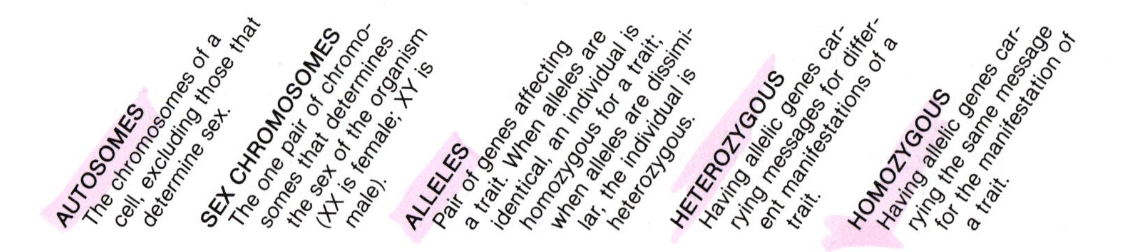

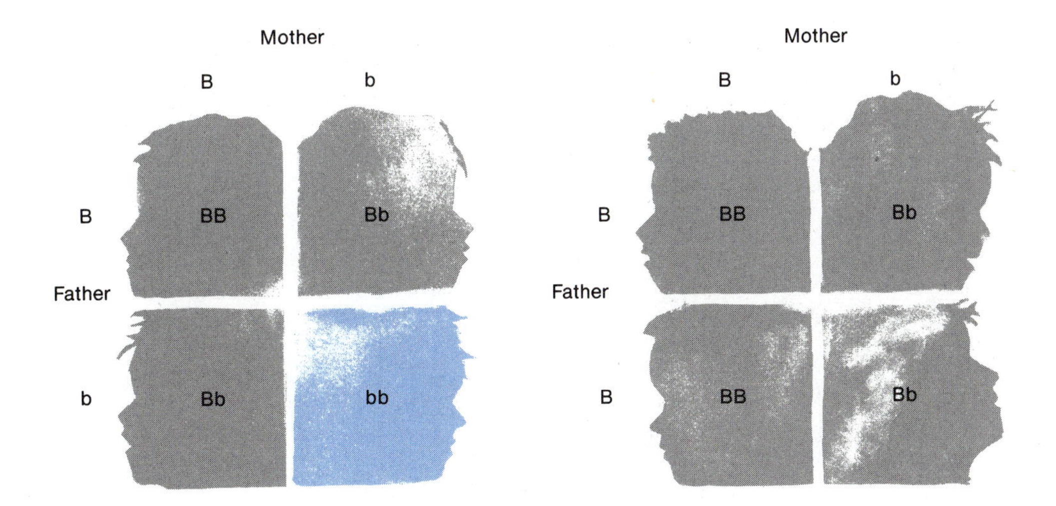

gous and carry genes for both brown and blue eye color. Further, suppose you mate with a person who also carries genes for both brown and blue eyes. In meiosis, 50 percent of your respective mature germ cells will carry genes for blue eyes. However, the chance of two phenotypically brown-eyed but genotypically heterozygous persons producing children with blue eyes is only 25 percent (see Figure 3–8). Any combination of a dominant brown gene pairing with a recessive blue gene will produce a brown-eyed child.

Suppose you, as a heterozygous brown-eyed individual, mate with a homozygous brown-eyed person (see Figure 3–9). All of your offspring will have brown eyes like their parents. However, theoretically 50 percent of them will carry recessive genes for blue eyes that they may pass on to their offspring.

What if you mate with a partner who is homozygous for blue eyes? What color will the eyes of your offspring be? Theoretically, half of them should have blue eyes and the other half will be phenotypically brown eyed but genotypically heterozygous for blue eyes as well (see Figure 3–10).

All of the offspring of parents who are both homozygous for blue eyes will be blue eyed (see Figure 3–11). There will be no opportunity for genes to work in a dominant-recessive relationship since the genes on both alleles will transmit the same information

Figure 3–8
(left) Possible genotypes of offspring produced by heterozygous parents carrying genetic information for both brown eye color (B) and blue eye color (b). Theoretically, 75 percent of the offspring will be brown-eyed; 25 percent will be blue-eyed.

Figure 3–9
(right) Possible genotypes of offspring produced by one parent carrying heterozygous genetic information for both brown eyes and blue eyes (Bb) and one parent carrying homozygous genetic information for brown eyes (BB). All offspring will have brown eyes, but 50 percent will carry genes for both brown and blue eyes.

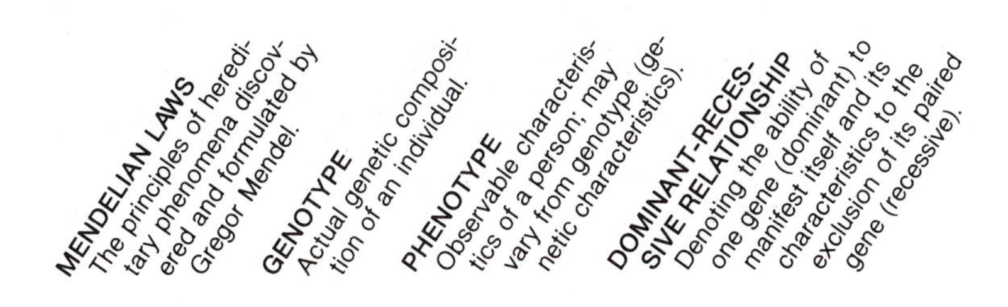

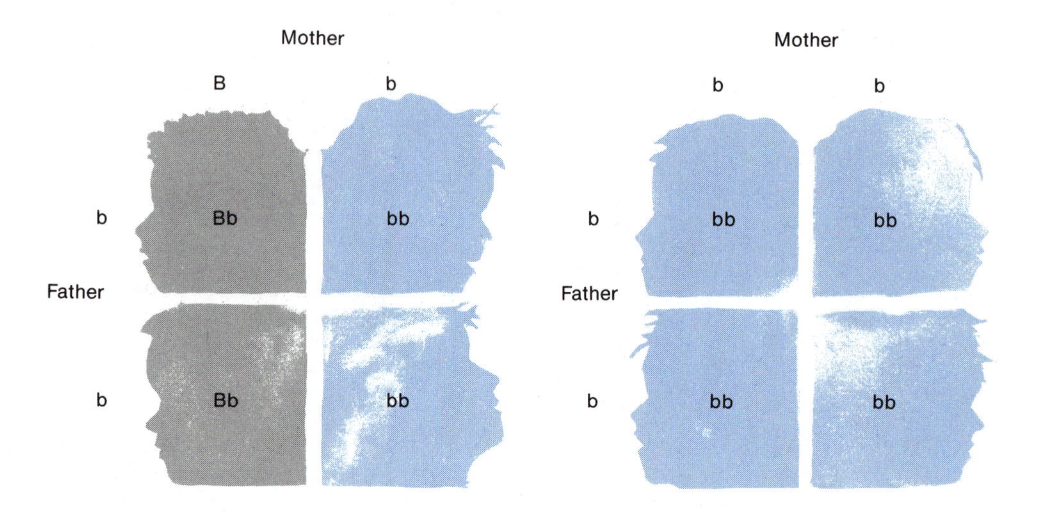

Figure 3–10
(left) Possible genotypes of offspring produced by one parent carrying heterozygous genetic information for both brown eyes and blue eyes (Bb) and one parent carrying homozygous genetic information for blue eyes (bb). Half the offspring will have brown eyes and half will have blue.

Figure 3–11
(right) Possible genotypes of offspring produced by both parents carrying the homozygous genetic information for blue eyes (bb). All offspring will have blue eyes.

determining the color of the pigment in the iris (colored portion) of the eye.

Dominance of one trait over another is common but not applicable in every case. In some situations of genetic inheritance, there is *incomplete* dominance: intermediate effects of messages from allelic regions of paired chromosomes are expressed. Recessive genes can and do exert some influence over dominant genes. In other cases, as with inheritance of AB blood type, there is *codominance.* Both the message for blood type A from one chromosome and the message for blood type B from the paired chromosome are expressed equally (Ayala and Kiger, 1984).

The sex of offspring is determined not by genes on alleles but by the sex chromosomes (see Figure 3–12). Every mature ovum contains one large X sex chromosome along with twenty-two autosomes. Every mature sperm contains either a large X or a small Y sex chromosome along with twenty-two autosomes. If a sperm containing an X chromosome fertilizes the ovum, the offspring of this conception will be female. Likewise, if a sperm containing a Y chromosome fertilizes the ovum, the offspring will be a male.

It has become possible to identify and differentiate sperm that will produce male and female offspring. Sperm that will produce male children (called **androsperm**) have longer tails and smaller heads and can be expected to reach the far ends of the oviducts most rapidly. However, the sperm that will produce female children (called **gynosperm**) can survive longer in the oviducts waiting for an ovum, due to the extra cytoplasm carried in their head region (see Figure 3–13). Knowledge of the differences in androsperm and gynosperm have allowed some parents to preselect the sex of their children (see Box 3–2).

ANDROSPERM
Y-carrying sperm.
GYNOSPERM
X-carrying sperm.

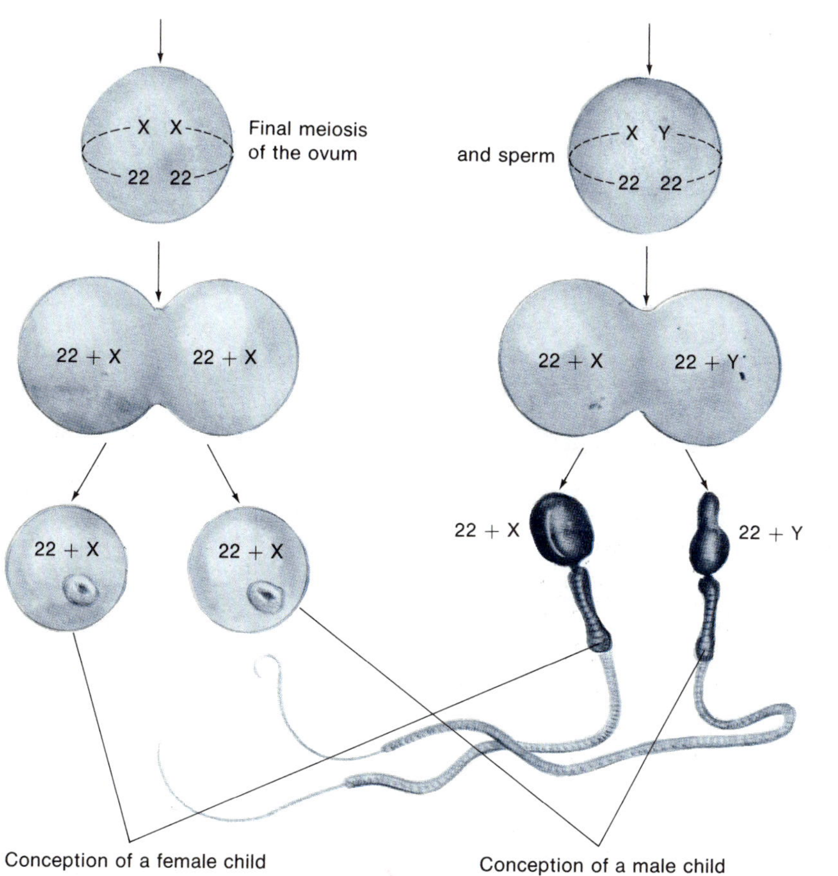

Figure 3–12
Sex of the offspring is determined by the sex chromosome contributed by the father's mature sperm (X or Y). All the mother's mature ova contain X chromosomes.

Figure 13–13
Schematic representation of mature spermatozoa. Those that will produce female offspring (carrying an X chromosome) have the larger, oval heads and shorter flagella.

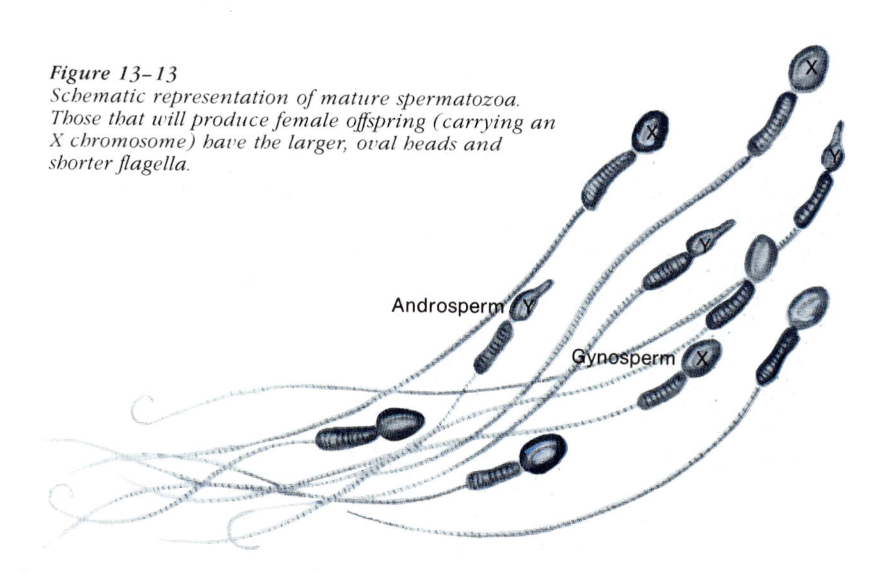

BOX 3–2

Would You Preselect the Sex of Your Child?

The ancient Greeks believed that tying a string around a man's right testicle would produce male sperm, while a string around his left testicle would create female sperm. By 1970 in the United States, with a knowledge of the differences between androsperm and gynosperm to guide him, Dr. Landrum Shettles proposed a more effective method of preselecting a child's sex. In order to produce a son, he advised alkaline (baking soda and water) douches for females prior to intercourse and deep penetration by the male during intercourse, which should occur at the time of ovulation. If a daughter is preferred, he advised acid douching (vinegar and water), intercourse with shallow penetration two to three days before ovulation, and sexual abstinence at ovulation.

By the 1980s, two California scientists had developed Gametrics, a laboratory method of preselecting offspring's sex for which they claim a success rate of about 75 percent (Ericsson and Glass, 1982). Gametrics separates X- and Y-bearing sperm in glass columns. If a son is desired, a solution of albumin is used. The androsperm, with their long tails and small heads, can swim through the sticky substance to the bottom of the column faster than gynosperm. If a daughter is preferred, a gelatinous powder is placed in the column instead. Gynosperm fall through the powder to the bottom of the tube more rapidly because they are heavier. Once the separation of X- and Y-bearing sperm is accomplished, the prospective mother is artificially inseminated with the sperm of her choice extracted from her mate's semen.

STAGES OF PRENATAL DEVELOPMENT

The growth and development of a single fertilized human ovum into a full-term infant in approximately 266 days is a truly remarkable phenomenon. From fertilization until birth, every day is marked by notable changes in the baby-to-be. Three stages of prenatal development will be highlighted first: (1) the changes leading up to and surrounding the implantation of the fertilized ovum into the wall of the uterus; (2) the embryonic stage; and (3) the fetal stage, with an emphasis on anatomy and physiology. A discussion of a multitude of environmental influences (both internal and external to the uterus) that can influence the developing organism will follow in the next section of this chapter.

MORULA The mass of cells resulting from early mitotic cell division of the zygote.

BLASTOCYST The cluster of cells resulting from cell division of a zygote in the first week after conception.

ENDOMETRIUM The mucous membrane lining the uterus.

Implantation As soon as fertilization and the merging of the twenty-three chromosomes of the ovum and sperm occur, the new cell becomes a *zygote.* This zygote is the prototype (model) for all body cells the new human being will produce throughout his or her lifetime. It undergoes mitotic cell division within a day and becomes a two-cell structure, each cell a carbon copy of the other: each containing twenty-three chromosomes identical to those contributed by the sperm paired with twenty-three chromosomes identical to those contributed by the ovum. Within another day it will reach a four-cell stage. By three days the four cells will have divided twice again to make a sixteen-cell structure called a **morula**. The nucleus of each cell will contain the full complement of forty-six paired chromosomes. These chromosome pairs are always identical to those formed at fertilization.

This early cell division occurs in the oviduct as the zygote-changed-to-morula is slowly transported by peristaltic (rhythmic, wavelike) movements toward the uterus. The ball of cells floats free in the uterus, growing progressively larger through mitotic cell division until about the sixth to tenth day after fertilization. At this time it reaches a stage called the **blastocyst** and contains over one hundred cells. It will now burrow into the well-prepared mucosal lining, called the **endometrium**, on the wall of the uterus (see Figure 3–14).

Figure 3–14
Schematic representation of the changes from zygote to implanted blastocyst in the first two prenatal weeks.

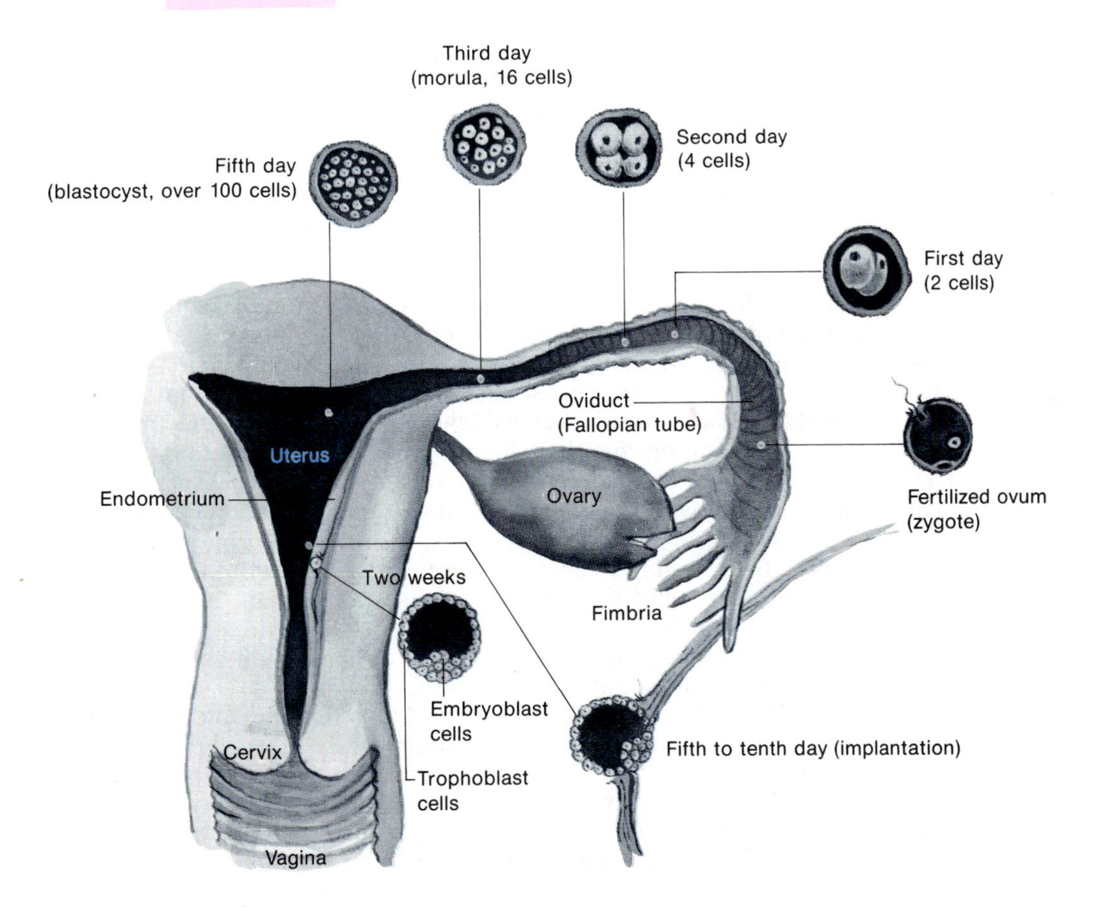

BOX 3–3

In Vitro Fertilization

There is new hope for many couples faced with infertility problems (see Chapter 8). Conception can now be accomplished in vitro (in a glass dish) and the fertilized ovum can be placed in the prospective mother's uterus for a normal nine-month course of pregnancy. Success rates for in vitro fertilization continue to climb as techniques improve.

The in vitro mother-to-be is induced to produce more than one ovum per month by using hormonal stimulant drugs (Marshall, 1984). When the Graafian follicles are fully mature, just prior to ovulation, the ova are surgically extracted from the follicles. The surgeon uses either a laparoscope or ultrasound imaging to guide the extracting needle. The ova are placed on a glass petri dish containing a nutrient-enriched culture medium and are allowed to incubate for several hours. Sperm from the father-to-be are then added to the special culture medium and the ova are again incubated, at body temperature, for at least a day. When fertilized ova have undergone two or three mitotic cell divisions, they can be inserted into the mother's uterus through the cervix.

The chance of implantation of at least one fertilized ovum is improved if more than one is transferred. This also creates the possibility of multiple births, a chance most parents are willing to take. While the majority of in vitro fertilized ova are spontaneously aborted in the first three months of pregnancy, some are carried to term.

In vitro fertilization vastly improves the chances of conception for men with low sperm counts or low sperm motility. It also allows women with blocked oviducts to become pregnant. If a man cannot produce viable sperm, his wife's ova can be fertilized by a donor's sperm. Likewise, a husband's sperm can fertilize a donor's ova and the fertilized ova can be inserted into the wife's uterus. It is also possible for a donor to carry the husband's and wife's sperm-ovum conceptus to term for them if the wife cannot sustain a pregnancy in her own uterus.

This endometrium, you will remember, was stimulated to grow by the hormones estrogen and progesterone produced by the corpus luteum after the ovum escaped from its Graafian follicle. The process by which the blastocyst burrows into the endometrium is called **implantation.** Some women will shed a little blood at this time. The bleeding may be mistaken for a menstrual period since it occurs about five to ten days after ovulation. It usually differs from a normal menstrual flow, however, in both amount and duration of bleeding.

If infertile couples opt for in vitro fertilization (a test-tube baby), the union of ovum and sperm and the growth from zygote to morula is accomplished in a petri dish rather than in the oviduct. The morula is then inserted into the uterus to continue growing and, it is hoped, to implant itself into the endometrium as a blastocyst within five to ten days (see Box 3–3).

The cells of the blastocyst will differentiate into embryonic cells and trophoblastic cells as it implants into the uterus. The embryonic cells become the future baby. The *trophoblastic* cells will form the **placenta** (vascular organ in which the embryo/fetus develops), the *amniotic sac* (innermost membrane enveloping the embryo/fetus), and the *amniotic fluid* to surround, support, and

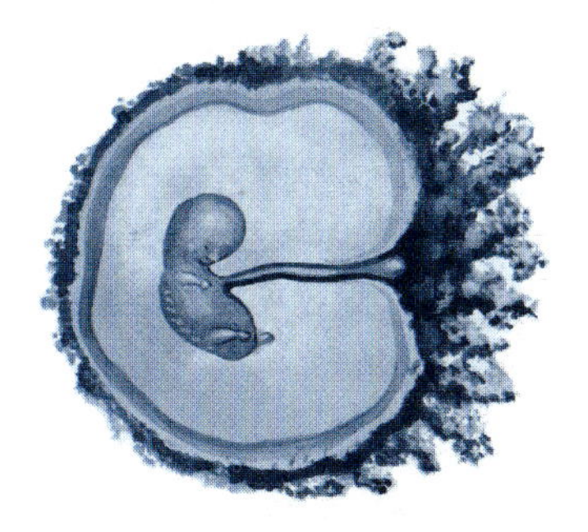

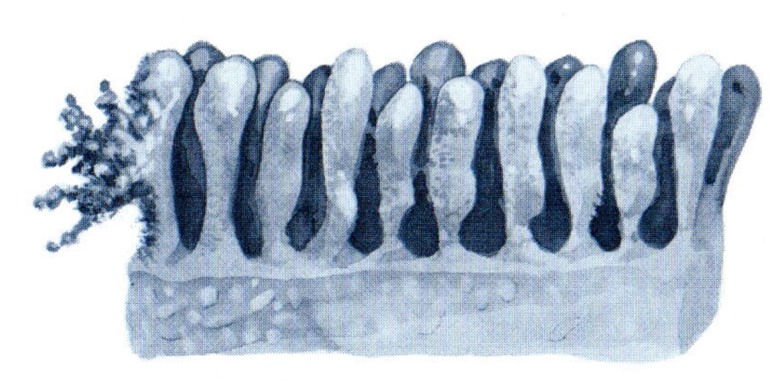

Figure 3–15
Early placental villi are short, stout projections into the uterine mucosa. Later they develop treelike side branches.

protect the developing baby. Cells containing identical chromosomal and genetic materials in their nuclei can assume different roles because only limited numbers of genes are allowed to function in each cell. The factors responsible for turning genes on and off as the process of differentiation proceeds are, as yet, unknown (Sutton, 1980).

Some of the trophoblastic cells form minute projections called **villi** (see Figure 3–15), which burrow into the endometrial lining of the uterus. The circulation of the baby-to-be is never directly connected to the blood vessels of the mother. Rather, exchange of oxygen, nutrients, and wastes is accomplished through diffusion between the blood vessels of the mother on the inner side of the uterus and the villi of the developing embryo/fetus on the outer side of the placenta. Nutrients, oxygen, and other products obtained from the mother by diffusion to the placental villi are then transported to the baby through the umbilical cord.

Other cells, especially the *chorion* (the outermost membrane enveloping the embryo/fetus), begin excreting chorionic hormones within a few days after implantation, or about two weeks after fertilization. These hormones, *human chorionic gonadotropin* (HCG), *human chorionic somatomammotropin* (HCS), *thyrotropin, estrogen,* and *progesterone* cause the "morning sickness" that about 50 percent of women experience in their pregnancies. Other signs and symptoms of pregnancy are listed in Table 3–1.

The fact that the placenta is tolerated in the uterus presents an immunological mystery that has inspired a great deal of investigation but has not as yet been solved. The problem revolves around the fact that the embryo/fetus is not immunologically identical with the mother. It carries **antigens** (materials that stimulate the body to produce a rejection response) from both the mother and the father. The lack of rejection is, consequently, puzzling.

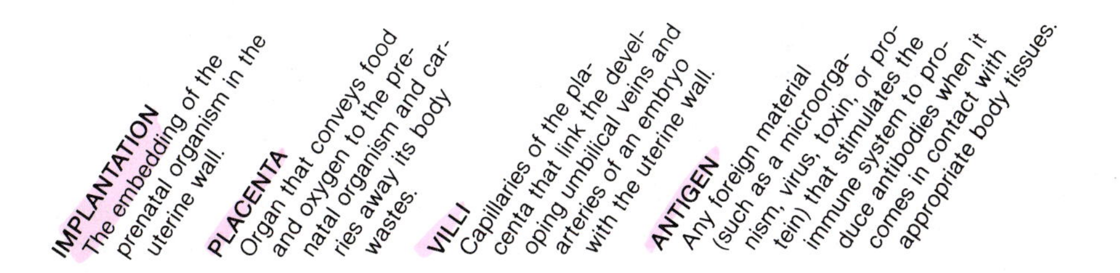

Table 3–1 Early Physical Signs and Symptoms Used to Diagnose Pregnancy

Presumptive	Probable	Positive
Missed menstrual period Nausea and vomiting Tingling sensation or pain in the breasts Fluttering sensation in the lower abdomen (caused by movement of the fetus) Urinary frequency and/or bladder irritability Constipation Weight gain Fatigue Thinning and softening of the fingernails Elevation of basal body temperature Darkening of skin over the forehead, nose, and cheekbones Darkening of nipples and areolas Increased facial or body hair Enlargement of the breasts Protuberance of the lower abdomen Increased vaginal discharge Softening of the cervix	Abdominal enlargement Uterine contractions (painless or with a sensation of tightening and pressure) Ballottement: when the physician taps the uterus from within the vagina the fetus bounces back, giving the sense of a floating object within the uterus Uterine souffle: a rushing sound heard through the abdomen caused by movement of maternal blood filling placental vessels and sinuses	Hearing fetal heartbeats Feeling the outline of the fetus through the abdomen X-ray film showing fetus Ultrasound picture of fetus (safer than x-ray) Electrocardiogram of fetal heart Radioimmunoassay for HCG (very precise and accurate test)

SOURCE: Adapted from R. W. Hale. Diagnosis of pregnancy and associated conditions. In R. C. Benson (Ed.), *Current Obstetric and Gynecologic Diagnosis and Treatment,* 5th ed. Los Altos, Calif.: Lange, 1984. Reprinted by permission.

Embryonic Period The placenta expands to form a balloonlike sac, the amniotic sac, inside which the embryo grows. The sac fills with amniotic fluid that serves as a shock absorber for the floating embryo and keeps the walls of the uterus from cramping continued development.

As the embryo grows, its cells differentiate into the *ectodermal, mesodermal,* and *endodermal layers.* These layers eventually become, respectively, the outer structures (skin, hair, teeth, nails, nerves), the middle structures (muscles, bones, heart, blood vessels), and the inner structures (intestines, endocrine glands, liver, pancreas, lungs, and respiratory tract). The second to eighth weeks following fertilization make up the **embryonic period.** In the fourth week the embryo appears C-shaped. The brain and heart experience the most rapid growth. Thickened areas that will become eyes and ears are visible (see Figure 3–16). In the fifth week the buds that will become arms and legs appear. Growth of the brain is faster than that of all the other body parts. By the sixth week the head is bent over the abdominal bulge. The heart and circulatory system, lungs and respiratory system, liver, pancreas, kidneys, genitals, nervous system, mouth, stomach, and intestines are all undergoing very rapid development. A hand with webbed fingers appears. By the seventh week the fingers are all clearly visible and toes appear. The head becomes more rounded and is supported by a neck area. The abdominal bulge and the umbilical cord become smaller. By the eighth week the embryo has a distinctly human appearance. At this point it is called a fetus rather than an embryo.

Figure 3–16
Development of the embryo week by week.

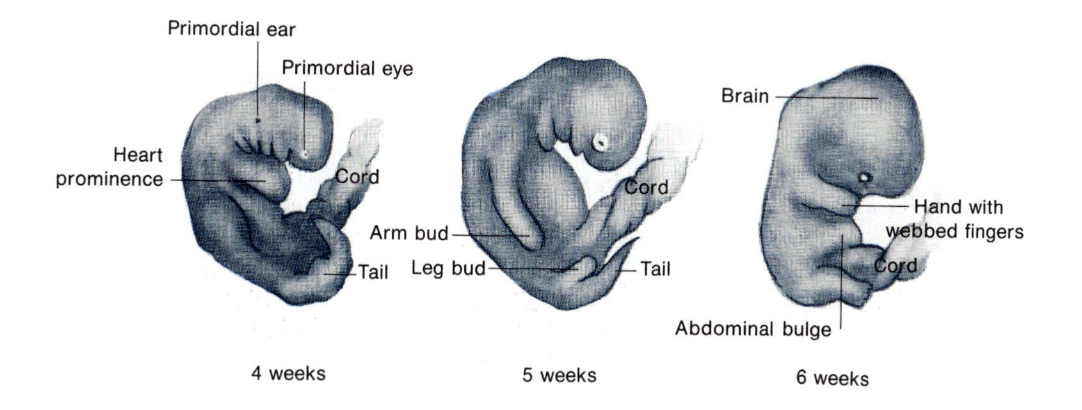

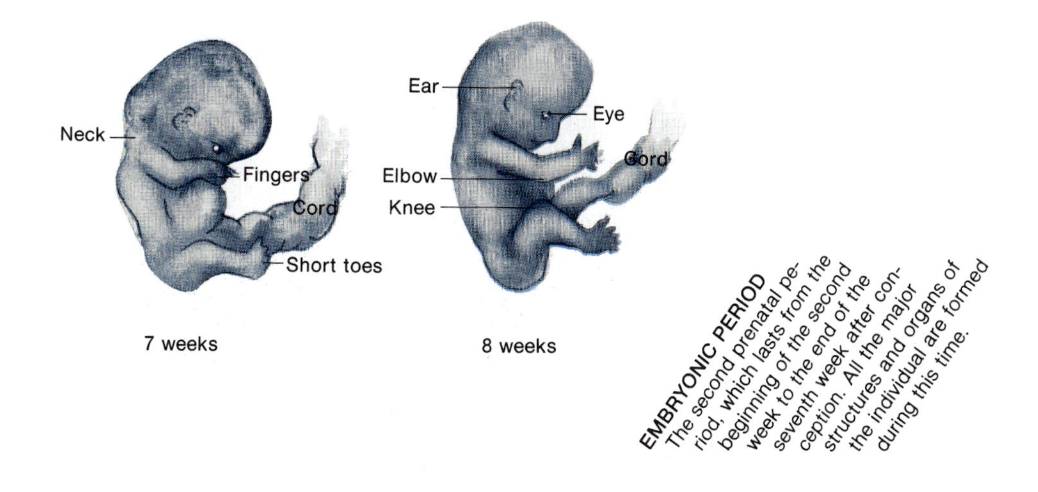

EMBRYONIC PERIOD The second prenatal period, which lasts from the beginning of the second week to the end of the seventh week after conception. All the major structures and organs of the individual are formed during this time.

The period of most rapid growth or development of any structure or function is called its **critical period**. The developing human is especially vulnerable to disruptive influences to the organ systems during the period of embryonic development. It is remarkable that, in spite of the enormous complexity of the organ systems, most babies are born with faultless or nearly faultless, well-integrated, functional systems. One of the reasons for this final perfection rate is the early rejection rate. Many blastocysts with imperfections die before they implant or are not allowed to implant into the uterine lining. Many blastocysts that implant incorrectly are likewise rejected. And embryos that are damaged by various extrinsic causes (to be discussed shortly) are also frequently naturally aborted. Many of these spontaneous abortions are simply experienced as a normal or slightly delayed menstrual period, the mother having had no prior knowledge of her pregnant state.

Fetal Period The fetus is less susceptible than the embryo to internal and external injuries to its organs, since the vital organs are usually past their critical period. The **fetal period** is characterized by growth and maturation of all of the structures that were formed in the embryonic stage. Weight gain is extraordinary. By the third month the sex of the fetus is distinguishable externally. By the fourth month the head is erect on the neck, the external ears are formed and stand out from the head, and the legs are well developed. By the fifth month the mother has usually felt movements of the legs or arms (called **quickening**). At first the movements can be compared to a bubble rising to the surface, or a butterfly fluttering, but later they are very noticeably kicks or punches that may cause the mother to wince. By six months the fetus has hair, eyebrows, and fingernails. Occasionally a fetus born around the twenty-sixth to twenty-ninth week (approximately seven months) can survive (see Figure 3–17). However, the mortality rate is high because the respiratory system is not yet adequately developed to support life outside the uterus. In the last two months the fetus rapidly accumulates fat deposits to use in the neonatal period. The fatty tissues give the fetus a plump, smooth look. The fetus is usually expelled from the uterus about thirty-eight weeks (nine months, or 266 days) after conception. In spite of intensive research, the exact mechanisms that trigger hormone changes in the mother's blood, the fetus, and in the uterus and cause the uterus to contract and push out the fetus are not yet well understood (Ganong, 1983). However, if these trigger mecha-

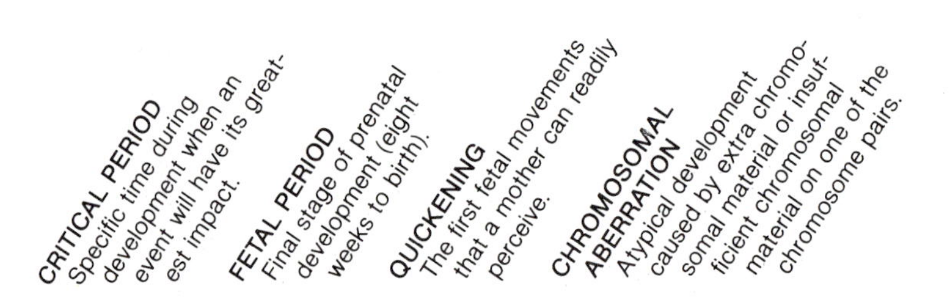

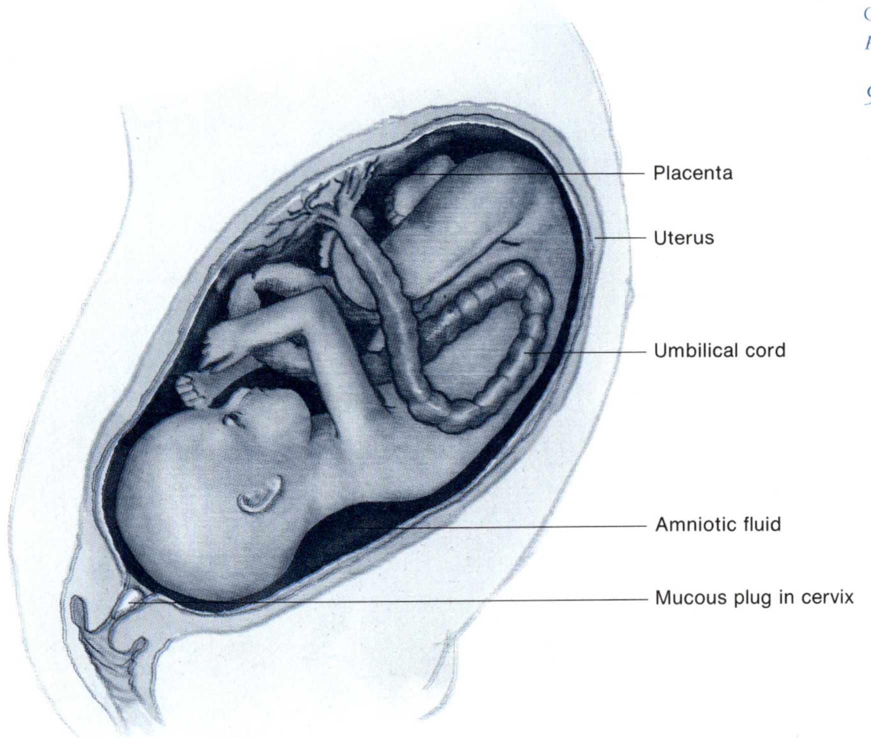

Figure 3–17
Fetus in utero at about seven months, with vital organ systems formed but immature. Visibly lacking are the protective layers of fat that are added during the last few weeks of the third trimester of pregnancy.

nisms fail and the fetus is retained in the uterus longer than about forty-two weeks, the delay can be dangerous rather than helpful to the baby-to-be.

Possible Prenatal Defects

Considering the enormously complicated mechanism of the development of an individual from germ cells to birth, and considering the multitude of factors that can affect the organism in intrauterine life, it is amazing that abnormalities are not more frequent. In this section of the chapter we will describe various reasons for some of the imperfections that do occur. Remember, however, that these are not all risks in every pregnancy. The norm is for full-term, healthy infants, not defective ones, to emerge from the nine months of prenatal existence.

CHROMOSOMAL ABNORMALITIES

When a fertilized ovum contains abnormal chromosomes, extra chromosomal materials, or absent chromosomes, it is usually spontaneously aborted, thus preventing the birth of defective infants. However, some disorders due to **chromosomal aberrations** (deviations from normal) are more viable than others and occur with enough frequency to merit description here. Down syndrome, for example, is the most frequent type of serious birth disorder.

At present the chromosome pairs are designated by numbers

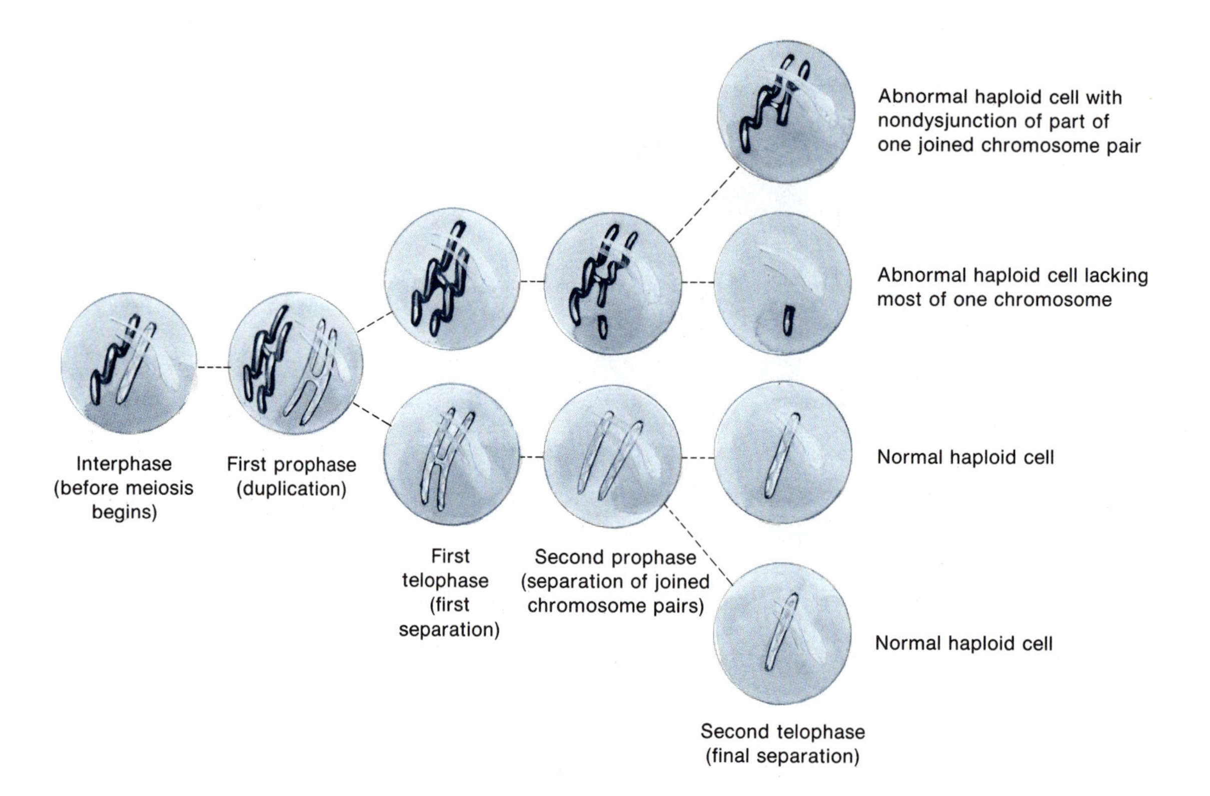

Figure 3–18
Schematic representation of meiosis showing only one chromosome pair and showing nondysfunction of one of the duplicated chromosomes. If a cell with extra material on a 21st chromosome is present in the zygote, it will develop into a person with Down syndrome. The abnormal haploid cell may be either an ovum or a sperm.

from 1 to 22, and the sex chromosomes are designated as XY or XX. **Down syndrome** results when one of the chromosomes on the twenty-first pair contains extra chromosomal material. Down syndrome is also called *21 trisomy* for this reason. The most common cause of the extra material is a failure of the members of the twenty-first chromosome pair to separate during the process of meiosis (see Figure 3–18). This failure to separate is known as *nondisjunction.* The cell that receives no member of the twenty-first chromosome pair, or only a fragment of the chromosome, cannot survive. However, the cell with the extra chromosome material, if present in the zygote, will develop into a person with Down syndrome. Another cause of extra material can be *translocation.* In this aberration one portion of a number 21 chromosome breaks off and attaches to a chromosome of another pair.

Nondisjunctions and translocations probably occur in all the autosomal chromosomes, but the recipients of the errors are usually aborted. Infants with trisomies of the eighteenth pair (*E trisomy* or *Edward syndrome*) sometimes survive prenatal development but die in early infancy. Infants with trisomy of the thirteenth chromosome pair (*D trisomy* or *Patau syndrome*) have also been born. They die by the second year of life (Robinson, Goodman, and O'Brien, 1984).

Children with Down syndrome may survive into adulthood.

Figure 3–19
Down syndrome children can usually be recognized by their characteristic facial features.

They have a variable degree of mental retardation, dwarfed physical height, eyes that slant upward, small ears, slightly protruding lips, and an enlarged tongue (see Figure 3–19). Congenital heart defects are common, and nystagmus (involuntary movement of the eyes) and enlarged liver and spleen also occur (Tjossem, DeLaCruz, and Muller, 1984).

The risk of chromosomal trisomies increases dramatically as a woman ages. For example, the incidence of Down syndrome is less than 1 in 1,000 in women under age 30, about 1 in 100 at age 40, and about 1 in 10 at age 50. Researchers have demonstrated, however, that in about 25 percent of cases, an older father contributed the extra chromosome (Tjossem, DeLaCruz, and Muller, 1984). These age-related risks of trisomy formation may stem from aged ova and sperm or from exposure of the gonads to radiation, viral infections, or other toxic agents.

Trisomies of the sex chromosomes are more common than autosomal trisomies because they are seldom lethal. A male born with extra X chromosomal material (XXY), produced by nondisjunction of the X chromosome pair during meiosis (refer back to Figure 3–18), is said to have **Klinefelter syndrome**. He typically has very small testes with near absence of seminiferous tubules, some breast development, and scant body hair. He is sterile. Bancroft, Axworthy,

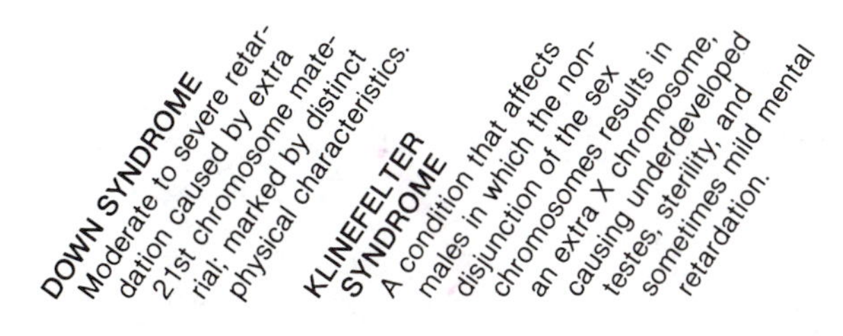

and Ratcliffe (1982), in a longitudinal study of Klinefelter boys, found that psychosocially they are timid, are lacking in self-confidence, and relate poorly with their peers.

A male with an extra Y chromosome (XYY) is typically taller and more muscular than the average man, with reduced fertility. Males with this sex chromosome trisomy were first identified as aggressive members of prison populations. Certainly not all XYY males are aggressive, criminal, or antisocial. Ratcliffe and Field (1982) reported that four out of fourteen young XYY males they studied had emotional disorders: severe temper tantrums, stealing, and enuresis.

Some males with multiple sex chromosomes (more than trisomies) have also been seen: XXXY, XXXXY. They have small testes, are sterile, and are mentally retarded.

Super- or meta-females with XXX or XXXX sex chromosomes may appear normal and sometimes have normal ovaries but diminished fertility. They are predisposed to a variety of psychological and intellectual abnormalities such as poor verbal skills and memory and information-processing deficits (Rovet and Netley, 1983).

A monosomy is occurrence of only one of a pair of homologous chromosomes in the zygote. A person with a monosomy can survive only if the unpaired chromosome present is an X sex chromosome. An X0 recipient is said to have **Turner syndrome**. She has abnormal ovarian development, sterility, short stature, and cardiac and skeletal deformities (see Table 3–2).

Table 3–2 Approximate Number of Children Born with Various Chromosomal Abnormalities

Abnormality	Incidence	
Down syndrome (trisomy 21)	15 per 10,000	
Klinefelter syndrome (XXY)	10 per 10,000	
Double Y (XYY)	9 per 10,000	
Trisomy X (XXX)	8 per 10,000	
Turner syndrome (X0)	1 per 10,000	
Edward syndrome (trisomy 18)	3 per 10,000	do not survive infancy
Patau syndrome (trisomy 13)	1 per 10,000	do not survive infancy

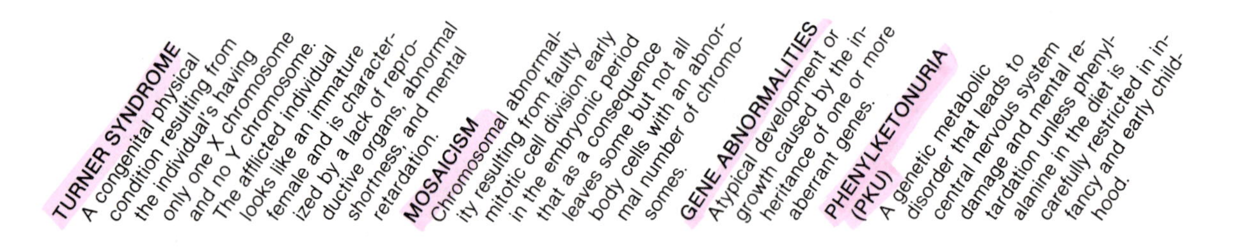

Mosaicism is another type of chromosomal abnormality. It results from chromosomal nondisjunction in some cells postmeiotically, during mitotic cell division early in embryonic development. The consequence is that abnormal daughter cells produce tissues of the body that have abnormal complements of chromosomes. The remaining normal daughter cells produce normal tissues with normal chromosome numbers. If this event occurs early in the development of embryonic cells, the person may show some departure from normal. If it occurs later, the person with chromosomal mosaicism may appear normal and function normally, since most body cells have the normal number of chromosomes. Abnormal organs, or even small islands of tissue within organs, may have abnormal chromosomal numbers. Mosaicism accounts for about 1 percent of Down syndrome individuals. They may only have some of the features of the disorder.

GENETIC ABNORMALITIES

Many diseases existing from birth are caused by deleterious recessive genes rather than whole chromosomal errors. Single **gene abnormalities** tend to be less lethal than those of whole chromosomes. The recessive disease-producing genes must be inherited from both mother and father in order to be manifest (see Figure 3–20).

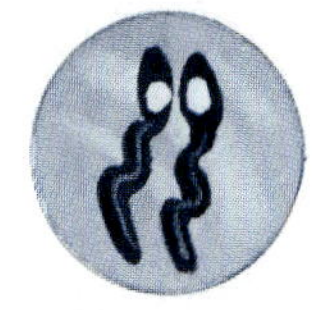

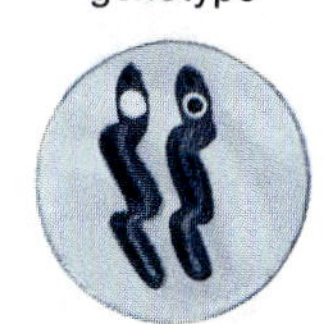

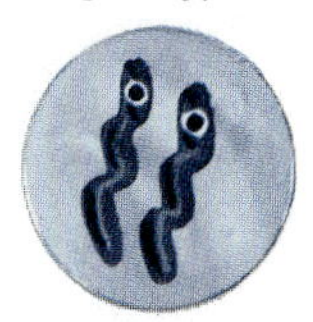

Figure 3–20
Schematic view of one pair of chromosomes per cell showing a normal allele, a carrier allele, and a disease-producing pair for any of the genetic recessive diseases.

Certain genetic defects have become quite well known, not so much because of their frequency but because of publicity surrounding them: phenylketonuria (PKU), galactosemia, Tay-Sachs disease, sickle-cell anemia, and cystic fibrosis. Because people tend to marry persons from their own cultural groups, many of these disorders are more prevalent in one population group than in others (see Table 3–3).

With **PKU** the recessive deleterious inheritance is manifest by lack of the enzyme *phenylalanine hydroxylase,* needed to convert phenylalanine, one of the essential amino acids found in most protein foods, into tyrosine. Without this conversion, phenylpyruvic acid and other substances are produced through alternate pathways. They accumulate to toxic levels in the central nervous system, causing mental retardation, tremors, seizures, and weak muscles. A nationwide screening program is now in effect to test newborns for PKU. A few drops of blood from the infant's heel can be used to determine the presence or absence of the disease. If it is present, treatment can be instituted before damage occurs. Special diets containing only a minimal amount of phenylalanine are given to PKU children to prevent accumulation of the toxic substances.

Table 3–3 Some Mendelian Disorders and the Population Group They Most Frequently Affect

Disorder	Extraction of most affected persons
Cystic fibrosis	Northwestern European
Phenylketonuria	Northwestern European
Tay-Sachs disease	Ashkenazic Jewish
Dysautonomia	Ashkenazic Jewish
Glucose dehydrogenase deficiency	African and Chinese
Mediterranean fever	Armenian, Italian, Greek
Adult lactose deficiency	Chinese
Adrenogenital syndrome	Eskimo
Sickle-cell anemia	African

Galactosemia is a genetic recessive disorder resulting in an inability to produce an enzyme involved in converting galactose (found in milk) to glucose. The affected person, unless given a diet free of galactose, develops liver disease, cataracts, mental retardation, and bouts of severe hypoglycemia (insufficient sugar in the bloodstream). A screening test is available for early diagnosis. Affected persons can lead normal lives if all sources of galactose are eliminated from the diet from earliest infancy.

Tay-Sachs disease is caused by a deficiency of another enzyme, which causes accumulation of certain lipids found in cells throughout the body, causing progressive degeneration of the central nervous system. Symptoms usually appear by about three to six months of age, and affected children seldom survive past age 3. The disease can be detected in the recessive form by simple blood tests, a valuable aid in genetic counseling.

Sickle-cell anemia differs from the genetically inherited diseases described so far because no enzyme deficiency is involved. Rather, the approximately 1 in every 600 blacks in the United States who have the disease have structurally abnormal hemoglobin-S, which tends to crystallize and precipitate in red blood cells that become slightly deoxygenated. This produces peculiar, elongated, sickle-shaped red blood cells. These sickled red cells can clump together and impede flow of blood to any part of the body. Affected persons may have severe anemia, frequent respiratory infections, and heart disease. Inheritance of only one recessive gene for sickle-cell disease (being a carrier of the trait) may cause some sickling to occur during oxygen depletion at high altitudes or after hard exercise. It does not, however, cause the crisis of obstructed blood vessels associated with sickle-cell disease. Sickle-cell tests are available to diagnose the disease in infants and to identify carriers of the sickle-cell trait. Bone marrow transplant is the only curative therapy currently available for sickle-cell disease.

Cystic fibrosis affects the pancreas, respiratory system, and sweat glands. It is usually manifested in infancy by *meconium ileus* (intestinal obstruction caused by a mass of thick feces), frequent or chronic respiratory infections, susceptibility to heat prostration, malabsorption from the bowel, and a general failure to thrive. Pancreatic enzymes, needed for digesting fats, proteins, and starches, are insufficient or absent. Thick mucous secretions and frequent mucous plugs develop in the airways and result in severe, recurrent, and often fatal respiratory infections. Because of the unusually high content of sodium in the sweat of these patients, they must consume extra salt to compensate for their losses. Treatment is directed at preventing infections and obstructions. Meconium ileus is treated with immediate surgery. Airways are kept open with inhalation therapy, postural drainage, physical therapy, and breathing exercises. Immunizations against some infections and rapid antibiotic therapy for others are used. In addition to salt, cystic fibrosis patients' diets are supplemented with vitamins, extracts of animal pancreas, and increased protein and calories, and are kept low in fat. Mortality, although still high, has been greatly reduced by these procedures.

If only one chromosome in males, the X, contains a recessive gene for certain diseases, the male will be affected. This is because the smaller Y chromosome of the pair does not have a corresponding gene to cancel out the effects of the recessive gene on the X (see Figure 3–21). These diseases (hemophilia and color blindness are

Male with one X-linked recessive gene (abnormal)

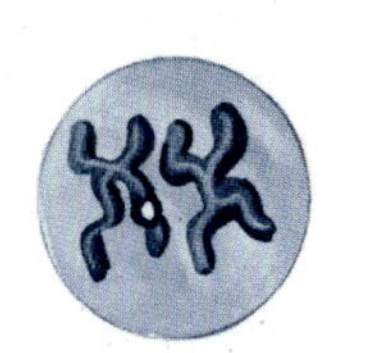

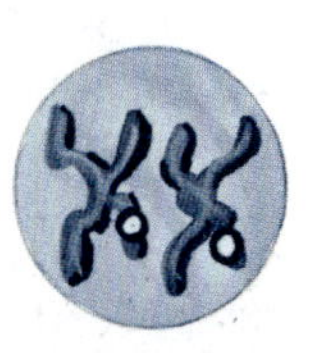

Figure 3–21
Schematic representation of the sex chromosomes in a cell showing the effects of possession of a recessive X-linked gene on males and females.

GALACTOSEMIA
An inborn error in the metabolism of galactose, which accumulates in abnormally large amounts in the blood. It can lead to nutritional defects and mental and physical retardation.

TAY-SACHS DISEASE
A genetic disorder of lipid metabolism.

SICKLE-CELL ANEMIA
A form of anemia in which abnormal red blood corpuscles of crescent shape are present.

CYSTIC FIBROSIS
Genetically inherited disease of children characterized by increased fibrous connective tissue, malfunctioning of the pancreas, and frequent respiratory infections.

the best known) are called **sex-linked** or X-linked. They appear in females only if both X chromosomes contain the recessive disease-producing gene, a rare occurrence.

In addition to chromosome and single-gene disorders, there are many diseases that are considered polygenic or multifactorial in inheritance. The second most common birth defect, **spina bifida**, covers a range of deformities of the spinal column (Sadler, 1985). In the spina bifida occulta form the dorsal portions of one or two vertebrae fail to fuse. A small tuft of hair can be seen growing externally over the affected area. The spinal cord and nerves are usually normal. In the meningocele form, more than two vertebrae fail to fuse, leaving a cleft in the spinal column through which the meninges of the spinal cord bulge. Externally, one sees a sac covered with skin. In the meningomyelocele form, the sac contains not only the meninges but also the spinal cord and its nerves. Neurological symptoms will be present but will depend on the amount and portion of the spinal column that lies within the sac. Symptoms may range from sensorimotor dysfunction to complete paralysis of the portion of the body below the meningomyelocele. In the myelocele (or rachischisis) form, the neural groove fails to close and neural tissue remains exposed to the surface. In the Arnold-Chiari malformation form, portions of either the medulla or the cerebellum project into the spinal column into a meningocele or meningomyelocele. This is a common cause of hydrocephalus (enlarged brain due to an excessive amount of cerebrospinal fluid). Heredity is obscure, but since spina bifida is several times more frequent in relatives of affected individuals than in the general population, genes are a factor. Other factors associated with occurrence of spina bifida include exposure to radiation, maternal age, maternal diabetes, and exposure to environmental toxins during pregnancy. Other common polygenic diseases are asthma, cleft lip and palate, pyloric stenosis, and congenitally dislocated hips (Robinson, Goodman, and O'Brien, 1984).

Parents who, through genetic counseling or past experience, know that they are at high risk for producing offspring with one of the chromosomal or genetically transmitted disorders may elect to have an **amniocentesis** done after the fourteenth week of pregnancy. This procedure is described in Box 3–4.

ENVIRONMENTAL HAZARDS

The influences of heredity and environment cannot be separated. From conception, when the genes on the chromosomes contributed by the ovum and the sperm first provide instructions for the development of structures and functions of the new organism, the environment provided by the mother begins to exert an influence. The genetic make-up of the organism provides a potential for functioning for any given trait in that individual. This potential may or may not then be utilized, depending on certain environmental influences. One tends to think of environment as something experienced after birth; however, there are many areas in which environment before birth can effect changes in the growing, developing embryo/fetus.

BOX 3–4

Amniocentesis

Amniocentesis, the removal of less than an ounce of amniotic fluid from the placental sac after the fourteenth week of pregnancy, allows medical technicians to diagnose Down syndrome, Klinefelter or Turner syndrome, and hundreds of other chromosomal aberrations and hereditary disorders before birth. In this procedure a sterile needle is inserted through the abdominal and uterine walls into the placental sac in order to withdraw amniotic fluid. Ultrasonography—locating the fetus with sound waves—is used to prevent injuries to the developing organism. The amniotic fluid, containing fetal cells, is then cultured and incubated to allow for analysis.

This relatively safe procedure is usually recommended to all pregnant women over age thirty-five and to those who have previously borne children with serious hereditary disorders. Questions have been raised about the use of amniocentesis, since some parents will choose to terminate a pregnancy through abortion if confirmation of a severe fetal disorder is made. In the majority of cases, however, amniocentesis simply confirms that no disorders are present. In those cases where a serious defect *is* confirmed, most parents choose to bear the child. They use the remainder of the pregnancy to make realistic plans for life with their child.

The mother's own health has a pronounced influence on the developing baby. In addition there are several agents, called *teratogens,* that can cause abnormal development if the mother is exposed to them while pregnant.

Prenatal Maternal Illness Women with certain preexisting health problems who become pregnant may have difficulty providing good uterine environments. Diabetes, heart defects, high blood pressure, drug addiction, alcoholism, and malnutrition are examples of preexisting maternal health problems that often adversely affect the developing organism.

Diabetic mothers, even when controlled before pregnancy, may have difficulty adjusting to their increased needs for both nutrients and insulin during pregnancy. Kitzmiller (1984) reported that the incidence of major congenital anomalies in infants of diabetic women is 6 to 12 percent, compared to 2 percent for infants of nondiabetic mothers. Major congenital malformations are most common in children of women whose diabetes is poorly controlled in the first weeks of pregnancy and in those who also have vascular disease.

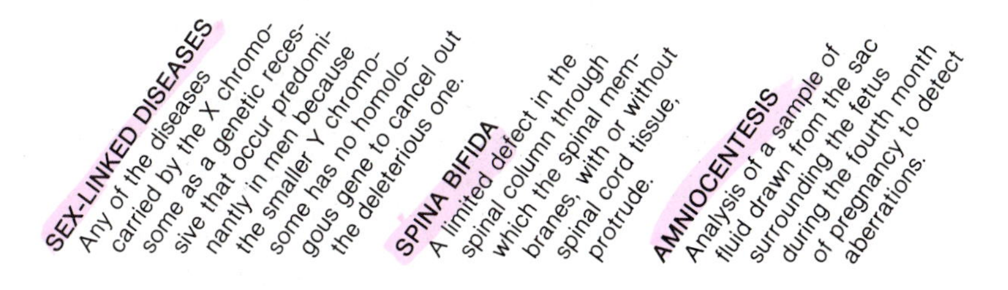

Later in pregnancy, polyhydramnios (an excess volume of amniotic fluid) is also common in poorly controlled diabetics. In the third trimester, fetal distress may develop if control is inadequate. Careful monitoring is necessary to prevent sudden fetal death. Insulin-dependent diabetics are usually hospitalized by thirty-six weeks of pregnancy in order to monitor both fetus and mother. If the fetus is macrosomic (large for date) a Caesarean section delivery is performed (see p. 118) at the thirty-eighth week or later. Risks to the newborn baby include respiratory distress syndrome (see p. 142), hypoglycemia (insufficient sugar in the blood), and poor feeding. Despite these risks, Kitzmiller (1984) reported that diabetic women have a 98 percent chance of delivering a healthy baby if they control their disease carefully during pregnancy. Yogman and colleagues (1982) reported that even healthy newborns of diabetic mothers have lower scores on the Neonatal Behavioral Assessment Scale (see Chapter 4). However, these problems are limited to infancy, and childhood development is usually normal.

Women with heart defects may have difficulty in tolerating the marked rise in blood volume associated with pregnancy. The amount of increase varies but averages 1500 milliliters (about three pints) (Ganong, 1983). If defective hearts cannot adequately handle the extra work load, infants are apt to show growth deficiencies and be born prematurely. Women with high blood pressure are also more apt to deliver prematurely, possibly due to impaired circulation to the uterus and placenta.

Narcotics (heroin, morphine, codeine) addiction in pregnant women can be transmitted to the fetus and leads to withdrawal symptoms in the neonate. Such babies suffer from tremendous irritability, seizures, tremors, diarrhea, and vomiting. They need tranquilization and intensive medical and nursing care in the neonatal period. Householder and colleagues (1982) found that even after the neonatal withdrawal was complete, infants born to narcotic-addicted mothers continued to have cognitive and psychomotor deficits and an impaired health status.

Pregnant women who have malnutrition, alcoholism, or alcohol abuse problems are at high risk for miscarriages or stillbirths. If they carry their fetuses to term, they typically produce poorly nourished newborns who are small for gestational age (SGA). Women who consume too much alcohol may also produce babies with fetal alcohol syndrome (see Box 3–5). SGA infants are known to be at risk for a variety of problems, including small head circumference and cognitive deficits (Simopoulos, 1983).

Pregnancy is never a good time for a woman to try to lose weight. Weight gains should be from 22 to 27 pounds, most of which should be added during the last six months of pregnancy when fetal weight gain is rapid. The National Research Council's Committee on Maternal Nutrition has recommended an approximate intake of 2100 calories per day in the first two months of pregnancy and 2400 calories per day thereafter. These calories should be obtained from the four basic food groups, not from "junk" foods, which can leave even obese mothers malnourished and at risk for SGA infants. Dwyer

BOX 3–5

Fetal Alcohol Syndrome

The third most common type of birth defect (after Down syndrome and spina bifida), and one that is completely preventable, is fetal alcohol syndrome (FAS). It is characterized by a thin upper lip, flat upturned nose, short, fissured eyelids, small head, growth deficiency during childhood, joint abnormalities, and, in many cases, a heart or other major organ defect. While physical appearance is unique, it is relatively minor compared to the major outcome of FAS: mental retardation (Darby, Streissguth, and Smith, 1981).

How much alcohol can a mother-to-be consume during her pregnancy without the risk of FAS and mental retardation? The answer is unknown. Chronic consumption of six or more alcoholic drinks per day places any woman at high risk. However, even regular intake of one to two ounces of absolute alcohol per day may result in FAS, especially if a woman is malnourished, smokes heavily, or uses other drugs. The danger of regular alcohol consumption is greatest during the first trimester, when fetal development of brain tissue and other vital organs is most rapid (Holzman, 1982). No "safe" lower limit for alcohol consumption at any time during pregnancy has been set (Furey, 1982). Alcohol use later in pregnancy, below that needed to produce FAS, has been associated with low arousal and poor habituation in newborn infants (Streissguth, Barr, and Martin, 1983).

(1984) suggested daily intake of two small (two-ounce) servings of meat, eggs, fish, or poultry; two small servings of legumes, nuts, or dried beans; four servings of milk or milk products; four servings of bread or cereal; and one serving each of a vitamin C-rich fruit or vegetable, a dark green vegetable, and another fruit or vegetable. Multiple-vitamin supplements are probably not necessary with a good diet. However, the need for folic acid (one of the B vitamins) doubles during pregnancy, and iron requirements increase to about 50 mg per day. It is difficult to obtain these amounts in a normal diet, so they are usually prescribed. Self-prescribing of megadoses of other vitamins may be unsafe to both mother and fetus. It is now possible to identify slow-growing fetuses early in pregnancy with ultrasonography. Compliance with nutritional counseling for the remainder of pregnancy can prevent, or ameliorate, most fetal malnutrition and SGA birth.

Women who have been generally well fed prior to their pregnancies and then suddenly eat poorly (as may happen with acute nausea and vomiting accompanying morning sickness) usually have adequate reserves to tide the developing embryo over until they can return to better eating habits. This was well illustrated by a classic study made of teenagers in the Netherlands (Stein et al., 1975). All subjects had well-nourished mothers who experienced a wartime famine (1944–1945) while they were pregnant. The children had below-normal birth weights but were otherwise normal throughout childhood and adolescence. The researchers suspected that other mothers who were not well nourished prior to the famine were not able to draw on reserves and therefore had miscarriages or stillbirths or lost their babies in infancy.

Illnesses that the mother contracts during her pregnancy may be as detrimental as preexisting illnesses. Especially dangerous are viral infections in the first trimester and eclampsia in the last trimester.

Viruses are small enough to diffuse through the villi of the placenta (called the **placental barrier**). Consequently, any viral infection during pregnancy can be dangerous to the embryo/fetus. This is especially true during the first trimester (first three months) of pregnancy, when most organ systems are in their critical period of development.

The virus that causes rubella or (German measles) is rarely found in populations of the United States and Canada anymore because of a successful vaccination program against it. Rubella continues to be a significant problem in many other areas of the world, however. If a mother has had rubella (or an immunization against it) prior to her pregnancy, her body will contain **antibodies** (proteins produced by the body in response to foreign substances, or *antigens,* that have the capacity to neutralize the antigens) that will destroy the virus whenever she comes in contact with it. However, if she is unprotected and contracts the disease, the virus is likely to infect the embryo as well. Common effects of rubella in the first trimester on the infant are heart defects, deafness, cataracts, mental retardation, and growth retardation. Maternal rubella is not as dangerous if it occurs later in pregnancy, by which time the organ systems are past their most susceptible critical period for development.

Viruses causing chickenpox, shingles, cytomegalovirus infection, and herpes (skin eruptions) can also cross the placental barrier and disrupt development during the critical embryonic period.

Chickenpox and shingles are caused by the same virus, varicella-zoster. If a mother-to-be has either of these diseases early in pregnancy, she is at high risk for spontaneous abortion or stillbirth, especially if she has a high fever with either disease. Infection with the varicella-zoster virus late in pregnancy may result in infection of the fetus that may last into the newborn period with characteristic pox lesions.

Cytomegalovirus infection early in pregnancy is associated with a high risk of abortion or stillbirth. Other effects of severe maternal infection that are observed in the newborn include chorioretinitis (inner eye inflammation), deafness, microcephaly (small, underdeveloped brain), hydrocephaly, or learning disability (Sever, 1983). At least 10 percent of infants who survive will be mentally retarded (Benson, 1984). However, it is estimated that over 30,000 cytomegalovirus-infected infants are born in the United States each year, yet fewer than 1,000 have severe disease (Benson, 1984).

The herpes simplex virus has two strains. Type 1 produces oral infections and type 2 produces genital infections. Intrauterine infection of the fetus with either strain can occur, usually close to the time of delivery. Generally, however, the newborn acquires the virus from the mother during the birth process (Sadler, 1985). Infections of neonates are due to the type 2 strain in about 95 percent of cases (Sever, 1983). If the virus is acquired, an inflammatory reaction occurs during the first three weeks of life. If the reaction is systemic, it is usually fatal. If there is central nervous system involvement, possibly with convulsions, the neurological sequelae may be mild to severe mental retardation, learning disabilities, or seizure disorder. If it is localized to vesicular lesions of the skin, the prognosis is good in about 50 percent of the cases. In the remainder of cases, a localized reaction may proceed to nervous system or total systemic involvement with the aforementioned sequelae. The best way to prevent perinatal herpes simplex infection is to perform Caesarean section deliveries on all women manifesting genital herpes simplex if lesions occur within two weeks of their due dates. This surgery should be initiated before rupture of the membranes of the amniotic sac.

The microorganism causing syphilis is also able to pass through the placental barrier. It can cause death, deafness, mental retardation, hydrocephalus, microcephalus, and osteitis (inflammation of bones) in the embryo or fetus.

The parasite that causes toxoplasmosis, contracted from exposure to cat feces or from eating poorly cooked meat, can also cross the placental barrier. It can also cause hydrocephalus or microcephalus of the newborn with resulting mental retardation or learning disabilities.

Preeclampsia (also called toxemia of pregnancy) is a condition in which a woman's blood pressure increases, her body accumulates salt and water, and she develops swelling (edema). Proteins can be found in her urine. *Eclampsia* is a more severe form of the disease in which convulsions or coma may develop. These, in addition to endangering the life of the mother, may result in fetal oxygen depriva-

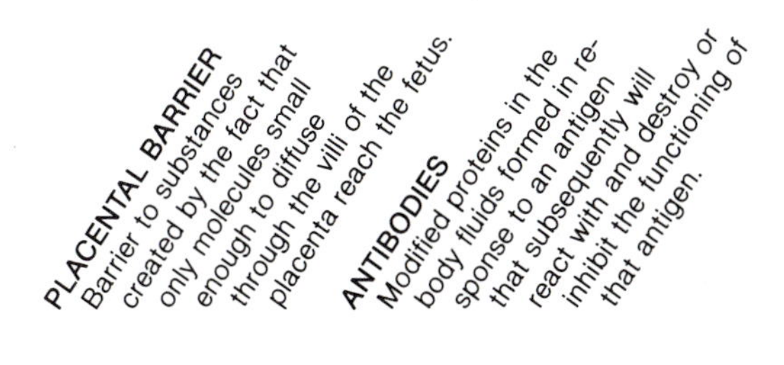

tion. DeAlvarez (1984) reported an incidence of preeclampsia in about 5 percent of pregnant women in the United States. It usually develops late in the third trimester. About 65 percent of all cases occur in primigravidas (women pregnant for the first time), most commonly in those under age seventeen or over age thirty-five. The cause of preeclampsia-eclampsia remains unknown. Suggested causes include maternal immune response to foreign fetal antigens, nutritional deficiencies, metabolic disturbances, idiosyncratic vascular reactivity, and maternal smoking. Preeclampsia-eclampsia was once believed to result from excessive weight gain or excessive salt intake, but these are no longer considered causative factors. The problem is generally relieved by the delivery of the baby and the placenta.

Drugs Some drugs are able to pass from mother to embryo (or fetus) through the placental barrier. Those that interfere with development and lead to birth defects in the child are called **teratogenic drugs**. The U.S. Food and Drug Administration (FDA) established five categories of drug teratogenicity, based on the drug's potential for causing birth defects: *A* indicates no risk, *B* indicates some small risk in animals, *C* indicates defects in animals (no evidence in humans), *D* indicates human fetal risk but the benefits to mother may outweigh the risk to fetus, and *X* indicates that the drug should never be given in human pregnancy. Drugs with an FDA *X* classification include ethyl alcohol, amphetamines, LSD, streptomycin, tetracycline, diethylstilbestrol (DES), iodide, methotrexate, podophyllin, phenotoin, and trimethadione.

Many "social drugs" are highly dangerous prenatally, especially in early pregnancy. Ethyl alcohol is associated with fetal alcohol syndrome. Amphetamines can cause cleft palate and blood vessel transposition in developing fetuses. Lysergic acid diethylamide (LSD) is associated with chromosomal abnormalities, limb abnormalities, and malformations of the central nervous system (Sadler, 1985).

No pregnant woman should take antibiotic drugs without the prescribed consent of an obstetrician. Streptomycin is classified *X* because it causes eighth cranial nerve damage (deafness) and multiple skeletal anomalies. Tetracyclines, if taken during the critical period for tooth development (six to ten weeks), stain the enamel irreparably and cause defective tooth formation. There is evidence that they may also inhibit the growth of long bones.

Diethylstilbestrol (DES), a synthetic estrogen that was formerly given to women to protect them during threatened miscarriages and that is still found in some birth control pills, produces changes in the vaginas of female fetuses. DES has been linked to a form of vaginal cancer in women whose mothers were given the drug during their pregnancies.

Iodides cause the thyroid gland to enlarge, which in turn can lead to congenital goiter, hypothyroidism, mental retardation, or respiratory difficulties postnatally. Many cough suppressants and expectorants that are sold over the counter without prescription contain iodides that, although safe normally, can be potentially dangerous during pregnancy.

TERATOGENIC DRUGS
Any drugs that cause malformations of the embryo/fetus within the uterus.

Methotrexate, found in psoriasis preparations, and podophyllin, found in laxatives, can cause multiple fetal anomalies. All over-the-counter, nonprescription drugs should be approved by an obstetrician before use in pregnancy.

Phenytoin and trimethadione, drugs used by epileptics, cause fetal hydantoin syndrome (FHS). Symptoms of FHS include heart abnormalities, facial clefts, and microcephaly.

While nicotine (in cigarettes) is not an FDA teratogenicity-rated drug, there is increasing evidence that it has teratogenic properties. Pregnant women who smoke may risk giving birth to SGA infants. On the average, smokers' babies weigh 200 grams less than the babies of nonsmoking mothers. This reduction in birth weight is directly related to the number of cigarettes smoked per day, with heavy smokers producing smaller infants than light smokers do. Sexton and Hebel (1984) reported that a counseling program to help one group of women quit smoking for the duration of their pregnancies increased their infants' birth weights significantly over those of a matched control group. Meyer (1978) reported that this weight differential is not a function of a pregnant woman's food intake. In smokers and nonsmokers matched for weight gains with a range from 5 to 40 pounds during pregnancy, the percentage of low-birth-weight infants born to the smokers was still higher. This lower birth weight is also associated with a decrease in length and head circumference. Cigarette smoking has also been linked to an increased risk for spontaneous abortion, stillbirth, and premature delivery (Johnston, 1981). It is associated with problems of delivery such as abruptio placentae and placenta previa (see p. 123), and premature rupture of the amniotic membranes (Naeye, 1981). Additionally, infants born of smoking mothers have lower Apgar scores (see p. 134) (Garn et al., 1981), increased respiratory problems, and poor orientation and habituation responses (Streissguth, 1983). Even at age 4 children born to smoking mothers had significantly poorer attention and orientation responses than matched offspring of nonsmokers (Streissguth et al., 1984).

The reasons why smoking has adverse affects on fetuses is not fully understood. The nicotine in cigarette smoke is a vasoconstrictor—it causes blood vessels to narrow. This may elicit both a rise in maternal blood pressure and a reduction of blood supply to the placenta. Nicotine also crosses the placenta and may serve as a toxic agent to inhibit growth. Growth retardation is also probably a result of the toxic effects of the carbon monoxide in the blood of smoking mothers, which also crosses the placental barrier (see Figure 3–22). Carbon monoxide reduces the amount of oxygen in the tissues and may lead to a state of fetal hypoxia (oxygen deprivation).

Immunologic Factors Our immune systems are very sensitive to the presence of foreign substances (for example, foreign proteins, viruses, bacteria, some pollens, or blood cells with proteins on their surfaces that are different from those on our blood cells) that enter our bodies. We produce antibodies to attack and help eliminate these antigens. We continue to manufacture specific antibodies to

1. Inhaled smoke, containing tar, nicotine, carbon monoxide, and other toxins, enters a mother's lungs via mouth and windpipe. The lungs serve as a way station where blood excretes carbon dioxide and takes in fresh oxygen.

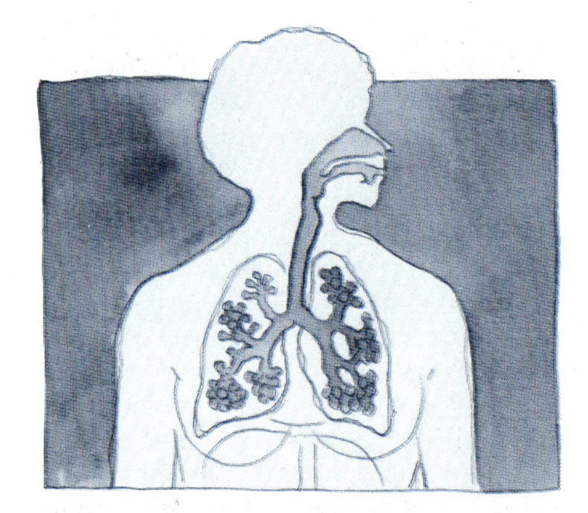

2. Inside the lungs tobacco smoke releases small particles of toxic substances into the maternal bloodstream and prevents the blood from taking up enough fresh oxygen.

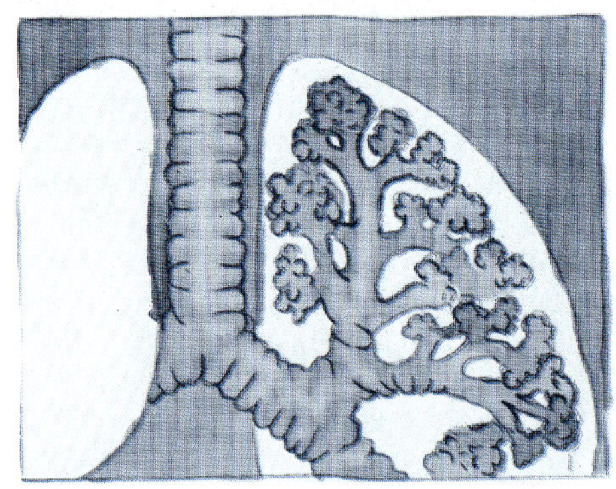

Figure 3–22
Effects of prenatal maternal smoking.

3. Arteries, arterioles, and capillaries carry the mother's blood to different organs, including her uterus, which holds the fetus (unborn baby) and placenta.

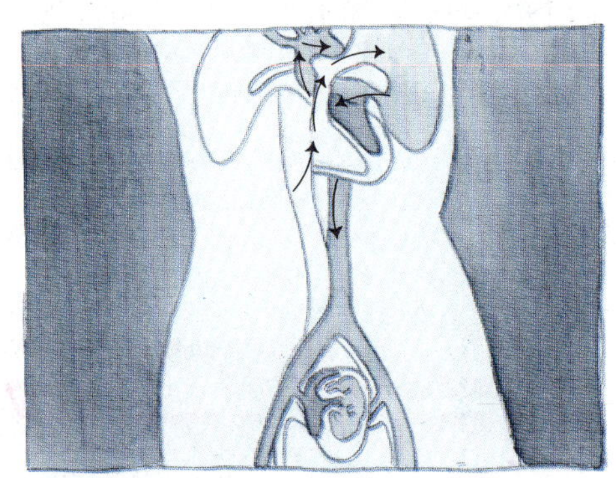

4. In the placenta the mother's blood delivers wholesome nutrients to the fetus as well as any harmful substances she eats, drinks, or inhales. The umbilical cord carries the blood to the fetus where it is distributed to its developing tissues and organs. Research evidence suggests that the higher levels of carbon monoxide in the blood of a smoking mother may deprive her fetus of the oxygen it needs for full growth.

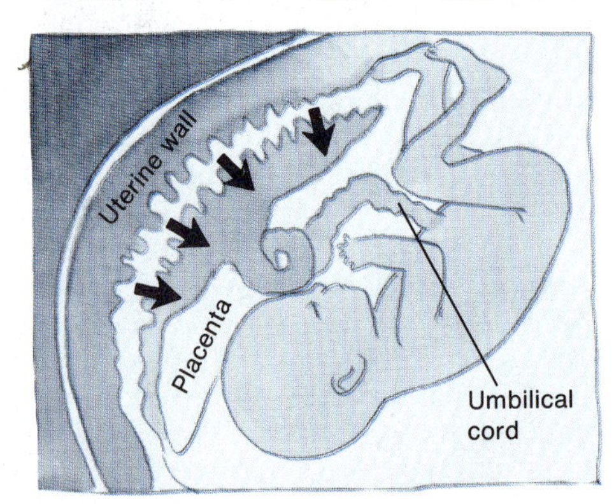

RH FACTOR An agglutinizing factor present in the blood of most humans; when it is introduced into blood lacking the factor, antibodies form in the blood.

guarantee us some degree of "immunity" against specific antigens for varying periods of time after our initial exposure to them. There are some fetal products (such as red blood cells) that can act as antigens if they get into the mother's circulatory system. These "antigens" can stimulate the mother's immune system to produce antibodies that can then pass back through the placental barrier and attack the fetal antigens. Antibodies directed against fetal cells may produce serious diseases in the fetus or neonate. An outstanding example of this is Rh sensitization.

The **Rh factor** is a substance found on the surface of red blood cells of approximately 85 percent of the Caucasian population and nearly 100 percent of the black and Asian population. People are said to have Rh-positive blood if they have this substance on their cells and Rh-negative blood if they lack it.

If a mother with Rh-negative blood has a child with Rh-positive blood, there is a possibility that some of the red blood cells of the fetus containing the Rh-positive factor may enter the mother's bloodstream at the time of labor or delivery. If so, the mother will build up antibodies against the Rh factor. If second or subsequent children inherit Rh-positive blood, there is a chance that the antibodies in the mother's bloodstream will pass through the placenta and destroy some of the fetus's red blood cells (see Figure 3–23). If this happens

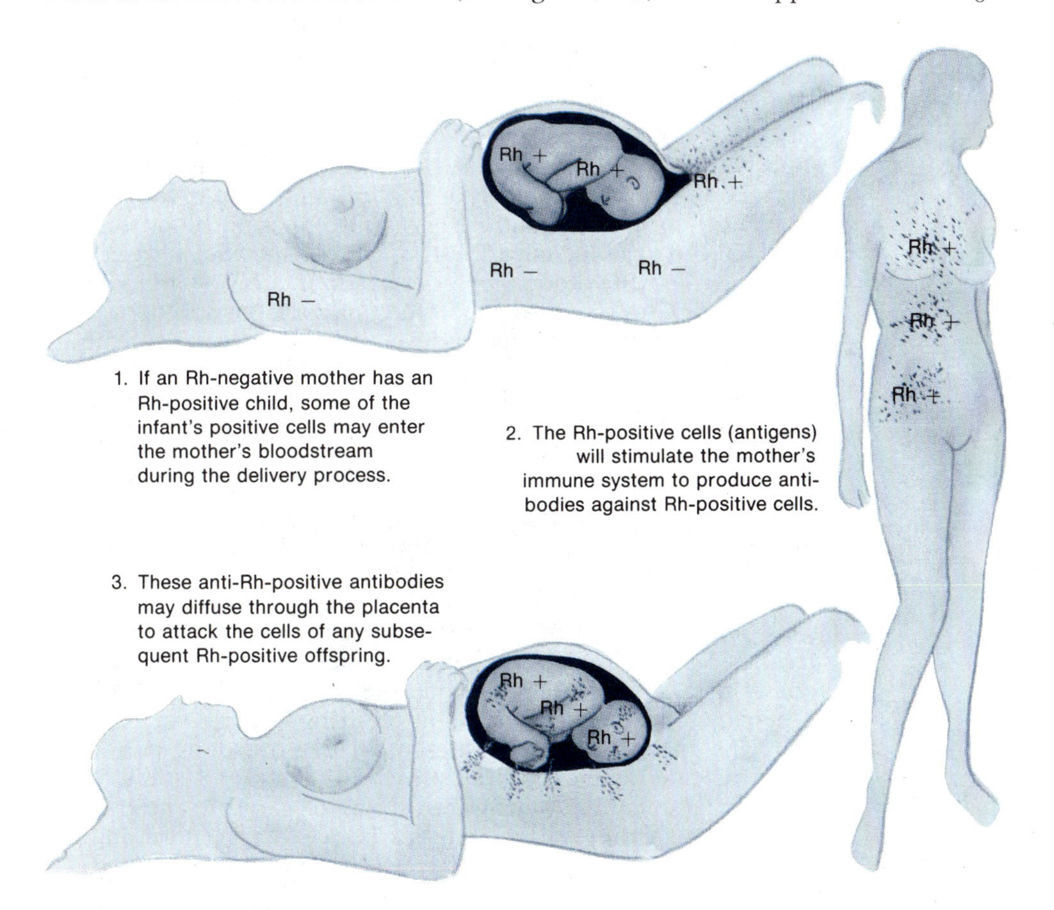

Figure 2–23
Antigen-antibody response to Rh factor in an Rh-negative mother.

the baby may be born with erythroblastosis fetalis (severe jaundice and anemia and ongoing destruction of red blood cells) and must have exchange transfusions to survive.

Not all Rh-negative mothers will have problems with Rh incompatibility in their children. If the father is also Rh-negative, all the children will inherit Rh-negative blood and no problems will arise. If the father is heterozygous for the Rh trait and carries one gene for Rh-negative blood, there is a 50-50 chance that offspring will inherit negative blood. However, if the father is homozygous for the Rh-positive blood trait, all the children will have Rh-positive blood. But Rh-positive blood from the fetus does not always enter the mother's bloodstream at the time of delivery, so an Rh-negative mother may bear several normal Rh-positive children without ever developing anti-Rh antibodies. Now, thanks to a new anti-Rh immunoglobulin (Rho GAM), an Rh-negative mother can be protected from developing antibodies with a simple intramuscular injection given within seventy-two hours after delivery or miscarriage. Rho GAM prevents the formation of Rh antibodies by the mother by combining with and destroying any fetal red blood cells that enter the mother's circulation. It must be administered following every delivery or miscarriage to ensure continued protection.

Another example of an immunological problem that may occur prenatally is **ABO incompatibility**. Blood cells can be divided into various types according to differences in glycoproteins (sugar-containing proteins) on their membranes. The most commonly known blood types are A, B, AB, and O; and Rh+ and Rh−. (There are many others, including Kidd, Kell, Duffy, Hr, M, N, P, and Lewis.) People who have cells with A-type antigens on their membranes spontaneously produce anti-B antibodies, and people with B-type antigens produce anti-A antibodies. Thus, crossing blood types A and B can lead to an incompatibility. If a pregnant woman has a fetus with a different ABO blood type from her own, it is possible for fetal blood entering her bloodstream at the time of labor and delivery to stimulate her immune system to develop antibodies against the incompatible type. Future offspring may be affected by these antibodies. However, A or B isoimmune disease of the newborn is usually milder than Rh incompatibility disease and only rarely requires exchange transfusion.

Radiation Too much exposure to radiation, whether from x-rays, radioactive materials used in some industries or research, leaks from nuclear power plants, or ultraviolet light from the sun or sunlamps, can have teratogenic effects on the developing embryo/fetus. These effects were made evident after the 1945 bombings of Hiroshima and Nagasaki, Japan. Surviving pregnant women who were within the fallout range experienced abortions (28 percent), neonatal deaths (25 percent), and mutations to their offspring (25 percent) (Sadler, 1985). The effects of radiation are cumulative over long periods of time so repeated overexposure during childhood and adolescence can produce mutations (changes in form) of genes in the ova and sperm stored in the ovaries or testes before they mature and are discharged for possible fertilization. For this reason, the gonads of both sexes should always be protected by a lead shield during irradiation with x-rays. It is not known how much exposure to radiation can safely be allowed during pregnancy. Dental x-rays are usually considered safe, although many doctors prefer pregnant women to avoid all exposure to radiation. The need for pelvic x-rays to diagnose obstetrical problems sometimes outweighs the possible radiation risks to the unborn child.

Birth

Approximately 266 days after conception, the new human being is ready to emerge into the outside world. Most obstetricians give mothers-to-be an expected due date, or EDC (expected day of confinement), figured from the last menstrual period by means of Nägele's rule. Seven days are added to the first day of the last menstrual period. Then three months are subtracted from this date. For example, if Mrs. K.'s last period began October 13, adding seven days would bring one to October 20 and subtracting three months would bring one to July 20. Mrs. K.'s due date would then be given as July 20. Only a very small fraction of women actually deliver on their due dates. Most deliver within ten to fifteen days before or after the estimated date.

The uterine contractions that expel the fetus and the placenta are known as **labor**. Labor differs for every woman and with every succeeding child, so statements about what it is like are at best generalizations. Jacklin and Maccoby (1982) reported that the length of labor for mothers giving birth to boys is longer than that for mothers giving birth to girls.

The average length of labor for a first child is thirteen hours; for subsequent children, it is eight hours. Labors lasting twenty-four to thirty-six hours and precipitous deliveries that occur on the way to the hospital are not uncommon.

Mild contractions called **Braxton-Hicks contractions** occur throughout pregnancy, preparing the uterus for its eventual task of expelling the fetus and placenta. They are painless early in pregnancy but may be felt as "false labor" toward term. These contractions may begin thinning out and enlarging the uterine floor prior to actual labor. False labor may also accompany the dropping of the baby's head into the brim of the pelvis. This is variously known as "lightening," "settling," or "dropping" (see Figure 3–24). With first babies it may occur as much as two weeks before labor. Lightening usually makes a pregnant woman feel more comfortable and allows a

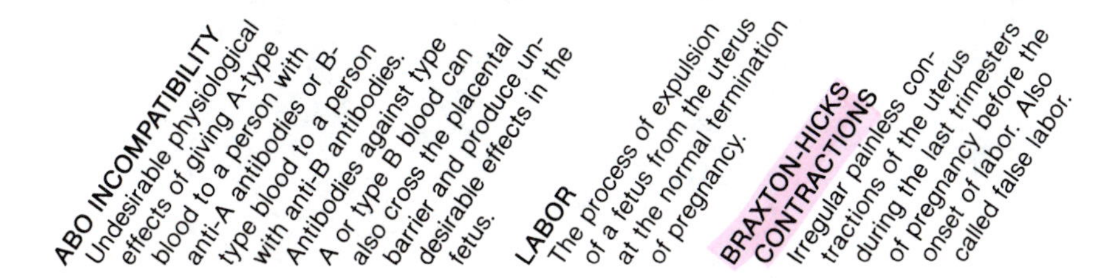

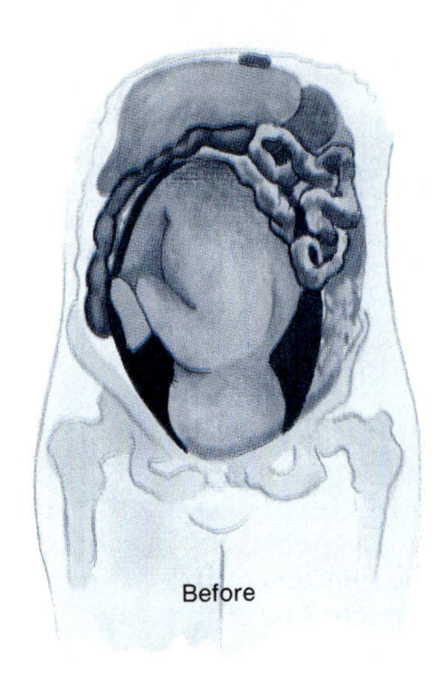

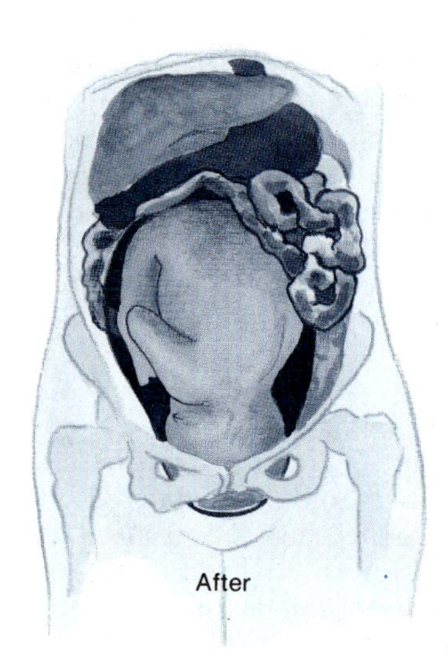

Engagement of the head late in pregnancy
(known as lightening, settling, or dropping)

Figure 3–24
When the baby's head descends into the pelvis, the mother feels some relief from the pressure on her abdominal organs.

Figure 3–25
Effacement and dilation of the cervix in the first stages of labor.

greater range of motion. Since the pressure on her diaphragm is relieved, she can breathe more easily.

As the baby's head presses down on the uterine floor, **dilation** and **effacement** occur. The pressure against the opening to the uterus (the cervix) causes it to *dilate* (enlarge) and *efface* (thin out). As this happens, a mucous plug, which has been keeping the cervix closed during pregnancy, is expelled. It is frequently streaked with blood and is referred to as the "bloody show." Real labor usually begins after the mucous plug is discharged and when the cervix begins dilating and effacing more rapidly (see Figure 3–25).

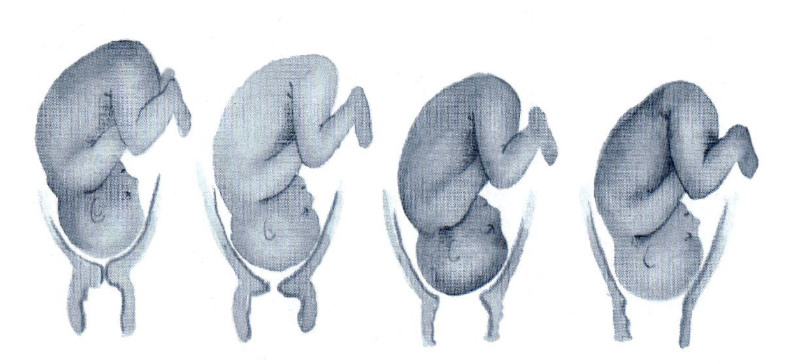

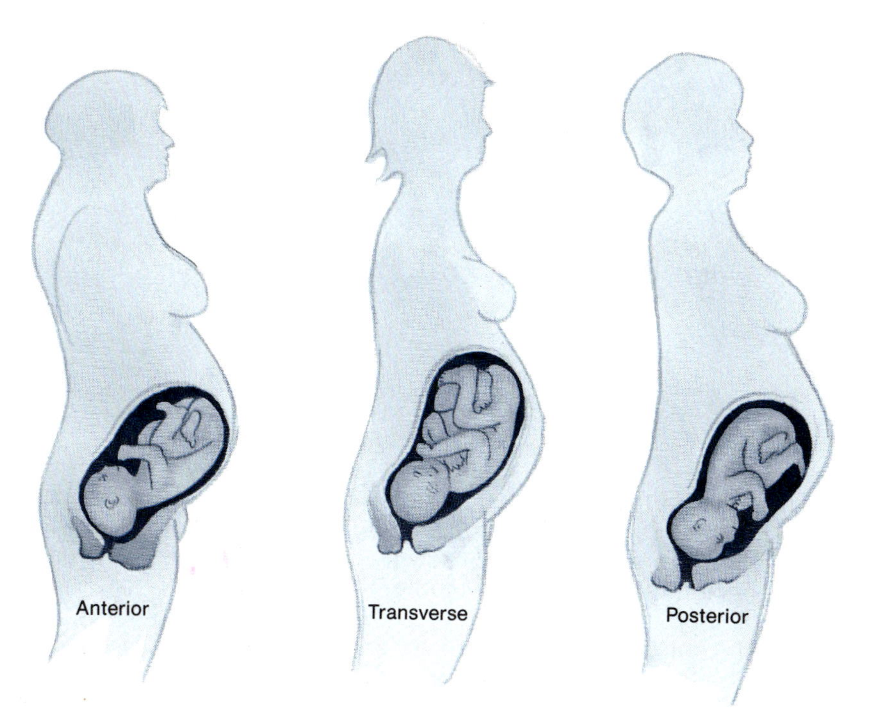

Figure 3–26
Commonly seen positions of the fetus in the uterus.

STAGES OF LABOR

There are four **stages of labor**. In the first stage the cervix effaces and dilates until it is wide enough (usually around four inches or ten centimeters or five fingers) to allow the baby's head to pass into the vagina. In the second stage the baby emerges from the vagina. In the third stage the placenta (or afterbirth) is expelled. The fourth stage is the immediate postpartum period.

PRESENTATIONS

Presentations refer to how the fetus is positioned during delivery. In about 96 percent of pelvic deliveries the top of the skull (vertex) comes out first. If the spine of the fetus is adjacent to the abdomen of the mother, the head descends in an *anterior lie* (see Figure 3–26). If the spine of the fetus is to the left or right side, the head descends in a *transverse lie*. With anterior or transverse lies the back

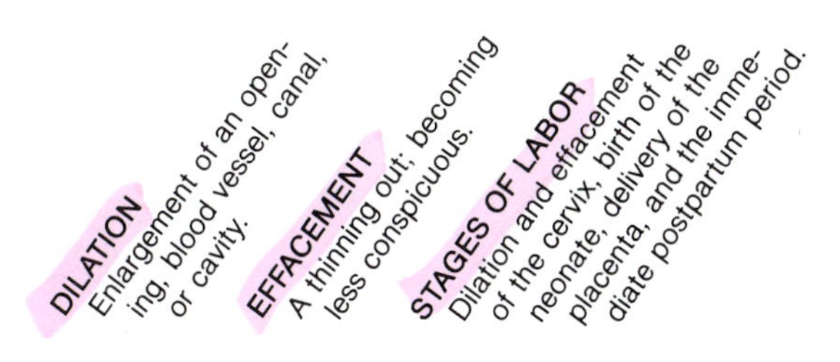

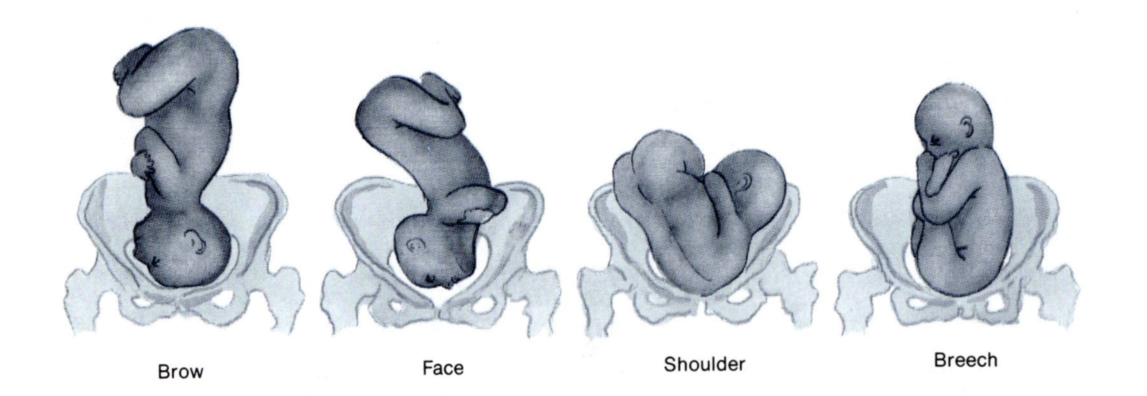

Figure 3–27
Atypical presentations of the fetus.

part of the skull emerges first, which is the easiest way for it to pass the pubic arch bones. If the spine of the fetus is lined up against the spine of the mother, the head descends in a *posterior lie*. The posterior position often causes the mother to experience more back pain during labor.

The bones of the fetal skull are not yet completely formed. The membrane-covered spaces between the bones, known as *fontanels* and *sutures,* allow the bones to overlap each other slightly, making it possible for the head of the fetus to slip past the tightest portion of the maternal pubic arch. This overlap process is known as **molding**.

Fetuses who do not emerge with the top of their skulls first may present with their brow, their face, their shoulder, or in a breech (buttocks or feet first) position (see Figure 3–27).

Brow and face presentations are difficult because the diameter of each of these presenting parts is too large to slip through the pelvic arch. If the fetus is very small or the bone structure of the mother is very large, or if the obstetrician can turn the head before it becomes engaged and molded, pelvic delivery may be possible. Frequently, the baby is delivered surgically through an incision made through the abdomen and uterus of the mother (**Caesarean section**). A fetus cannot be delivered shoulder first. When the abnormal position is recognized before or in early labor, the fetus can be turned. If the position cannot be changed, delivery must be by Caesarean section.

Breech presentations occur in about 3 percent of all deliveries. There are variations of breech where buttocks alone, buttocks and

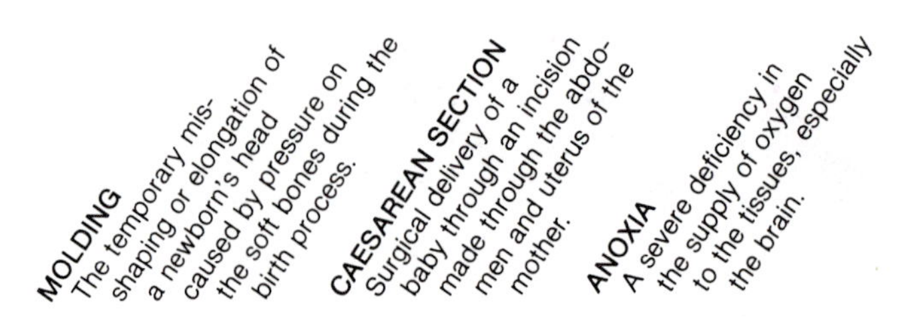

both feet, single foot, or knee may descend first. Breech babies may be delivered vaginally or through Caesarean section.

As the fetus is forced down through the cervix by uterine labor contractions, the membranes of the amniotic sac rupture and the amniotic fluid gushes out. This is commonly referred to as "the water breaking." It occurs during the first stage of labor for most women but occasionally may precede it.

AIDS TO DELIVERY

Analgesics such as sedatives, narcotics, and barbiturates help the laboring mother relax and lose her sensitivity to the pain of the uterine contractions. Anesthesia takes away all the sensations either by putting the mother to sleep (general inhalation anesthesia) or by deadening the sensations in a portion of her body (regional anesthesia).

Some anesthesia must be used for Caesarean section deliveries. Once uncommon (4 percent of deliveries in the 1950s), "C-section" births are now rising precipitously. Brody (1981) reports that in some hospitals they constitute about 40 percent of deliveries, whereas the overall percentage of deliveries by Caesarean in 1980 was 18 percent. This increase reflects many physicians' desire to deliver the baby as quickly as possible if any signs of fetal distress occur.

All of the anesthetic agents and many of the analgesics share a common hazard. They can be transmitted across the placental membrane to the fetus. This transmission can contribute to a drowsy infant at birth who may have difficulty initiating his or her own respiration. Failure to initiate respirations can lead to hypoxia (less than normal content of oxygen in the tissue) or **anoxia** (severe deficiency of oxygen in the tissues). Without oxygen, brain tissue will be damaged or destroyed. Sedatives (Valium, scopolamine), narcotics (Nisentil, Demerol), barbiturates (secobarbital, thiopental), and inhalation anesthetics (halothane, nitrous oxide) are all known to cause respiratory depression in the newborn. Other negative outcomes associated with these drugs include alternations in heart rate, lowered Apgar scores, and abnormal patterns of sleep and wakefulness (Broman, 1983). The baby may also have trouble feeding and lose more weight than normal in the first two to three days after delivery. Some evidence exists that the effects of hypoxia or anoxia at birth may be long lasting. Hollenbeck and colleagues (1984) found that both smiling and infant-parent touching were diminished at one month when labor medication doses produced a depressed neonate. Inhalation anesthetics especially are associated with poor neuromotor and psychomotor functioning throughout the first year (Broman, 1983). Another drug, oxytocin, which may be used to induce labor or stimulate contractions during labor, is also associated with long-lasting infant psychomotor deficits. Another class of drugs used to induce labor, the prostaglandins, do not seem to have such adverse effects. No form of obstetric anesthesia is 100 percent safe and satisfactory. The obstetrician and anesthesiologist must weigh the dangers of drugs against the need for relieving pain in Caesarean births and difficult labors and deliveries and for promoting relaxation in very tense, anxious mothers. The American Academy of Pediatrics (1978) recommended that obstetricians administer the lowest effective doses of the safer obstetric medications for pain relief during labor and avoid the use of those shown to have more adverse effects on neonates.

The use of analgesics and anesthesia during labor may necessitate the use of obstetrical **forceps**, double-bladed tongs that fit over the head of a fetus and grasp it without compressing it. They are used if a mother loses her ability to push the fetus out during the second stage of labor due to the effects of medication or exhaustion. Forceps frequently leave marks showing where they were placed on the infant's head, but such marks disappear in a few days.

Many women have an **episiotomy** whether or not they use obstetrical medications. In this procedure a straight cut is made from the vagina back toward the anus but slightly to one side to enlarge the vaginal opening just before the head delivers. In some hospitals it is done routinely while in others it is performed only if it seems likely that the perineal tissues will tear without an episiotomy. A straight cut is easier to sew up and heals better than a ragged-edged tear. An episiotomy also gives the baby more room to get out.

Odent (1982) reported that the best way to prevent tearing of perineal tissues, and to reduce the need for an episiotomy, is to allow the mother to assume a squatting position during her labor contractions. This not only reduces the pain of contractions but also allows the best relaxation and optimum stretch of the perineal muscles. It is also a good posture to help the rotation of the fetus's head inside the pelvis. Many hospitals now have birthing chairs, which allow the mother to be supported in a vertical, semi-squat posture during contractions. Other maternity units allow laboring women to find for themselves a posture in which they are most comfortable during contractions, whether a squat, a kneeling position, or simply bending forward.

Natural childbirth is the delivery of an infant without the use of inhalation anesthetics or instrumentation. Prepared childbirth, often called natural childbirth, refers to the fact that the mother (and usually a partner, such as the father) has been educated beforehand about the physiology of pregnancy, knows what to expect during labor and delivery, and has been taught perineal muscle-building exercises, relaxation techniques, breathing mechanisms, and comfort aids to assist in the birth (see Figure 3–28). In the United States such preparation usually comes from a series of six to eight evening meetings conducted in the **Lamaze method of delivery** or the Grantly Dick-Read method. Prepared childbirth encourages parents to expe-

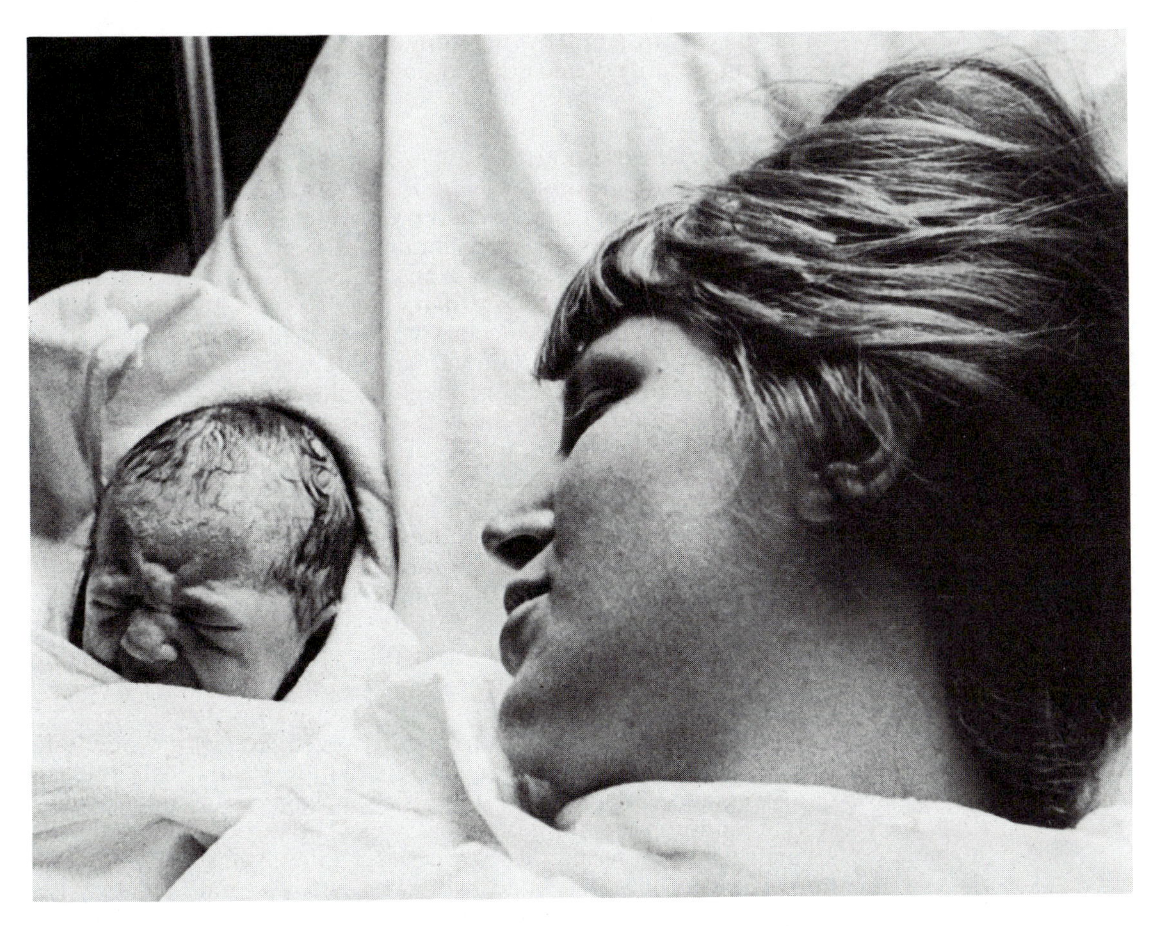

Figure 3–28
Prepared childbirth allows a mother to experience the incredible moment of her child's birth while fully awake. It minimizes the danger that the infant will be drowsy or have difficulty breathing due to the effects of obstetric medications.

rience birth together. It discourages the mother from relying on drugs or tranquilizers that will cause her to sleep through or be hazy during childbirth. It does not necessarily mean drugless labor. Use of drugs may be initiated at the discretion of the doctor or midwife (a nurse specially trained to help women in childbirth) in attendance. Usually the program greatly alleviates tensions and pain and reduces the need for analgesia or anesthesia.

The **LeBoyer method of delivery** was developed to reduce fetal birth trauma. The delivery room is kept dark and quiet. Sometimes soft music is played. The newborn is immediately immersed in warm water to simulate amniotic fluid. Then the infant is gently stroked and massaged as it adjusts to the external world. Finally, the neonate is put to the mother's breast while she is still in the delivery room, to facilitate an immediate bonding between mother and child. The father is also encouraged to be present and to participate in holding, cuddling and stroking his newborn.

Evidence exists that the presence of the father, or some other companion, to support a mother during labor and delivery has many beneficial effects. Kennell (1982), in a comparative study of twenty mothers with and without a supportive companion during labor, found that labor was shortened a mean time of ten hours for mothers with support. In addition, the supported mothers remained awake longer after delivery and stroked, talked to, and smiled at their infants more.

POSSIBLE BIRTH DIFFICULTIES

Precipitous (sudden or abrupt) deliveries can be dangerous to the fetus, as can prolonged labors and premature, low-birth-weight, and postmature deliveries. Other common problems affecting labor and delivery are prolapsed cord, placenta previa, and abruptio placentae.

Precipitous deliveries are dangerous to the fetus because the skull bones are compressed rapidly rather than molding gradually. This can cause tears in the membranous coverings of the brain or even occasionally the severing of veins in these membranes. Another danger of precipitous labor is that the fetus may be born unattended. Such labors last less than three hours.

Prolonged labor may occur as a result of weak uterine contractions, a large fetus, abnormal fetal position, abnormally shaped pelvic arch, multiple fetuses, or emotional reasons. With prolongation of labor (over twenty-four hours) the fetus runs the risk of acquiring an infection, suffocating, or sustaining brain damage from insufficient oxygenation. Obstetricians may provide rest for the mother with weak contractions in order to improve them, may stimulate contractions with drugs, or may perform a Caesarean section to deliver the infant.

Preterm infants (born before thirty-seven weeks) are now differentiated from low-birth-weight (less than five and a half pounds or 2,500 grams) infants. Small-for-gestational-age (SGA) infants have weight inappropriate for gestational age, whether born before thirty-seven weeks or at term. These variables, while given separate definitions, may occur in the same baby (that is, a preterm baby may be both SGA and low birth weight) but do not necessarily do so (a preterm baby may weigh 6 pounds when born at thirty-six weeks of gestation and simply be preterm, with neither SGA nor low birth weight labels added). These infants have respiratory problems, poor temperature control, limited nutrient deposits, difficulties with sucking and swallowing, poor resistance to infection, fragile blood vessels that may hemorrhage, poor kidney function (which may lead to electrolyte imbalance), and poor liver functioning (which may lead to jaundice). Their survival may be endangered by respiratory distress syndrome (also called hyaline membrane disease) (see Chapter 4). Factors associated with low birth weight, preterm delivery,

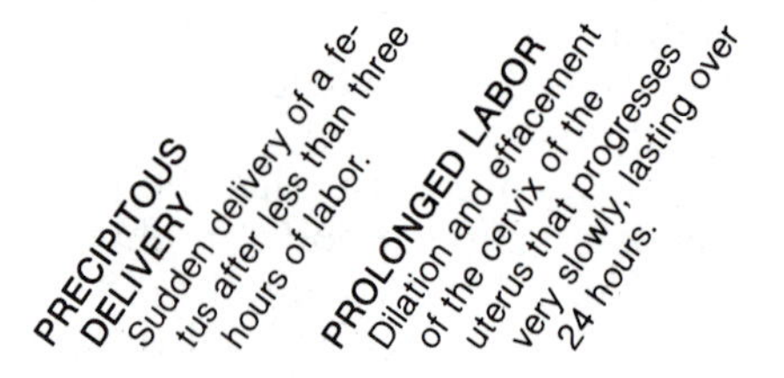

and SGA include youthfulness or agedness of the mother, maternal prenatal infections, maternal prenatal alcohol abuse, malnutrition or heavy smoking, emotional traumas during pregnancy, multiple fetuses, and placental malfunctioning. Saco-Pollitt (1981) found that high-altitude living (14,000 feet above sea level) also contributes to low birth weight. Infants born with low birth weights may also have a genetic inheritance for small size. Whatever the reason for their size, these babies need extra warmth, nutrition, protection, and possible assistance with cardiac and respiratory functioning.

The survival rates for preterm, low-birth-weight, and SGA babies all increased dramatically in the 1980s due to new drugs and techniques such as intermittent positive pressure ventilation, transcutaneous oxygen monitoring, and ultrasound scanning.

Another serious threat to the life of a fetus is prolapsed umbilical cord. The pressure of the descending head of the fetus on a cord that has partially slipped through the cervix ahead of the fetus cuts off the oxygen supply. Unless the doctor can intervene to relieve the pressure on the cord, the fetus can suffocate. If the cervix is fully dilated, the doctor will speed the delivery with forceps. If it is not, the doctor will deliver the baby by Caesarean section.

With *placenta previa* and *abruptio placentae,* the placenta separates from the uterus early in labor. The fetus can suffocate if enough placenta separates to interfere with oxygen exchange via the uterus. In placenta previa the placenta is implanted abnormally low in the uterus. The dilation and effacement of the uterus that occur with labor cause it to separate too soon. In abruptio placentae the placenta is implanted normally but separates too soon, sometimes because of eclampsia or high blood pressure in the mother, or for unknown reasons. In both placenta previa and abruptio placentae, the chances of saving the life of the fetus depend on the amount of speed of the separation and the prompt diagnosis and intervention by trained medical personnel.

Summary

The uterus of the mother is the primary prenatal environment. As protected as the developing infant seems floating in amniotic fluid within an amniotic sac, within a placenta, within a uterus, within a mother, he or she is dependent on and influenced by other things external to this setting. The greater world of the mother's environment—nutrition, health, rest, shelter, and stress—affects the embryo/fetus. The progress of prenatal development also depends on the interactions of the genetic materials inherited from each parent.

The ovum and sperm divide by meiosis as they mature and become haploid cells, with only half (twenty-three) the normal (forty-six) cell complement of chromosomes. Conception of a new human being occurs with the union of one ovum with one sperm. This union produces one new cell with forty-six chromosomes, twenty-three inherited from each parent. This new cell proceeds to divide again and again, by mitosis, reproducing other cells, each with forty-six chromosomes and each carrying the same genetic materials inherited from the parents. The period of growth and development in the uterus lasts for nine months. The cells differentiate. Some become the placenta, others become the baby-to-be (skin, muscles, bones, heart, lungs, and so on). From the second to the eighth week the developing organism is called an embryo. By the end of the embryonic period, he or she has a distinctly human appearance. For the remainder of the prenatal period, the baby-to-be is called a fetus.

While the norm is for fetuses to survive the nine prenatal months and emerge as healthy full-term infants, some imperfections do occur. Down, Klinefelter, and Turner syndromes are examples of problems caused by chromosomal defects. Tay-Sachs disease, PKU, and sickle-cell anemia are examples of problems caused by genetic defects. Spina bifida is an example of a problem that is multifactorial in inheritance.

The embryo/fetus also can be harmed by teratogenic factors in the mother's environment such as excessive radiation, some infections, poor nutrition, many drugs, alcohol abuse, and malnutrition.

At the end of nine months the fetus is ready to emerge into the external world as a truly unique individual. The contractions that push the baby out are called labor. Most babies are expelled head first. The average length of labor for a first child is thirteen hours, and for subsequent children it is eight hours. If either the fetus or mother has problems that might interfere with a normal vaginal delivery, the baby may be delivered by surgery through the abdomen of the mother (Caesarean section).

Considering the enormously complicated process from formation of mature haploid ova and sperm to fertilization to implantation to delivery of a full-term baby, it is amazing that normal infants are the rule, not the exception. Never again in the life span will growth and change be quite as rapid or quite as comprehensive.

Questions for Review

1. Outline the fertilization process by stating what, where, when, and how it occurs.
2. Discuss the pros and cons of in vitro fertilization.
3. If a mother is homozygous for brown eyes and a father is heterozygous, what is the likelihood of their child having brown eyes?
4. Galactosemia is inherited as a recessive condition. If a couple, neither of whom has galactosemia, give birth to a child with galactosemia, what ratio of children also affected would be expected among their subsequent offspring?
5. Many women have diets that consist of high amounts of carbohydrates and low amounts of protein. Discuss what effects this kind of diet may have on a developing fetus.
6. Prepare a possible program of intervention to improve the nutritional status of pregnant women that might be culturally acceptable and economically feasible.

7. Write a short letter persuading a friend to give up smoking during her pregnancy. Give her reasons for doing so.
8. The LeBoyer method of childbirth is one in which every effort is made to reduce the trauma of the birth for the infant. What might be some of the arguments against using the LeBoyer method from the point of view of the hospital staff?
9. The United States is the highest user of anesthesia and analgesics during childbirth. Why do you think this is so?
10. More women today are considering the option of having their children at home, with the assistance of a midwife. List the pros and cons of this type of delivery.

References

American Academy of Pediatrics Committee on Drugs. Effect of medication during labor and delivery on infant outcome. *Pediatrics,* 1978, *62,* 402–403.

Ayala, F. A., and Kiger, J. A. *Modern genetics,* 2nd ed. Menlo Park, Calif.: Benjamin/Cummings, 1984.

Bancroft, J.; Axworthy, D.; and Ratcliffe, S. G. The personality and psychosexual development of boys with 47 XXY chromosome constitution. *Journal of Child Psychology and Psychiatry,* 1982, *23,* 169–180.

Benson, R. C. Medical and surgical complications during pregnancy. In R. C. Benson (Ed.), *Current obstetric and gynecologic disorders and treatment,* 5th ed. Los Altos, Calif.: Lange, 1984.

Brody, J. E. As Caesarean birth rate grows, so does a debate. *The New York Times,* Sunday, September 20, 1981, p. E22.

Broman, S. H. Obstetric medications. In C. C. Brown (Ed.), *Childhood learning disabilities and prenatal risk: Pediatric round table series; 9.* Skillman, N. J.: Johnson & Johnson Baby Products Company, 1983.

Darby, B. L.; Streissguth, A. P.; and Smith, D. W. A preliminary follow-up of eight children diagnosed with fetal alcohol syndrome in infancy. *Neurobehavioral Toxicology and Teratology,* 1981, *3,* 157–159.

DeAlvarez, R. R. Preeclampsia-eclampsia and other gestational edema-proteinuria-hypertensive disorders. In R. C. Benson (Ed.), *Current obstetric and gynecologic disorders and treatment,* 5th ed. Los Altos, Calif.: Lange, 1984.

Dwyer, J. T. Impact of maternal nutrition on infant health. *Resident and Staff Physician,* 1984, *30*(8), 19–30.

Ericsson, R. J., and Glass, R. H. *Getting pregnant in the 1980s: New advances in infertility treatment and sex preselection.* Berkeley: University of California Press, 1982.

Furey, E. M. The effects of alcohol on the fetus. *Exceptional Children,* 1982, *49,* 30–34.

Ganong, W. F. *Review of medical physiology,* 11th ed. Los Altos, Calif.: Lange, 1983.

Garn, S. M.; Johnston, M.; Ridella, S. A.; and Petzold, A. S. Effect of maternal cigarette smoking on Apgar scores. *American Journal of Diseases of Children,* 1981, *135,* 503–506.

Hale, R. W. Diagnosis of pregnancy and associated conditions. In R. C. Benson (Ed.), *Current obstetric and gynecologic diagnosis and treatment,* 5th ed. Los Altos, Calif.: Lange, 1984.

Hartland, E. S. *Primitive paternity.* Washington, D.C.: American Medical Society, 1909.

Hollenbeck, A. R.; Gewirtz, J. L.; Sebris, S. L.; and Scanlon, J. W. Labor and delivery medication influences parent-infant interaction in the first postpartum month. *Infant Behavior and Development,* 1984, *7,* 201–209.

Holzman, I. R. Fetal alcohol syndrome (FAS)—A review. *Journal of Children in Contemporary Society,* 1982, *15,* 13–19.

Householder, J.; Hatcher, R.; Burns, W. J.; and Chasnoff, I. Infants born to narcotic-addicted mothers. *Psychological Bulletin,* 1982, *92,* 453–468.

Jacklin, C. N., and Maccoby, E. M. Length of labor and sex of offspring. *Journal of Pediatric Psychology,* 1982, *7,* 355–360.

Johnston, C. Cigarette smoking and the outcome of human pregnancies: A status report on the consequences. *Clinical Toxicology,* 1981, *18,* 189–209.

Kennell, J. H. The physiologic effects of a supportive companion (doula) during labor. In M. H. Klaus and M. O. Robertson (Eds.), *Birth, interaction and attachment: Pediatric round table series; 6.* Skillman, N.J.: Johnson & Johnson Baby Products Company, 1982.

Kitzmiller, J. L. Diabetes mellitus: Medical and surgical complications during pregnancy. In R. C. Benson (Ed.), *Current obstetric and gynecologic diagnosis and treatment,* 5th ed. Los Altos, Calif.: Lange, 1984.

Marshall, J. R. Infertility. In R. C. Benson (Ed.), *Current obstetric and gynecologic diagnosis and treatment,* 5th ed. Los Altos, Calif.: Lange, 1984.

McAuliffe, K., and McAuliffe, S. Keeping up with the genetic revolution. *New York Times Magazine,* November 6, 1983, 41–44, 93–97.

Meyer, M. B. How does maternal smoking affect birth weight and maternal weight gain? Evidence from the Ontario Perinatal Mortality Study. *American Journal of Obstetrics and Gynecology,* 1978, *8,* 888–893.

Naeye, R. L. Nutritional/nonnutritional interactions that affect the outcome of pregnancy. *American Journal of Clinical Nutrition,* 1981, *34,* 727–731.

Needham, J. *A history of embryology,* 2nd ed. Cambridge: Cambridge University Press, 1959.

Odent, M. The milieu and obstetrical positions during labor: A new approach from France. In M. H. Klaus and M. O. Robertson (Eds.), *Birth, interaction and attachment: Pediatric round table series; 6.* Skillman, N. J.: Johnson & Johnson Baby Products Company, 1982.

Ratcliffe, S. G., and Field, M. A. S. Emotional disorder in XYY children: Four case reports. *Journal of Child Psychology and Psychiatry,* 1982, *23,* 401–406.

Robinson, A.; Goodman, S. I.; and O'Brien, D. Genetic and chromosomal disorders, including inborn errors of metabolism. In C. H. Kempe, H. K. Silver, and D. O'Brien (Eds.), *Current pediatric diagnosis and treatment,* 8th ed. Los Altos, Calif.: Lange, 1984.

Rovet, J., and Netley, C. The triple X chromosome syndrome in childhood: Recent empirical findings. *Child Development,* 1983, *54,* 831–845.

Saco-Pollitt, C. Birth in the Peruvian Andes: Physical and Behavioral consequences of the neonate. *Child Development,* 1981, *52,* 839–846.

Sadler, T. W. *Langman's medical embryology,* 5th ed. Baltimore: Williams & Wilkins, 1985.

Sever, J. L. Perinatal infections and damage to the central nervous system. Paper presented at the Symposium on Prenatal and Perinatal Factors Relevant to Learning Disabilities. ACLD 20th International Conference, Washington, D.C., February 1983.

Sexton, M., and Hebel, J. R. A clinical trial of change in maternal smoking and its effect on birth weight. *Journal of the American Medical Association,* 1984, *251,* 911–915.

Simopoulos, A. P. Nutrition. In C. C. Brown (Ed.), *Childhood learning disabilities and prenatal risk: Pediatric round table series; 9.* Skillman, N. J.: Johnson & Johnson Baby Products Company, 1983.

Stein, Z.; Susser, M.; Saenger, G.; and Marolla, F. *Famine and human development: The Dutch hunger winter of 1944/45.* New York: Oxford University Press, 1975.

Streissguth, A. P. Smoking and drinking. In C. C. Brown (Ed.), *Childhood learning disabilities and prenatal risk: Pediatric round table series; 9.* Skillman, N. J.: Johnson & Johnson Baby Products Company, 1983.

Streissguth, A. P.; Barr, H. M.; and Martin, D. C. Maternal alcohol use and neonatal habituation assessed with the Brazelton Scale. *Child Development,* 1983, *54,* 1109–1118.

Streissguth, A. P.; Martin, D. C.; Barr, H. M.; Sandman, B. M.; Kirchner, G. L.; and Darby, B. L. Intrauterine alcohol and nicotine exposure: Attention and reaction time in 4-year-old children. *Developmental Psychology,* 1984, *20,* 533–541.

Sutton, H. E. *An introduction to human genetics,* 3rd ed. Philadelphia: Saunders, 1980.

Tatum, H. J. Contraception and family planning. In R. C. Benson (Ed.), *Current obstetric and gynecologic diagnosis and treatment,* 5th ed. Los Altos, Calif.: Lange, 1984.

Tjossem, T.; DeLaCruz, F. F.; and Muller, J. Z. *Facts about Down syndrome.* (NICHD Fact Sheet 421–948:15.) Washington, D.C.: U.S. Government Printing Office, 1984.

Yogman, M. W.; Cole, P.; Als, H.; and Lester, B. Behavior of newborns of diabetic mothers. *Infant Behavior and Development,* 1982, *5,* 331–340.

Infancy

4

Chapter Outline

Physical Development
- *Physical Status of the Neonate*
- *Physical Status of a Low-Birth-Weight Neonate*
- *Physical Changes in the First Two Years*
- *Health Maintenance*
- *Common Health Problems in Infancy*

Cognitive and Language Development
- *Maturation of the Infant Brain*
- *Piaget's Sensorimotor Stage*
- *Language Beginnings*
- *Fostering Learning*

Psychosocial Development
- *Social Milestones*
- *Attachment*
- *Separation*
- *Erikson's Concept of Trust Versus Mistrust*
- *Individual Differences*
- *Fathering*
- *Discipline*

Key Concepts

accommodation
Apgar scale
assimilation
attachment
bonding
brain stem
Brazelton Neonatal Behavioral Assessment Scale
central nervous system
cephalocaudal development
cerebellum
cerebrum
colic
colostrum
contact comfort
croup
discipline
eczema
equilibrium
expressive language
failure-to-thrive syndrome
general to specific development
holophrases
hospitalism
individual differences
infant daycare
iron deficiency
lactation
language acquisition device (LAD)
lanugo
motor milestones
myelin
neurons
neurotransmitters
phonetic drift
pincer grasp
proximal-distal development
pyloric stenosis
receptive language
respiratory distress syndrome
reflex abilities
rooming-in
schemas
secure attachment
sensorimotor intelligence
soft spots
stranger anxiety
sudden infant death syndrome (SIDS)
synapses
telegraphic speech
trust versus mistrust
vernix caseosa

♥

Case Study

Bernadette's son had been delivered at term, weighing 8 pounds and receiving a five-minute Apgar score of 10. He was healthy! He continued to be healthy, eating and sleeping well and displaying an "easy" temperament. He hit his developmental milestones on schedule: sat alone at twenty-six weeks, crawled at twenty-eight weeks, and pulled himself up to a standing position at thirty weeks. Then, one Sunday morning in his thirty-sixth week of life, he vomited after his morning feeding. "So what?" said Bernadette's husband when she told him. She persisted. Mikey's cry didn't "sound" normal. He was like a ragdoll in her arms between his crying attacks. Something was really wrong! Besides, he had never vomited—spit up, yes; vomited, no. Against her husband's advice, Bernie called the pediatrician (it *was* a Sunday morning). The doctor agreed to meet the family in his office in half an hour. When the pediatrician examined little Mikey, he found a hard abdominal mass. The party of four trekked to the nearest hospital, where an x-ray revealed an intussusception (a telescoping of one portion of the intestine into another). Because the obstruction was found so early after its inception, it was possible to cure it (separate the telescoped parts) with a barium enema, nullifying any need for surgery. The pediatrician complimented Bernadette for her wisdom in calling him, even on a Sunday morning. "How did you know?" he asked. "It was the way he cried," she answered. What do you think?

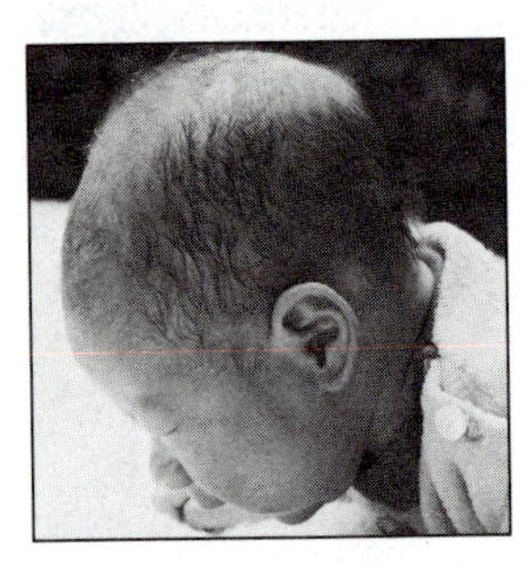

Specialists in the study of infancy differ in their opinions as to when infancy ends and childhood begins. The word *infant* derives from the Latin *infans,* which literally means "without language." The consensus of many experts is that the advent of meaningful verbal communications marks the end of infancy. However, some babies begin using language before they reach their first birthdays, walk, or perform other childlike activities. Other babies walk, run, are toilet trained, and reach their third birthdays before they talk. Consequently, this delineation of infancy is often invalid. Other experts use motor milestones such as toddling, toilet training, or self-feeding with a spoon to mark the end of infancy. Such boundaries also have too much variability from infant to infant to make a good case for using them for all babies. Age is a third frequently used demarcation for the end of infancy. Again, controversy exists. Do infants become toddlers at eighteen months? At two years? At three years? This chapter will be eclectic. All three boundaries (language, motor tasks, age) will be encompassed to some extent. Infants will be defined as human beings who have not yet learned to communicate in short sentences or walk around with ease and who are below the age of two. This admits that some babies will be toddlers before age two, whereas others will still be babies up through their second birthdays. The term *neonate* is much easier to define. It will be used to describe newborn infants in their first month of life (see Figure 4–1).

Physical Development

Physical development in infancy exceeds that of any other time period of the postnatal life span. Although the brain stem, heart, and lungs of a neonate are mature enough to make extrauterine life possible, many other organ systems are still quite immature. They undergo rapid changes during infancy, resulting in a metamorphosis from a totally dependent, reflex-bound neonate to a toddling, talking, thinking child. Let us examine the physiological status of the neonate first. This will provide a base from which to compare the physical growth and development that take place in the months of infancy.

Figure 4–1
All babies, from the helpless newborn to the rather independent, self-propelling, self-feeding 18-month-old, are in the stage of life called infancy.

PHYSICAL STATUS OF THE NEONATE

At birth the infant must literally conform to a whole new world. The era of floating in a warm amniotic sac filled with fluid, and of having all the necessities of life supplied through the umbilical cord, ends. The transition from intrauterine to extrauterine existence requires many adaptations (see Figure 4–2).

Survival of newborns requires that circulation be transferred from the umbilical cord to the heart and lungs. The blood vessels that pass from the placenta to the baby through the cord must be obliterated. First the cord is "stripped." This means that the doctor or midwife pushes all the extra blood and nutrients from the cord toward the baby's body. Then the cord is clamped shut close to the infant's abdomen. Finally, the cord is cut between the clamp and the placenta (afterbirth). The cord stump remaining on the baby dries up and falls off in one or two weeks. The umbilical site heals to form the navel. Further adjustment of the circulatory system to extrauterine life requires that two openings, present prenatally, now close. The *ductus arteriosus,* which is a shunt between the aorta and the pulmonary artery, obliterates by about age three months. The *foramen ovale,* which is the connection between the right and left atria of the heart, functionally closes with a flap by the first week of life, although it may not obliterate completely until much later (Wolfe and Wiggins, 1984).

Survival of newborns also requires that the exchange of oxygen and carbon dioxide be immediately transferred from the placenta

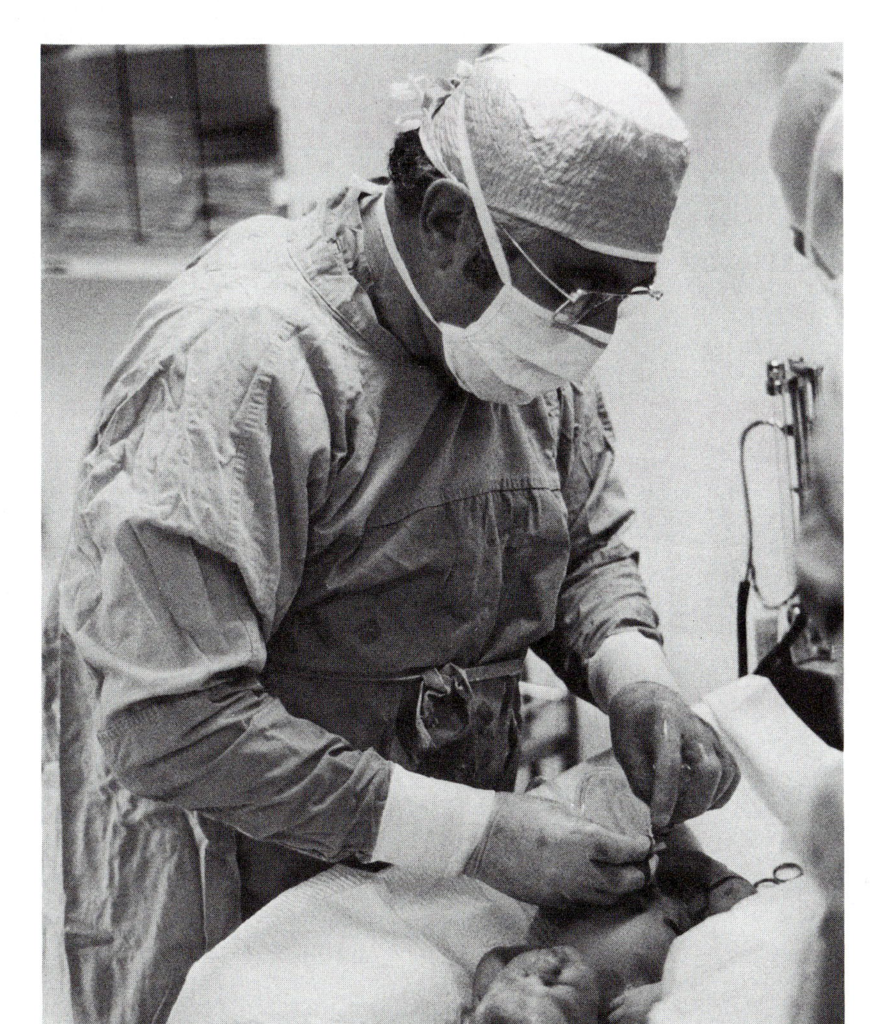

Figure 4–2
The transition from intrauterine to extrauterine life is a giant step for a neonate. Breathing and blood circulation are transferred to the infant's lungs and heart as the tie to the mother through the placenta and umbilical cord is severed.

to the lungs. Respiratory reflexes become activated within seconds after birth. Due to pressure changes infants usually breathe spontaneously and cry, thereby allowing full expansion of their lungs with air. However, if the respiratory passages are still blocked with residual amniotic fluid and mucous from intrauterine life, breathing will be impaired. For this reason babies are held upside down after delivery, allowing such materials to drain out. Some hospitals also routinely suction the respiratory tracts of newborns to hasten free breathing. The belief that the doctor or nurse must slap the baby to start it crying is false. Spanking, back slapping, or immersing the baby in cold water can be dangerous. Instead, the newborn's heels are flicked or tapped lightly. If infants do not begin to breathe within sixty seconds of birth, resuscitation measures may be instituted to prevent cell damage or death from lack of oxygen (called anoxia). Failure of the respiratory control center to maintain adequate breathing, whereby oxygen is taken into the body, is a leading cause of mortality and brain damage in infants. Likewise, any failure of the cardiovascular system to pump the oxygen-carrying blood to the brain can result in brain injury or death. The demands of the neonate's brain for oxygen are huge. The blood vessels that supply the brain with oxygen are still quite fragile. In addition, the muscular support system for the head is weak. Consequently, the baby's head must be supported carefully and not be allowed to jerk rapidly from side to side. Such careless handling, which can cause blood vessels to break, can lead to brain damage or death.

At one minute and again at five minutes after hospital deliveries, babies are given a score for heart rate, respiratory effort, muscle tone, reflex irritability, and color, on the **Apgar scale** developed by Dr. Virginia Apgar (see Table 4–1). Very few infants are oxygenated enough to have pink extremities at one minute. Therefore, the highest score an infant usually first receives is 9. By five minutes healthy infants may receive a score from 5 to 10. Infants scoring lower than 4 need some form of prompt diagnosis and treatment. Very low-birthweight infants generally have low Apgar scores.

Table 4–1 The Apgar Scoring Method

Sign	0	1	2
Heart rate	Absent	Below 100	Over 100
Respiratory effort	Absent	Minimal; weak cry	Good; strong cry
Muscle tone	Limp	Some flexion of extremities	Active motion; extremities well flexed
Reflex irritability (response to stimulation on sole of foot)	No response	Grimace	Cry
Color	Blue or pale	Body pink; extremities blue	Pink

SOURCE: From Virginia Apgar, *Anesthesia and Analgesia,* 32:260, 1953. Reprinted by permission.

Survival of the neonate requires that warmth be provided in the period immediately following birth. The temperature-regulating mechanism of the nervous system is not yet sufficiently mature to allow the baby to shiver or sweat to raise or lower his or her own temperature. Also, the intrauterine environment is much warmer than room temperature. The neonate is accustomed to an environment closer to 32°C (90°F). In this so-called neutral thermal environment the baby finds it easier to breathe. Therefore, in hospitals infants are dried and placed in warm basinettes or carriers as soon as respirations are established and the cord is clamped. Many times stockinette caps are put on the infants' heads to help retain body heat.

The immune system at birth is in an immature stage, requiring that neonates be protected against infections. Silver nitrate drops or a penicillin ophthalmic solution are instilled in the neonate's eyes immediately after birth as a prophylactic (protective or preventive) measure against gonorrheal infection. Various hospitals may separate mother and infant for the period of hospitalization (usually two to three days) or may allow modified or complete **rooming-in**. With separation the infant remains in a nursery and is taken to the mother only for short visiting hours (usually coinciding with feeding for both breast- and bottle-fed infants). With modified rooming-in the baby stays in a crib at the mother's bedside when she so desires and is taken back to the nursery when she wants to rest or at night. Complete rooming-in permits the infant to remain with the mother constantly. Each plan has its immunological advantages and disadvantages. When an infant resides in a nursery with other babies, there is a danger of cross-infant spread of infections. When an infant stays in the mother's room, there is a danger of infections being carried in by visitors or the mother. Actually, neonates have some immunity to a few organisms at birth, despite their own limited ability to produce antibodies. They have received certain antibodies through the placenta prenatally. These will protect them for several months while their own immune system is developing. Neonates who are breast-fed receive additional antibodies, first through **colostrum**, a sweet, thin fluid that is produced by the breast for two to three days before milk comes in, and then later through the mother's milk. The merits of both breast- and bottle-feeding will be discussed in more detail later in this chapter.

The neonate is seldom very hungry immediately following delivery. First feedings are either sweetened water or colostrum. Milk is generally not tolerated well for a day or two. Then milk becomes the only necessary source of nutrients for the remainder of the neonatal period.

Sleep is the infant's major activity throughout the first few weeks of life. Some babies stay awake for an hour and a half before their first sleep; others fall asleep after delivery and do not wake up fully until the second or third day.

The bones of the skull ossify (change from cartilage to bone) prenatally and protect the neonate to some extent, although there

APGAR SCALE
Medical technique to measure adjustment of neonate at birth; measures appearance, pulse, grimace, activity, and respiration.

ROOMING-IN
Hospital policy that permits the neonate to remain with the mother in her room during the postdelivery days.

COLOSTRUM
The fluid secreted by the breasts for several days after childbirth preceding the secretion of milk.

are still spaces between the bones of the head at birth (called anterior and posterior fontanelles and sutures—see Figure 4–3). These unossified areas allow the skull bones to overlap each other slightly during labor (molding) to ease passage through the pubic bones of the mother. The fontanelles are commonly referred to as the **soft spots** in infants. The two largest ones, on the top back and top front of the skull, will not close until about four months and nine to eighteen months respectively. Although they are covered with a tough membrane, care must be taken to prevent injury to the baby's brain and the underlying brain vessels through these fontanelles.

Although the skull bones, clavicle, spine, and long bones of the extremities partially ossify (change from cartilage into bone) before birth, many other bones-to-be in the neonate are still completely cartilage. Ossification continues rapidly in infancy. Therefore, an abundant supply of calcium from milk is a vital part of the diet. The

Figure 4–3
Schematic representation of sutures and fontanelles (soft spots) of a neonate's skull.

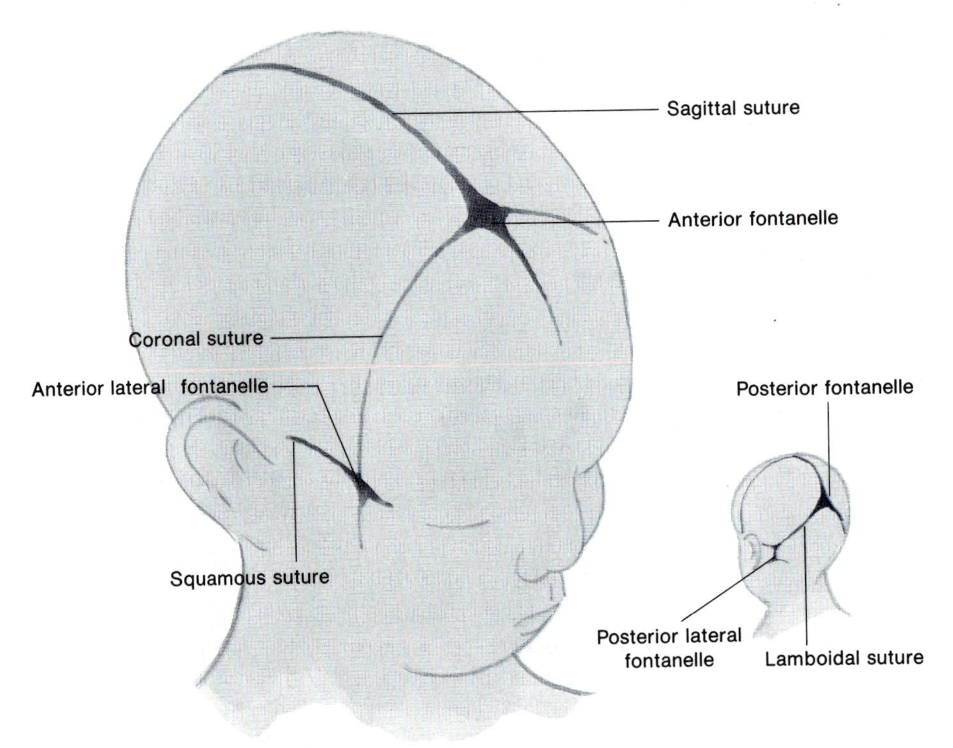

SOFT SPOTS The spaces (fontanelles) between the bones of the neonate's skull that have not yet changed from cartilage to bone.

LANUGO Fuzzy prenatal and neonatal body hair.

VERNIX CASEOSA Waxy covering of the fetus in utero; also found on the neonate.

rate and sequence of bone development has been established by means of x-ray visualization. There is an established correlation between the degree and location of ossification and the age of the infant. This relationship is the basis of the concept of *bone age*. Factors that interfere with bone maturation, such as severe disease, endocrine disorders, and malnutrition, all influence the development of bones in a way that can be measured by comparing the infant's bone age to chronologic age. At birth, as well as throughout infancy, childhood, and adolescence, girls are more advanced in skeletal development than boys.

The neonate's physical appearance is quite different from that of older infants and much different from adults. The head accounts for about one-fourth of the newborn's length. It may be misshapen for a day or two postnatally due to pressures on it during birth. The trunk is small with rounded shoulders and narrow hips. Skin is generally thin and wrinkled and may be covered with downy hair called **lanugo**. A rosy red flushing is common when the neonate cries. Some of a white, waxy substance composed of fatty materials that "varnished" and protected the fetus in the uterus may still be present on the skin. This is called **vernix caseosa** (see Figure 4–4). The hands remain curled and both arms and legs remain in their fetal posture for several days. Both male and female neonates have slightly enlarged breasts and genitals due to the prenatal influence of maternal hormones.

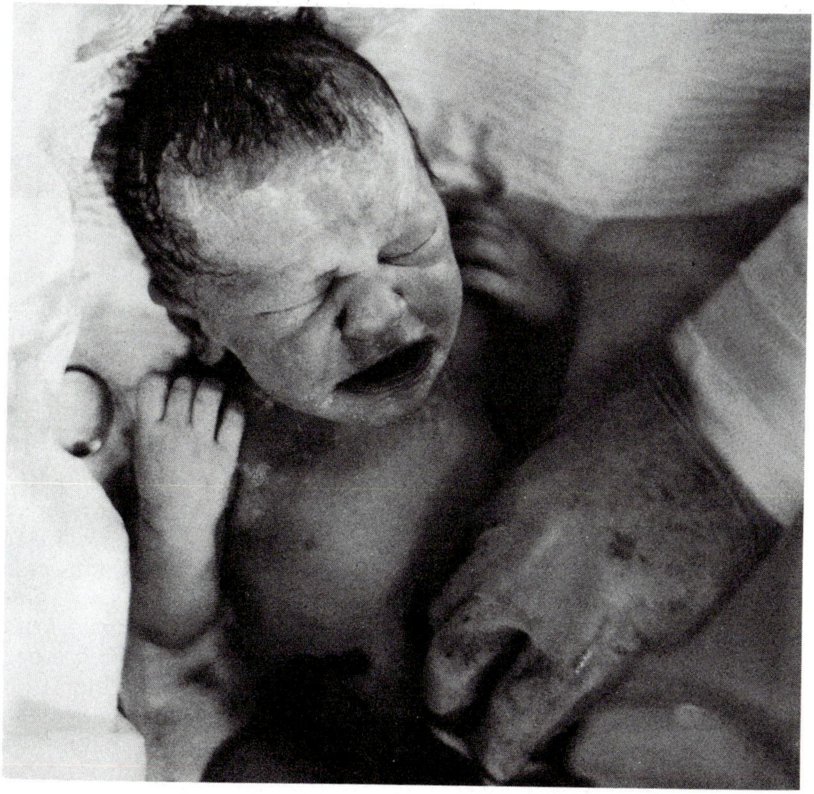

Figure 4–4
A healthy newborn may appear to have been dusted with flour. White splotches of vernix caseosa contrast with the deep flush that develops when the baby cries. A newborn's head appears disproportionately large compared to the trunk and curled up arms and legs.

Physiologic jaundice, a yellow coloration of the skin and of the sclera (white area) of the eyes, appears in about one-third of full-term neonates in the first three days of life (Koops and Battaglia, 1984). It is caused by the presence of high levels of bilirubin (a bile pigment that is a breakdown product of fetal hemoglobin) in the bloodstream. It usually disappears by one week of age, at which time the infant's liver has developed the ability to metabolize the bilirubin, allowing its excretion. If the bilirubin level climbs too high, the neonate may be placed under full-spectrum lights (phototherapy), which enhances the breakdown of bilirubin. The baby's eyes must be covered for their protection if phototherapy is used. The prognosis (probable course and chance of recovery) is excellent.

All babies' eyes are dark smoky blue at birth. The lids are puffy and the tear ducts do not function. Eye muscles are weak and the eyes may occasionally drift to a crossed or wall-eyed position. Visual focus is fairly rigid, and the neonate will see objects best if they are from 9 to 20 inches from the eyes, the distance the mother's face is from the baby when she is breast-feeding. Visual acuity is poor in the newborn: a neonate can see at 20 feet what an adult with normal visual acuity can see approximately at 300 feet (Banks and Salapatek, 1983). Neonates attend mainly to large objects close to their faces.

Hearing is developed in the uterus where the fetus is exposed to the sounds of the mother's heartbeat, intestinal rumblings, and blood flow. The neonate will turn to sounds immediately after birth and often sleeps more peacefully when background noises are present. For this reason many newborn nurseries play a tape or record of the "Lullaby of the Womb."

The average neonate weighs about 3000 to 4000 grams (6½ to 8½ pounds). Length is about 50 to 53 centimeters (20 to 21 inches). The average here means the ranges where considerable numbers of infants fall. An infant may be above or below average and still be normal. Although girls are more advanced in bone maturation at birth, boys tend to be, on the average, 5 ounces heavier and one-half inch longer than girls.

Survival of neonates is enhanced by several **reflex abilities** that exist at birth (see Table 4–2). When placed face down on a surface, they will draw their knees up under their abdomens and turn their heads to one side (head-turning reflex). Normal infants are thus protected from smothering unless someone puts a soft pillow or loose cloth under their heads that will block the respiratory tract or prevent the lifesaving head turn. When placed on their backs, infants turn their heads to one side and extend their arms and legs on the corresponding side (tonic neck reflex). This reflex is often referred to as the fencing reflex, as the neonate looks ready to practice the art of fencing, minus the foil. When the cheek is touched, infants will turn their heads toward the contact (the rooting reflex—see Figure 4–5). This reflex may assist them to find food. They begin to make sucking movements (sucking reflex) followed by swallowing movements (swallowing reflex), on any suckable object with which their mouth makes contact. Neonates cough, sneeze, and yawn reflexively and frequently. These actions help them clear their respiratory tract

REFLEX ABILITIES Those actions performed involuntarily by infants in response to some external excitation.

Table 4–2 Reflex Abilities of the Neonate

Reflex	Description	Age range
Protective head turning	Place infant face down. Infant will turn head to one side.	Birth on
Tonic neck	Place infant on back with head to one side. Arm and leg on head side will be extended.	1–10 months
Moro	Startle baby. Arms and legs will first extend, then flex close to body.	0–6 months
Stepping	Hold baby upright with feet touching surface. Infant will alternately raise and lower feet.	0–6 months
Smiling	Watch infant when well fed and sleepy. Lips curl up in smile posture.	0–1 month
Grasp	Place fingers in infant's palms. Lift. Infant will hold tight enough to be raised from surface.	0–4 months
Babinski	Stroke bottom of baby's foot. Toes will extend out and apart.	1–12 months
Rooting	Stroke baby's cheek. Head will turn toward stimulus.	0–12 months
Sucking	Place clean finger in infant's mouth. Infant will suck it.	0–12 months

Figure 4–5
To root means to poke around in order to find something. The rooting reflex of human infants leads to a turning of nose and mouth in the direction of any facial touch. This may lead to food or to other novel encounters.

and get more oxygen. When they are startled by sudden noises or bumps on their beds, they throw out their arms and legs, then hug them together over their bodies (startle or Moro reflex). This reflex is frequently used to assess the neurological status of the baby, as is the Babinski reflex. This latter refers to an extension of the toes when the bottom of the foot is stroked. Consistent absence of both reflexes suggests a delay in central nervous system maturation. When supported upright, neonates will appear to take steps (stepping or dancing reflex). When well fed and about to drop off to sleep, they will smile (smiling reflex). (This reflexive smile precedes the social smile that occurs in response to interactions with people some weeks later.) When an object or finger is placed in the neonate's hand, he or she will grasp it so tightly that often the baby can be lifted off the surface (grasping reflex).

The **Brazelton Neonatal Behavioral Assessment Scale** (NBAS) (Brazelton, 1973) is used to assess neurological integrity by examining twenty reflex items. It is also used to assess twenty-seven behavioral items to see how a neonate (under thirty days old) responds to the environment while he or she is in six states:

quiet sleep	active alertness
active sleep	animated alertness
drowsy alertness	irritable or crying awake state

Observations are made on lability (instability) of state, skin color, rapidity of build-up, self-quieting activity, startle responses, irritability, and spontaneous movement. Psychophysiological reactions can be judged on motor, cognitive, social, and temperamental modalities. The assessment takes about a half hour and must be done by a trained examiner who has established an acceptable level of observer reliability. Many pediatricians, nurses, and researchers are so trained and routinely assess the behavioral status of neonates with the NBAS. It provides valuable information about a neonate's current status.

A modification of the NBAS was developed by Als, Brazelton, and other colleagues (1982) to assess preterm infants. Called the Assessment of Preterm Infant Behavior (APIB), it examines a series of interactions between the observer and the preterm infant to determine the preterm's abilities to organize and integrate behaviors.

Another new form of neonatal assessment is brain electrical activity mapping (BEAM). It utilizes neurological techniques that have been available for years, the electroencephalogram and sensory evoked potentials, and makes the data more visible with computerized topographic mapping. An extension of BEAM, called significance probability mapping (SPM), can be used to delineate regions of difference between two BEAM images. This neurophysiological technique may make it possible to distinguish between neurologically normal and abnormal neonates in the first three days of life (Duffy, Burchfiel, and Lombroso, 1979).

BRAZELTON SCALE Scale used to assess the neurological integrity and behaviors of neonates.

Perhaps the simplest way to assess a neonate's status is to listen to his or her cry. Zeskind and Lester (1978) demonstrated that biological differences between normal and abnormal neonates can be detected with spectrographic analyses of cry features, even in babies who show no other abnormal physical, behavioral, or neurological signs (see Box 4–1).

What's in a Neonate's Cry?

In a study conducted by Zeskind and Lester (1978), college students, male as well as female, all inexperienced with infants, listened to the taped crying of neonates, some of whom were normal while others were not. Could the naive students tell the difference? Yes! They rated the cries of the high-risk neonates as sounding more urgent, grating, arousing, piercing, distressing, discomforting, aversive, and sick. In another study, Zeskind (1980) asked parents and nonparents to report what response they would make to each neonate that they heard crying on tape. Some of the cries were of normal infants, others were of high-risk infants. Parents (experienced with crying babies) were better than nonparents in choosing appropriate responses. Parents consistently responded with tender and caring behaviors for "sick" sounding cries and with behaviors that would effectively terminate crying for "distressing" and "urgent" sounding cries. No parent chose a "wait and see" or "give a pacifier" response for the cry of a high-risk neonate. The urgent, distressing, sick sounds identifiable even by nonparents are functional in eliciting appropriate caregiving behaviors.

When the cries of normal and high-risk neonates are subjected to sound analysis using a spectrograph, a machine that can plot the threshold, latency, activity, and frequency of cries, the differences heard by students, parents, and everyone else can clearly be seen. Neonates who appear "normal" on routine physical and neurological examinations may be spotted by spectrum analysis to really be "at risk." This is especially true of neonates who have been malnourished prenatally (Zeskind, 1983). The cries of malnourished infants are harder to elicit, are shorter in duration, and contain more high-pitched sounds than the cries of normal infants. Different types of cries can signal different risks. Down syndrome babies, for example, tend to have unusually low-pitched cries. There are indications that infants at risk for sudden infant death syndrome may also produce low-pitched cries. Preterm babies produce "urgent" high-pitched cries.

Adult listeners are aroused by high-pitched cries. While the initial effect is functional and produces caregiving, over time the effect may be paradoxical. An aroused adult may feel inadequate, frustrated, and hostile toward the crying baby, especially if all attempts to soothe the infant fail. There is some evidence that high-pitched crying may be associated with later parental abuse. Can a parent intervention program for preterm neonates with high-pitched cry sounds reduce the risk of eventual abuse? Zeskind and Iacino (1984) began a program to help mothers of hospitalized high-risk neonates increase their contact and familiarity with their babies and develop more realistic expectations for their behaviors. Unexpectedly, this intervention had another positive effect. It facilitated the recovery and reduced the hospitalization time of the preterm high-pitched-crying neonates in the study.

PHYSICAL STATUS OF A LOW-BIRTH-WEIGHT NEONATE

As defined in Chapter 3, low-birth-weight infants weigh less than 5½ pounds (2500 grams). They may be preterm (born before thirty-seven weeks) or small for gestational age (SGA) with weight inappropriate for gestation, whether born before thirty-seven weeks or at term. While there has been a dramatic reduction in neonatal deaths in all birth-weight and gestational-age categories with expanding technology and neonatal expertise, the United States still loses more than 45,000 neonates each year. Two-thirds of these are low-birth-weight infants. About 7 percent of all neonates fit into the low-birth-weight category.

While low-birth-weight neonates differ, depending on their weight, gestation, and possibility of other major problems (such as chromosomal or genetic anomalies), most of them face problems with respiration, temperature maintenance, susceptibility to infection, feeding, nutrition, and liver functioning.

A common sequela to low birth weight is **respiratory distress syndrome** (also known as hyaline membrane disease of the lung). The immature lungs lack *surfactant,* a substance that lubricates the air sacs and helps them inflate. The neonate consequently suffers hypoventilation and hypoxia. The airways must be kept open mechanically. This is now accomplished with continuous positive airway pressure (CPAP) delivered into an incubator, a hood, or tubes in the infant's nose or mouth. The pressure is continued several days or until the neonate produces the surfactant that will allow his or her own air sacs to remain inflated. When CPAP is in use, measurements of the oxygen and carbon dioxide mixture of the neonate's blood can be taken continuously through a method that allows for monitoring across the skin. When CPAP is discontinued, another monitoring device is attached to the incubator or bassinet to emit a warning signal should respirations cease.

Another common sequela to low birth weight is a large ductus arteriosis, which allows too much blood to be pumped into the lungs with each heartbeat. In the majority of cases, this problem can be alleviated by injecting an intravenous drug that triggers closure of the duct. If drug therapy doesn't work, surgery may be required to prevent heart failure.

Low-birth-weight infants have both limited fat deposits to protect them from heat loss and poor temperature-regulating abilities. Consequently, they must be kept warm in incubators or bassinets set to maintain skin temperature between 96.8° and 98.6°F. Provision of adequate heat allows the neonate to use the calories taken in for growth rather than heat expenditure. Incubation also helps prevent infection, to which low-birth-weight and preterm infants are particularly susceptible. Parents may still visit their tiny baby, and even help care for him or her through incubator windows.

Nutritional status is often poor in low-birth-weight, preterm, and SGA infants. Many have limited stores of iron, which can lead to later anemia. Many are also hypoglycemic (which can lead to brain damage) and deficient in folic acid. Correcting nutritional imbalances and maintaining adequate postnatal nutrition are extremely

RESPIRATORY DISTRESS SYNDROME
Breathing disorder of the newborn caused by underexpansion of the lungs and reduced lung volume, resulting in hypoventilation and hypoxia.

critical, yet difficult. Low-birth-weight infants have weak sucking, swallowing, gag, and cough reflexes, leading to difficulty in feeding and danger of aspiration (Koops and Battaglia, 1984). They often require intravenous feeding. Frequent blood samples are taken to assure that the neonate has sufficient glucose, which can be added to the special low-birth-weight-infant formula. The formula has extra protein as well as vitamins, carbohydrates, and fats balanced to support rapid postnatal growth.

Since liver functioning is not well developed, many low-birth-weight neonates develop jaundice caused by an excess of bilirubin, which is normally broken down by the liver. This can be corrected with phototherapy.

Low-birth-weight and preterm babies often have poor muscle control, poor orienting responses, and abnormal reflex activity. Researchers have discovered that incubators equipped with oscillating waterbeds and rhythmic sounds (Burns et al., 1983) and that provision of manual rocking and sound stimulation to incubator babies (Barnard and Bee, 1983) will improve their motor responsiveness.

Transfer from an incubator to a bassinet is usually accomplished when the neonate reaches 4 to 4½ pounds (1800 to 2000 grams). Weaning from intravenous to oral feedings is accomplished gradually when the neonate is strong enough to suck and swallow without aspirating the formula. When milk is substituted for the special formula, some vitamin and glucose supplementation may still be necessary to prevent the development of rickets or hypoglycemia. Discharge from the hospital is accomplished when the neonate eats well, has all existing medical problems under control, and weighs from 4½ to 5½ pounds (2000 to 2500 grams).

The long-term prognosis for surviving low-birth-weight, preterm, or SGA infants is much better than it used to be. Still, most will show some delays in sensorimotor, personal-social, and gross motor abilities for one to two years (Ungerer and Sigman, 1983). Some will have permanent neurological difficulties, learning and perceptual problems, or mental retardation. Permanent handicaps are greatest in the neonates who are the tiniest or the most neurologically abnormal at birth (Pape and Fitzhardinge, 1981).

The physical status of a low-birth-weight neonate is improved when parents become actively involved in care during the time while the neonate is still hospitalized (Holmes, Reich, and Pasternak, 1984). Most intensive-care nurseries not only allow this but also teach mothers and fathers how to relate to their babies and to be sensitive to their special needs. Even with this interaction encouraged, parents report many negative reactions to having an infant in intensive care. They are often emotionally upset, uncertain about the prognosis, inconvenienced by the distance and expense of traveling to the hospital and the need to find babysitters for the children at home, alienated from the tiny baby who seems to belong more to the hospital staff than to them, resentful that they cannot take their baby home, and worried about the special care their infant will need after discharge (Jenkins and Pederson, 1985).

The prognosis for the neonate is improved when parents are

able to adjust to all the problems and expenses manifest by their high-risk infant. This is easier if parents are not economically disadvantaged and if they have strong social support networks available to them in their extended families and communities. Substantial differences in performance have been found between surviving low-birth-weight infants depending on their advantaged or disadvantaged status (Koop, 1983). Home intervention programs following discharge can help improve performance in high-risk babies of low-income families (Ross, 1984).

PHYSICAL CHANGES IN THE FIRST TWO YEARS

The physical changes of infancy progress in an orderly, sequential way for all normal babies, although the exact age at which each change occurs varies widely from baby to baby. In the 1930s and 1940s, Arnold Gesell, Frances Ilg, and other colleagues at the Yale University Clinic for Infant and Child Development filmed and studied sequences of infant development. Out of this "cinemanalysis" came the conclusion that growth proceeds in certain developmental directions: from head to foot (called the **cephalocaudal** direction), from the center outward (called the **proximal-distal** direction), and from **general to specific** movements. Cephalocaudal direction is well illustrated by the fact that the head accounts for about one-fourth of the infant's length at birth. By about one year of age the circumference of the chest finally becomes equal to that of the head. Proximal-distal development can be illustrated by the fact that the first bones to ossify are those close to the central, vital organs (skull, clavicle, spine) that are formed before birth. The bones of the fingers and toes do not begin to ossify until the end of the first year of life. General to specific development can be illustrated by grasping. Infants can crudely hold larger objects with both hands by about four months. They can hold smaller objects in one hand between thumb and forefinger (**pincer grasp**) by about one year of age. Figure 4–6 illustrates these developmental directions.

Gesell's studies also indicated that development does not progress in a straight line but rather oscillates back and forth between periods of rapid and slower maturing.

Visual acuity develops rapidly during infancy. While a neonate's vision is about 20/300, a six-month-old's acuity is closer to 20/100 (Banks and Salapatek, 1983). Color perception is similar to that of adults by three months, and infants can discriminate among forms. The ability to accommodate the curvature of the lens to shift

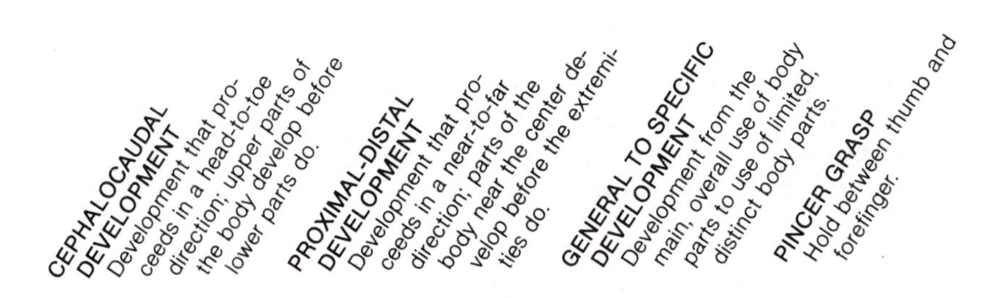

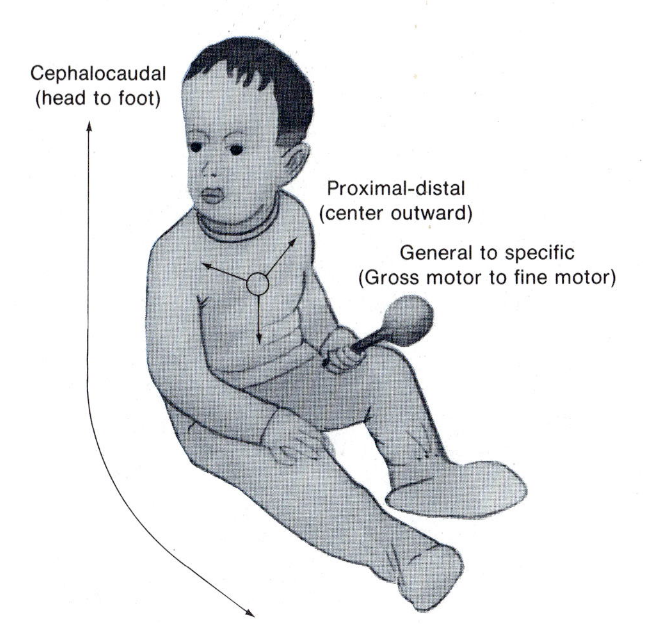

Figure 4–6
Principles of developmental direction.

from focusing on near objects to far objects (or vice versa) also becomes rapid and quite accurate by age three months.

Hearing is well developed at birth. Recent research suggests that infants are relatively more sensitive than adults to high-frequency sounds (Aslin, Pisoni, and Jusczyk, 1983). Many infants seem to prefer the higher pitched, slower rhythm voices of women (*motherese*) to the lower pitched voices of men (Fernald, 1985). Infants will attend to changes in rhythm or pitch of voice. If you want a baby to listen, change your voice.

Weight and height both progress at faster rates in infancy than at any other postnatal time of life. Infants tend to gain an average of about 20 grams (two-thirds of an ounce) per day during their first five months and 15 grams (half an ounce) per day for the remainder of their first year of life. They can be expected to double their birth weight by about five months, triple it by one year, and quadruple it by two years of age. Length during infancy can be expected to increase by about 25 to 30 centimeters (10 to 12 inches) during the first year and by about 12 more centimeters (5 inches) at the end of the second year. The weight of the brain multiplies most rapidly during infancy as cells enlarge, acquire longer, branched processes, and gain myelin sheathing. At the end of the first eighteen months, the brain is 75 percent of its adult weight. By the fourth year, 90 percent of its final weight has been attained (Holt, 1977).

Nutrition is vitally important to an infant's physical development. Milk is sufficient to meet all of a baby's needs for the first six months of life. It may be mother's milk, formula, or a combination of both. Breast-feeding is encouraged because the mother's milk, in addition to being less expensive, always ready, sterile, and at the right temperature, has many nutritional advantages. It contains higher levels of lactose (milk sugar), vitamin C, and cholesterol than cow's milk and less protein. It also has a more efficient nutritional balance between iron, zinc, vitamin E, and unsaturated fatty acids (O'Brien and Hambidge, 1984). Human milk seems particularly suitable for rapid brain cell development. The additional cholesterol in early infancy may induce the production of enzymes required for cholesterol breakdown in adulthood. Another advantage of breast milk is that it also contains anti-infectious agents, which protect the baby from gastrointestinal and upper respiratory disease. Still another benefit of breast-feeding is that the process of **lactation** (secretion of milk by the mammary glands) stimulates the uterus to shrink more rapidly to its normal nonpregnant size.

Some mothers are discouraged from breast-feeding because it takes so long. Neonates require long feeding times, whether breast- or bottle-fed. By one month of age, all babies speed up their feeding time as they become more proficient at sucking, even though the amount they consume remains the same (Pollitt, Consolazio, and Goodkin, 1981).

Breast-feeding may be discouraged if the mother is in poor health, malnourished, or taking medicines that may diffuse into the breast milk. On occasion a nursing mother may also develop an abscess (swollen, inflamed area) on one of her nipples that prevents her from feeding her infant. Breast-feeding may also be discouraged if the infant is low birth weight, SGA, or premature or suffers from a cleft (split) lip or palate. Sometimes the mothers' milk may be pumped and fed to these infants until they are able to suckle.

Many commercial formulas are available for bottle-feeding infants, fortified with vitamins and modified in protein content. Formulas may also be prepared at home from milk, water, and sugar in a ratio determined by the physician. Many infants have adverse reac-

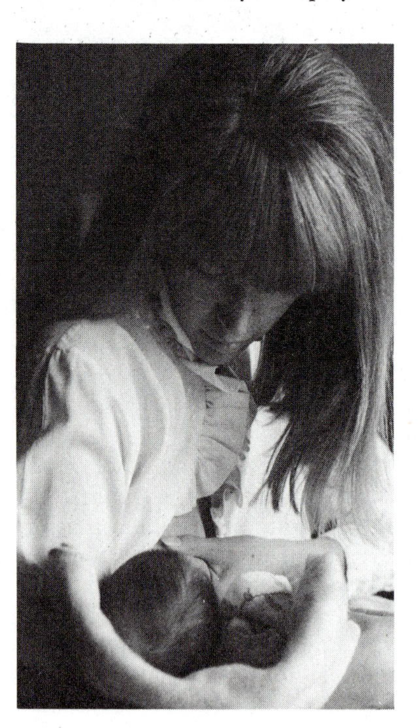

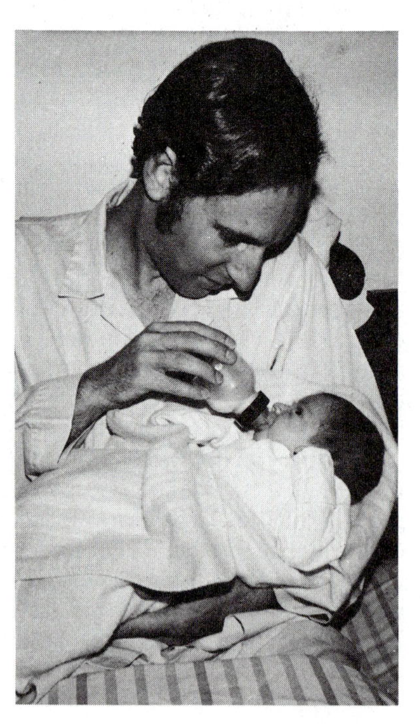

Figure 4–7 *Both breast- and bottle-feeding can satisfy an infant's need for contact comfort when adults cuddle the baby close and provide face-to-face attention.*

LACTATION Secretion of milk by the mammary glands; breast-feeding.

tions to cow's milk (rashes, wheezing, diarrhea). For this reason many of the commercial formulas substitute vegetable proteins for cow's milk.

Care must be taken when bottle-feeding to assure cleanliness and psychological stimulation (touching, fondling, and so on). Either breast- or bottle-feeding can be satisfying if accompanied by tender loving care (see Figure 4–7). Infants should not, however, be fed by propping bottles in their cribs. In addition to emotional deprivation, this may cause obstruction of the eustachian tube and lead to repeated ear infections.

Figure 4–8
Solid foods may be introduced by about the sixth month, depending on the size and appetite of the infant and the advising physician's preferences. Feedings should be calm, pleasant, uninterrupted times for both baby and caregiver.

Solid foods are generally started after six months, depending on the advising physician. Rice cereal is usually introduced first, followed by strained fruits, vegetables, and meats. Foods with soy or wheat flour and egg products are postponed until the infant is older. Only a few teaspoons of all new foods are introduced at a time. They are continued for a day or two to ensure that no allergic manifestation (such as rash) or other evidence of intolerance (for example, diarrhea) will occur before starting another new food (see Figure 4–8).

Infants generally show a decrease in appetite between about twelve and eighteen months of age. They are no longer growing as rapidly and therefore do not need as many calories. "Empty calories," such as those found in candy and soda pop, should especially be avoided.

Sleep needs and patterns oscillate greatly during the first two years, both in the life of a single child and between infants. Immediately after birth, infants spend most of their nonfeeding time in sleep. The sleep-wake cycles are short, however, and the neonate can be expected to wake up every two to three hours (even during

Table 4–3 Motor Milestones of Infant Development

Achievement	Approximate age
Chin up off mattress	1 month
Chest up off mattress	2 months
Rolls over	3 months
Sits with support	4 months
Sits without support	6 months
Stands holding on	6 months
Crawls	7 months
Gets to sitting position alone	8 months
Pulls self to stand	8 months
Walks	12 months
Walks backward	15 months
Climbs stairs	18 months

SOURCE: Adapted from *Denver Developmental Screening Test* (DDST), by W. K. Frankenburg and J. B. Dodds. Copyright 1967 by the University of Colorado Medical Center. Reprinted by permission.

the night). The age at which the baby sleeps through the night (considered about a six- to eight-hour stretch) varies greatly from infant to infant. Some babies surprise their parents by doing this shortly after birth whereas others cannot do so until they are from four to seven months old. Between ages one and two most infants adapt their schedules to a morning and an afternoon nap and a full night (about twelve hours) of sleep.

The ages at which infants reach the **motor milestones** of development (see Table 4–3) vary considerably. The "average" baby, determined by mean values of hundreds of infants, does not exist. Normative data can often be misleading. Each idiosyncratic human being develops at his or her own rate on each of the motor tasks. In general, girls are ahead of boys in skeletal maturity and motor development (Silver, 1984).

Motor milestones depend on genetic factors, biologically rooted tempos of growth, maturation of the central nervous system, skeletal ossification, nutrition, physical health, environmental space, freedom, stimulation, and even psychological well-being. On the "average," one-month-old infants can lift their chins up above a surface when lying on their stomachs. By two months they can probably lift their chests up slightly and gaze around. At about three months their locomotor maturation is sufficient to hold their heads steady as they are pulled up. They also begin to roll over and reach for objects that attract their attention. At about four months they may grasp and hold objects for which they reach. By age five months infants can generally bear weight on their legs. They can also hold their heads steady when in an upright position. At six months many infants sit alone. Their grasp is also now practiced enough to allow them to

MOTOR MILESTONES The major developmental tasks of a period (such as infancy) that depend on muscular movements.

transfer objects from hand to hand. Between seven and nine months, babies may assume a sitting posture on their own, learn to creep or crawl, or stand with support. Between nine and twelve months they begin to take walking steps when their hands are held or to walk along by holding the furniture or some other support (see Figure 4–9). They also have practiced their grasp enough to have moved from a whole-hand carry to a pincer grasp (opposition of thumb and forefinger). Most babies can walk alone by twelve to fifteen months. By eighteen months they may be able to run stiffly and even climb stairs.

The ages for eruption of teeth in infancy are far more varied than the ages for accomplishing motor tasks. An occasional infant is born with a tooth. Other infants show no signs of teething until well past their first birthday. Early and late teething patterns appear to have genetic bases and are not related to other aspects of physical and motor development. Infants are often irritable and have increased salivation while teething, but teething does not cause diarrhea, fever, or other systemic disturbances (Beedle, 1984). Teething rings may hasten tooth eruption as well as provide pain relief when infants chew on them.

Figure 4–9
There is considerable variation in the age at which each infant begins to walk independently. Most babies require some support in their early attempts to step out into the world.

HEALTH MAINTENANCE

One of the most important ways in which caregivers provide for infant health is by supplying appropriate nutrients in a loving manner. Neonates' initial requirements are for colostrum or sweetened water. By the third or fourth postnatal day, they will require breast milk or formula. Caregivers have the responsibility to ensure that the nutrients provided are appropriate, germ free, available in sufficient supply when requested, and offered concurrently with tender loving care. As infants mature, it is necessary to add to the diet to ensure that they have sufficient protein, iron, calcium, and vitamins for the most active growth period of their life.

A second important aspect of infant health maintenance is protection from infection. Although some degree of immunity from pathogens to which the mother is immune is passed on prenatally, the infant is still highly susceptible to disease. Persons with known active infections (such as colds, influenza, "strep" or "staph" infections) should not kiss and cuddle or even be in contact with infants. If the primary caregiver becomes ill, another person should help meet the baby's needs until the caregiver can recover.

Because infant skin is sensitive, it must be protected, especially from sun and from diaper rash. Diaper rashes are common, even in the cleanest homes, because of the ammonia that forms from urine in wet diapers. It burns and irritates the skin, causing redness or blisters. Protective powders or creams may be necessary to supplement prompt changing of wet diapers if rashes are common. If blisters become infected, further diagnosis and treatment are required.

Many infections that contributed in the past to high infant mortality and morbidity in later life can now be prevented through a series of immunizations: diphtheria, whooping cough (pertussis), tetanus, poliomyelitis, measles, mumps, and rubella (also typhoid fever—depending on locale). All babies should be taken to a physician or well-baby clinic for immunizations. The usual schedule is for three immunizations for DPT-OPV (diphtheria, whooping cough, tetanus, and oral polio vaccine) in the first year and one for measles, mumps, and rubella after the first birthday at approximately two-month intervals as decided by the physician. Injections are not given at the scheduled visit if an infant has an active upper respiratory infection or other illness. The exact timing of immunizations is not nearly as important as the fact that the infant receive all of the injections eventually. Records of immunizations should be carefully kept with other important documents. Some school systems will need to see these records before enrolling the child in kindergarten or first grade. They are also important to alert physicians to the timing of later booster injections.

Some parents fear the whooping cough (pertussis) vaccination because it has been associated with adverse side effects. It is somewhat more risky than the other immunizations (Harding, 1985). Tenderness at the inoculation site and fever are common sequelae. Many physicians routinely advise aspirin for two to twelve hours following immunization. A more serious risk is central nervous system (CNS) reactivity. This has been estimated to occur in 1 out of 310,000 vaccinations (Fulginiti, 1984). The occurrence of any CNS symptoms is an absolute contraindication to giving further doses. Parents must be sure to report convulsions or collapse after injections to the physician. Even when CNS symptoms have developed after the whooping cough vaccine, few infants have suffered permanent sequelae. In contrast, the disease itself is most severe in infancy. Of all the deaths from pertussis in the United States in a recent seven-year period, 72 percent were infants (McIntosh and Lauer, 1984). Several hospitals have reported outbreaks of the disease in the last decade. Unfortunately, no immunity for pertussis is passed from mother to child prenatally. Antibiotics are also ineffective during the paroxysmal "whooping cough" stage of the disease, when deaths may occur from apnea or lung collapse.

Regularly scheduled visits to a doctor's office or clinic during infancy are important for health maintenance above and beyond immunizations (see Figure 4–10). Trained personnel are able to weigh, measure, and examine the baby and determine the presence of any possible disorders or developmental delays that may require attention. They can provide nutritional guidance and counseling for questions and problems. Establishing contact with medical supervisors also aids the supervisors' capacity to respond appropriately to patients in emergencies.

An additional important aspect of health maintenance for infants is accident prevention. Motor vehicle accidents are the leading cause of accidental death in infancy. Even in nonaccidents infants may be killed because of sudden stops and turns (Agran, 1981). Safety seats for infants are now mandated in the United States and must be used. They should comply with the Federal Motor Vehicle Safety Standard and contain the words "dynamically tested," which means they have passed a crash situation test. Infants must be pro-

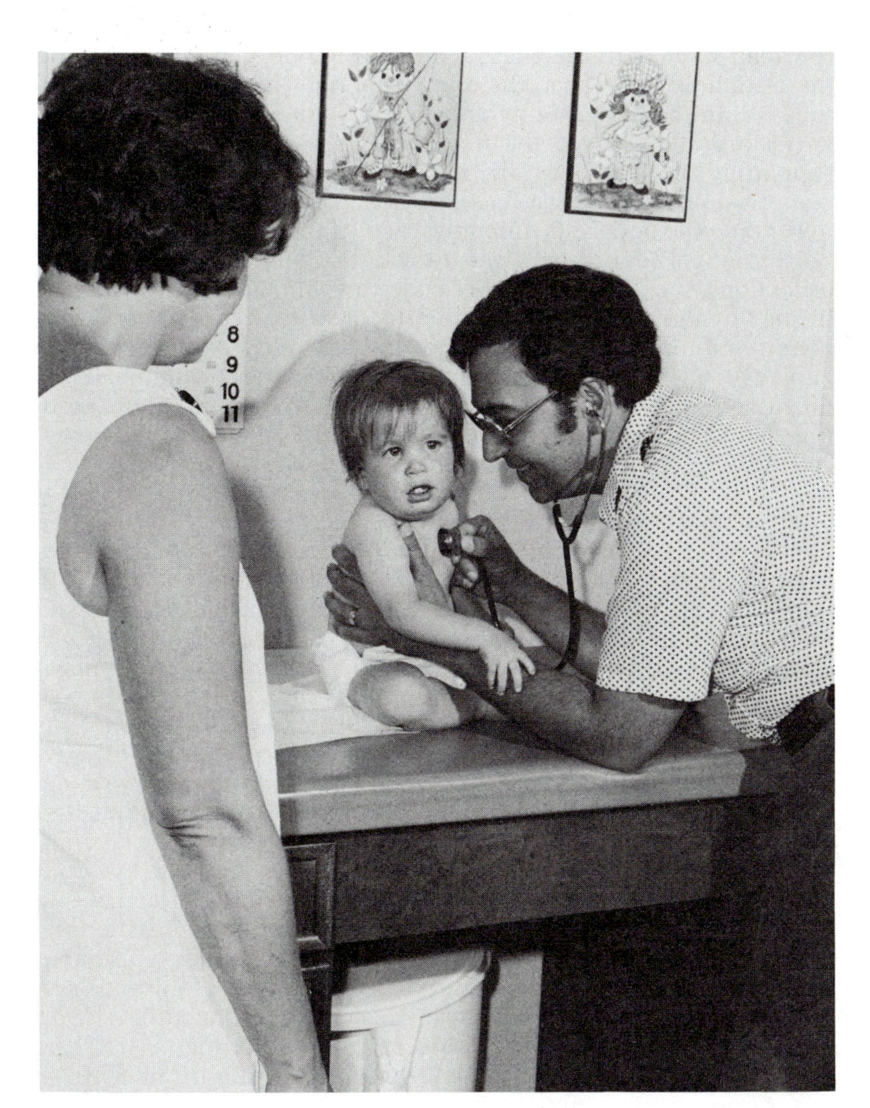

Figure 4–10
Regular well-baby examinations by a pediatrician, family practitioner, or nurse practitioner prevent many health problems and identify others before they become serious.

tected from falls: out of cribs with bars down or full-sized beds, down open stairwells, off furniture. They must be kept away from open heaters and fireplaces, matches, hot stoves and hot radiators and out of the paths of motor vehicles. Poisons, pins and nails, and any small objects that could be swallowed must be kept out of reach. Any large containers of water (bathtubs, swimming pools) are especially hazardous. Infants must be watched carefully and taught to respect and obey the command "no" for their own safety's sake.

Should a serious accident occur, the baby should be rushed to the nearest doctor's office or hospital with someone telephoning ahead. If the problem is aspiration and the infant chokes and turns blue, a caregiver should first try to dislodge the object blocking the airway with blows to the back. The infant should be straddled over the adult's arm with the chest and head supported in one hand and the head held lower than the body. Then four rapid blows are carefully administered between the baby's shoulder blades. If the object is not dislodged, place the free hand over the infant's back and, supporting the head, turn the baby over and administer four chest compressions in rapid succession. Again, be certain the head is lower than the body and supported well. If the infant has swallowed poison, water should be given to dilute it and then, with certain exceptions, the baby should be made to vomit. This can be accomplished by tickling the back of the throat with a blunt object, or by giving 1 tablespoon (15 milliliters, ½ ounce) of syrup of ipecac in a cup (240 ml, 8 ounces) of water. Do not induce vomiting if the infant has swallowed strong acid, lye, or petroleum products (gasoline, kerosene, paint thinner) or if the infant is unconscious or convulsing. (See First Aid for Poisoning in Chapter 5.)

COMMON HEALTH PROBLEMS IN INFANCY

Infections of the digestive system and the respiratory tract are the leading health problems of infants from birth to one year of age in the United States. From one to two years of age, accidents and respiratory infections cause the most difficulty. Colic is not a disease. It affects about 20 percent of all infants for reasons as yet unknown (see Box 4–2).

Diarrhea and vomiting are frequent problems in young infants and are usually related to irritation of the digestive system by new foods, contaminated foods, or nonfood substances. Infants become dehydrated very quickly when they lose fluids because of the smaller total fluid volume of their bodies. Consequently, diarrhea and vomiting must be treated promptly.

Projectile vomiting (hurled forward) and constipation may signal *pyloric stenosis,* a narrowing of the end of the stomach called the pylorus. This occurs in 1 out of 500 infants, three to four times more commonly in males, for unknown reasons. If the problem is untreated, the baby will lose weight and become dehydrated. Treatment is to surgically enlarge the pyloric opening as well as to treat the complications of the disease (dehydration, stomach inflammation).

Severe malnutrition problems can arise if infants are weaned too early from milk and are not given other foods equally rich in proteins. The effects of malnutrition on brain cell development are profound. During the first eight months of life the proliferation of brain glial cells (supporting cells) can be curtailed by insufficient dietary protein (Winick, 1976). In addition, lack of protein reduces branchings of the nerve fibers, enlargement of cell bodies, and myelinization of the cell processes. Infancy is a critical period for growth of brain tissue. This can be illustrated by outcomes of kwashiorkor (a disease caused by severe protein deficiency) in humans. Babies acquiring the disease in early infancy suffer irreparable brain damage. Those who are affected by kwashiorkor in later infancy (from fifteen to twenty-nine months) have a milder residual mental retardation. Children who do not get the disease until they are beyond age three can recover normal intelligence once they are cured

COLIC Unexplained, explosive, inconsolable crying in infants from two weeks to three months of age.

BOX 4–2

What's in a One- to Three-Month-Old's Cry?

While neonatal cries may indicate developmental risk (see Box 4–1), periods of explosive, inconsolable crying between the ages of approximately two weeks and three months of age may betoken nothing more serious than **colic**. Nothing more serious? To parents of the 20 percent of babies who have colic, such a phrase will seem a gross understatement. Yet colicky babies are healthy, not diseased. They gain weight and are physically robust. They also have screaming attacks, often accompanied by twisting, turning, or stiffening, which last more than three hours a day, occur at least three days a week, and continue for at least three weeks (usually into the third month of life), for which no physical cause can be found.

Almost all infants have some periods of unexplained fussiness in this age period. You will hear descriptions such as "a touch of colic," or "very colicky" offered to explain cries. There are gradations of colic. What sets colicky crying apart from plain crying is the severity and inconsolable nature of the colic attacks, their persistence and repetition, the abrupt and dramatic mood shifts seen in the colicky baby, and the failure to explain the attacks with any medical or environmental causes.

Many myths exist about colic. Weissbluth (1985) assures that colic is *not* due to: birth order, sex, intelligence of the baby, educational status of the mother, the nursing mother's diet, the failure to breastfeed, any food allergy, any gastrointestinal problem, fresh air, or maternal anxieties. While its fundamental cause is still unknown, research suggests that colic is set off by one or more physiological factors, including disordered regulation of breathing, temperament, sleep cycles that are not synchronized with other body rhythms, or abnormal levels of naturally occurring substances such as prostaglandins or progesterone (Weissbluth, 1985). The problem usually corrects itself by about three to four months.

Can anything be done to "cure" colic and stop the screaming attacks before three or four months? No. Some physicians advise changing formulas, burping the baby more frequently, applying warm water bottles, giving herbal tea, or tranquilizing the baby (or tranquilizing the mother!). The colic continues to wax and wane until it has run its natural course. Some physicians will prescribe dicyclomine (Bentyl), a smooth muscle relaxant that eases spasms of the intestines. (It may also calm the baby's central nervous system.) It does not cure colic, but it makes crying spells shorter and less frequent and makes the baby more consolable. Most experts advise soothing the colicky infant with rhythmic motions until each attack abates: rocking, jiggling, bouncing, patting, massage, a ride in the car—whatever works to console him or her.

The baby should never be left alone to cry it out for hours at a time in the first few months of life. A classic study by Bell and Ainsworth (1972) demonstrated that ignoring early cries leads to more crying later, while responding to early cries leads to less crying later. Parents must be patient and loving during the many trying hours of colic attacks in the first few months. The baby is normal and the colic will disappear.

of their malnourished, diseased state. Even infants who are protein deficient but not severely affected enough to have kwashiorkor show developmental lags three years after hospital discharge (Grantham-McGregor et al., 1982) and learning disabilities during the elementary school years (Galler, Ramsey, and Solimano, 1984).

Another danger of inadequate infant nutrition is **iron deficiency**. As many as 35 percent of infants in the one to three age range are iron deficient, according to the National Health Nutrition Examination Survey (Honig and Oski, 1984). Infants have iron stores from prenatal development sufficient to last them for about four months. After that, they must obtain additional iron from breast milk, iron-fortified infant formulas, or iron-containing solid foods. Signs of iron deficiency in infants include irritability, weakness, poor appetite, and susceptibility to infections (Githens and Hathaway, 1984). Iron-deficient infants also perform poorly on mental development scales. Their performance dramatically improves within two weeks of treatment with intramuscular iron injections (Honig and Oski, 1984). Honig and Oski noted that iron-deficient infants also manifest very solemn faces. All efforts to establish positive rapport and obtain smiles fail. This is an important behavioral index that can be used to alert caregivers, nurse practitioners, and physicians to which babies need medical screening for potential iron deficiency.

Less profound, but also unfortunate, are the effects of overfeeding in infancy. Babies who are fed many more calories than they can utilize from day to day develop additional fat cells to store the unused energy sources (Fomon, 1974). The hypothesis that these fat cells that proliferate in infancy continue to replace themselves as children grow has not been supported with research evidence. However, adults who were fat as infants are frequently obese or have difficulty holding their weight down to their normal range for size. Some researchers believe there is a genetic predisposition for obesity (Stunkard et al., 1986). An alternate theory about the effect of infant overfeeding on adult obesity is that eating patterns become habits that are difficult to change. Whatever the explanation, fat babies tend to become overweight adults. Fat babies also appear to be at increased risk of infection (O'Brien and Hambridge, 1984).

Eczema is a manifestation of allergic reaction to one or more foods or other substances. It is characterized by inflammation of the skin and intense itching. It may be a wet form with running sores or a dry form with bright red patches of scales. The tendency to become allergic is inherited, but an infant's eczema may be a reaction to

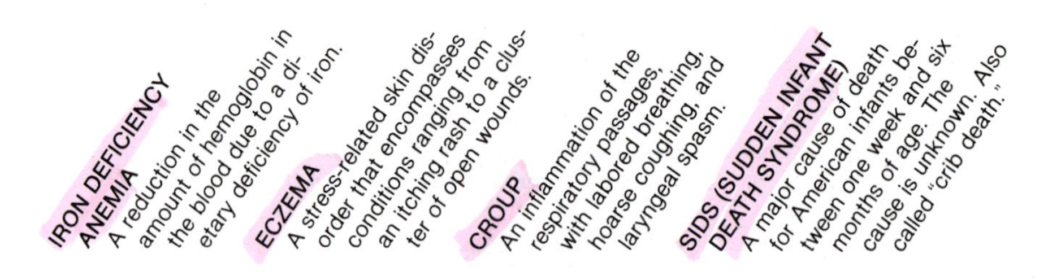

a food or substance quite different from the food or substance to which the parents are allergic. Doctors frequently have difficulty determining the source(s) of the infant's problem. Until they do, medication is usually prescribed to help relieve the itching.

Respiratory illnesses common in infancy include croup, influenza, colds, bronchitis, pneumonia, and ear infections. **Croup** is a mild form of tracheobronchitis common in infancy. It produces a loud spasmodic cough that can be very frightening (it may sound like the infant is strangling). Croup is rarely serious unless complicated with other infections. Humidifiers generally liquefy the secretions in the air passages making it easier for them to be coughed up and expelled. Colds are more common between age six months and two years. The nasal passages may become filled with thick, bubbling mucus. Decongestants are usually prescribed promptly to help reduce secretions from the airways and to shrink congested mucous membranes. Ear infections are common sequelae to colds. Ear infections must be treated promptly to prevent rupture of the eardrum and possible hearing loss from tissue destruction and to prevent chronic ear infections that can lead to meningitis (inflammation of the membranes of the brain or spinal cord).

An increasing amount of heart surgery is being done to repair congenital anomalies in early life, for compelling reasons. Some of the other frequent reasons for surgery in infancy are repair of cleft lip or palate, relief of intussusceptions, hernias (passage of a portion of the intestine through spaces between muscle layers in the abdominal wall), and hydroceles (accumulation of fluid around the testes). Any surgery that can wait until a baby is three years or older is usually delayed because of the traumatic effects of hospitalization and separation from caregivers during infancy.

The heartbreaking syndrome of crib death, called **sudden infant death syndrome (SIDS)**, in which apparently normal, well-developed infants suddenly stop breathing, is being studied intensively in many centers. Although the fundamental cause of SIDS is unknown, Steinschneider (1972) suggested apnea (the spontaneous interruption of respiration) as an etiological (causative) factor. Guilleminault and colleagues (1979) found increased episodes of apnea in sleep recordings of infants who later suffered from SIDS.

SIDS can occur anytime in the first six months, but it is most common between two and four months. It can occur in any infant but is more common in urban, low-income, male, twin, low-birthweight, and premature babies and in infants born to teenage parents, especially those who discontinued school early. Frequently, there is a history of maternal smoking during pregnancy.

In most major hospitals today infants who have frequent apnea episodes are watched closely. At discharge from the hospital their caregivers may be instructed to monitor breathing with the help of a machine that signals apnea occurrences. Usually just moving the baby is enough to restore breathing. Someone from the baby's family must always be close enough to hear the monitor in case it should give its warning ring. As the infant matures, the apnea episodes become less frequent and are eventually outgrown. The monitor can then be returned to the hospital for use with another potential SIDS baby.

The **failure-to-thrive syndrome (FTTS)**, in which apparently normal infants fail to maintain established patterns of weight gain and growth and fall below the third weight percentile, (97 percent of same-age infants weigh more—see Figure 4–11) without evidence of systemic disease or abnormality, is another condition whose fundamental cause is unknown. Barbero (1982) found FTTS infants to have characteristically bizarre eye behaviors. They gaze wide-eyed, scan the environment continuously, and avoid eye contact. Neurologically, they are either spastic and rigid or floppy with a near absence of muscle tone. Many other researchers have suggested that FTTS involves a deprivation of loving, sensitive caregiving. Drotar and Malone (1982) reported that mothers of FTTS infants have diminished responsiveness to their infants and tension-laden interactions when they do respond. Further, they found that the mothers have strained relationships with others. Gagan, Cupoli, and Watkins (1984) presented research evidence that mothers of FTTS infants do, indeed, have poor social support networks. Frequently, they are financially disadvantaged or teenage mothers. Barbero (1982) reported that mothers of FTTS infants have histories of some emo-

Figure 4–11
Standard measurement percentiles for length and weight of infants in three-month increments. Each healthy infant tends to stay near his or her same line of trajectory (e.g., 50th length percentile) over time as he or she grows and develops.

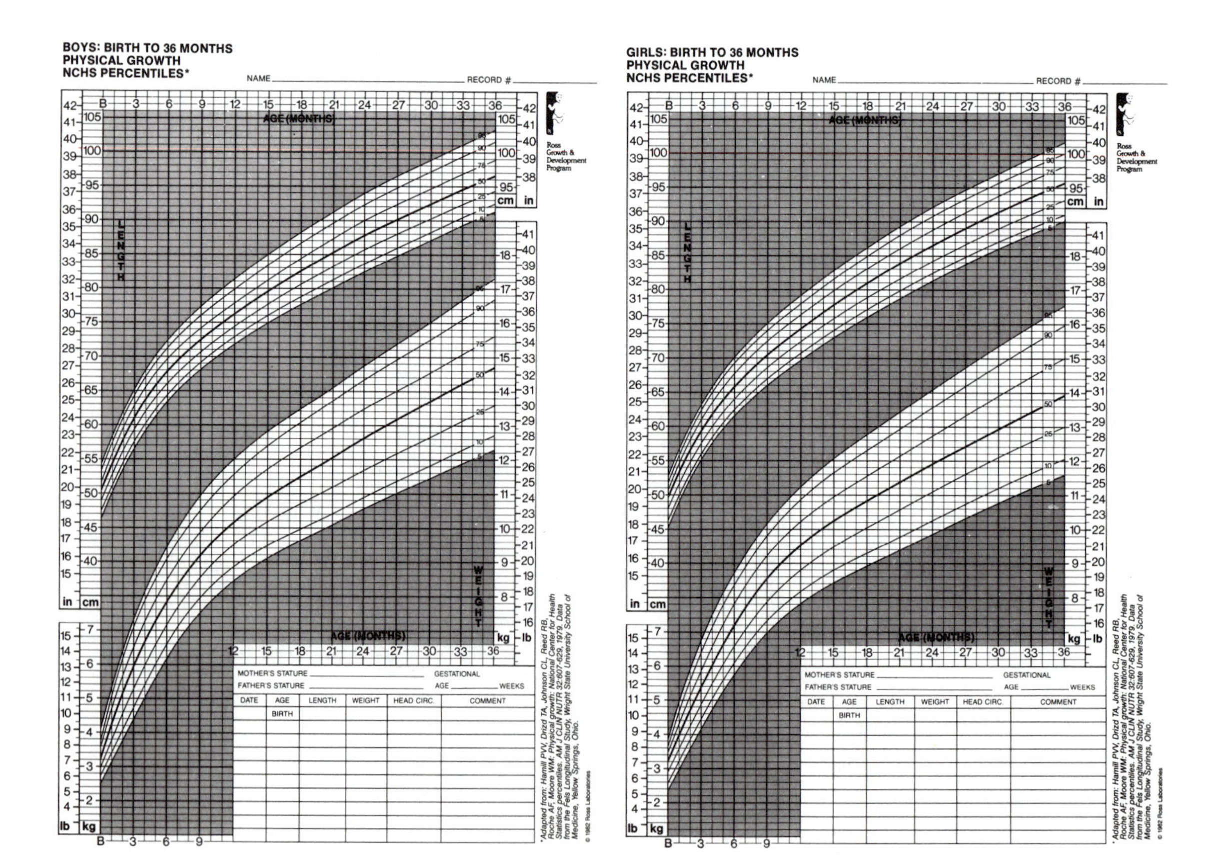

tional trauma: deprivations in their own childhoods, problems during their pregnancies, disruptions in the neonatal period (prematurity, complications of parturition, acute illness in mother or infant), or current life stresses (marital strains, mental or physical illnesses, drug abuse, financial crises). FTTS babies do gain weight when hospitalized and appropriately nurtured. This, however, is only the beginning of treatment. Therapy must be continued in the home to counter patterns of dysfunction and encourage a more adaptive, high-quality attachment between mother and infant to assure future nurturing and sensitive caregiving.

Cognitive and Language Development

There are some people who believe that infant behaviors (shaking a rattle, drinking from a cup, walking) are taught by sensitive caregivers or are learned in environments carefully supplied with the right kinds of stimuli. There are many other people who feel that infant behaviors will gradually evolve, regardless of caregivers or environmental stimuli, as the baby matures physiologically. Before reading on, consider these two explanations of the acquisition of new behaviors by infants. Which one seems more correct to you? There is an inseparable element of interplay between maturation and experience in infant learning. However, in recent years educators have accepted the idea that many aspects of cognitive behavior are controlled by maturation. Until the central nervous system reaches the necessary degree of maturity for any given skill (such as shaking a rattle), no amount of teaching or assisted practice will enable the infant to accomplish the task. Only after the appropriate neuroanatomical development has occurred will experience be of value.

MATURATION OF THE INFANT BRAIN

The **central nervous system** (brain and spinal cord) is one of the more immature organ systems in the neonate. The three major regions of the brain are the cerebrum, the cerebellum, and the brain stem (see Figure 4–12). The **brain stem** consists of four structures: the pons, medulla, midbrain, and diencephalon. At birth these structures are developed more fully than the rest of the brain. They help regulate respiration, heartbeat, blood pressure, coughing, sneezing, swallowing, vomiting, postural reflexes, and some motor coordination. Less well developed in the neonate is the **cerebellum**, through which motor activities, balance, and joint position sense are coordinated. It develops very rapidly in the early months of life. The **cerebrum** consists of two large lobes called cerebral hemispheres. Each

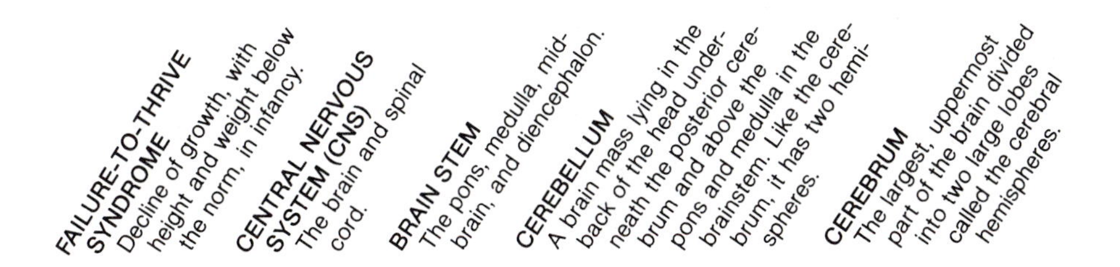

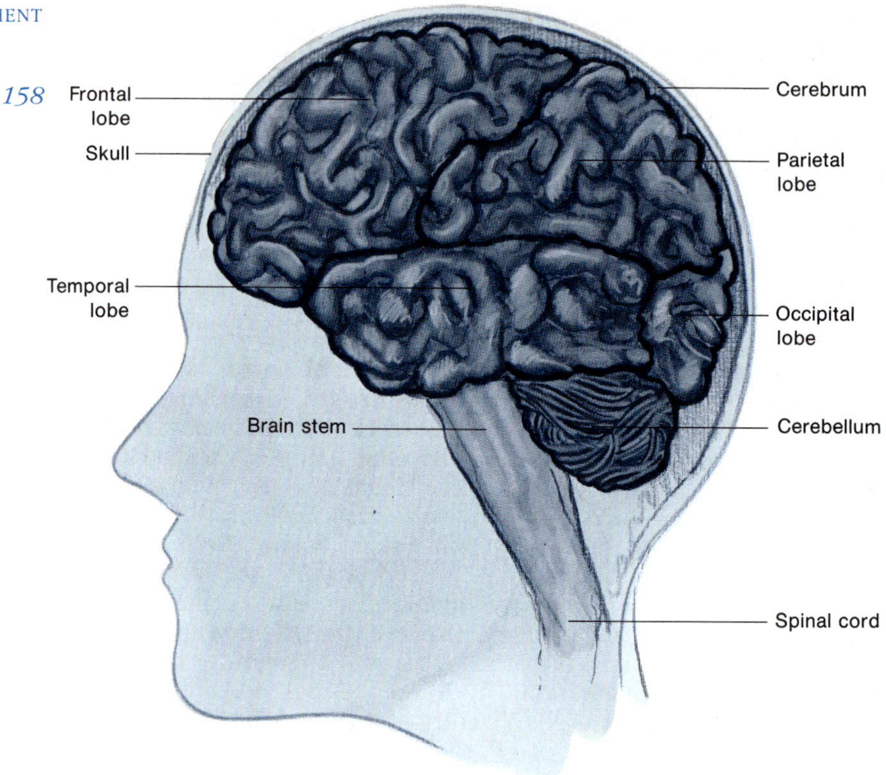

Figure 4–12
The major structures of the human brain are the cerebrum, the cerebellum, and the brain stem. Each cerebral hemisphere is described by a frontal lobe, parietal lobe, occipital lobe, and temporal lobe.

is subdivided into four smaller regions. They control learning, thought, and memory as well as sensory and motor functions. However, the growth of the cerebrum is slow, spanning infancy, childhood, and adolescence.

All portions of the brain are composed of nervous tissue that consists of three elements: nerve cells, nerve fibers, and the supporting structure of cells and fibers known as the neuroglia, or simply glia. The glial cells (named from the Greek word for glue) account for approximately 90 percent of all brain tissue. Nerve cells, called **neurons**, in spite of their relatively small number (10 percent of brain cells, or approximately 10 billion cells) sustain life, control thought, consciousness, and memory, direct all our involuntary and voluntary muscle movements, and are in charge of all the senses. When we speak of brain cells, we usually mean neurons, the workers, rather than glia, the supporters.

At birth most, if not all, of the neurons that the brain will ever have are present. Glial cells may continue to be added during infancy. The rapid growth and development of the brains of infants consist mainly of additions of glial cells and increases in size of existing neurons rather than proliferation of new cells. The increase in size of neurons is due to increases in materials in the nuclei and cytoplasm of the cell bodies, increases in the length and branchings of the cell processes (the axon and dendrites), and the growth of a layer of fatty, insulating material called **myelin** around the axon and dendrites (see Figure 4–13). This myelin allows messages to be transmitted with greater speed and ease across the cell processes. Brain growth during infancy also involves an increase in the number of **synapses** (junctions between cell processes of neurons) and the development of **neurotransmitters** (chemical substances that facilitate the transmission of impulses across the synapses between nerve endings). Eventually each neuron will be able to receive input from about 1,000 other neurons and send messages to hundreds more. There are indications that the more activity a region of neurons gets, the greater the chance of creating extra dendritic branches and synaptic connections (Lenard, 1983). There is also evidence that environmental stimulation in infancy will produce the neuronal activity that, in turn, will increase dendritic growth (Greenough and Juraska, 1979). The same is probably true of environmental stimulation in childhood, adolescence, and even adulthood, although the dendritic branching occurs at increasingly slower rates.

At birth the brain weighs, on the average, 335 grams (about 11 ounces). Its adult size will eventually be 1300 to 1400 grams (about 3 pounds). The neurons and glia are massed together in a fashion different from their final organization and relative positions; they await rearrangements, growth and interconnections among fibers, and the development of myelin sheathing.

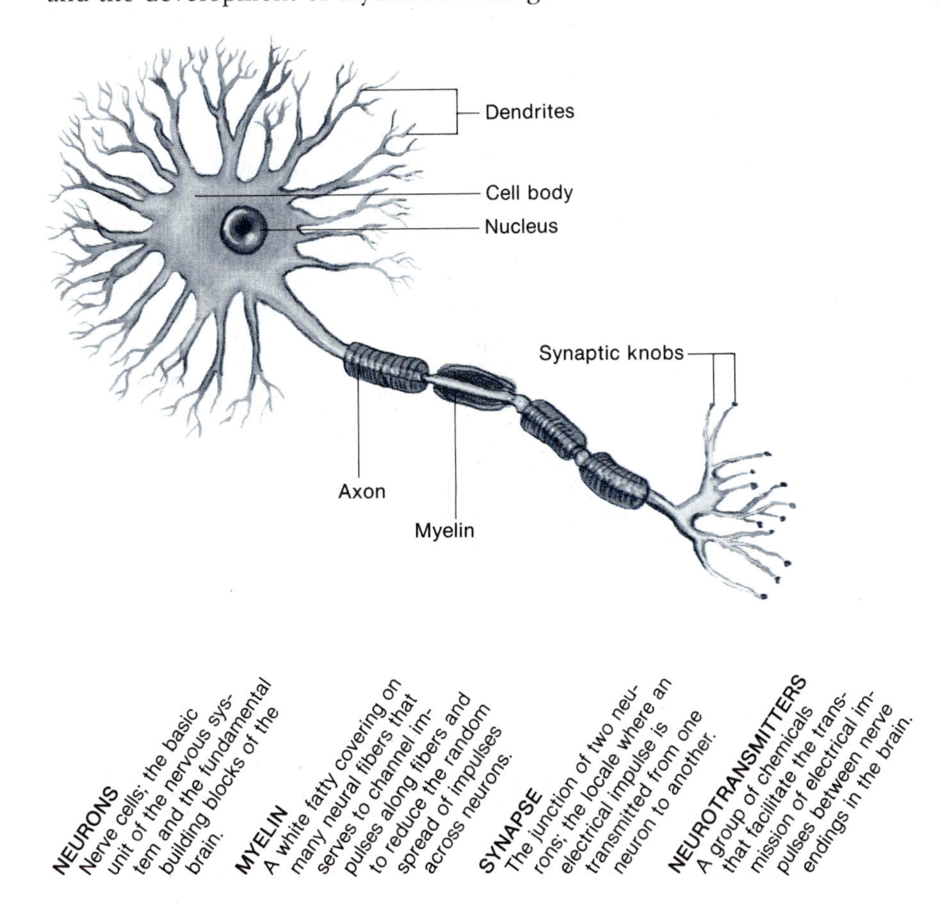

Figure 4–13
Schematic view of a neuron, showing multiple layers of myelin around the axon.

NEURONS Nerve cells; the basic unit of the nervous system and the fundamental building blocks of the brain.

MYELIN A white fatty covering on many neural fibers that serves to channel impulses along fibers and to reduce the random spread of impulses across neurons.

SYNAPSE The junction of two neurons; the locale where an electrical impulse is transmitted from one neuron to another.

NEUROTRANSMITTERS A group of chemicals that facilitate the transmission of electrical impulses between nerve endings in the brain.

While neurons can generate new fibers, the cell body of a central nervous system neuron cannot be replaced once it is destroyed. For this reason, the infant brain must be carefully protected.

PIAGET'S SENSORIMOTOR STAGE

Piaget's theory of cognitive development stresses four major periods through which humans pass in the course of intellectual maturation: (1) sensorimotor intelligence in infancy; (2) preoperational thinking in early childhood; (3) concrete operational thought in later childhood; and (4) formal or logical intelligence in adolescence and adulthood. Piaget's period of sensorimotor intelligence is easily juxtaposed with knowledge of the organization of the infant brain. As various cells of the cerebral cortex mature, develop interconnections among one another, and acquire myelin sheathing, infants become able to learn increasingly more about the world around them. The cells of the cerebral cortex are of three types: *motor, sensory,* and *associational.* Motor cells are concerned with muscle activities. Sensory cells are concerned with input from the senses (vision, smell, hearing, taste, touch). These cells mature, and become myelinated before the associational cells. The activities of memory, reason, and judgment are believed to take place in the cells that mature last, the associational cells. In Piaget's first stage of **sensorimotor intelligence** an infant's knowledge of the world comes about primarily through sensory impressions and motor activities. In Piaget's second stage of learning, ages three to seven, the preoperational stage, learning coincides with increased maturation of the associational cells.

Piaget saw the sensorimotor period of intellectual development as having six distinctive substages, each of which will be described separately. Remember, however, that there is no overt end to any substage nor any obvious entrance into the next. Even if infants are watched very carefully for each new substage to emerge, more often than not they are discovered to have acquired new abilities and practiced them enough to be proficient before the observer quite realizes what has happened. Some infants pass through the succession of stages earlier than others. The rate of intellectual development depends not only on maturation but also on genetic factors, physical and emotional health, nutrition, and the kinds of stimulation that the infant receives from the environment. Piaget and his colleagues were hesitant to attach an expected age of emergence on any of the substages of sensorimotor development. Instead they provided a wider approximate range for going through the stages. More impor-

tant than when the behaviors emerge is the fact that they develop in the sequence that he described.

As infants experience various phenomena through their motions and senses, the complementary processes of assimilation and accommodation are put into play. **Assimilation** refers to the process of absorbing new information from the environment and using current structures to deal with the information (see Chapter 2). Any objects or events that elicit exploring behaviors from the infant are being assimilated. **Accommodation** refers to the process whereby the infant (or child, or adult) alters his or her behaviors and adjusts existing schemas to the requirements of the object or event just assimilated. Assimilation and accommodation allow infants to integrate new learning with old and thus adapt to their ever-expanding environments. As a result of infants' encounters with new stimuli, they acquire increased numbers of **schemas**. These, according to Piaget (1952), are mentally organized patterns of behaviors. When any structures are assimilated and accommodated into sharply defined schemas, a state of relative **equilibrium** is said to exist. However, this equilibrium sows seeds for its own destruction. Inconsistencies and gaps previously ignored become salient. They usher in a new disequilibrium requiring more exploration of newly perceived phenomena that, in turn, become new schemas.

The first stage of sensorimotor intelligence is the *reflex stage*. Piaget felt that for the first month of life the primary learning that takes place is bound to naturally occurring reflexes: rooting, sucking, swallowing, head turning, crying, startling, grasping, and smiling (refer back to Table 4–2). He proposed that assimilation to these reflexes occurs because the infant has a basic tendency to exercise any available behaviors and make them function. In exercising reflexes such as sucking, the infant learns more about the environment. The infant will suck not only a nipple but also a finger, a blanket, someone else's cheek, a piece of clothing, or anything that happens to come in contact with the mouth. During the first stage the infant learns to differentiate among various stimuli and accommodate to them. Thus, before long, when hunger is strong, only a nipple will elicit strong sucking movements. A blanket may elicit a loud howl of protest instead.

The second stage of sensorimotor intelligence commences at about one month and may last two to three months. Piaget called it the stage of *primary circular reactions*. Examples of some primary circular reactions (behaviors that become habits) are sucking on a specific object (thumb, finger, pacifier, toy), turning to look in the direction of a sound, reaching for and grasping an object, smiling at a friendly (later at a familiar) face, and showing anticipatory behaviors before routine procedures (feeding, diapering, being put to bed) (see Box 4–3).

Infants begin to pay more attention to moderately novel stimuli during this stage. Both events totally assimilated and accommodated and events with no elements of familiarity are given less heed. Infants seem to prefer the challenge of attending to environmental events that almost, but not quite, fit into their preexisting schemas. These events require assimilation and accommodation work.

During the phase of primary circular reactions, infants seem content to focus on objects or events that occur naturally. Soon, however, maturational processes make it possible for them to begin experimenting with novel actions of their own. At first such experi-

BOX 4–3

Observations of P. Through the Sensorimotor Substages

1. *Reflexes* At 0;0(21)* P. is lying quietly in her crib when K. begins to cry. P. also begins to cry. K.'s cry seems to be the stimulus which set her reflex in motion.

2. *Primary circular reactions* At 0;3(1) P. has been amused by a mobile above her with a string from it loosely attached to her foot so that the mobile moves when her foot moves. The string slips off her foot. She continues moving her foot and watching the mobile. When nothing happens, she moves her foot more vigorously, causing the mobile to move slightly as the crib jiggles.

3. *Secondary circular reactions* At 0;4(16) P. is lying on her back in her crib watching a mobile move as she shakes her right hand. A string is attached from the mobile to her right wrist. Her babysitter removes the string from her right wrist and slips it over her left wrist. She shakes her right wrist vigorously. She then shakes both hands as she becomes agitated. When the mobile moves, she continues shaking both hands more calmly. Soon she shakes only her left hand. (On following days, she becomes accustomed to shaking only the hand to which the loop is attached after one or two trials when the mobile-string game is presented.)

On 0;4(20) when the mobile is presented without a string, she shakes both hands until she fortuitously hits the side of her crib, making the mobile move. In time she discovers that kicking the side of the crib to which the mobile is attached causes greater movement. Thereafter she experiments with the rate and rhythm of her kicks to make this interesting sight last.

4. *Coordination of secondary schemas* At 0;8(13) P. spends a great deal of time walking in a walker. Her mother has tied a string to the walker to "haul it in" when P. gets too close to forbidden objects. On this day, P. is sitting on the floor and she spies the walker. She does not know how to crawl but after looking around a while she notices the string. She picks it up and "reels in" the walker.

5. *Tertiary circular reactions* At 1;2(5) P. accidentally drops a toy on the floor of the car. She cannot retrieve it herself because she is buckled into a carseat. After only a moment's hesitation, she reaches for her mother's hand and pushes it down toward the object, indicating her desire to have the toy returned.

6. *Mental combinations* At 1;6(18) P. has tired of K. as he has been at her house for 5 hours. When K.'s mother comes in the door, P. immediately goes and gets K.'s coat and happily waves "bye-bye," without a word being said about K.'s need to leave.

---

*Numbers indicate years, months, and days in age at time of observation.

The observations were made of a friend's child by the author.

mental actions occur by accident. Serendipity! Infants arrive at the fortunate discovery that they can alter stimuli themselves and thus make them slightly novel and more interesting (see Table 4–4).

Piaget called the third substage in infancy the period of *secondary circular reactions.* "Secondary" refers to the idea that habits developed in the preceding substage can now be embellished with new, more advanced actions. "Circular" refers to the fact that these behaviors are continually repeated by the baby in play. This stage is believed to characterize infants ranging in age from about four months through about eight to ten months. During this phase of development an intentionality in the babies' behaviors becomes apparent. Infants show evidence of finding ways to make interesting events last or be repeated.

Table 4–4 Some of the Achievements in the Six Substages of Piaget's Sensorimotor Stage

(1) Reflexes	**(2) Primary circular reactions**	**(3) Secondary circular reactions**	**(4) Coordination of secondary schemas**	**(5) Tertiary circular reactions**	**(6) Mental combinations**
Exercises innate behavioral patterns	Causality: Repeats simple, pleasurable activities	Causality: Shows abbreviations of intentional actions; combines actions	Causality: Shows expectations of events or behaviors	Causality: Recognizes causes other than self and differentiates cause and effect	Causality: Perceives causality mentally and is able to detour simple forms of action
	Imitation: Imitates actions that have previously been performed	Imitation: Successfully imitates modeled activity if it is familiar	Imitation: Begins to imitate novel behaviors	Imitation: Adept at novel imitations including sounds	Imitation: Can imitate model who is no longer present
	Object concept: Coordinates looking and hearing	Object concept: Conducts visual or tactile search if behavior was ongoing	Object concept: Employs a variety of searching behaviors	Object concept: Correctly follows visual sequence of object's movements	Object concept: Can infer object's location even when tricks of perception are introduced
			Means to ends: Uses purposive behavior to attain goal	Means to ends: Experiments with new means to attain ends	Means to ends: Uses short cuts by thinking rather than groping
			Object in space: Discovers perspective in depth	Object in space: Sees spatial relationships enough to fit different shapes into corresponding openings	Object in space: Has symbolic representation of space; can solve detour problems

During the phase of secondary circular reactions infants will imitate models if the patterns of behavior being demonstrated are familiar. They cannot yet copy novel actions.

Prior to this stage infants behave as if objects out of their sight no longer exist. Now they begin to have a rudimentary object concept or sense of object permanence. If an object with which they are involved disappears, they will attempt a visual or tactile search for it. However, they will not search for things that have been out of sight for a moderate period of time.

Piaget called the fourth stage the *coordination of secondary schemas.* It commences when indications arise that babies have a definite goal in mind for what they are doing and so is often referred to as the "means-ends stage." The approximate age range is from eight to twelve months. By this time infants are more mobile. They can sit alone, reach and grasp, creep, crawl, or perhaps even pull to stand or take a few unassisted steps. "Coordination of schemas" refers to the idea that infants now need and use more than one pattern of behavior to attain the goal they have in mind. They may use one schema as a means for attaining the goal and a second for dealing with the goal. Active experimentation is also evident as infants combine schemas in different ways. New skills and awarenesses emerge as they discover means to ends that were not originally intended.

Infants come to show anticipation of events that do not depend on their own actions. They expect people to act in certain ways, indicating their appreciation of the laws of causality independent of their own behavior. They also acquire the ability to imitate novel behaviors of models. Further, they now show a variety of searching behaviors for vanished objects.

During the first year and a half of life infants basically know their worlds through sensory impressions and motor activities. In substage five, *tertiary circular reactions,* the roots of some rudimentary associational cell functioning are seen. By this time, approximately twelve to eighteen months, babies can incorporate both the actions of combining schemas and the memory of results of their experiments into their mental storehouse. Piaget located the beginnings of rational judgment and intellectual reasoning in this cyclic behavior. When confronted with obstacles, infants can now invent new means for handling them. They do not have to rely solely on schemas that were successful before. They now also recognize causes and effects quite clearly. Object concept is so well established that, even if something is hidden in a succession of places, the infant will search for it in the place where it was last seen.

The last of the sensorimotor phases postulated by Piaget is the *invention of new means through mental combinations.* By this time, approximately eighteen months to two years, infants can understand and use some language. Language is used to form mental symbols for patterns of behavior. Infants now try to think about problems and develop solutions on a mental rather than a physical level. Simple new forms of behavior are initiated and carried through without the past trial-and-error steps. Imitations of behaviors that occurred some time earlier can be seen. Mental representations of the behaviors are copied. A physically present model is not needed. This last substage precedes the beginning of a whole new period of symbolic thought: that of preoperational intelligence.

Some people feel that infants may be able to generate mental hypotheses about their experiences sooner than Piaget surmised. J. J. Gibson (1966) and E. Gibson (1969) have argued that infants can pick up perceptual information from birth and attend to the relevant features of the objects they perceive. Bruner (1973) argued that even from the outset the infant's motor activities show the control of an intention. And Kagan (1979) reported that infants develop a concept of an object's permanence (as evidence by searching behaviors) much sooner than Piaget predicted. However, the sequence in which Piaget described development has been well replicated. In addition, we are convinced that maturation of the central nervous system must precede the acquisition of increasingly intelligent behaviors.

LANGUAGE BEGINNINGS

The word *infans* means "without language." In their early development, even though infants are not yet speaking, they are actively learning about the words that they hear spoken. As their neurological development progresses, they become increasingly able to organize and comprehend language. As their cognitive development progresses, they find increasing numbers of things for which some representational symbol would be helpful (such as objects that are out of sight but not out of mind).

From birth the auditory nerve tracts are myelinated enough to allow infants to hear. They attend to sounds of distress and respond in like fashion. (Any newborn-nursery nurse can verify the fact that when one baby starts crying, others will do the same.) They soon differentiate caregivers' voice sounds from strangers' voices. They show signs of actively listening and trying to tune in sounds that caregivers make to them in feeding, diapering, and soothing situations. A discrimination of different intonations is soon followed by the ability to differentiate between various vowel and consonant sounds. The ability to understand the spoken word is called passive or **receptive language**. It is, in fact, the primary or fundamental aspect of language. The ability to produce meaningful utterances is called **expressive language**. Understanding of language becomes increasingly proficient as babies get older. Whether they are responding to intonation, to actual words, or both, it is possible to either soothe or upset a baby of three to four months with comforting or

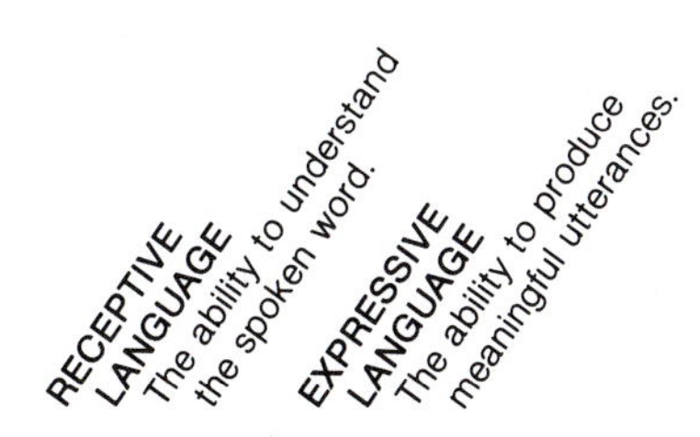

frightening statements. By nine to eleven months receptive language is organized to the point that infants will obey spoken commands (such as give me the ball, play pat-a-cake, finish your juice).

There are many precursors of expressive language also present at birth. The sensory nerve fibers that control speech are myelinated enough to allow infants to utter sounds, although neither the associational nor the motor cells are mature enough to allow the production of meaningful words. It is interesting that the speech center in the brain, which is located in the temporal lobe on the left side of the cerebrum (the right side in some left-handed persons), borders on the areas of the motor cortex that control both mouth-tongue movements and hand movements. This proximity of the speech center to the hand-control area of the motor cortex is important. We all express ourselves with our hands as well as with our mouths. Infants especially use many gestures in association with simple words as they acquire language.

The earliest sounds made in infancy are cries (Table 4–5). By about two weeks of age infants add new sounds to their repertoire such as coos and goo sounds, which are easily formed in the back of the mouth with undifferentiated tongue movements. These sounds have a nasal quality because the epiglottis is very close to the soft palate. The larynx and nasopharynx are short and high in the neck, the mouth is flat because of the absence of teeth, and the tongue more nearly fills the mouth (see Figure 4–14). The tongue moves mostly back and forth as for sucking and swallowing. In fact, the earliest sounds seem to be passive reflex activity, by-products of feeding and sucking (Oller, 1980).

These sounds are soon practiced whenever the baby is alert and contented. It seems, as Piaget (1952) stated it, that infants have a basic tendency to exercise any available reflexes and make them function. Infants' sounds change dramatically from week to week.

Table 4–5 Progression of Expressive Language with Approximate Ages of Emerging Ability

Language characteristic	Age of emergence
Crying	Birth
Wide variety of meaningless speech sounds	1–6 months
Babbling syllables (such as "mamamama")	3–12 months
Phonetic drift (frequent practice of phonemes of "mother tongue")	6–18 months
Echolalia (imitation of simple sounds such as "dog, hat, pop")	6–18 months
Holophrasing (one-word sentences)	12–24 months
Telegraphic speech (subject-verb-object communiques)	18–36 months
Complete sentences	24–48 months

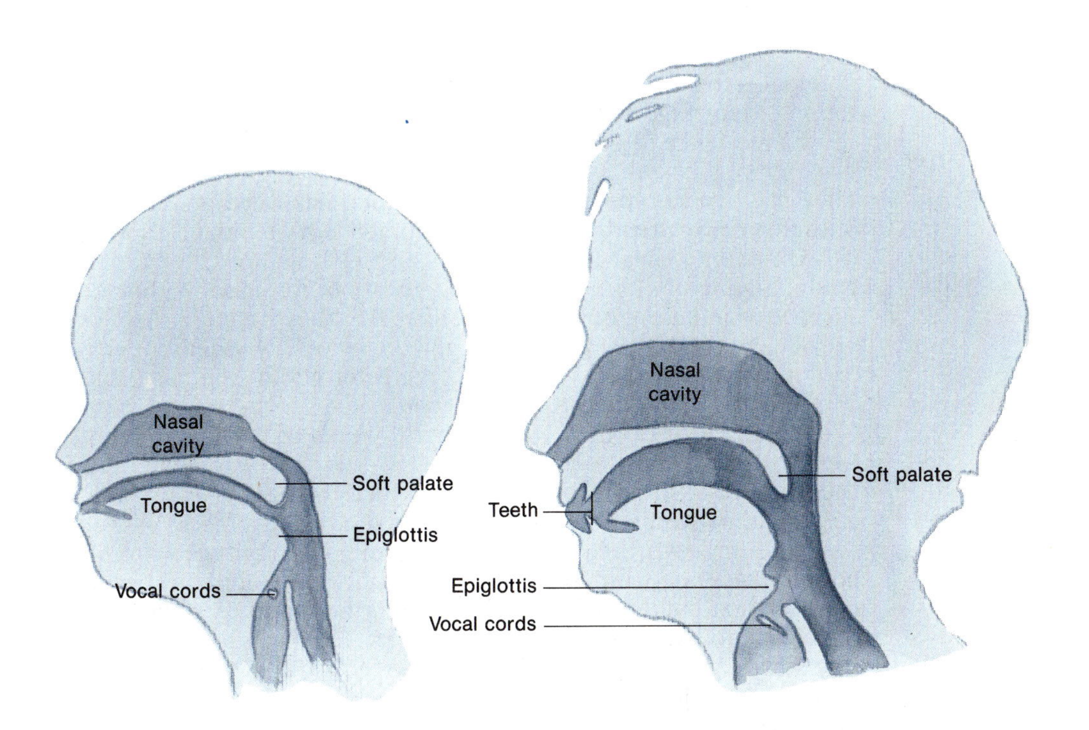

Figure 4–14
Comparison of the vocal tracts of an infant and an adult.

They can be elicited by stimuli such as smiling adult faces, voices, or touch. Frequently, infants will try to reproduce one particular sound on which they focus their attention. This can no longer be viewed as reflexive activity. As the oral and nasal cavities separate, grow, and change in shape, and as the tongue becomes more mobile, infants become capable of producing more vowels and consonants. They particularly babble the consonants p, b, m, t, d, and n, all formed in the front of the mouth. They seem to enjoy exercising their speech organs and newly acquired phonetic abilities. Many cultures use labels for primary caregivers that are combinations of these consonants plus vowels, perhaps because they are the sounds infants make (such as papa, mama, dada, nana). Infants in all cultures produce all the possible sounds of every known language. They try out whistles, snorts, chuckles, squeals, gutteral sounds, trills, glides, even "Bronx cheers." These "babbles" reflect anatomical changes. Raspberry sounds are evidence of increased air pressure in the mouth as the larynx and nasopharynx disengage, squeals and growls become possible as the larynx descends into the neck, and noncry yelling becomes possible as motor coordination between the respiratory system and the larynx improves (Kent, 1980).

At about six months babies begin to display what linguists call **phonetic drift**. They concentrate on producing only the sounds (phonemes) that they hear spoken. We have forty-five phonemes in English. Some languages have as many as seventy-one. Phonemes are fundamental sounds. Some of our alphabetical letters represent several different phonemes (for example, *ō*, *ô*, *o͞o*, *oo*, *oi*, *ou* are all represented by *o*). The infants' drift toward reproducing only the

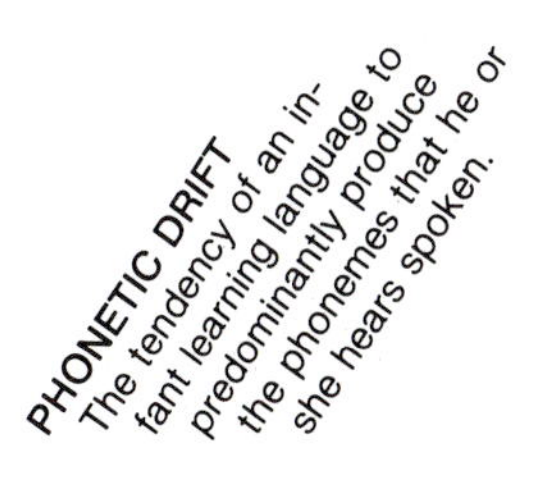

phonemes that they hear spoken indicates the beginnings of organization of language in the brain. As early as age three months infants show a tendency to babble more when they are alone. When a caregiver approaches, they quiet to attend to the caregiver's utterances. They may try to repeat sounds that adults make if the sounds are ones they have already produced and practiced. At about six months it is common to hear infants string syllables together (for example, gumgumgumgum, duhduhduhduh). By about ten months they may succeed at imitating novel sounds, ones that they have not practiced previously. The imitation of words spoken by others, whether perfect or imperfect, is called *echolalia.* Finally, at about one year, many babies make their first meaningful words.

In all cultures the first meaningful words of babies are nouns. Next to emerge are verbs, the action words. The early utterances usually refer to subjects or objects on which the baby acts, not familiar things to which caregivers attend (such as diapers). Between ages one and two infants generally acquire the ability to produce **holophrases**, one-word sentences that convey a complete message ("Up"). Next, infants learn to expand their holophrases by attaching them back-to-back to other nouns or verbs. They thus form two-word sentences ("Mommy milk, Daddy come"). The one- and two-word sentences used by babies are generally accompanied by many gestures that, together with the context, make them comprehensible. For example, the single word *chair* might mean "I want to get up in the chair," "I want the chair moved," or "I want you to get out of the chair," or it may simply be a label ("This is a chair") or a question ("Is this a chair?"). The words produced by infants are usually the most salient portions of any message. Early speech is often referred to as **telegraphic speech** because, as in telegram messages, the articles, pronouns, prepositions, conjunctions, and auxiliary verbs are omitted. In organizing and coding receptive language, infants acquire an understanding of the most meaningful units of speech. No one teaches them to use nouns and verbs first. They learn this sequence on their own.

Noam Chomsky (1968) proposed that humans (as opposed to other species) have an innate capacity to learn language, which he labeled the **language acquisition device (LAD)**. During infancy, as neurological maturation proceeds, the unique ability of the brain to sort out basic sounds and to extract from sentences the most meaningful elements becomes apparent. During early childhood the brain's language acquisition ability becomes even more sophisticated. Children will implicitly use underlying rules for constructing

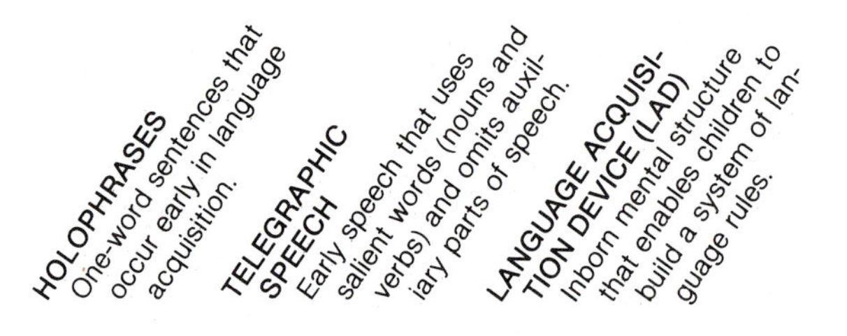

HOLOPHRASES One-word sentences that occur early in language acquisition.

TELEGRAPHIC SPEECH Early speech that uses salient words (nouns and verbs) and omits auxiliary parts of speech.

LANGUAGE ACQUISITION DEVICE (LAD) Inborn mental structure that enables children to build a system of language rules.

sentences in their native tongue. They will invent new sentences that they have never heard adults utter. They will also make rule-based mistakes (for example, adding *-er* and *-s* incorrectly to irregular verbs and nouns). These aspects of language development will be discussed further in Chapter 5.

FOSTERING LEARNING

Parents and other caregivers can have an enormous influence on their infants' intellectual development, within the limits set by heredity. Language can be stimulated in many verbal ways. Cognition can be enhanced by providing exciting stimuli to be assimilated and accommodated, as well as through various behaviors of the caregivers.

The more infants are talked to directly, the better. From birth on, caregivers should talk to their babies when they are awake and alert, as during bathing and diaper changes. This early practice helps

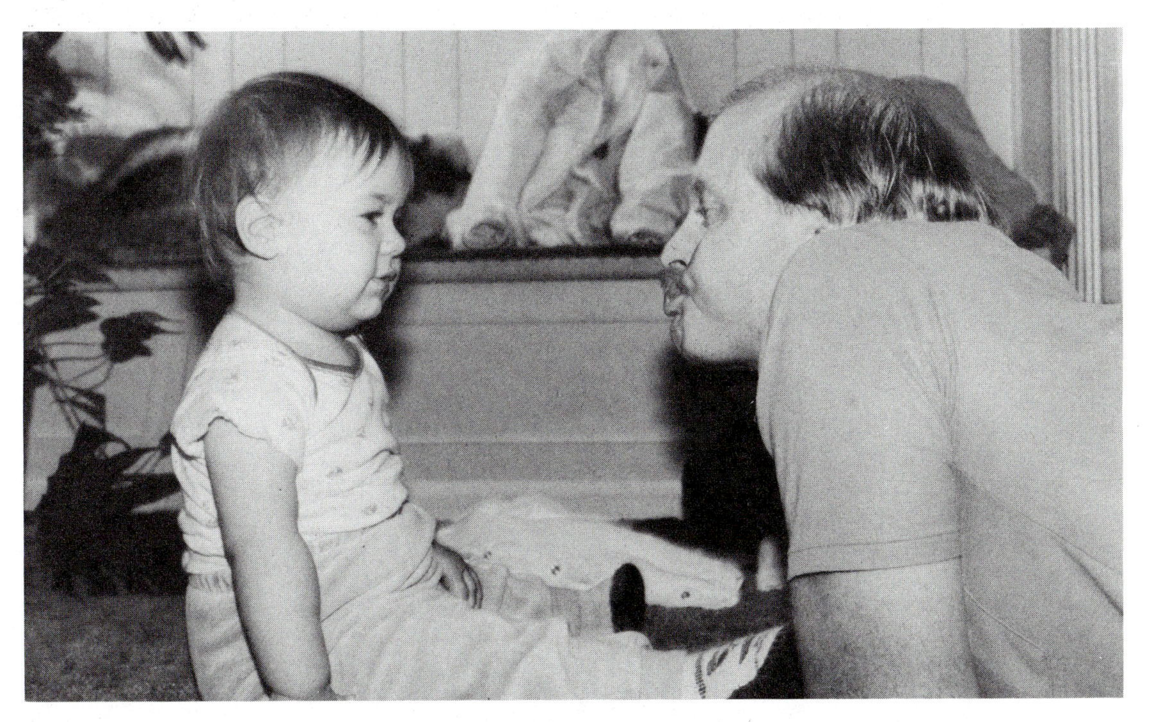

Figure 4–15
Language development is enhanced by face-to-face vocal interactions between infant and adult. The infant enjoys both hearing the adult's words and hearing her own sounds imitated back to her.

form habits of talking to infants that are invaluable later. This early talking also helps soothe babies and allows them to become familiar with and form attachments to their caregivers. Many parents are not sure what they should say to their babies. It is not important what is said, only that speaking occurs. Infants can be told the plans for the day. They can be sung to or told nursery rhymes. There is an old Norwegian saying that a much nicknamed baby is a much loved baby. Infants can be called all kinds of affectionate things and can be complimented over and over without any damage (see Figure 4–15).

At about three or four weeks, when infants begin cooing, caregivers can increase the rate at which babies play with sounds by imitating some of them. These vocal games should continue right up to and beyond the time when infants produce their first words. After about six months babies will begin to imitate some of the caregivers' sounds. From two or three months on, when infants begin working to understand words with their receptive language capabilities, caregivers should begin labeling things clearly (bottle, dog, blanket). They should show enthusiasm and encourage infants to label things as well, even if the baby's word for an object is incomprehensible. With infants' increasing development, explanations can be given for all the activities of infant care. Questions may be asked and, after a few moments, answered. All of these verbal activities will stimulate language development in infants.

Once babies begin making holophrases (one-word sentences) and gestures in attempts to communicate, every attempt should be made to understand these messages. If caregivers cannot comprehend, they can ask the baby "Show me." Sometimes, if another child is present, the child will understand what the baby is saying. When caregivers grasp what infant vocalizations mean, they should expand the holophrases or gestures into longer sentences, both encouraging the baby's word and modeling additional language (for example, "Up."—"Up? Very good! Billy wants to go up. Up Billy goes. Up into the chair").

The same kinds of verbal stimulation that are useful in fostering language also encourage cognitive growth. Caregivers should be responsive to infants' calls for information, for encouragement, for a change of play materials, or for a change of scenery.

In earliest infancy babies spend most of their time either in their cribs or with a caregiver—being fed, changed, bathed, or comforted. While in their cribs, infants can be placed on their stomachs a portion of the time. Their head-turning reflex will prevent smothering. They will try to lift their heads and eventually develop enough strength to also lift their chests. The view from this position can be stimulating. While on their backs, infants should not have to stare at a blank ceiling. A mobile, toys strung on elastic, or a friendly face is infinitely more interesting. Caregivers can foster intellectual growth during feedings, baths, and diaper changes by talking, singing, showing toys, and even just holding babies so they can view the caregivers' faces. Facial features are very potent stimuli to infants and hold their attention longer than any other configurations. Other games are also stimulating (touching baby's nose, bouncing baby on a knee, blowing on baby's stomach).

By about three months infants begin to reach for things and should be given opportunities to do so. Mobiles or toys strung on strong elastic and placed over the crib or across infant recliner seats are good. When infants can grasp the toys, it is important to keep safety in mind. Although expensive toys are not necessary, babies should not be given objects small enough to swallow, wooden toys with slivered, rough surfaces, or peeling or paint-chipped toys. Even some new toys can be dangerous (hard, thin plastic objects that break into sharp pieces, breakable rattles filled with BB pellets, stuffed animals with button or prong eyes that pull off easily). Babies explore everything with their mouths as well as with their hands (see Figure 4–16). Safer infant toys include beads on strong cords, sturdy rattles, large soft balls, blocks with rounded corners, push and pull toys, and stuffed animals with felt faces.

The right learning toy for a baby at any age is one that produces pleasure and excitement and is safe. Hunt (1964) called finding such toys "the problem of the match." If the toy is too familiar, the baby will be bored. If it is too novel, the baby will ignore it. The best learning toys are ones with both familiarity and challenge (something new to be assimilated and accommodated). Toys with various colors or sounds or textures or shapes should be provided. Several safe, manipulatory objects can be kept in one place to present to babies at playtimes. This allows them to select their own "match."

When babies acquire the motor coordination to hold their own heads steady, they can be propped or carried in upright positions more of their waking hours. This opens up whole new horizons of interesting sights for them. Caregivers who provide many social stim-

Figure 4–16
When babies can sit without support, their hands become freer for exploration. They pick up everything in reach. They examine what they grasp with hands, mouth, and any new teeth. Be sure that objects near baby are safe toys, not valuable adult properties.

ulations for their infants go a long way in fostering intellectual development. Babies learn a great deal from walks, from visits to stores or other people's houses, and from being with adults and other children in their own homes. Although playpens have their uses, it is good to give babies freedom to explore more than just a small square space from day to day. Back, front, or shoulder carriers are available that allow caregivers to take infants wherever they go with minimal difficulty.

During the latter half of the first year of life, babies become quite mobile. They will roll over, sit up alone, scoot, creep, crawl, pull themselves up to stand, and eventually walk. They will explore anything they can touch, which necessitates putting unsafe things out of reach. This is a time of active learning. Caregivers should respond to babies' bids for help, movement, encouragement, and the like. Play objects that no longer attract attention should be removed and new ones provided (see Box 4–4). It is also good to

BOX 4–4

Choosing Infant Toys

Burton White (1975) divided the first thirty-seven months of life into seven progressive developmental phases and provided detailed lists of instructions for toys, strategies, and parental practices aimed at enhancing cognitive (also physical and psychosocial) development during these phases. He wrote that most families do quite well for the first six to eight months, but few provide adequate mental stimulation during the especially crucial phases between eight to fourteen months and fourteen to twenty-four months. He suggested taking the child out of restrictive devices such as playpens, jump seats, and gated areas by these ages and accident-proofing the living area instead so the infant can roam freely and explore. He suggested toys such as graduated-size containers to fit into each other, hinged objects, stiff-paged books, surprise boxes (such as Jack-in-the-box), busy boxes, collections of safe small objects, balls, dolls, and water-play objects for eight- to fourteen-month-olds. Ridenour (1982) warned that the common practice of putting infants in this age range in walkers is inherently dangerous. The infant walker does not accelerate the onset of walking and the infant must be constantly supervised to prevent accidents.

Fourteen- to twenty-four-month-old infants can benefit from four-wheel devices on which they can sit and move, smaller toy cars and trucks with movable (but not removable) wheels to spin, doll carriages, swings and small slides, safe stairs to climb, pull toys, Ping-Pong balls and footballs, small plastic human and animal figures, pots and pans, pails, boxes, plastic jars and containers, small chairs and tables, full-length door mirrors, paper, crayons, simple puzzles, stacking toys, toy telephones, plastic pop beads, and stuffed animals. DeStefano and Mueller (1982) suggested that when two infants in this age group are playing together, smaller toys foster conflict while larger toys encourage more positive interactions. White (1975) discouraged such common activities as placing the infant in front of the television (even the educational programs) at these ages or doing any forced teaching (ABC's, numbers). He also warned against giving the child expensive but unstable tricycles, wind-up toys that require an adult to wind them, and popular but dangerous metal soldiers or small cars with parts that can be pulled off.

organize days so that changes of play objects or scenery occur frequently enough to prevent boredom.

One additional important way in which families can foster intellectual development in infants is to provide diets with enough protein (including milk and milk products, meat, poultry, fish, beans) for maximum neurological development. Diets also need to be balanced out with other essential foods (fruits, vegetables, cereals, and grains). They should not be so calorie laden, however, that infants become too fat to move about freely.

Psychosocial Development

Many of the things families do to foster learning also foster psychosocial development. The most important person in any infant's life is the primary caregiver, usually the mother or mother-substitute. Fathers, other family members, and community members are also very important persons to infants' psychosocial development.

SOCIAL MILESTONES

As babies grow, their waking social periods lengthen. Table 4–6 shows how "social" human beings develop during infancy. By about two months babies will smile spontaneously at any human face. If the recipient responds, babies will usually make happy noises. Between two and five months babies begin reaching for and grasping objects, including human noses, glasses, hair, and clothing. Whatever they successfully grasp, they bring to their mouths. By five or six months (or whenever the first teeth emerge), babies will also bite any objects brought to their mouths. At about the half-year mark babies begin trying to feed themselves. They also hold on to objects that they have grasped tightly and resist the efforts of caregivers to take them away. If objects are put out of reach, they will try hard to get them. By seven to nine months social games such as peek-a-boo and drop and fetch are fun. Babies may insist on continuing such sport to the annoyance of caregivers. At about this time babies also

Table 4–6 Selected Social Milestones of Infancy with Approximate Average Age of Emergence

Milestone	Age of emergence
Regards face	1 month
Social smile evoked by face	2 months
Selective smiles to familiar faces	5 months
Plays social games (for example, peek-a-boo)	7 months
Shy with strangers	8 months
Resists separation from caregiver	10 months
Holds own cup for drinking	12 months
Uses spoon to feed self	18 months
Removes own clothing	18 months
Plays interactive games (such as hide and seek)	24 months

become shy or anxious around strangers (called **stranger anxiety**). They may cling, cry, and vigorously protest any separation from the primary caregiver(s). When caregivers are present, babies may explore everything. As mobility increases, their curiosity leads them to every nook and cranny and through every door possible. Safety precautions become vital.

By about one year of age babies begin indicating what they want with gestures or holophrases. If babies do not receive gratification when they seek it, or if they are thwarted or frustrated in obtaining some goal, they may have angry outbursts (temper tantrums). Mealtimes may be chosen as prime time for demonstrations of self-help skills: drinking from a cup, using a spoon. They also may be chosen as opportunities to test the word *no.* Food likes and dislikes may be exaggerated in the process. By age eighteen months most babies are walking and finding additional ways to satisfy their curiosity. They imitate many of the behaviors of their caregivers, in psychosocial behaviors as well as physical actions. They will also attempt to undress and dress themselves—the first rather well, the second with many frustrations. They may not have as much stranger anxiety by this time, but they still desire a close relationship with their primary caregiver(s). Jealousies become evident—of other adults, other children, even of pets or time-consuming activities such as housework or phone conversations. The word *no* enters into what appears to be virtually all requests made of them by parents and others. As you can see, by the end of infancy babies have already experienced a fairly wide range of social-emotional reactions (among them, distress, delight, anxiety with strangers, attachment to caregivers, curiosity, pleasure, frustration, anger, jealousy, and negativism).

ATTACHMENT

In his ethological research (the study of the behavior of animals through direct observation), the Nobel Prize winning naturalist Konrad Lorenz (1935) labeled the characteristic following and imitation of the mother animal by the young that is necessary for survival as *imprinting.* In the human species motor coordination is not sufficiently developed at birth to allow for much following or imitating. Psychologists believe the dual processes of **bonding** and **attachment** are the human equivalent of imprinting.

Historical Overview In the 1940s and 1950s a few studies were made of infants raised in overcrowded orphanages. In general, the foundlings received only physical care. They were kept in cribs with

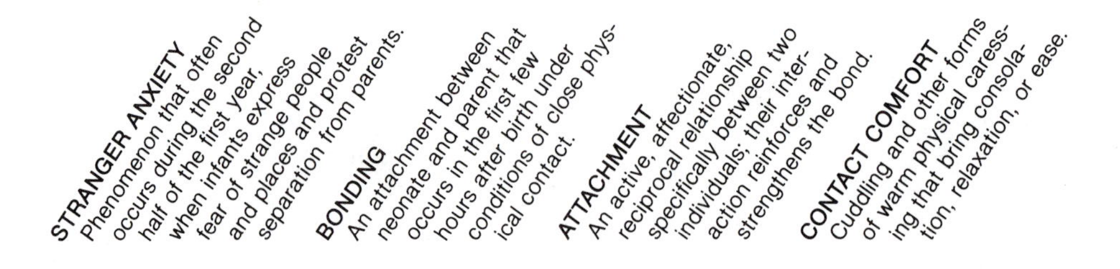

covered sides, possibly to prevent the spread of infections, and were picked up only on alternate days for baths. Bottles were propped in the cribs. Visual contact with the caregivers and other babies was minimal due to the covered crib sides. Verbal stimulation was limited, and no toys were provided. Rene Spitz (1946) found that after two years of such care 37 percent of the infants had died, and the survivors seemed emotionally starved. In another study, Wayne Dennis found that over 60 percent of such deprived infants could not sit up alone at age two, and 80 percent could not walk by age four (Dennis and Najarian, 1957). Today many of these orphans would be diagnosed as having failure-to-thrive syndrome and the suggested cause would probably be a failure of attachment.

John Bowlby (1951), a psychoanalyst working in England with parents and children separated by long-term illnesses or the havoc worked by World War II, described the serious sequelae in children separated from adults to whom they had already become attached. (These reactions are discussed further on p. 180.) In his writings, Bowlby emphasized the importance of an infant's developing a primary attachment to a caring, responsible adult. This attachment figure becomes a source of security from whom the infant can move out to explore the world and to whom he or she can retreat when danger threatens.

Harry and Margaret Harlow (1965) demonstrated the importance of cuddling (**contact comfort**) for attachment and emotional well-being when they raised newborn rhesus monkeys with either chicken wire-covered or terry cloth–covered "surrogate" (substitute) mothers. Infant monkeys spent much more time with the warm, cloth-covered mothers even when they were fed from bottles placed in the wire mothers. When a frightening object like a moving toy was placed in their cages, they ran to the cloth-covered mother for comfort, even though in some cases the Harlows had fashioned these mothers with fright-producing faces. If the warm, cloth-covered mothers were absent, the baby monkeys became extremely agitated. They did not seem to be comforted by the presence of the wire mothers, even with milk or kindly faces. The Harlows attributed this to *contact hunger.* Human infants also have a psychological need for close contact with a warm, soft person to whom they can become attached.

William Butler Yeats wrote "Love comes in at the eye." When mothers are given their babies following birth, they move in order to align their eyes with their infant's for a mutual gaze. This is followed by tentative touching, much like that between a man and a woman falling in love. Reva Rubin (1963) described the unfolding of this contact, so important between mother and child. The new mother, at first tentatively, very gently explores her neonate with a fingertip, stroking a small area (hair, profile). She behaves very much like a person in the first throes of courtship. Gradually, as she is not rejected, she braves a whole-hand stroke. Using palm as well as fingers she makes contact with larger areas of her baby (back, head, buttocks). Finally she uses her arms as an extension of her whole body. Rodholm and Larsson (1982) reported similar behavior patterns when fathers first approach their neonates. They begin by touching the infant's extremities, then the trunk, and finally the face, using their fingertips first, then proceeding to their palms.

Bonding Klaus and Kennell (1982) believe there is a sensitive period in the first few hours after birth when the parents should have close contact with their neonate in order for bonding to occur. This, they believe, is necessary for optimizing later attachment. Kennell and associates (1979) reported that mothers who had extensive contact with their neonates in the first three days after birth had more secure attachments with their infants at one year than did mothers who held their neonates only at feedings in the postpartum period. They also showed a greater commitment to the infant, greater confidence in their mothering abilities, and better caregiving skills. Brodish (1982) found that early neonatal bonding improved early feeding and generally was beneficial to infant health status. Keller, Hildebrandt, and Richards (1985) found that encouraging a group of fathers to spend one hour privately with their infant during each of the first two days of life, and another two hours with both the mother and the infant, resulted not only in a strong father-infant bond but also in improved paternal self-esteem. Compared to control fathers, they provided more care for their infants and had a more positive overall affect in their relationship with their infants.

There is some suggestion (Leifer et al., 1972) that parents whose infants are kept in incubators for weeks after delivery have difficulties forming later attachments. They make less eye contact, smile less, and hold their own babies less closely. Klaus and Kennell (1982) reported that if the separation is prolonged some parents think of their baby as belonging to someone else. It takes these unbonded mothers a considerable time to learn to do the simplest tasks such as feeding a bottle and changing a diaper. They are noticeably hesitant and clumsy in their movements.

While studies suggest benefits from early parent-infant bonding, they do not prove that bonding must occur in the neonatorium. Many infants who have been incubated have later become securely attached, especially if the mother has spent time with her baby in spite of the incubator. Chess and Thomas (1982) in a review of the literature on bonding suggested that the attachment of parent and infant in the early neonatal period is not crucial to the child's later psychosocial development. Goldberg (1983) argued that we do not yet know how important (or unimportant) bonding is due to a lack of appropriate research. She warns that while early contact should be supported, parents must not be made to feel that they are already failures if, for some reason (such as maternal postpartum problems, preterm infant problems), they cannot establish an early bond.

An attachment bond between an infant and a primary caregiver (or caregivers) nearly always has developed by the second half of the first year. The exceptions are babies who have failure-to-thrive syndrome without organic cause. Call (1984) substituted the psychiatric term "reactive attachment disorder of infancy" for these infants if, as they grow older, they continue to remain detached from their primary caregivers. Various attachment behaviors of the very young infant toward the caregiver that have been associated with bonding include grasping, visual following, clinging, reaching, and smiling. One can be quite sure attachment has occurred when certain behaviors are observed later in the first year, such as separation anxiety, fear of strangers, repeated visual or tactile contacts with the caregiver while exploring unfamiliar territory, and the kinds of following and imitative behaviors that characterize imprinting in other species.

Secure Attachment Ainsworth and her colleagues (1978) used brief infant separations from mothers, in a laboratory situation, to assess degrees of security of attachment. In this study, which has become a prototype for many other studies of the mother-infant attachment bond, the infants were briefly separated from their mothers twice. In the first instance, the mother left the baby in a laboratory room with a stranger. In the second separation, the mother left the baby in the room alone. Of special interest to Ainsworth was the baby's behavior in the two reunion episodes following separation. About two-thirds of the babies sought to be close to their mothers on reunion, and if picked up they tended to resist being put down again too soon (Ainsworth, 1982). They were labeled as having **secure attachment**. About one-quarter of the babies reacted by avoiding their mothers on reunion. They had not seemed to be very disturbed by the separation. They were called *insecurely attached of the anxious/avoidant type.* Less than 10 percent of the babies reacted by seeking proximity to their mothers on reunion but behaving in an ambivalent or even angry fashion. These babies had been acutely distressed in the separation episodes. They were labeled *insecurely attached of the anxious/ambivalent type.*

Ainsworth and her colleagues had made sixteen visits to each infant's home at three-week intervals since their subjects were three weeks of age, observing and recording behaviors. Consequently, they were able to correlate mothering behaviors with infant attachment types. Mothers of securely attached infants were most sensitive to their babies' needs and communications. Mothers of insecurely attached infants were less emotionally expressive, felt more aversion to close bodily contact with their babies, and were more frequently overwhelmed with irritation, resentment, and anger. Durrett, Otaki, and Richards (1984) reported that in a study they did, mothers of securely attached infants perceived fathers as being more supportive than did mothers of insecurely attached infants. In a similar study, Egeland and Farber (1984) found that mothers of anxious/avoidant babies tended to have negative feelings about motherhood, to be more tense and irritable, and to treat their babies with more careless indifference.

Other researchers have followed securely and insecurely attached infants in other situations on a longitudinal basis. By age two, securely attached infants are more independent and have better spatial abilities than insecurely attached infants (Hazen and Durrett, 1982). They also are more playful, have longer attention spans, and have larger vocabularies (Main, 1983). By ages four to five, securely

SECURE ATTACHMENT
Affectionate reciprocal relationship between parent and infant in which infant resists being separated from caregiver.

attached infants become more independent and positive in nursery school settings, while insecurely attached infants show signs of high dependency as well as resistant and avoidant behaviors (Sroufe, Fox, and Pancake, 1983).

Kagan (1978) advised caution in concluding that insecure attachment in infancy produces irreversible consequences in the adult lives of all persons. He studied infants in an isolated subsistence farming village in Guatemala who typically spend their first year of life confined in a small dark hut. They are hardly spoken to, not played with, and poorly nourished. When they emerge from the hut in their second year of life, attachment is minimal and they are socially and intellectually retarded. However, by adolescence most of these slow starters can perform nearly as well as children in the United States and Canada on standardized tests. He suggested that the mind and the personality are elastic in early childhood: easily deformed by shearing forces but able to rebound when and if the deforming forces are removed. Sroufe (1979) wrote that the lasting consequences of early inadequate care may be subtle and complex, taking the form of increased vulnerability to repeated stress in later life.

SEPARATION

What happens to an attachment bond when an infant is separated from the primary caregiver(s) for an extended period of time, such as when hospitalized or when placed in a daycare setting for several hours a day? The answers depend on the infant, the caregiver(s), and the reasons for separation. We will look at infant daycare first.

Infant Daycare While approximately 40 percent of mothers of infants work, most of them provide for infant care in the homes of relatives or friends, or in their own homes with babysitters, husbands, or other relatives. Only about 10 percent use infant daycare centers (Klein, 1985). In **infant daycare** situations, a few adults provide care for several infants. This poses certain limitations on the establishment of close, one-to-one relationships between infants and caregivers. Administrators of most infant daycare centers are aware of the importance of attachment in an infant's life and will make efforts to have one or two adults care for the same babies every day to provide a measure of continuity. However, there are problems. Infants are brought to daycare centers early in the morning before the hour that the parent must be at his or her own job. They are picked up to be taken home late in the afternoon or early in the evening, after the parent has finished work and perhaps run a few errands. An infant's time at the daycare center averages eight and a half to ten hours a day. Employees of daycare centers usually work eight-hour days and, in addition, may leave for holidays, vacations, lunch or coffee breaks, and sick days. Thus, in spite of efforts to have just one or two adults provide most of the care for any given infant, that infant will in fact experience an array of caregivers over time.

INFANT DAYCARE A situation in which several infants are cared for all day by a few adults in a setting specifically designed for infant care.

Infants will also experience the need to share their favorite caregiver (the one to whom they become most attached) with other infants. They will have experiences of waiting and watching while other infants are being held, cuddled, fed, changed, and entertained by their adult. If they become sick, their caregiving is either taken over by a relatively unfamiliar nurse in the daycare center or they must remain at home, often with an equally unfamiliar babysitter. However, in spite of some of these roadblocks to establishing caregiving characterized by consistency, continuity, and sameness, many daycare centers do a good job of providing sensitive attention to each infant's individual needs (see Figure 4–17).

The best centers are those with fewer children and more staff who work shorter hours and do not have housekeeping as well as childcare responsibilities (Howes, 1983). A special program in Syracuse, New York tied routine caregiving activities such as feeding and diapering into joyful emotional encounters aimed at providing the maximum in a living, loving environment for its babies (Lally and Honig, 1977). This center, with an enriched cognitive stimulation program as well, is a model for many of the excellent infant daycare centers being run at present. It trains caregivers in workshops and through its manual, *Infant Caregiving: A Design for Training* (Honig and Lally, 1972).

In an extensive review of research studies of infants in daycare compared to home-reared infants, Clarke-Stewart and Fein (1983) reported that infants do form close relationships with daycare staff.

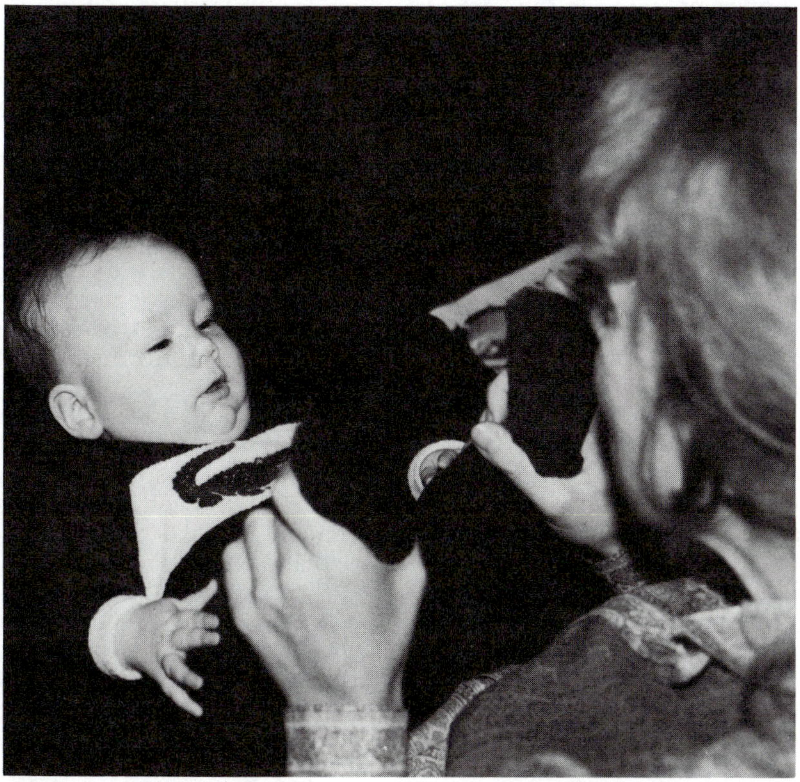

Figure 4–17
High-quality infant daycare can meet a baby's needs for tender loving care, verbal stimulation, and sensory and perceptual arousal as well as for feedings and diaperings.

These attachments, however, do not replace nor weaken the mother-infant primary attachment bond. A host of research reports have documented the idea that attached infants, despite long hours in daycare, still prefer their mother, go to her for help, stay closer to her, approach her more often, go to her when distressed or bored, and interact with her more. They do not greet the daycare staff with the same joy in the morning as they do their mother at night. Clarke-Stewart and Fein also cited many studies suggesting that infants who have experienced daycare have more social competence and do as well or better on measures of intellectual development. The studies Clarke-Stewart and Fein reviewed were of infants who attended high-quality programs.

Honig (1983), in an attempt to evaluate various infant programs, pointed out that it is difficult to compare programs and rate their effectiveness because each has different staff, methods, goals, durations, and populations served. Although research demonstrates that good infant daycare is not detrimental, it does not suggest that it is necessarily better than care by parents. Many experts on infant and early child development feel that working mothers should be allowed extended leaves of absence from their jobs during the early months of their infants' lives. Many experts also support government subsidies to infant daycare facilities, as well as development of more centers at the mothers' working places. The United States lags behind the countries of Western Europe as well as China, Japan, and the Soviet Union in the support it provides for families with young children.

Hospitalization Research reports on the effects of infant-caregiver separation for hospitalization are in contrast to the findings of little or no emotional trauma engendered by daycare separations. The situations, however, differ in many respects. A healthy baby goes to a cheerful, familiar daycare center every day. A sick baby goes to a busy, unfamiliar hospital for the first time in his or her memory. It is far easier to adjust to a new environment where substitute caregivers behave like primary caregivers (holding, talking, entertaining with toys) than to accept a setting where a multitude of strange adults poke, pry, and administer painful or distasteful treatments and medicines.

Research on the effects of hospitalization show a steady change in a baby's behavior over time, related to the length of the hospital stay. Bowlby (1973) described a syndrome called **hospitalism** with progressively worse symptoms. He classified these changes into three progressive stages: protest, despair, and denial. They are illustrated in the vignette in Box 4–5. The denial stage involves a deterioration of the primary attachment bond.

The behaviors described in Box 4–5 can be alleviated to a great extent by unrestricted visiting privileges or rooming-in for the primary caregiver during the baby's hospitalization. However, factors such as the nature of the child's illness, the necessities of caring for other children at home, or the primary caregiver's emotionality may make frequent visiting imprudent. The worst age period for hospital-

BOX 4–5

Hospitalism

An eighteen-month-old baby enters the hospital in his mother's arms, sick and frightened. A stranger in white takes him from his mother and carries him to a strange room, past other staring or crying or busy people. She places him in an unfamiliar, cage-like crib and undresses him. His mother is requested to leave the room. His clothes disappear. Other strangers approach, stick cold things in him, hold him down, tell him to stop crying. When they finally go away, they pull bars up around him. Mother eventually reappears and is crying. She says she must leave, and does. He feels sick, hurt, rejected, frightened. He shakes the crib bars and screams (*protest*).

For a day or two the protesting is vigorous. He calls for his mother when she is gone and cries when she is present. He is not only uncomfortable from the illness (or other reason for hospitalization) but also angry at everyone who approaches. He is grief stricken each time his mother or other family members depart. Gradually his behavior becomes subdued, passive, more obedient. He is "settled in." He starts to keep his emotions bottled inside rather than exhausting himself with futile protests. He may regress in feeding and/or toileting practices (*despair*).

For the next few days he becomes increasingly more pleasant toward the hospital staff and more and more hostile toward his family. If he is hospitalized for an extended period, he withdraws attachment behaviors from all adults. He appears self-sufficient, even happy. He hardly notices when adults, especially his primary caregiver(s) come and go (*denial*). If a baby is separated long enough to begin denying his attachments to loved adults, his normal social-emotional development may be in jeopardy. It may take an extended period of time for him to learn to trust his caregiver(s) again.

When the baby is discharged from the hospital, another series of behaviors are typical. At first he is withdrawn. He warms up to auxiliary family members first. For hours, or days, he demonstrates hostility toward his primary caregiver. He demands things and is jealous of his siblings. Then for several days he clings to his primary caregiver. He also is defiant, destructive, disobedient, and negativistic. He must relearn feeding, dressing, and any bowel or bladder control. For weeks he remains babyish, anxious, and fearful, has sleep difficulties and nightmares, is easily angered, and has temper tantrums. If his parents treat him with sympathy and love, in a few months he will probably recover from his trauma.

based separation from the primary caregiver(s) is from about six months to three years. Prior to six months, attachments may not be fully developed. After age three, a child begins to understand reasons for the separation and hospitalization and may not suffer the progressive stages of protest, despair, and denial that characterize hospitalism.

ERIKSON'S CONCEPT OF TRUST VERSUS MISTRUST

Erik Erikson, the neo-Freudian psychodynamic theorist, postulated that humans go through eight stages of psychosocial development in the life span. At each stage there exists a nuclear conflict that is a critical period for the development of certain attitudes and the resolution of certain conflicts. Although nuclear conflicts are never completely resolved during the stage in which they emerge, a satisfactory resolution of most of the conflict leaves a person with the judgment and ability to handle repeated upsurgings of the basic conflict throughout life. A person who has not passed through a stage successfully experiences a feeling of insecurity related to that "sense" throughout life.

Erikson proposed that, in infancy, the most crucial sense to be formed is a sense of **trust**, a feeling of confidence and reliance that caregivers will provide adequate care. Erikson's sense of trust does not explicitly deal with attachment; however, it is implicit in the theory that an attachment does form between the infant and the caregiver as trust develops. Erikson (1963) suggested that when a primary caregiver meets the infant's needs consistently, with sameness and continuity, the infant learns to rely not only on the external provider but also on himself or herself. Trust is learned very early through situations such as feedings and diaperings. If infants are fed when hunger first appears rather than being allowed to develop strong hunger pangs, they are more comfortable in their surroundings. Likewise, if they are diapered when they first wet rather than being left cold and dirty, a feeling that the world is a good place is more apt to develop. **Mistrust**, the alternative, develops when infants are allowed to experience frequent bouts of overwhelming hunger or are left dirty, uncomfortable, irritated, and irritable. Effective caregiving for the development of a sense of trust involves far more than just taking care of an infant's bodily needs. It prospers with warm, tender touching, verbal soothing, verbal stimulation, smiles, and sensory arousal with toys and changes of scenery. It flounders with careless holding, inconsistent caregiving, and an absence of things to see and do while awake. Table 4–7 presents some of Erikson's descriptions of the sense of trust versus the sense of mistrust and some adult behaviors that foster each attitude.

INDIVIDUAL DIFFERENCES

Few people believe today, as John Locke postulated in the seventeenth century, that the infant is born a "tabula rasa" (clean slate) or a moldable piece of clay. The kind of adults that infants become is most certainly influenced by caregivers and care received, but behaviors of caregivers do not create the whole physique, intelligence, and personality of adults. Each infant comes into the world as sculptor as well as clay to be molded. Each has his or her own unique characteristics for self-shaping; for seeking or avoiding stimuli; for organizing, interpreting, and adapting to experiences. Normal babies differ remarkably. Some infants, when held, cling. Others need considerable support and behave more like rag dolls. Some infants sleep a great deal more than others. **Individual differences** in babies have been noted in their activity, rhythmicity, adaptability, approach, threshold, intensity, moods, distractibility, and persistence (Thomas and Chess, 1977). Babies also differ in the amount of their crying, soothability, and capability for self-comforting behavior as well as in their capacity to assimilate and accommodate sensory stimuli.

In a longitudinal study of more than 200 infants in New York City, Thomas and Chess (1977) identified three different types of

Table 4–7 Erikson's First Nuclear Conflict: Trust Versus Mistrust

Sense	Eriksonian descriptions	Fostering adult behaviors
Trust	Ease of feeding, depth of sleep, relaxation of bowels	Caregiving techniques with consistency, continuity, and sameness
	Feeling of inner goodness	Sensitive attention to the baby's individual needs
	Reliance on outer providers	Firm sense of personal trust-worthiness; conviction that there is meaning to what one is doing
	Trust in the capacity of one's own organs to cope with urges	
versus		
Mistrust	Sense of having been deprived, divided, abandoned	Inconsistent, neglectful caregiving
	Anxiety, rage, lack of control of urges	Withdrawal from situations where baby tests the relationship and demands attention to his or her needs

SOURCE: Based on material from "Eight Ages of Man" from *Childhood and Society* (2nd ed., revised) by Erik H. Erikson, with the permission of W. W. Norton & Company, Inc. Copyright 1950, © 1963 by W. W. Norton & Company, Inc.

temperaments in infants. They defined *temperament* as behavioral style, the "how" of behavior in contrast to the "what" (abilities) and the "why" (motivation). Others prefer to define temperament as individual differences in the expression of emotionality and arousal (Campos et al., 1983). The three types of temperament classified by Thomas and Chess (difficult, easy, and slow-to-warm-up) account for the way about 60 percent of all infants behave emotionally. Other infants' behaviors can be classified as intermediate to these three patterns. *Difficult* babies do not have predictable rhythms; they react intensely, approach slowly, adapt slowly, and are more frequently negative in mood. *Easy* babies do have predictable rhythms, react mildly, approach and adapt quickly, and are more frequently positive in mood. *Slow-to-warm-up* babies approach and adapt slowly and are

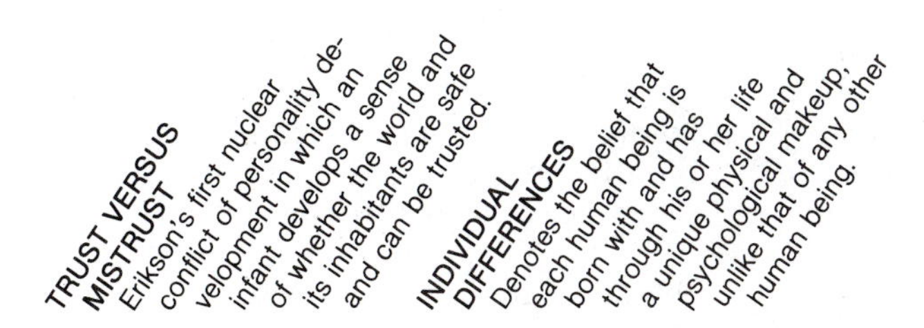

more negative in mood. However, they are not as intense as difficult infants. They react mildly and have a relatively low activity level. Infants with difficult temperaments are much more fearful of strangers (Berberian and Snyder, 1982). They are, in Erikson's polar conflict scheme, least trusting.

The differences in uniquely individual infants can be detected in the first three days of life and tend to be stable over time. For example, Korner and her associates (1985) compared activity electronically monitored in infancy with activity levels in the same children at four to eight years of age, as measured by ambulatory microcomputers, and found that the least vigorous infants became the least active children. Twin studies demonstrate that temperamental styles are not simply a result of parenting practices. Identical twins are much more alike on temperament ratings than same-sexed fraternal twins, both at nine months (Torgersen and Kringlen, 1978) and at four and seven years (Goldsmith and Gottesman, 1981).

Many studies have shown sex differences in infant temperaments. Boy babies have been shown to be more active, both awake and in sleep, to fuss and cry more, and to soothe less easily, which may eventually lead to fewer attempts to soothe them by caregivers. Girl babies have been shown to babble to faces and voices more and in turn to receive more face-to-face vocal exchanges. Carey and McDevitt (1978), who developed a reliable and widely used infant temperament questionnaire, found another sex difference in temperament: girls are less approaching toward strangers at four to eight months than boys are. As infants move into their second year of life, boys have been observed to show more aggression, and independence, whereas girls have been observed spending more time close to familiar adults, vocalizing. Arguments as to whether observed sex differences are genetically or environmentally induced, or both, are unresolved. Most research suggests socialization factors heavily affect the outcome.

Ethnic differences in temperament have also been seen as early as a few days after the birth, suggesting genetic as well as environmental effects of race. Kagan, Kearsley, and Zelazo (1978) reported that Chinese American infants are less vocal, less active, more socially inhibited, and more negative in mood than non-Chinese infants. When Hsu, Soong and their associates (1981) used the Carey Infant Temperament Questionnaire in Taiwan to assess Chinese babies, they found them to be more withdrawing, less adaptable, less distractible, and more negative in mood than Chinese American babies. Weissbluth (1982), also using Carey's Infant Temperament Questionnaire, likewise found Chinese American babies to be less adaptable, more withdrawing, and more negative in mood. However, he also found that they have, on the average, a shorter total sleep duration than non-Chinese infants. Weissbluth associated decreased duration of night sleep with more negative mood. Ethnic differences in temperament may exist, but they may be associated with environmental as well as genetic differences. They also can change, even though they usually do not.

Weissbluth and Liu (1983) reported that temperament may be

associated with active sleep patterns (movement throughout the night in a restless fashion with many small movements of the hands, feet, or eyes). Active sleep is more frequently found in male infants and is correlated with difficult temperament and shorter attention spans. Weissbluth and Green (1984) also found that difficult temperament may be associated with low plasma progesterone levels in infancy, reflecting individual differences in neurophysiology.

Is it useful to label infants by temperament type, especially when research cannot prove that all temperaments remain stable or clearly explain why differences exist? Carey (1981) gave many reasons why assessing temperament and discussing results with parents are useful. One need not use the labels "difficult" or "slow-to-warm-up." However, parents will not feel as inadequate, guilty, apprehensive, or angry if they realize that their infant's slow approach to foods or people, intensity, or negative moods, for example, are temperamental characteristics that are inborn. Temperament assessments can help parents anticipate their infant's probable behaviors and emotional expressions. With this insight, they can shift their interaction patterns to mesh with their baby's reactions and prevent potentially serious family problems.

FATHERING

How large a role do fathers play in infant caregiving? Carey (1981) reported that about 10 percent of the fathers in one of his studies were not well enough acquainted with their babies to complete an Infant Temperament questionnaire. Nevertheless, the general consensus is that fathers are more involved with their infants today than they were in past decades. The current encouragement for fathers to attend prenatal childbirth classes, participate in labor and delivery, and participate in the care of the neonate in the hospital may be one of the reasons. Another may be the fact that about 40 percent of mothers return to work shortly after their infant's birth and need the father's help with infant care. Keyes and Scoblic (1982) found that if mothers work or attend school, fathers do spend more time interacting with their infants. However, they do not spend as much time feeding and changing diapers as the mothers (Parke, 1982). They are more likely to play with their infants, and their play is more physical and rousing, especially with sons (Power and Parke, 1982). The vigorous physical play between fathers and male infants may have a beneficial effect. The boys may learn to regulate their physical aggression as a result of experiencing competent rough and tumble play in which nobody gets hurt (Parke, 1982).

Fathers engage in more essential caregiving (feeding, diapering) if a mother undergoes Caesarean section delivery (major surgery) and truly needs help in the postpartum period (Entwisle and Doering, 1980). Fathers are also more apt to feed and diaper if their infant is premature, low-birth-weight, or small for gestational age. There is some evidence that an infant's development is enhanced by more paternal participation (Pederson, Rubenstein, and Yarrow, 1980). Other evidence suggests that a mother's competence in caregiving is enhanced by paternal support of her efforts (Pederson,

1981). While the research does not suggest that a father's participation is essential to infant care, it does suggest that it is very beneficial.

DISCIPLINE

Discipline refers to limit setting, to training that develops self-control. It is a very important part of helping babies feel safe and secure. It helps them gain the ability to "trust oneself and the capacity of one's own organs to cope with urges; . . . to consider oneself trustworthy enough so that the providers will not need to be on guard lest they be nipped" (Erikson, 1963, p. 248). This feeling of emerging inner goodness and self-control is best brought about through consistent, predictable caregiving routines.

At birth infants' wake-sleep-hunger cycles are irregular. Initially, it is a good idea to offer food each time babies awaken from sleep, although before long babies will develop their own feeding schedules, based on their intestinal absorption patterns and the amount of milk they can tolerate at each feeding. In general, periods of hunger vary from two and a half to four hours apart. Soon caregivers will recognize hunger cycles and will discover the approximate rhythm of their own infants. At this point they can begin to modify their infants' demands to fit a pattern. Knowing the approximate length of time between hunger cycles will help establish another rhythm, the day-night sleep pattern. Most neonates will automatically have one longer nap (ranging from four to eight hours) per day and several shorter ones (ranging from one to three hours). The longer nap does not necessarily occur at night, but a knowledge of the approximate time between hunger pangs during the short nap periods will enable caregivers to "move" the longer nap to night. They simply can wake the baby for feedings every two and a half (or three, or four) hours during the day and provide a warm, dark quiet atmosphere conducive to sleep at night. The "longer" night nap may only be four hours initially, but it gradually lengthens until babies "sleep through the night." The age at which an eight-hour-plus night occurs varies widely from infant to infant. A few will do this from birth. Others may not oblige until five to seven months of age. Other ways that caregivers can demonstrate consistency and continuity and help establish disciplined, patterned days for their infants are by bathing the baby at approximately the same time each day and by making feedings routine (feeding, burping, feeding, burping, diapering, talking).

As babies get older, naps get shorter, although most infants still require a morning as well as an afternoon nap and an uninterrupted sleep through the night. Babies who are consistently put to nap at certain times are less likely to resist being put to bed than babies who are put down irregularly.

DISCIPLINE Training that develops self-control, character, orderliness, and efficiency.

Food likes and dislikes are common in infancy. To avoid confrontations over foods, caregivers should try a new food when the baby is neither tired nor painfully hungry. If babies resist, caregivers should stay calm. They can try the food again later. Food jags are also normal. Infants may resist all but one food for a few days, but these jags seldom last long. It is more important to have regular, pleasant meals than to insist on eating exactly what is on a plate at every sitting. If meals are kept pleasant and patterns of interpersonal relationships include mutual affection and trust, "imitative" older infants will soon want to eat the same foods that the caregivers eat. When self-help urges surface, finger foods can be supplied (such as soft fruit slices, cooked and cooled vegetables, soft cheeses, cereals, bread cubes).

Biting usually follows teething. Infants learn what not to bite (caregivers, siblings, furniture) and what is acceptable (teething toys, dry toast, biscuits) through the caregiver's prompt substitution of appropriate teething materials for inappropriate ones.

Mobility requires a certain amount of household and outdoor engineering and an acceptance of the word *no* as voiced by caregivers. Household engineering consists of such things as moving dangerous items out of reach or locking them away, using gates as necessary on stairwells or off-limits areas, putting safety plugs in electrical outlets, and keeping pins, scissors, nails, or other sharp or poisonous items off the floors. Outdoor engineering consists of providing a fenced-in play yard, teaching the baby never to cross a street without holding an adult hand, checking toys and playground equipment for sharp edges or toxic paint flakes, and never leaving the baby alone in a wading pool or around open water (see Figure 4–18). Infants also learn to respect and obey the word *no* if it is used only when needed and reinforced by removing the infant from the temptation (or vice versa). However, if it is voiced often and indiscriminately with no follow-up, it will eventually be ignored.

Figure 4–18
Babies love to play in water. Once infants can walk, they become adept at finding water sources. Never leave an infant unattended in a tub, a wading pool, or around other bodies of water. Even a shallow puddle can be hazardous to an infant.

Temper tantrums are often avoidable if caregivers have not rewarded temper cries (see Box 4–6) and are alert to signs of hunger, exhaustion, boredom and/or frustration in their babies. Angry outbursts are much more likely to occur when babies are hungry and tired than when they are fed and rested. Likewise, they occur more frequently when toys, scenery, or social contacts thwart or frustrate some goal than when babies are able to observe or explore phenomena that interest them. If an infant desires something forbidden and must be thwarted with a *no,* it is a wise caregiver who can quickly find an acceptable substitute goal for the baby or in some way distract attention to another interesting stimulus.

Summary

The neonatal period (birth to one month) is a high-risk period in the life of the new human being. Neonates are dependent on others to meet their needs for food, warmth, tender loving care, stimulation, and protection from infectious agents.

Growth and change are rapid in the first two years of life. Babies double their birth weight by about five months, triple it by one year, and quadruple it by two years. They add about 10 inches to their length in the first year and about 5 more inches in the second year.

BOX 4–6

What's in a Four- to Twelve-Month-Old's Cry?

By four to six months of age, the cry sounds of an infant are not indicative of a high-risk neonate (see Box 4–1) or of colic (see Box 4–2). Do all older babies' cry sounds become similar? Not at all. Each unique baby has a repertoire of cries by this age to communicate such things as hunger, pain, sleepiness, or boredom, and the primary caregiver(s) can tell which cry is which. Nevertheless, some babies' cries are generally more irritating and "spoiled" sounding than the repertoire of cries of other babies. Lounsbury and Bates (1982) asked mothers of four- to six-month-old infants to rate the taped cries of twelve unrelated same-aged infants in terms of their emotional responses to the cries. Would the crying induce sadness, a desire to mother the baby, irritation, or a perception that the infant on the tape was "spoiled"?

Each infant whose crying had been taped had previously been classified by temperament type, difficult to easy. Unrelated mothers, not knowing the temperament ratings, were expert at picking out the "difficult" babies. They were the babies who sounded "spoiled" and whose cries were most irritating. Spectrum analysis of the cries of difficult babies reveal that they have longer pauses within and between cry sounds.

Are babies with "difficult" temperaments spoiled? Weissbluth (1985) advised that, while the temperament is inborn, the caregivers' reactions to the irritating crying can either reduce its frequency or exacerbate the screaming ("spoil" the baby) after four to six months. His recommendation: "During the first few months, always respond promptly to your baby's crying. After four, five, certainly six months, it is time to change your tactics" (p. 135). Weissbluth focused on night crying or crankiness between midnight and 5 A.M., which he called "trained night crying." A baby who fusses all night will not be pleasant the next day (just like adults!). The caregiver should shift the baby's sleep schedule. Weissbluth recommends waking the baby early, allowing no naps after 11 A.M., and imposing bedtime at 9 P.M. The bedroom should be kept dark (no night light) and quiet, and the baby should be left alone, even if he or she cries. Within a few days, the crying will occur less frequently, for shorter and shorter durations, and soon the baby will "learn" to sleep through the night. Parents need to be consistent with this imposition of discipline. Weissbluth adds a word of caution. If the baby habitually snores and appears to have difficulty breathing at night, parents should consult a pediatrician. If the baby awakens agitated and frightened from nightmares or night terrors, parents should check to see if he or she has a fever. If night terrors are common, parents should consult a pediatrician or perhaps a child psychologist.

Health maintenance requires providing good nutrition, adequate sleep, and a clean environment lovingly. Infections must be treated promptly; many can be prevented through a series of immunizations. Safety precautions to prevent accidents are essential. Common health problems include digestive and nutritional disorders and respiratory infections. Sudden infant death syndrome and failure-to-thrive syndrome affect a smaller number of infants each year. Efforts are actively being made to both understand and prevent these syndromes.

Piaget described a sensorimotor stage to explain the cognitive changes that occur during infancy. In six progressively more complex phases, babies move from predominant exercise of preexisting reflexes to actions that require remembering and planning.

Learning to talk has its roots in receptive language. The baby listens to communications of others and begins to sort out meanings of words. Expressive language (talking) usually begins with nouns (mama, dada, nana, papa). Verbs are usually added next. Holophrases, which are one-word sentences, may consist of nouns or verbs ("go"; "bottle"). Eventually the baby begins to string nouns and verbs together ("Daddy come"). Early speech is described as telegraphic because, as in telegrams, only the most salient words are reproduced.

Tender loving care is a crucial ingredient in the make-up of emotionally healthy infants. Babies need contact comfort as well as feeding and diapering. They become attached to their primary caregivers and show joy at reunion and anxiety at separation. Erikson described the most crucial social learning in infancy as trust, a sense of confidence and reliance in the caregivers. Trust develops when caregivers are consistent with schedules and discipline as well as sensitive to the baby's needs.

Babies show individual differences in temperament early. Babies shape their caregivers' behaviors as well as respond to them.

Fathers play with infants more than assist with care but can also be competent at feeding and diapering. All caregivers involved with an infant should be sensitive to the baby's needs and temperament and should be consistent with schedules and discipline.

Questions for Review

1. What might be the developmental consequences for an infant who receives little or no stimulation?
2. The number of women in the labor force has been increasing over the past twenty years. Many of these women are placing their infants in daycare centers. Discuss the pros and cons for families. In your discussion, consider the needs of all family members—mother, father, and child.
3. Infants require a great deal of attention to both their physical and psychological needs. Describe some of their psychological needs.

4. Describe the process that occurs immediately after birth, when the baby adapts from intrauterine to extrauterine life.
5. Discuss how families can foster cognitive development in infancy.
6. Describe a normal and an abnormal cry of a neonate, a two-month-old, and a ten-month-old, discussing probable reason(s) for the abnormal cry you have described at each of the ages.
7. Your neighbor's infant has been diagnosed as having non-organic failure-to-thrive syndrome. What does this tell you? Can you help your neighbor?
8. Infants may reside in foster homes for a number of months, often being moved from one home to another. Consider the future implications of this lifestyle on such children. Speculate as to the resolution of Erikson's trust-mistrust conflict in these situations. Also, discuss the consequences in terms of formation of the attachment bond.
9. Your friend has written to ask you whether you think she should breast-feed or bottle-feed her infant. Write a letter in response giving pros and cons of each.

References

Agran, P. F. Motor vehicle occupant injuries in noncrash events. *Pediatrics,* 1981, *67,* 838–840.

Ainsworth, M. D. S. Early caregiving and later patterns of attachment. In M. H. Klaus and M. O. Robertson (Eds.), *Birth, interaction and attachment: Pediatric round table series; 6.* Skillman, N.J.: Johnson & Johnson Baby Products Co., 1982.

Ainsworth, M. D. S.; Blehar, M. C.; Waters, E.; and Wall, S. *Patterns of attachment.* Hillsdale, N.J.: Lawrence Erlbaum, 1978.

Als, H.; Lester, B. M.; Tronick, E.; and Brazelton, T. B. Towards a research instrument for the assessment of preterm infants' behavior. In H. E. Fitzgerald, B. M. Lester, and M. W. Yogman (Eds.), *Theory and research in behavioral pediatrics,* vol. 1. New York: Plenum, 1982.

Apgar, V. Proposal for a new method of evaluation of the newborn infant. *Anesthesia and Analgesia,* 1953, *32,* 260.

Aslin, R. N.; Pisoni, D. B.; and Jusczyk, P. W. Auditory development and speech perception in infancy. In M. M. Haith and J. J. Campos (Eds.), *Infancy and developmental psychobiology.* Vol. 2 of P. H. Mussen (Ed.), *Handbook of child psychology,* 4th ed. New York: Wiley, 1983.

Banks, M. S., and Salapatek, P. Infant visual perception. In M. M. Haith and J. J. Campos (Eds.), *Infancy and developmental psychobiology.* Vol. 2 of P. H. Mussen (Ed.), *Handbook of child psychology,* 4th ed. New York: Wiley, 1983.

Barbero, G. Failure-to-thrive. In M. H. Klaus, T. Leger, and M. A. Trause (Eds.), *Maternal attachment and mothering disorders: Pediatric round table series; 1.* Skillman, N.J.: Johnson & Johnson Baby Products Co., 1982.

Barnard, K. E., and Bee, H. L. The impact of temporally patterned stimulation on the development of preterm infants. *Child Development,* 1983, *54,* 1156–1167.

Beedle, G. L. Teeth. In C. H. Kempe, H. K. Silver, and D. O'Brien (Eds.), *Current pediatric diagnosis and treatment,* 8th ed. Los Altos, Calif.: Lange, 1984.

Bell, S. M., and Ainsworth, M. D. S. Infant crying and maternal responsiveness. *Child Development,* 1972, *43,* 1171–1190.

Berberian, K. E., and Snyder, S. S. The relationship of temperament and stranger reaction for younger and older infants. *Merrill-Palmer Quarterly,* 1982, *28,* 79–94.

Bowlby, J. *Maternal care and mental health.* Geneva: World Health Organization, 1951.

Bowlby, J. *Attachment and loss: Separation.* New York: Basic Books, 1973.

Brazelton, T. B. Neonatal behavioral assessment scale. *National Spastics Society Monograph.* London: Heineman, 1973.

Brodish, M. S. Relationship of early bonding to initial infant feeding patterns in bottle-fed newborns. *Journal of Obstetric, Gynecologic and Neonatal Nursing,* 1982, *11,* 248–252.

Bruner, J. S. Organization of early skilled action. *Child Development,* 1973, *44,* 1–11.

Burns, K. A.; Deddish, R. B.; Burns, W. J.; and Hatcher, R. P. Use of oscillating waterbeds and rhythmic sounds for premature infant stimulation. *Developmental Psychology,* 1983, *19,* 746–751.

Call, J. D. Child abuse and neglect in infancy: Sources of hostility within the parent-infant dyad and disorders of attachment in infancy. *Child Abuse and Neglect: The International Journal,* 1984, *8,* 185–202.

Campos, J. J.; Barrett, K. C.; Lamb, M. E.; Goldsmith, H. H.; and Stenberg, C. Socioemotional development. In M. M. Haith and J. J. Campos (Eds.), *Infancy and developmental psychobiology.* Vol. 2 of P. H. Mussen (Ed.), *Handbook of child psychology,* 4th ed. New York: Wiley, 1983.

Carey, W. B. The importance of temperament-environment interaction for child health and development. In M. E. Lewis and L. A. Rosenblum (Eds.), *The uncommon child.* New York: Plenum Press, 1981.

Carey, W. B., and McDevitt, S. C. Revision of the infant temperament questionnaire. *Pediatrics,* 1978, *61,* 735–739.

Chess, S., and Thomas, A. Infant bonding: Mystique and reality. *American Journal of Orthopsychiatry,* 1982, *52,* 213–222.

Chomsky, N. *Language and mind.* New York: Harcourt, Brace & World, 1968.

Clarke-Stewart, K. A., and Fein, G. G. Early childhood programs. In M. M. Haith and J. J. Campos (Eds.), *Infancy and developmental psychobiology.* Vol. 2 of P. H. Mussen (Ed.), *Handbook of child psychology,* 4th ed. New York: Wiley, 1983.

Dennis, W., and Najarian, P. Infant development under environmental handicap. *Psychological Monographs,* 1957, *71* (Whole No. 436).

DeStefano, C. T., and Mueller, E. Environmental determinants of peer social activity in 18-month-old males. *Infant Behavior and Development,* 1982, *5,* 175–183.

Drotar, D., and Malone, C. Family-oriented intervention with the failure-to-thrive infant. In M. H. Klaus and M. O. Robertson (Eds.), *Birth, interaction and attachment: Pediatric round table series; 6.* Skillman, N.J.: Johnson & Johnson Baby Products Co., 1982.

Duffy, F. H.; Burchfiel, J. L.; and Lombroso, C. T. Brain electrical activity mapping (BEAM): A method for extending the clinical utility of EEG and evoked potentials data. *Annals of Neurology,* 1979, *5,* 309–321.

Durrett, M. E.; Otaki, M.; and Richards, P. Attachment and the mother's perception of support from the father. *International Journal of Behavioral Development,* 1984, *7,* 167–176.

Egeland, B., and Farber, E. A. Infant-mother attachment: Factors related to its development and changes over time. *Child Development,* 1984, *55,* 753–771.

Entwisle, D. R., and Doering, S. G. *The first birth.* Baltimore: Johns Hopkins University Press, 1980.

Erikson, E. *Childhood and society,* 2nd ed. New York: Norton, 1963.

Fernald, A. 4-month-old infants prefer to listen to motherese. *Infant Behavior and Development,* 1985, *8,* 181–195.

Fomon, S. J. *Infant nutrition,* 2nd ed. Philadelphia: Saunders, 1974.

Frankenburg, W. K., and Dodds, J. B. *Denver Developmental Screening Test Manual.* Denver: University of Colorado Medical Center, 1967.

Fulginiti, V. A. Immunization. In C. H. Kempe, H. K. Silver, and D. O'Brien (Eds.), *Current pediatric diagnosis and treatment,* 8th ed. Los Altos, Calif.: Lange, 1984.

Gagan, R. J.; Cupoli, J. M.; and Watkins, A. H. The families of children who fail to thrive: Preliminary investigations of parental deprivation among organic and nonorganic cases. *Child Abuse and Neglect, The International Journal,* 1984, *8,* 93–103.

Galler, J. R.; Ramsey, F.; and Solimano, G. The influence of early malnutrition on subsequent behavioral development: III. Learning disabilities as a sequel to malnutrition. *Pediatric Research,* 1984, *18,* 309–313.

Gibson, E. J. *Principles of perceptual learning and development.* New York: Appleton-Century-Crofts, 1969.

Gibson, J. J. *The senses considered as perceptual systems.* Boston: Houghton Mifflin, 1966.

Githens, J. H., and Hathaway, W. Hematologic disorders. In C. H. Kempe, H. K. Silver, and D. O'Brien (Eds.), *Current pediatric diagnosis and treatment,* 8th ed. Los Altos, Calif.: Lange, 1984.

Goldberg, S. Parent-infant bonding: Another look. *Child Development,* 1983, *54,* 1355–1382.

Goldsmith, H. H., and Gottesman, I. I. Origins of variation in behavioral style: A longitudinal study of temperament in young twins. *Child Development,* 1981, *52,* 91–103.

Grantham-McGregor, S. M.; Powell, C.; Stewart, M.; and Schofield, W. N. Longitudinal study of growth and development of young Jamaican children recovering from severe protein-energy malnutrition. *Developmental Medicine and Child Neurology,* 1982, *24,* 321–331.

Greenough, W. T., and Juraska, J. M. Experience-induced changes in brain fine structure: Their behavioral implications. In M. E. Hahn, C. Jensen, and B. C. Dudek (Eds.), *Development and evolution of brain size: Behavioral implications.* New York: Academic Press, 1979.

Guilleminault, C.; Ariagno, R. L.; Forno, L. S.; Nagle, L.; Baldwin, R.; and Owen, M. Obstructive sleep apnea and near miss for SIDS: Report on an infant with sudden death. *Pediatrics,* 1979, *63,* 837–843.

Harding, C. M. Whooping cough vaccination: The case presented by the British national press. *Child: Care, Health & Development,* 1985, *11,* 21–30.

Harlow, H. F., and Harlow, M. The affectional systems. In A. Schrier, H. Harlow, and F. Stollnitz (Eds.), *Behavior of nonhuman primates,* vol. II. New York: Academic Press, 1965.

Hazen, N. L., and Durrett, M. E. Relationship of security of attachment to exploration and cognitive mapping abilities in 2-year-olds. *Developmental Psychology,* 1982, *18,* 751–759.

Holmes, D. L.; Reich, J. N.; and Pasternak, J. F. *The development of infants born at risk.* Hillsdale, N.J.: Lawrence Erlbaum, 1984.

Holt, K. *Developmental pediatrics.* Reading, Mass.: Butterworths, 1977.

Honig, A. S. Evaluation of infant/toddler intervention programs. *Studies in Educational Evaluation,* 1983, *8,* 305–316.

Honig, A. S., & Lally, J. R. *Infant caregiving: A design for training.* New York: Media Projects, 1972.

Honig, A. S., and Oski, F. A. Solemnity: A clinical risk index for iron deficient infants. *Early Child Development and Care,* 1984, *16,* 69–84.

Howes, C. Caregiver behavior in center and family day care. *Journal of Applied Developmental Psychology,* 1983, *4,* 99–107.

Hsu, C.; Soong, W.; Stigler, J. W.; Hong, C.; and Liang, C. The temperamental characteristics of Chinese babies. *Child Development,* 1981, *52,* 1337–1340.

Hunt, J. M. How children develop intellectually. *Children,* 1964, *11*(3), 83–91.

Jenkins, S., and Pederson, D. R. Maternal responses to premature birth. Poster paper presented at the meeting of the Society for Research in Child Development, Toronto, April 1985.

Kagan, J. The baby's elastic mind. *Human Nature,* January 1978, *1*(1), 66–73.

Kagan, J. Structure and process in the human infant: The ontogeny of mental representation. In M. H. Bornstein and W. Kessen (Eds.), *Psychological development from infancy: Image to intention.* Hillsdale, N.J.: Lawrence Erlbaum, 1979.

Kagan, J.; Kearsley, R. B.; and Zelazo, R. P. *Infancy: Its place in human development.* Cambridge, Mass.: Harvard University Press, 1978.

Keller, W. D.; Hildebrandt, K. A.; and Richards, M. E. Effects of extended father-infant contact during the newborn period. *Infant Behavior and Development,* 1985, *8,* 337–350.

Kennell, J. H.; Voos, D. K.; and Klaus, M. H. Parent-infant bonding. In J. Osofsky (Ed.), *Handbook of infant development.* New York: Wiley, 1979.

Kent, R. D. Articulatory and acoustic perspectives on speech development. In A. P. Reilly (Ed.), *The communication game: Pediatric round table series; 4.* Skillman, N.J.: Johnson & Johnson Baby Products Co., 1980.

Keyes, C. B., and Scoblic, M. A. Fathering activities during early infancy. *Infant Mental Health Journal,* 1982, *3,* 28–42.

Klaus, M. H., and Kennell, J. H. *Parent-infant bonding.* St. Louis: Mosby, 1982.

Klein, R. P. Caregiving arrangements by employed women with children under 1 year of age. *Developmental Psychology,* 1985, *21,* 403–406.

Koop, C. B. Risk factors in development. In M. M. Haith and J. J. Campos (Eds.), *Infancy and developmental psychobiology.* Vol. 2 of P. H. Mussen (Ed.), *Handbook of child psychology,* 4th ed. New York: Wiley, 1983.

Koops, B. L., and Battaglia, F. C. The newborn infant. In C. H. Kempe, H. K. Silver, and D. O'Brien (Eds.), *Current pediatric diagnosis and treatment,* 8th ed. Los Altos, Calif.: Lange, 1984.

Korner, A. F.; Zeanah, C. H.; Linden, J.; Berkowitz, R. I.; Kraemer, H. C.; and Agras, W. S. The relation between neonatal and later activity and temperament. *Child Development,* 1985, *56,* 38–42.

Lally, J. R., and Honig, A. S. The family development research program. In C. Day and R. Parker (Eds.), *The preschool in action.* Boston: Allyn & Bacon, 1977.

Leifer, A.; Leiderman, P. H.; Barnett, C. R.; and Williams, J. A. Effects of mother-infant separation on maternal attachment behavior. *Child Development,* 1972, *43,* 1203–1218.

Lenard, L. The dynamic brain. *Science Digest,* 1983, 65–67, 118–119.

Lipsitt, L. P. Critical conditions in infancy: A psychological perspective. *American Psychologist,* October 1979, *34*(10), 973–980.

Lorenz, K. Z. The companion in the environment of the bird. *Journal Ornithology,* 1935, *83,* 137–215, 289–413.

Lounsbury, M. L., and Bates, J. E. The cries of infants of differing levels of perceived temperamental difficultness: Acoustic properties and effects on listeners. *Child Development,* 1982, *53,* 677–686.

Main, M. Exploration, play, and cognitive functioning related to infant-mother attachment. *Infant Behavior and Development,* 1983, *6,* 167–174.

McIntosh, K., and Lauer, B. A. Infections: Bacterial and spirochetal. In C. H. Kempe, H. K. Silver, and D. O'Brien (Eds.), *Current pediatric diagnosis and treatment,* 8th ed. Los Altos, Calif.: Lange, 1984.

O'Brien, D., and Hambridge, K. M. Normal childhood nutrition and its disorders. In C. H. Kempe, H. K. Silver, and D. O'Brien (Eds.), *Current pediatric diagnosis and treatment,* 8th ed. Los Altos, Calif.: Lange, 1984.

Oller, D. K. Patterns of infant vocalization. In A. P. Reilly (Ed.), *The communication game: Pediatric round table series; 4.* Skillman, N.J.: Johnson & Johnson Baby Products Co., 1980.

Pape, K. E., and Fitzhardinge, P. M. Perinatal damage to the developing brain. In A. Milunsky, E. A. Friedman, and L. Gluck (Eds.), *Advances in perinatal medicine,* vol. 1. New York: Plenum, 1981.

Parke, R. D. The father's role in family development. In M. H. Klaus and M. O. Robertson (Eds.), *Birth, interaction and attachment: Pediatric round table series; 6.* Skillman, N.J.: Johnson & Johnson Baby Products Co., 1982.

Pedersen, F. A. Father influences viewed in a family context. In M. E. Lamb (Ed.), *The role of the father in child development,* 2nd ed. New York: Wiley, 1981.

Pedersen, F. A.; Rubenstein, J. L.; and Yarrow, L. J. Infant development in father-absent families. *Journal of Genetic Psychology,* 1980, *135,* 51–61.

Piaget, J. *The origins of intelligence in children.* (M. Cook, trans.) New York: International Universities Press, 1952.

Pollitt, E.; Consolazio, B.; and Goodkin, F. Changes in nutritive sucking during a feed in two-day- and thirty-day-old infants. *Early Human Development,* 1981, *5,* 201–210.

Power, T. G., and Parke, R. D. Play as a context for early learning: Lab and home analysis. In L. M. Laosa and I. E. Sigel (Eds.), *The family as a learning environment.* New York: Plenum Press, 1982.

Ridenour, M. V. Infant walkers: Developmental tool or inherent danger. *Perceptual and Motor Skills,* 1982, *55,* 1201–1202.

Rodholm, M., and Larsson, K. The behavior of human male adults at their first contact with a newborn. *Infant Behavior and Development,* 1982, *5,* 121–130.

Ross, G. S. Home intervention for premature infants of low-income families. *American Journal of Orthopsychiatry,* 1984, *54,* 265–270.

Rubin, R. Maternal touch. *Nursing Outlook,* 1963, *11,* 828–831.

Silver, H. K. Growth and development. In C. H. Kempe, H. K. Silver, and D. O'Brien (Eds.), *Current pediatric diagnosis and treatment,* 8th ed. Los Altos, Calif.: Lange, 1984.

Spitz, R. Anaclitic depression. *Psychoanalytic Study of the Child,* 1946, *2,* 313–342.

Sroufe, L. A. The coherence of individual development: Early care attachment and subsequent developmental issues. *American Psychologist,* October 1979, *34*(10), 834–841.

Sroufe, L. A.; Fox, N. E.; and Pancake, V. R. Attachment and dependency in developmental perspective. *Child Development,* 1983, *54,* 1615–1627.

Steinschneider, A. Prolonged apnea and the sudden infant death syndrome. *Pediatrics,* 1972, *50,* 646–654.

Stunkard, A. J.; Sorensen, T. I. A.; Hanis, C.; Teasdale, T. W.; Chakraborty, R.; Schull, W. J.; and Schulsinger, F. An adoption study of human obesity. *The New England Journal of Medicine,* 1986, *314*(4), 193–198.

Thomas, A., and Chess, S. *Temperament and development.* New York: Brunner/Mazel, 1977.

Torgersen, A. M., and Kringlen, E. Genetic aspects of temperamental differences in infants: A study of same-sexed twins. *Journal of the American Academy of Child Psychiatry,* 1978, *17,* 433–444.

Ungerer, J. A., and Sigman, M. Developmental lags in preterm infants from one to three years of age. *Child Development,* 1983, *54,* 1217–1228.

Weissbluth, M. Chinese-American infant temperament and sleep duration: An ethnic comparison. *Developmental and Behavioral Pediatrics,* 1982, *3*(2), 99–102.

Weissbluth, M. *Crybabies. Coping with colic: What to do when your baby won't stop crying!* New York: Berkley, 1985.

Weissbluth, M., and Green, O. C. Plasma progesterone concentrations and infant temperament. *Developmental and Behavioral Pediatrics,* 1984, *5*(5), 251–253.

Weissbluth, M., and Liu, K. Sleep patterns, attention span, and infant temperament. *Developmental and Behavioral Pediatrics,* 1983, *4*(1), 34–36.

White, B. L. *The first three years of life.* Englewood Cliffs, N.J.: Prentice-Hall, 1975.

Winick, M. *Malnutrition and brain development.* Oxford: Oxford University Press, 1976.

Wolfe, R. R., and Wiggins, J. W., Jr. Cardiovascular diseases. In C. H. Kempe, H. K. Silver, and D. O'Brien (Eds.), *Current pediatric diagnosis and treatment,* 8th ed. Los Altos, Calif.: Lange, 1984.

Zeskind, P. S. Adult responses to cries of low and high risk infants. *Infant Behavior and Development,* 1980, *3,* 167–177.

Zeskind, P. S. Production and spectral analysis of neonatal crying and its relation to other biobehavioral systems in the infant at risk. In T. Field and A. Sostek (Eds.), *Infants born at risk: Physiological, perceptual and cognitive processes.* New York: Grune & Stratton, 1983.

Zeskind, P. S., and Iacino, R. Effects of maternal visitation to preterm infants in the neonatal intensive care unit. *Child Development,* 1984, *55,* 1887–1893.

Zeskind, P. S., and Lester, B. M. Acoustic features and auditory perceptions of the cries of newborns with prenatal and perinatal complications. *Child Development,* 1978, *49*(3), 580–589.

♥

Early Childhood

5

Chapter Outline

Key Concepts

animism
artificialism
associative play
astigmatism
authoritarian-autocratic parenting
authoritative-reciprocal parenting
autonomy versus shame and doubt
basic food groups
birth order
catch-up growth
centration
child abuse
child neglect
cooperative play
Dreikurs's stages
egocentrism
eidetic memory
Electra complex
growth trajectory
hearing defects
hyperopia
indifferent-uninvolved parenting
inductive discipline
indulgent-permissive parenting
initiative versus guilt
intuitive phase
leukemia
limbic system
long-term memory
love-withdrawal discipline
mental images
Montessori program
motor norms
myopia
nursery schools
Oedipal complex
onlooker play
otitis media
parallel play
pica
preconceptual phase
preoperational thought
power-assertive discipline
realism
secular growth trend
sexual abuse
sex typing
short-term memory
sibling rivalry
solitary play
strabismus
strep throat
symbolic play
syncretism
syntax
toilet training
transductive reasoning
Wilms' tumor

♥

Case Study

Elise brought Ami into her life when she was three years old, shortly after meeting a real Ami at a picnic. Elise resisted the efforts of others to convince her that Ami lived at Ami's own house with Ami's own parents. Elise knew better. Ami lived in her head.

Sometimes Ami played nicely with Elise, but more often she was mean. For example, Elise liked to look at pictures and then close her eyes and continue to look at them in her mind. Whenever she did this, Ami would change the picture or make it go away. Elise liked to take her doll for walks in a doll carriage. However, whenever she did so, Ami would make the doll carriage bounce over bumps and send the doll flying. One day Elise pulled the head off her favorite doll. When her Mom asked why, Elise answered, "Ami made me do it." Another day, Elise smeared and broke her mother's lipstick when trying to put the top back on the lipstick tube. She then hid the whole thing. When her Mom found it and asked her whether she did it and why she hadn't brought it to her and confessed, Elise answered, "Ami hid it."

Elise and Ami often had long conversations about Elise's real friends. Ami did not like Elise to play with Betsy. One day, a new imaginary friend appeared in Elise's head. The new friend was Rebecca, Ami's older sister. During the day, Rebecca would "go to school." She would come "home" in the late afternoon. Then Elise would tell Rebecca about Ami's mischief. In particular, she would report that Ami wouldn't let her play with Betsy. Rebecca would then intercede and Elise could go over to Betsy's house. Rebecca would make Ami leave Elise alone. Sometimes Rebecca brought her own friends "home" from school. None of Rebecca's friends liked Ami but they all liked Elise. However, Rebecca would never allow her friends to tease or be mean to Ami.

Shortly after Rebecca appeared, both sisters went away.

During the extended time that Elise had one imaginary friend, Ami, and the short time that Elise had two imaginary friends, Ami and Rebecca, Elise's mother was patient and tolerant of the fantasy. Once when a neighbor overheard Elise blame Ami for something, the neighbor suggested that Elise was "disturbed" and should be taken to a child psychologist. Elise's mother disagreed. She felt it was a normal part of Elise's development. What do you think?

Somewhere between eighteen and thirty months of age infants develop motor coordination and speech skills to such an extent that their parents and other adults view them as toddlers rather than as babies. They walk, they have increased success with talking, and they strive for a measure of self-reliance and independence. This chapter will describe the various aspects of development separately. These boundaries serve only to help the reader visualize the changes in each area more clearly. Try, if you can, to visualize children you know (or have known) who are approximately ages two, three, four, and five. Keep each whole child in mind as you read about development in the separate areas. The two-year-old is quite different from the five-year-old. In each section the contrasts between younger and older children will be stated.

Early childhood in this text will include the period of time between infancy and elementary school. (Elementary does not include nursery school or daycare.) Late childhood (to be described in the next chapter) will encompass the years from beginning elementary school to adolescence. Within early childhood two other phase parameters will occasionally be used: toddlers and preschoolers. Toddlers are younger children, ages two and three. Preschoolers are children ages four and five.

Physical Development

Physical development can be defined as that which primarily affects the body as contrasted with the mind. The two are truly inseparable because of constant interactive effects. Nevertheless, before examining the development of cognition, language, and psychosocial changes, which depend heavily on mental development, we will look at the body changes of early childhood. Nutrition, health status, physical disabilities, and accidents play an important part in determining how each unique child grows physically. Remember that feelings of love, belonging, and self-esteem can also affect physical growth.

GROWTH PROGRESSION

The rate of growth in the early childhood years is slower than in infancy but follows the same general principles. Development proceeds from head to foot (cephalocaudal direction), from the center outward (proximal-distal direction), and from general to specific movements (see Figure 4–6, page 145). To illustrate cephalocaudal development, let us look at head and leg circumferences of five-year-olds. Mean head circumference is approximately 50 centimeters, or about 90 percent of its adult size. In contrast, leg circumference is only about 22½ centimeters, less than 45 percent of its adult size (see Figure 5–1). As an example of proximal-distal development, the young child will develop the ability to use the muscles of the upper arms (pushing) and upper legs (jumping) long before the ability to use the finger muscles (molding clay) or toe muscles (picking up marbles) develops. To illustrate the principle of development from general to specific, consider finger muscles again. First, the young child will have the gross motor ability to push a lump of clay from

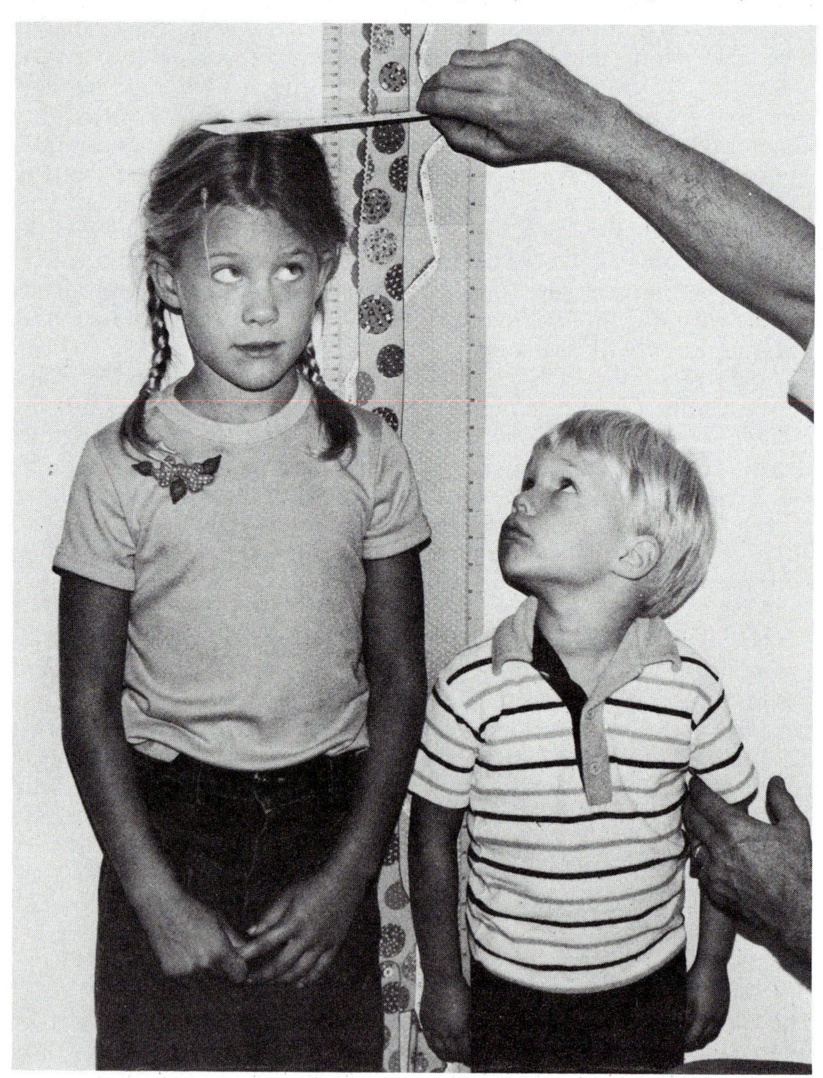

Figure 5–1
The cephalic (head) development of these seven- and four-year-olds is nearly identical as measured by the circumference of their heads. It is the length of the arms, trunk, and legs that give the seven-year-old her larger appearance.

round to flat then to round again. Later, the fine muscles will acquire the necessary coordination to push the clay into a shape roughly resembling what the child desires (such as a dog or a car).

Physical changes occur at a relatively slow, even, and continuous pace. Only occasionally will starts and stops be noticeable. Sometimes a child will concentrate so intently on one aspect of development (for example, tricycle riding) that other emerging abilities will seem to falter or even regress (for instance, bladder sphincter muscle control). Most children will also show some seasonal variations in the speed of their growth. Increases in height occur more rapidly in the spring, whereas increases in weight occur more rapidly in the fall. Illnesses and periods of malnutrition can temporarily slow children's rates of growth, but when diseases are cured or missing nutrients supplied, a catch-up phenomenon usually occurs. The **catch-up growth** may be as much as 400 percent above normal and will continue until appropriate norms are achieved, at which point growth slows to normal (Ganong, 1980). Exceptions depend on the length and severity of the disease or malnourished state, on the organs affected, and on the emotional state of the child. The brain is more susceptible to permanent injury because its cells are not replaced once they are destroyed. (To some extent, other brain cells may take over the functions of the missing cells.)

To help you appreciate some of the physical changes that occur in early childhood, **motor norms** (the average motor achievements by age for a large number of children) are presented in Table 5–1. Gross motor accomplishments are those that involve the use of the large muscles (arms and legs). Fine motor tasks are those that require use and coordination of smaller muscles (especially fingers).

Figure 5–2 depicts the rating sheet used in the Denver Developmental Screening Test (DDST), a reliable and widely accepted normative schedule that allows an examiner to ascertain how much a child has matured in various areas. It has bars that represent the age span between which 25 percent and 90 percent of children perform each item. It alerts professionals to the possibility of developmental delays so that appropriate diagnostic studies may be pursued. Any screening test must be quickly and easily administered for the purpose of differentiating normal from abnormal. Screens are not meant to diagnose. The DDST does not give an intelligence quotient (IQ) or developmental quotient (DQ). Recently, an even shorter form of the DDST has been developed (Frankenburg, 1981). It is a prescreen, to use before the full DDST. If parents have a high school

CATCH-UP GROWTH
A period of rapid growth, following a period of illness or malnutrition, that continues until the child has attained the height of his or her previous normal growth curve.

MOTOR NORMS
Standards for achievement of activities related to muscle use set for various ages of childhood.

Table 5–1 Gross and Fine Motor Accomplishments of Early Childhood

Gross motor accomplishments	Age range in years	Fine motor accomplishments	Age range in years
Throws ball overhand	2–2½	Imitates vertical line	2–3
Balances on one foot 1 second	2–3¼	Dumps raisin from bottle	2–3
Jumps in place	2–3	Builds tower of eight cubes	2–3¼
Pedals tricycle	2–3	Copies circle	2¼–3¼
Broad jumps	2–3¼	Imitates bridge with three cubes	2¼–3½
Balances on one foot 5 seconds	2½–4¼	Picks longer line, three of three	2½–4¼
Balances on one foot 10 seconds	3–6	Copies plus sign	2¾–4½
Hops on one foot	3–5	Draws man, three parts	3¼–5
Heel-to-toe walk	3¼–5	Imitates square (demonstrated)	3½–5½
Catches bounced ball	3½–5½	Copies square	4½–6
Backward heel-to-toe walk	3¾–6	Draws man, six parts	4½–6

SOURCE: Adapted from Frankenburg & Dodds, Denver Developmental Screening Test (DDST), University of Colorado Medical Center, 1967. Reprinted by permission.

education, the prescreen consists of a ten-item questionnaire. If the parents are not educated, the prescreen consists of twelve performance items for the child to accomplish. If one or more of the prescreen items indicate delay, the full DDST is given. Together, the two forms of the new prescreens are known as the Prescreening Developmental Questionnaire (PDQ).

Frankenburg, in collaboration with Bettye Caldwell, well known for the development of HOME (Home Observation for Measurement of the Environment), also constructed a screening test for potential school problems, to be given to children between birth and age six in their own homes. Known as the Home Screening Questionnaire (HSQ), it may be useful in determining which preschoolers need a longer, more comprehensive evaluation for potential school problems. With early diagnosis and early intervention, difficulties may be alleviated.

The average North American two-year-old stands about 81 to 84 centimeters high (32 to 33 inches) and weighs about 11 to 13 kilograms (26 to 28 pounds). By age five the average child stands about 109 to 111 centimeters high (40 to 44 inches) and weighs about 18 to 19 kilograms (39 to 41 pounds). Remember that "average" refers to the sum of many quantities divided by the number of quantities

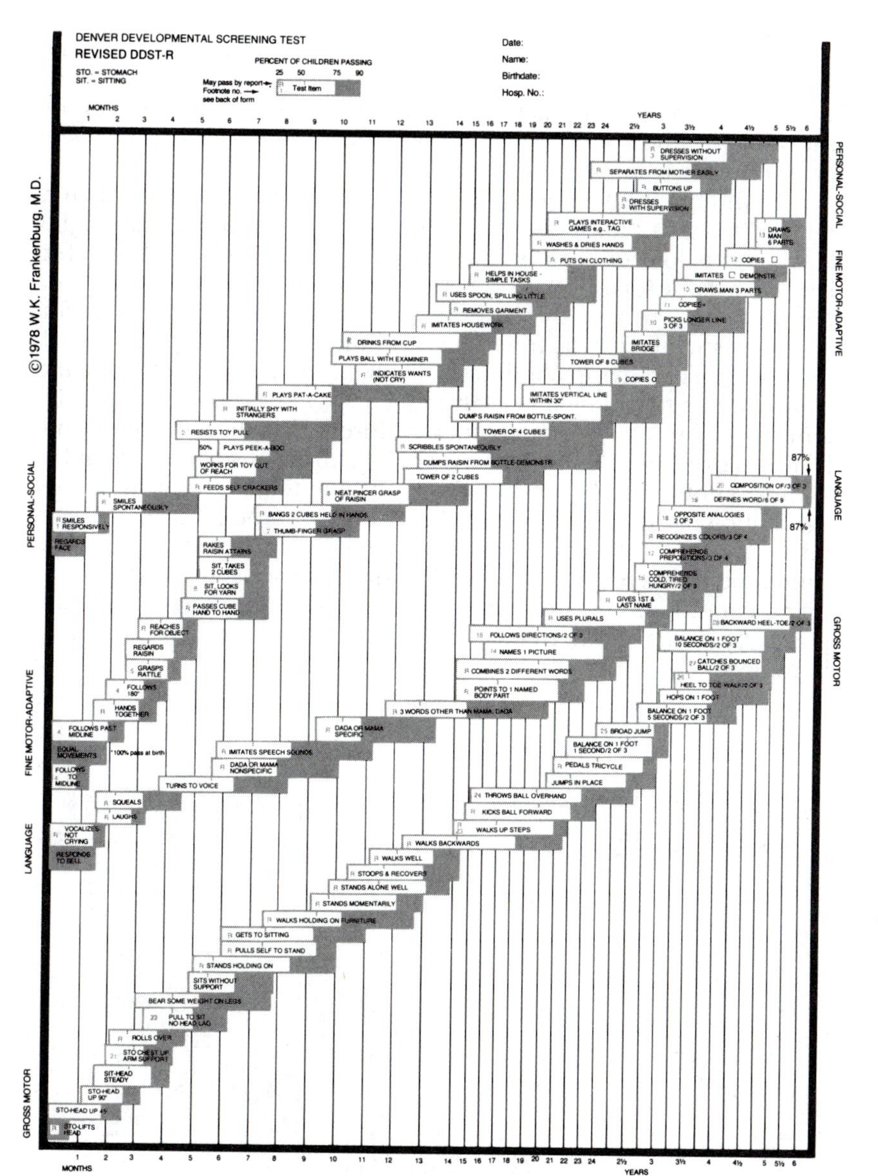

that were added together. No one child should be expected to be average in every respect. Weight gain in early childhood averages about 2 kilograms (4 pounds) per year, and height increases range from 6 to 8 centimeters (2½ to 3½ inches) per year.

Studies of well-nourished European and American children over the past one hundred years have revealed a trend toward increased height each decade up through the 1960s. In countries such as England and Norway and in the upper socioeconomic strata of the United States, this growth seems to have leveled off (Roche, 1979).

Figure 5–2
Denver Developmental Screening Test (DDST-R). (Reprinted by permission of Dr. William K. Frankenburg, University of Colorado Health Sciences Center.)

In many countries where food supplies are scarce, this growth trend (called the **secular growth trend**) has not been seen. When you next visit a museum with artifacts from the early days of American colonization, notice how much smaller the people must have been to wear or use the items on display.

Although the average two-year-old may be about 82 centimeters tall and weigh 12 kilograms, there is a wide range of normal heights and weights for children of every age. These can be found by referring to a table of standard measurement percentiles for age (see Figure 5–3). Individual variations in height and weight are influenced by such factors as inheritance, ethnicity, sex, hormones, nutrition, health, and living conditions. Each child's growth in height and weight tends to follow a **growth trajectory** (curved path) that maintains its place in relation to other children's height and weight trajectories over time. Thus, a two-year-old child whose height is in the 90th percentile can be expected to continue to be tall for his or her age, following the 90th percentile trajectory throughout childhood into adulthood. (Weight is more susceptible to dietary factors and cannot be predicted as successfully.) Improvements in nutrition and living conditions can increase height only within certain limits imposed by genetic inheritance and secretions of growth-promoting hormones.

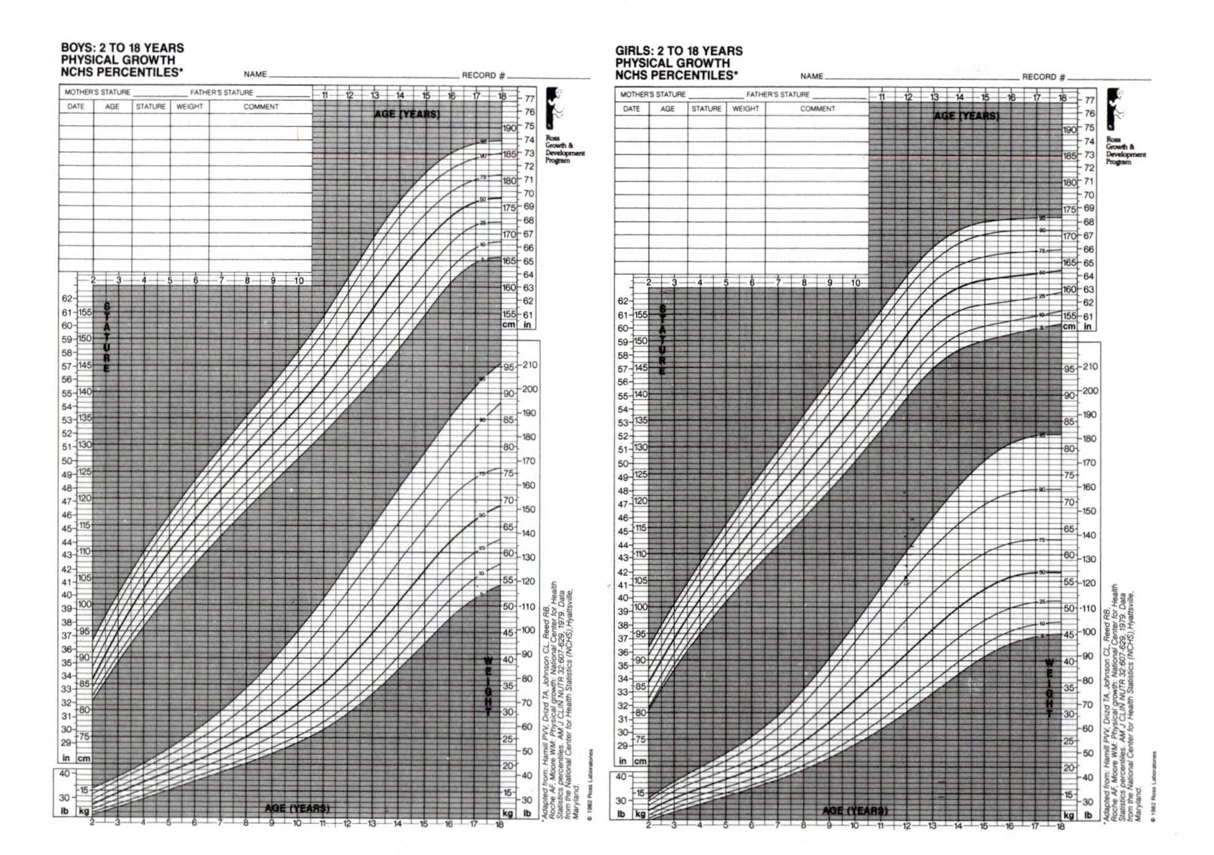

Figure 5–3
Standard measurement percentiles for height and weight of children. Data from the National Center for Health Statistics.

Arms and legs grow fastest during early childhood, trailed by trunk growth. Head growth follows at a much slower pace, mainly because brain weight has already reached approximately 75 percent of its adult size by age two. The top-heavy, short-legged appearance of babyhood changes to proportions more nearly resembling those of adults between ages two and five. When two-year-olds bend over, they can usually touch their heads to the floor without bending their knees. By age five they can simply place their hands flat on the floor without bending their knees. When two-year-olds wave good-bye, their arms go straight up without any elbow bending, yet their hands reach only slightly above their heads (see Figure 5–4). Five-year-olds' hands reach considerably higher up into the air. Potbellies are typical of toddlers, due to the forward placement of the bladder in their relatively short trunks and to the lordosis (curvature) of their as yet unelongated spines. By age five to six potbellies and the lordosis disappear.

Visual acuity improves dramatically during early childhood. Remember from Chapter 4 that vision at birth is about 20/300. By age one, it may improve to about 20/100, by age two to about 20/40, by age three to about 20/30, and finally by age four to "normal" acuity of 20/20 (Silver, 1984). Accommodation of the lens to shift from near to far focus on different objects is rapid, and convergence—bringing the image seen by each eye to a central point where it appears the same—is smooth.

Consider the following comparisons of neuromuscular skills of toddlers and preschoolers. At age two children walk upstairs holding onto a hand, rail, or wall. They place both feet on each step before proceeding to the next. By age three they begin to alternate their feet, one to a step. By age four they have usually ceased to hold on to anything and alternate steps going both down and up. By age five they may well run up the stairs. For many two-year-olds, holding a glass of liquid and drinking without spilling is a feat. By age three children can drink well and feed themselves complete meals with very little assistance. At age three children undress (quite successfully) and attempt to dress (less successfully). By age five, children dress without assistance, including washing face and hands and brushing teeth. Five-year-olds may even be able to tie the laces of their shoes in neat bows. Three-year-olds learn to jump over or off objects and maintain their balance. Five-year-olds learn to jump rope in rhythm. Two-year-olds may or may not be toilet trained for daytime (there will be more on toilet training later in this chapter). If they are, there are still generally some accidents. Three-year-olds

Figure 5–4
The arms of two-year-olds are so short that only their hands extend above their heads when they raise their arms. The arms reach the top of the head at midpoint on the forearm. Adult arms are long enough so that the elbow usually extends above the top of the head.

SECULAR GROWTH TREND
A trend toward increased height in each new generation of children seen in well-nourished children over the past 100 years.

GROWTH TRAJECTORY
The curved path that a normal child follows in terms of growth in height and weight.

may or may not be night trained. Five-year-olds generally take care of their own toilet needs, unannounced and unassisted.

NUTRITION

Adequate nutrition consists of a diet balanced with proteins, carbohydrates, and fats plus sufficient vitamins and minerals. It can be achieved by giving children two or more servings every day from each of the four **basic food groups**: (1) milk, (2) vegetables and fruits, (3) meat (or protein), and (4) breads and cereals. However, a well-balanced diet may be more difficult to attain during early childhood than it was during infancy. Appetites decrease and fluctuate from day to day. Children use mealtimes to test their independence with food refusals, food jags, or dawdling. They discover "junk" foods and demand them. Between-meal snacks are common. They are also sometimes necessary because active young children burn up their intake of mealtime calories rapidly. Nevertheless, snacks may interfere with intake of more nutritious foods at mealtimes. Caregivers can help assure adequate nutrition in early childhood by providing snacks with substantive nutritive value (milk, cheese, nuts, whole grain cereals or bread, fruit, vegetables). If young children have not been allowed to spoil their appetites with junk foods and if their food refusals do not receive a great deal of attention, they will eat most of the foods served at meals in imitation of their caregivers. Food dislikes are generally related to highly seasoned or strong flavors or the foods that the caregivers also dislike. Vitamin supplements are rarely necessary if young children have balanced diets. Some health care professionals recommend vitamins with iron if the child shows signs of iron deficiency anemia. Others may recommend fluoride pills to help prevent tooth decay.

Some children develop an abnormal craving for certain unnatural "foods" (dirt, laundry starch, play-dough, chalk, chips of paint, or plaster). This craving is called **pica**. The ingestion of flakes of paint or plaster with lead in it or other leaded substances (e.g., solder, brass alloys, home-glazed pottery) or inhalation of lead (as from fruit tree sprays and fumes from burning batteries) can lead to moderate to severe anemia with weakness, irritability, and weight loss. If more than 0.5 mg of lead per day is ingested, the child can have a toxic reaction (lead poisoning) causing peripheral neuritis (inflammation of the outer nerves) or lead encephalopathy (inflammation of the brain leading to delirium, convulsions, coma, brain damage, or death) (Rumack, 1984). Even relatively low levels of body lead in children are associated with poorer cognitive functioning

BASIC FOUR FOOD GROUPS Milk; vegetables and fruit; meat or protein foods; and breads and cereals.

PICA Hunger for nonfood substances.

than expected (Bellinger and Needleman, 1983), often before any signs of gross motor impairment are seen (Thatcher et al., 1983). About 4 percent of American children have elevated lead levels in their blood and should receive medical treatment. Pica may also be related to psychosocial problems that need investigation and treatment.

COMMON ILLNESSES

Young children are more susceptible to the common cold than their elders are because they have had less opportunity to build up antibodies against the numerous viral organisms that cause it. Their shorter eustachian tube and shorter respiratory tract also contribute to more cold sequelae, or invasions by other pathogenic organisms. Otitis media (ear infection), pharyngitis (sore throat), strep throat, tonsillitis, bronchitis, and pneumonia are common sequelae in young children. The sooner colds are treated, the less chance of more severe outcomes. They should not be treated with aspirin if they are caused by a viral agent (see Box 5–1). Other common health problems in early childhood include impetigo and boils (skin eruptions), conjunctivitis (pink eye), chickenpox, gastroenteritis (vomiting/diarrhea), and urinary tract infections.

BOX 5–1
Aspirin and Reye's Syndrome

The common cold, gastrointestinal flu, and chickenpox are the result of infection with one or more viral agents. There is some suggestion that acetylsalicylic acid (aspirin) and other salicylate drugs may interact with viruses causing illnesses in children to produce the rare condition known as Reye's syndrome. Other causes have also been suggested (insecticides, herbicides, metabolic defects), but none has been conclusively linked (together with a virus) as the cause of the disease. Until more is known about Reye's syndrome, however, aspirin bottles now carry a warning not to give the product to children nineteen years and under with chickenpox or flu without first consulting a physician. Vulnerability seems to be limited to childhood.

Symptoms suggestive of Reye's syndrome are vomiting, irrational behavior, restlessness, convulsions, progressive stupor, and coma (Silverman and Roy, 1984). The diagnosis of Reye's syndrome can be confirmed or denied with laboratory tests. Treatment is supportive and life-saving in about 70 percent of affected children. However, some residual neurologic damage is common, especially in children whose Reye's syndrome was preceded by chickenpox.

There is no indication that giving aspirin to children for the pain or fever accompanying bacterial infections, toothaches, or headaches or the pain accompanying minor accidental injuries carries any risk when given as directed.

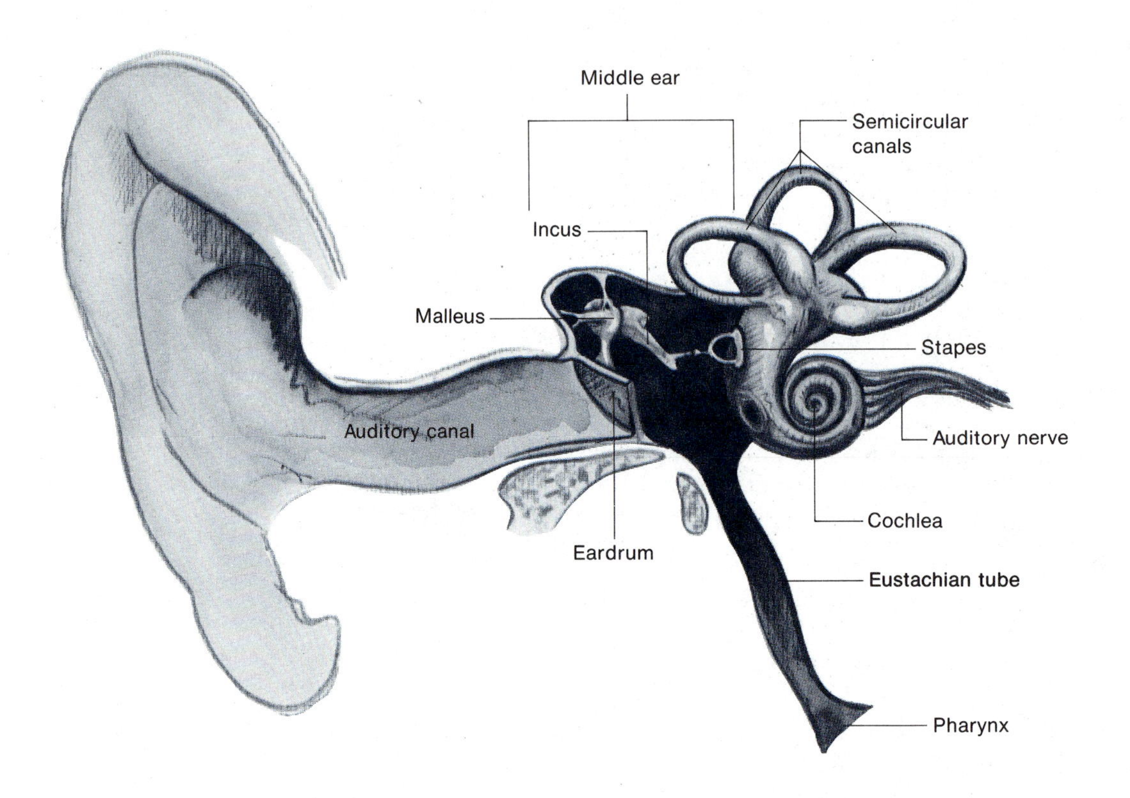

Figure 5–5
Middle ear infections (otitis media) are common sequelae to upper respiratory infections in early childhood. Eustachian tubes are shorter and less vertical in toddlers than in adults. Middle ear effusions cause the eardrum to bulge painfully and sometimes to rupture.

Prevention of illness in young children rests on good nutrition, sufficient sleep, exercise, freedom from emotional stress, avoidance of close contact with persons with contagious diseases, and immunizations against the preventable diseases (measles, mumps, diphtheria, whooping cough, tetanus, and polio). These measures will not assure that a young child will be able to resist all infections, however. The second line of defense against illness is prompt diagnosis and treatment of the symptoms of any illness.

Let us briefly review the symptoms and treatment of some of the more common health problems already mentioned.

Earaches can develop quickly in young children as the viral or bacterial organisms causing colds or upper respiratory tract infections (URIs) travel up the eustachian tube and create ear infections (see Figure 5–5). Schmitt and Berman (1984) estimate that one-third of the pediatrician's time is spent in the diagnosis and management of **otitis media**, a condition in which the middle ear, the portion just behind the eardrum, is filled with fluid. Symptoms include complaints of earache, fever, irritability, other signs of a URI, sleep difficulty, or tugging at the ear. In some cases, the eardrum will have ruptured and caregivers will see watery fluid or pus draining from the ear. Treatment is a full ten- to fourteen-day course of antibiotics. Caregivers must give the medicine for the complete time prescribed even if the child appears completely well in two to three days. If the child does not show improvement in forty-eight hours, the physician should be notified. Many of the organisms responsible for otitis

OTITIS MEDIA Inflammation of the middle ear.

media have become resistant to some antibiotics and another medication may need to be prescribed. Inadequate treatment of earaches can lead to ruptured eardrums, chronic ear infections, hearing loss, mastoiditis, or meningitis. Children with persistent otitis media may have temporary tubes placed in their eardrums to prevent ear damage.

Strep throat, caused by streptococcal bacteria, is another illness for which caregivers must provide the full ten-day course of antibiotics, even if the child's sore throat clears up in one or two days. Inadequately treated strep may lead to more serious illness, such as scarlet fever, pneumonia, or meningitis. A strep infection may travel far from the respiratory tract to the skin (causing impetigo) or to the heart or kidneys. Meningitis, rheumatic heart disease, and acute glomerulonephritis (kidney infection) are strep sequelae that are often either fatal or result in permanent disabilities in young children. The symptoms of strep throat are similar to the symptoms of viral sore throats (fever, irritability, throat redness and pain, and often some exudate on the tonsils or pharynx). Consequently, a throat culture should be taken whenever a young child has a persistent sore throat.

Impetigo, also caused by streptococcal bacteria, is a rapidly spreading, highly infectious skin rash characterized by honey-colored crusts, which form over vesicles that swell and break. Young children with impetigo may develop high fevers and be acutely ill or may simply have the skin lesions. Regardless, they must have a complete ten-day course of antibiotics to prevent the strep complications previously discussed. On occasion, impetigo is caused by staphylococcal bacteria. These bacteria are more frequently associated with two other common problems of young children: boils and conjunctivitis.

Boils are referred to as furuncles, and the condition of having one or more boils is called furunculosis. The boil or furuncle originates in a hair follicle. It becomes red, swollen, hot, and painful and fills with pus. Treatment may include incising and draining the abscessed area. It also includes a course of antibiotics effective against staphylococci.

Conjunctivitis is the most common of all pediatric ocular disorders (Ellis, 1984). Symptoms include redness of the conjunctiva ("pink eye"), a purulent discharge, and sticking together of the eyelids in the morning. It is usually treated with antibiotic ointment instilled into the eye several times a day.

Chickenpox is caused by a viral organism of the herpes family: herpes zoster. (Other herpes lesions will be discussed in Chapter 7.) Also called varicella, chickenpox is characterized by crops of skin vesicles that break and become crusted with the escaping fluid. It is usually a mild disease in young children. Some parents have "chickenpox parties," intentionally exposing their children to a child with chickenpox in hopes that they will contact the disease while young. (It has more serious sequelae in adolescents and adults.) Systemic varicella infection confers an immunity to the disease in the future. Itching is usually the only symptom in early childhood, and treatment consists of cool baths or calamine lotion to

relieve the itching. Aspirin should not be given to a child with chickenpox (see Box 5–1).

Symptoms of gastrointestinal infections such as vomiting, cramps, or diarrhea should always be reported to a medical practitioner. It is not safe to assume that the child has "flu," a stomach upset, or an intestinal virus. The problem may be one of a magnitude requiring surgery, such as appendicitis or intussusception (an obstruction of the bowel caused by a telescoping of one portion of the intestine into another). Even if the gastrointestinal upset is due to a virus, young children can become seriously dehydrated quite quickly. They may also go into shock from loss of fluids and salts after a prolonged bout of vomiting and diarrhea. Aspirin should not be given for gastroenteritis caused by viruses (see Box 5–1).

Urinary tract infections are uncommon in boys after infancy but quite frequent in preschool girls. They usually result from fecal contamination of the short female urethra situated so close to the anus, and from infrequent voiding. They are usually caused by bacteria, not viruses, and can be successfully treated with antibiotics. If relief from symptoms (abdominal pain, strong-smelling urine, increased urinary frequency, painful urination, fever, vomiting) is not obtained in forty-eight hours, the physician should be notified and another antibiotic should be prescribed. Some bacterial organisms are resistant to certain antibiotics. Inadequate treatment may result in a more serious kidney infection. Little girls should be taught to wipe themselves carefully from front to back after every bowel movement to prevent infection.

PHYSICAL DISABILITIES

Some young children are born with or develop physical disabilities that are not life threatening but that remain with them for a lifetime. These include sensory defects (visual and hearing problems) and motor defects (cerebral palsy or congenital-orthopedic impairments).

About five to 10 percent of preschool children have some kind of visual impairment (Schmitt, 1984). Visual defects frequently remain undetected until toddlerhood. Parents may begin to suspect that vision is faulty if they see their child hold an object very far from or very close to the face while examining it. Friends or neighbors may point out problems such as squinting or crossed eyes that parents have grown so accustomed to viewing that they ignore.

Most young children are normally hyperopic (farsighted). When the globe of the eye expands to rounder, fuller, more adultlike proportions at about ages four to six, this farsighted condition will correct itself. However, some children with a higher grade of **hyperopia** in early childhood may experience headaches and eyestrain from this refractive error. If so, they should be fitted for glasses to relieve such symptoms and to improve their near vision. Glasses will not interfere with the progressive development of the eye and the accompanying decrease of hyperopia. Two other refractive errors of the eyes that young children may experience are **myopia** (nearsightedness) and **astigmatism**. In astigmatism one or more of the refractive surfaces of the eye have unequal curvature, which interferes with clear focusing. Glasses will correct the errors of both myopia and astigmatism and should be prescribed for and worn by young children as soon as the defects are discovered.

Strabismus (crossed eyes, walleyes) is difficult to diagnose in infants because their eye muscles normally allow the eyes to drift occasionally. However, by age six months eyes should move in parallel (conjugate eye movement). Many parents ignore strabismus, believing it will correct itself in time. Although some cases of strabismus lessen with the passage of time, the more usual course of events is for the child to habitually use only one eye. This may eventually cause the turning eye to lose its visual function. Uncorrected strabismus is one of the leading causes of monocular (one-eyed) blindness. It is preventable with early diagnosis and treatment. In general, vision can be saved if the deviating eye is brought back into binocular (two-eyed) functioning as early as possible. Children should be taken to an ophthalmologist for correction of strabismus whenever it persists beyond infancy. Not only will treatment prevent monocular blindness, but it will protect the child from the psychologically damaging effects of being cosmetically different from others.

Mild or moderate **hearing defects** occur in approximately 1 percent of young children (Schmitt, 1984). They are frequently unrecognized during early childhood. Caregivers or physicians may suspect a hearing defect if children do not attempt to speak by age two, if they fail to respond to out-of-sight noises, or if they tilt their heads or assume straining-forward postures while listening to communications. Although some hearing defects are congenital, the majority are acquired in early life as a result of inadequately treated otitis media or accidental ear injuries. As soon as hearing defects are suspected, they should be given medical attention for confirmation, diagnosis, and treatment, as hearing is very important to the development of language, cognition, and social skills in young children. If a hearing defect cannot be corrected with a hearing device, a child may be started in lip reading and sign language instructions as early as two to three years of age, the period when language development occurs in hearing children. Caregivers should also learn to communicate in sign for the benefit of the child.

Motor defects, such as spasticity (increased muscle tone with sudden, burstlike movements), other forms of involuntary movements, difficulty walking, and lack of coordinated muscle move-

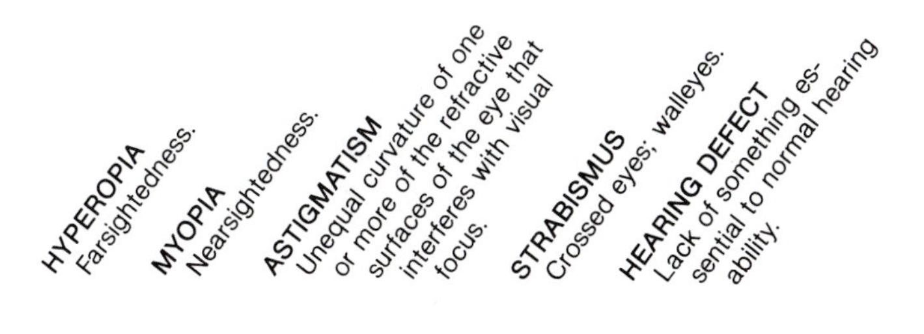

ments may not become obvious until early childhood. Motor defects in young children are most frequently the result of cerebral palsy, spina bifida (see Chapter 4), or some congenital orthopedic problem.

Cerebral palsy (CP) is a condition characterized by nonprogressive disturbances of control of motor power and coordination. It is related to developmental defects, intrauterine degeneration of some parts of the brain, anoxia, or brain trauma at the time of birth. In addition, it can sometimes result from postnatal factors (accidents, meningitis, encephalitis). Spastic hemiplegic CP is now appearing as the most common cerebral palsy syndrome (Wing and Roussounis, 1983). It is not usually detected until the end of infancy and has no clear-cut prenatal or perinatal injury as a causative factor (see Table 5–2). With early diagnosis and physiotherapeutic treatment, speech training, or corrective orthopedic devices, children with CP can be helped to gain some degree of control of their muscle movements.

Table 5–2 Classifications of Cerebral Palsy Based on Predominant Area of Involvement and Type of Motor Disability

Type	Characteristics
Spastic	A condition of the muscles in which there is hypertension, involuntary contraction, and rigidity, causing stiff and awkward movements
Spastic hemiplegia	One-sided body involvement, arms more than legs. Right-sided hemiplegia is far more common than left-sided involvement
Spastic quadriplegia	All four limbs involved
Spastic diplegia	All four limbs involved, legs more than arms
Spastic double hemiplegia	All four limbs involved, arms more than legs
Spastic paraplegia	Legs only involved
Spastic monoplegia	Only one extremity involved
Spastic triplegia	Three extremities involved
Choreoathetosis (dyskinesia)	Slow, repeated, involuntary, purposeless, writhing muscular movements, especially of hands and fingers
Ataxia	Uncoordinated muscle movements
Flaccidity	Lack of tone in muscle movements; overly relaxed, flabby muscles
Mixed types	Combinations of choreoathetosis or ataxia with a spastic CP

ACCIDENT RISKS

Accidents are both the leading cause of death in early childhood and the most frequent nonfatal problem for which medical services are sought. Young children are especially vulnerable to hazardous conditions in their environments because of their cognitive immaturity and their strivings for autonomy and initiative. They cannot be trusted to remember and obey safety rules. Explanations of rules may be poorly understood. Young children also often act impulsively to gratify their immediate egocentric needs or desires, rather than considering their knowledge of rules and regulations.

Motor vehicle mishaps account for nearly one-half of the fatal accidents of early childhood. Caregivers have a dual teaching responsibility in relation to automotive safety: precautions for riding in a motor vehicle and precautions while playing near areas where motor vehicles travel.

Many in-car fatalities can be avoided if children are required to stay in seatbelts or restraining devices. It is the law! In addition, children must be taught to keep their hands and feet away from all the instruments used by the driver. The back seat is, in general, a safer place for children than the front, and the center seat is safer than the window seats. A child riding on an adult's lap is not safe.

Young children should not be allowed to play in streets or busy driveways at any time and should be supervised while playing in areas accessible to roadways. In addition, toddlers and preschoolers should cross streets only with adult supervision. Young children should be taught the correct cautious "stop, look, and listen" procedures whenever they are near traffic.

The fourth leading cause of death in young children is ingestion of poisonous substances (Rumack, 1984). The four household areas from which toddlers and preschoolers most often take and taste poisonous materials are the cabinet under the kitchen sink, the medicine cabinet, the bedroom dresser, and the garage (or storage area for shop, automotive, and garden supplies). Mothers frequently fail to realize that detergents and other cleaning aids can seriously injure or kill children when eaten or drunk in relatively small quantities. All household cleaning products should be stored on a high shelf out of children's sight and reach. Likewise, beauty products, paints, paint removers, petroleum products, fertilizers, insecticides, and the like should be high and hidden from view. Medicine cabinets should be kept locked, and medicines should never be stored with foods (see Figure 5–6). If cough syrups or liquid antibiotics must be refrigerated, they should be kept separate from foods in covered containers. Safety caps should be replaced securely (Table 5–3 lists some common poisonous substances that children ingest).

All potentially poisonous materials should be kept in the original, labeled containers in which they are purchased. If a supply must be transferred to another box, can, or jar, caregivers should never choose one that has been associated with food (such as a soda bottle or soup can). The new container should also be clearly labeled as to contents and precautions. Young children should be taught not to

Figure 5–6
Small children can be expert at climbing to reach attractive items they see their parents use. All household poisons and medicines should be kept out of sight and out of all possible reach.

Table 5–3 Some Common Household "Poisons" That Young Children May Ingest

- Alcohol-based hair tonic
- Alcoholic beverages
- Ammonia
- Antihistamines
- Antiperistaltics (especially Lomotil)
- Aspirin
- Bath oil
- Bleach
- Boric acid
- Cement and glue
- Charcoal lighter fluid
- Clinitest tablets
- Cologne, toilet water
- Contraceptive pills
- Deodorizers
- Depilatories
- Detergents
- Diet pills
- Disinfectants
- Diuretics
- Drain cleaners
- Eyedrops
- Fabric softener
- Fingernail polish
- Fingernail polish remover
- Floor wax
- Fumigants
- Fungicides
- Gasoline
- Hair spray
- Heart medicine
- Insecticides
- Iodine
- Iron pills
- Kerosene
- Laxatives
- Leaded paints
- Leakage from batteries
- Lighter fluid
- Metallic hair dye
- Moth balls
- Nitroglycerin
- Permanent wave lotion
- Permanent wave neutralizer
- Phisohex
- Photographic solutions
- Rodenticides
- Rubbing alcohol
- Sedatives
- Shampoo
- Shaving lotion
- Shoe polish
- Silver polish
- Sleeping pills
- Solder
- Toilet bowl cleaners
- Toxic houseplants
- Turpentine/paint thinner
- Vitamins
- Washing soda
- Weed killers

play with any supplies belonging to adults. Cosmetics such as hair spray or cologne can cause fatalities. Children should especially be taught not to taste or eat anything unless it has been given to them by a parent or known caregiver. When medicines are necessary, caregivers should caution the child that the substance is a medicine, not a candy, and should only be taken when given by a known adult. The chart in Figure 5–7 may be copied and given to parents of young children to remind them of safety precautions and first aid measures. Caregivers should fill in all the phone numbers at the bottom of the chart. If they do not know their nearest poison control center, they can call the toll free information number 1-800-555-1212 and ask for the number of the poison center in their city or state.

Other accidents that are seen frequently in early childhood include burns and scalds, animal bites, drowning, electrocution, head injuries, and sprains from falls. If a preschooler experiences a brief loss of consciousness, a seizure, irritability, drowsiness, vomiting, or an unsteady gait following a head injury, he or she may have a concussion. Medical attention should be sought immediately. Frequently the child will be hospitalized for observation for more serious complications of the head injury. Seizures or symptoms of internal bleeding may not be apparent until twenty-four to forty-eight hours after the injury. Caregivers must assume the responsibility for protecting toddlers and preschoolers from sources of flame or heat, from strange animals, from open water, from falls, and from sources of electrical current. Active, curious young children cannot be expected to remember all rules and protect themselves from environmental hazards.

Figure 5–7 (opposite) *This chart* should *be reproduced and placed in a prominent place in homes with young children. Be sure to fill in the local phone numbers and keep them near the telephone.*

TERMINAL ILLNESSES

The second leading cause of death in early childhood is from neoplasms (cancers). The third leading cause of death is from complications of severe congenital anomalies of the central nervous system

PREVENTING CHILDHOOD POISONINGS

Each year, thousands of children are accidentally poisoned by medicines, polishes, insecticides, drain cleaners, bleaches, household chemicals, and garage products. It is the responsibility of adults to make sure that children are not exposed to potentially toxic substances.

Here are some suggestions:

(1) Insist on safety closures and learn how to use them properly.
(2) Keep household cleaning supplies, medicines, garage products, and insecticides out of the reach and sight of your child. Lock them up whenever possible.
(3) Never store food and cleaning products together. Store medicine and chemicals in original containers and never in food or beverage containers.
(4) Avoid taking medicine in your child's presence. Children love to imitate. Always call medicine by its proper name. Never suggest that medicine is "candy"—especially aspirin and children's vitamins.
(5) Read the label on all products and heed warnings and cautions. Never use medicine from an unlabeled or unreadable container. Never pour medicine in a darkened area where the label cannot be clearly seen.
(6) If you are interrupted while using a product, take it with you. It only takes a few seconds for your child to get into it.
(7) Know what your child can do. For example, if you have a crawling infant, keep household products stored above floor level, not beneath the kitchen sink.
(8) Keep the phone number of your doctor, Poison Center, hospital, police department, and fire department or paramedic emergency rescue squad near the phone.

FIRST AID FOR POISONING

Always keep syrup of ipecac and Epsom salt (magnesium sulfate) in your home. The former is used to induce vomiting and the latter may be used as a laxative. These drugs are used sometimes when poisons are swallowed. Only use them as instructed by your Poison Center or doctor, and *follow their directions for use.*

Inhaled Poisons If gas, fumes, or smoke have been inhaled, immediately drag or carry the patient to fresh air. Then call the Poison Center or your doctor.

Poisons on the Skin If the poison has been spilled on the skin or clothing, remove the clothing and flood the involved parts with water. Then wash with soapy water and rinse thoroughly. Then call the Poison Center or your doctor.

Swallowed Poisons If the poison has been swallowed and the patient is awake and can swallow, give the patient only water or milk to drink. Then call the Poison Center or your doctor. *CAUTION:* Antidote labels on products may be incorrect. Do not give salt, vinegar, or lemon juice. Call before doing anything else.

Poisons in the Eye Flush the eye with lukewarm water poured from a pitcher held 3–4 inches from the eye for 15 minutes. Call the Poison Center or your doctor.

DOCTOR____________ POISON CENTER____________ AMBULANCE____________

POLICE____________ FIRE DEPARTMENT____________ HOSPITAL____________

(This page may be reproduced for purposes of education in poison prevention. Courtesy of Rocky Mountain Poison Center, Denver, Colorado.)

(such as spina bifida and hydrocephalus) or defects in the heart and circulatory systems, respiratory system, genitourinary system, or musculoskeletal system.

Parents are usually aware of the possibility that a cancer or congenital anomaly will make their child's life expectancy short. Most still hope that surgery, improved medical techniques, protection from accidents and infections, or a miracle will give their child additional years of life. Toddlers and preschoolers themselves have only a vague concept of what it will mean to die. Death may be viewed simply as a transient phenomenon, similar to sleep. (In Chapter 12 we will present an in-depth discussion on children's concepts of death, related to their levels of cognitive maturity.) Although death itself may not be overly fear provoking, young children do experience stress, anxiety, and the fears and anxieties observed in their caregivers. As difficult as it may be, caregivers should treat terminally ill children much the same as they treat normal children. Overindulgence can be more upsetting than helpful. Children feel more secure knowing that there are limits to the things they may request and the actions they may perform.

Cancers unfortunately claim the lives of many preschoolers each year. The predominant forms of cancer of early childhood are leukemia and Wilms' tumor. **Leukemia** of early childhood involves a proliferation of abnormal white blood cells. It nearly always has an acute onset with anemia, bruises, limb pain, spleen and liver enlargement, and fever. The cause remains unknown. Treatment involves antileukemic chemotherapy (use of chemical substances or drugs), cranial irradiation if the central nervous system becomes involved, antibiotics for any infections, blood platelet transfusions if internal bleeding occurs, red blood cell transfusions, and antiemetics for nausea and vomiting related to the chemotherapy. Some leukemia can be treated with bone marrow transplants if a suitable donor is available. With aggressive chemotherapy about half of leukemic children can now be expected to survive free of disease for five years or longer (Tubergen, 1984). The word "cure" as applied to leukemia is difficult to define, but there are increasing numbers of long-term survivors.

Wilms' tumor is a cancerous mass that develops in the kidneys of some young children. There are usually no symptoms before the mass is felt by a caregiver while washing or dressing the child, or by a pediatric nurse practitioner or physician during a routine physical examination. Treatment involves surgical removal of the tumor and chemotherapy (and sometimes also radiation) postoperatively. The cure rate is excellent, from 80 to 90 percent.

Cognitive and Language Development

It is far easier to explain how physical abilities emerge and become refined during early childhood than it is to give reasons for human acquisition of knowledge and understanding. We do not yet completely comprehend how or where the brain stores memories. Karl Lashley (1960), a respected neuropsychologist who spent thirty years searching for the location of memory, stated, tongue-in-cheek, that it is so elusive it is nearly impossible to accept the fact that we learn. This section will discuss the elusive neuroanatomical roots of learning, Piaget's view of the preoperational stage of intellectual development, and the further acquisition of language by toddlers and preschoolers. Next comes a consideration of the role of play in fostering cognitive and language development. The meaning of early schooling in young children's lives will be reviewed, and finally, the family's role in fostering learning will be discussed.

MEMORY DEVELOPMENT

Elusive though memory is, a few hypotheses about it have persisted. The most frequent view is that it has two stages: immediate, **short-term memory** (often called "telephone-number" or "scratchpad" memory) and distant, **long-term memory**. Some evidence also suggests that there may be an intermediate third stage, or even several intervening stages. Still other research suggests an asymmetry of cerebral hemisphere functions that accounts for the stagelike character of memory (Springer and Deutsch, 1981).

The first phase of an answer to the question "How is memory stored?" is given a suggestive answer by the short-term/long-term hypothesis. Short-term memories are believed to consist only of electrical impulses. The electrical activity that is generated by the millions of bits of incoming material recorded by the brain every minute may be so brief and shallow as to be dissipated rather quickly, before there is a chance to imprint anything in a long-term memory storage system. On the other hand, some bits of incoming information may excite a great deal of electrical activity, enough to overflow the boundaries of the short-term registration system and to generate some longer-term retention of knowledge with potential for retrieval.

Research suggests that a pair of small structures in the brain play a vital role in pushing some information into a retention process rather than allowing it to dissipate and be forgotten. These are the hippocampi, small seahorse-shaped structures located within both temporal lobes. The hippocampi are a part of the **limbic system**, a major function of which appears to be the regulation of emotions. Bits of information registered in the brain that excite some emotional reactions are more apt to be converted into long-term memories through the hippocampi than are such things as telephone numbers. The role of the hippocampi has been made clearer through brain surgery. Neurosurgeons discovered that if the hippocampus in only one side of the brain is accidentally or unavoidably destroyed,

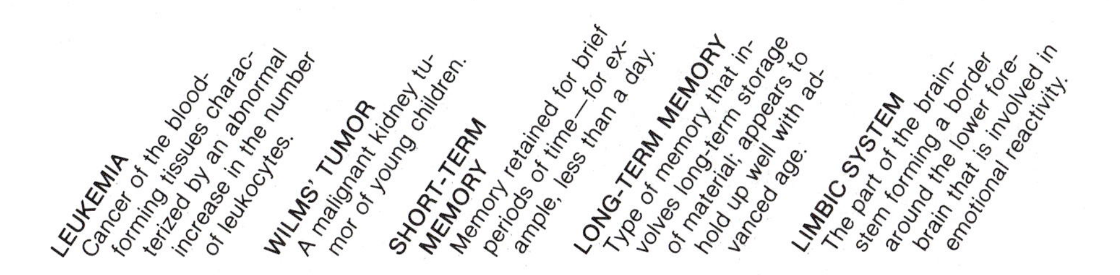

memory functions are only temporarily disrupted. However, if both hippocampi are destroyed, memory functions are severely disrupted.

The role of the hippocampi in discriminating between trivial information and relevant material to commit to memory is very important to intellectual functioning. The Russian cognitive psychologist A. R. Luria (1968) (see Chapter 2) reported a case study of a man whose hippocampi sent too many things into his long-term memory bank. Although he could reproduce lengthy series of numbers or words or describe minute details of any given scene, he appeared to be dull-witted. He remembered whole situations rather than singling out key points. Consequently, his understanding of the meaning of any given subject matter was poor. He also could not forget images he no longer needed. New impressions collided with old in a chaotic manner.

About 5 percent of children also have the ability to remember minute details of a situation (Haber, 1969). We call this **eidetic memory**. It is sometimes also referred to as photographic memory since the whole picture rather than important material seems to be retained. Eidetic imagery is usually lost in the maturation process, since intelligence requires the ability to blot out irrelevant stimuli and to focus on only the most salient elements of a situation.

A second phase of an answer to the question "How is memory stored?" concerns the mechanisms by which the brain changes short-term impressions into long-term memories. Here neuroscientists are puzzled. For years memories were believed to be laid down in electrical circuit patterns called *engrams.* For the most part, searches for the location of engrams were frustrating. Karl Lashley (1960) taught rats to run mazes, then systematically cut away portions of their cortex to see where the information had been stored. Yet, even with 90 percent of their cortex removed, the rats still staggered through the mazes. Memory seemed to be everywhere rather than localized.

One rather recent hypothesis is that human memories may be stored as holograms. Holograms are three-dimensional pictures that are made on photographic film without the use of a camera. They consist of interference patterns caused by splitting a laser light into two parts: one is bounced off an object or scene, and the other interacts with this reflection. Several interference patterns can be superimposed on a single holographic plate so efficiently that millions of bits of material can be stored and retrieved with ease. If even a small part of a hologram is illuminated with an appropriate stimulus, a whole object or scene will be reproduced in detail. It may be years before we understand whether brain cells store information as holograms, or if some other sophisticated mechanisms are responsible for memories.

Research studies have given us reason to believe that chemical changes and the synthesis of new protein materials by the brain cells are involved in the transfer of bits of incoming information from short- to long-term memory. Will and colleagues (1977) found increases in glial cells (supportive cells), in cholinesterase (an enzyme present in glial cells), and in acetylcholinesterase (an enzyme involved in synaptic transmission of neural impulses) in rats after they had been raised in separate, well-equipped cages and taught new tasks. Furthermore, rats raised in stimulating environments or trained in various skills have heavier brains than unstimulated rats raised in common barren cages. Likewise, rats fed high-protein diets learn more efficiently than malnourished rats.

A variety of drugs have been found that improve the brain's ability to fix memories. However, these drugs have side effects: some are poisonous; some cause convulsions; some are addictive. It may be possible that in the future safer drugs will be manufactured that will help push more bits of information into a retention system and help synthesize the necessary chemicals to store them permanently. In the meantime, we are left with the knowledge that sufficient dietary intake of protein is essential to the neurophysiology of remembering. We can also be comforted by the knowledge that too much remembering is chaotic, and our ability to focus on some things and forget others is essential to intelligent behavior.

PIAGET'S PREOPERATIONAL STAGE

Preoperational thought, which characterizes children from about ages two to seven, is advanced over sensorimotor intelligence because children have the emerging ability to remember and use their memories. Piaget explained the new ability to think about situations, rather than just behave in them, in terms of symbolic functions, or the formation of **mental images**. Although the acquisition of expressive language marks the beginning of the preoperational stage of intellectual development, mental images are not always words. A word is one limited kind of mental sign. It refers to an object or situation, the meaning of which the child shares with all others in the environment (up, milk, Daddy). Other mental images are very personal (the feeling of cold and fright that accompanies being put on an adult toilet seat rather than on a child's seat or potty chair). Although mental symbols may have words associated with them, they are also partly nonverbal. You can appreciate this if you consider your own thoughts. When you think in words, your mental progress is considerably slower than when you allow whole images to flash by at a time.

Piaget gave specific names to both the mental images that a child forms to represent an object or situation and the objects or situations themselves. The mental images (words, nonverbal symbols) are called signifiers. The objects or events to which they refer are called the significates. In the dual processes of assimilation (taking in and consolidating information) and accommodation (chang-

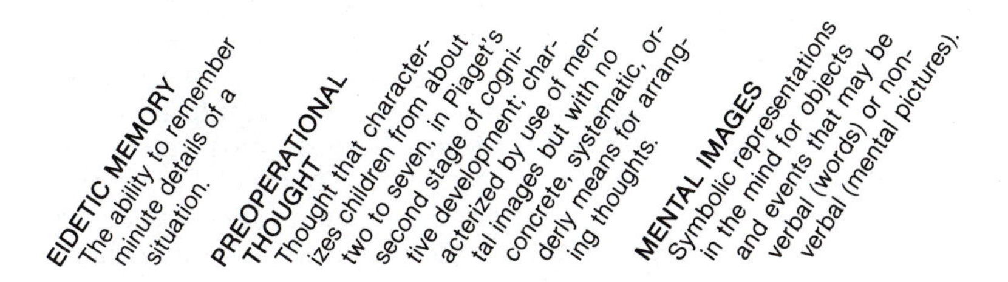

Figure 5–8
A signifier can be both a word (for example, "Muffin") and all the nonverbal, sensory, perceptual feeling tones that accompany a child's mental image of a certain dog.

ing concepts to fit with the newly assimilated information), young children apply signifiers to more and more diverse phenomena (significates). (See Figure 5–8.)

The ability to use mental images, though a major achievement in cognitive development, does not become refined for a number of years. For example, young children sometimes have difficulties differentiating between the real and the unreal, or between actual events and their fantasies. They also have a limited appreciation of the point of view of anyone but themselves. They try to find explanations for things based on their own personal needs and experiences. Their receptive and expressive language abilities, though expanding daily, also leave them puzzled about many things they see, hear, or experience. They are unable to phrase adequate questions to clarify all of their confused thoughts. In many cases, they do not feel confused but simply "understand" events in illogical ways that make sense to them. Piaget referred to this early phase of mental reasoning as *preoperational.* Young children's logic does not function according to "operations." Piaget defined operations as systematic, orderly ways of mentally turning an action over to its starting point again and again and integrating it with other reversible actions (Piaget and Inhelder, 1956). This reversible, thoughtful operating does not begin to emerge until about age seven.

Piaget subdivided the preoperational period into two phases: preconceptual and intuitive. **Preconceptual** refers to the young child's tendency to group together facts as they are acquired, not separating the real from the fantasized and not classifying or defining objects and events in a systematic manner. **Intuitive** refers to the emerging ability of young children to separate disparate objects and events into some rudimentary classification. However, the classifications are still faulty by standards of adult logic. The preconceptual phase of reasoning characterizes children from about two to four. The intuitive phase begins to be evidenced in children from about four to seven.

Piaget saw imitation as a bridge between sensorimotor intelligence and preoperational thought. In infancy imitations are first manifested as repetitions of ongoing behaviors—if such behaviors are encouraged and copied by caregivers. Eventually, infants become able to repeat novel behaviors demonstrated by caregivers. In the final stage of sensorimotor development, infants begin to imitate absent models. In the preoperational period, the character of imitations is advanced further. Long periods of time may elapse between observations and imitations. When imitations occur, they may be very abbreviated forms of the original actions. The child is no longer bound to physical repetitions of behaviors. The imitations can be performed mentally. In other words, imitations are internalized.

Piaget described the thoughts of young children as characterized by **egocentrism**. They view everything in relation to themselves. It is difficult for them to see fully any other point of view. When they take the role of another person in symbolic play, they make some strides toward such an understanding. The ability to think more systematically about the needs and experiences of others helps mark the transition into a higher level of cognitive functioning.

The egocentricity of young children often leads them to believe that their thoughts and actions are shared by others, especially by the primary caregivers. It therefore seems unreasonable to them for an adult to say, "I don't understand." They are unable to reconstruct their thought processes to show how they arrived at their conclusions and help others understand. Because they think that saying or thinking something makes it so, they frequently feel no need to justify their conclusions anyway.

Communication difficulties arise not only because young children are just learning to label things, phrase requests, ask questions, and make statements but also because they presume that much is understood without its being voiced. Furthermore, a small portion of young children's speech is not aimed at communicating. It is practice speech and resembles a monologue. Although it may be directed toward another person, the child will make little or no effort to be sure the listener hears or understands. Like adults, children may frequently hear only what it seems they want to hear. However, there may be a difference between the selective listening of a child and that of an adult: A young child may lack the competence to understand a message that is adequately imparted; adults may simply lack the will to understand.

Egocentricity leads to another problem in reasoning that Piaget called **centration**, or centering. Young children focus on one point of view, their own. They center on one need, desire, or aspect of a problem and find it difficult to shift their attention elsewhere. This cognitive limitation can make them appear persistent, single-minded, and sometimes stubborn.

Preoperational children deal with things exactly as they appear to them in their here-and-now egocentric perception. Piaget called this perception **realism**. Psychological events such as thoughts, dreams, and names are things of substance to the child. For example, names are believed to be inherent in a thing. When Piaget (1933) asked a young child how people knew that the sun was called the sun, he got this answer: They saw it was called the sun because they could see it was round and hot. For another example of realism, Piaget discussed preschoolers' views of dreams. They are believed to come from outside, sometimes from God; they can be made of wind and sometimes can punish their viewer for misdeeds of the day.

Facts acquired by young children can be described as animistic, artificialistic, and syncretistic (see Table 5–4). **Animism** refers to the child's endowment of life, consciousness, and will to physical objects and events (pincushions feel the prick of a pin, clouds feel rain, and grass feels hurt when it is pulled). However, animistic notions may be fleeting (the pincushion may not be sentient until stabbed; grass may not feel the child playing on it but only another person's pulling on it). Dolgin and Behrend (1984) found animism strongest in five-year-olds. Physical similarity to animates (stuffed animals, dolls) and apparent self-movement of inanimates (vehicles) contributed to their animistic beliefs. **Artificialism** refers to the child's tendency to believe that all objects and events in the world

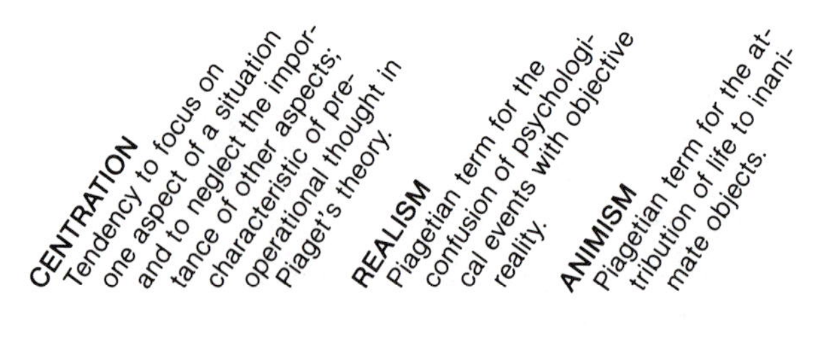

Table 5–4 Some Characteristics of Preconceptual Cognition Described by Piaget

Characteristic	Definition
Animism	Belief that inanimate objects (such as wind, fire, and water) have life, feelings, and purpose
Artificialism	Belief that all objects and events exist to serve needs of humans, especially the self
Centration	Focus on one and only one aspect of an object or event; blindness to others' points of view
Egocentrism	Self-centeredness; consideration of oneself and one's own interests to the exclusion of others
Fantasm	Perception of things that have no physical reality, such as an imaginary friend
Irreversibility	Inability to reconstruct mental processes to see what led to conclusions reached
Morality of constraint	Belief that rules are sacred and unalterable (yet unable to really follow rules)
Realism	Belief that names, thoughts, and so on have objective reality; thinking something so makes it so
Syncretism	Belief that co-occurring events belong together; for example, good china on table means that Grandma will come
Transductive reasoning	Reasoning from particular to particular; assignment of co-occurring events as cause and effect

were made by humans for humans (clouds exist to give us shade, rain comes to make splashy mud puddles, and lakes were put on earth for swimming or boating. Night comes so we can sleep). **Syncretism** refers to the child's tendency to fuse a multitude of diverse phenomena together. If you ask a young child the question "Why does the water flow downstream?" the answer is apt to be based on some co-occurring phenomenon ("The rocks push it down").

Reasoning in the preoperational period is most apt to be from particular to particular. Piaget called this **transductive reasoning**. It is midway between the two forms of reasoning that adults use—inductive and deductive. When we induce something, we use a few

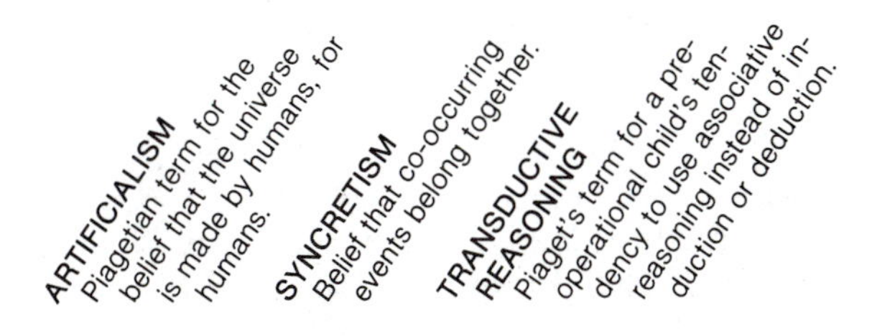

particulars to arrive at generalizations. For example, after meeting several persons from Denmark, all of whom have blond hair, we might generalize that all Danes are blond. When we deduce something, we reason from general to particular. For example, after observing many college students wearing bluejeans, we might word an invitation to a student "Wear your bluejeans," assuming that he or she must own a pair. The results of young children's transductive reasoning may or may not be logical by adult standards (they are always logical to the child) but they are usually fascinating. Piaget (1962) provided these examples:

> J. at 2 years, 9 months: "She hasn't got a name" (a little girl a year old).—Why?—"Because she can't talk."
> J. at 2 years, 10 months (showing a postcard): "It's a dog."—I think it's a cat.—"No, it's a dog."—Is it? Why?—Why do you say it's a dog?—Why do you think it's a dog?—"It's grey." (p. 232)

The ability to separate objects or events into classes in an ordered, systematic way is absent in the preoperational stage, although some children begin to see relationships among diverse phenomena from about ages four through seven. Piaget felt that they may be unaware of why they do what they do. Therefore, he called this phase *intuitive.* When young children are asked to classify a set of materials, they either make small, partial alignments (an array of materials with some groupings but no overall guiding plan) or complex objects (arrangement of the materials into an interesting form or picture).

Time and number concepts are hazy in early childhood since a genuine understanding of them requires the operation of classification. Toddlers can learn to "wait a minute" and to recognize times for bodily activities like eating and sleeping. The concepts of yesterday and tomorrow gradually become meaningful, but even though preschoolers may voice the words *next month* or *next year,* they have little appreciation for the length of time involved. Thus they may wait impatiently for Christmas in September, regardless of their knowledge of the three-month wait. They may count numbers fluently yet be unable to select correctly which array has the most candies if one is piled and the other is spread out. Even after counting the candies, they are apt to choose the array that looks bigger to them, although it actually has fewer candies.

Piaget referred to preoperational thought as bad thought or thought akin to daydreams, because it lacks the sophistication of adult logic. However, it is advanced over sensorimotor intelligence and improves continually as the child experiences more and increasingly adapts to his or her environment.

LANGUAGE ACQUISITION

Noam Chomsky's proposition that humans have an innate ability to acquire language was presented in Chapter 4. Although imitation is also considered an important factor in language acquisition, most linguists are now paying more attention to the relationship between cognition and language. Children are known to dissect rules and structural principles (**syntax**) from whatever language they hear. The special information-processing abilities in their brains make it possible for them to invent their own sentences as they begin speaking. Although not always grammatically correct, such sentences adequately convey their intended messages ("All gone nap"; "Mommy up teddy"). Children can interpret from the massive input of language because they hear what is relevant. They learn first to impart back to others only what is necessary to be understood. Later they learn to embellish their sentences with the auxiliary parts of speech. Although language is first acquired spontaneously and rules of construction operate automatically, language acquisition continues to span both early and late childhood. Many adults, in fact, are still attempting to make their word usage more polished, precise, and correct.

The first words used by children are *holophrases* (entire ideas expressed by one word). The words used are generally nouns or verbs. It requires mental planning by children to move to the next stage of speech, selecting appropriate nouns and verbs from their repertoire to make an understandable two-word sentence. Piaget, for this reason, used acquisition of language as the end point of sensorimotor intelligence and the beginning of preoperational thought, which requires the use of mental processes.

Toddlers embellish their two-word, telegramlike utterances with additional meanings through gestures and intonations. "More milk," for example, may be a meager, bashful request made with head lowered and hands clutching a glass close to the body. It may be a loud demand voiced in an insistent tone with the glass thrust out in both hands. It may also be a question, evidenced by a rise in pitch at the end of the utterance.

Many of the first words of toddlers become overextended in their meanings. For example, a child may call several four-legged real and stuffed animals "Fluffy" because they resemble a cat that is so named.

Three-word sentences generally emerge around the middle of the second year. They usually take the form of subject-verb-object (for instance, "Billy feed kitty"; "Daddy go car"). Some of the common constructions involve requests for actions, requests for repetitions, labeling of phenomena, claims of possession, signs of disagreement or negation, descriptions of location, descriptions of absent objects, and questions. In addition to forming three-word sentences, children begin voicing nearly incessant "whys." They seem more interested in keeping adults talking than in receiving answers. For example, an adult begins: "I'm going to make lunch." The child responds, "Why?" "Because I'm hungry." "Why?" "Because I haven't had anything to eat since breakfast." "Why?" "Because I've been busy." "Why?" . . . And so it goes.

A short while after children master two- and three-word sentences, they begin constructing longer messages. At first they often put two short ideas together end to end. For example, a child may

SYNTAX
The rules for combining morphemes to form words and sentences.

say, "Candy is gone, on the table." It will be a while before they can embed one idea within another ("The candy on the table is gone").

During the early phases of language acquisition, pronunciation of words may be understood only by regular caregivers. It is not uncommon for a child to impart a message to an unfamiliar person and have it received as gibberish. However, if the recipient of the message calls on the caregiver, he or she can usually decipher it quite easily.

Between ages two and a half and three, children begin to use pronouns, especially *I, me, you,* and *it.* They often confuse *I* and *me.* The possessives *my* and *mine* emerge soon after the noun-verb stage, as do the articles *a, an,* and *the.* Children also begin to add inflections to their nouns and verbs about this time, especially plurals and past tenses. Idiomatic nouns are pluralized by the same rule as regular nouns, and idiomatic verbs are given the same past tense as regular verbs (see Table 5–5). This is because children learn the syntax (rules and structural principles) of the language before they learn the exceptions to the rules.

Table 5–5 Examples of Language Errors Made by Children Applying "Rules" to Irregular Words

Plurals	Past tenses			Contractions
sheeps	seed	*or*	sawed	amn't
foots	goed		wented	donen't
gooses	braked		broked	willn't
tooths	singed		sanged	she've
mices	gived		gaved	he've
oxs	doed		dided	they's
serieses	knowed		knewed	we's
deers	beginned		beganned	
loafs	bited		bitted	
childs	swimmed		swammed	
mans	bringed		broughted	
halfs	choosed		chosed	
mooses	camed		comed	
leafs	doed		dided	
trouts	writed		wroted	
knifes	ringed		ranged	
lifes	rided		roded	
thiefs	shaked		shooked	

Word usage multiplies rapidly once a child begins speaking. The average number of words used may sprout from 50 at age two to 1000+ by age three. Although average refers to a number representing the mean of a wide range of observed children, it does not necessarily suggest what a normal child should accomplish. In the area of language development, where individual differences are vast, average is not synonymous with normal. Some children learn to speak earlier than the ages indicated in the preceding descriptions. Other normal children acquire language much later. Many persons who became quite famous in their adult lives are known to have been slow to speak in childhood (among them Albert Einstein, Winston Churchill, Virginia Woolf). Some children learn more than one language at the same time (see Box 5–2).

The speech control center is located in the dominant cerebral hemisphere. For most persons this is the left side. Girls show advanced development of their left cerebral hemispheres during infancy and early childhood, which possibly accounts for some of their accelerated verbal fluency. However, the right hemisphere can take over control of speech if the speech area of the left hemisphere is injured in childhood. Lenneberg (1967) reported that left-sided in-

BOX 5–2
When Should a Child Learn a Second Language?

If the sensitive period for speech center development is from birth to puberty, shouldn't children learn second (or third, or fourth) languages while young? Yes. Research and experience have shown that adults learn foreign languages slowly and speak with accents, whereas children are able to learn languages rapidly and speak without accents.

Should you wait until a child has learned the syntax and a large number of words in one language before you teach another? Not necessarily. If parents are bilingual (or multilingual), they can teach their children two or more languages from infancy. In fact, this is what many experts recommend.

Won't children mix up the words and syntax of two languages if they learn them simultaneously? In many countries where the population is bilingual or multilingual (Czechoslovakia, Switzerland), children learn two or more languages with ease. Garcia (1980) cautioned that there may be a temporary use of interlanguage (incorporation of two languages in speech). However, between the ages of three and seven, children sort out the separate grammars and by age seven they can usually keep the languages totally separate.

Is there a way to help children avoid interlanguage? Yes. Parents can make learning two or more separate grammars easier by using the languages in different situations. Some examples: Spanish in the home, English outside the home; Spanish at meals, English except at meals; Spanish before 3 P.M., English after 3 P.M.; Spanish when Mom is present, English when Mom is absent. There are many other possibilities. Parents must also be careful not to model interlanguage usage.

Children learning two languages simultaneously may not acquire fluency in either one as rapidly as monolingual children do. However, they may have less difficulty learning subsequent foreign languages because they develop larger, more flexible cerebral speech centers (Albert and Obler, 1978).

juries that occur before speech is learned have little or no effect on language acquisition. The right hemisphere simply develops a speech center, and learning progresses on schedule. If a speech-center injury occurs after language acquisition but prior to adolescence, the child temporarily loses language. When it is relearned, it is acquired in much the same way as it first developed (by holophrases and telegraphic sentences and so on) but at a more rapid rate. The sensitive period for speech-center development in the nondominant hemisphere, however, seems to end at puberty. After this, speech-center injuries leave people with a limited or nonexistent ability to reacquire language.

PLAY AND COGNITION

Play is a very important activity of young children. It is a major means of coming to know and understand the world around them. As children acquire the ability to think and to use language, they can play at new levels. They are no longer limited to here-and-now explorations and imitations. They can transcend time and space and remember and think about things in the past as well as about things that are out of sight. Their play becomes more symbolic, filled with fantasy figures and elements of "let's pretend."

Young children use play as a means of assimilating and accommodating themselves to numerous bits of information. Symbolic functions lead to a new kind of play: **symbolic** (also called dramatic) **play**. This is one of the chief synthesizers of ego, emotions, cognition, and socialization in early childhood. Internalized imitations can be repeated, not verbatim, but with the stress where the child desires (for example, on words or behaviors that were not well understood, on words or behaviors that brought pleasure), or with the elimination of words or behaviors that caused pain or frustration. Piaget saw symbolic play as a very important aspect of a child's emotional life as well as of his or her cognitive development (see Figure 5–9). Children use symbolic play to adjust to reality and to get reality to conform to their own personal needs and desires. They can act out conflicts. They can acquire power. They can project their shortcomings or limitations onto others. They can make their wishes come true. Play serves a multitude of functions. When you observe children in play, you will learn a great deal about what they are trying to comprehend. Consider for example, the following vignette:

> Signe, age two and a half, was left with a babysitter while her parents attended a concert. The following day she announced that she would attend a concert. She went to the closet for the shoes her mother had worn. She then went to stand next to her crib, the last place she had seen her mother stand prior to the concert. Her father, observing this, asked her if she knew what a concert was. He then explained that it is a musical performance. She immediately began singing, "Mary had a little lamb . . ."

SYMBOLIC PLAY Piagetian concept to describe play in which child makes something stand for something else.

Signe was not only trying to understand about a concert, she was reenacting a situation that had frustrated her. She had been quite

Figure 5–9
Play is a child's richest learning medium. It contributes to physical, cognitive, language, and social-emotional development.

unhappy about her parents' departure the previous evening. In her play she remained at home and gave herself a concert.

Chance (1979) identified four kinds of play of young children: symbolic play (as just described), manipulative play, physical play, and games. Manipulative play relies on sensory and motor exploration as well as on the cognitive task of trying to understand (such as how a broken toy came apart, how to assemble Lego blocks or puzzles). It may also be viewed as skill-mastery play. Physical play emphasizes motor actions, such as running, jumping, hopping, swinging, and riding on vehicles. It enhances a child's preception of his or her body in space. It is more apt to be done with other children, but it can also be a form of solitary play. Game playing is social play. The child needs to interact with another child, a parent, or another adult according to some rules or conventions. While games are more common in older children, any interaction between a young child and either another child or an adult that has a routine, turns-taking, sharing quality can be described as a game. All of these forms of play enhance cognitive and language development. Play also aids physical development, emotional development, and sociability.

Although young children are encouraged to play with other children, it takes several years to learn to consistently play cooperatively. At all ages, most children are keenly interested in the activities of other children. Two-year-olds usually engage in **solitary play**, **onlooker play**, or **parallel play** (see Table 5–6). Confrontations over desired toys are common. With increasing age, the child moves to more **associative play** in which some shared activity and communication occurs. Finally, around school age, children move to more **cooperative play** in which there are more rules and goals to their activities.

Many young children develop imaginary playmates. In a review of the literature on these companions, Schilling (1985) reported that they typically make their appearance by age two-and-a-half to three years. While common in both sexes, they are slightly more frequent

Table 5–6 Changes in the Social Quality of Play with Age

Type of play	Characteristics
Onlooking	Observation of play of others with no actual participation beyond some communication (such as question/answer).
Solitary	Play of a self-contained nature; examining and manipulating toys, other objects, or people without reference to the other(s) in the vicinity.
Parallel	Play with one or more other companions but without any real exchange of interests or materials beyond a possible tug-of-war for some coveted object.
Associative	Play with one or more other companions loosely, doing the same thing with the same or similar materials, but separately, without a great deal of social interchange.
Cooperative	Play of relatively long duration and complexity with various roles and goals and sometimes leader(s) and follower(s).

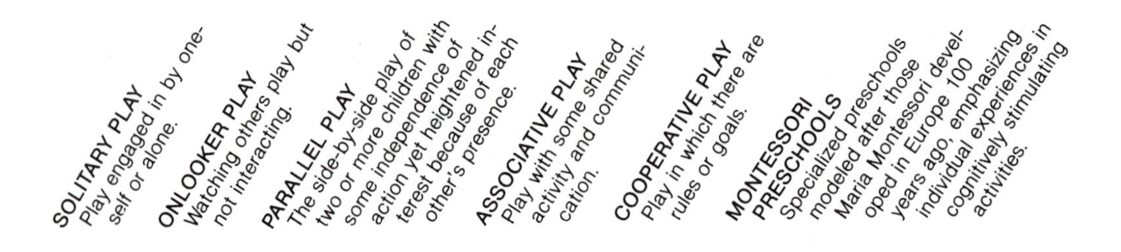

in girls. The evidence suggests that children who have imaginary companions may be more creative, more intelligent, and more verbal than other children. The imaginary friend can suggest activities to the child, be blamed when mistakes and accidents occur, compliment the child in the absence of adult approval, and participate (silently) in a great deal of symbolic play. Imaginary friends are part of the normal development of imaginative play behavior in some young children and should be treated with tolerant sensitivity and respect by adults.

EARLY SCHOOLING

Preschool programs have been available to American parents for over one hundred years. Susan Blow began the first kindergarten in the United States in St. Louis in 1873. It was modeled after the teaching styles of Friedrich Froebel, a German educator who gave the name ***kindergarten*** (meaning "child's garden") to his preschools in Europe. It stressed sitting down and learning. Within a few years other preschools opened that stressed play and group interactions rather than rigid lessons for very young children.

John Dewey (1916), an American educator, espoused the need for young children to receive individual experiences in meaningful activities geared to teach societal adjustment (such as handicrafts, gardening, nature studies). The activities were guided by the teacher but were not forced. His theories had a profound influence on American progressive education during his long (ninety-three-year) lifetime.

Maria Montessori, an Italian physician-turned-educator, developed preschools in Europe (first in Rome, later in Holland) that stressed individual experiences in meaningful activities of a cognitive nature (for example, sandpaper letters, graduated cylinders to teach size concepts, numbers' games) (Montessori, 1912). Like Dewey's activities, they were guided by the teacher but not forced. Her schools emphasized quiet, good manners, respect for others, and a regard for the upkeep of all the educational materials. Montessori schools spread to the United States in the early 1900s. In the 1920s they were eclipsed by schools following Dewey's philosophies. However, in recent years modified **Montessori programs** have again become very popular in the United States. Although her rigid schedules and strict insistence on quiet have been somewhat relaxed, many of her teaching materials are used just as she developed them at the turn of the twentieth century.

Daycare centers grew out of the work programs established by Franklin Roosevelt to help the United States out of the depression of the 1930s. They expanded during the 1940s to provide care for children whose mothers went to work during World War II. They practically became extinct during the 1950s and early 1960s. The economic situation, and women's return to out-of-home jobs, however, have catapulted daycare centers into an increasingly popular position today. Clarke-Stewart (1982) in a review of contemporary daycare programs, pointed out that they appear to accelerate children's peer relationships while not hindering parent-child relationships.

Some programs may also accelerate children's development of independence, knowledge, curiosity, inventiveness, and problem-solving abilities. (See also the discussion of daycare in Chapter 4.)

Nursery schools emphasizing socialization, group play, and creativity (and marked by permissive discipline) became popular after World War II. They were frequently attached to colleges or universities to provide a setting for training student teachers and for doing child development research, as well as for early schooling of young children whose parents could afford the tuition. The postwar era also saw the development of free half-day kindergartens that were attached to elementary schools. In some areas of the country, attendance at kindergarten before entering first grade even became mandatory.

Emphasis on socialization became partially displaced by an emphasis on cognitive development in the 1960s. The nation became especially concerned about its underprivileged children. Compensatory (to make up for deprived conditions) programs were developed such as the U.S. government's project Head Start, Peabody College's DARCEE program (which emphasized improving environmental factors), and the Bereiter-Engelmann program in Champaign, Illinois (which emphasized language learning through programmed drills). Follow-up research on the graduates of many of the 1960s compensatory programs, completed when the students were in high school, demonstrated their long-lasting benefits. For example, Lazar and Darlington (1982) found graduates of twelve compensatory programs, as compared to controls, to more often meet their school's basic requirements, to be proud of their achievement and accomplishments, and to score higher on achievement and IQ tests. Swartz and Pierson (1982) found that graduates of compensatory programs consistently scored higher on cognitive tests than their older siblings who had not attended programs. Miller and Bizzell (1984) found DARCEE and Montessori graduates higher on achievement and IQ tests when compared to students who had attended Bereiter-Engelmann programs or traditional nursery schools.

Many experts worry that public schools do not continue to support children who are given good starts in enriched preschool programs. In many cases, the transition from a high-quality preschool to a public school with a low adult-to-child ratio has produced student problems in personal-social adjustment (Honig, Lally, and Mathieson, 1982).

Other experts worry that early schooling can make children prematurely anxious about failure and deprive them of the wonderful world of play (Ames and Chase, 1974; Suransky, 1982) or will create "hurried children." Hurried children (those who grow up too fast, too soon) make up a large portion of the older troubled children experiencing school failure, drug dependence, trouble with the law, or chronic psychophysiological illnesses (Elkind, 1981).

Today caregivers may be able to choose (depending on their locale) from a vast array of specialized preschool programs. How and if these programs will enhance a child's intellectual and social-emotional development depends on many factors. The child's unique

NURSERY SCHOOLS Programs for young children emphasizing socialization attended before enrolling in grammer school; usually before age five.

personality must be considered. Some programs serve the needs of shy children better than others. Likewise, some programs and teachers are more effective with active children. Some stress only the acquisition of cognitive or language skills. Others stress social adjustment. Many are eclectic (composed of material gathered from various sources). Some children enjoy a program more if it is not too long or too frequent (perhaps half-days two or three times a week). Others enjoy full-day programs five days per week. The reasons why caregivers send children to early childhood education programs can also influence the children's enjoyment and progress in the centers. If parents believe in the programs and cooperate with the teachers, their children are more apt to find the atmosphere pleasant than are children whose parents resent the intrusions of the daycare center or the school or feel guilty for leaving their children a part of the day (see Figure 5–10).

ROLE OF THE FAMILY IN FOSTERING LEARNING

Experiences—rich and varied opportunities to personally observe, hear, taste, smell, or touch a multitude of objects and circumstances—are the *sine qua non* for fostering intellectual growth. Without experiencing phenomena children cannot assimilate and accommodate them into their mental structures. In addition, the emotional climate in which a child learns about his or her world is

Figure 5–10
Preschool programs can provide an opportunity for children to interact with other children and to participate in many activities not readily available in their own homes.

important. An insecure, frightened, unhappy child seldom evidences much curiosity or interest in the events taking place around him or her other than those things that relate directly to his or her emotional problems. Nutrition is also important to learning. A hungry child is often lethargic and seldom very curious. Severe malnutrition can limit a child's ability to form new memories as well as interfere with interest and attention. Chronic health problems also interfere with energy and interest levels.

The right learning environment is one that produces pleasure and excitement in children. In such a setting parents and others involved in the child's care will not have to push. The child's interests will propel him or her to explore and discover more and more. The stimuli in the environment must satisfy what Hunt (1964) called "the problem of the match." Stimuli must not be too familiar, or they will bore children. But if stimuli are too novel, they will be ignored. Outings, trips, toys, games, and household articles that have elements of both familiarity and novelty will best capture and hold children's interests (see Figure 5–11).

Figure 5–11
Parents can do much to enrich their children's environments and stimulate their cognitive development through outings, books, games, plenty of one-to-one interaction, and attention to needs for nutritious food, rest, safety, and love.

Verbal stimulation is also important to children's intellectual development. Parents should spend time every day talking and listening to their children in face-to-face communication. It is not the amount of speech heard by children that fosters language but the way in which communications occur. A child may be bombarded with human vocalizations throughout the day (from television, other

children, adults), but unless words are spoken to the child directly, they will have little meaning. The most meaningful communications are those that come from the adults to whom the child has the greatest attachments. Parents and other significant adults can foster language by encouraging children's efforts to communicate. Holophrases and short sentences can be repeated approvingly and then expanded into longer sentences with the addition of the auxiliary parts of speech. Adults can explain objects or events to a child as they interact together and help the child move toward a greater comprehension of the world and its properties. It is also important that adults attend to children's questions. Although a steady stream of "whys" can be exasperating, children who ask them want either information, attention, or both. The best time to provide both is when children are open and receptive. Caregivers who repeatedly say "Not now" or "Wait" and then fail to find a few minutes to spend with the child will discourage further attempts by the child to obtain explanations or information from them.

Reading aloud to children will entertain and stimulate them. Children can ask for and get instant replays of scenes. In addition to the pluses of being able to turn pages back, ask questions, and discuss interesting aspects of the pictures in books, reading provides a warm emotional climate between children and caregivers (see Figure 5–12).

Some adults have a decided impact on cognitive achievements.

Figure 5–12
Reading pleasure is enhanced when the child can sit close to the reader, can "read" the pictures, and can tell by the tone of voice of the reader that he or she is enjoying the book as well.

Effective caregivers more often give specific, well-organized instructions, encourage children to ask questions or talk about activities, engage children in interactions, and praise their progress. Burton White (1971) found that caregivers of competent children provided safe environments with a variety of toys and interesting household objects where children had a great deal of freedom for exploration. He also found that these caregivers set definite limits to dangerous behaviors and were available as consultants to answer questions, give directions, or provide encouragement when children needed it.

Psychosocial Development

The process of socialization is no longer viewed as a one-way street with family and community members acting on the child to bring about socialization. Children's unique characteristics and ways of interacting with the world influence the ways in which others react to them. Socialization is thus seen as a reciprocal process in which children influence adult behaviors as certainly as adults influence children's behaviors. Children's patterns of behavior range from shy to bold, passive to active, cuddly to distant, lethargic to alert, serious to carefree. The same adult behaviors aimed at socialization of each child can have different effects.

Havighurst (1972) suggested that there are nine developmental tasks to be accomplished during infancy and early childhood:

1. Achieving physiological stability.
2. Learning to take solid foods.
3. Learning to talk.
4. Learning to walk.
5. Forming simple concepts of social and physical reality.
6. Learning to relate emotionally to parents, siblings, and other people.
7. Learning to control the elimination of body wastes.
8. Learning to distinguish right and wrong and develop a conscience.
9. Learning sex differences and sexual modesty.

The first five tasks were discussed in Chapter 4. In this section we will discuss the last four tasks, as well as Erikson's suggested nuclear conflicts of early childhood.

ERIKSON'S NUCLEAR CONFLICTS OF EARLY CHILDHOOD

Given a sense of trust in the caregivers to whom they are attached, children move into what Erik Erikson described as the second nuclear conflict of life, that of achieving **autonomy versus shame and doubt** (see Table 5-7). Erikson, as a psychodynamic theorist, followed Freud's lead in viewing toilet training and anal functioning as a foremost concern of the child at about age two. As young children develop the sphincter control necessary for holding on and letting go, caregivers begin to ask them not to wet or mess their pants.

Table 5–7 Erikson's Second Nuclear Conflict: Autonomy Versus Shame and Doubt

Sense	Eriksonian descriptions	Fostering adult behaviors
Autonomy	Sense of inner goodness	Encouragement to stand on own feet
	Sense of self-control	Firm control of child's anarchy due to lack of sense of discrimination
	Sense of good will and pride	Gradual and well-guided experiences of free choice
versus		
Shame	Sense of premature or foolish exposure	Shaming as punishment technique
	Sense of being too visible	Suppression of self-expression
	Desire to sink out of sight	Overcontrol of actions
		Little free choice
Doubt	Sense of inner badness	Critical of self-help efforts
	Secondary mistrust with a need to look back or behind	Overprotective

Children have the power to obey or disobey, which, for them, can be a heady feeling. If accidents are cleaned up without much fuss, if children are encouraged to hold on long enough to get to a toilet, and if praise is given for toileting successes, children will gain a sense of self-control, of inner goodness, and of pride. These feelings are basic to autonomy as Erikson described it. However, if children are punished and made to feel foolish for their accidents, a sense of shame will be encouraged. If they are kept in diapers and given no opportunities to control their urges, a sense of doubt will be fostered. Feelings of shame and doubt are not healthy personality attributes. As Erikson (1963) stated: "Too much shaming does not lead to genuine propriety but to a secret determination to try to get away with things, unseen—if, indeed, it does not result in defiant shamelessness" (p. 253). Doubt is the brother of shame. It is a sense of

inner badness and secondary mistrust with a need to look back or behind. Erikson believed that many adult persecution complexes may have their origins in the compulsive doubting that begins in early childhood.

Although toilet training methods (to be discussed later in this chapter) play a role in the nuclear conflict of learning autonomy versus learning shame and doubt, they are by no means the only caregiving techniques involved. During the second and third years of life, toddlers are actively trying to stand on their own two feet, literally as well as figuratively. Increased neuromuscular development gives them the capabilities for walking, running, climbing, pushing, pulling, holding on tight, and exploring their worlds in ways heretofore impossible. Likewise, cognitive and language development make it possible for them to think about their actions and make their wills known to their caregivers. "No!" and "Me!" (meaning "Let me do it myself") are oft-repeated holophrases of two- and three-year-olds. Caregivers do not have an easy time helping children develop a sense of autonomy in all the activities of daily living. Toddlers need many experiences of being able to choose among alternatives (to play inside or outside, to wear the blue pants or the brown pants, to have a peanut butter or a cheese sandwich). However, caregivers should carefully phrase questions for children to allow situations in which either choice will be acceptable. When a particular behavior is necessary (such as going to bed, holding hands to cross a street, letting go of another child's hair), caregivers should not give a choice. Rather, they must exercise firm control over the toddler's wish to be in command. Erikson stressed that young children do not have the wisdom or the sense of discrimination to know what behaviors are acceptable or unacceptable, healthy or unhealthy. If caregivers give in too often to children's stubborn streaks and insistent demands, children may develop long-lasting personality conflicts. They may become fearful of their own powers. If their willfulness leads them into too many unfortunate accidents, they may develop what Erikson describes as a sense of shame and self-doubt. Autonomous behaviors are best nourished in a climate where both experiences of free choice and firm control of anarchy are continuously available.

Remember that, although Erikson's theory describes a primary time of life for resolving each of the ego's eight conflicts, he stressed that personality development is a continuous process. All of the conflicts of the ego can be present even at times when they are not of central importance. Children who develop a strong sense of trust in infancy will still experience feelings of suspicion and fear in later life. However, they will be able to resolve these conflicts of trust versus mistrust more easily if their infant experiences are predominantly trust promoting. Likewise, children who develop a strong sense of autonomy as toddlers should be able to emerge from conflicts of autonomy versus shame and doubt without undue difficulty throughout life.

In the preschool period (about ages four to five) children begin the work of resolving Erikson's third nuclear conflict, that of

Table 5–8 Erikson's Third Nuclear Conflict: Initiative Versus Guilt

Sense	Eriksonian descriptions	Fostering adult behaviors
Initiative	Quality of undertaking, planning, and "attacking" a task	Provide opportunities for child to plan and carry out own activities
versus		
Guilt	Anxiety about own behavior being bad Fear of wrongdoing leading to overcontrol and overconstriction of own activities	Inhibit child from starting own activities Deride child's efforts at doing for self

acquiring a sense of **initiative versus guilt** (see Table 5–8). Having discovered that they can do for themselves, they become curious about how much they can do, and when, and where. They also wonder about who else they can be. They explore answers to these questions by pretending to be other people and by scrutinizing more and more diverse phenomena. Acquiring a sense of initiative involves thrusting out into a wider world of childhood and assuming new interests and activities. Energy levels are high, curiosity is profound, and explorations are vigorous.

Erikson, following Freud, saw elements of genital interest and sexual conflicts accompanying the nuclear conflict of acquiring initiative versus assuming guilt. Curiosity and cognitive maturity lead young children to take note of the differences between males and females. Their role-playing imitations of other people include pretending to be husband and pretending to be wife. Children most frequently assume the roles of the adults in their lives whom they recognize as sharing sexual sameness. According to psychoanalytic theory, **Oedipal** and **Electra complexes** are common at this time. Oedipus was a legendary Greek king who killed his father and married his mother. A boy with an Oedipal complex presumably wants to replace his father as his mother's husband. Many boys become very attached to the females in their lives during the preschool period. Electra was a legendary Greek woman who harbored an intense love for her murdered father and bitterly hated her mother, whom she blamed for her father's death. A girl with an Electra complex

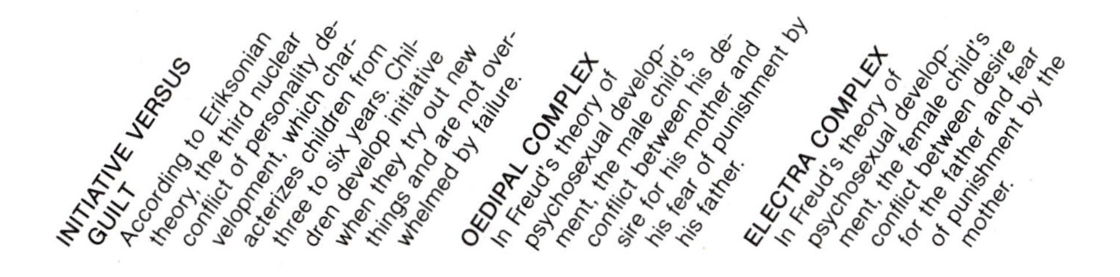

supposedly wishes to replace her mother as her father's wife. Many preschool girls do become very attached to the men in their lives during early childhood.

Erikson (1963) wrote that jealousy and rivalry directed at the same-sex parent is part of the conflict of initiative versus guilt. As he stated it, "Initiative brings with it anticipatory rivalry with those who have been there first and may, therefore, occupy with their superior equipment the field towards which one's initiative is directed" (p. 256). The jealousy and rivalry that accompany the nuclear conflict of initiative versus guilt can also be directed at other children (siblings, friends) and at other adults.

The development of a sense of initiative involves a great deal more than sex-role jealousies and rivalries. Erikson described initiative as a quality of undertaking, planning, and attacking a task. It involves creativity and assertiveness. Children need to be given opportunities to plan for and carry out their own activities during the preschool years. They need to begin to develop a feeling that they are, to some extent, masters of their own fates.

During this period of childhood a rudimentary sense of conscience, or the *superego* as the psychodynamic theorists call it, becomes apparent. Preschoolers begin to show signs of anxiety about their misbehaviors, even while they find it difficult to control their impulsive, assertive, and/or aggressive actions. Erikson (1963) calls this feeling of concern about misdeeds a sense of guilt. Preschoolers try to take control of their own behaviors as well as to manipulate the behaviors of other objects or persons. Sometimes they are harsher critics of their own behaviors than are their caregivers. Erikson warned that children who are frequently criticized, derided, and inhibited from initiating behaviors may learn to overcontrol and overconstrict themselves to the point of self-obliteration. They may develop deep regressions (returns to earlier development stages). They may also form lasting resentments of caregivers who do not live up to the high standards they exact for themselves. No child emerges from early childhood without some feelings of guilt for initiating behaviors beyond the realms of what the caregivers or the larger social order find permissible. Learning the rules, regulations, and limits of behavior involves exploration of confines and occasional overstepping of bounds. Adults need to help children establish a sense of right and wrong without laying on their shoulders an unduly heavy burden of guilt for all the mistakes that naturally accompany active, curious exploration of people (through role playing), places, and things.

PARENT-CHILD SOCIALIZATION INTERACTIONS

Chapter 2 introduced identification as a major element in Robert Sears's social-learning theory of child development. Sears believed that identification with caregivers follows attachment bonding (Sears, Rau, and Alpert, 1965). Children learn to enjoy and look forward to contact with the caregivers who are their primary sources of nurture and affection. This attachment bond motivates children to imitate the behavior of the loved adult(s). Imitations are performed not only in the caregivers' presence but also in their absence. Sears pointed out that imitating absent caregivers has a secondary reinforcing value. It makes the caregivers seem closer. Children's incorporations of adult behaviors into their own acts can give them a sense of autonomy and initiative and can aid them in controlling their own emotions and behaviors.

The strength of the identity bond is related to factors such as affection bonds, responsiveness of others to imitations, recognition of similarities between self and others, and perceptions of power. Albert Bandura (1977) and his associates found that adults who are recognized as wielding power and controlling status are more salient as identification models than are those adults or children who are simply rewarding. Knowledge of this perception-of-power component in strong identification bonds may bring comfort to some parents who fear the effects of peer models, television models, or super-rewarding (spoiling) adult models on their children. Although children practice many different roles in their play, they are most apt to take on the characteristics of the most loved, nurturing, and powerful persons in their lives.

Rudolf Dreikurs, a neo-Freudian therapist, stressed the importance of parent-child interactions on children's social development (see Figure 5–13). Dreikurs wrote that unless children feel worthwhile and protected by their family unit, they may go through one or more of four progressive stages aimed at capturing attention and a sense of being wanted (Dreikurs and Soltz, 1964). **Dreikurs's stages** involve behaviors that are usually unpleasant for the adult(s) involved in parenting: (1) seeking attention directly through loud talking, interrupting, annoying, being cute; (2) seeking power through stubbornness, assertiveness, rebelliousness; (3) seeking revenge through aggressive acts aimed to injure or inflict pain; and, finally, (4) giving up altogether and retreating into a silent, passive stance. Most children have times when they feel the need to overcome feelings of inferiority by seeking attention or power or revenge. Parents who reassure their children that they are loved and wanted can reduce these attention-seeking behaviors. Dreikurs feared that if children cannot achieve a feeling that they are worthwhile and protected by their families they may give up efforts to interact in a social way with others.

Many people are concerned that the increased percentage (about 40 percent) of mothers of preschoolers working outside the home will weaken identification bonds and children's feelings of being loved and wanted. Pederson and colleagues (1982) found that

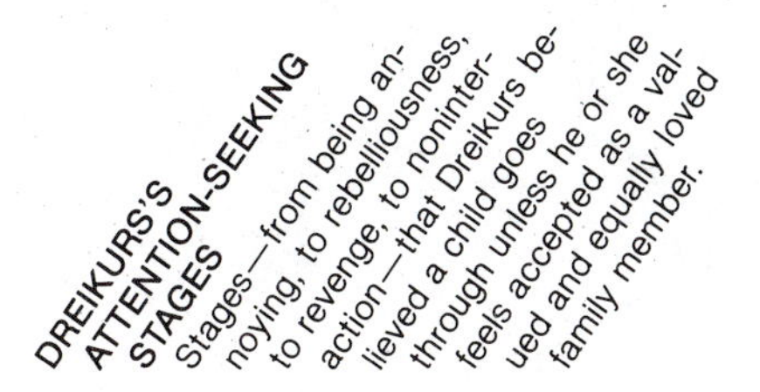

Figure 5-13
Dreikurs's theory of interfamilial social interaction patterns.

employed mothers spend more time interacting with their children in the evening hours than do nonworking mothers. Easterbrooks and Goldberg (1985) found that maternal employment outside the home had little effect on the security of the toddlers they studied. They warned, however, that there may be "sleeper effects" of early maternal employment. If they exist, they can only be discerned with longitudinal research. Bronfenbrenner, Alvarez, and Henderson (1984) found that employed mothers perceive their three-year-old daughters in much more favorable ways than they do their sons. Fathers exhibit the same pattern. Further research is needed to discern what differential effect, if any, full-time maternal employment has on the socialization of young sons and daughters. Such research must also consider the socialization interactions of the child and the substitute caregiver(s) in the absence of the parents.

Let us look at specific ways in which parent-child interactions foster the accomplishment of the developmental tasks of early childhood as set forth by Havighurst (1972).

Toilet Training **Toilet training** should not begin until the child is both physically and mentally ready to accept it. In many cases unpleasant toilet training is associated with negative attitudes toward the whole pubic area and ultimately toward the genitals and sexuality. Physical readiness (ability to voluntarily control the sphincter muscles) is seldom achieved before thirteen to eighteen months for the bowels (anal sphincter) and even later for the bladder. Starting earlier than this will only result in disappointment for the parents and a sense of shame and doubt for the toddler. Mental readiness suggests that the child is old enough to find a mess in the diaper a nuisance. Toddlers are seldom as interested in cleanliness as their caregivers. In fact, they are often curious about their feces and may take them out of their diapers to examine. Wet diapers, which become cold and irritating, often bother them more. However, around thirteen to eighteen months children can understand that dry diapers (or training pants) are, in fact, more comfortable than dirty ones. They also have a desire to imitate loved caregivers and model them, and they appreciate the pleasure they give caregivers when they deposit their feces (and later urine) in a potty or toilet. Mental readiness, then, suggests understanding of the benefits of dry over wet and motivation to please the caregivers by modeling them and by using the toilet. Caregivers should be patient and casual about the training process. Sitting on a toilet should be kept under ten minutes and should be a pleasant time. Successes should be rewarded with a hug, applause, or some positive feedback. No negative feedback should be given for failures. It is emotionally easier for a toddler to be trained later than earlier. Ultimately all normal children do become toilet trained, and the process should not be allowed to damage the loving relationship between caregiver and child.

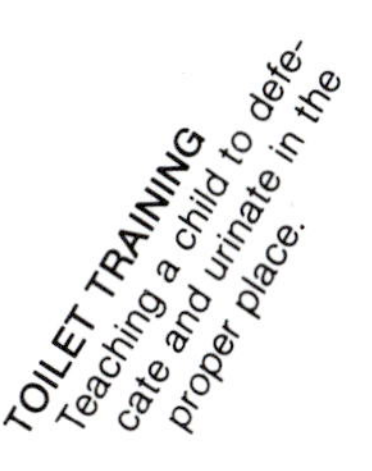

Discipline and Conscience Development It would be much easier to be a parent if experts in child development could produce an easy formula for stopping all children's bad behaviors and quickly inculcating high moral standards and pleasing behaviors. They cannot. There is an abundance of advice available on how to discipline children, but it is not simple or straightforward. It is full of ifs, ands, and buts. This is because children are not all alike. And their behaviors change from day to day, even from minute to minute. In addition, parents are not all alike, and parental behaviors are subject to mood swings just as children's actions are. The long-range goal of discipline is to help children develop a conscience that will guide their behavior toward the positive and away from the negative. This conscience (or superego) formation involves gaining an understanding of the moral codes of the social order (learning right from wrong). It also involves gaining mastery over one's own behavior (learning self-control). The short-term goal of discipline is to stop behaviors that are dangerous, destructive, pain producing, or annoying. Some forms of discipline are more effective in bringing about the long-range goal of helping children guide their own behaviors according to society's moral standards. Other forms of discipline better serve the short-term goal of stopping an ongoing set of bad behaviors.

Power-assertive discipline (physical punishment, threats of punishment, commands backed by force, shouting), and also **love-withdrawal discipline** (scolding, refusing to look, listen to, or speak to a child, isolating the child, giving dirty looks, or explicitly stating dislike or disapproval) are more effective for bringing about immediate compliance with parental directives. However, **inductive discipline** (explaining reasons why behavior should change and explaining the consequences of the behavior) is more effective for bringing about the long-term goals of internalization of moral standards and self-regulation of behaviors.

Identification with the powerful and nurturant parents or other caregivers helps the child model his or her behaviors, and also adopt attitudes about right and wrong, as long as this brings the child security and/or external reinforcements. Beyond identification comes *internalization,* a process whereby beliefs about right and wrong (and other attitudes) become so deeply inculcated that they continue to exist regardless of any external reinforcements. This internalization process can be equated with conscience development, the long-range goal of discipline.

Over the past twenty years, experts have been advising parents to avoid power-assertion and love-withdrawal discipline and use reason (inductive discipline) to foster internalization of moral values and development of a conscience to regulate behavior. However,

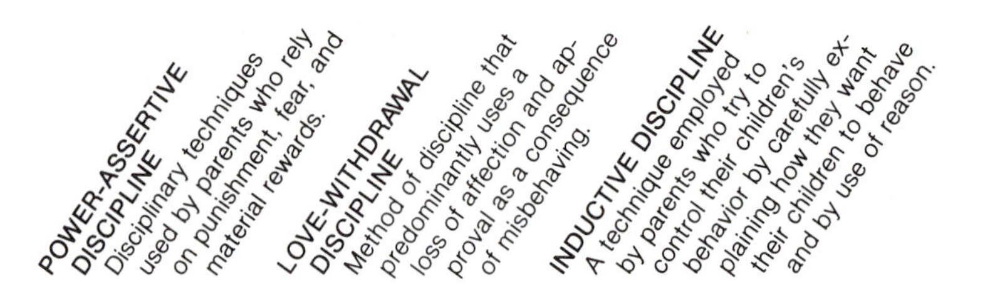

"reason" can be applied in many ways, and the ways which it is used are of utmost importance to its outcome.

Bearison and Cassel (1975) compared person-oriented reasoning (directed toward the feelings, thoughts, needs, or intentions of a person) with position-oriented reasoning (directed toward rules or status). Person-oriented reasoning is preferable. Children whose mothers used person-oriented reasoning fostered more conscience development as it related to seeing the perspectives of others. Hoffman (1975) found that parents who use person-oriented reasoning do have more altruistic children, but only if they are concurrently loathe to use much power assertion.

Most parents who have learned about inductive discipline try to use it. However, because they must also stop ongoing bad behavior when it occurs, a situation in which "reason" is seldom effective, they also rely on either power assertion or love withdrawal, or both, at times. Overuse of power assertion negates the value of reasoning.

Kuczynski (1982) compared other-oriented reasoning (the same as person-oriented) with self-oriented reasoning (directed toward the consequences of noncompliance on the self). Other-oriented reasoning resulted in more compliant behavior in children, but only in the absence of an adult. Grusec and Kuczynski (1980) reported that mothers differ in their use of power assertion or induction depending on whether they want to influence the child's behavior when they are present, or when they are absent. They use more inductive reasoning when they want children to be able to regulate their own behavior appropriately in situations where the mother will not be present to assert her own control.

Many people firmly believe that children behave better when their parents or other adults are absent. While the Kuczynski research suggests this, it may only be true if parents use person-oriented induction frequently in their interaction time with their children. They must spend time with the children to accomplish this!

Love withdrawal, like power assertion, is ineffective as a technique to enhance the development of a conscience and self-regulation of behavior in children (see Figure 5–14). Furthermore, frequent use of love withdrawal is correlated with high anxiety levels in children and avoidance of the parent who withdraws the love.

Parents are not alike. They differ in the ways they assert power, or withdraw love, or use inductive techniques of discipline. They differ in the relative amounts of each type of discipline they use. They also differ along two other important dimensions:

1. Demandingness (high to low).
2. Responsiveness (high to low).

High demandingness can be defined as a large amount of regulation and control of a child's behavior by a caregiver, while low demandingness suggests few attempts to regulate or control a child's activities. High responsiveness can be defined as child-centered parenting with a great deal of time and attention spent engaging the child in activity or responding to the child's ongoing activities. Low re-

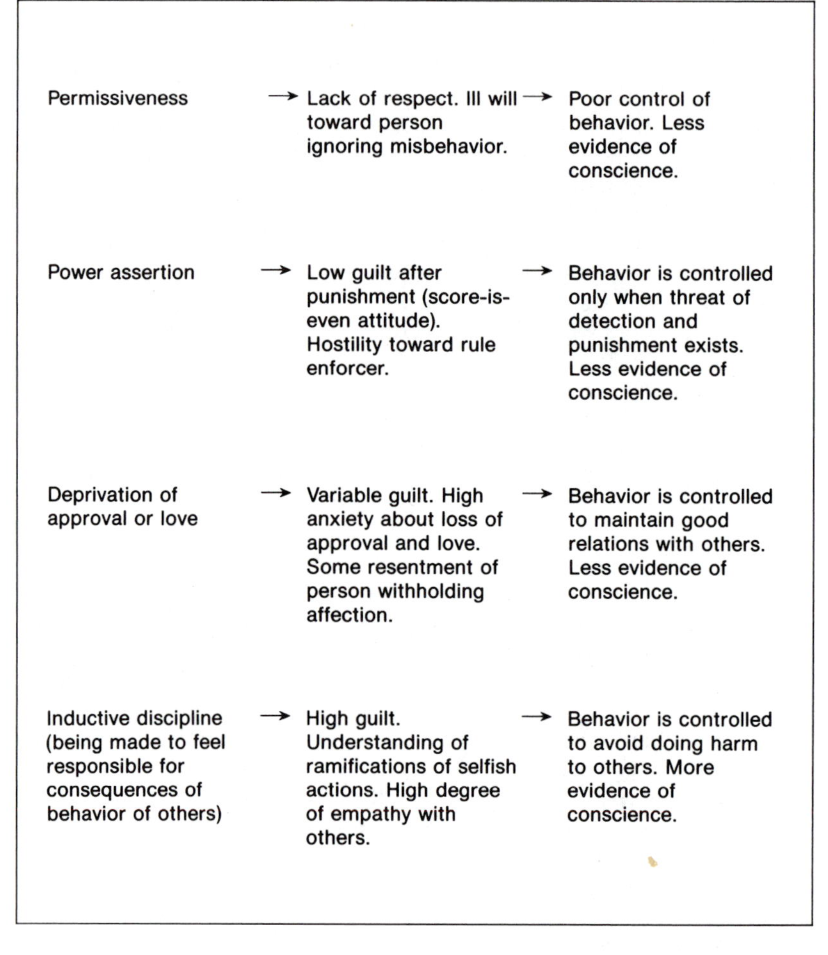

Figure 5–14 *Consequences of the predominant use of various forms of discipline.*

sponsiveness suggests parent-centeredness with little time and attention spent engaging or answering the child.

Maccoby and Martin (1983) presented a fourfold scheme that characterizes caregivers who differ on these dimensions. Caregivers high on both demandingness and responsiveness fit the **authoritative-reciprocal parenting** pattern. Caregivers high on demandingness but low on responsiveness fit the **authoritarian-autocratic parenting** pattern. Caregivers low on demandingness but high on responsiveness

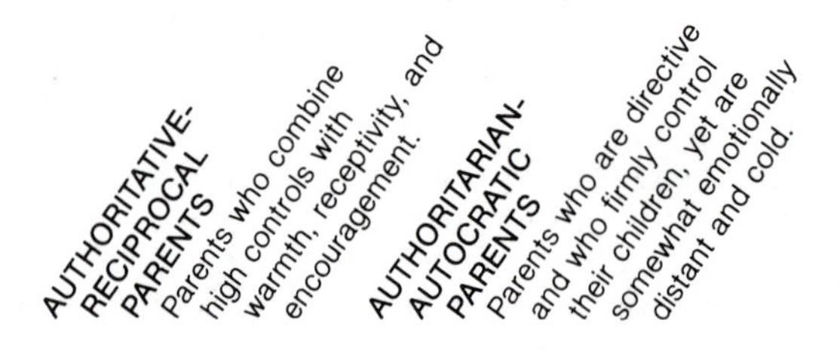

fit the **indulgent-permissive parenting** pattern, and caregivers low on both characteristics fit the **indifferent-uninvolved parenting** pattern.

The most positive parenting pattern in terms of conscience development (internalizing society's moral standards and gaining mastery over one's own behavior) is the authoritative-reciprocal pattern. Authoritative parents are most apt to raise children who are high in both self-esteem and self-control. They are also more independent and more socially responsible. Authoritarian-autocratic parents are more apt to raise children who are low in self-esteem and self-control. The children are, however, highly controlled and obedient when in the presence of their autocratic parents or other adults. They tend to be shy and socially withdrawn with their peers as well as with adults.

The less positive parenting patterns in terms of conscience development are those that involve low demandingness from parents. Indulgent-permissive parents are apt to raise children who are immature. The children feel a need to rely on others for advice or help, and they tend to be socially irresponsible, impulsive, and often aggressive. The least positive pattern of parenting is the indifferent-uninvolved pattern. Children raised by neglecting parents are apt to become aggressive, impulsive, and delinquent. They show little evidence of internalization of moral standards or the ability to regulate their own behaviors. The consequences of the four patterns of parenting are less clear-cut when parents use a mixture of all four patterns of behavior as their children mature.

Many conflicts between adults and children are avoidable. Adults should be cognizant of normal behaviors of growing children at different ages. They should not become so responsive and so demanding as to intrude constantly in the lives of even their youngest children. Such overcontrol will interfere with the acquisition of autonomy and initiative. Nor should parents insist on rules and regulations that are beyond the child's ability to understand or obey.

Discipline is not easy, but it is necessary. Children want and need help in controlling their impulsive actions in early childhood. Parents who are prepared to give this help in an inductive manner, without resorting to an overuse of power assertion or love withdrawal, and with a relatively high level of responsiveness and demandingness, can be successful disciplinarians.

Gender-Role Learning **Sex typing** (also called gender identification) does not occur all at once but continues gradually throughout

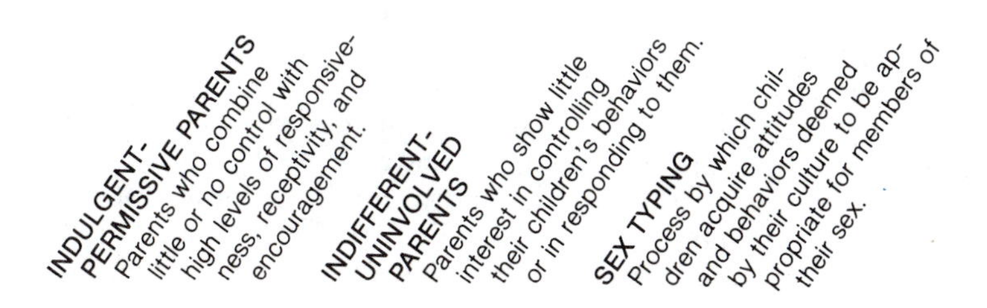

the course of development. It is not an all-or-none phenomenon at any time. Many factors are related to the way in which young children (and adults) perceive themselves as gender typed. Huston (1983) defined sex typing as multidimensional. It involves many constructs (including beliefs about sex typing, self-perceptions, preferences, and behavioral adoptions) across many content areas (including biological sex, activities, personal-social attributes, gender-based social relationships, and style characteristics). Masculine gender typing and feminine gender typing are not polar opposites. One can be more masculine in some constructs or content areas and more feminine in others and still be appropriately gender typed as male or female.

Freud (1933/1965) felt that cross-sex identifications (boys with their mothers and girls with their fathers) occur during early childhood and are the cause of tremendous personality conflicts. He not only posited Oedipal and Electra complexes but also believed that castration fears and penis envy take place in boys and girls respectively. Freud believed that it is not until about age six that emotionally healthy boys and girls identify with the same-sex parent. Contemporary theories of the acquisition of gender typing have moved away from Freud's position, both because of the conflicts he posited and because of the rather late age at which he felt appropriate gender identification takes place.

Social-learning theorists have proposed that gender typing occurs much earlier (Mischel, 1966). Soon after attachment bonds are formed, boys and girls begin to imitate the behaviors of adults they perceive as nurturant, powerful, and bearing strong resemblance to themselves. Gender identity grows stronger with the passing of time. Boys are rewarded and praised for modeling masculine behaviors, whereas girls are rewarded and praised for modeling feminine actions. Both may be punished or ignored for inappropriate sex behaviors. Thus, according to social-learning theory, boys and girls learn appropriate sex behaviors both because they imitate the behaviors of all the adults whom they find nurturant and powerful and because of the rewards and attention that these appropriate behaviors bring.

The cognitive-developmental theory of Kohlberg (1966) emphasizes cognitive aspects over reinforcement aspects of gender typing. Kohlberg views role taking more as a developmental phenomenon. Children model like-sexed persons because, first, they perceive their resemblance to those persons, and second, they want to imitate the persons whom they most closely resemble.

Martin and Halverson (1981) proposed a schematic information-processing model as a variation of the cognitive explanation of gender typing. As children process information, they simplify it by trying to fit it into categories: "things for me," and "things not for me." Eventually their schemas will include the categories of gender, as gender differences are easily discriminated even by young children. "Things for me" will be broadened to include "things for (my sex)." They will change, modify, and refine their stereotypes over time and adopt activities and attributes that they perceive as being appropriate for their sex, or for which they are given reinforcements.

Toddlers and preschoolers now usually all sport pants and shirts, regardless of sex. Many caregivers make a conscious effort to teach their sons to avoid fights, to express emotions, to nurture dolls, animals, or other children, and to help with cooking and housework. Daughters are rewarded for doing the same things but are also taught to assert themselves, to run, climb, and roughhouse, and to help with garbage collection and mechanical fix-it tasks. However, efforts to raise non-gender-typed children do not succeed. Even if parents studiously avoid labeling any activities, interests, personality attributes, or social behaviors as more appropriate for one gender than another, children will learn their culture's expectations from television, books, peers, other family members, and persons in the surrounding community.

Marsha Weinraub and her colleagues (1984) found that a significant number of two-year-olds knew their own gender identity. When shown pictures of men, women, boys, and girls, they could also verbally label them by gender. Many could also classify adult tasks, adult clothing, and adult possessions by stereotypic gender appropriateness. While some two-year-olds showed preferences for toys "appropriate" to their own sex, none could classify children's toys into gender categories. By three years of age, the majority of children knew that men wear suits, shirts, and men's hats and shave, while women wear dresses and blouses, carry pocketbooks, and wear makeup. However, less than one-third of them had stereotyped concepts about children's toys yet. Perhaps this speaks to the efforts of many contemporary parents and other caregivers to encourage non-gender-stereotypic play: boys are given dolls (such as "My Buddy," male Cabbage Patch Kids, action figures); girls are given racecars and trucks.

Although classifications by gender appropriateness were rare, toy preferences in the Weinraub et al. study did show gender typing. McLoyd and Ratner (1983) offered preschool children their choice of play with a panel containing household appliances or a panel containing car appliances. Toy preferences in this study also indicated gender typing: girls preferred the house, boys the car. Ashton (1983) found that preschool children's toy preferences were directly related to the stereotypes found in picture books that they had just seen and heard read. The Weinraub group found that, in general, boys showed more awareness of gender roles than girls. O'Keefe and Hyde (1983) found that preschool boys chose gender-stereotyped occupations for themselves more frequently than did girls. This may reflect the current efforts to encourage little girls to pursue any options without concurrent efforts to encourage boys to pursue traditionally female occupations.

While acquiring gender identity and sex-typed schemas about their worlds, children also became curious about the reasons for gender differences. In the process of being toilet trained, they learn that society deems modesty to be appropriate—they should not run around without pants, at least not outside their homes. Many preschoolers must also accept a new sibling. Questions arise about pregnancy, birth, and sex of the baby. Sex education should begin in situations such as these and continue whenever children ask questions about gender differences, sex, or sexually related topics (see Box 5-3).

SIBLING RELATIONSHIPS

The presence of one or more brothers or sisters in a family adds new dimensions to social development beyond the caregiver/child

BOX 5-3

When and Where Should Sex Education Begin?

Sex education should begin in the home as soon as each child asks his or her first question about sex: "Does Mommy have a penis?" "Why doesn't Daddy have breasts?" "Where do babies come from?" "Why can't I touch my [genitals] in public?"

The child's major source of sex education should be the parents or primary caregivers. Most children acquire the bulk of their knowledge about sex in the preschool years. Sex education courses taught in schools, even if they are offered at the elementary school level, are usually too little, too late. Even extreme efforts to shelter children from a knowledge of sex differences usually end in failure. Sex differences are too obvious to remain unnoticed. If children learn that it is taboo to ask their caregivers questions about sex, they ask their siblings or friends or invent their own answers. Often they acquire a great deal of emotionally upsetting misinformation in this manner.

If parents and other caregivers are open and honest about sexual matters and answer all questions in a simple manner when and where they occur, children usually go back about their play. Most experts believe that sex education is best accomplished by such simple direct information given inside the home. However, it is not easy for all families to give on-the-spot, simple answers to questions, especially if they have had a lifetime's practice in keeping sexual matters hidden. Their embarrassment may be acute. Children can sense their parents' timidity and may quit asking questions.

Because about 25 percent of child sexual abuse occurs in early childhood (and 75 percent of victims are girls), parents need to teach youngsters that they have the right to keep their body private and that they must tell someone right away if anyone tries to touch them in private areas. This aspect of sex education should not be delayed until children are older. Many books for preschoolers are available to help families explain sex and gender roles. In order to help caregivers provide sex information, writers of such books present answers to the questions that they know preschoolers ask. These books are especially useful for parents who are too embarrassed to answer children's questions directly. Often the interaction required to read the book to the child will break the ice and allow the parent(s) to continue giving factual information about sex, when the child desires it, in the future.

behaviors discussed so far. Children may form affectionate bonds to each other in spite of the fact that the outward demonstrations of their relationship seem to consist mainly of fights. Younger children imitate their older siblings as well as their caregivers. They may play with them, learn from them, and receive some measure of help and nurturance from them. Young children may also experience the conflicting and painful emotions of fear, jealousy, anger, or hatred directed toward another sibling, especially a new arrival or one who consistently receives favored treatment from the caregivers. **Sibling rivalry** varies from family to family. Several interacting variables can affect its relative intensity, strength, and endurance. One factor that plays an important role in the formation of jealousies is the quality and quantity of time and attention that caregivers bestow on each individual child. The greater the differential treatment and favoritism shown, the more intense are the rivalries that develop. Although it is not always possible for caregivers to treat children equally, explanations should be given to children about the reasons for any preferential treatment that is protested as unfair. As much as possible, caregivers should try to be fair with the time, attention, and favors they give each child.

Two other factors that, along with parental treatment, have an interactive effect on sibling rivalry are sex and age spacing. The stresses of sibling rivalry are greater between opposite-sex than between same-sex children. (Minnett, Vandell, and Santrock, 1983). Children who are spaced closer together have more rivalry than children who are born further apart. The most stressful spacing is under two years apart. The one- to two-year age range is the period of time when toddlers are beginning to establish a sense of self. They have strong attachment and dependency bonds with their primary caregiver(s). They struggle for a sense of autonomy and yet are vulnerable to feelings of shame and doubt. Their cognitive immaturity prevents their understanding why the caregiver must spend time with the baby. They resent the intruder who has usurped their favored status.

Lynch (1982) found that mothers of children spaced less than two years apart also experienced the most stress. The toddlers' competition for her attention was difficult for her to handle. She had less time for herself and was constantly fatigued. Maternal stress was least when children were spaced four to six years apart. By about age four preschoolers are better able to understand reasons for sharing the caregivers' time with a new sibling. They are also more independent of adults and can initiate and carry through more of their own activities without assistance.

Parents can help prepare young children for the arrival of a new sibling by explaining what life will be like with an eating, sleeping, crying baby in the house. When the infant arrives, parents should expect negative reactions from the toddler or perhaps some regression toward more infantile behaviors. Field and Reite (1984) found decreased activity, decreased sleep, and increased negative emotions in preschoolers whose mothers had just returned from the hospital with another child; these reactions are suggestive of depres-

SIBLING RIVALRY
Competition between brothers and sisters.

sion. Lynch (1982) found that mothers frequently expected their toddlers to regress in toilet training or eating habits, but they were unprepared for the behavior changes—the "different" way their children acted. Children should be helped to understand that such feelings are normal but cannot be expressed in acts of physical aggression toward baby, mother, or self. Parents should spend as much time as possible with the toddler when the baby is asleep and reassure him or her of their love. Sibling rivalry can be further reduced by allowing the older child to help the adult nurture and protect the baby (holding, feeding, diapering, entertaining, and so on). (See Figure 5–15.)

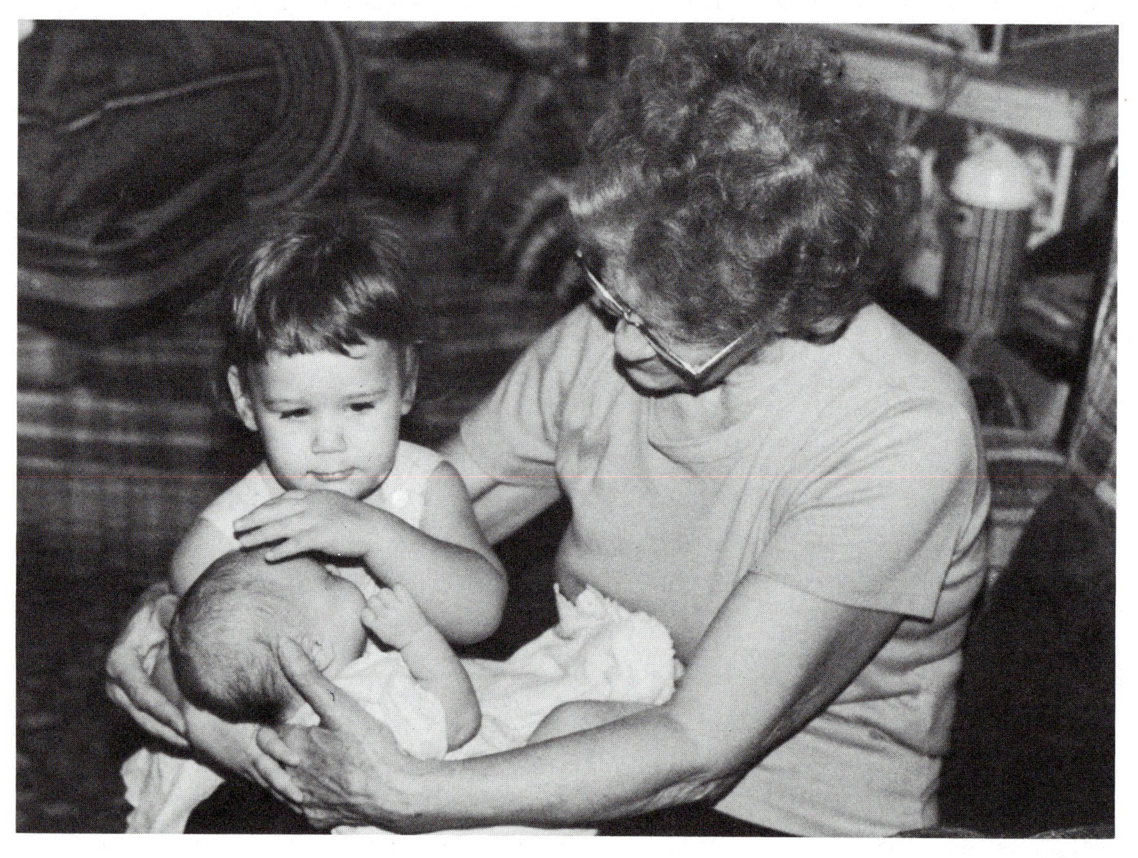

Figure 5–15
Sibling rivalry is generally more intense when babies are born close together. Although some rivalry is unavoidable, it can be minimized by encouraging and praising the nurture of one sibling by the other and by giving each child individual attention.

Studies of large groups of firstborn, middle-born, and youngest children have revealed certain personality traits that are apt to characterize people of different **birth orders**. One must be cautious in interpreting these studies too literally. Although a significant number of persons of a particular birth order may share similarities, any one person can be quite different from the group and still be normal. Firstborn children have been shown to begin talking sooner than later borns. They model and identify more with their parents. By age thirty-three months, firstborns are more sociable with unfamiliar peers (Snow, Jacklin, and Maccoby, 1981). In school, they are superior to later-born children in both reading ability and achievement motivation (Glass, Neulinger, and Brim, 1974). They score higher on National Merit Scholarship exams (Breland, 1974), and more than a fair share of them achieve high-status employment for their adult years (doctors, lawyers, professors, engineers, architects, business executives).

Parents tend to spend more time with firstborn children. Although every developmental task accomplished is new and unique, parents pressure firstborns to perform more. They also have higher expectations for them, and give them more positive ratings (Baskett, 1985). If a firstborn female is followed by a second-born male, however, and a family places a great deal of emphasis on male achievement, the firstborn daughter may not experience as much parental involvement or support (Thomas, 1983). Sutton-Smith and Rosenberg (1970) described second- and middle-born children as more easygoing and cheerful or, occasionally, as neglected. The neglect of middle-born children has occasionally been documented (Kidwell, 1982) but is probably not as pervasive as folk wisdom suggests. Many middle-born children experience a strong sense of being loved, wanted, and attended. Kalliopuska (1984) found middle-born children more prone to empathize with the feelings of others than firstborns.

Youngest children have been variously described as more immature and self-conscious or as more popular and outgoing. When families are very large, children usually have fewer pressures on them in the cognitive and social realms. Zajonc (1976) suggested this as a reason for the lower intelligence of later-born children in large families. However, not all later-born children in large families have lower IQs than their older siblings. Again, one must be cautious in interpreting "norms" as representative of all children. Families with six or more children usually insist on a handful of very firm rules but thereafter give the children a lot of freedom. In some families this freedom may have a positive effect on intelligence.

TELEVISION'S EFFECT ON CHILDREN

Television is considered an influential, extrafamilial agent of socialization in North American culture today. Its role is similar to that of radio and movies in past generations. Programs such as *Sesame Street* and *The Electric Company* bring a great deal of enrichment to young children's lives. It is not uncommon to see preschoolers incorporate parts of these programs or characters from them into their fantasy play. These programs were developed collaboratively by aca-

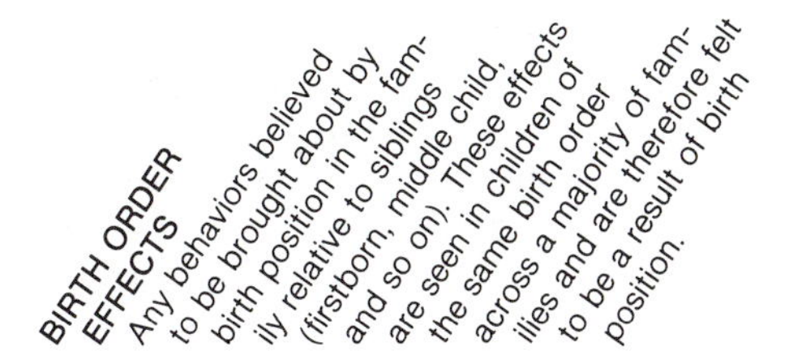

demic researchers and television producers to provide both intellectual stimulation and entertainment for young children. However, they constitute a small proportion of the programming that young children watch each day. Television viewing can become habitual, an unstimulating, passive, escapist activity. Children may remain in front of televisions in spite of their boredom because the alternatives to watching it are even less stimulating. Viewing time varies from child to child, day to day, and home to home, but may average about three or four hours a day and more on cartoon day (Saturday).

Professionals are concerned about the amount of television young children watch and also about the content of TV programming. There are few worthwhile children's programs besides those produced by the Corporation for Public Broadcasting. Many of the newer "children's programs" are in fact little more than program-length commercials. They are released concurrently with a line of toys associated with the program's theme or characters, to promote the toys. In some cases arrangements are made with broadcasters to share in the profits generated from the sales of toys when these program-length commercials are purchased by TV stations (Kunkel and Watkins, 1985).

Many of the cartoons made for children are full of violence, criminal themes, and an unrealistic view of death/resurrection. Gerbner (1972) took the time to count violent episodes in children's cartoons and got an average rate of 25.1 per hour. In a classic study in 1963, Bandura and his associates demonstrated that preschool children remember and subsequently perform violent acts that they see on television. It does not matter whether the models are human or cartoon characters, children imitate whatever situations or characters they watch. Efforts to demonstrate that TV viewing and modeling of violence work in conjunction with other factors (age, sex, family interactions) have been successful. Individual children react differently to the violence they watch. However, it is difficult to predict how any given child will react to any given episode of TV violence. It has a large effect on a small percentage of youngsters and a small but significant effect on a large percentage of youngsters (Liebert, Sprafkin, and Davidson, 1982).

The content of the programs that children watch also teaches gender-role stereotypes and ethnic stereotypes. Males outnumber females in television roles. They have prestigious jobs. They are powerful, aggressive, assertive, and intelligent. In contrast, women more often play romantic or domestic roles. Commercials for boys' toys tend to be fast paced and full of action. Those for girls' toys are slow paced and accompanied by pastel colors and soft music. Except for blacks, ethnic minorities are practically ignored on television. If seen, they are often cast as victims or villains.

The content of commercials directed at young children is of special concern to child development professionals because preschoolers do not have the necessary cognitive abilities to defend themselves against skillful persuasion. In many cases, they cannot tell when programming ceases and advertising begins. This is especially true when program characters promote products. Children often see commercial messages as sacred, unalterable truths. It is difficult to persuade young children that the information imparted by advertisements may be misleading—that certain toys and certain foods are *not* essential to their happiness or well-being.

Efforts are being made to improve the quality of children's television content and the quantity of programs available for children but, because such programming is not very profitable, change comes slowly. Most early childhood educators and child psychologists feel that preschool television viewing should be limited to those programs designed to capture and hold the attention of the very young children. It should not be used as a baby-sitter. Nor should children be placed in front of programs such as *Sesame Street* with the command "learn," especially if they are actively involved in some other creative activity. Viewing time should reflect their genuine interest time, no more.

CHILD ABUSE AND NEGLECT

Child abuse is defined as the intentional, nonaccidental physical or sexual abuse of a child by a parent or other caregiver entrusted with his or her care. It is not a new phenomenon. However, research studies and media reporting have only recently brought it out from behind closed doors and into national awareness. In the early 1970s many states passed laws making it easier for doctors, nurses, social workers, and law enforcement officers to report cases of child neglect or abuse. By the mid-1980s, over one million cases of child abuse were being reported annually.

Serious physical abuse—that which results in permanent injury (or death)—is given the clinical name *battered child syndrome*. Physicians have a duty to fully evaluate battered children and to guarantee that no repetition of trauma will be permitted to occur (Kempe et al., 1985). However, repetitions are common. Fatal battering is most apt to occur in children under age five.

Research indicates that abusive adults can be found in all socioeconomic levels in our society and in all ethnic groups. Many factors, singly or in combination, contribute to high-risk situations in which child abuse is more likely to occur. Parents under stress, as from factors such as unemployment, poor marital relations, social isolation, depression, or health problems, may be at high risk for abusive behavior. They may be more physiologically aroused in response to any child behaviors, and this hyperresponsivity may mediate aggression (Wolfe, 1985).

Certain child attributes or behaviors may also be very aversive to some parents and contribute to the incidence of abuse. Frequent crying, clinging, vomiting, sleep difficulties, bedwetting, aggressive outbursts, hyperactivity, or a difficult temperament may frustrate parents and lead to their aggressive outbursts. There is also evidence that some abusive parents have misconceptions about children, little knowledge of normal child development, and an orientation toward punishment as a remedy for problems (Steele, 1975).

Many of the other factors contributing to child abuse are poorly understood because the data are sparse. Starr (1978) saw the need

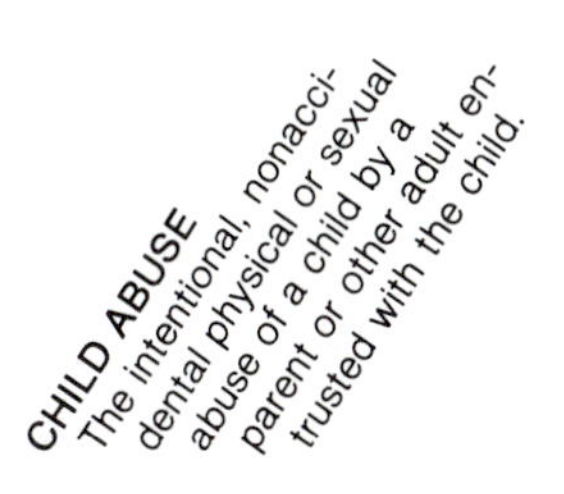

for research that considers the total ecology of the family: the abused child, the psychological characteristics of the parent(s), the social forces impinging on the family, institutions, and the total character of the society.

Our relatively recent awareness of the prevalence of battered children in our society has led to new programs to help abusive parents and to intervene on behalf of injured children. Programs for adults should be multidisciplinary to provide aid in the many areas in which they suffer their seemingly insurmountable stresses and to educate them in areas of self-control and child-rearing techniques. Programs for parents should also not be incriminating. Self-help groups and "parents anonymous" programs can be especially effective. Programs for children, such as specialized daycare centers, should emphasize the affectionate, trustworthy side of adults. They should also help abused children overcome their feelings of unworthiness and develop more positive self-concepts. With increased understanding of the factors involved in child abuse, with increased efforts to provide intervention and treatment facilities in our communities, and with concern for psychological as well as physical health maintenance practices in our homes, it is hoped that child abuse (as well as other forms of violence in our society) will soon begin a downward spiral.

Sexual abuse, which was rarely reported just ten years ago, now accounts for about 20 percent of reported child abuse cases in some areas of the country. The incidence has probably not risen dramatically; rather, public awareness and willingness to report it has. A retrospective (looking backward) study of 200 prostitutes revealed that 60 percent of them had been unreported victims of sexual abuse as children (Silbert and Pines, 1981). About two-thirds of the prostitutes reported that the abusive figure was a father or father substitute. Goodwin, Cormier, and Owen (1983) reported that grandfather-granddaughter incest accounts for about 10 percent of intrafamilial sexual abuse. Other perpetrators reported quite frequently are uncles, cousins, brothers, stepbrothers, or close friends of the family. The victims can be either male (25 percent) or female (75 percent). Experts feel that children almost never make up stories about being sexually abused (Faller, 1984). They more commonly keep their victimization a secret. The trauma has extremely negative emotional, physical, and attitudinal impacts, especially on very young children. About 25 percent of sexual abuse is perpetrated on preschoolers.

Prevention of sexual abuse begins with early sex education, including information about the possibility of molestation, given by a loved and trusted source. Finkelhor (1984) suggested discussing touching with children, differentiating good, bad, and confusing touches. Confusing touches are those that make a child feel "funny" or "mixed up," those that go under the pants, in private places, or all over. Children should be encouraged to tell someone right away about both bad touches (hitting, bullying, trapping) and confusing touches. They should understand that they have the right to refuse inappropriate touching and the right to keep their bodies private.

Sexual abuse prevention also includes parental follow-up on any communications from children that molestation has been attempted or perpetrated. Adults are often loathe to believe such messages from children, especially when the accused abuser is a family member, relative, or close friend. Caregivers must help frightened children defend themselves. The issue should be confronted head on, and appropriate steps should be taken to guarantee that no further repetition of the abuse will be permitted. Any child who has been sexually abused should be provided with psychological counseling to ameliorate the negative emotional impact of the trauma.

Child neglect is far more prevalent than child abuse (Wolock and Horowitz, 1984). The consequences can be just as long lasting—physically, mentally, and emotionally. Child neglect is defined as being remiss in attending to any aspect of child care (nutrition, health care, accident prevention, cognitive stimulation, toilet training, discipline, or provision of a sense of being loved and wanted). Failure-to-thrive infants are frequently victims of neglect (see Chapter 4). Egeland and Sroufe (1981) documented the high incidence of neglect among parents who fit the indifferent-uninvolved pattern of parenting. These parents are detached, emotionally uninvolved, often depressed, and uninterested in the child they neglect.

Lally (1984) presented a view of child neglect that goes beyond the characteristics of the neglecting parent(s). Many such parents feel psychologically isolated from other adults. They may be caught up in nonnurturant social systems. Primary prevention aimed at teaching them how to care for the children they neglect is not enough. They need to be helped to find family and social support networks to decrease their feelings of isolation and helplessness. Links with community service agencies should be established, maintained, and strengthened so the parents will be able to find services (psychological, economic) when they need them. Beyond parent-child relationships and parent–social support system networks, Lallay saw an association between child neglect and cultural beliefs. If a society espouses the concepts of each person for himself or herself, survival of the fittest and achievement through aggressive competition, the weaker, less "fit" (by reasons of health, sex, ethnicity, age, economics) and less aggressive parents will be more apt to feel powerless to succeed and more apt to give up trying. On the other hand, if a society embraces the concepts of cooperative efforts, good Samaritanism (willingness to help fellow beings in distress), and the fellowship of humankind, parents will be more apt to give as well as receive nurture.

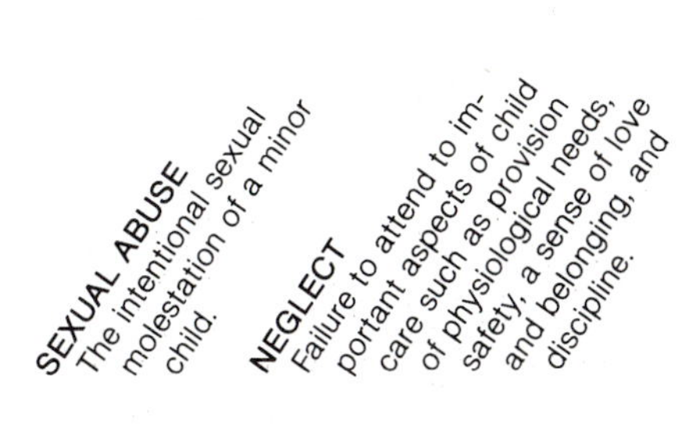

Summary

Physically, children stretch out in early childhood. Arms, legs, and trunks grow fastest, helping to alter the top-heavy, short-legged appearance of babyhood. Motor skills emerge and improve (running, stair climbing, tricycle riding, ball throwing).

Health maintenance requires adequate nutrition, sleep, exercise, affection, and protection from accidents. Regular check-ups, immunizations, and prompt diagnosis and treatment of infections can control many serious diseases and childhood disabilities.

Piaget described the cognitive developmental stage of early childhood as *preoperational.* Children explore, manipulate, and remember. The new dimension of improved mental functioning and use of mental images makes possible rapid acquisition of language and stimulates symbolic (dramatic) play. Play is a major synthesizer of ego, emotions, cognition, and socialization used to assimilate and accommodate more and more information. However, children in play and in other situations do not yet classify and systematically define their thoughts. They have difficulty separating fantasy from reality. They view their worlds from an egocentric perspective. Quite often they are very persistent and self-centered about felt needs or desires. By the end of the preoperational period, children begin to use rudimentary classifications. They also recognize some rights and needs of others and become less egocentric.

Language development reflects the egocentricity of early childhood. *I, me,* and *mine* are often heard. *Why?* is also common. Word usage multiplies rapidly, going from the few holophrases of beginning speech to more than a thousand words in a year. The age of beginning speech varies greatly from child to child, some starting shortly after their first birthday, others waiting until after their third. One-to-one verbal stimulation fosters language usage.

Many young children attend preschools oriented toward cognitive and language development, toward socialization, or both. Day-care centers usually combine a range of preschool-type activities with caregiving: meals, naps, nurturing. Neighborhood play groups help children relate to children and adults outside the family as well.

Erikson described the nuclear conflicts of early childhood as first achieving a sense of autonomy versus shame and doubt and then thrusting out into the wider world with initiative rather than guilt. Learning self-help skills (feeding, dressing, toileting) and having freedom to explore and communicate are basic to autonomy. Shame and doubt, and then guilt, may arise if children's strivings toward independence bring continued negative feedback.

Discipline is essential in early childhood. Three kinds may be used alone or in combination: power assertion, love withdrawal, and induction. Induction has the greatest potential for directing conscience formation.

Questions for Review

1. Parents are often told that having only one child will be harmful to that child. Do you agree or disagree with this? If you feel it would be harmful, how old do you think the first child should be before parents have another child? Discuss.
2. Imagine you are the physician or nurse practitioner working with a family who has a child with a severe physical handicap. What would you advise the family in terms of coping with this handicap? Consider the coping needs of the child and the parents in such a situation.
3. How does the preoperational stage described by Piaget differ from the sensorimotor stage that occurs in infancy?
4. Why do you think the age of two is commonly called the "terrible twos"?
5. What gifts (toys, games, and so on) do you think would be good for a child in early childhood? In your answer consider the developmental needs of a child at this time in the life span.
6. Children as young as the age of five—possibly younger—have been diagnosed as being emotionally disturbed. Do you agree or disagree that it's possible for a child at this age to be severely emotionally disturbed?
7. Do you agree or disagree that the violence seen on television and in other parts of society is producing more violence in children? Discuss.
8. Jamie is an only child who is very active, bright (but not precocious), healthy, and extremely inquisitive. His parents are looking for a preschool setting for him to attend mornings. Describe a nursery setting that would be most appropriate for Jamie. What qualities should the parents look for in the teachers? In the physical environment?

References

Albert, M. L., and Obler, L. K. *The bilingual brain.* New York: Academic Press, 1978.

Ames, L. B., and Chase, J. A. *Don't push your preschooler.* New York: Harper & Row, 1974.

Ashton, E. Measures of play behavior: The influence of sex-role stereotyped children's books. *Sex Roles,* 1983, *9,* 43–47.

Bandura, A. *Social learning theory.* Englewood Cliffs, N.J.: Prentice-Hall, 1977.

Bandura, A.; Ross, D.; and Ross, S. Imitation of film-mediated aggressive models. *Journal of Abnormal and Social Psychology,* 1963, *66,* 3–11.

Baskett, L. M. Sibling status effects: Adult expectations. *Developmental Psychology,* 1985, *21,* 441–445.

Bearison, D. J., and Cassel, T. Z. Cognitive decentration and social codes: Communication effectiveness in young children from differing family contexts. *Developmental Psychology,* 1975, *11,* 29–36.

Bellinger, D. C., and Needleman, H. L. Lead and the relationship between maternal and child intelligence. *Journal of Pediatrics,* 1983, *102,* 523–527.

Breland, H. M. Birth order, family configuration, and verbal achievement. *Child Development,* 1974, *45,* 1011–1019.

Bronfenbrenner, U.; Alvarez, W. F.; and Henderson, C. R., Jr. Working and watching: Maternal employment status and parents' perceptions of their 3-year-old children. *Child Development,* 1984, *55,* 1362–1378.

Chance, P. *Learning through play: Pediatric round table; 3.* Skillman, N.J.: Johnson & Johnson Baby Products Co., 1979.

Clarke-Stewart, A. *Day care.* Cambridge, Mass.: Harvard University Press, 1982.

Dewey, J. *Democracy and education.* New York: Macmillan, 1916.

Dolgin, K. G., and Behrend, D. A. Children's knowledge about animates and inanimates. *Child Development,* 1984, *55,* 1646–1650.

Dreikurs, R., and Soltz, V. *Children: The challenge.* New York: Hawthorn, 1964.

Easterbrooks, M. A., and Goldberg, W. A. Effects of early maternal employment on toddlers, mothers, and fathers. *Developmental Psychology,* 1985, *21*(5), 774–783.

Egeland, B., and Sroufe, L. A. Attachment and early maltreatment. *Child Development,* 1981, *52,* 44–52.

Elkind, D. *The hurried child.* Reading, Mass.: Addison-Wesley, 1981.

Ellis, P. O. Eye. In C. H. Kempe, H . K. Silver, and D. O'Brien (Eds.), *Current pediatric diagnosis and treatment,* 8th ed. Los Altos, Calif.: Lange, 1984.

Erikson, E. H. *Childhood and society,* 2nd ed. New York: Norton, 1963.

Faller, K. C. Is the child victim of sexual abuse telling the truth? *Child Abuse and Neglect, The International Journal,* 1984, *8,* 473–481.

Field, T., and Reite, M. Children's responses to separation from mother during the birth of another child. *Child Development,* 1984, *55,* 1308–1316.

Finkelhor, D. The prevention of child sexual abuse: An overview of needs and problems. *SIECUS Report,* 1984, *13*(1), 1–5.

Frankenburg, W. F. Early screening for developmental delays and potential school problems. In C. C. Brown (Ed.), *Infants at risk: Pediatric round table; 5.* Skillman, N.J.: Johnson & Johnson Baby Products Co., 1981.

Freud, S. *New introductory lectures on psychoanalysis.* New York: Norton, 1965. (Originally published 1933.)

Ganong, W. F. *Review of medical physiology,* 8th ed. Los Altos, Calif.: Lange, 1980.

Garcia, E. E. Bilingualism in early childhood. *Young Children,* 1980, *35,* 52–66.

Gerbner, G. Violence in television drama: Trends and symbolic functions. In G. A. Comstock and E. A. Rubenstein (Eds.), *Television and social behavior, vol. I: Media content and control.* Washington, D.C.: U.S. Government Printing Office, 1972.

Glass, D. C.; Neulinger, J.; and Brim, O. G., Jr. Birth order, verbal intelligence, and educational aspiration. *Child Development,* 1974, *45,* 807–811.

Goodwin, J.; Cormier, L.; and Owen, J. Grandfather-granddaughter incest: A trigenerational view. *Child Abuse and Neglect, The International Journal,* 1983, *7,* 163–170.

Grusec, J. E., and Kuczynski, L. Direction of effect in socialization: A comparison of the parent vs. the child's behavior as determinants of disciplinary techniques. *Developmental Psychology,* 1980, *16,* 1–9.

Haber, R. N. Eidetic images. *Scientific American,* 1969, *220,* 36–44.

Havighurst, R. *Developmental tasks and education,* 3rd ed. New York: McKay, 1972.

Hoffman, M. L. Altruistic behavior and the parent-child relationship. *Journal of Personality and Social Psychology,* 1975, *31,* 937–943.

Honig, A. S.; Lally, J. R.; and Mathieson, D. H. Personal-social adjustment of school children after 5 years in a family enrichment program. *Child Care Quarterly,* 1982, *11,* 138–146.

Hunt, J. M. How children develop intellectually. *Children,* 1964, *11,* 83–91.

Huston, A. C. Sex typing. In E. M. Hetherington (Ed.), *Socialization, personality and social development.* Vol. 4 of P. H. Mussen (Ed.), *Handbook of Child Psychology,* 4th ed. New York: Wiley, 1983.

Kalliopuska, M. Empathy and birth order. *Psychological Reports,* 1984, *55,* 115–118.

Kempe, C. H.; Silverman, F. N.; Steele, B. F.; Droegemueller, W.; and Silver, H. K. The battered-child syndrome. *Child Abuse and Neglect, The International Journal,* 1985, *9,* 143–154.

Kidwell, J. S. The neglected birth order: Middleborns. *Journal of Marriage and the Family,* 1982, *44,* 225.

Kohlberg, L. A. A cognitive-developmental analysis of children's sex role concepts and attitudes. In E. E. Maccoby (Ed.), *The development of sex differences.* Stanford, Calif.: Stanford University Press, 1966.

Kuczynski, L. Intensity and orientation of reasoning: Motivational determinants of children's compliance to verbal rationales. *Journal of Experimental Child Psychology,* 1982, *34,* 357–370.

Kunkel, D. L., and Watkins, B. A. Children and television. *Washington Report,* 1985, *1*(4), 1–8.

Lally, J. R. Three views of child neglect: Expanding visions of preventive intervention. *Child Abuse and Neglect, The International Journal,* 1984, *8,* 243–254.

Lashley, K. *The neuropsychology of Lashley: Selected papers.* (F. Beach, Ed.) New York: McGraw-Hill, 1960.

Lazar, I., and Darlington, R. Lasting effects of early education: A report from the consortium for longitudinal studies. *Monographs of the Society for Research in Child Development,* 1982, *47*(2–3), Serial No. 195.

Lenneberg, E. *Biological foundations of language.* New York: Wiley, 1967.

Liebert, R. M.; Sprafkin, J. N.; and Davidson, E. S. *The early window: Effects of television on children and youth.* New York: Pergamon Press, 1982.

Luria, A. R. *The mind of a mnemonist.* (L. Solotaroff, trans.) New York: Basic Books, 1968.

Lynch, A. Maternal stress following the birth of a second child. In M. H. Klaus and M. O. Robertson (Eds.), *Birth, interaction and attachment: Pediatric round table; 6.* Skillman, N.J.: Johnson & Johnson Baby Products Co., 1982.

Maccoby, E. E., and Martin, J. A. Socialization in the context of the family: Parent-child interaction. In E. M. Hetherington (Ed.), *Socialization, personality, and social development.* Vol. 4 of P. H. Mussen (Ed.), *Handbook of child psychology,* 4th ed. New York: Wiley, 1983.

Martin, C. L., and Halverson, C. F., Jr. A schematic processing model of sex typing and stereotyping in children. *Child Development,* 1981, *52,* 1119–1134.

McLoyd, V. C., and Ratner, H. H. The effects of sex and toy characteristics on exploration in preschool children. *Journal of Genetic Psychology,* 1983, *142,* 213–224.

Miller, L. B., and Bizzell, R. P. Long-term effects of 4 preschool programs: 9th and 10th grade results. *Child Development,* 1984, *55,* 1570–1587.

Minnett, A. M.; Vandell, D. L.; and Santrock, J. W. The effects of sibling status on sibling interaction: Influence of birth order, age spacing, sex of child, and sex of sibling. *Child Development,* 1983, *54,* 1064–1072.

Mischel, W. A social-learning view of sex differences in behavior. In E. E. Maccoby (Ed.), *The development of sex differences.* Stanford, Calif.: Stanford University Press, 1966.

Montessori, M. *The Montessori method.* New York: Stokes, 1912.

O'Keefe, E. S. C., and Hyde, J. S. The development of occupational sex-role stereotypes: The effects of gender stability and age. *Sex Roles,* 1983, *9,* 481–492.

Pederson, F. A.; Cain, R. L.; Zaslow, M. J.; and Anderson, B. J. Variation in infant experience associated with alternative family roles. In L. Laosa and I. Sigel (Eds.), *Families as learning environments for children.* New York: Plenum, 1982.

Piaget, J. Children's philosophies. In C. Murchinson (Ed.), *A handbook of child psychology.* Worcester, Mass.: Clark University Press, 1933.

Piaget, J. *Play, dreams and imitation in childhood.* (C. Gattegno and F. Hodgson, trans.) New York: Norton, 1962.

Piaget, J., and Inhelder, B. *The child's conception of space.* (F. Langdon and J. Lunzer, trans.) London: Routledge & Kegan Paul, 1956.

Roche, A. F. Secular trends in human growth, maturation and development. *Monographs of the Society for Research in Child Development,* 1979, Serial No. 179.

Rumack, B. H. Poisoning. In C. H. Kempe, H. K. Silver, and D. O'Brien (Eds.), *Current pediatric diagnosis and treatment,* 8th ed. Los Altos, Calif.: Lange, 1984.

Schilling, L. S. Imaginary companions: Considerations for the health profession. *Early Child Development and Care,* 1985, *22,* 211–223.

Schmitt, B. D. Ambulatory pediatrics. In C. H. Kempe, H. K. Silver, and D. O'Brien (Eds.), *Current pediatric diagnosis and treatment,* 8th ed. Los Altos, Calif.: Lange, 1984.

Schmitt, B. D., and Berman, S. Ear, nose and throat. In C. H. Kempe, H. K. Silver, and D. O'Brien (Eds.), *Current pediatric diagnosis and treatment,* 8th ed. Los Altos, Calif.: Lange, 1984.

Sears, R. R.; Rau, L.; and Alpert, R. *Identification and child rearing.* Stanford, Calif.: Stanford University Press, 1965.

Silbert, M. H., and Pines, A. M. Sexual child abuse as an antecedent to prostitution. *Child Abuse and Neglect, The International Journal,* 1981, *5,* 407–411.

Silver, H. K. Growth and development. In C. H. Kempe, H. K. Silver, and D. O'Brien (Eds.), *Current pediatric diagnosis and treatment,* 8th ed. Los Altos, Calif.: Lange, 1984.

Silverman, A., and Roy, C. C. Liver and pancreas. In C. H. Kempe, H. K. Silver, and D. O'Brien (Eds.), *Current pediatric diagnosis and treatment,* 8th ed. Los Altos, Calif.: Lange, 1984.

Snow, M. E.; Jacklin, C. N.; and Maccoby, E. E. Birth order differences in peer sociability at 33 months. *Child Development,* 1981, *52,* 589–595.

Springer, S. P., and Deutsch, G. *Left brain, right brain.* San Francisco: W. H. Freeman, 1981.

Starr, R. H., Jr. The controlled study of the ecology of child abuse and drug abuse. *Child Abuse and Neglect,* 1978, *2,* 19–28.

Steele, B. F. *Working with abusive parents from a psychiatric point of view.* U.S. Department of Health, Education & Welfare Publication OHD75–70. Washington, D.C.: U.S. Government Printing Office, 1975.

Suransky, V. P. *The erosion of childhood.* Chicago: University of Chicago Press, 1982.

Sutton-Smith, B., and Rosenberg, B. G. *The sibling.* New York: Holt, Rinehart & Winston, 1970.

Swartz, J. P., and Pierson, D. E. Cognitive abilities of sibling pairs: Evaluating the impact of early education. *Psychological Reports,* 1982, *51,* 171–176.

Thatcher, R. W.; Lester, M. L.; McAlaster, R.; Horst, R.; and Ignatius, S. W. Intelligence and lead toxins in rural children. *Journal of Learning Disabilities,* 1983, *16,* 355–359.

Thomas, J. H. The influence of sex, birth order, and sex of sibling on parent-adolescent interaction. *Child Study Journal,* 1983, *13,* 107–114.

Tubergen, D. G. Neoplastic disease. In C. H. Kempe, H. K. Silver, and D. O'Brien (Eds.), *Current pediatric diagnosis and treatment,* 8th ed. Los Altos, Calif.: Lange, 1984.

Weinraub, M.; Clemens, L. P.; Sockloff, A.; Ethridge, T.; Gracely, E.; and Myers, B. The development of sex role stereotypes in the third year: Relationships to gender labeling, gender identity, sex-typed toy preference, and family characteristics. *Child Development,* 1984, *55,* 1493–1503.

White, B. *Experience and psychological development.* Englewood Cliffs, N.J.: Prentice-Hall, 1971.

Will, B. E.; Rosensweig, M. R.; Bennett, E. L.; Herbert, M.; and Morimoto, H. Relatively brief environmental enrichment aids recovery of learning capacity and alters brain measures after postweaning brain lesions in rats. *Journal of Comparative and Physiological Psychology,* 1977, *91,* 33–50.

Wing, E., and Roussounis, S. H. A changing pattern of cerebral palsy and its implications for the early detection of motor disorders in children. *Child: Care, Health and Development,* 1983, *9,* 227–232.

Wolfe, D. A. Child-abusive parents: An empirical review and analysis. *Psychological Bulletin,* 1985, *97,* 462–482.

Wolock, I., and Horowitz, B. Child maltreatment as a social problem: The neglect of neglect. *American Journal of Orthopsychiatry,* 1984, *54,* 530–543.

Zajonc, R. B. Family configuration and intelligence. *Science,* 1976, *160,* 227–236.

♥

Late Childhood

6

Chapter Outline

Physical Development
- *Health Maintenance*
- *Accidents*
- *Nutrition*
- *Chronic Illnesses*
- *Emotional Disorders*

Cognitive and Language Development
- *Piaget's Concrete Operations Stage*
- *The Learning Process*
- *Language*
- *The Concept of Intelligence*
- *Creativity*
- *Problems of Learning*

Psychosocial Development
- *Erikson's Concept of Industry Versus Inferiority*
- *Parent-Child Socialization Interactions*
- *The School's Role in Socialization*
- *Peer Interactions*
- *Neighborhood Interactions*
- *The Effect of Technology*

Key Concepts

androgyny
asthma
attentional deficit disorder
authoritarian parenting
authoritative parenting
behavior modification
character education
concrete operations
conservation
conventional morality
creativity
dental caries
diabetes mellitus
encoding
encopresis
enuresis
epilepsy
expressive orientation
giftedness
hay fever
Heimlich maneuver
indulgent parenting
industry versus inferiority
instrumental orientation
intelligence quotient (IQ)
Jensenism
learning disabilities
mediation
mental retardation
meritocracy
metamemory
morality of constraint
morality of cooperation
morality training
motor tics
negligent parenting
numbering
peer group
physical causality
premorality
prosocial behaviors
psychological causality
psychophysiologic disorder
reciprocity
reversibility
Rosenthal effect
school phobia
serous otitis media
sex-role models
single-parent family
specific reading disorder
stepparenting
stuttering

♥

Case Study

Arthur was the new kid in school, a fourth grader. He was having trouble finding a place in the well-established boys' cliques. His mother made him walk to and from school with his little sister, a first grader. He was teased about this, about his short stature, about the way he talked, about his clothes—about everything that nine-year-old boys find to tease each other. Two boys in particular made Art's life miserable. Both worked as crossing guards before and after school. One afternoon, they persuaded Norman, a boy who attended a different school, to meet them at their school for the purpose of beating Arthur up. They showed Norman who Art was, then left their crossing guard posts to follow Norman, who was in turn following Art and his little sister, so they could watch the fight. When Norman attacked Art, Art fought back. In fact, Art was a good fighter and quickly got the best of Norman. Then he and his sister ran home. They barely had time to tell their mother what had happened when a group of five boys knocked on their door: Norman, the two crossing guards, and two boys they had picked up along the way. The boys wanted Art to come out and fight some more. Art's mother called the school principal to report that the crossing guards were at her house asking for a fight rather than helping other children cross the street. The principal responded by calling a meeting of the two crossing guards and Art the next day. The crossing guards of course denied any part in the fight between Norman and Arthur. They said they had left their posts only because they heard the fighting. The principal believed them. She scolded Arthur for fighting on his way home from school. She warned the crossing guards not to leave their posts early again. She was satisfied that justice was done. What do you think?

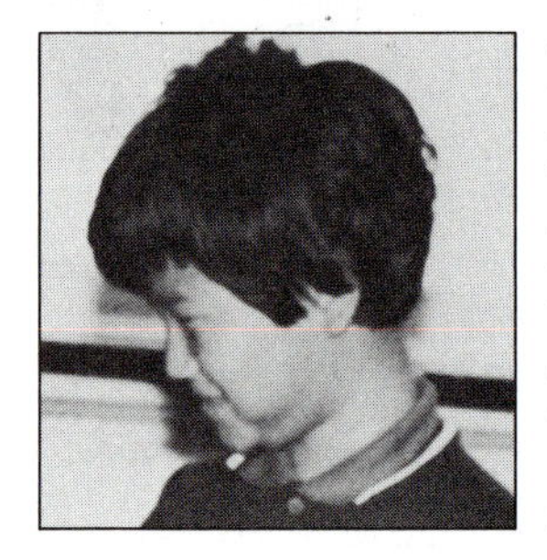

The years from six to twelve are sometimes called *middle childhood* as well as *late childhood.* The latter rubric will be used here as an inclusive term, as "middle childhood" suggests that the early-adolescent age range comprises late childhood, which most adolescents would consider a misnomer. The term *middle childhood* will be used occasionally to describe children of ages six, seven, and eight.

The most vivid memories of childhood are usually those from the late childhood period covered in this chapter: best friends (the "at home" ones such as neighbors and siblings and the "at school" ones who probably changed frequently); school; lunch hours at school; teachers and classes; after-school activities; homework; chores; sports; clubs; summer vacations; secrets; places to hide when one wanted to be alone.

During late childhood, peers become extremely important. For this reason some writers call it the "gang age." It is also called the "dirty age" for reasons of both language and laundry. It is a time of considerable development physically, socially, emotionally, cognitively, and linguistically. The old adage that a child's personality is set by age seven has not been supported by research. Personality is a dynamic phenomenon, influenced throughout life by all kinds of environmental pressures. One should not become locked into the notion that the late childhood years are simply a latent passage of time in which a child grows bigger but otherwise emerges unchanged. Early childhood experiences do have a profound impact on the continuous development of children in the physical, cognitive, and psychosocial realms as they grow older. For example, the manner in which the Eriksonian conflicts of trust, autonomy, and initiative are resolved in the early years influences the way children react during their school years. However, new conflicts arise, new solutions are generated (interpreted by the accumulation of past experiences), and children change as they grow.

Physical Development

Between the ages of six and twelve, children's proportions become more adultlike. Arms and legs get longer, giving more gangliness, the abdomen flattens, and the shoulders, chest, and trunk broaden. The lordosis (spinal curvature) of early childhood disappears, and the back appears straighter. Although the size of the head changes very little, facial proportions are altered considerably. The forehead broadens, the nose grows larger, the lips get fuller, and the jaw juts out from the chin. All of this takes place at a gradual but steady pace (see Figure 6–1). Height increases by about 5 to 7 centimeters (2 to 3 inches) a year, and weight gains vary from 1½ to 2½ kilograms (3 to 6 pounds) a year. Medical professionals may examine children carefully if they do not grow taller in a year's time, but they do not get as concerned about failure to gain weight. This is because weight is influenced by exercise and environment as well as by diet. A well-fed, well-nourished child may get taller but not add kilograms (pounds) over a one-year time span and still be well within a normal range.

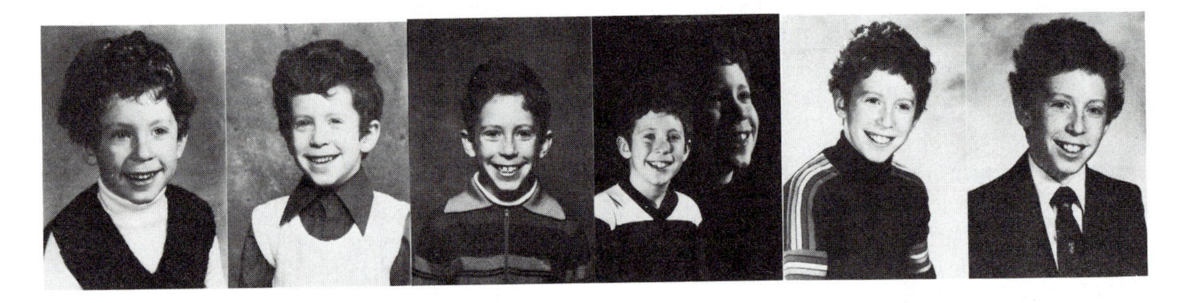

Figure 6–1
During the grade-school years facial features lose their babyish look and become more adultlike.

Normal, healthy children engage in plenty of activities that enhance large-muscle (gross motor) development—walking, running, jumping, climbing, throwing, catching, skating, swimming, dancing, and riding bicycles, skate boards, or horses. They also may be required by their families to engage in other less pleasant physical activities like daily chores. In general, "baby fat" decreases and muscle mass increases in both sexes during the elementary school years.

Small-muscle (fine motor) movement—eye-hand coordination—and manual dexterity become considerably more precise with each advancing year of childhood, especially when opportunities exist for the exercise of the small muscles. Printing and cursive writing in the schools provide such exercise. So do model building, sewing, weaving, painting, coloring, clay modeling, other arts and crafts projects, stamp collecting, dressing dolls, playing of musical instruments, taking photographs, and other such hobbies.

The physical appearances of elementary school children vary considerably within each age group. Consider the differences of your own peers when you were in the fifth or sixth grade, not only by sex and race but also by such factors as height, weight, and physical build. Girls are generally slightly taller and heavier than boys of the same age until the mid-teens. Body types (short-fat, tall-thin) tend to remain stable even as height and weight are added. Sheldon (1940) identified three basic physiques: *endomorphy, mesomorphy,* and *ectomorphy* (see Figure 6–2). Endomorphs are short and round, meso-

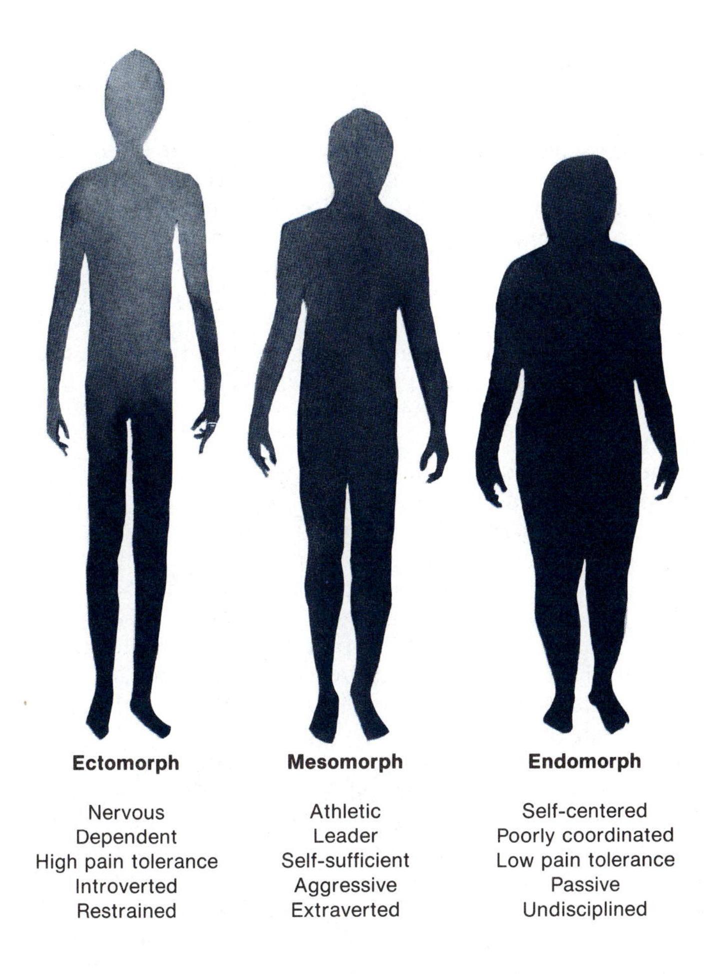

Figure 6–2
Studies correlating body build and temperament have had a profound impact on social stereotyping but remain controversial. Can you think of children who are exceptions to these characterizations?

morphs are muscular with broad shoulders and narrow hips, and ectomorphs are long and lean. Sheldon also assigned basic personality types to each body type. Research on body typology as it relates to behavior has waxed and waned over the last 40 years, with interest rekindled in the mid-1980s with a book by Wilson and Herrnstein (1985) suggesting that a mesomorphic body build is associated with aggression and criminal activity. However, the assumption that body build determines personality remains highly criticized. There are too many exceptions to prove the rule (see Figure 6–2 and try to think of exceptions you know). A strong belief in a stereotyped personality to go with physical build may lead to a self-fulfilling prophecy effect. Current research leads us to suspect that genetic factors and hormone balances do make body build (especially bone structure, less so weight) relatively stable from childhood to adulthood. The pace of maturation may affect one's self-concept. Concentrations of hormones in the blood may also affect behavior, but there is not an invariant relationship between physique and personality.

Around the age of six or seven every child experiences the

seemingly toothless grin caused by the shedding of deciduous teeth in preparation for the permanent ones. Throughout late childhood deciduous teeth are lost and permanent teeth are gained, until by age twelve most youngsters have twenty-eight out of the eventual thirty-two permanent teeth. (Lacking are the wisdom teeth that may or may not manage to arrive during adolescence.) These larger teeth make the mouth appear bigger and add character to the face. However, these teeth often grow in crooked. Numbers of children learn to cope (and sometimes even feel part of the gang) with their orthodontist visits and braces. Straightening of teeth not only gives a more pleasing appearance but also allows for a more satisfactory development of the lower part of the face and for a better bite.

HEALTH MAINTENANCE

The leading noninfectious health problem in the Western world is **dental caries**. In spite of fluoridated water (which retards tooth decay) in many communities, approximately 98 percent of North American children have some cavities and fillings in their teeth. The high consumption of sugar is held primarily responsible. Every year each North American child consumes an average of 125 pounds of sugar. Most of this is invisible, a part of processed foods (soups, ketchup, fruit drinks, sodas, salad dressings, cured meats, canned or frozen fruits and vegetables, breads, cereals). Children should be trained to brush their teeth with a fluoridated toothpaste regularly. They should also be encouraged to drink milk or water rather than sugared drinks and to have vegetables, cheese, or nuts rather than pastries or candy for snacks. Beedle (1984) recommends that children be given sweets only at mealtimes, when the tooth-destroying acidogenic bacteria that act on sugars are buffered by other foods and saliva. He cautions that sucrose, especially in forms that cling (taffy, caramel) or have prolonged contact (hard candy, lollipops, chewing gum) should be eliminated.

Adequate sleep for school-age children decreases with age. A six-year-old should probably get from ten to twelve hours of sleep. A twelve-year-old should probably get from eight to ten hours of sleep. Adequate sleep varies from child to child. If children go to sleep easily, sleep soundly, and wake refreshed, they are getting enough rest. Sleep disturbances such as restlessness, periods of wakefulness, and nightmares may be symptoms of emotional upset rather than excess sleep. School-age children may find sleep difficult for reasons such as unhappy relationships with teachers or friends, poor academic progress, quarreling between parents or siblings, fears, feelings of low worth, or guilt. Parents can help alleviate sleep disturbances by having a set bedtime hour that is followed firmly with few exceptions, by avoiding severe bedtime punishments, by providing an atmosphere conducive to discussing problems that have occurred in the course of the day, by reassuring the child that he or she is a valued and loved family member, and by soothing away fears.

DENTAL CARIES Decay of teeth.

Roughly 25 percent of children need to wear glasses by late childhood. Most children are normally hyperopic (farsighted) until after about age six when their eyes begin to reach more adultlike proportions. Some children, however, are myopic (nearsighted), a problem that is not outgrown. Others have problems with astigmatism (a situation in which images are improperly focused on the retina due to faulty curvature of the cornea or lens) or strabismus (crossed-eyes, walleyes; refer back to Chapter 5). Glasses can correct hyperopia, myopia, astigmatism, and strabismus and should be worn to prevent eyestrain or progressive loss of visual function (see Figure 6–3).

A school-age child's health should be safeguarded with an annual routine physical examination. Some school systems provide this service free of charge to parents. If they do not, the family should make arrangements to have each child visit a physician or pediatric nurse practitioner yearly. Although it may seem unnecessary for apparently healthy children, there are conditions that may go unrecognized by parents (for example, heart murmurs, high blood pressure, anemia, hearing losses, developmental delays). In addition, parents should call or take children to a medical facility for treatment whenever they have prolonged illnesses. The more common illnesses of late childhood include the common cold, acute otitis media, viral pharyngitis and tonsillitis, gastroenteritis, bronchitis, conjunctivitis, strep throat, impetigo, chickenpox, urinary tract infections, pneumonia, and boils (Schmitt, 1984). These diseases were discussed in more detail in Chapter 5.

Infections account for approximately 70 percent of school

Figure 6–3
Children with visual defects such as myopia and astigmatism should wear glasses. Failure to correct vision can result in blurred or distorted images that confuse children and interfere with learning.

absences. Many children continue to go to school when sick and spread their disease to other children. Infections are less threatening than they were a generation ago because of the availability of antibiotics and immunizations, but untreated infections still may lead to serious complications (such as strep infection leading to rheumatic heart disease and glomerulonephritis) that occasionally result in the death of a child.

ACCIDENTS

Accidents are the main cause of death and physical handicaps during childhood. School-age children most commonly get hurt in motor vehicle accidents. They also swallow or come in contact with poisons, burn themselves, have serious falls from bicycles, trees, or buildings, have water-related accidents, get frostbite, or choke.

Child victims of major accidents should be considered seriously injured until their condition is proved stable. The first adult to arrive at the accident scene must be sure that the child can breathe. This may involve clearing debris (chewing gum, food) from the airway and positioning the head to the side. If there is any question of head or spinal injuries, the child should not be moved until trained medical personnel arrive. If the child is unconscious and pulses are not palpable, cardiopulmonary resuscitation (CPR) may be required. While it would be ideal for all adults to be trained in CPR, no adult who lacks sufficient knowledge of the procedure should attempt it. It is better to call for help. Water-related accidents may also require CPR. Near-drowning victims are usually hospitalized overnight for observation.

If a child is bleeding, direct pressure should be applied over the hemorrhage site. If the child has been burned, a clean cool cloth should be placed on the burned area until help arrives. First aid for poisoning is presented in Chapter 5. Frostbite requires a prolonged (20-minute) rewarming of the affected area(s) with warm (38–40°C/100°F) water, not hot water, snow, vigorous rubbing, or dry heat. If a child is choking, any older child or adult in the vicinity should immediately apply the **Heimlich maneuver**:

1. The rescuer stands behind the victim and wraps his or her arms around the victim's waist.
2. The rescuer makes a fist with one hand and places the thumb side of the fist against the victim's abdomen, slightly above the navel and below the rib cage.
3. The rescuer grasps the fist with his or her other hand and presses it into the victim's abdomen with a quick upward thrust.
4. This thrust is repeated several times if necessary.

HEIMLICH MANEUVER
A method of dislodging objects blocking the airway by applying sudden pressure just under the rib cage.

Although adults cannot always watch school-age children, they can repeatedly warn them against accident hazards and help them find safe play areas (see Figure 6–4) and teach safety consciousness.

Figure 6–4
Normal, healthy school-age children delight in attempting daring deeds that demonstrate their gross and fine motor coordination and muscular agility. Adult supervisors should remind children to play in safe areas and obey safety precautions. Most of the 19 million annual accidental injuries to school-age children are preventable.

NUTRITION

Good nutrition in middle and late childhood does not necessarily mean eating the same foods as adults. For example, adult breakfasts are often toast and coffee or just coffee. What should a growing child have for breakfast? Many television advertisements would have them eat a sugary, vitamin-fortified, degermed, bleached, artificially flavored and colored, packaged cereal. Most cereals are at least 20 percent sugar; many are over 50 percent sugar. When taken with milk, their nutritional benefits are improved. However, many are eaten dry in the rush to get off to school, or as snacks on the bus, or while walking to school. Other "fast" breakfasts taken in similar fashion are toast and jam or breakfast pastries. These meals of highly refined carbohydrates and sugar are digested rapidly and leave the child hungry again long before lunch. Many children skip breakfast altogether. This is especially hard on the mind and the body because of the absence of any nutrients for an eight- to sixteen-hour period since dinner the night before. Pollitt, Leibel, and Greenfield (1981) found that skipping breakfast significantly decreased the accuracy of responding to a number of problem-solving tasks in nine- to eleven-year-old well-nourished children.

Some protein food should certainly be part of breakfast (such as a glass of milk, an egg, nuts, cheese, meat, soybeans, peanut butter, or high-protein cereal). If the protein foods chosen for breakfast are low in carbohydrates (egg, nuts, meat), some other higher carbohydrate food should also be provided (milk, orange juice, toast, rice, or cereal). Fat, such as provided by butter or margarine, is also necessary for the normal functioning of the body. However, it is usually found in protein foods in sufficient quantity to make it unnecessary to add separately. Sugar is not necessary to good nutrition and contributes to problems of dental caries and obesity. Starches are changed by digestion into glucose (blood sugar), which the body needs. They do not contribute as much to tooth decay, and, because they are digested more slowly, they delay "hunger pangs" longer.

In a normal day a growing child should have three servings of milk or milk products (skim milk, yogurt, cheese), two servings of protein foods (fish, poultry, meat, peanut butter, nuts, dried beans and peas, lentils, eggs), three or four slices of bread or its equivalent (rice, pasta, cereal), one fruit or fruit juice, and at least two vegetables.

Children do not usually need vitamin supplements if they balance their diets from the four basic food groups (milk, meat or protein foods, breads and cereals, vegetables and fruits). Some physicians and pediatric nurse practitioners may recommend iron supplementation if proteins and other foods preferred are iron poor. Some school-aged children become vitamin and mineral conscious and shop for extra sources of "fad" nutrients in health food stores. They should be warned that the fat-soluble vitamins (A, D, E, and K) are stored in the body and can build up to toxic levels when ingested in excess amounts. Vegetarian diets can provide adequate nutrition if they are supplemented with milk and cheese, or also with eggs. Pure vegetarian (vegan) diets are not recommended for children because they do not supply the adequate complete proteins necessary for growth.

Obesity affects from 5 to 10 percent of preadolescent children (O'Brien and Hambidge, 1984). Many obese children are poorly nourished because they overeat predominantly high-carbohydrate foods. Compulsive eating may reflect a disturbed parent-child relationship (for example, the child uses food to compensate for lack of attention, or the parent forces food on a child to alleviate his or her guilt or insecurities about parenting). Obesity may also have a genetic basis, or result from learned family behaviors (as when all family members overeat) or from some body malfunction (such as underactive thyroid). Obesity can handicap a child physically and socially. Peers are often extremely cruel in their ridicule of an overweight child.

Undereating, or improper eating, is less easily spotted but is also hazardous to good health. A malnourished child may have growth retardation, weakness, fatigue, anemia, tooth decay, poor posture, and a lowered resistance to infection. In some cases food deprivation may be a form of child neglect. More commonly, the malnourished child is fed, but for reasons of economics, ignorance, or overuse of convenience foods, the child is given insufficient protein, vitamins, and iron. High "empty-calorie" diets also increase the need for certain vitamins (such as thiamine) necessary for the metabolism of sugars. Some properly fed children are also malnourished because they refuse to eat the foods placed before them. Lack of appetite may be due to filling up on empty-calorie snacks (soda, candy, cookies), poor health, lack of exercise, emotional tensions, or a desire to be thin.

CHRONIC ILLNESSES

Chronic illnesses affect between 5 and 10 percent of the late-childhood population (Haggerty, 1984). Children with chronic illnesses may have to miss school or have special school arrangements made for them. When possible, it is preferable to keep the child within a normal school setting for the benefits to social and emotional development. Handicapped children need love, laughter, and a sense of independence as surely as do other children. Some of the more common chronic illnesses of childhood are diabetes, epilepsy, serous otitis media, hay fever, asthma, and enuresis.

Diabetes mellitus, sometimes referred to as sugar diabetes, may begin in childhood. In the juvenile form of diabetes, with rare exceptions, insulin-producing cells in the pancreas are destroyed. For this reason, it is also called insulin-dependent diabetes, since insulin must be given by injection once or twice a day (Silver, Gotlin, and Klingensmith, 1984).

A small percentage of black children who develop diabetes may have a rare diabetic syndrome that at first appears to be an insulin-dependent disease but is biochemically different. Pancreatic insulin-producing cells are not completely destroyed and, as the child grows older, they once again begin to produce insulin. These children will initially need insulin injections but are able to be controlled with diet and pills to lower blood sugar levels during the course of their disease. They then resemble persons with non-insulin-dependent diabetes, a less severe form of the disease (see Chapter 10).

Acute symptoms of insulin deficiency include increased thirst, abdominal or leg cramps, loss of weight, frequent urination (sugar can be found in the urine), lassitude, emotional disturbances, collapse, and finally a comatose condition if treatment is not initiated. Treatment consists of giving the child insulin (by injection) and regulating the child's diet and activity. A child may go into a hypoglycemic state (state of low blood sugar level, also referred to as insulin shock) when he or she fails to consume enough carbohy-

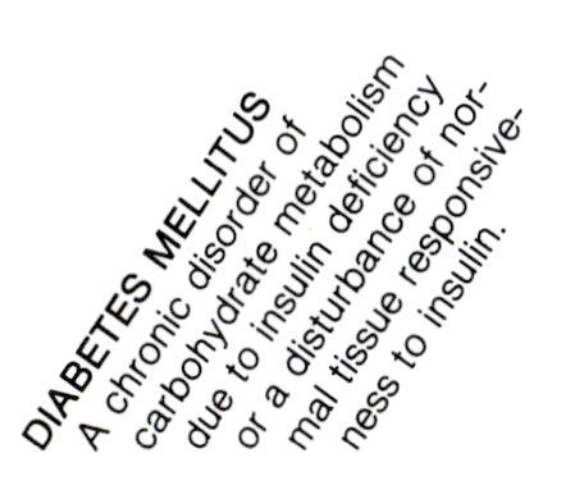

drates, exercises too much (exercise facilitates uptake of sugar by the body's muscles so not as much insulin needs to be administered to maintain the same blood sugar level), or takes too much insulin. A hypoglycemic state may first be noted by behavior changes such as inattention, confusion, sleepiness, or irritability. If the child does not ingest some form of sugar, loss of consciousness or convulsion may occur. For this reason diabetic children usually carry candy with them or ask for sweet food or drink whenever they experience any of the symptoms of hypoglycemia.

Having diabetes mellitus may make children feel angry and frustrated. They have difficulty achieving autonomy due to their dependence on others caused by the disease. They must always remember their medicine, watch their diet and their exercise, and be concerned about infections or suffer consequences that hardly if ever affect their nondiabetic friends. Diabetic children must also learn how to safely give themselves insulin injections. Their disease is not curable, only controllable with constant care.

Hagen, Anderson, and Barclay (1985) found that diabetic children have more problems than other children in a variety of areas beyond those of disease complications. While diabetic children have normal intelligence, they often underachieve and have social adjustment problems in school. They are at an increased risk for special education class placement and for repeating a grade in their early school years. Parents have a responsibility to monitor their entrance to school and their adaptation to the classroom environment. Each year, the parents must educate the child's teachers and other school staff (secretaries, administrators, food service personnel, coaches) about their diabetic child's special concerns and needs. Diabetic children should be treated as normally as possible and should be encouraged to perform up to their capabilities, without jeopardizing their medical disease management.

Epilepsy is a condition characterized by recurring seizures or loss of consciousness accompanied by convulsive muscle movements. Convulsive movements that are associated with loss of consciousness may be due to a recognized cause such as electric shock, high temperatures, brain infections or tumors, low blood sugar, or inflammatory disease of blood vessels in the brain. Such seizures are called symptomatic seizures. The underlying disorder in the brain causing grand mal or petit mal epilepsy is not known.

In grand mal epilepsy the attack most often begins with a warning called an aura. The aura may take a variety of forms, from the hallucination of a smell, taste, vision, or sound to an abnormal feeling in some part of the body. Immediately after that, consciousness is lost. The child falls to the ground and may suffer injuries in so doing. There then appears rigidity of the muscles, followed by sharp, short, interrupted jerking movements. When the convulsions stop, the child may remain unconscious for only a few seconds or for a long period of time. He or she is generally sleepy and may have headaches for a few hours following each seizure.

In petit mal seizures children suddenly lapse into blank stares. They are unaware of their surroundings and may blink or smack their lips. The spell may only last ten to fifteen seconds. Children do not fall, have convulsive movements, or feel sleepy during or after petit mal seizures.

Once epilepsy has been diagnosed, the child can usually be kept symptom free with daily medication. As with diabetic children, parents of epileptic children must educate the child's teachers and other relevant school personnel each year about the medical management and special concerns of their epileptic child. While seizures probably will not occur in a child controlled with medications, school staff should know how to recognize them and what to do in the event that one should take place in school.

About 5 percent of school-age children have some degree of temporary or permanent hearing loss. Even a mild, fluctuating hearing impairment such as that resulting from **serous otitis media** (condition in which the middle ear is filled with fluid) can interfere with learning and should not be ignored. Children with chronic serous otitis media, or cycles of serous otitis media alternating with acute otitis media (see Chapter 5), may eventually sustain a permanent conductive hearing loss from damage to the hearing structures. Other causes of hearing loss include accidental ear injuries and congenital defects. Hearing tests should be conducted annually on school-age children. Significant hearing loss is often an overlooked cause of learning difficulties. If schools do not do routine hearing screening, parents should ask their physician or nurse practitioner to include this in the child's annual check-up.

Some form of allergic reaction is normal in everyone who comes in contact with certain antigens (such as viruses or bee stings) for which they have no antibodies (see Chapter 3, p. 000). Some forms of antigens (also called *allergens*) cause reactivity only in certain members of the population. **Hay fever**, also known as allergic rhinitis (inflammation of the nasal mucous membrane) occurs seasonally as a result of exposure to specific wind-borne pollens (Pearlman, 1984). Symptoms include itching, sneezing, eye tearing, swollen nasal passages, and nasal obstruction leading to mouth breathing, nasal speech, inability to clear the nose with blowing, difficulty eating, and difficulty sleeping. In addition, children may develop headaches, sore throats, laryngitis, and serous otitis media as an extension of the allergic reaction. The acute symptoms can be treated with antihistamines, decongestants, and short-term use of steroid drugs. On occasion, the polyps (obstructing, bulging masses of tissue) that form in the nasal passages in response to allergens may have to be removed surgically.

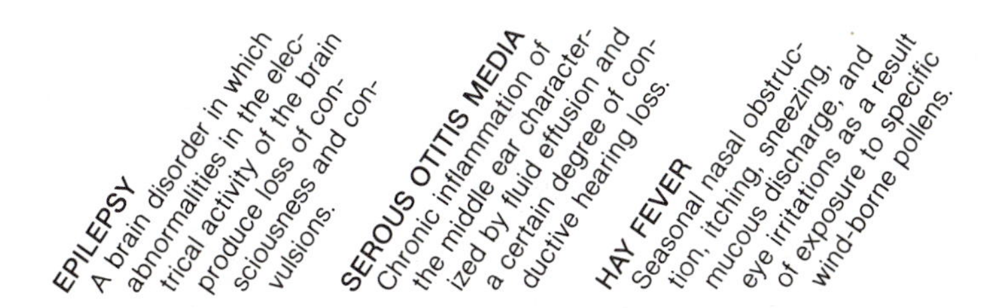

If the inciting antigens can be identified, children may be given a series of injections of extremely small amounts of the substances to make them hyposensitive to those allergens in the future. Some physicians and parents may try environmental control instead, protecting the child from exposure to whatever substances incite the allergic reaction. In many cases, this may involve a move to another area of the country where the offending seasonal pollen does not exist. However, children prone to hay fever may develop new allergies to substances that exist in that area of the country after a few years. Severe symptoms may cause prolonged school absences and the need for home-bound instruction, make-up work, or special scheduling considerations. Seasonal bouts of illness should not be allowed to become an excuse for underachievement. School personnel need to be sympathetic and understanding of the special considerations necessitated by the onset of symptoms of hay fever.

Asthma is a reaction of the bronchial tubes to allergens (house dust, plants, molds, insects, feathers, furs, pollens, foods, cooking odors, medicines, chemicals, smoke, paint fumes), to changes in temperature or barometric pressure, to overexercise, to viral upper respiratory infections, or to psychologic factors (Pearlman, 1984). A child does not have to be allergic to have asthma, and even when the child does have a hypersensitivity to certain allergens, exposure to the allergens is seldom the only precipitating factor in an asthma attack.

An attack involves bronchial wheezing that may become so serious that one wonders whether the child will be able to catch his or her next breath. Persons witnessing such attacks are often frightened. The asthmatic child becomes flushed and his or her skin becomes moist. Emergency care may involve aerosolized drug administration and transportation to a hospital for further supportive care. The child's own fear, together with onlookers' fears, only serves to make the attack worse. It is more helpful if persons attending an asthmatic attack remain calm and keep onlookers away. It is difficult not to overprotect and be anxious about asthmatic children. Long-term asthma treatment has many facets: drugs, control of suspected or known environmental allergens, allergen desensitization programs, relaxation exercises, parental counseling, and child counseling. Parents can help their children most by understanding the underlying causes of attacks, reassuring the child of his or her continued worth and loved position in the family, setting definite limits on the child's behavior rather than "spoiling" him or her after attacks, and remaining calm and reassuring during and after the at-

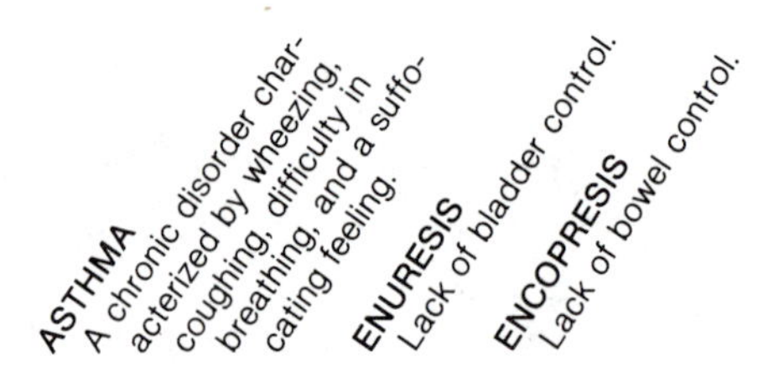

tacks. Meijer (1981) found that asthmatic children typically become very dependent on their mothers. The children need to be helped to assume more and more responsibility for their own care without unreasonable fears of initiating asthma attacks. Parents need to be helped to allow the child and others outside the family (such as school personnel) to assume some of the burden of care for disease management as well. Many children outgrow or have a lessening in severity of their asthma after adolescence.

By late childhood normal children should have day and night bladder and bowel control. Lack of bladder control, which most commonly occurs at night, is called **enuresis**. One in five children wets the bed occasionally at age seven, and one in fourteen still does so at age ten. Some of the possible reasons for this delayed control are given in Table 6–1. The treatment of enuresis depends on the cause of the problem. Medicines, behavior modification procedures, and family counseling have all been beneficial to some children.

Table 6–1 Factors That May be Associated with Bedwetting in Late Childhood

Genetic: tends to run in families
Delayed physical maturation
Stresses of early childhood may have interfered with acquisition of control
Functional disorder of bladder (associated with frequent urination during the day as well)
Inadequate toilet training (neglected or excessively punitive)
Emotional or behavioral difficulties
Current environmental stress
Physical abnormality of urinary tract (rare)

SOURCE: Adapted from M. Rutter, *Helping Troubled Children* (New York: Plenum, 1975).

Lack of bowel control, called **encopresis**, is a rarer problem in late childhood. Children who still soil their pants in the school years are usually suspected of having some emotional disturbance. Smearing the feces frequently accompanies the "accident." Some form of psychotherapy is usually necessary before encopresis is stopped.

EMOTIONAL DISORDERS

There is increased recognition today of the interactive effects of physical illnesses and mental disturbances in the etiology (causes) of emotional disorders. Consequently, it is becoming less and less clear whether abnormal behaviors should be labeled emotional disorders or physical disorders or should not be differentiated in this way. Children with enuresis and encopresis are often classified as having mental disorders. So are children with attention deficit disorder, mental retardation, and school phobia (discussed respectively on pages 298, 299, and 321 of this chapter) and children with the eating disorder pica (discussed in Chapter 5). Children with stereotyped movement disorders, psychophysiologic disorders, depression, childhood schizophrenia, and autism are also classified as having mental disorders. We will not discuss the latter two disorders in this chapter because they occur rarely.

Many boys (less commonly girls) develop some minor **motor tics** between the ages of six and twelve. These tics are involuntary, repetitive, nonpurposeful movements of a part of the body, such as eye blinks, facial twitches, tongue clicks, or jerking movements of the shoulders, arms, or legs. They occur at irregular intervals, increasing when the child is tense, decreasing during nonanxious concentration, and disappearing during sleep. These stereotyped motor tics may last from two weeks to usually not more than one year (Shapiro and Shapiro, 1980). No treatment is necessary unless the motor movement becomes severe and does not fluctuate and wane. Adults are advised not to criticize the child, call attention to the tic, or punish him or her for not stopping the behavior. Tics disappear most readily when the child's anxieties decrease and his or her self-confidence improves.

Tense, anxiety-disordered children quite frequently develop symptoms of a **psychophysiologic** (psychosomatic) **disorder**. This term refers to any health problem that is suspected to result from emotional stress too great for a person to handle in the course of everyday living. Developing a physical illness gives the person "time off." Most psychophysiologic disorders are not recognized as such. Allergies, rashes, headaches, stomachaches, cramps, indigestion, and backaches are often psychophysiological, although these conditions can also be caused by physical problems or a combination of factors. As many as one-fifth of the children taken to pediatricians, pediatric nurse practitioners, or family practitioners may be suffering from purely psychophysiologic disorders. When parent-child (or, more rarely, peer-child) relationships are improved, the disease fades. Drug treatment is at best superficial and usually wasteful. Unless the emotional disturbance can be alleviated, the disease will persist or one illness will be substituted for another.

Depression is estimated to occur in as much as 5 percent of the late-childhood population. Depressed children share the common denominator of feeling unloved or misunderstood by their families. Many are products of the indifferent-uninvolved pattern of parenting described in Chapter 5 or of child abuse or neglect. Others may be reacting to the loss of a parent through divorce, death, hospitalization, or institutionalization. Symptoms of childhood depression are not the same as for adults. Children more often mask their

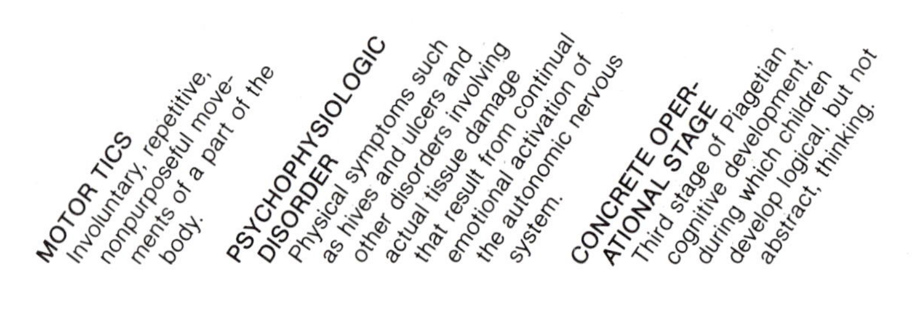

deep depressions behind temper tantrums, irritability, violent acts, running away from home, psychophysiologic symptoms, excessive weight gain or loss, accident-proneness, academic problems, or sexual escapades. In many cases, depressed children attempt suicide (Kosky, 1983). If a child is despondent enough to talk about or undertake suicide, protective hospitalization may be warranted. Actual suicides do occur in childhood. Parents, as well as the child, need psychiatric counseling and some environmental manipulation to prevent further attempts. Antidepressant medications do not appear to be effective in children and can have severe side effects (Nurcombe, 1985).

Cognitive and Language Development

Piaget's descriptions of cognition in late childhood will be described first in this section, followed by examination of the various processes involved in learning, language development, intelligence testing, the debate over how much of intelligence is inherited, factors involved in creativity, and problems of learning in late childhood.

Cognitive and language development move from the realm of parental concerns to the sphere of school interests during the years from six to twelve. An assortment of teachers help children learn increasingly complex cognitive skills and try to stimulate their drives for mastery and achievement. School becomes the center of children's extrafamilial lives, occupying about one-half of their waking hours Monday through Friday nine months of each year. As we shall see in the social-development section of this chapter, school plays an important role in instilling many attitudes not related to academics.

PIAGET'S CONCRETE OPERATIONS STAGE

Piaget postulated that the period of **concrete operations** begins somewhere between ages six and eight, depending on each particular child and his or her maturation, physical experiences, and social interactions. Some children may begin some concrete operations as early as age four.

As we discussed in Chapter 5, the preoperational child (approximately age two to seven) gradually acquires the ability to see a few relationships between things and to handle some simple classifications. Piaget felt such successes were more intuitive than reasoned, however. The period of concrete operations is differentiated from preoperations by the fact that children now learn to reason about what they see and do. In fact, psychologists and educators sometimes call the years between six and eight the age of reason because children do acquire the ability to think things through (see Figure 6–5).

During concrete operations children develop the ability to apply rules to the new things they see and hear. Such rules help them understand and classify new phenomena. The rules constantly undergo revisions and expansions. For example, a child in the concrete operations stage who sees a baby alone in a room will reason that a mother must be near because babies cannot take care of themselves. When a group of adults enter a room, the child will search out the most probable mother. If the mother is not present and a father attends the baby instead, the child will have to revise the rule to include the fact that adults other than mothers also take care of babies.

Figure 6–5
Children in Piaget's stage of concrete operations find games like chess challenging. They utilize their abilities to apply rules, comprehend ordering and classifications, reverse their mental calculations, and conserve equivalency.

Children who perform concrete operations can reverse their thoughts (called **reversibility**) and mentally imagine things as they were before any actions were taken. In the previous example the child who had expected to see a mother can think back to the moment the baby was discovered alone and realize that, although a mother was expected, what he or she actually knew was that some adult must be near.

The ability to consider reciprocal relations also develops during the concrete operations stage. **Reciprocity** refers to the corresponding or interchangeable action or relation that one person or thing has on another. Using the example of the baby in the room, the child knew that some adult must take care of the infant. Once the child identified the caregiving adult, knowledge of reciprocity would lead the child to understand that taking care of a baby requires changes in an adult. The inverse of the baby being fed, diapered, and carried is that the adult has learned how to nurture and protect an infant. Further observations of the infant and the adult will lead the child to other reciprocal associations—baby fusses, adult produces bottle; baby continues to fuss, adult rocks baby; baby quiets, adult talks to baby; baby coos and smiles, adult smiles and talks more to baby.

The ability to use and reverse mental operations makes possible an important concrete skill described by Piaget, that of **conservation**. Conservation involves the ability to understand that a quantity does not change simply because the form it takes varies. For example, an amount of water does not change when it is poured from a tall thin glass into a short fat glass. Likewise, an amount of clay does not change when it is rolled from a ball shape into a snake shape. In order to conserve, children must be able to look over the whole field of a problem. They must pay attention to more than one characteristic at a time. For instance, they must be able to perceive changes in width as well as changes in height. They also need to be able to reverse the operation mentally and visualize what the properties were like before the change in form occurred. Preoperational children find conservation difficult because they center on only a limited aspect of the task and do not reverse their thoughts.

Conservation of mass or quantity develops first. Conservation of weight, where children need to have a relative knowledge of the weights of different materials as well as attend to size, develops more slowly. For example, if a child were given a scale and asked to measure two equal amounts of playdough and two much smaller but equivalent-weighted amounts of lead, the first surprise would be that the lead is the same weight. However, a nonconserving child could be made to guess that one lump of playdough would be heavier than the other if its shape were changed. Likewise, it would confuse the child if one lump of playdough plus one-half of the other lump of playdough were compared to one lump of lead plus two-thirds of the other lump of lead. A nonconserver would attend to size and guess that the larger-appearing lumps of playdough would weigh more. By ages nine to twelve (depending on cognitive ability and experience), children will learn to attend to weight as well as size and know that the lead will weigh more. Conservation of volume seldom is achieved before age twelve. For this the child will have to realize that two equivalent amounts of material will both displace the same amount of water even though one is ball shaped and the other some other shape (such as flat as a pancake).

Another demarcation separating preoperational from concrete-operational children is the ability to comprehend **numbering**. Piaget explained that children become more adept at handling number correspondence and ordering problems during the concrete operations period because they use rules, reverse their mental calculations, and conserve. When preschoolers are shown a bouquet of flowers bunched together and an equal-sized bouquet spread out,

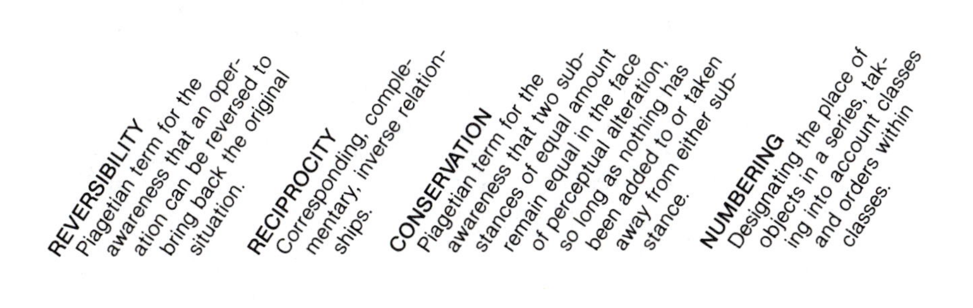

they will choose the separated bouquet as larger because it looks larger. Even when they count the flowers and determine that the two bouquets have an equal number of blossoms, they will again be deceived about size when one bouquet is spread out. In contrast, concrete-operational children handle problems of judging more, less, and the same by using a numerical count whenever possible. They will then conserve equivalency or nonequivalency no matter what form the rearrangement takes (see Figure 6–6).

Figure 6–6 *Which box has more flowers? Preschoolers will base their answer on their perceptions of* more *and* less. *Children in the concrete-operational stage of cognitive development will count the flowers before giving an answer.*

Piaget (1941/1965) viewed numbering as a synthesis of two other operations: ordering and classifying. If preoperational children are asked to place in order a group of articles that come in a range of sizes, they will order them haphazardly, often letting them represent some figure such as a house or train. Concrete-operational children develop the ability to order them from largest to smallest or vice versa. If new gradients are introduced, the concrete-operational children will insert them correctly into their series. In the concrete operations period children will also ascertain what objects belong to the same class, what objects differ, and how. They do this concurrently with ordering and ultimately with numbering. When presented with a box containing ten chocolates, five jelly beans, and five marbles and asked "Which is most—chocolates, candies, or toys?" a preoperational child will probably say that there are more chocolates than candies. They will not see that the chocolates are also in the class of candies. Concrete-operational children would not have this difficulty. They can handle many classifications and subclassifications while ordering a series of objects.

One of the hallmarks of children's ability to order and classify is their appreciation of where they are in relation to the rest of the universe (for example, Ten Hills section, City of Baltimore, State of Maryland, eastern seaboard, mid-Atlantic States, United States of America, northern hemisphere, continent of North America, planet Earth, solar system, Milky Way galaxy, cosmos). Their fascination with collections of various kinds (such as cards, rocks, shells, bottle caps, or stamps) also reflects this interest in classification.

Time conceptions (days, months, years) become increasingly well differentiated during this stage, but ideas of historical time and far future may remain vague until ages ten to twelve, depending on maturation, physical experiences, and social interactions. Many an adult has been horrified to learn that a school-age child may equate the days of the adult's youth with the days of the Civil War, for example, or fail to understand that some of the projections of science fiction might be possible by his or her own adult years. Space perceptions also develop gradually in the concrete operations period. Perception of self in space develops early. However, an understanding of geographical distances, such as those between states and countries, or of celestial distances, such as those between the earth, sun, moon, and stars, develops considerably more slowly.

Another notable achievement during the concrete operations period is an understanding of the differences between **physical** and **psychological causality**. Young children with their animistic and artificialistic ideas (see p. 222) listen for or invent psychological causes for events. For example, many preschoolers who ask about the flame in a cigarette lighter would not be satisfied with an explanation of flints and lighter fluid. They would feel better answered if it were explained that the flame shot up so you could light your cigarette. As children grow older, they want to understand physical causes, although an interest in and knowledge of physical causality may not entirely replace ideas of psychological causality. Older children (even adults) may understand and be able to explain the physical cause for an occurrence (for example, a higher flame on a cigarette lighter resulting from more oxygen intake) yet speak of a psychological cause for the same occurrence as well ("Aha, the flame is shooting up to get you!").

The concrete-operational abilities do not all appear at once. They develop gradually as children have more and more experience manipulating objects and discovering rules about sizes, shapes, weights, classes, volume, time, space, and causality. All the operations become more refined and sophisticated with age and experience until children finally bridge the gap between concrete operations and the formal operations that characterize the reasoning of most literate adolescents and adults (see Table 6–2).

THE LEARNING PROCESS

Children use various processes to assimilate and accommodate their experiences and newly acquired information into their range of

Table 6–2 Abilities found in the Concrete Operations Stage

Ability	Explanation
Classifying	Sorting a group of objects into related divisions (such as same size, same shape, same color)
Ordering	Placing a group of objects in succession (for example, tallest to shortest, smallest to largest, by rank)
Numbering	Designating the place of objects in a series, taking into account classes and orders within classes
Conservation	The ability to understand that a material system remains unchanged while internal changes of any kind occur
of quantity	The amount of a material is neither increased nor diminished by changing its form
of weight	The heaviness of a material is neither increased nor diminished by changing its form
of volume	The space occupied by a material is neither increased nor diminished by changing the form of parts of the material
Reversibility	The ability to effect a change and then go back to the original condition by a physical or mental reversal of the change
Reciprocity	The ability to understand corresponding complementary, inverse relationships (for example, A to B is the equivalent of B to A)
Seeking physical causality	Looking for physical causes for events rather than believing in fantasy-based psychological causes
Applying rules	Use of established guides and regulations for actions and conduct rather than meeting egocentric desires
Spatial awareness	Knowing where one is in geographical distances
Time consciousness	Knowing where one is in relation to past, present, and future

knowledge. An appreciation of how information is processed is central to a comprehension of cognitive development.

Some facts are stored only briefly (short-term memory), whereas others are retained over a longer period of time (long-term memory). The hippocampi, structures that are part of the limbic system of the brain, seem to play a vital role in pushing some information into a retention process rather than allowing it to dissipate. The major function of the limbic system is the regulation of emotions. It does not seem merely coincidental that new information that arouses a personal emotional reaction is more apt to be remembered longer than information that is emotionally neutral.

An awareness of memory processes by themselves is called **metamemory**. Young children are unlikely to think about trying to remember things for future retrieval. They have what Flavell (1970) has called a production deficiency. However, with increasing age, children develop strategies for remembering subject matter that they deem important. They learn to categorize incoming information in various ways: **encoding** stimuli with a word or series of words, with pictorial images, or with concepts. Encoding refers to the use of symbols or a code to represent something else. Some children develop more effective strategies for remembering than others. The more experience children have in finding ways to remember subjects, the more adept they become at fixing memories. However, the amount a given person can memorize at any one time is limited by the development and functioning of the brain cells. An average six-year-old can usually remember and repeat five digits presented verbally. By age ten it is usually possible to remember seven digits (a telephone number). Some adults may have immediate recall of ten digits. A six-year-old may remember and repeat two unrelated words presented verbally. A ten-year-old may remember three or four unrelated words. An adult may be able to handle up to six. Short-term memory of adults is generally longer lived than short-term memory of children.

The ability to recall and retrieve various objects and events from the memory is greatly enhanced by the comprehension and use of a large vocabulary. For example, an adult would be hard pressed to remember more than two or three words presented verbally in a foreign language. Or, if children were asked to draw ϕ ψ Ω from memory, they would have difficulty unless they had previously learned the Greek alphabet. If children know the Greek alphabet, they can use the names *phi, psi, omega* to mediate the image of the Greek symbols they see.

Mediation refers to a middle step between perceiving and remembering that aids memory. Verbal mediators, in the case of the Greek letters, would help children remember and respond with a correct drawing. This is much simpler than trying to remember the descriptions "circle with a vertical line through it, vertical line with a half circle open on top attached at midpoint," and so on. Mediation involves using skills already acquired to help develop new skills or acquire greater dexterity at old skills. Mediators are frequently verbal (that is, words we tell ourselves between taking in new information and responding to it). Reading increases school-age children's use of words as mediators (see Figure 6–7). Sometimes nonverbal mediators like pictures are used. The ability of deaf children to respond to new information quickly and accurately illustrates the use of some nonverbal mediators.

METAMEMORY Conscious or intuitive knowledge about memory.

ENCODING Categorizing incoming information.

MEDIATION Using skills already developed to help acquire new skills or greater dexterity at old skills.

Figure 6–7
Reading contributes to competency in all cognitive and learning skills, such as assimilating and accommodating new concepts, attending, encoding, mediating, remembering, and increasing the total range of knowledge. Children prefer books with humor, action, sustained suspense, and new information.

School-age children learn new information best when they are aroused to attention and have some definite verbal or physical examples with which to encode and mediate their perceptions. Abstractions are still difficult for the child in Piaget's stage of concrete operations. Invisible, intangible concepts such as the grace of God, infinity, and algebraic equations cannot be readily grasped during late childhood.

LANGUAGE

Growth in vocabulary and use of language continue throughout adulthood. Children not only gain larger vocabularies during their late childhood years, they also correct grammatical and pronunciation errors and learn subtle meanings of old words. Puns and figures of speech finally become meaningful. Inflections allow children to alter the meaning of what they say. Language becomes fun! School-age children usually enjoy the play of words and invent or adopt word games that allow them to show off and improve their speech proficiency. Jokes based on double word meanings, slang, colloquialisms, curse words, feigned accents, secret languages, and ciphered messages abound between ages six and twelve. Children also try to use words that they do not fully understand (such as radar, communist, light year) and they often misinterpret expressions according to what they believe the words should mean.

School-age children acquire new sophistication in their sentence constructions. Sentences may be compound or complex and may contain compound subjects, compound verbs, and relative clauses. Children's understanding of certain syntactical forms (arrangements of words) in sentences also improves. Noam Chomsky (1972), who postulated the language acquisition device (LAD) in humans (see Chapter 4), believed there are two levels of language:

surface and deep structures. For example, the sentences "Helen is easy to please" and "Helen is eager to please" have the same surface structure, but their deep structures, their underlying meanings, are very different. School-age children come to understand that Helen is the object of the first sentence (others try to please her), whereas she is the subject of the second and will try to please others. Younger children may have difficulty with the first sentence. They may be confused about what Helen is to do.

Some school-age children have problems with articulation, vocal quality, or the rhythm of their language. Common articulation problems are omissions of sounds (such as "at" for "hat"), substitutions ("dreth" for "dress"), distortions (such as /s/ in "sink" articulated like /z/), and additions ("go-a to the store-a"). Parents, teachers, and speech therapists can usually help the child correct the incorrect utterances as long as the cause of the disorder is not due to brain or nerve damage or oral-facial abnormalities (for example, cleft lip). Problems with vocal quality can be characterized as hypernasality (too much nasal emission during speech), hyponasality (too little nasal emission), or disorders of pitch (too high or too low), intensity (too loud or too soft), or flexibility (monotone). Children with such problems can usually be helped to find an acceptable voice in their repertoire and reinforce it until it is consistently used. Disorders of rhythm are either **stuttering** (the interruption of speech fluency through blocked, prolonged, or repeated words, syllables, or sounds) or cluttering (rapid, garbled, disorganized speech). Cluttering is often considered a form of stuttering. Stuttering is about four times more common in boys and usually begins before age eleven (Silver, 1984). Both organic and psychogenic factors are suspected causative factors. Gemelli (1982) stressed psychogenic factors such as ego functioning and parental patterns of behavior in cases of persistent stuttering. Many therapists feel that psychotherapy for both parents and child is a necessary adjunct to speech therapy for persistent stutterers.

THE CONCEPT OF INTELLIGENCE

Pause for a moment from your steady reading to answer each of these questions in your own words: What is intelligence? What is common sense? What is street sense? What is rote memorization ability? What is speed reading? What is creativity? What is logic? Are these terms related? Which concepts are most applicable to you?

Many scholars have struggled to define the concept of intelligence in a manner that would appeal to others. The dilemma continues. Spearman (1927) proposed that intelligence consists of a "g" factor, for general abilities related to deduction and logical analysis, and an "s" factor, for specific abilities such as spatial reasoning and arithmetic skills. Thurstone (1938) proposed that intelligence could be defined by seven primary abilities: verbal comprehension, word fluency, number, space, memory, perceptual speed, and reasoning. Guilford (1967) proposed 120 factors of intelligence related to the dimensions of mental operations, contents, and products. Cattell (1971) proposed that intelligence consists of fluid intelligence (for

STUTTERING
The interruption of speech fluency through blocked, prolonged, or repeated words, syllables, or sounds.

innate capacities) and crystallized intelligence (for what one has learned). Sternberg (1982) proposed four general components of intelligence: the ability to learn and profit from experience, the ability to think or reason abstractly, the ability to adapt to the vagaries of a changing and uncertain world, and the ability to motivate oneself to accomplish expeditiously the tasks one needs to accomplish.

Regardless of which definition of intelligence (if any) appeals to you, can you propose a way to test for the concepts defined? Many scholars today feel that the intelligence tests we now have are, at best, achievement tests. They only test what a child (or adult) has learned. Let us briefly review why some of them were developed.

Alfred Binet and Theophile Simon standardized the first widely accepted intelligence test in France in 1905 for the purpose of determining which children were normal and could attend public schools and which ones were retarded and could not. Binet shunned the notion of formulating levels of intelligence beyond normal and subnormal. The concept of an intelligence quotient (IQ), giving degrees of mental difference, was proposed by a German, William Stern, in 1911 (Wolf, 1973). Lewis Terman at Stanford University revised the Binet-Simon test for use in the United States in 1916, changed its name to the Stanford-Binet test, and used Stern's concept of **intelligence quotients**. An IQ is obtained by dividing tested mental age by chronological age and multiplying by 100. Thus if a six-year-old child answered questions on an intelligence test to the expected mental level of an eight-year-old, his or her IQ would be 133 ($MA/CA \times 100 = 8/6 \times 100$). If another six-year-old answered questions only at the expected mental level of a four-year-old, his or her IQ would be 66 ($4/6 \times 100$). The average IQ is 100; for example, a six-year-old answering questions up to and not beyond the level expected for a six-year-old would have an IQ of 100 ($6/6 \times 100$). Most IQs fall in a normal distribution range from 70 to 130 (see Figure 6–8). With a knowledge of where a child's tested IQ falls relative to a normal distribution of scores, one can determine whether a child belongs in a gifted class (IQ above 130), a special education class (IQ below 70), or a regular class (IQ between 70 and 130).

David Wechsler developed an alternative IQ test that scored verbal and performance abilities separately to get an idea of each unique individual's particular strengths and weaknesses. This approach allowed for special educational attention to be directed to areas of weakness. Both the Stanford-Binet and the Wechsler tests have to be administered to one person at a time by a trained tester and are therefore considered more trustworthy than IQ tests admin-

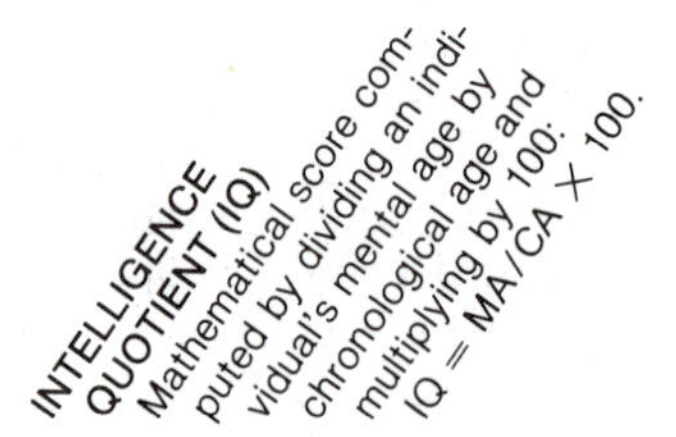

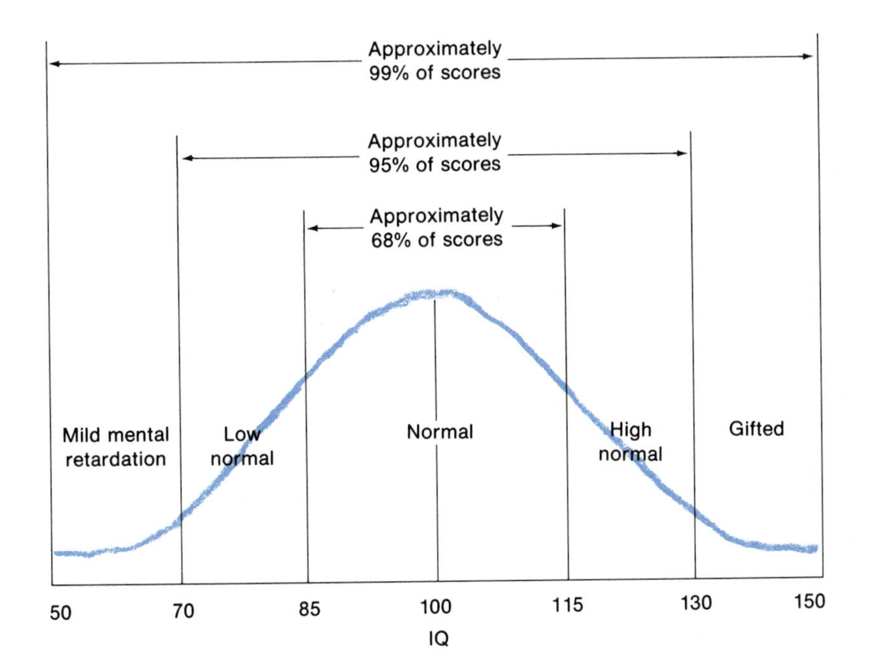

Figure 6–8
The normal frequency distribution of IQ scores.

istered to a whole class (or other group) by an untrained tester (such as a classroom teacher). However, group IQ tests are easier and far less expensive to administer and are, consequently, more prevalent. Today there are hundreds of IQ tests: individual tests, group tests, long forms, short forms, so-called culture-fair tests, even tests based on Piaget's studies in genetic epistemology. They have been developed to enable educators and others to classify students (or adults) by intelligence level and to place them in appropriate classes or jobs.

As the number, complexity, and uses of IQ tests have grown, so have the criticisms of them. Some advocates of IQ testing hail them as one of psychology's greatest achievements. More important, IQ examinations can gauge how well or poorly a person has learned academic materials in comparison with others of the same age group. In addition, IQ tests can also predict, to some degree, how well or poorly a person will continue to learn academic subjects in the future. Critics of IQ tests see them as one of psychology's most shameful bequests and often call for an abolition of their use (Scarr, 1978). In Washington, D.C., California, New York City, Philadelphia, and Minneapolis, IQ testing has been terminated.

One of the problems of IQ tests is that they can measure only a rather small set of the abilities known to be a part of intelligence. Different tests also give pictures of different factors of intelligence and, in effect, look at different "intelligences." No IQ test is known to measure the best (in theory) construct of intelligence (Carroll and Horn, 1981).

For a test to fairly gauge and predict the relative learning abilities of a group of people, the members of the group should all have had equal exposure to the kinds of information required on the test. One seldom finds so homogeneous a group. The testing of linguistic minorities as if they had had equal exposure not only to the test materials but also to the language of the tester is an especially sensitive issue. In 1981 there were an estimated 3.5 million U.S. school-age children with limited English proficiency (Olmedo, 1981). These included Hispanics, Asians and Pacific Islanders, Native Americans, Eskimos, and recent immigrants from many other parts of the world. Many of these children, after IQ testing, have been incorrectly labeled "learning disabled" or "mentally retarded."

There are numerous conventional IQ tests translated into minority group languages, including nearly one hundred tests in Spanish (Samuda, 1975). However, they are most often translated into the formal form of the language rather than into the dialects used by the populations to which they are given. Translated items of tests also frequently lose the equivalent meaning of the original English items. To be fair, children with limited English proficiency should be tested by examiners of their own ethnic and linguistic background or have a bilingual interpreter present. Even here problems arise. In which language should the directions be given first? There are many culture fair/culture free IQ tests available that are also used to test children who have not had an equal exposure to the kinds of information found on the standard IQ tests. Research evidence indicates, however, that these tests are neither culture fair nor culture free, and they are not considered viable alternatives to the standard tests. See Figure 6–9 for an example of cultural biases on standardized tests.

Test takers should all be relaxed during an examination so they can think clearly. They should be motivated to answer the questions correctly to the best of their ability. Test anxiety or lack of motivation to perform can severely mask real abilities. Terrell and Terrell (1983) suggested that for a test to fairly predict the relative learning abilities of black children with a high level of mistrust of whites, the test should be administered by a black examiner.

Children may have tremendous fluctuations in their IQ scores from test to test, as much as from 20 to 30 points, depending on the circumstances surrounding the test administration. For example, one child's tested IQ dropped 50 points after the death of his mother, and another's IQ score increased 30 points after psychotherapy relieved some anxieties. When such variations due to environmental conditions are possible, the accuracy of any one measurement is necessarily open to question.

The one thing on which test advocates and critics agree is that a child's success in learning school subjects can sometimes be predicted by IQ tests. However, the reasons for this are unclear. Rosenthal and Jacobson (1968) demonstrated how influential expectation can be on actual performance in their widely known study *Pygmalion in the Classroom.* After intelligence tests were given to all the children in a class, teachers were told that certain children with just average test scores were going to bloom and show unusual academic growth during the school year. At the end of the year the children were retested. Those who were singled out as bloomers

1. Which of these two objects go together?

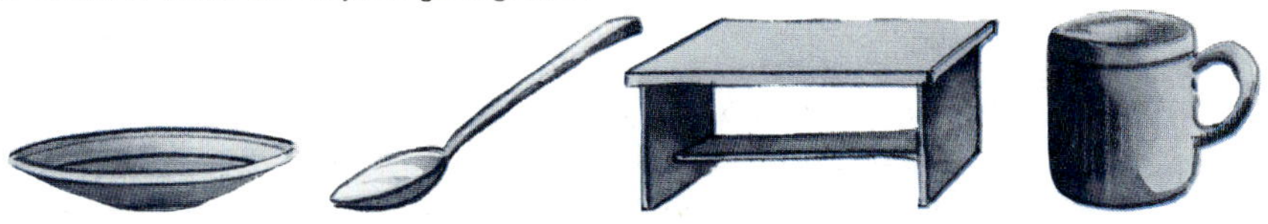

"Correct" answer = Cup and saucer

Culture-bound possibilities that would be judged "incorrect" =

- spoon (for example, for soup) from saucer/cup
- mix (as for chocolate) in cup
- place cup on table
- place spoon on table

2. Pick the object that does not belong.

"Correct" answer = Oven (other objects are living)

Culture-bound possibilities that would be judged "incorrect" =

- rose (other objects relate to foods)
- clam (other objects exist on land; clam resides in water)
- tree (other items can be found inside a house)

Figure 6–9
Samples of IQ test items that have culture-bound "correct" answers.

actually had improved and showed higher retest scores. The teachers' belief that they were going to bloom was held responsible for their higher scores. The treatment of a student according to some preconceived notion of how he or she will learn, which has the effect of eliciting the expected academic performance, has come to be known as the **Rosenthal effect**.

Contemporary IQ tests are far less effective as predictors of behaviors outside of schools (job performance, social intelligence, motivation, and so on). Some research suggests that the testing of

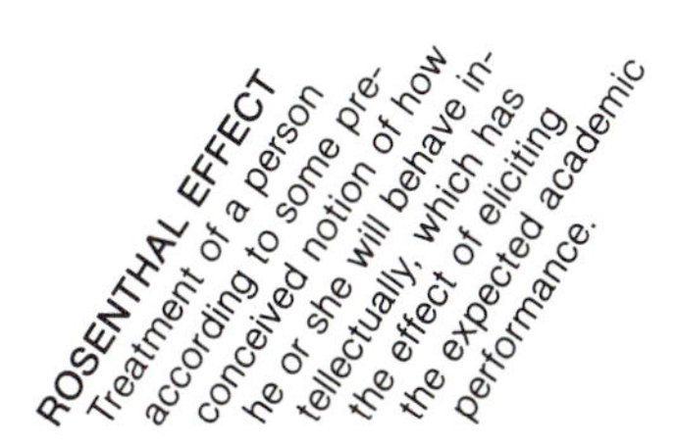

intelligence will be more accurate in years to come. Sternberg (1981) gave an optimistic view of the effects on testing practice of looking at automatic and controlled processes in cognition, time-sharing abilities, abilities to deal with novel tasks, practical intelligence, and executive processes. The ability to reliably test for these components of intelligence may be years away, however.

Arguments wax and wane about whether children inherit their intelligence from their parents (grandparents, distant relatives) or whether parents and teachers inspire and motivate them to learn well. Most people believe that both heredity and environment influence intellectual ability. The argument became heated in the early 1970s after Berkeley psychologist Arthur Jensen (1969) published a review of several studies of intelligence. He concluded that about 80 percent of intelligence is inherited. He suggested further that "genetic factors are strongly implicated in the average Negro-white intelligence difference" (p. 82). He placed this difference at about 15 IQ points. This belief became known as **Jensenism**. He testified before a U.S. Senate Committee on Education that with only about 20 percent of intellect influenced by environment, the vast sums of money spent on compensatory education, in his opinion, were wasteful. (Compensatory education programs, such as Head Start, were trying to provide intellectually stimulating environments to preschool children from deprived backgrounds before they began formal schooling.) Tempers flared, especially among educators trying to provide better educational opportunities for poor children. Jensen was accused of being a racist. Richard Herrnstein (1971) wrote in support of Jensen and added fuel to the fire of controversy. He suggested that since different people inherit different intellectual abilities, we should educate people for jobs according to their merit, or practice a system of **meritocracy**. A Nobel Prize-winning physicist, William Shockley (1972), further suggested that persons with inherited low IQs might be monetarily rewarded for being sterilized to halt the production of more low-IQ children.

Counterarguments were abundant. Supporters of Head Start and other compensatory education programs emphasized that children given enriched environmental stimulation through early schooling could enhance the extent to which their genetic potential was realized. Others questioned whether the impoverished and suppressive environments of these children were not more responsible for failing to evoke and nurture their IQs. In France in 1982, Schiff and his colleagues showed that infants of unskilled workers adopted by families from the top of the socioprofessional scale showed an in-

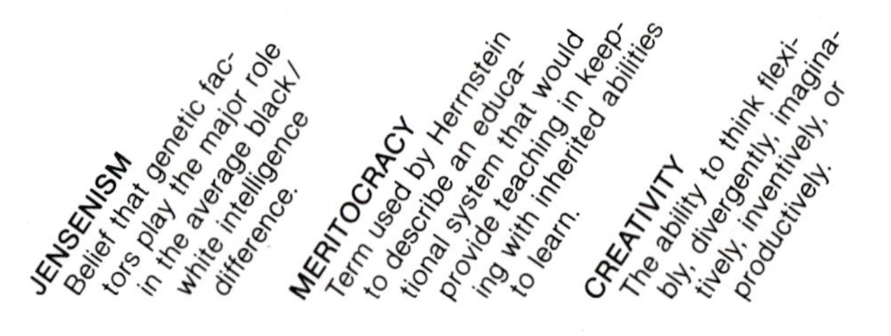

crease in their IQ scores over time when compared to their siblings who remained in their natural homes.

In 1980 Jensen published a book defending his thesis. He put less emphasis on the data of the British psychologist Sir Cyril Burt, which he had cited in his 1969 article, because Burt's research has been revealed to be fraudulent (Hearnshaw, 1979). In his 1980 writing, Jensen put forth a massive technical analysis of the merits and validity of those IQ tests that have shown inherited intellectual differences.

No valid answer can really be provided to the question of how much heredity and environment affect intelligence until the concept of intelligence itself is clearly defined and until some way is found to both validly and reliably measure the agreed upon components of intelligence.

CREATIVITY

School-age children display varying degrees of creativity that are not at all synonymous with academic ability. **Creativity** is variously defined as the ability to think flexibly, divergently, imaginatively, inventively, or productively. Tests to measure creativity have been developed, but their validity and merits have not been widely accepted. Is it good to be creative? A paradox exists in contemporary schools about creativity. Although its merits are extolled, its development is frequently inhibited. Examinations, for example, usually have only one right answer. Any novel, unconventional ideas, even though practical and correct in their own way, as are those answers to the IQ questions in Figure 6–9, are penalized. Children who have been fond of freestyle drawing, sculpturing, and building in their preschool years may be encouraged to make school products look like the models in workbooks or those that their teachers have made. Creative children can become embarrassed or ashamed of work that differs from the norm. They learn to inhibit their originality and their urges to strike out on their own. A child's attendance at school five days a week learning conventional school subjects also reduces the amount of time he or she has for creative enterprises. A young Mozart would be hard pressed to write two operas and a large number of arias, serenades, symphonies, masses, and divertimenti during late childhood today.

Creative development is fostered by adults who are tolerant of unconventional products or responses, who encourage free use of materials and free flow of conversation, who accept children's own best efforts, and who allow children to proceed on tasks without a lot of specific directions (see Figure 6–10). Special rooms are needed where messy projects are permitted and time is unlimited. Such places are not conducive to maintaining clean, quiet, well-ordered households or to teaching large classrooms of children certain required subjects in short school terms. Teachers who are firmly committed to developing creativity in their students might also find their efforts inhibited by administrators or parents who value conventional academic performance more highly than originality and imagination.

Figure 6–10 *Creative children are able to express their experiences or dreams and fantasies in original, unique ways. They use materials and ideas freely rather than repeating tried and true procedures. Adults can foster creativity by encouraging and praising new productions and refraining from giving specific "how-to-do-it" directions.*

PROBLEMS OF LEARNING

Attempts to classify children with learning disabilities have been hampered by the controversies surrounding IQ testing. Minority children (particularly Blacks, Hispanics, and Native Americans) have been overassigned to special-education classes due to test scores, as, to a lesser extent, have been other children from poverty backgrounds and boys in general (Reschly, 1981). An additional problem in classifying learning disabilities centers on labels. Terms such as *dyslexia* (difficulty in reading) and *hyperactivity* have been overused and are being replaced with more precise labels such as **specific reading disorder** (for dyslexia) and **attentional deficit disorder** (for hyperactivity). The term *borderline mental retardation* has also been dropped. Children must now score two standard deviations below the mean on a standardized intelligence test and show significant impairment of adaptive skills to be labeled even mildly retarded.

Specific Developmental Disorders **Learning disabilities** not involving mental retardation or the symptoms of attentional deficit disorder are now labeled according to the specific disability observed:

> Specific reading disorder (basic reading skills and reading comprehension)
> Specific arithmetical disorder (mathematics calculation and mathematics reasoning)
> Developmental language disorder (oral expression and listening comprehension)
> Developmental articulation disorder (oral expression)
> Coordination disorder (written expression; other fine or gross motor skills)

Many learning disabled (LD) children do not fit even into these neater, more well-defined categories. There are many variations of reading problems or speech problems, for example. Some children have developmental disparities in two or more areas or concurrently have attentional deficit disorders. A learning-disabled child has an imbalance with one (or two) delayed or impaired skill(s) (for example, language) while other perceptual, cognitive, or motor abilities are achieved normally. Learning-disabled children can benefit greatly from special-education instruction in the area of their delay or impairment. In some cases children "outgrow" their disability by, or during, adolescence. For example, children who have persistent problems perceiving letter or word reversals in reading and writing (see Figure 6–11) may gradually "see" the differences in directionality.

There are many hypotheses about the causes of learning disabilities. One is that prenatal abnormal hormone secretions may have permanent effects on brain structure and function (McEwen, 1983; Marx, 1983). In males, excess prenatal androgens appear to cause the right and left cerebral hemispheres to develop asymmetrically. Their differential growths, interacting with other hormone levels, may explain why so many more boys than girls show learning disabilities and anomalous dominance patterns. Other suspect factors include neuroanatomical disorders (Hynd and Hynd, 1984), starting school too early (Ames, 1983), childhood depression (Colbert et al., 1982), genetic inheritance (Moser, 1983), malnutrition during pregnancy (Simopoulos, 1983), prenatal maternal infections (Sever, 1983), prenatal drug use (Gray and Yaffe, 1983), obstetric

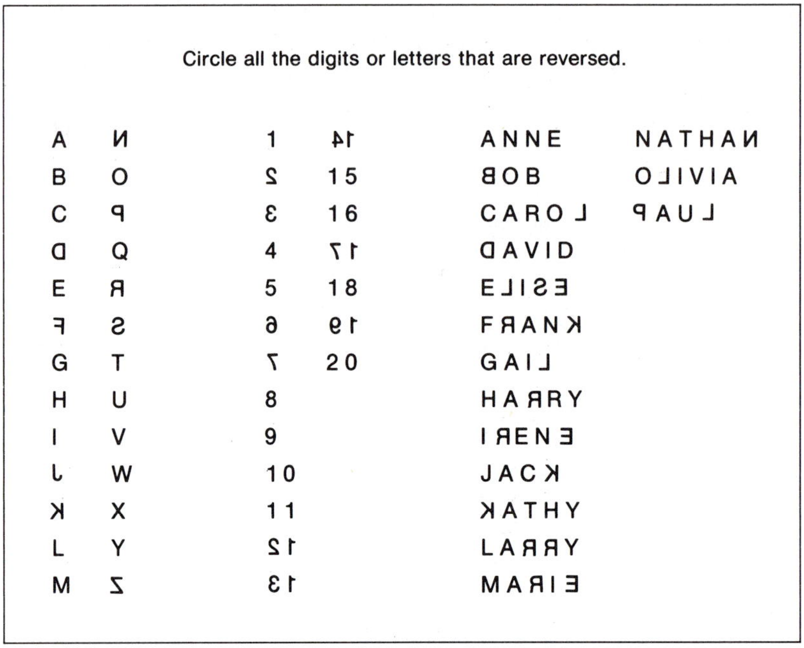

Circle all the digits or letters that are reversed.

A	И	1	ᔭɿ	ANNE	NATHAИ
B	O	Ƨ	15	ᗺOB	O⅃IVIA
C	ꟼ	Ɛ	16	CARO⅃	ꟼAU⅃
ᗡ	Q	4	٢ɿ	ᗡAVID	
E	Я	5	18	E⅃IƧƎ	
ꟻ	Ƨ	ð	eɿ	FЯANꓘ	
G	T	٢	20	GAI⅃	
H	U	8		HAЯRY	
I	V	9		IЯENƎ	
⅃	W	10		JACꓘ	
ꓘ	X	11		ꓘATHY	
L	Y	Ƨɿ		LAЯЯY	
M	ᘔ	Ɛɿ		MAЯIƎ	

Figure 6–11
A child with a specific reading disability (dyslexia) will have difficulty perceiving the errors on this list.

medications (Broman, 1983), obstetric trauma (Creevy, 1983), low birthweight (Cohen, 1983), socioeconomic factors (Robbins, 1983), and environmental pollutants (Needleman, 1983). In most cases of LD the exact cause cannot be determined with current diagnostic tools. The incidence of LD is unknown, due to problems with diagnosis and classification. Silver (1983) estimated that approximately 3–7 percent of the school-age population is affected.

Attention Deficit Disorders The new classification of attentional deficit disorder (ADD) for what in the past had variously been called *hyperactivity, hyperkinesis, minimal brain injury,* and *minimal brain dysfunction* reflects current knowledge that a high level of physical activity and neurological impairment are not universals for all affected children. Although ADD children may have some sort of brain dysfunction, its nature cannot be precisely defined or located. Because some affected children show *hypoactivity* (are underactive), this classification now has two subcategories: attentional deficit disorder with hyperactivity, and attentional deficit disorder without hyperactivity.

The problem in ADD revolves more around the quality of the child's activities than the amount. Most typically, affected children have short attention spans. This inability to focus on a task leads to many irrelevant motor behaviors that are especially noticeable when other children are task involved (as in school classrooms). In the "silent" form of ADD (without hyperactivity) the affected child may sit rather quietly through a task, unable to concentrate, and consequently learn little. In the more common hyperactive form the child may have temper outbursts, rapid mood changes, frequent crying, or explosive and unpredictable behaviors, and may be demanding, easily frustrated, aggressive, restless, and fidgety.

The causes of ADD, like the causes of LD, may lie in genetic inheritance, prenatal factors, postnatal factors (premature birth, birth trauma, anoxia at birth), factors related to an early childhood accident (a blow to the head, ingested poison), factors related to an early childhood disease (encephalitis, meningitis), some disorder in neurotransmitter metabolism in the brain, or dietary factors. Hyperactive behaviors can also be caused by some emotional disturbances and have no relation to ADD. The relationship between ADD children and their parents may also become emotionally taut due to the ADD problem. The afflicted child is difficult to handle. Parents, especially if they do not know the diagnosis, may feel frustrated and guilty about the supposed inadequacies of their child-rearing procedures.

Two classes of drugs are used to treat ADD: stimulants and antidepressants, both of which "quiet" affected children and allow them to attend to tasks and control their motor behaviors. Both groups of drugs work by affecting neurotransmitters (chemicals that transmit signals across the synaptic gap between nerve cell processes). In some communities physicians have overlabeled and overdrugged active, difficult children who are actually free of ADD symptoms. This dangerous practice and its exposure by the mass media with resultant lawsuits has recently made many health professionals wary of any quick diagnoses and drug prescriptions for ADD.

Many other therapies are currently popular as well: hypoglycemic (low-sugar) diets, megavitamin therapy, the Feingold diet (natural foods without artificial colors, flavors, or other additives), use of strong black coffee, and "patterning" exercises of muscles. The overall effectiveness of these later therapies is questionable, although they have been successful in alleviating ADD in some children and have made it possible for them to attend regular school classes without taking drugs.

Mental Retardation The American Association on Mental Deficiency's (AAMD) definition of **mental retardation** has three requirements (Grossman, 1977):

1. Significantly subaverage general intellectual functioning.
2. deficits in adaptive behavior.
3. symptoms manifest in the developmental period.

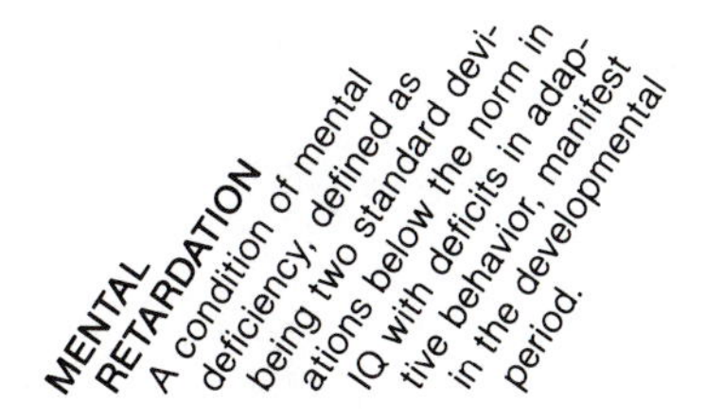

Significantly subaverage IQ means two standard deviations below the mean or an IQ score lower than that obtained by 97 to 98 percent of children the same age. Many children formerly labeled retarded are now considered low normal IQ, or slow learners. *Mildly retarded* children are considered "educable" to the extent that in special classes they can learn some elementary school subjects, adjust socially to a point of future independent living, and learn some occupational skills to allow future partial or total self-support. *Moderately retarded* children are considered "trainable." They can be expected, after special education, to learn some self-help skills, some social adjustment within a family or neighborhood setting, and some occupational skill that will be useful in a residential institution or sheltered workshop. *Severely* and *profoundly retarded* children are not able to learn many self-care skills, or social or occupational skills, and consequently need to be supervised and cared for throughout their lives.

There are several known causative factors of mental retardation. In many cases, however, a causative factor cannot be found. Known etiologies include prenatal causes such as biochemical disorders, chromosomal abnormalities, prenatal disruption of the developing brain due to diseases of the mother, radiation, malnutrition, maternal use of some medicines or drugs, and maternal-fetus blood incompatibilities; perinatal causes such as premature birth, birth injuries, and birth anoxia; and postnatal causes such as meningitis, encephalitis, and nutritional deficiencies. In a Swedish study of the causes of mental retardation, Blomquist, Gustavson, and Holmgren (1981) found that the etiology could not be traced in 45 percent of cases. Prenatal causes were considered relevant in 43 percent, perinatal in 7 percent, and postnatal in 5 percent of the other children analyzed. Similar results would probably be found in North America, with unknown etiology and prenatal factors accounting for the majority of the cases of mental retardation. There are no cures for mental retardation, only supportive services for the training and care of these special children.

Giftedness Children with significantly above-average IQ scores (two standard deviations above the mean) are considered intellectually **gifted**. They make up about 2 to 3 percent of the population. Although giftedness may seem out of place in a discussion of problems of learning, many such children become bored and frustrated in regular school classrooms and fail to be educated to their learning

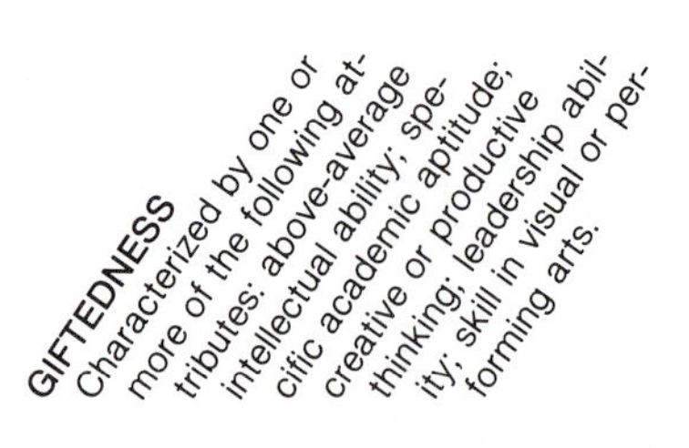

potential. A definition of giftedness at the federal level (Marland, 1972) includes five areas in which the children can be gifted:

1. General intellectual ability.
2. Specific academic aptitude.
3. Creative or productive thinking.
4. Leadership ability.
5. Visual and performing arts.

Marland (1972) reported to Congress that talented children in the United States were a neglected national resource and suggested that their identification and nurture should be a national priority. He suggested that methods in addition to IQ testing should be used to find these children. Gifted children with limited English proficiency or from cultures that put them at a disadvantage on standardized IQ tests (among them blacks, Hispanics, Native Americans, and Asians) are often underrated when judged only by a single intelligence test. In addition some parents in poor areas regard it as a stigma to have an intellectually gifted child and try to make the child appear less brainy. Identification is further hampered by the tendency of many child prodigies to learn with zeal from respected adults yet become frustrated, restless, and resistant to any motivational techniques of other adults (often teachers) for whom they have little respect. The challenge in educating intellectually gifted children is to find a good fit between child and teacher and between the child's needs and the academic offering (Vail, 1979). Although gifted children have been found to be superior in physical and health characteristics, they are also frequently socially uncomfortable and extremely sensitive. Many need help to achieve a balance in their social and emotional as well as intellectual lives (see Box 6–1). Many states and cities now appropriate money for special educational programs for gifted students—in groupings with regular classes, in special classes, in special schools, or with advanced curricula. Others meet the challenge by accelerating the child through school. Then they must adapt to being with children older and larger than themselves, who may resent the little "egghead" in their midst. While many accelerated gifted students are lonely, Ludwig and Cullinan (1984) reported that few have behavioral repercussions. They seem to have the ability to accept the jealousy and resentment of others for what it is and phase it out. The failure to identify or stimulate highly intelligent, creative, or talented students to their potential is a problem of learning yet to be solved by many schools.

Psychosocial Development

The American family has changed profoundly in the last few years. Approximately 60 percent of the mothers of school-age children work outside the home each day. Child-rearing responsibilities are being shared by neighbors, extended family members, and increasingly by fathers who arrange flexible work schedules to allow them to spend more time at home with their children. Persons in the fam-

BOX 6–1

A Gifted Eleven-Year-Old Boy's Attempt to Explain His "Education"

They laughed at me.
They laughed at me and called me names,
They wouldn't let me join their games.
I couldn't understand.
I spent most playtimes on my own,
Everywhere I was alone,
I couldn't understand.

Teachers told me I was rude,
Bumptious, over-bearing, shrewd,
Some of the things they said were crude.
I couldn't understand.
And so I built myself a wall,
Strong and solid, ten foot tall,
With bricks you couldn't see at all,
So *I* could understand.

And then came Sir,
A jovial, beaming, kindly man,
Saw through my wall and took my hand,
And the bricks came tumbling down,
For *he* could understand.

And now I laugh with them,
Not in any unkind way,
For they have yet to face their day
And the lessons I have learned.
For eagles soar above all birds,
And scavengers need to hunt in herds,
But the lion walks alone,
And now I understand.

SOURCE: "The Wall," from *The World of the Gifted Child,* by P. L. Vail. Copyright © 1980 by Penguin Books. Reprinted by permission.

ily's social network, the school, and the children's peer group join the parents as major socialization influences. How potent the school and the peer group can become depends in part on how strict or relaxed the family members are in exerting their own forces on shaping behaviors and on the stoutness of the family's affectionate bond. Children from ages six to twelve still identify with and model their behavior after the people they perceive as most nurturant and powerful. Strong family ties relegate peers and school to a position of lesser importance, whereas weak family ties escalate the influence of significant others in the community.

ERIKSON'S CONCEPT OF INDUSTRY VERSUS INFERIORITY

Erik Erikson's fourth age in his concept of "The Eight Ages of Man" is the nuclear conflict of **industry versus inferiority**. True to his Freudian background, Erikson saw the middle to late childhood period as one characterized by a latent interest in sex. Oral, anal, and genital concerns are supposedly sublimated. As the child becomes master of many of the concerns of early childhood (autonomy, walking, running, talking, toilet training, initiative, sex-role identification), he or she moves beyond the womb of the family. As Erikson (1963) put it, "The inner stage seems all set for 'entrance into life,' except that life must first be school life, whether school is field or jungle or classroom" (p. 258). The nuclear conflict between developing a sense of industry and acquiring an uncomfortable sense of inferiority remains paramount throughout the elementary school years.

School, with its objectives, methods of instruction, discipline, and system of rewards and punishments, is an influential force in shaping or negating a sense of industry. In school, children concentrate on the important but impersonal tools of the adult world: reading, writing, arithmetic, science, and social studies. They must apply themselves to tasks and persist in the work involved until some satisfactory completion point is reached. This ability to persevere is fostered by increasing cognitive maturity, by social reinforcements, and by children's own drives for industry. The ability to bring a productive situation to completion carries with it a sense of pleasure and pride in accomplishment. The more experiences children have winning recognition from others and feeling inner pride, the more anxious they become to finish projects and produce or accomplish things. The mere exploration of or experimentation with materials as practiced in early childhood becomes less rewarding.

Schools foster industry by requiring that tasks be finished (completing a reader; writing sentences, paragraphs, compositions; learning to spell; learning arithmetic computations). They often teach children to work together toward some stated goal (the division of labor principle). Two or three children may be assigned a joint project, or the whole class may work together on some undertaking that will then be displayed to other classes or to parents. In some foreign countries children actually begin working together during their elementary school years to produce articles for sale to the outside community. The recognition that comes to a whole group for their production serves to bolster the sense of industry of each participant.

Although the school has the potential to foster a sense of industry in each pupil, this possibility is not always realized. Standards used by various schools or teachers to measure adequacy, such as tests, praise for classroom performance, and report cards, can leave some children well recognized and others reproached for mediocrity or inadequacy. When children fail to win recognition for their efforts to accomplish things or for the products they make, they are in danger of developing a sense of inferiority.

Erikson (1963) wrote that too many experiences of being made to feel inadequate and inferior in the early years of elementary school may cause a child to revert back to the more isolated, less industry-conscious stage of the Oedipal and Electra attachments. He

INDUSTRY VERSUS INFERIORITY According to Eriksonian theory, the fourth nuclear conflict of personality development, which occurs during middle childhood. Children must learn the skills of their culture or face feelings of inferiority.

warned that families have a responsibility to prepare children for the realities of school life with its competitiveness and its emphasis on accomplishments. Likewise, the school has a responsibility to sustain the promises of worth and respectability made by families to children. Each child should be helped to feel industrious and adequate no matter how his or her skills compare with others. Children who are continually made to feel inferior during their elementary school years may be doomed to feeling mediocre or inadequate throughout their lives. Just as babies who fail to develop a sense of trust in infancy may have lifelong difficulties in trusting others, children who fail to develop a sense of industry in the elementary school years may have lifelong underlying weaknesses when it comes to taking their productive places in the world.

The family is an important source of feedback for children to assure (and reassure) them that their industry is both recognized and approved. Children who have learned the pleasure of work completion in school will try to win approval at home through the demonstration of various achievements and productions. It is important that parents give children this needed approval. It is also important that children be given projects or work to do at home to help foster their growing sense of industry (see Figure 6–12). Although boys and girls may bemoan the addition of chores to their daily activities, such household tasks provide a sense of contributing to the work of the family unit. They also give children a sense of pride in accomplishment when their completion is recognized, approved, or rewarded with praise or some other tokens of appreciation. Favoritism is one way in which families detract from the child's sense of industry and contribute to his or her sense of inferiority. Sensitive parents will be careful not to continuously praise or reward one sibling over another. They will avoid the use of comparisons to motivate behaviors. Instead, they will help all their children feel adequate with their own special skills and accomplishments.

Figure 6–12
Daily chores can help children develop a sense of industry, as described by Erikson, if family members approve their efforts. Chores can also help children feel that they are contributing to the work of the family unit.

There are many activities in which elementary-school-age children can participate outside of the home and school that serve to enhance their sense of industry. Many organized clubs offer individual recognition in such forms as uniforms, badges, pins, award ribbons, trophies, or even monetary prizes. Depending on age, children can join the Cub Scouts or Brownies, Boys or Girls Club, 4-H, Boy or Girl Scouts, Demolay or Rainbow, swim teams, other sports teams, drama groups, dance or music groups, religious groups, and many other organizations where they can work together with others learning new skills of the adult world. Erikson (1963) pointed out that it is by no means always in schools with special teachers that children receive systematic instruction. Many adults teach their specialized skills to children simply by dint of gift and inclination. Often the most effective teachers are older siblings, neighbors, and friends, both within and outside of organized groups.

Computer activities and games provide still another avenue for developing a sense of industry in childhood. In individual competitions a child often has the prerogative of choosing a rival of equal or similar abilities so winning happens in a balanced ratio with losing.

In discussing inferiority Erikson (1963) warned of the danger that threatens individuals and society when the schoolchild is made to feel that his or her worth is related to skin color, parental background, or fashionable clothes. Parents, teachers, neighbors, or club leaders can be guilty of conscious or unconscious prejudicial judgments about some children's accomplishments. Such adult prejudices often make children feel inferior. And, unfortunately, at this time in children's lives self-concept is usually not well defined. During the elementary school years children are apt to believe adult evaluations rather than recognize prejudice for what it is. It is important for parents and friends to reassure children of their worth if and when the children report the prejudicial comments of others. Although some experiences of feeling inadequate are bound to occur during late childhood, a sense of industry can be fostered by sensitive adults who counterbalance their criticisms with constructive suggestions, recognition of completed work, and praise (see Table 6–3).

Table 6–3 Erikson's Fourth Nuclear Conflict: Industry Versus Inferiority

Sense	Eriksonian descriptions	Fostering adult behaviors
Industry	Application of self to given skills and tasks	Give systematic instruction in skills and tasks
	Effort to bring productive situation to completion	Give recognition of things produced
	Attention and perseverence at work with pleasure obtained from effort	Specialized adults and older children outside family also instruct and recognize progress
	Work beside and with others brings sense of need to divide labor	
versus		
Inferiority	Disappointment in own tool use and skills	Family failed to prepare child for life in school and with other adults and unrelated children
	Sense of inadequacy among tool partners	School failed to sustain promises of earlier stages
	Lost hope of association in industrial society	Adult makes child feel that external factors determine worth rather than wish and will to learn
	Sense of being mediocre	

Moral Development The gradual growth of moral and ethical standards is influenced by identification with and modeling of parents, teachers, siblings, and other significant persons; cognitive growth; and the predominant kind of discipline used to correct a child's unacceptable behaviors (power assertion, love withdrawal, induction; see Chapter 5). The goal of **morality training** is to produce a person whose conscience will direct his or her behaviors toward the good and away from the bad without continual external reminders.

Freud (1935/1953) believed that children learn early moral standards through two kinds of identification with their caregivers: by anaclitic identification (based on a fear of losing love) and by aggressive identification (based on a fear of punishment by the aggressor or authority figures). He believed anaclitic identification is used more by girls in identifying with their mothers, whereas aggressive identification is used more by boys in identifying with their fathers. Identification of girls with their fathers and boys with their mothers can be either anaclitic or aggressive. Freud saw these identifications as a basis for children's knowledge of right and wrong. The fear of losing love and the fear of punishment are the forces that help children avoid unacceptable behaviors.

Research has amply demonstrated that principles of stimulus-response learning apply to the development of moral behaviors. When a response to a given behavior is attention, approval, praise, or some material reward, that behavior will be repeated. When the response is negligence or punishment, the behavior is less apt to be repeated.

Social-learning theorists such as Bandura and Walters (1963) emphasize the fact that children model the behavior of adults with whom they most strongly identify. Children's modeling takes the form of striving to incorporate everything about the loved or powerful adults into their own lives, including moral standards and values. Learning theorists claim that children receive a sense of security and approval when they emulate caregivers' behavior. This modeling of behavior then becomes a need because of the emotional rewards it brings to the child.

Although identification with and modeling of the loved caregivers within the family is felt to be the *sine qua non* for beginning moral development, children also learn moral rules by modeling the behavior of friends, neighbors, teachers, and television characters, especially during late childhood.

Piaget (1932/1965) distinguished two moral levels in children. The earliest is a **morality of constraint** that puts wrongdoing in terms of damage done and emphasizes submission to authority. The second moral level is a **morality of cooperation** that judges wrongdoing by the intent of the doer and establishes rules by mutual agreement.

In the period of morality of constraint, or moral realism, children feel an obligation to comply with rules because they see them as sacred and unalterable. They measure the degree of wrongness of an act by the amount of damage done, without consideration of motivation. Here are two sample stories that Piaget (1932/1965) used to question children to help him determine their level of morality. Which of the two children is naughtiest?

> A. Alfred meets a little friend of his who is very poor. This friend tells him that he has had no dinner that day because there was nothing to eat in his home. Then Alfred goes into a baker's shop, and as he has no money he waits until the baker's back is turned and steals a roll. Then he runs out and gives the roll to his friend.
>
> B. Henriette goes into a shop. She sees a pretty piece of ribbon on a table and thinks to herself that it would look nice on her dress. So while the shop lady's back is turned she steals the ribbon and runs away at once. (p. 123)

A child in Piaget's stage of morality of constraint would judge the first offense as more serious because the roll was bigger and more expensive than the ribbon. In effect, the child would ignore the intent of the transgressor. The sin would simply be judged by its magnitude. In this case the worse crime would be stealing the larger, more expensive item. At this stage children also believe that punishment wipes away the sin. The more severe the punishment is, the fairer it is. During this stage they see punishment as inherent in the external world, a view that Piaget called a ***belief in immanent justice***. The following question asked by Piaget (1932/1965) illustrates a belief in immanent justice:

> PIAGET: In a class of very little children the teacher had forbidden them to sharpen their pencils themselves. Once, when the teacher had her back turned, a little boy took the knife and was going to sharpen his pencil. But he cut his finger. If the teacher had allowed him to sharpen his pencil, would he have cut himself just the same?
>
> SIX-YEAR-OLD: He cut himself because it was forbidden to touch the knife.
>
> PIAGET: And if he had not been forbidden, would he also have cut himself?
>
> SIX-YEAR-OLD: No, because the mistress would have allowed it. (p. 252)

Children in this level of morality believe that all misdeeds are ultimately punished. If adults fail to punish, nature somehow intercedes. It is common for children to believe that accidents, bad dreams, or illnesses are the just punishments for their sins of the day.

In late childhood a morality of cooperation, or moral relativism, replaces the morality of constraint. As children begin to help make the rules for their games, they see them as less absolute and immutable. Children begin to see others' points of view and realize how different motivations underlie different actions. Justice comes to be viewed in a social context and in terms of equity and equality.

Kohlberg (1969) expanded Piaget's ideas of moral development, postulating six stages rather than two. Brief summaries of his stages are presented in Chapter 2, page 60, and in Chapter 7, page 369. Kohlberg saw the first two stages, which characterize early childhood, as **premorality**, where hedonistic, self-serving urges are paramount. He labeled stages 3 and 4 as **conventional morality**, where the social context is paramount. School-age children are concerned with maintaining the good relations and approval of others (stage 3) and grow cognitively aware of the need to show respect for authority and maintain the social order for the sake of having order (stage 4). Some children will reach what Kohlberg called postconventional morality by late childhood. Here the conscience is paramount. Laws may be altered or disobeyed under extenuating circumstances. These last two stages of postconventional morality will be further discussed in Chapter 7. Although a few cognitively superior children (with excellent teachers of highest-level moral development to learn from, identify with, and model) may reach postconventional stages, it is more common to find adolescents and adults who are still functioning at the stages of conventional morality.

Kohlberg asserted that Piaget placed too much emphasis on the child's respect for rules and authority in the morality of constraint. He said young children's obedience was instead a recognition that the parents are more powerful. He emphasized both cognitive growth and social experience as requirements for moving to conventional moral stages. Kohlberg stated that children advance in sequence from one stage to another, rather than by leapfrogging any of them. However, some older children may be morally behind due to lack of experience or cognition, whereas younger children may advance to higher levels than might be expected for their ages. Research by Turiel (1966) supported Kohlberg's theory. Turiel gave his subjects stories like the one below and asked them to choose an ending and explain the reasons for their choices:

> In Europe, a woman was near death from a special kind of cancer. There was one drug that the doctors thought might save her. It was a form of radium that a druggist in the same town had recently discov-

> ered. The drug was expensive to make, but the druggist was charging ten times what the drug cost him to make. He paid $200 for the radium and charged $2000 for a small dose of the drug. The sick woman's husband, Heinz, went to everyone he knew to borrow money, but he could only get together about $1000, which is half of what it cost. He told the druggist that his wife was dying and asked him to sell cheaper or let him pay later. But the druggist said, "No, I discovered the drug and I'm going to make money from it." So Heinz got desperate and broke into the man's store to steal the drug for his wife. Should the husband have done that? (Kohlberg, 1963, pp. 18–19)

Answers such as "You really shouldn't steal the drug" or "The druggist should get some profit from his business" are conventional role-conforming answers, stages 3 and 4. A stage 6 answer would reflect the acceptability of changing a conventional rule when unusual circumstances make it desirable.

In Piaget's and Kohlberg's views, moral development can be influenced by identification and modeling, but it is mainly tied to cognitive development. To reach higher moral levels children need many experiences with choosing right and wrong and opportunities to reason out the whys and wherefores of the choices.

How then do parents, teachers, and other adults help provide for optimal moral development? In order to develop a mature moral orientation and the ability to guide their own behavior toward the positive and away from the negative, children need (1) to know right from wrong, (2) to be able to control their own urges to do wrong, and (3) to consider the rights and needs of others before acting. By middle to late childhood children can be expected to understand most of the things that family and society define as unacceptable, but they cannot always be expected to have control over their urges toward selfish or aggressive acts. Nor can they or anyone be expected always to consider the rights and needs of others. Adults can help children learn both self-control and consideration of others.

In helping school-age children learn to control their own behaviors, adults should be mindful of the child's own needs for industry, approval, and respect. An involved, happy child is more apt to be in control of his or her behaviors than a frustrated, unhappy, or bored child. Adults can help school-age children find challenging work and play activities that are neither too simple nor too difficult. They can remain close by to supervise children's activities and be ready to suggest alternatives if frustration or boredom seems imminent. When fighting occurs, adults can let children share in the process of deciding right from wrong and the way to correct the situation. Adults should not impose all rules and regulations on school-age children. Rather, children should have many opportunities to discuss rules in a democratic fashion and to help decide on the final regulations. As children mature, they can be given more and more adultlike responsibilities, both in their work and play and in their decision-making activities. When they achieve, they can be given ample respect, approval, and praise.

In helping school-age children learn to respect the rights and needs of others, adults can use inductive reasoning. They can tell children as often as possible how misbehaviors make them feel and how they suppose the wrongful actions make others feel. During middle and late childhood it becomes possible for children to see others' points of view and to consider others' motivations, although they often need to be reminded. The more practice children have considering how others feel, what motivates other people's actions, and what other people need, the more apt they are to consider these factors before acting in the future.

An extension of considering the rights and needs of others is engaging in **prosocial behaviors**, altruistic actions directed at satisfying the wants or needs of others, such as giving or sharing. Eisenberg-Berg (1979) suggested that prosocial moral reasoning develops by stages similar to Kohlberg's moral stages. Very young children are hedonistic and seek only their own pleasure. In stage 2, children learn to share because this behavior will bring attention and approval from adults. In stage 3, children become able to empathize with others and they want to help others feel good, happy, warm, or less hungry. Finally, people can achieve the stage in which they not only empathize with others but also realize that they will feel good about themselves if they help, or, conversely, will feel at conflict with their own internalized sense of right and wrong if they refuse to care and share.

Research on how children acquire prosocial behaviors and the extent to which they engage in actions that aid or benefit others suggests the interplay of many factors: societal norms (organization of work, community life, religious practices, family structure, sex roles), cultural norms (competition/cooperation), the effects of media presentations of prosocial behaviors, cognitive maturity, learning by modeling, child-rearing variables (discipline, affection, involvement), peer interactions, the child's personality, and the child's personal life experiences (Radke-Yarrow, Zahn-Waxler, and Chapman, 1983).

Discipline Adults responsible for the care of older children are still obliged to provide some form of control over misbehavior, although the discipline may take many different forms from that used with younger children. Physical punishment leads to hostility toward the punishing parent, as well as a low sense of guilt for the misbehavior once "the score is settled." School-age children also model physical punishment and may strike back at the parent or strike out at a scapegoat (such as a sibling, friend, or pet). Love-withdrawal discipline leads to some resentment of the parent withholding affection and to a high degree of anxiety about the loss of approval and love. Permissiveness leads to a lack of respect for the parent.

Inductive techniques of discipline are most effective as the child matures cognitively. These are attempts to control the child's actions by explaining reasons for a change of behavior or explaining the consequences of the undesirable deed in terms of its effects on others (see Chapter 5). Some psychologists also recommend programs of **behavior modification** to help parents alter single unacceptable behaviors. Interestingly, parents must usually change their own behavior before they can modify their child's behavior. Parents need to realize that they may be subtly rewarding a child for a behavior that they find unpleasant, since this behavior persists along with the good behaviors they knowingly reward. Parents must learn to reinforce an alternative behavior that they consider more desirable.

An example will help make a behavior modification program more understandable. Parents who came to a child development clinic complained that their six-year-old son continually pestered them. An interaction between the parents and the boy was videotaped for about fifteen minutes and then played back for the parents. They were able to see that while their son entertained himself with toys, they read their magazines or newspapers and ignored him. However, when he threw a toy at them or grabbed at their reading materials, they quit reading and tried to discipline him. The parents saw that they paid attention to their son only when he pestered them. In effect, they were positively reinforcing the attention-seeking behavior they hated. To eliminate pestering the parents first agreed to set aside time to talk to, play with, or read children's books to their son. They then decided what to use as a negative or neutral stimulus when the undesirable behavior occurred. (A negative stimulus is a punishment such as social isolation. Neutral is nothing, an ignoring of the behavior.) They also agreed on a positive reinforcer for a desired behavior to replace the aversive act. (Positive reinforcers may be material rewards such as money, candy, or toys; privileges such as watching television or going to favorite places; or social reinforcers such as attention, affection, praise, or reassurance.) At first the child was rewarded for short periods without pestering behavior. Later the child had to complete a longer segment of good behavior to get a reward. Still later the child had to complete several days of desirable behavior to get the reward. Material possessions were used as reinforcers at first, and social reinforcers were later substituted for the material rewards. Finally the child's behavior maintained itself without continual rewards. The parents were consistent, firm, and calm during the program. When they gave positive reinforcers, they told the child what the behavior was that they were rewarding to ensure that their son knew what they wanted and liked.

Often when behavior modification fails, the parents' motivation is at fault. They may not really want their child to change. Lack of consistency is also a problem. When a child is rewarded for a bad behavior, he or she is apt to try the bad behavior again and again to get another reward.

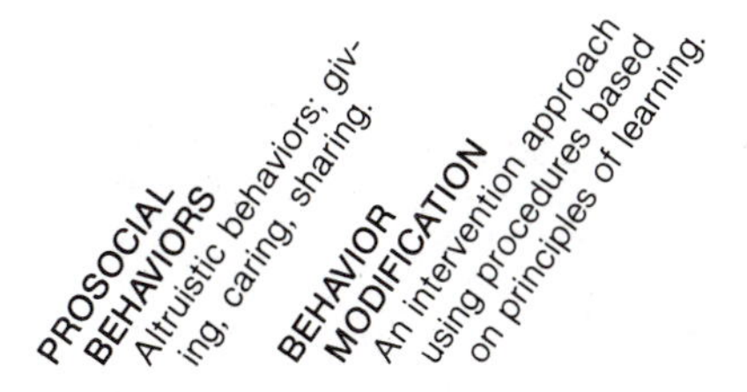

Critics of behavior modification feel that such engineering of human behavior is disrespectful of free will and somehow lowers children to the level of trained animals. Proponents, especially B. F. Skinner (1971) in his book *Beyond Freedom and Dignity,* argue that conditioning of human behavior occurs all the time anyway. In Skinner's opinion, it is better to carefully think through desirable behaviors for children than to let behaviors contributing to the hostile, aggressive side of human nature be taught or learned without any attempts to modify or control them.

Rudolf Dreikurs felt that discipline will succeed most if parents avoid any kind of power play with their children (Dreikurs and Stolz, 1964). He proposed that children be left to discover the negative consequences of bad behavior for themselves. If they fail to put clothing in the hamper, they will soon find themselves with no clean clothes to wear. If they annoy the cook, their dinner will be delayed. If they leave toys where they do not belong, the toys will disappear. This is actually a form of inductive discipline. The inducement to behave is more felt than heard. (Parents can give reasons for the child's suffered consequences based on the child's own actions whenever they are not immediately apparent.) Dreikurs advocated a family democracy where all family members feel loved, wanted, worthwhile, and equally important. He feared that when parents set themselves up as the exclusive lawmakers and law enforcement agents, they assume a position of being "more equal" than the children.

Maccoby and Martin (1983) looked at the overall social climate that parents establish in the home. They identified four major patterns of parental behavior: authoritarian, authoritative, indulgent, and neglecting. The **authoritarian parent** assumed complete control over rules and regulations, used physical punishment and stayed somewhat detached and cool toward the children. The **authoritative parent** was in control but allowed feedback from the children about rules and regulations, used inductive techniques of discipline (occasionally backed up by physical punishment), and was receptive and warm toward them. The **indulgent parent** seldom asserted control, made few demands about rules and regulations, and was warm to the children. The **neglecting parent** made few demands on children and was also characteristically detached, cool and unresponsive. (See Chapter 5 for a review of the consequences of such social climates.)

Social responsibility is greatest in children with authoritative parents. Too much authority holds back the socialization of independence and responsibility taking, lax control sometimes leads to so-

cially disruptive, immature behaviors, and neglecting leads to multiple problems with emotions and behaviors. When Siegal and Cowen (1984) asked school-age children to evaluate the disciplinary climates created by their mothers, they found that authoritative mothering was not only the most effective in terms of positive socialization but was also what children preferred. Induction was rated very favorably as a disciplinary technique. Contrary to what you might expect, children did not advocate permissiveness. Physical punishment was given mild approval.

Siegal and Cowen did not look at disciplinary techniques used by fathers. Evaluations of their role in discipline are worthy of further examination. More mothers now work for economic and psychosocial reasons and must rely on the father's help with child-rearing. Hoffman (1977) found that fathers are more actively involved with their school-age children when the mother is employed. Carlson (1984) found that boys whose fathers share responsibility for child care hold fewer sex-role stereotypes. Warm and accepting fathers who play with and listen to their children have children who are more motivated to achieve and sons who perform better on intellectual tasks (Epstein and Radin, 1975). In contrast, father-absent boys have been shown to decline in both achievement motivation and ego strength over time (Fry and Scher, 1984). Father absence seems to affect boys more than girls. Wallerstein and Kelly (1976) found that both boys and girls whose fathers are absent had poorer impulse control and lower levels of moral development.

Weinraub (1978) pointed out that the old practice of assigning fathers the roles of breadwinner and ultimate disciplinarian (as in "Wait till your father gets home!") may have caused men to feel conflict about their parenting roles. Today's increased reliance on the father's assistance as a warm, nurturing, accepting second parent results in more adequate child care. Two competent individuals are better able to meet a growing child's needs for physical, emotional, and intellectual stimulation than one.

Sex Role Differentiation Chapter 5 presented various theories about how and why children learn to adopt behaviors appropriate to their own sex. Freud believed that by about age six boys renounce their Oedipal complexes and identify with their father and girls renounce their Electra complexes and identify with their mother, primarily through repression. Social-learning theorists state that boys identify with their fathers and girls with their mothers by age six because of the rewards and attention such sex-appropriate behaviors bring. Cognitive theorists claim that children model the like-sexed parent because they perceive their resemblance to that parent and want to imitate the person whom they most closely resemble.

When children are old enough to join clubs, stay overnight with friends, and spend at least half of their waking hours outside the home, they develop other **sex-role models** besides their parents. Pitcher and Schultz (1984), after an extensive review of interdisciplinary literature, presented a strong case for peers becoming major agents of sex-role socialization. The sex-appropriate behaviors children learn at home are tested against a wide sample of men and women, boys and girls. When confusion develops, children are most apt to model the persons who are nurturant and powerful and who dole out the rewards for sex-appropriate behavior, which may mean learning a double or triple standard. For example, a girl may learn that her liberated mom wants her to fight her own battles, that her more traditional schoolteacher believes girls should never fight, and that peers think she should fight verbally but not physically. Or a boy may be told at home that it is appropriate to cry only when physically hurt, by his peers that it is never appropriate to cry, and by a teacher that even in situations where the hurt is emotional, it is still all right to cry.

In the past and in some cultures today girls are expected to develop an **expressive orientation** to life. They are supposed to be nurturant, sympathetic, dependent, and emotional. Boys are expected to develop an **instrumental orientation** to life. They are supposed to be task-oriented, brave, independent, and unemotional. Edwards and Whiting (1980) studied three contemporary cultures that socialized expressive and instrumental orientations in their girls and boys: India, Mexico, and Okinawa. Girls were frequently asked to do household tasks, take care of infants, and do work that kept them home in the company of female adults. In these cultures, a great many statistically significant sex differences were found. In contrast, in three cultures that emphasized equal treatment of boys and girls, the United States, the Philippines, and Kenya, fewer sex differences were found (see Figure 6–13).

Figure 6–13 *Sex-role stereotypes are not clear-cut in North American society. Many boys and girls can feel comfortable demonstrating needs for dependence and independence, pursuing both instrumental and expressive tasks, dressing in similar fashion, playing the same games, and eventually choosing the same careers. This photo shows two girls and one boy.*

Why should a girl be taught that she is weak and needs to depend upon a man? Why should a boy be taught that he must be strong and never depend on anyone? Would it not be better to allow girls and boys to develop according to their own capabilities? Mental health professionals, child psychologists, feminists, educators, and many other concerned persons in the United States and Canada have done much to have sex-role stereotypes reduced. Many textbooks and books for parents stress the advantages of androgynous child-rearing. An **androgynous** child incorporates positive aspects of both expressive and instrumental characteristics into his or her personality. Research indicates that androgynous personalities are healthier than those who are highly sex-typed. Both high femininity and high masculinity have been correlated with higher anxiety and lower self-esteem (Bem, 1976). Nevertheless, while American culture in general, and some families in particular, may encourage non-sex-stereotypic behaviors, American children are not free of sex-role learning and sex-typed behaviors. The most striking sex differences are in the greater rough-and-tumble aggressive behaviors of boys, and the greater verbal behaviors of girls (Maccoby, 1980).

Beliefs that American girls are more submissive and less assertive than American boys have not been borne out by research studies. Submissiveness is defined as nonhostile, noncoercive behavior that involves taking into consideration the power, authority, or feelings of others while denying or not standing up for one's own feelings and beliefs. Assertiveness is defined as the direct, nonhostile, noncoercive expression of thoughts, desires, beliefs, or feelings (Deluty, 1979). After observing school-age children over an eight-month period in a wide variety of naturally occurring school-related activities, Deluty (1985) found very few submissive behaviors in either sex and a great many assertive behaviors in both sexes. Both boys and girls were consistently assertive in situations that varied in structure from sitting passively to moving about more actively (such as art and music classes). American girls as well as boys know how to express their thoughts and feelings in a nonhostile manner, make requests for behavior, stand their own ground in arguments, and assert themselves both positively and negatively in many other ways. They all do so frequently. Deluty's research also confirmed the finding of many other studies: boys are more aggressive than girls. However, they assert themselves much more frequently than they aggress.

Single Parenting About 50 percent of American children will spend a portion of their childhood in a **single-parent family** for rea-

sons of divorce, desertion, death, unwed mothering, or single-parent adoption. About 40 percent of marriages end in divorce. In about 90 percent of divorces involving children, the mother becomes the custodial single parent. (See Box 6–2 for a discussion of single-parent fathering.) Children without fathers generally do less well in school and have poorer inpulse control, less ego strength, and lower levels of moral development.

One frequently hears the phrase "Kids are resilient" or "Chil-

BOX 6–2

Single-Parent Fathers

About 10 percent of fathers are awarded physical custody of their children in divorce proceedings or become single-parent fathers for reasons of death, desertion, or single-parent adoption. While single-parent fathers are statistically uncommon, there are in fact about 900,000 of them in the United States (Meredith, 1985). Like single mothers, they must combine jobs outside the home with cooking, cleaning, laundry, chauffeuring of children, handling repairs, paying bills, shopping, balancing the budget, functioning as both father and mother, nurturing, disciplining, protecting their children's health and safety, assisting with schoolwork, and being solely in charge inside the home. The physical and emotional stresses of so doing may exhaust them (just as it does single mothers).

Although caring for younger children might be easier for a father, most children awarded to their father's custody are older (Glick and Norton, 1978). Older children, while they can lend a hand with housework, are harder to discipline, more vocal about their anger, fears, griefs, and displeasures, and more interfering with any attempts on their father's part to develop his own social life.

Single fathers often find housework simple compared to parenting interactions. Fathering of daughters can be especially problematic. Girls without mothers have been rated as less feminine, less independent, and more demanding than those without fathers. In contrast, boys without mothers are more sociable and more mature (Santrock and Warshak, 1979). Santrock and his colleagues (1982) have suggested that father custody may be more beneficial for some boys and mother custody for some girls, although both sexes do better with frequent contact with both parents.

dren bounce back quickly" to assuage fears about the effects of separation and divorce on them. Are they true? Generalizations from one situation to another are of questionable validity, but researchers are finding that a great deal of childhood anguish accompanies the breaking up of the two-parent family, the primary foundation of security. The aftermath is long-lived. In fact, children usually do better during the crisis of the actual break-up and have their worst problems with guilt, anger, fear, and depression later. The "bouncing back" seldom occurs until at least a year after the divorce.

When a mother first becomes a single parent, she experiences shock, fears and anxieties, anger, bitterness and resentment (especially if the husband now enjoys his freedom and/or a new sex partner), and a loss of self-esteem. Less than one-third of departing fathers make significant monetary contributions to support their families, so most single-parent mothers must work. Most experience a degree of downward economic mobility. Many must move to more modest housing and give up luxuries such as a car. The physical and emotional stresses of working, balancing a budget, running the house, functioning as both father and mother, and being solely in charge can exhaust even the best organized and most emotionally stable women. Weinraub and Wolf (1983) found that many of them work longer hours, withdraw from their social networks, and become socially isolated. They also become more authoritarian and erratic in the ways they discipline their children (Hetherington, 1979).

Single parenting usually becomes easier with time. Mothers can learn how to cope by talking to other single parents in organizations such as Parents Without Partners. Counseling by mental health professionals is also desirable. Single parenting is not as difficult if the mother has sufficient money, has help from her ex-husband, parents, neighbors, relatives or friends, and has people with whom to communicate openly and honestly about her physical, social, and emotional concerns.

When children first lose one of their parents, they commonly experience anger, fears and insecurities, guilt, and depression (Hetherington, 1979). They worry about whether their needs will be met, they worry about the absent parent, and they worry whether they are still loved. Kalter and Plunkett (1984) found that about one-third of all children believe their behaviors caused the parents' divorce. Most fantasize about, or actively try to negotiate, a parental reunion. If they must move away from old friends, neighbors, and classmates, or if they see less of their mother as she takes on extra work, they often feel extreme loneliness and depression. Depression in children often takes the form of school setbacks, uncontrolled tempers, destructive acts, and antisocial behaviors. Boys seem to have many more problems adjusting to a father's loss than do girls (Hetherington, Cox, and Cox, 1982). They refuse to comply with many of their mother's requests, adding to her already stressed existence. In turn, mothers often view their sons in a more negative light than their daughters. There are several things the parents can do to help tensions subside and bring children back to a sense of well-being again:

- Assure each child that the break-up was not his or her fault.
- Assure each child that he or she is still loved.
- Communicate with each child about his or her feelings of anger, fear, guilt, and depression.
- Be honest about your own feelings of anxiety, fear, and so on.
- Be honest about the financial situation.
- Don't tear down the reputation of the absent parent.
- Keep quarrels with the absent parent hidden.
- Encourage frequent contact with the absent parent (if he or she is emotionally well adjusted).
- Discourage fantasies and hopes for reconciliation (unless feasible).
- Discuss and reach agreement with the absent parent on rules of child-rearing and discipline and keep them consistently.
- Keep consistent schedules for meals, school, bedtime, and the like.
- Look to others (parents, siblings, friends, competent housekeeper) for support when stressed.
- Seek professional counseling or therapy for self or children if stressed.

Mothers are frequently concerned about how their sons will achieve a stable masculine self-concept without a live-in male role model. They should encourage their ex-husband or a trusted uncle, grandfather, brother, neighbor, or other male to interact with the boy(s). They should encourage independence and task mastery, praise their sons' strengths and capabilities, and present maleness in a positive light.

Some mothers overprotect their sons and reward dependence. This may be due to the mothers' own needs for love and attention. Such mothers may encourage physical contact, discourage any signs of aggression, and interfere with the boys' attempts at task mastery. Such boys may become timid and retiring underachievers and have a low sense of worth in terms of their maleness. Some mothers who engage in overprotective behaviors also add a barrage of negative comments about men to their child-rearing tactics. The mother's comments about the inadequacy, incompetence, or worthlessness of men further decreases the son's sense of masculine worth.

In some cases of father absence the mother withdraws from or neglects her children. This may be especially true of sons, who she expects can take care of themselves and maybe even her. Boys so treated may dramatize their masculinity, toughness, and independence, yet they may still have an underlying low sense of worth as males.

STEPPARENTING Taking on the responsibilities for protecting and raising the children of one's spouse.

Stepparenting Within a period of five years, most children who experienced the breakup of their biological parents will have to learn to adjust to a **stepparent** (new spouse of one's biological parent). Approximately one out of eight children in the United States today has a stepparent (Einstein, 1979). It is hard to make generalizations about how children react to a new adult in the home since each situation is different. However, some jealousy and resentment is common. Blending new families is easier if the stepparent does not try to replace the biological parent but rather becomes a third (or fourth) parent. Relationships develop slowly and should not be rushed. The stepparent and real parent should agree on child-rearing and discipline and be consistent in their use, supporting each other when necessary. Although it is normal for one child to warm up to the stepparent before others, showing favoritism should be studiously avoided. Boys often accept a stepfather sooner than girls (Santrock et al., 1982). Daughters tend to be angry with their mothers for remarrying and anxious about the new interactions required.

Stepfathers are usually more acceptable to children than are stepmothers, and younger children accept stepparents more easily than do older children (Stapleton and MacCormack, 1981). While initially children may have more behavioral problems in the reconstituted family than in the single-parent family (Nunn, Parish, and Worthing, 1983), over an extended period of time remarriage can mitigate the negative effects of the divorce and single-parent experience if the mother's (or father's) new partner is supportive, caring, and willing to work at becoming a parent (Rutter, 1979). Designing workable stepparent/stepchild(ren) relationships can seem overwhelmingly difficult at times. Many families cannot meet the challenge. Over 40 percent of newly blended families end in divorce within five years (Einstein, 1979).

THE SCHOOL'S ROLE IN SOCIALIZATION

A well-liked schoolteacher, especially one who resembles a child in some way (sex, race, religion, ethnicity) may be taken on as a role model by a child. Sometimes a teacher will be aware of the child's modeling; often he or she will not. Consider these examples of teachers' impacts. James Conant (1970), a scientist who helped develop the atom bomb, attributed his early interest in chemistry to a teacher. In his autobiography he wrote "I doubt if any schoolteacher has ever had a greater influence on the intellectual development of a youth than Newton Henry Black had on mine" (p. 160). Helen Keller (1954) paid an even greater tribute to her teacher. She wrote "All the best of me belongs to her—there is not a talent, or an aspiration or a joy in me that has not been awakened by her loving touch" (p. 46).

A classic study by Lewin, Lippitt, and White (1939) helped make educators aware of the ways in which a teacher can influence the social behavior of the members of a group. They compared autocratic (dictatorial), democratic, and laissez-faire (let people make or do what they choose) teaching styles. They discovered that, although the autocratic teachers ostensibly had good classes, the democratic teachers actually had the better ones. When the autocratic teachers left their groups, fighting broke out immediately. The laissez-faire teachers had fighting occur even in their presence. In the democratic atmosphere policies were established by the group, and the children felt some responsibility for the rules they helped make. They showed little aggression in either the teacher's presence or absence. They also liked the democratic leaders and worked harder for them.

The findings of this classic study are not unlike recent research pointing to both greater effectiveness and popularity of authoritative (democratic) parenting styles (see Chapter 5). More recently, Deci and his colleagues (1981) showed that autonomy-oriented (democratic) teachers contributed to their pupils' desires to learn and to their self-esteem much more than did control-oriented (autocratic) teachers. Autocratic teachers are more apt to engage in stereotypic labeling, which can have a Rosenthal effect by eliciting expected classroom performance. Tom, Cooper, and McGraw (1984) found that authoritarian (autocratic) teachers have higher grade expectations for girls than for boys, for Asians than for whites, and for middle-class than for lower-class students. Ball and his colleagues (1984) also found that teachers lower their performance expectations for children from divorced single-parent families, especially boys.

While many persons believe that schools do not make a difference, research suggests differently. In an extensive review of school effects on pupil progress, Rutter (1983) concluded that effective schooling needs to be measured not only by scholastic attainment but also by attitudes toward learning, classroom behavior, social functioning, absenteeism, continuation in education, and ultimate employment. School features that may contribute to beneficial effects include resources, size of class, composition of student body, degree of academic emphasis, classroom management, pupil participation, discipline, and staff. One good teacher, especially in the first grade, can have remarkably persistent positive effects on students (Pedersen, Faucher, and Eaton, 1978).

The effect of a teacher on in-classroom social behavior is better known than the carryover effect of socialization practices from class to home life. Much carryover modeling depends on the child's perception of the importance of the teacher's nurturance and power. Much also depends on how well caregivers understand, agree with, and are willing to adopt different school socialization practices in their homes. A parent's negative attitude toward a school may contribute to a child's school phobia (see Box 6–3).

Carryover of socialization practices from school to home has been demonstrated more clearly in the Soviet Union than here (Bronfenbrenner, 1970). **Character education** is considered one of the most important functions of Soviet schools. Parents and community members are duty bound to uphold the socialization procedures begun in the schools. Children are taught to help each other and work together for the good of the class and the school and, ultimately, for the good of the community and their country. They learn to praise and criticize their own and others' shortcomings rather than hide them. The group then decides on rewards or punishments for behaviors. Selfishness is one of the most serious offenses. A. S. Makarenko is the Soviet equivalent to America's Dr. Spock. But, whereas Spock deals predominantly with physical health, Makarenko

BOX 6–3

School Phobia

A fairly prevalent (and disabling) childhood phobia is **school phobia**. It is a condition where a child actually develops physical symptoms of illness when left at school (headache, vomiting, cramps, diarrhea, hysteria, crying). It is most common when a child first starts school but may also occur after a trauma, such as the loss of a parent by death or divorce. School phobic children are typically more dependent, immature, and anxious than nonphobic children (Trueman, 1984). The root of the child's anxiety is not usually the school or the teacher but a fear of being separated from the parent(s). However, over time the child may also become fearful of teasing or rejection by teacher or peers or of being called on to recite. Symptoms decrease if a parent stays in the classroom. However, this is not a good solution to the problem. The immediate goal is to get the child to remain in school without the parent (usually the mother) on whom he or she has become overly dependent.

To help children overcome school phobia, parents need to be made aware of the ways in which they consciously or unconsciously convey the impression to their child that all will not be well during the separation. School phobic children may sense that the parent will be lonely without them or that the parent does not believe the school environment is as safe, healthy, and loving as the home environment. Parents differ in their readiness to accept their own roles in their child's problem. Many need help in adjusting to their child's being in school all day. In some cases, the mother's entry into some out-of-the-home employment will allow her to "untie the apron strings" and more willingly send her child to school. In some cases one or both parents need psychotherapy to work out their underlying conflicts about allowing their child autonomy, and about the safety of the school. This can help to prevent recurrences of the child's perceived need to remain at home.

When a school phobic child does experience physical symptoms in school, he or she should be sent to the nurse's office, not home. As soon as possible, the child should be gently but firmly returned to the classroom. Home teaching should not be prescribed. It creates more psychological invalidism for the school phobic child and encourages continuation of the conscious or unconscious parental domination.

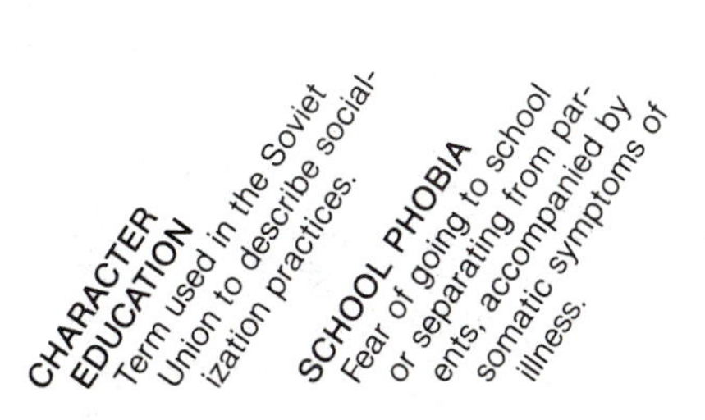

deals with character education and informs parents how to raise socially upright citizens.

PEER INTERACTIONS

Just as a schoolteacher may have an impact on social development, so too can friends. But peers' influence on a child's behavior can be weak or strong depending on several factors, including:

- The age of the child.
- The sex of the child.
- The self-esteem of the child.
- The intelligence of the child.
- The amount of time spent with peers.
- The stoutness of the parent-child affectionate bond.
- The amount of time spent in positive interactions with the family grouping.
- The family's acceptance of various peer group members.
- The values and activities of the peer group.
- The child's position in the peer group.

Children use their friends as sounding boards and testing grounds for the values and attitudes they have learned at home. In many cases the **peer group** (group of persons of equal rank or status) can be more democratic than the home. Instead of rules being laid down by authority figures, they are debated, with some or all of the group having a say in what they should be. Home values and attitudes may be upgraded or watered down, depending on the participants in the group.

Children increasingly turn to their peers for assistance. Nelson-LeGall and Gumerman (1984) asked children whom they would seek help from in academic and social contexts and found that their preferences for parental help decrease with age. Peer academic tutoring or collaboration brings with it unique motivational and cognitive benefits for the participating children (Damon, 1984).

During late childhood, friendships become more stable. Friends are usually of the same sex and often of the same race, religion, culture, or socioeconomic standing in the community. Organized activities such as scouting, sports, and religious group projects tend to strengthen friendships and add a cohesive element to a group. Little League participation for a season, for example, has been shown to enhance the self-esteem of each player involved (Hawkins and Gruber, 1982). When children work together, they learn new respect for each other. Occasionally after a group failure or defeat members will turn against each other, but in general teamwork strengthens friendships.

PEER GROUP
The group of persons who constitute one's associates, usually of the same age and social status.

Children in groups will often do things they would never do on their own. This can take the form of increased altruism (for example, visiting nursing homes) or increased delinquency (such as destroying property). Many a quiet, well-behaved child has joined fights or used vulgar language along with a supporting peer group. Peers have the potential to influence each other for good or for bad. Some children are more powerful in influencing others, and some children are more influenced by powerful peers.

Being a member of a minority group within a larger community frequently presents special problems for school-age children. If the minority group is large enough so a child can find a cohesive group of same-sex friends, the going is easier. Without the benefit of other minority group members to have as friends, it is often difficult for a child to feel worthwhile or acceptable to the mainstream peer groups. By the elementary school years children have learned many prejudices of their parents. Prejudice is an insidious thing. Even members of minority groups may adopt the mainstream culture's prejudice against themselves. Fu and Fogel (1982), for example, found that both black and white children from the South had a white-positive/black-negative bias.

Prejudice can be leveled at religious, language, cultural, and economic groups as well as different racial groups. Obese children, chronically ill children, even homely children may find themselves the objects of peer ridicule. Communities and neighborhoods differ in what they find acceptable and unacceptable. Children may be ostracized if they come to school in old fashioned, small, or ragged clothes or without spending money for candy or other things purchased by the majority of the school children.

Children who lack English language proficiency or speak with accents are especially vulnerable to rejection by their classmates. The "melting pot" ideology is not appreciated by many children during late childhood. Unless a child can master English quickly, he or she is usually denied the opportunity to participate fully in the social activities of the English-speaking core group. Unless steps are taken to teach the English language before other academic subjects are introduced, children fall farther and farther behind and may be classified as retarded in spite of very adequate intelligence. A proposed Bilingual Education Act, to fund school projects to teach the English language to children before attempting to teach them other academic subjects, was withdrawn by the Reagan administration in 1981. Schools themselves must now set up programs and find teachers to conduct English language classes in the children's native languages, or else the teaching of English must be done at home or in the community before the child can learn in school.

NEIGHBORHOOD INTERACTIONS

Increasingly, child developmentalists are looking beyond the family, the school, and the peer groups for other socialization influences on growing children. Neighbors are emerging as important sources of support in late childhood. As more mothers work outside the home, more at-home neighbors are being asked to watch the children after school until a parent returns. Research on latchkey children (those who are home alone after school) has suggested that while they tend to be more independent, they may also suffer more fears, loneliness, boredom, and depression (Long and Long, 1983).

Because of increased fears of child abduction and child molestation, many parents make a conscious effort to introduce their children to several trusted neighbors to whom they may run when and if a stranger or older child bothers them. Many nonworking parents also want their children to know about safe houses, where trusted neighbors will be home, along the route to and from school. If children walk through business areas, parents may also want their children to know about stores, libraries, and the like where the child will be recognized, sheltered, and perhaps allowed to use the telephone. Bryant (1985) found that children who had more identifiable sources of support in their neighborhoods scored higher on several measures of social and emotional functioning. Helping school-age children develop a social network in their neighborhoods has many beneficial effects for safety, socialization, and mental health.

THE EFFECT OF TECHNOLOGY

The influences of parents, schools, peers, and neighbors are modified, somewhat, by machines of technology available to children. In addition to the standard television, children now increasingly have videocassette recorders (VCRs). These enable them to tape television programs for late replay or to play movies, games, or music videos on their television sets. Children also have easy access to video games in video arcades or other public places. Many of them are exposed to computers in their schools. And many children now have personal home computers. The effects of television on children are better known than the effects of computers.

The average North American child is a heavy consumer of television, watching two to four hours a day on school days and more on weekends and school vacations. By completion of high school most children will have spent about twice as much time in their lives watching television as learning in school classrooms. Researchers have focused more on the negative concepts they acquire (aggression, junk food preferences, and highly sex role-typed stereotypes) than on positive behaviors (rule obedience, empathy, altruism) they might learn (see Figure 6–14).

The aggressive content of television programming is high. Television networks have kept prime-time evening programs less violent until after 9 P.M. However, many children are allowed to stay up to watch the more violent programs. Pearl (1984) reported that over the past ten years there has been more violence on children's weekend programs than on prime-time television.

In the past some social scientists felt that watching televised violence would have a cathartic effect and displace or dissipate a child's need to be aggressive in the real world. This theory has not been supported by research. Instead, it has been found that the more violent programming children watch, the more aggressive they become in all aspects of life: conflicts with parents, fighting, and delinquent behaviors. They also view the world with more suspicion and distrust and perceive violence as an effective solution to conflict (Pearl, 1984). (See also the discussion of modeling TV violence in Chapter 5.) Cordes (1985) reported that children who watch more

TV are more likely to believe it is real. This is true not only for violence but also for the stereotyped behaviors of men and women.

Children also believe in the enormous presence, in all homes except their own, of certain toys and foods. Commercial time on TV is higher than most people imagine, about 22 percent of the broadcast day. Commercials rarely last longer than 60 seconds, but they are skillfully designed to stay on their viewers' minds for much longer. Many commercials hire celebrities to endorse foods or toys for children. Ross and her colleagues (1984) found that eight- to fourteen-year-olds prefer products promoted by celebrities and fail to see that the ads are staged, especially when the commercial includes live action. Ads for highly sugared or salted foods with low nutritional value ("empty calories" or "junk food") are flashed at children frequently, as often as eight times per hour on Saturday mornings. Children in turn beg to be allowed to eat such foods and consider their parents particularly cruel if they deny them the "pleasures" the ad kids have consuming these foods.

Figure 6–14
Television influences the social development of children with its own views of reality and pictures of how the "rest of the world" acts and reacts.

Feldstein and Feldstein (1982) analyzed televised toy commercials and found that they not only have more males than females per commercial but also are more likely to put females in passive roles. Prime-time adult-oriented commercials do the same (Mackey and Hess, 1982).

Wright and Huston (1983), and Greenfield (1984) made it

clear that television has a rich (although as yet largely untapped) potential for enhancing the cognitive and psychosocial development of children. Information-processing skills and retention of information can be increased and improved by television. Children can also acquire many positive concepts such as the efficacy of moral and prosocial behavior. It is unfortunate that so few programs have been developed to exploit these positive growth possibilities.

Parents can enhance the television viewing time of their children by watching programs with them, asking questions, explaining concepts, and discussing outcomes and alternatives to outcomes. When used wisely and sparingly, television can be an interactive, educational, and constructive family activity. Heavy viewing is potentially harmful, especially with the paucity of age-specific educational programming for children and the abundance of aggressive, violent, sex-stereotyped programs.

Greenfield (1984) described computer technology as another potentially great tool for enhancing children's cognitive and psychosocial development. While children watching television are passive, children playing video games are active. They must plan moves ahead and develop their fine motor skills to play competently. However, the natures of the video games children play have an effect on their behaviors, just as TV programming does. Some video games are terrifyingly violent, with ultimate goals of destruction and annihilation. Ascione and Chambers (1985) reported that children who play violent video games are less likely to help other children.

Some people have worried that computer games are addictive—that children who play a great deal may feel a compulsive need to continue playing. There is little research evidence to support this. Most video games encourage interactions among competitive players. Rather than creating obsessive-compulsive socially isolated children, they often stimulate social involvement. Playing may improve the social acceptance and self-esteem of children who are competent at video games even though they are not good athletes or good scholars. Levin (1985) wrote that mastering a new computer game has much in common with solving a math problem. It absorbs the child's whole attention. The child must modify, augment, delete, or transform behaviors in order to succeed.

Many children now have personal computers in their own homes. They use these home computers not only to play games but also to create new games or other programs. Many adults who have tried to learn one or more computer languages are amazed at how quickly children learn these languages relative to their own slow progress. Progamming requires that the user provide the machine with a carefully thought through series of instructions. This requires patience and a willingness to correct one's mistakes. The computer provides a precise feedback of what it has been told to do. It cannot be intimidated into doing something else. The user must take full responsibility for any "bugs" in the program and find out what he or she has done wrong. Levin (1985) writes that this process teaches children humility and personal responsibility. They cannot displace the blame for mistakes onto others. Many schools have incorporated computer programming courses into their curriculum. It will be years before we have a clear idea of the effects of computers on education and on the intellectual, social, and emotional development of children. However, there is an optimistic sense that this technology can be beneficial. It will not be a substitute for parents and teachers, but, used wisely, it can enhance the work of adults in rearing children.

Summary

Middle and late childhood is a time when the community begins to play a much greater part in children's lives. Peers, neighbors, and schoolteachers often spend as much or more time with children as children spend with their families.

Physical growth is slower than it was in early childhood and slower than it will be in adolescence.

During this age, as in previous ages, good health is maintained through good nutrition, adequate sleep, and satisfying family, school, peer, and community relationships. Children continue to get many respiratory infections. Tooth decay reaches epidemic proportions. Some children have chronic illnesses or are impeded by psychological problems during late childhood. Accidents, although not as numerous as for younger children, also occur frequently.

Cognitive growth proceeds in what Piaget called the concrete operations stage. With experience children develop abilities to apply rules, reverse mental operations, see reciprocal relations, conserve, order, classify, conceive of distances in time and space, and understand physical and psychological causation. Learning processes that children may use to assimilate and accommodate new information include mediation and encoding.

Language is fun in late childhood. Children practice accents, secret languages, foreign words, slang, and words with double meanings. Sentences acquire sophistication as understanding of complex syntactical forms develops.

Intelligence is a concept not well understood. IQ tests give an inexact estimate of children's abilities to learn school subjects but do not accurately predict abilities to learn other life-situation skills. Although both heredity and environment affect intelligence and IQ scores, the relative strength of each influence is unknown. Creativity can adversely affect test scores when only one answer is acceptable. Although creativity is extolled, its development may be inhibited in many structured situations.

Problems of learning may take the form of specific disorders of reading, arithmetic, language, coordination, attentional deficits, mental retardation, or giftedness.

Erikson described the nuclear conflict of this age group as that of developing a sense of industry versus developing feelings of inferiority. Family members, the school, peers, and other community members all have power to enhance or defeat a child's sense of industry. Children need many experiences of applying themselves to tasks and persisting through their completion. They also need positive feedback that their efforts are worthwhile.

Moral development is enhanced by identification with and modeling of adults with high ethical standards. It is also tied to cognitive development and disciplinary techniques. To reach higher moral levels, children need many opportunities to deal with choices of right and wrong and to reason out the whys and wherefores of the choices.

Parents may use authoritarian, authoritative, permissive, or neglectful patterns of discipline. The authoritative pattern is the most effective and is also preferred by children.

Appropriate sex-role differentiation for both boys and girls is enhanced by both the father's and mother's warm and active participation in child-rearing. Sex roles may be moving toward androgyny (each human acquiring both male and female traits) with more mothers working outside the home and more fathers participating in child care.

More children are being raised in single-parent families and/or learning to adjust to stepparents as separations, divorces, and remarriages become more common. Separation from a loved parent is traumatic for a school-age child, and behavioral effects are seen for at least a year after the loss.

The influence of schoolteachers, the peer group, and neighbors on children is related to their self-concepts, parent-child affection bonds, and the amount of time spent with these significant others. Children in groups will often do things they would not do on their own. Minority-group children may find social acceptance especially problematic.

Televisions and computers can be called electronic family member(s). Most children grow up spending more time in front of the TV set than in school. Many children now spend a considerable amount of time playing with computers as well. These "family members" can be used wisely and can enhance interactions between parents and children. They can also impede family communications, teach their own messages about violence and stereotyped sexual behaviors.

Questions for Review

1. Outline a day's nutritional program for a school-age child suggesting what you will serve for breakfast, pack in the school lunch, and provide for dinner. Will snacks be provided? Desserts? If so, what will they be?
2. What side would you take on the inheritance of intelligence question? Do you believe that most of intelligence is determined by inheritance and that only a minimal amount of intelligence can be affected by environmental stimulation, *or* do you believe that intelligence is almost equally affected by inheritance and environment? Discuss. If possible, use some real-life examples to support your answer.
3. Learning involves an ability to select or discard certain amounts of irrelevant material. How do you think the following affect the learning process: depression, unhappy home atmosphere, poor health, feelings of inferiority? Be specific in your answer. Consider what has been discussed in previous chapters on cognitive development and emotional development.
4. There is often a tug-of-war between the values children are exposed to in the home and those of their peers. How can parents be consistent and adhere to their values without creating greater conflict within the child by putting down the child's friends?
5. Discuss particular ways in which the family can foster the positive resolution of what Erikson described as the *industry versus inferiority* conflict during middle-late childhood. Also describe the ways the family can negatively influence this resolution.
6. Some individuals argue that male and female children are "born different" and that most sex-role behavior is only minimally influenced by environment. Others argue that "male" and "female" behavior is generally socialized into children and is not innate. Which viewpoint do you subscribe to? Discuss.
7. Children who belong to a minority group are constantly faced with feeling different from or feeling inferior to those in the mainstream of society. Project what these feelings might do to the ambition and motivation of these children and their ability to meet challenges in the future.

8. Some adults fear that frequent playing of computer games will result in socially withdrawn, unrealistic, physically unfit children who believe in violence over mediation for problem solving. Other adults feel that computers will enhance children's abilities to take an active, realistic, honest involved role in life. What do you think children are learning from computers?

References

Ames, L. B. Learning disability: Truth or trap? *Journal of Learning Disabilities,* 1983, *16,* 19–20.

Ascione, F. R., and Chambers, J. H. Videogames and prosocial behavior: The effects of prosocial and aggressive video games on children's donating and helping. Paper presented at the meeting of the Society for Research in Child Development, Toronto, April 1985.

Ball, D. W.; Newman, J. M.; and Scheuren, W. J. Teachers' generalized expectations of children of divorce. *Psychological Reports,* 1984, *54,* 347–353.

Bandura, A., and Walters, R. *Social learning and personality development.* New York: Holt, Rinehart & Winston, 1963.

Beedle, G. L. Teeth. In C. H. Kempe, H. K. Silver, and D. O'Brien (Eds.), *Current pediatric diagnosis and treatment,* 8th ed. Los Altos, Calif.: Lange, 1984.

Bem, S. L. Probing the promise of androgyny. In A. Kaplan and J. Bean (Eds.), *Beyond sex-role stereotypes: Readings toward a psychology of androgyny.* Boston: Little, Brown, 1976.

Blomquist, H. K., Jr.; Gustavson, K. H.; and Holmgren, G. Mild mental retardation in children in a northern Swedish county. *Journal of Mental Deficiency Research,* 1981, *25,* 169–186.

Broman, S. H. Obstetric medications. In C. C. Brown (Ed.), *Childhood learning disabilities and prenatal risk: Pediatric round table; 9.* Skillman, N.J.: Johnson & Johnson Baby Products Co., 1983.

Bronfenbrenner, U. *Two worlds of childhood: US–USSR.* New York: Russell Sage Foundation, 1970.

Bryant, B. K. The neighborhood walk: Sources of support in middle childhood. *Monographs of the Society for Research in Child Development,* 1985, *50*(3, Serial No. 210).

Carlson, B. E. The father's contribution to child care: Effects on children's perceptions of parental roles. *American Journal of Orthopsychiatry,* 1984, *54,* 123–136.

Carroll, J. B., and Horn, J. L. On the scientific basis of ability testing. *American Psychologist,* 1981, *36*(10), 1012–1020.

Cattell, R. B. *Abilities: Their structure, growth, and action.* Boston: Houghton Mifflin, 1971.

Chomsky, N. *Language and mind.* New York: Harcourt Brace Jovanovich, 1972.

Cohen, S. L. Low birthweight. In C. C. Brown (Ed.), *Childhood learning disabilities and prenatal risk: Pediatric round table; 9.* Skillman, N.J.: Johnson & Johnson Baby Products Co., 1983.

Colbert, P.; Newman, B.; Ney, P.; and Young, J. Learning disabilities as a symptom of depression in children. *Journal of Learning Disabilities,* 1982, *15,* 333–336.

Conant, J. *My several lives.* New York: Harper & Row, 1970.

Cordes, C. Kids who watch more found more likely to believe TV. *APA Monitor,* 1985, *16*(11), 35.

Creevy, D. C. Obstetrical trauma. In C. C. Brown (Ed.), *Childhood learning disabilities and prenatal risk: Pediatric round table; 9.* Skillman, N.J.: Johnson & Johnson Baby Products Co., 1983.

Damon, W. Peer education: The untapped potential. *Journal of Applied Developmental Psychology,* 1984, *5,* 331–343.

Deci, E. L.; Schwaratz, A. J.; Sheinman, L.; and Ryan, R. M. An instrument to assess adults' orientations toward control versus autonomy with children: Reflections on intrinsic motivation and perceived competence. *Journal of Educational Psychology,* 1981, *73,* 642–650.

Deluty, R. H. Children's action tendency scale: A self-report measure of aggressiveness, assertiveness, and submissiveness in children. *Journal of Consulting and Clinical Psychology,* 1979, *47,* 1061–1071.

Deluty, R. H. Consistency of assertive, aggressive, and submissive behavior for children. *Journal of Personality and Social Psychology,* 1985, *49*(4), 1054–1065.

Dreikurs, R., and Stolz, V. *Children: The challenge.* New York: Hawthorn Books, 1964.

Edwards, C. P., and Whiting, B. B. Differential socialization of girls and boys in light of cross-cultural research. In C. M. Super and S. Harkness (Eds.), *New directions for child development: Anthropological perspectives on child development.* San Francisco: Jossey-Bass, 1980.

Einstein, E. Stepfamily lives. *Human Behavior,* April 1979, 63–68.

Eisenberg-Berg, N. Development of children's prosocial moral judgment. *Developmental Psychology,* 1979, *15,* 128–137.

Epstein, A. S., and Radin, N. Motivational components related to father behavior and cognitive functioning in preschoolers. *Child Development,* 1975, *46*(4), 831–839.

Erikson, E. H. *Childhood and society,* 2nd ed. New York: Norton, 1963.

Feldstein, J. H., and Feldstein, S. Sex differences on televised toy commercials. *Sex Roles,* 1982, *8,* 581–587.

Flavell, J. H. Developmental studies of mediated memory. In H. Reese and L. Lipsitt (Eds.), *Advances in child development and behavior,* vol. 5. New York: Academic Press, 1970.

Freud, S. *A general introduction to psychoanalysis.* (J. Rivere, trans.) New York: Permabooks, 1953. (Originally published, 1935.)

Fry, P. S., and Scher, A. The effects of father absence on children's achievement motivation, ego-strength, and locus-of-control orientation: A five-year longitudinal assessment. *British Journal of Developmental Psychology,* 1984, *2,* 167–178.

Fu, V. R., and Fogel, S. W. Prowhite/antiblack bias among southern preschool children. *Psychological Reports,* 1982, *51,* 1003–1006.

Gemelli, R. J. Classification of child stuttering: Part II. Persistent late onset male stuttering, and treatment issues for persistent stutterers—psychotherapy or speech therapy, or both? *Child Psychiatry and Human Development,* 1982, *13,* 3–34.

Glick, P. G., and Norton, A. J. Marrying, divorcing and living together in the U.S. today. *Population Bulletin,* 1978, *32,* 3–38.

Gray, D. B., and Yaffe, S. L. Prenatal drugs. In C. C. Brown (Ed.), *Childhood learning disabilities and prenatal risk: Pediatric round table; 9.* Skillman, N.J.: Johnson & Johnson Baby Products Co., 1983.

Greenfield, P. M. *Mind and media: The effects of television, video games, and computers.* Cambridge, Mass.: Harvard University Press, 1984.

Grossman, H. (Ed.). *Manual on terminology and classification in mental retardation.* Washington, D.C.: American Association on Mental Deficiency, 1977.

Guilford, J. P. *The nature of human intelligence.* New York: McGraw-Hill, 1967.

Hagen, J. W.; Anderson, B. J.; and Barclay, C. R. Issues in research on the young chronically ill child. *Topics in Early Childhood Special Education,* 1985, *5*(4).

Haggerty, R. J. Forward: Chronic disease in children. *Pediatric Clinics of North America,* 1984, *31,* 1–2.

Hawkins, D. B., and Gruber, J. J. Little League baseball and players' self-esteem. *Perceptual and Motor Skills,* 1982, *55,* 1335–1340.

Hearnshaw, L. S. *Cyril Burt, Psychologist.* Ithaca, N.Y.: Cornell University Press, 1979.

Herrnstein, R. IQ. *Atlantic Monthly,* September 1971, 43–64.

Hetherington, E. M. Divorce: A child's perspective. *American Psychologist,* 1979, *34*(10), 851–858.

Hetherington, E. M.; Cox, M.; and Cox, R. The aftermath of divorce. In J. Stevens and M. Mathews (Eds.), *Mother/child, father/child relationships.* Washington, D.C.: National Association for the Education of Young Children, 1978.

Hetherington, E. M.; Cox, M.; and Cox, R. Effects of divorce on parents and children. In M. Lamb (Ed.), *Nontraditional families.* Hillsdale, N.J.: Erlbaum, 1982.

Hoffman, L. W. Changes in family roles, socialization, and sex differences. *American Psychologist,* 1977, *32,* 644–657.

Hynd, G. W., and Hynd, C. R. Dyslexia: Neuroanatomical/neurolinguistic perspectives. *Reading Research Quarterly,* 1984, *19,* 482–498.

Jensen, A. How much can we boost IQ and scholastic achievement? *Harvard Educational Review,* 1969, *39,* 1–123.

Jensen, A. *Bias in mental testing.* New York: Free Press, 1980.

Kalter, N., and Plunkett, J. W. Children's perceptions of the causes and consequences of divorce. *Journal of the American Academy of Child Psychiatry,* 1984, *23,* 326–334.

Keller, H. *The story of my life.* New York: Doubleday, 1954.

Kohlberg, L. The development of children's orientations toward a moral order. I: Sequence in the development of human thought. *Vita Humana,* 1963, *6,* 11–33.

Kohlberg, L. The cognitive-developmental approach to socialization. In D. Goslin (Ed.), *Handbook of socialization.* Chicago: Rand McNally, 1969.

Kosky, R. Childhood suicidal behavior. *Journal of Child Psychology and Psychiatry,* 1983, *24,* 457–468.

Levin, G. Computers and kids: The good news. *Psychology Today,* 1985, *19*(8), 50–51.

Lewin, K.; Lippitt, R.; and White, R. Patterns of aggression in experimentally created social climates. *Journal of Social Psychology,* 1939, *10,* 271–299.

Long, T., and Long, L. *The handbook for latchkey children and their parents.* New York: Arbor House, 1983.

Ludwig, G., and Cullinan, D. Behavior problems of gifted and nongifted elementary school girls and boys. *Gifted Child Quarterly,* 1984, *28,* 37–39.

Maccoby, E. E. *Social development: Psychological growth and the parent-child relationship.* New York: Harcourt Brace Jovanovich, 1980.

Maccoby, E. E., and Martin, J. A. Socialization in the context of the family: Parent-child interaction. In E. M. Hetherington (Ed.), *Socialization, personality, and social development.* Vol. 4 of P. H. Mussen (Ed.), *Handbook of child psychology,* 4th ed. New York: Wiley, 1983.

Mackey, W. D., and Hess, D. J. Attention structure and stereotype of gender on television: An empirical analysis. *Genetic Psychology Monographs,* 1982, *106,* 199–215.

Marland, S. P. *Education of the gifted and talented.* Report to the Congress of the United States by the U.S. Commissioner of Education. Washington, D.C.: U.S. Government Printing Office, 1972.

Marx, J. L. The two sides of the brain. *Science,* 1983, *220*(4595), 488–490.

McEwen, B. S. Hormones and the brain. In C. C. Brown (Ed.), *Childhood learning disabilities and prenatal risk: Pediatric round table; 9.* Skillman, N.J.: Johnson & Johnson Baby Products Co., 1983.

Meijer, A. A controlled study on asthmatic children and their families: Synopsis of findings. *Israel Journal of Psychiatry and Related Sciences,* 1981, *18,* 197–208.

Meredith, D. Dad and the kids. *Psychology Today,* 1985, *19*(6), 62–67.

Moser, H. W. Genetics. In C. C. Brown (Ed.), *Childhood learning disabilities and prenatal risk: Pediatric round table; 9.* Skillman, N.J.: Johnson & Johnson Baby Products Co., 1983.

Needleman, H. L. Environmental pollutants. In C. C. Brown (Ed.), *Childhood learning disabilities and prenatal risk: Pediatric round table; 9.* Skillman, N.J.: Johnson & Johnson Baby Products Co., 1983.

Nelson-LeGall, S. A., and Gumerman, R. A. Children's perceptions of helpers and helper motivation. *Journal of Applied Developmental Psychology,* 1984, *5,* 1–12.

Nunn, G. D.; Parish, T. S.; and Worthing, R. J. Perceptions of personal and familial adjustment by children from intact, single-parent, and reconstituted families. *Psychology in the Schools,* 1983, *20,* 166–174.

Nurcombe, B. The search for childhood depression. In L. P. Lipsitt (Ed.), *The Brown University human development letter,* Whole Special Report, 1985.

O'Brien, D., and Hambidge, K. M. Normal childhood nutrition and its disorders. In C. H. Kempe, H. K. Silver, and D. O'Brien (Eds.), *Current pediatric diagnosis and treatment,* 8th ed. Los Altos, Calif.: Lange, 1984.

Olmedo, E. L. Testing linguistic minorities. *American Psychologist,* 1981, *36*(10), 1078–1085.

Pearl, D. Violence and aggression. *Society,* 1984, *21*(6), 17–22.

Pearlman, D. S. Allergic disorders. In C. H. Kempe, H. K. Silver, and D. O'Brien (Eds.), *Current pediatric diagnosis and treatment,* 8th ed. Los Altos, Calif.: Lange, 1984.

Pedersen, E.; Faucher, T. A.; and Eaton, W. W. A new perspective on the effects of first grade teachers on children's subsequent adult status. *Harvard Educational Review,* 1978, *48,* 1–1.3.

Piaget, J. *The moral judgment of the child.* (M. Gabain, trans.) New York: Free Press, 1965. (Originally published, 1932).

Piaget, J. *The child's conception of number.* New York: Norton, 1965. (Originally published, 1941.)

Pitcher, E. G., and Schultz, L. H. *Boys and girls at play: The development of sex roles.* New York: Praeger, 1984.

Pollitt, E.; Leibel, R. L.; and Greenfield, D. Brief fasting, stress, and cognition in children. *American Journal of Clinical Nutrition,* 1981, *34,* 1526–1533.

Radke-Yarrow, M.; Zahn-Waxler, C.; and Chapman, M. Children's prosocial dispositions and behavior. In E. M. Hetherington (Ed.), *Socialization, personality, and social development.* Vol. 4 of P. H. Mussen (Ed.), *Handbook of child psychology,* 4th ed. New York: Wiley, 1983.

Reschly, D. J. Psychological testing in educational classification and placement. *American Psychologist,* 1981, *36*(10), 1094–1102.

Robbins, F. C. Forum. In C. C. Brown (Ed.), *Childhood learning disabilities and prenatal risk: Pediatric round table; 9.* Skillman, N.J.: Johnson & Johnson Baby Products Co., 1983.

Rosenthal, R., and Jacobson, L. *Pygmalion in the classroom: Teacher expectation and pupils' intellectual development.* New York: Holt, Rinehart & Winston, 1968.

Ross, R. P.; Campbell, T.; Wright, J. C.; Huston, A. C.; Rice, M. L.; and Turk, P. When celebrities talk, children listen: An experimental analysis of chldren's responses to TV ads with celebrity endorsement. *Journal of Applied Developmental Psychology,* 1984, *5,* 185–202.

Rutter, M. *Helping troubled children.* New York: Plenum, 1975.

Rutter, M. Protective factors in children's response to stress and disadvantage. In M. W. Kent and J. E. Rolf (Eds.), *Primary prevention of psychopathology, vol. 3: Social competence in children.* Hanover, N.H.: University Press of New England, 1979.

Rutter, M. School effects on pupil progress: Research findings and policy implications. *Child Development,* 1983, *54,* 1–29.

Samuda, R. J. *Psychological testing of American minorities: Issues and consequences.* New York: Dodd, Mead, 1975.

Santrock, J. W., and Warshak, R. A. Father custody and social development in boys and girls. *Journal of Social Issues,* 1979, *35*(4), 112–125.

Santrock, J. W.; Warshak, R. A.; Lindbergh, C.; and Meadows, L. Children's and parents' observed social behavior in stepfather families. *Child Development,* 1982, *53,* 472–480.

Scarr, S. From evolution to Larry P., or what shall we do about IQ tests? *Intelligence,* 1978, *2,* 325–342.

Schiff, M.; Duyme, M.; Dumaret, A.; and Tomkiewicz, S. How much could we boost scholastic achievement and IQ scores? A direct answer from a French adoption study. *Cognition,* 1982, *12,* 165–196.

Schmitt, B. D. Ambulatory pediatrics. In C. H. Kempe, H. K. Silver, and D. O'Brien (Eds.), *Current pediatric diagnosis and treatment,* 8th ed. Los Altos, Calif.: Lange, 1984.

Sever, J. L. Maternal infections. In C. C. Brown (Ed.), *Childhood learning disabilities and prenatal risk: Pediatric round table; 9.* Skillman, N.J.: Johnson & Johnson Baby Products Co., 1983.

Shapiro, A. K., and Shapiro, E. S. *Tics, Tourette syndrome and other movement disorders.* Bayside, N.Y.: The Tourette Syndrome Association, 1980.

Sheldon, W. H. *The varieties of human physique.* New York: Harper & Row, 1940.

Shockley, W. Dysgenics, genecity, raceology: A challenge to the intellectual responsibility of educators. *Phi Beta Kappan,* 1972, *53,* 297–307.

Siegal, M., and Cowen, J. Appraisals of intervention: The mothers' versus the culprits' behavior as determinants of children's evaluations of disciplinary techniques. *Child Development,* 1984, *55,* 1760–1766.

Silver, H. K. Growth and Development. In C. H. Kempe, H. K. Silver, and D. O'Brien (Eds.), *Current pediatric diagnosis and treatment,* 8th ed. Los Altos, Calif.: Lange, 1984.

Silver, H. K.; Gotlin, R. W.; and Klingensmith, G. J. Endocrine disorders. In C. H. Kempe, H. K. Silver, and D. O'Brien (Eds.), *Current pediatric diagnosis and treatment,* 8th ed. Los Altos, Calif.: Lange, 1984.

Silver, L. B. Forum. In C. C. Brown (Ed.), *Childhood learning disabilities and prenatal risk: Pediatric round table; 9.* Skillman, N.J.: Johnson & Johnson Baby Products Co., 1983.

Simopoulos, A. P. Nutrition. In C. C. Brown (Ed.), *Childhood learning disabilities and prenatal risk: Pediatric round table; 9.* Skillman, N.J.: Johnson & Johnson Baby Products Co., 1983.

Skinner, B. F. *Beyond freedom and dignity.* New York: Knopf, 1971.

Spearman, C. *The abilities of man.* London: Macmillan, 1927.

Stapleton, C., and MacCormack, N. When parents divorce, DHHS Publication No. (ADM)81-1120. Washington, D.C.: U.S. Government Printing Office, 1981.

Sternberg, R. J. Testing and cognitive psychology. *American Psychologist,* 1981, *36*(10), 1181–1189.

Sternberg, R. J. Reasoning, problem solving and intelligence. In R. J. Sternberg (Ed.), *Handbook of human intelligence.* New York: Cambridge University Press, 1982.

Terrell, F., and Terrell, S. L. The relationship between race of examiner, cultural mistrust, and the intelligence test performance of black children. *Psychology in the Schools,* 1983, *20,* 367–369.

Thurstone, L. L. *Primary mental abilities.* Chicago: University of Chicago Press, 1938.

Tom, D. Y. H.; Cooper, H.; and McGraw, M. Influences of student background and teacher authoritarianism on teacher expectations. *Journal of Educational Psychology,* 1984, *76,* 259–265.

Trueman, D. What are the characteristics of school phobic children? *Psychological Reports,* 1984, *54,* 191–202.

Turiel, E. An experimental test of the sequentiality of the developmental stages in the child's moral judgments. *Journal of Personality and Social Psychology,* 1966, *3,* 611–618.

Vail, P. L. *The world of the gifted child.* New York: Walker, 1979.

Wallerstein, J. S., and Kelly, J. B. The effects of parental divorce: Experiences of the child in later latency. *American Journal of Orthopsychiatry,* 1976, *46,* 256–269.

Weinraub, M. Fatherhood: The myth of the second-class parent. In J. Stevens and M. Mathews (Eds.), *Mother/child, father/child relationships.* Washington, D.C.: National Association for the Education of Young Children, 1978.

Weinraub, M., and Wolf, B. M. Effects of stress and social supports on mother-child interactions in single- and two-parent families. *Child Development,* 1983, *54,* 1297–1311.

Wilson, J. Q., and Herrnstein, R. *Crime and human nature.* New York: Simon & Schuster, 1985.

Wolf, T. H. *Alfred Binet.* Chicago: University of Chicago Press, 1973.

Wright, J. C., and Huston, A. C. A matter of form: Potentials of television for young viewers. *American Psychologist,* 1983, *38,* 835–843.

♥

Adolescence

7

Chapter Outline

Physical Development
- *Pubescent Changes*
- *Nutritional Status*
- *Health Maintenance*

Cognitive Development
- *Piaget's Formal Operations Stage*
- *Sex Differences in Cognition*

Psychosocial Development
- *Identity*
- *Moral Development*
- *Family Interactions*
- *Peer Interactions*
- *School Interactions*
- *Community Interactions*
- *Sexual Interactions*
- *Delinquency*
- *Substance Abuse*
- *Alienation*

Key Concepts

achievement needs
acne
adolescent egocentrism
adolescent suicides
affiliation needs
alienation
androgens
anorexia nervosa
bulimia
combinatorial analysis
cultural universals
delinquent acts
depression
estrogens
fear of success
foreclosure of identity
formal operations
general adaptation syndrome
generation gap
genital stage
heightism
hypothetical-deductive reasoning
identity achievement
identity diffusion
identity versus role confusion
independence training
infectious mononucleosis
interpropositional thinking
iron deficiency anemia
juvenile crimes
late maturation
menarche
menstruation
moratorium
nearness training
obesity
peer pressures
precocious puberty
primary sexual characteristics
principled moral reasoning
progestins
puberty
pubescence
secondary sexual characteristics
sexually transmitted diseases
socialized conduct disorders
status offenses
substance abuse
teenage pregnancy
unsocialized conduct disorders

Case Study

Rory's high school had a strict policy about school absenteeism. If a parent or guardian did not call the school each day that a pupil was absent, the student would be given demerits. The demerits had to be worked off by helping to clean the school on Saturdays. Demerits were also acquired for other misdemeanors. Work details were posted each Thursday, with no explanation of how the demerits were earned.

Rory broke his right wrist during a gym period partially because of a negligent act by the coach. A cast was put on, which extended halfway down his hand, preventing his thumb and fingers from meeting. Consequently, he could neither hold a pen nor write. His mother called the school to explain his absence the first day after the accident. She wanted to ask that the school allow Rory to "take notes" with a tape recorder, take quizzes or exams orally, and forgo homework assignments until the cast was removed. With so many requests, she asked for the principal rather than the secretary to whom school absences were supposed to be explained.

The principal was very understanding. He assured Rory's mother that special arrangements would be made to enable Rory to keep up in his classes despite his shattered wrist. The principal did send notes explaining Rory's inability to write to the teachers of his academic subjects. However, he neglected to tell the coach that Rory would not be able to take gym for a while. He also forgot to put Rory's name on the absence list for that day. The gym teacher, noticing Rory's absence, checked the list then assigned him demerits for not attending gym class. The same thing happened several days in a row.

When Rory saw his name on the Saturday work detail, he went to ask why. The school secretary explained that it was for unexcused absences from gym. Rory tried to explain that he couldn't take gym with a cast and that he thought his mother and the principal had arranged for him to be excused. The secretary assured Rory that she made such arrangements, not the principal, and that his mother should have called her.

Rory became upset and asked to see the principal. The principal heard Rory's anger, not his logic, heard his disrespect, not his plea for justice. Rory tried to explain that the coach saw the accident happen, got the nurse, heard the nurse recommend an x-ray, and heard her call Rory's parents to come and take him to a hospital. How could the coach not remember and check to see whether Rory had broken any bones? Why hadn't the principal notified the gym teacher? The principal, in turn, got angry at Rory for daring to question the behavior of the gym teacher and the school principal. He assigned Rory still more demerits for disrespect. Rory's mother then had to call the school secretary to have her notify the gym teacher that Rory would not be participating in athletics until the cast was removed, to prevent more demerits from accruing.

Adolescents are supposed to learn to make principled moral judgments, respectful of the rights and obligations of all concerned parties. Cognitive maturity and examples of others enhance this moral maturity. What kind of examples did Rory's school coach, secretary, and principal set? What should a more mature Rory have done?

Most people cannot recall a particular time when they entered adolescence and left childhood behind. In our society the adolescent period often seems to be a holding pattern between childhood and adulthood (see Figure 7–1). In early adolescence it is easy to slip back into the role of a child to suit a particular purpose. In early adulthood many individuals choose to slip back into adolescent roles.

Adolescence is often defined by both physical and social hallmarks. It is a period of time marked by the biological changes of puberty, and it is a transitional time socially. The adolescent becomes identified as a person increasingly able to make his or her own decisions about school, leisure, employment, and friends. It ends with an independence from the family of origin, brought about

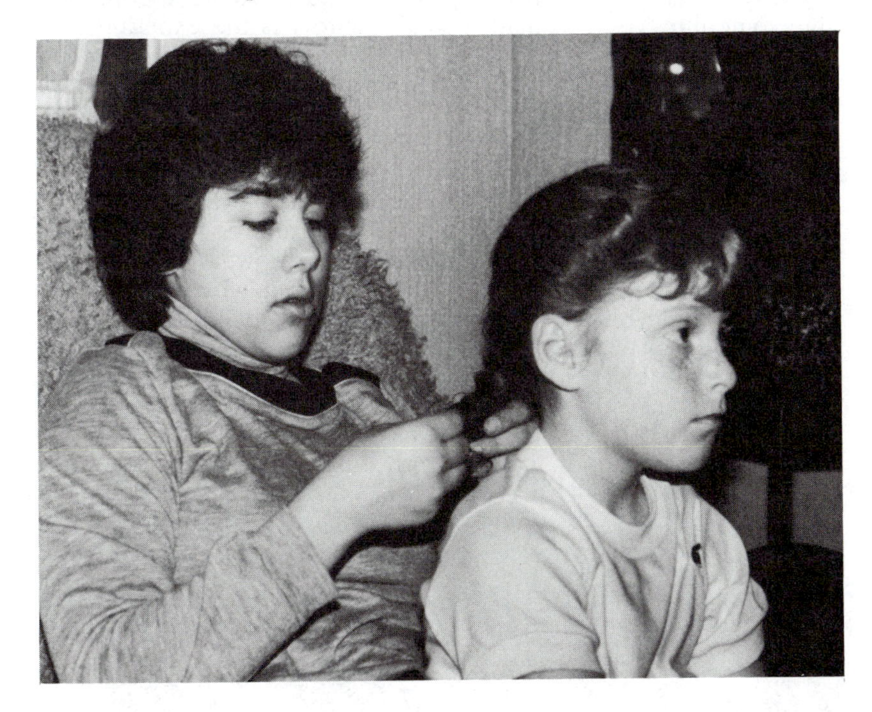

Figure 7–1
The self-concept of an adolescent often undergoes rapid changes—from child to adult, from dependent to self-sufficient, from unity to opposition with families. Old and new tasks are combined in the ever-changing challenge of the transition period between childhood and adulthood.

by marriage or a full-time job. For some people adolescence may begin at age eleven or twelve and go on through the middle to late twenties. For others it may start at age fourteen or fifteen and end within a year. Physical, cognitive, and social factors all help determine the length of the adolescent period.

Physical Development

The physical changes of adolescence lead to new problems for the growing, developing human being. It is not easy to adjust to all the bodily changes that occur. When an adolescent looks in the mirror, he or she is apt to constantly see changes in appearance, real or imagined. Increases in height and weight, budding sex organs, body hair, facial blemishes, oversized hands or feet, and in the male an enlarging Adam's apple are just some of the changes that may confront the adolescent. The lack of these things, if friends have them, may also worry the teenager.

PUBESCENT CHANGES

Puberty encompasses the one to two years of rapid growth and development during which individuals become capable of sexual reproduction. Sexual maturity for a girl is often defined as the time of the first menstrual period, but this is slightly inaccurate. Most girls are not immediately fertile after beginning to menstruate. They may have anovulatory cycles (not productive of mature ova) for one or two years before they can actually bear children.

It is harder to mark a point in time when boys reach sexual maturity or have the ability to produce and ejaculate sperm. One sign sometimes used to mark sexual maturity is the experience of wet dreams. However, wet dreams are environmentally influenced and may occur long after sexual maturity. A more scientific way to determine male fertility is to microscopically examine urine for evidence of sperm.

In both girls and boys a growth spurt accompanies puberty. If one feels the need to pinpoint sexual maturity, one can keep regular measurements of changes in height, weight, and body proportions during adolescence. The period of most rapid growth is a good indication of the time when sexual maturity is being achieved.

The physical hallmarks of **pubescence** (changes accompanying arrival of sexual maturity) vary from individual to individual. They are initiated by the periodic secretion of luteinizing hormone releasing hormone (LHRH) by the hypothalamus in the brain. These periodic bursts of LHRH cause the pituitary gland in the brain to release gonadotropic hormones, which in turn stimulate the gonads (ovaries

in the female and testes in the male), and to a lesser extent the adrenal cortex, to produce and release sex hormones. Both males and females produce "male" sex hormones called **androgens** (the most potent androgen is testosterone), "female" sex hormones called **estrogens** (the major estrogen is estradiol), and **progestins** (the major progestin is progesterone). Males, however, produce a higher amount of androgens whereas females produce a higher amount of estrogens and progestins.

Breast budding, characterized by firm nodularity, may be the first indication a girl has of increased sex hormone production and the approach of adolescence. Then her nipples become pigmented, her breasts swell, and her hips grow wider as the pelvic structures expand and fat is deposited over them. Pubic hair grows and eventually becomes abundant and curly.

The onset of menstruation is known as **menarche**. A female begins menstruating when her production of estrogen, in response to bursts of LHRH, becomes cyclic (see Figure 7–2). An increased production of estrogen occurs approximately once each month in the ovaries as Graafian follicles develop prior to the maturation and discharge of an ovum (*ovulation;* refer back to Chapter 3). After ovulation the immature Graafian follicles and the one mature follicle that sheds its ovum (now called the corpus luteum) degenerate unless fertilization of the ovum took place. The corpus luteum's degeneration causes a decreased production of estrogen and progesterone (meaning "for pregnancy"). Two or three days after the production of hormones stops, the uterus sheds the lining it had developed for the eventuality of a pregnancy. This discharge of the bloody lining is **menstruation**. The lowered estrogen and progesterone levels also trigger the ovaries to begin ripening more Graafian follicles, and thus hormone levels again rise until another ovulation. Rising estrogen levels stimulate the production of another deep, red, spongy endometrial lining on the inner uterine wall in readiness for the possibility of a fertilized ovum reaching the uterus after the next ovulation. If it doesn't, decreased hormone levels will again trigger the shedding of this lining in another menstrual period. The first several menstrual periods of puberty are often irregular in both amount of flow and timing. The interval between menses may be longer or shorter than is characteristic later in the woman's life (Silver, 1984).

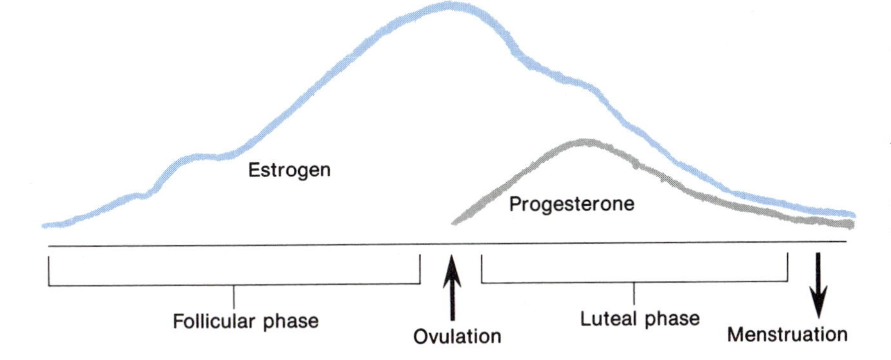

Figure 7–2
Schematic representation of approximate monthly estrogen and progesterone fluctuations in the female, with high and low secretions based on a 30-day cycle. (Increases or decreases in daily hormone production can also be influenced by heightened emotionality or illness.)

The age at which sexual maturity is reached in American girls varies from nine to seventeen years, with an average age of first menstruation now at twelve years, five months. The range for sexual maturity in American boys is estimated to be from eleven to nineteen. Girls, on the average, reach sexual maturity about two years ahead of boys. As noted in Chapter 5, there is a trend toward each generation growing larger than the last. Tanner (1962) reported a similar trend moving the age of puberty down with each successive generation. He reported that in 1840 the average girl's first menstrual period occurred in her seventeenth year. Bullough (1981), after examining ancient Roman, medieval, and nineteenth-century medical documents, questioned Tanner's theory. His data show an onset of menstruation between the ages of twelve and fourteen in the past. Tanner's claim of average menarche at seventeen in the 1840s was based on a small, isolated Norwegian population. Bullough found a slight decline in the twentieth century (about one year), probably due to improved nutrition and health care. Climate apparently has little effect on age of menarche. Nigerian and Eskimo girls begin menstruating at approximately the same ages. Likewise, race appears to have little effect when nutritional and health factors are similar (Silver, 1984).

Menstruation is accompanied by expectations of physical discomfort, increased emotionality, poor school performance, and the need to interrupt regular activities (Clarke and Ruble, 1978). These negative attitudes and expectations are culture bound and reportedly come from parents, friends, health classes, and books (for young girls), and TV and male friends (for boys). Such beliefs in menstrual distress can become a self-fulfilling prophecy for girls. Brooks-Gunn and Ruble (1982) found that girls who learned about menstrual distress from male sources rated menstruation as more debilitating and negative than girls who learned from female sources. Premenarcheal girls (those who have not yet reached menarche) usually expect to experience more distress (pain, water retention, irritability, poor concentration) than postmenarcheal girls (Brooks-Gunn and Ruble, 1982). Early maturers and girls who were unprepared for their menarches usually have more negative attitudes about their periods than well-prepared girls and girls who achieved menarche at later ages (Ruble and Brooks-Gunn, 1982). Initially, menarche creates confusion, ambivalence, and inconvenience. Most girls report being scared and upset at their first period (Whisnant and Zegans, 1975; Weideger, 1976) and more self-conscious afterward (Koff, Rierdan, and Jacobson, 1981). However, Greif and Ulman (1982) reported that some postmenarcheal girls later described menarche as a positive event that helped them "fit in" with other girls and reorganize and clarify their sexual identity.

Menstrual cramps are often given as reasons for school absence or refusal to participate in athletics. However, nonprescription medicines containing aspirin or acetaminophen can be used to alleviate cramping if and when it occurs, and few adolescent females need to curtail their regular activities because of menstruation.

Advertising campaigns for medicines to alleviate symptoms of *premenstrual syndrome* (PMS) have brought a new excuse for nonparticipation at school or home to the adolescent female population. However, premenstrual syndrome is more common in older women and is discussed as a reason for possible psychological debilitation in Chapter 9. Mood changes in teenagers related to menstruation are generally less pronounced and can usually be consciously controlled (Muller, 1985).

Increased production of sex hormones in boys causes the penis and testes to enlarge and pubic hair to grow. Some boys will also experience some breast development (gynecomastia) early in puberty. The changes in males and females that contribute to reproductive maturity (changes in the breasts, ovaries, and vagina or in the penis and testes) are called **primary sexual characteristics**. The increased androgens and estrogens of pubescence also bring about many other physical changes in males and females that are known as **secondary sexual characteristics**; they include changes in body shape, additional body hair, voice changes, organ changes, and even acne.

Females develop broader hips and more body fat as they grow older. Hair grows on their arms and legs, in their axillae (armpits), and, in some girls, on the upper lip or near the nipples of the breasts as well. Such facial and chest hair does not grow as coarse and thick as it does on males. Girls' voices become fuller and richer with the lengthening of their vocal cords. Sweat glands become active and body odors and acne appear.

Males also develop acne and increased overall body perspiration at puberty. They develop coarse facial and body hair as well as pubic and axillary hair. Testosterone acts on the receptor cells of their shoulders, causing them to broaden (Tanner, 1974). While females develop more body fat, males develop more muscle tissue. Males average 50 percent muscle and 15 percent fat, while females average 40 percent muscle and 25 percent fat after puberty.

Before adolescence girls and boys are similar in muscle strength. However, during pubescence boys have a great increase in muscle size and strength and develop more muscle mass and more force per gram of muscle (Tanner, 1974). This is especially true of the upper body. Boys develop proportionately longer limbs, while girls have proportionately larger trunks and wider pelvic girdles. The femurs of females are attached to the pelvis at more oblique angles than those of males (Tanner, 1962). Males' center of gravity is more medially located because of their wider shoulders and narrower hips. Males have still another advantage after puberty: they develop more aerobic power—greater ability to get oxygen to body cells and

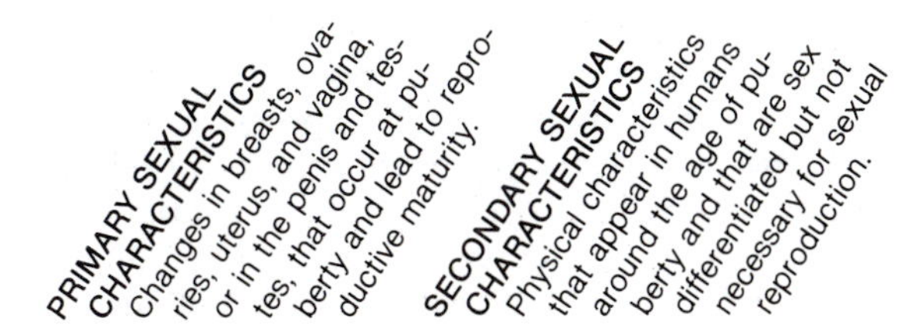

get rid of waste products. Many of these physical differences between the sexes give the average male an advantage over the average female in our traditional sports, although there are always exceptions. In long-distance swimming, for example, females have an advantage because of their narrower shoulders, lighter muscles, and more fat-cell-insulated and buoyant bodies.

The heart approximately doubles its weight in both sexes during adolescence, growing slightly larger in boys than in girls. Blood pressure also rises appreciably, with males eventually having higher systolic pressure than females.

Both boys and girls experience voice changes. Girls' voices become fuller and richer. Boys' voices become lower and louder. The deeper male voice results from enlargement of the larynx (the Adam's apple) and the lengthening of the vocal cords. In the process of acquiring a mature male voice, a boy may occasionally experience embarrassing voice breaks or squeaks in the middle of sentences. Although parents may be proud of this sign that their son is maturing, siblings and friends usually giggle.

Pubescent changes all take time. There may be from two to five years between the first notice of pubic hair and the final development of full sexual maturity. The extreme awkwardness that some teenagers experience during pubescence is a result of rapid and uneven development. Arms and legs or hands and feet may reach adult proportions before the rest of the body. Teenagers may discover that their big feet and longer legs are only two strides away from the snack table rather than three as they knock it over, or they may misjudge the distance they have to reach for the milk with the same results. Girls as well as boys may experience some of these judgment problems. Such problems probably result from failure to adapt quickly enough to changed body proportions. Some teenagers never go through an awkward stage. Others seem to learn to be clumsy and continue having difficulties long after any physical adjustment problems can be blamed.

When teenagers fail to develop signs of sexual maturity at the time when their friends are changing, their self-concept and self-esteem can sometimes be adversely affected. They may feel different, unacceptable, alone. Girls usually are less affected by **late maturation** than boys because a small or flat-chested female is not without sex appeal in our society. A late-maturing girl may also find that her lithe, slender body makes it easier for her to excel in sports. However, late-maturing girls may gravitate toward a younger peer group to find acceptance, or they may be excessively modest among their own age-mates. Many late-maturing girls are taken to physicians to ascertain whether an endocrine disturbance is contributing to the delays.

Late-maturing boys have been shown to be less popular, to feel inferior and less secure, to be more emotionally expressive (eager, animated, energetic, talkative), to resort to more attention-seeking behaviors, and to have more body and social acceptance concerns than their normally maturing peers (Clausen, 1975). Occasionally they may withdraw from social activities, possibly due to embarrassment about their small size, smooth faces, and high-pitched voices. Late-maturing boys tend to be more passive in boy-girl relationships. They usually do not date as early as their age-mates. Gillis (1982) discussed the prevalence of **heightism** in our society (similar to racism and sexism) directed against shorter people, especially against shorter men. He felt that late-maturing teenage boys may begin smoking to try to give an older appearance or may turn to drugs to ease anxiety and relieve depression about being smaller. He wrote that the male-taller, female-smaller rule has got to go.

Early-maturing teenagers generally have a much easier time than late maturers. Girls may temporarily seek out older girls for friends and usually begin dating earlier than their age-mates. Early-maturing boys are usually looked up to by their peers and rank high in dominance hierarchies. Early-maturing adolescents tend to be more independent, self-confident, and self-reliant than late maturers.

In rare cases, puberty may begin before age nine in girls or before age ten in boys. This is known as **precocious puberty**. It may be inherited in 5 to 10 percent of boys but its cause in girls remains a mystery. Occasionally, a benign tumor is found on the hypothalamus, causing early release of LHRH, the hormone that triggers puberty. Children with precocious puberty may feel isolated and socially rejected. Boys often become more aggressive and hyperactive than their peers. Adults often expect the child to act as old as he or she looks, rather than as old as he or she is, which is frustrating for the child. Many physicians try to delay sexual maturation with hormone therapy in precocious children to allow them to begin puberty later, at a more appropriate age (Johnson, 1983).

NUTRITIONAL STATUS

The nutritional needs of adolescents and the role of good nutrition in maintaining physical and emotional health cannot be overemphasized. Teenagers need to eat more during pubescence to provide their bodies with the nutrients necessary for rapidly accelerating growth. Appetites normally correspond with the need for more food. However, many teenagers get into trouble nutritionally. They eat the wrong kinds of foods, they overeat, or they refuse to eat to keep fashionably slim.

Calcium and iron have been identified as the most common deficiencies in adolescent diets. Protein deficiency also frequently accompanies iron deficiency. The substitution of sodas for milk and snack foods for meals helps to account for these deficits. Calcium

LATE MATURATION Puberty that occurs significantly later than the average.

HEIGHTISM Prejudice directed against shorter people.

PRECOCIOUS PUBERTY Puberty beginning before age nine in girls or before age ten in boys.

deficiency can lead to osteomalacia, better known as rickets, although this disorder is generally due to a deficiency of vitamin D that prevents the efficient absorption of calcium from the intestine. Most milk is fortified with vitamin D, but if teenagers do not get adequate amounts of milk or other dairy products and if they have minimal exposure to the ultraviolet light of the sun, which initiates production of vitamin D from precursors found in the skin, they may get symptoms of osteomalacia. These include softening and bowing of the long bones and subsequent problems such as backache and kyphosis (humpback).

Iron deficiency results in a reduced amount of blood hemoglobin, the substance that carries oxygen from the lungs to the tissues, and anemia (a reduced amount of red blood cells). The anemia may be manifested by lack of energy, quick fatigue on exertion, poor muscle tone, shortness of breath, and a pale appearance. Although both sexes may have **iron deficiency anemia**, it is much more common in teenage girls. They lose iron each month in their menstrual flow and may not eat enough of the iron-rich foods (red meats, eggs, fortified cereals and breads, green leafy vegetables) to replace it.

Newman and his colleagues (1986) suggested that teenage boys whose diets include a good deal of cholesterol and other saturated fats (as from pork, beef, egg yolks, butter, and cream) may already have fatty streaks of cholesterol and fibrous plaques in their blood vessels, an early sign of atherosclerosis. No fibrous plaques are found in females, who may be protected from early build-ups of plaque due to the estrogens in their blood vessels.

A lack of protein can lead to a malnourished state. Malnourished people can appear of normal weight or even overweight due to their intake of calories through carbohydrates, but they are more susceptible to infections, are mentally sluggish, have short attention spans, tend to be irritable, and have the same problems of fatigue and poor muscle tone that the anemic person has. Table 7–1 outlines adolescent caloric needs according to age, height, weight, and sex.

Obesity One out of every ten American teenagers suffers from **obesity** (is more than 20 percent above his or her ideal body weight;

Table 7–1 Caloric Needs of Adolescents

	Age	Height	Weight	Daily caloric needs
Females	9–12	4′7″	75 lbs.	2400
	12–15	5′1″	100 lbs.	2500
	15–18	5′4″	120 lbs.	2300
Males	12–15	5′1″	100 lbs.	3000
	15–18	5′7″	135 lbs.	3400
	18–21	5′9″	155 lbs.	2950

McAnarney and Greydanus, 1984). Many of these overweight individuals can also be malnourished because they support their eating habits with carbohydrates and fats rather than with proteins. Obese youths are prone to both health problems and personality problems.

Obese youths usually see themselves as unattractive in appearance and socially less acceptable than their thinner peers. They may spend a great deal of time and money on diets and slimming plans. Lacking signs of immediate improvement, they become discouraged and overeat as a way to assuage their feelings of failure. This vicious circle may repeat itself over and over. Meanwhile the obese youth's self-image remains low. Often one or both parents are also overweight. Munching on high-calorie snacks while watching television may be a part of their lifestyle and exercises may be shunned. There is evidence from adoption studies that obesity may have a strong hereditary component as well. Stunkard and his colleagues (1986) reported that 80 percent of the offspring of two obese parents became obese, as compared with no more than 14 percent of the offspring of two parents of normal weight, even when raised by adoptive parents.

Rodin (1978) reported that many overweight individuals are highly, sometimes uncontrollably, responsive to the sight and smell of food. Even after dieting they remain extremely responsive to food cues. Any kind of emotional arousal (excitement, distress, amusement) increases the likelihood that hyperresponsive persons will eat. Even nonemotional stimulants like coffee and tea increase the desire to eat. For this reason few overweight people manage to keep off the weight they lose through dieting. Rodin hypothesized that the unhappiness due to social pressures and discrimination against the obese and the frustrations surrounding failed diets are two factors that keep adolescents fat rather than cause their problems. She also found that less physical activity in the obese is a result of being overweight rather than the cause. Rand (1979) presented data to show that fear of sexuality, a reason commonly given for the adolescents' layer of protective fat, is also rarely a cause of obesity. For most overweight adolescents there is rather a resentment of the cultural stereotype that obesity decreases sexuality. They have sexual desires but less opportunity to find dates.

Overweight adolescents should be encouraged to start good diets and learn how to keep themselves within a normal weight range. Diets should begin with a complete medical examination. Because adolescents frequently rebel against having someone else supervise their eating, they should be given a great deal of nutritional information. If they can understand what they should eat and

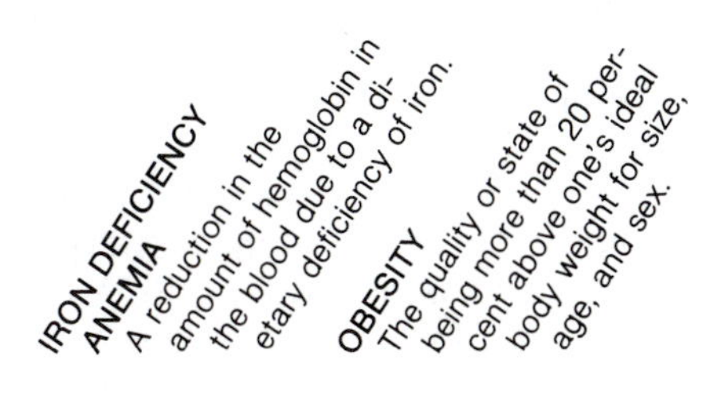

why, the diet has a greater chance for success. Drugs that suppress the appetite should be used only under the direction of a physician and only for a short time. The effects of drugs decrease rapidly with continued use. Self-control must be learned. In order to be effective, any diet plan should also include increasing physical activity with a structured and consistent exercise program. Adolescent diet programs should also include a generous amount of counseling for the psychological problems and negative self-concept that accompany obesity. Unless youths develop more positive self-images and learn to cope with their responsivity to food cues, they may regain any weight lost. Finally, an adolescent diet should allow the youth to take part in peer-group activities as much as possible, substituting low-calorie nutritious foods for the usual "empty" but high-calorie snacks.

Anorexia Nervosa and Bulimia When a person experiences a severe weight loss without the presence of a disease associated with weight loss (such as diseases of the bowel, tumors, or severe infections), **anorexia nervosa** may be suspected. This is often referred to as a diet gone out of control. About 85 percent of sufferers of anorexia nervosa (called *anorexics*) are teenage girls; the remaining anorexics are males or nonteenagers. Anorexics have a morbid fear of fatness, and a distorted attitude toward eating, food, or weight that overrides hunger, admonitions, reassurances, and threats. Encouraging an anorexic to eat and/or regain some weight usually generates a hostile response. The American Psychiatric Association (1980) has listed five diagnostic criteria for anorexia nervosa:

1. Intense fear of becoming obese, which does not diminish as weight loss progresses.
2. Disturbance of body image, such as claiming to "feel fat" even when emaciated.
3. Weight loss of at least 25 percent of original body weight or, if patient is under eighteen years of age, weight loss from original body weight plus projected weight gain expected from growth charts may be combined to make the 25 percent.
4. Refusal to maintain body weight over a minimal normal weight for age and height.
5. No known physical illness that would account for the weight loss.

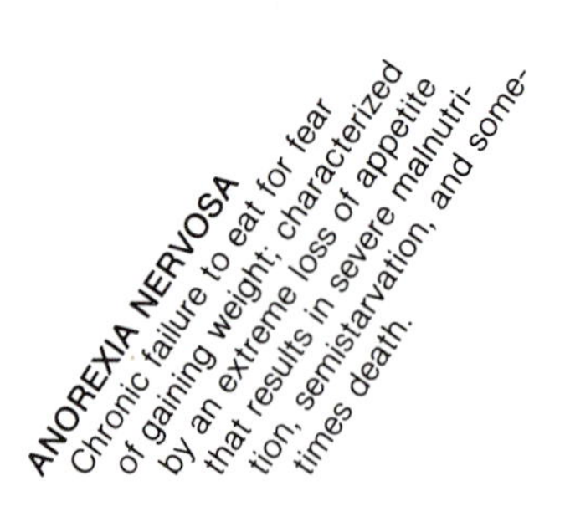

There is also frequently a cessation of menstrual periods, an abnormally slow heartbeat, growth of soft, downy body hair, vomiting (which may be self-induced), social withdrawal, shivering, loss of head hair, constipation, periods of overactivity, and a companion disorder, bulimia.

Anorexia nervosa is becoming increasingly common in our country, now affecting about 1 percent of middle class females. Typical anorexics are intelligent, ambitious, and anxious to do well. Anorexics may perceive that they are not meeting their family's high standards or that peer-group popularity is not what it should be. Frequently they have obsessive-compulsive personality characteristics—they are perfectionists, work oriented, and preoccupied with rules, schedules, and the like. Many other factors may contribute to the syndrome (see Figure 7–3). It is important that the patient and the patient's family be assured that nobody is to blame for the illness. Death rates from starvation have been reported as high as from 5 to 15 percent of patients with anorexia nervosa. However, with more accurate identification and treatment, current starvation mortality rates are reported as low as 0 to 2 percent (Andersen, 1985).

Treatment includes individual psychotherapy for the anorexic patient, nutritional therapy, and family counseling. When weight loss is rapid or severe, or if metabolic abnormalities are present, hospitalization is recommended. Many hospitals use behavioral contracts to encourage weight gains—granting privileges in exchange for eating a sufficient amount of food or reaching a certain caloric intake each day. A danger inherent in such behavior modification programs is that anorexia nervosa will be converted to bulimia. The patient needs to feel in control of the situation, understand the disease process and its effects, understand nutritional needs, and be reassured that she (or he) will not be coerced into becoming obese. Many patients with anorexia nervosa have difficulty in identifying

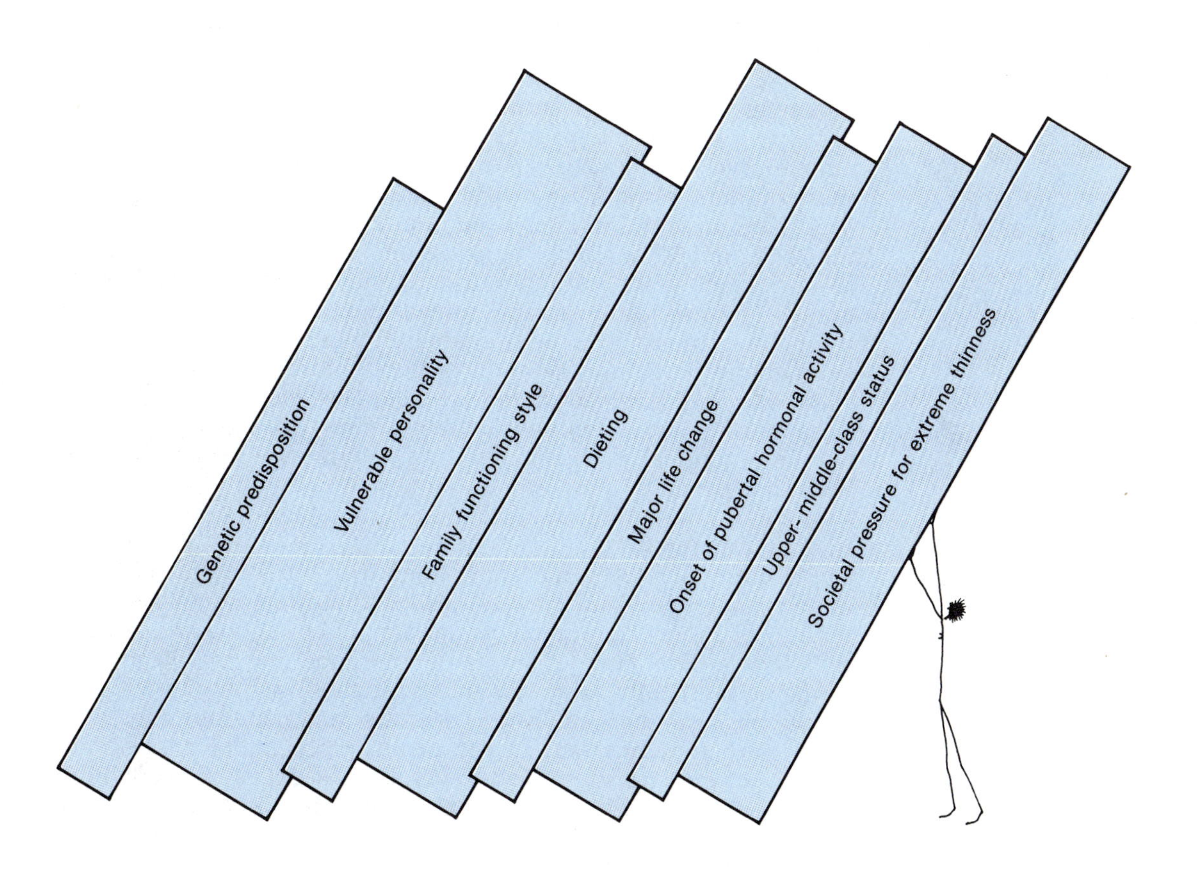

***Figure** 7–3*
Factors contributing to the development of anorexia nervosa. Source: A. E. Andersen, Practical Comprehensive Treatment of Anorexia Nervosa and Bulimia. *Baltimore: Johns Hopkins University Press, 1985), p. 51.*

their inner mood states and giving voice to their inner feelings. Families should be involved in helping patients identify and express their uncomfortable moods. Families should learn how to keep communication channels open after the treatment program ends. Many patients suffer relapses, especially when they have not learned how to relate more effectively with family members and work together to maintain well-being.

Bulimia is excessive binge eating. In bulimic episodes people may gorge themselves with several times their normal daily caloric intake. After such binges, however, they usually induce vomiting or take several laxatives to assure that the food is not digested.

Bulimia may occur in association with anorexia or may occur apart from it. It may be found in as many as 5 percent of high school females, 19 percent of college women, and 5 percent of college men (Halmi, Falk, and Schwartz, 1981). Bulimic patients without concurrent anorexia nervosa tend to be older, more extraverted, and more histrionic, self-dramatizing, and impulsive (Andersen, 1983). They may be of normal weight or even overweight. Their binge-purge episodes tend to occur during periods of stress, in unstructured time, or late in the evening. Food is usually bolted down in private, sometimes in extreme quantities. Then the bulimic usually induces vomiting or abuses laxatives or diuretics to counteract the effects of the eating. Bulimics may become distressed at their inability to control their abnormal eating behaviors. The syndrome may end in suicide or in death from physical complications such as cardiac arrhythmia or severe potassium depletion and blood alkalosis caused by the persistent vomiting and laxative-diuretic abuse. Treatment consists of helping the bulimic patient gain control of binge episodes, supportive psychotherapy, dietary therapy, and family therapy. In some cases antidepressant drugs may be used when features of depressive illness are present. Bulimics need to understand the events prompting their binges and learn to substitute other activities (walking, relaxation exercises, calling a friend) when they feel an urge to gorge. They should learn to spread their caloric intake evenly throughout the day to avoid overwhelming hunger. Families need to learn about the nature of the disorder and to blame neither themselves nor the patient for its occurrence. They can learn a variety of ways to support the bulimic patient's efforts to eat normally and handle emotional problems such as loneliness, anxiety, or anger without binging.

HEALTH MAINTENANCE

The majority of adolescents are blessed with good health, and their maintenance needs are minimal: proper nutrition, adequate sleep, mastery of stress. The infectious diseases of childhood are past and the chronic debilitating diseases of adulthood are still ahead. One of the weakest links in youths' health is their inability to adjust to all the physical changes (rapid growth, sexual maturity) and social changes that occur. Many of the health problems of adolescence are related more to stress than to germs (headaches, menstrual distress, depression, and so on). When stresses build to a level with which teenagers cannot cope, they are apt to get sick, have an accident, develop disturbed eating patterns, or escape through alcohol or drugs.

Selye (1974) differentiated between *eustress* (pleasant and stimulating stress) and *distress*. Our lives would be barren without any stressors. Both eustress and distress can be helpful in small doses and harmful in overloads. Selye described a **general adaptation syndrome (GAS)** to stress that has three stages: (1) alarm, (2) resistance, and (3) exhaustion.

If a person cannot resist the rigors of stress and becomes exhausted, he or she is apt to give up on life. Parents, teachers, and counselors can help adolescents "resist" the stresses they face by allowing them to discuss each one openly and honestly. Rather than telling a teenager how difficult or hopeless things are, one can concentrate on ways to adapt. Teenagers who feel that they can control outside events to some extent by their activities and who can control themselves and their emotional reactions to a large extent cope much better than those who feel powerless over life. Openness to change and involvement in life enhance adaptability.

In many cases adolescents can avoid persons who upset them, choose friends who provide warmth, acceptance, approval, and positive feedback, and spend as little time as possible in situations where they are made to feel uncomfortable. They can be helped to recognize their own signs of distress (such as pounding heart, dry mouth, headache, backache, neckache) and find ways to relieve the discomfort (talk to a trusted counselor, find help to accomplish the task). Selye (1978) emphasized the need to recognize signs of boredom as well. Life is less stressful when one is fully engaged in positive activities. Mastery of stress is also more easily accomplished when one gets sufficient exercise and is free of disease, well nourished, and adequately rested.

Acne Most teenagers have some skin eruptions surrounding puberty. **Acne** is characterized by blemishes, pimples, blackheads, whiteheads (pustules), or cysts of the skin (see Figure 7–4). These most often appear on the face, shoulders, back, and buttocks. Occasionally the pustules and cysts may become infected and result in semipermanent or permanent scarring of the skin below.

Acne is caused by an increase in the activity and secretions of the sebaceous glands of the skin. Oversecretion is thought to be related to increased production of the sex hormones during adolescence. The thick substance secreted by the glands, called sebum,

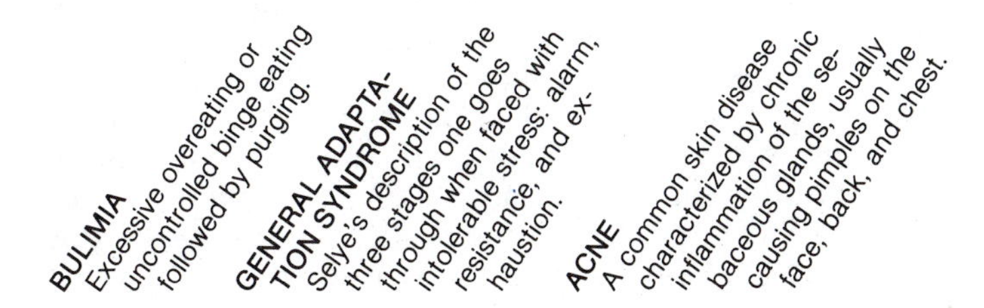

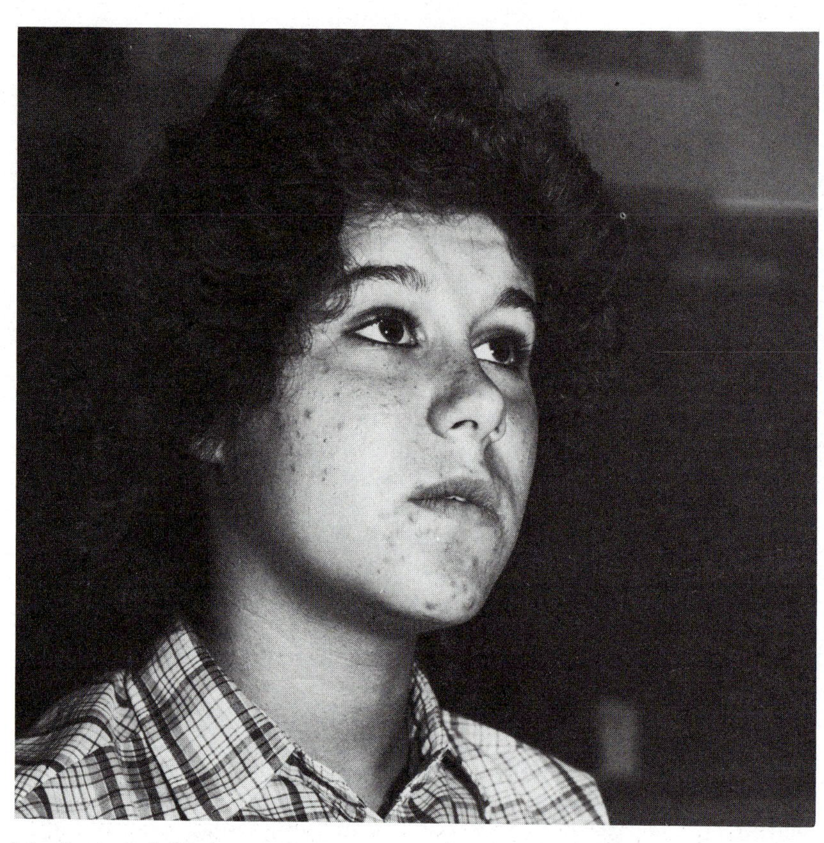

Figure 7–4
Acne is considered one of the worst curses of adolescence by many teenagers.

blocks hair follicles and interacts with other skin substances such as dust, dirt, and bacteria to cause blemishes.

Some teenagers seem to be able to avoid acne simply by normal cleansing of the skin. For others, frequent washing is essential. In addition, attention to diet (three nourishing meals a day with more protein and vegetables, less sugar and fat), sufficient sleep, and stress mastery all decrease the outbreak of blemishes. Although chocolate, nuts, and fried foods may aggravate acne, there is no good evidence that they cause it. Occasionally acne is induced by ingestion or exposure to some drugs (such as corticosteroids or hydantoins; Weston, 1980). Adolescents are cautioned not to squeeze or rub their pimples or to cover them with greasy cosmetics. Special topical acne preparations are available that help hide the blemishes without further blocking the pores. Some antibiotics are also available that help clear up the skin. A new drug called Accutane, which shrinks the sebaceous glands, is sometimes prescribed for cystic acne. However, it has unpleasant side effects: irritation of mucous membranes, nosebleeds, dry eyes, dry mouth, chapped lips, and thinning of the hair (Physicians' Desk Reference, 1986). Its use must be monitored carefully. It must not be given to females who might become pregnant while undergoing treatment because it is teratogenic (causes fetal abnormalities). For mild cases of acne the favored therapy is frequent cleansing of the skin and shampoos at least twice a week.

Infections Adolescents' resistance to infections is lowered when they are overtired, poorly nourished, or stressed. Many teenagers have frequent colds, episodes of acute respiratory and gastrointestinal problems, and skin infections (acne, warts), and teenage girls frequently have urinary tract or vaginal infections.

Warts are caused by infection with the human papillomavirus. They are irregularly shaped tumors usually found on the hands and fingers. If left alone, they usually disappear within six to nine months, although new ones may appear elsewhere. Because they have such an unpleasant appearance, most teenagers want them removed. Several therapies are available, including creams, freezing with liquid nitrogen, salicylic acid plasters, scalpel excision, electrosurgery, and radiotherapy. The latter three treatments often leave scars that may be as cosmetically unpleasant as the wart itself.

Vaginitis is often associated with the physiological changes in the flora of the vagina at puberty that leave the vagina more vulnerable to invasion by pathogenic organisms (McAnarney and Greydanus, 1984). Many (though certainly not all) cases of vaginitis are diagnosed as sexually transmitted diseases. The various kinds of vaginitis will, consequently, be discussed later in this chapter.

Infectious mononucleosis (usually referred to simply as "mono") is often referred to as "the kissing disease" because the organisms causing it have a low communicability—they must be spread by direct contact such as kissing or eating or drinking from the same utensils. Mono can occur at any age but is most frequently seen in teenagers and young adults (Fulginiti, 1984). A quality of exclusivity may link together the "have-had-mono" sufferers from the "have-nots." However, the disease has its dangers. The most characteristic symptom of mononucleosis is extreme tiredness. Sufferers discover themselves lacking their usual get-up-and-go. They also may have fever or sore throats, develop an enlarged spleen, develop faint skin rashes, or have enlarged lymph glands (especially in the neck, axillae, and groin). Many mononucleosis patients develop some temporary abnormal liver functioning. Some of them may also develop hepatitis, an inflammation of the liver. The treatment of infectious mononucleosis varies according to the symptoms that are manifested. It always includes rest. Usually the diet is supplemented with increased protein, vitamins, and iron. The period of convalescence from infectious mononucleosis may be quite prolonged depending on complications and the previous health of the patient. The most serious complication is rupture of the enlarged, fragile spleen with massive blood loss and shock. This requires surgery and removal of the spleen.

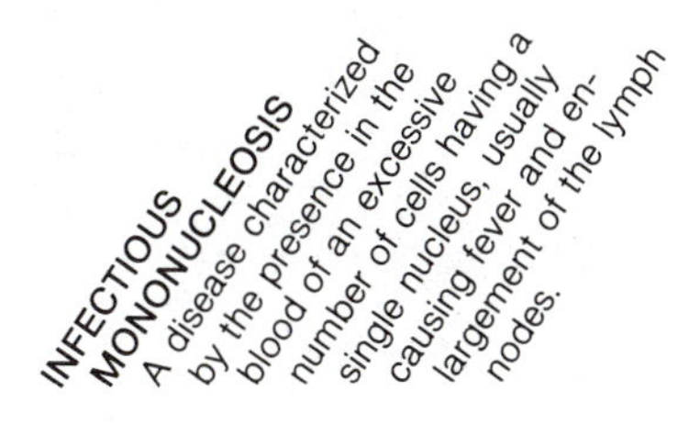

Accidents Adolescents often behave recklessly in their bids for identity and independence. Accidents are the leading cause of death among teenagers, and their incidence has been increasing over the last two decades, now accounting for over 25,000 deaths per year (McAnarney and Greydanus, 1984). Most of these are due to car accidents, many of which are precipitated by driving at excessive speeds under the influence of alcohol or other drugs. Other causes of death or injury that occur with some frequency in the adolescent population are homicide, suicide, sports injuries, drownings, poisonings, drug overdosing, and work-related injuries. The number of hospitalizations or long-term home convalescences from these accidents loom large (see Figure 7–5). Prolonged absence from school due to injuries not only affects educational progress but also disrupts social and emotional development. Many accidents are believed to have an emotional basis. Youths who feel depressed about such things as family strife, school failure, popularity problems, obesity, or poor health may be especially accident prone. Alcohol and drug abuse may also make adolescents more accident prone.

Cognitive Development

New levels of intellectual functioning usually emerge in adolescence. New cognitive abilities, along with the physical changes of puberty, profoundly influence the social development of teenagers. For this reason we will discuss changes in intelligence before describing the social advances of adolescence.

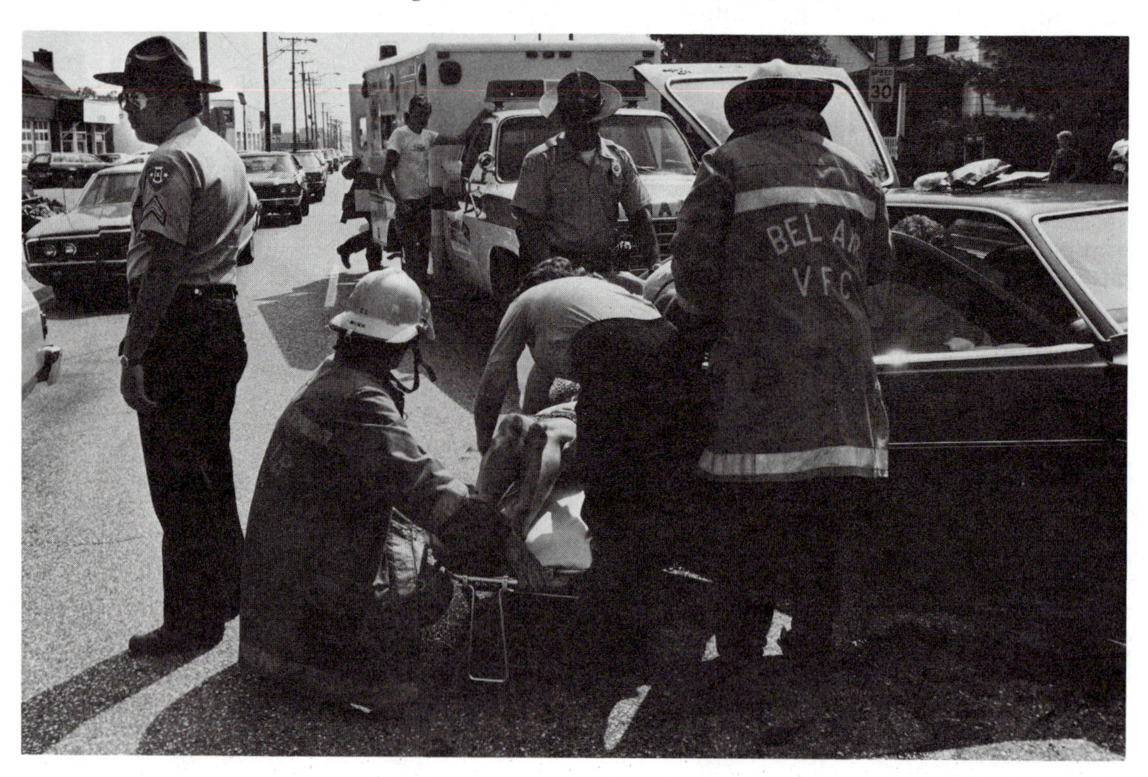

Figure 7–5 *Accidents involving motor vehicles are the leading cause of death and physical handicaps in American adolescents.*

Piaget saw **formal operations** as the last phase of cognitive development, the highest level of intellectual functioning. Although his descriptions of earlier intellectual stages have been well accepted and supported by research, the formal operations stage is frequently criticized—for going too far, for not going far enough, and for not adequately describing the new level of thought.

Those who criticize Piaget for going too far cite research studies that show a lack of formal-operational thought in many adults, both in this country and abroad. Children who have opportunities to manipulate relationships with concrete props, who learn written vocabularies with abstract words, and who are encouraged to develop ordering concepts usually shift from concrete- to formal-operational thinking as defined by Piaget somewhere around adolescence. Most North American children begin school early and are instructed with extensive written language. Written language (symbolic thought) helps children deal with ideas that go beyond concrete here-and-now perceptions of the world. Children who read are helped to think in terms of abstract possibilities as well as in terms of actuality. However, nonreaders, poor readers, and people in cultures that do not ordinarily use written language may not reach formal operations in the Piagetian sense.

Those who feel that a formal stage does not characterize adolescent thought include Keating (1980), who suggested that the new cognitive abilities of adolescents emerge gradually rather than abruptly as Piaget's stage theory supposed. He prefers to think of formal-operational reasoning as a general increase in memory span and ability to plan rather than as a new stage. Overton and Newman (1982) suggest that a lack of formal-operational thought in a test situation may reflect a lack of performance ability in the testing situation, or at the task presented, rather than an actual lack of formal-operational structural capacity. Overton and Mechan (1982) found that individuals perform inconsistently across various tasks used to judge a formal operations stage.

Some cognitive psychologists argue that the "formal" structural abilities Piaget described are too simplistic to describe the reasoning abilities of many adults. Commons, Richards, and Kuhn (1982), for example, found that many university undergraduate and graduate students use systematic and metasystematic reasoning, which is more complex and powerful than formal-operational reasoning. Other criticisms focus on the fact that Piaget took little account of the identity confusion, increased sexual libido, and other psychosocial concerns of adolescents in his analysis of their cognitive reasoning abilities.

PIAGET'S FORMAL OPERATIONS STAGE

In the period of concrete operations children become proficient at what Piaget (1958) called first-order operations—abilities such as classification, serialization, and one-to-one correspondence. In the formal operations stage, youth acquire second-order operations—abilities to see new kinds of logical relationships between classes or between and among several different properties. The reasoning be-

FORMAL OPERATIONS According to Piaget, the stage of cognition characterized by the ability to think abstractly.

hind first-order operations is called intrapropositional thinking, since it deals within the context of just one property (for example, weight). The more advanced reasoning of second-order operations is called **interpropositional thinking**, since it involves handling relationships among several different properties (density, size, and weight). The term *propositional thinking* simply refers to an ability to take objects that have been ordered into classes or correspondence in one way and then to see relations among the classes in some new, logical ways.

In formal operations a person also becomes proficient at what Piaget (1958) calls **combinatorial analysis**, which involves seeing all the possible variations of a problem, or separating out all the possible variables, and then testing them systematically. Imagine all the possible sums of money one can make from a quarter, a dime, a nickel, and a penny. Formal operators will systematically set about the task of combining the coins in some set order such as:

no coins	= 0¢	N + D	= 15¢
P alone	= 1¢	N + Q	= 30¢
N alone	= 5¢	D + Q	= 35¢
D alone	= 10¢	P + N + D	= 16¢
Q alone	= 25¢	P + N + Q	= 31¢
P + N	= 6¢	P + D + Q	= 36¢
P + D	= 11¢	N + D + Q	= 40¢
P + Q	= 26¢	P + N + D + Q	= 41¢

People still functioning with concrete operations will combine coins to make various sums but will not do so systematically. They will forget what combinations they may already have used. Piaget tested youths' level of operations by asking them to do such things as discover the right combination of chemical liquids to produce a yellow-colored liquid or to discover how rods of different composition, length, thickness, and form varied in their ability to bend when weights were attached. When people have the ability to think logically and work systematically, they often enjoy trying to solve such problems. They may even resent hitting the correct solution by chance on an early manipulation, thereby losing out on the fun of trying all the other possibilities.

When one sets out on a task with the idea of testing each possible variation to discover the correct solution(s), one is practicing what Piaget (1958) called **hypothetical-deductive reasoning**. Such reasoning involves solutions that are real (what can be seen, or felt,

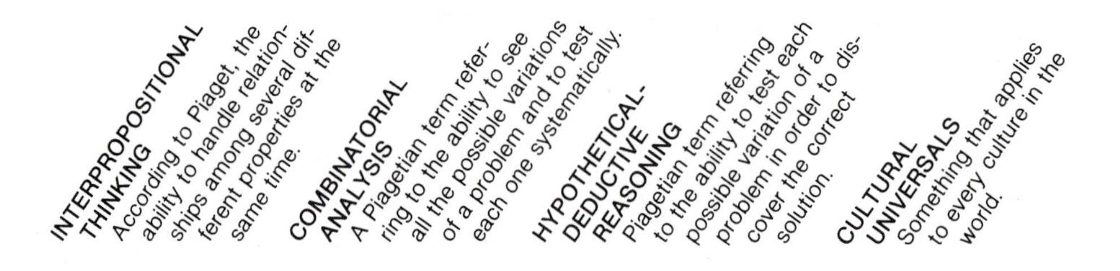

or experienced) and also encompasses abstract "what-if" solutions to the problem—in short, all the possible variations. A set of hypotheses is derived and then tested to deduce the correct answer(s). Piaget described formal operations as the combination of interpropositional thinking, combinatorial analysis, and hypothetical-deductive reasoning (Piaget and Inhelder, 1969). The total process, as he described it, is characterized by painstaking accommodation to detail and careful analyses and observations.

The abstract reasoning abilities of the formal operations period usher in both new creative skills and renewed creative urges in many teenagers. No longer limited to concrete perceptions of the world, they may express their new visions of possibilities in art forms such as painting, sculpture, music, dance, poetry, short stories, diaries, religious tracts, letters, science projects, or inventions. Many creative products of the teenage years show a certain raw-edged sensitivity and originality that is often missing from more mature works. As many creative preschoolers seem to give up original productions when they enter school, many creative teenagers also become inhibited about, or embarrassed by, their creative expressions when they enter the world of adults. Artists who continue to produce as they pass from adolescence to adulthood usually tone down and refine their works so they are more socially acceptable and salable, sometimes at the expense of the freshness of the product.

SEX DIFFERENCES IN COGNITION

By adolescence most girls show a preference for school subjects involving human relationships or verbal skills (literature, composition, foreign languages, history), whereas boys show a preference for subjects with numerical and spatial relationships (math, science). Many people argue that these academic preferences are learned; others contend that they have genetic links.

On the genetic-inheritance side, all cultures have sex-role stereotypes and divide activities into men's and women's work. Men are generally physically stronger and predominate in aggressive, protector, provider roles. Women have and care for children. Mead (1935/1949) found that two **cultural universals** characterize every human social order: The roles assigned men and women differ, and men have higher social status. Theoretically, there should be some genetic bases for such invariant findings.

We know that alterations in hormones neonatally can effect permanent, irreversible changes in postnatal behavior. These changes are called inductive hormone effects. Female fetuses exposed to synthetic forms of progestin, which can have androgenic properties, may be masculinized at birth (enlarged clitoris). They are also more tomboyish, assertive, independent, and self-reliant and often have higher than expected IQs. Behavioral changes related to increased secretions of estrogen and androgen at puberty are called activational hormone effects. (We are less sure that hormones rather than social learning are responsible for the sex differences that emerge at puberty.) Male fetuses born with a genetic recessive disease called androgen-insensitivity syndrome (all body cells are insensitive to androgen) develop the sexual characteristics of females at puberty (clitoris and eventually breasts) but have no gonads (neither ovaries nor testes). Their behavior is more feminine and their verbal abilities are better than their spatial abilities.

From about age eleven on normal girls consistently do better at tasks involving word fluency, grammar, spelling, creative writing, and other aspects of language. Boys begin to excel at tasks requiring spatial perception around age seven and accelerate to adult levels by age eleven. Larger sex differences on measures of mental rotation (ability to visualize how a two- or three-dimensional figure will appear when rotated) favoring boys can be detected across the life span (Linn and Petersen, 1985). Benbow and Stanley (1983) found that large sex differences in mathematical reasoning ability emerge in adolescence, especially among intellectually gifted students. Boys outnumber girls about 13 to 1 among those students who score 700 or above in the math section of the scholastic aptitude test (see Figure 7–6). Our IQ tests are carefully balanced with tests of math, verbal skills, and spatial abilities to assure that neither boys nor girls will have a scoring advantage over the other. Biologists suggest that these sex differences in mental abilities are best explained by genetically based organizational differences in the brain. In females the dominant cerebral hemisphere of the brain develops earlier than it does in males. The speech center is located in the dominant hemisphere (the left side of the brain for right-handed people). This may account for the earlier use of language by girls. Spatial ability is localized on the nondominant hemisphere. The minor hemisphere appears to become more highly developed in boys at puberty.

On the environmental learning side, most girls are consciously or unconsciously encouraged to like subjects that will make them better wives and mothers. If they voice a desire for a career, they are encouraged to choose one in which they can help others (such as secretary, nurse, teacher). Boys born with androgen-insensitivity syndrome and raised as girls are socialized in the same way. Boys, on the other hand, are encouraged to show off their own individual abilities, to excel as individuals, and to acquire status. Working at self-improvement, it is assumed, is as important for a boy as working to help others is for a girl. Boys are encouraged to pursue occupations where they can be competitive, assertive, and logical (business, law, electronics, mechanics). These are broad generalizations but nevertheless hold true for a great proportion of our society. Many people contend that the differential treatment of the sexes begins at birth. The pink and blue colors attached to babies cue in adults as to what to expect from and do for the infants. In childhood girls are **nearness trained** and kept close to home, whereas boys are **independence trained** and encouraged to behave more like Tom Sawyer. Girls are praised for making good conversation (verbal skills), and boys are attired in reinforced play clothes and expected to wander far afield (spatial skills). Grown-ups tend to worry more about boys who do not go exploring or irreparably stain or tear their clothing than about those who do.

NEARNESS TRAINING Education to remain dependent on others.

INDEPENDENCE TRAINING Education to become self-sufficient.

Psychologists have published research data to support the no-

A. **Vocabulary**

shrewd ochre flaunt piscatorial
espionage spangle belfry dilatory

Verbal analogies

Aspiration is to future as ______________ is to past.

(a) hope (b) regret (c) joy (d) ire

Scepter is to authority as ______________ is to weakness.

(a) crown (b) sickness (c) crutch (d) ruler

Spelling

seize voluble existence accession
facetious porcelain misdemeanor rhythm

Comprehension

The propaganda of a nation at war is designed to stimulate the energy of its citizens and their will to win and to imbue them with an overwhelming sense of the justice of their cause. Directed abroad, its purpose is to create precisely contrary effects among citizens of enemy nations and to assure to nationals of allied or subjugated countries full and unwavering assistance.

The title below that best expresses the ideas of this passage is:

(a) Propaganda's failure
(b) Designs for waging war
(c) Influencing opinion in wartime
(d) The propaganda of other nations
(e) Citizens of enemy nations and their allies .... ( )

B. **Math**

Which of the following fractions is greater than $\frac{2}{5}$?

(a) $\frac{2}{7}$ (b) $\frac{3}{8}$ (c) $\frac{9}{25}$ (d) $\frac{1}{4}$ (e) $\frac{5}{12}$

Of the following, the value closest to that of $\frac{42.10 \times .0003}{.002}$ is

(a) .063 (b) .63 (c) 6.3 (d) 63 (e) 630

Find the value of a in the equation $5a - .5a = 9$

(a) 0 (b) 2 (c) 9 (d) 4 (e) 6

Spatial Relations

Which one does not belong?

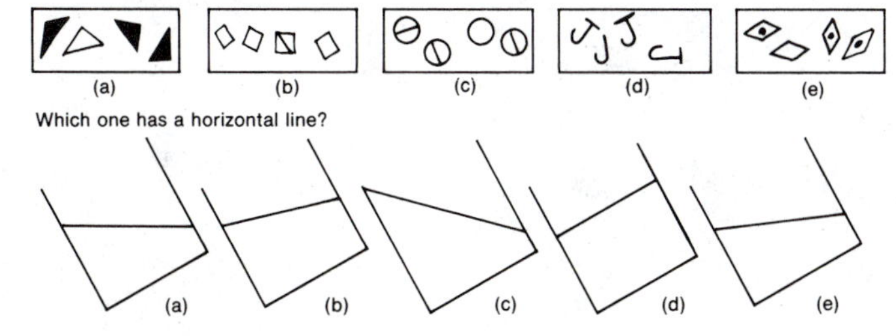

Which one shows the standard in a different orientation?

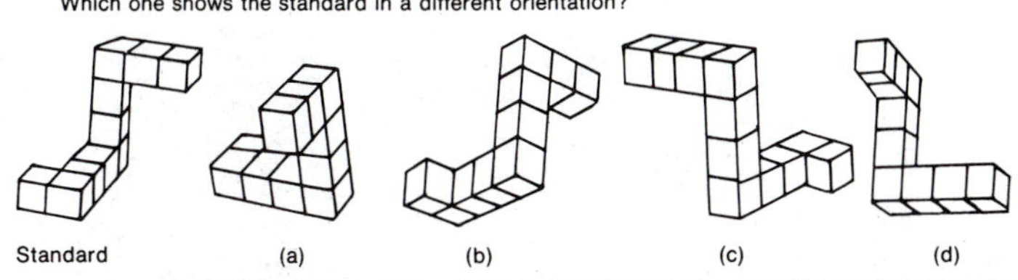

Figure 7–6
Examples of intellectual tasks on which (A) females excel and (B) males excel.

tion that the intellectual differences that become more pronounced after puberty are due to some females' **fear of success** or concern with the negative consequences of success. Some parents still warn daughters not to be "too smart" or they will never have dates (or get married). In a series of studies, Horner (1972) asked women to finish this story: "After first term finals, Anne finds herself at the top of her medical school class. . . ." About 65 percent of them wrote endings such as: "Anne is ugly, studies all the time, and will never get married" or "A mistake has been made. Anne is not really at the top of the class" or "Anne feels guilty. She will finally have a nervous breakdown, quit medical school, and marry a successful young doctor." In Horner's studies only about 10 percent of men wrote similar endings to the story: "After first term finals, John finds himself at the top of his medical school class. . . ." In a replication of Horner's research, Hoffman (1974) found that fear of success was slightly higher in men. Donelson (1977) found different causes for men's fear of success. They may question the value of success, the goals, or the emptiness of achievement. Women's fear of success is more often fear of social rejection. In fact there are more negative social consequences for intellectual achievement for women.

Hoffman (1972) suggested that women's **affiliation needs** may be higher than their **achievement needs** because they are socialized to value acceptance by others more than a position of being the brightest and the best (see Figure 7–7). High school girls often express more concern about finding the right man to marry than about making the right career choice. Erb (1983) found that in early

Figure 7–7 *The need to achieve versus the need to affiliate is often problematic for females who may be torn between their desires to pursue careers and their concerns about wifehood and motherhood.*

adolescence girls typically show an interest in traditional feminine career areas such as service while boys show an interest in high technology. Elder and MacInnis (1983) found that the most achievement-oriented adolescent girls are those whose mothers are well educated, are employed, and have a strong sense of their own personal worth as women. While there are currently more women starting college than men, they are more apt to drop out before graduation to get married. Men are more apt either to graduate or to continue to professional schools, and men eventually earn about three-quarters of all doctoral degrees.

Psychosocial Development

Families are often stunned when their obedient older children become egocentric, argumentative teenagers. They remember the **generation gap** between themselves and their own parents, but they never thought it would happen with their own offspring. All their very best efforts to raise understanding, understandable children seem to have been for naught. Why do teenagers argue? Adolescent developmentalists can offer many explanations. The hormonal changes associated with puberty affect moods. The physical changes in the body are confusing. Suddenly the more adult-appearing individual feels social pressures to become more independent, make a career choice, make sexual choices, take on more adult responsibilities, and "grow up." In addition, the new level of formal-operational thought can substantially alter the adolescent's outlook on his or her own life.

While not all adolescents experience a generation gap (a parting of the ways or breach with their parents) or an identity crisis (a crucial or decisive turning point in their self-concepts), most social scientists agree that an identity search and new self-discoveries characterize psychosocial development. The quest for identity affects moral development, family and peer interactions, and sexual explorations. Identity confusion can also lead to delinquent acts, drug abuse, and alienation.

IDENTITY

Havighurst (1972) proposed eight developmental tasks to which adolescents must address themselves in their identity quest:

1. Accepting one's physique and using the body effectively.
2. Achieving new and more mature relations with age-mates of both sexes.

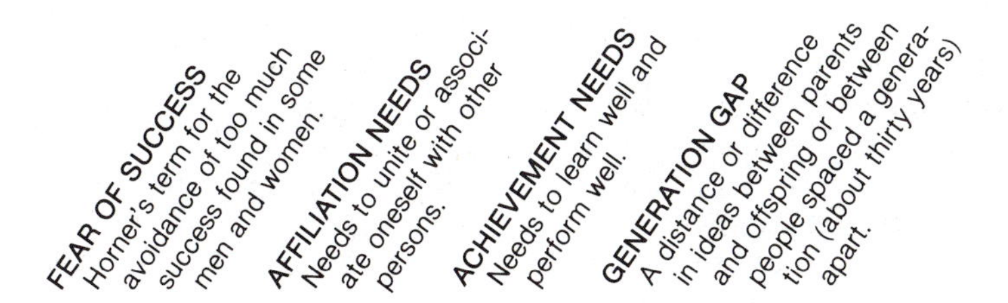

3. Achieving a masculine or feminine gender role.
4. Achieving emotional independence from parents and other adults.
5. Preparing for a career.
6. Preparing for marriage and family life.
7. Desiring and achieving socially responsible behavior.
8. Acquiring a set of values and an ethical system as a guide to behavior—developing an ideology.

Theorists early in this century characterized adolescence as a time of great turmoil (see Box 7–1). Sigmund Freud emphasized the sexual over the cognitive turmoil of childhood and adolescence. He called pubescence the **genital stage** and saw in it a second Oedipal conflict. Youth, he felt, need to free themselves once again from love for their opposite-sex parent. He commented that in so doing adolescents frequently choose parent-substitutes as their first outside love objects: Girls have crushes on older men, and boys have crushes on older women. Normal adolescents, Freud said, finally move away from self-love, same-sex friend attachments, and parental dependency and place their libidinal interest in opposite-sex persons. In Freud's opinion, a healthy self-concept is achieved along with the change from infantile sexual impulses to mature heterosexuality. Several years of confusion pass between a teenager's desire for and achievement of an extrafamilial heterosexual relationship.

Freud's daughter, Anna (1946), also emphasized the ego's response to the increased sexual/libidinal interests of the genital stage. She believed that in some cases the adolescent is id dominated: impulsive and hedonistic. In other cases the adolescent is superego controlled: rigid and rejecting of all things sexual. The dominant superego may lead to an overuse of the defense mechanisms of denial, intellectualization, and sublimation. Anna Freud also proposed that some adolescents use a new defense: asceticism. They practice strict self-denial, refuse to engage in any pleasurable activities, and may even engage in subtle forms of self-torture. In Anna Freud's view, the struggle of the adolescent years is to find a balance (ego) between the extremes of self-pleasuring and self-denial.

Let us look at how more recent theorists have viewed psychosocial development in adolescence, especially as it pertains to a search for an identity. We will begin with Piaget, who felt that the ability to perform formal operations brings in its wake a period of egocentricity, which then leads to new self-discovery.

Egocentricity Piaget (1958) suggested that there are three periods of great egocentrism in the process of development: infancy, early childhood, and early adolescence. He linked **adolescent egocentrism** to cognitive development. In this new form of egocentricity, where all the realms of possibility are viewed, adolescents fail to distinguish between their own conceptualizations and those of the rest of society. They believe everyone should come to terms with

BOX 7–1

Is Storm and Stress Characteristic of the Adolescent Years?

The American psychologist G. Stanley Hall (1904) wrote two volumes about adolescent development at the turn of the century. His studies led him to conclude that the biological changes of puberty inevitably usher in a period of *Sturm und Drang* (storm and stress). He described adolescents as emotionally unstable, with wide mood swings between elation and depression, extraversion and introversion, energy and lassitude. The American anthropologist Margaret Mead (1928) wrote that storm and stress may characterize adolescents living in Western industrialized nations but not teenagers in cultures such as Samoa who make a smooth, swift transition from child status to adult status after a tribal initiation ceremony. An Australian anthropologist, Derek Freeman (1983), criticized Mead's data and data collection methods and reported that Samoans also experience turmoil during their adolescent years.

What accounts for these different findings? Does the answer lie in cohort differences? Were Samoan teenagers less stressed in the 1920s than in the 1970s? Did Mead and her interpreters look only at the cooperative pleasant Samoans and neglect those who were rebellious, hostile, and alienated? Is turbulence universal between the years of thirteen and nineteen? Jerome Dusek and John Flaherty (1981) conducted a longitudinal study of over 500 adolescents in Syracuse, New York and judged that it is not. Their subjects, who were from a wide range of social class levels and ethnic backgrounds, showed continual and gradual growth characterized by continuity and stability. They concluded, "The person who enters adolescence is basically the same as that who exits it" (p. 39).

their own idealistic schemes. They fail to take into account the fact that others may not like or want what they want.

As teenagers see many alternatives to their parents' directives, they become unwilling to accept commands without questioning them. However, they see so many alternatives that they become confused. Frequently teenagers egg their parents into debate just for the sake of discussing alterations of decisions. This may occur even when they approve of the decision. They also turn to their peers to discuss modifications of nearly everything. Using, in Piaget's terms, interpropositional thinking, combinatorial analysis, and hypothetical-deductive reasoning, adolescents may construct an ideal society, an ideal religion, an ideal school, or an ideal family. They compare their ideals with reality and find the adult ways of constructing soci-

GENITAL STAGE Freudian term for the adolescent stage of mature sexuality.

ADOLESCENT EGOCENTRISM Belief held by the adolescent that other people are preoccupied with his or her appearance and behavior.

ety lacking. Teenagers, however, seldom make any great efforts to bring about their own best of all possible worlds. As Elkind (1970) put it, "The very same adolescent who professes his concern for the poor spends his money on clothes and records, not on charity."

Although moral idealism may be high in adolescence, the behavior of teenagers is unpredictable and generally quite self-centered. This egocentricity often takes the form of believing that everyone is extremely concerned with adolescents' appearance and behavior. When young teenagers walk into a crowd, they feel that all eyes are on them, a perception that accounts for the many hours they spend grooming and preening before a mirror. They believe everyone will admire what they admire, which helps explain their failure to understand why adults criticize their faddish clothes, their music, and/or their friends (see Figure 7–8).

Figure 7–8
Piaget wrote that adolescents' egocentrism is a symptom of their ability to conceptualize not only their own thoughts but the thoughts of other people as well. They often believe that others are as preoccupied with their appearance and behavior as they are. This belief causes them to be both vain and self-conscious at the same time.

Youthful egocentrism is also frequently turned into self-criticism. When young people find themselves lacking in some way, they believe the whole world will see the same deficiency. They may plunge into self-improvement exercises or write diaries full of resolutions to change. They may also shroud their efforts to better themselves in secrecy. This desire for privacy in adolescence often leads to lying. In the elementary school years children tell lies that are generally tall tales. They may even come to believe their own stories. In early adolescence young people hide behind deceitful screens they erect to confuse others. However, they know the truth full well in their own minds.

Piaget felt that the egocentricity of early adolescence is modified with age as a consequence of hypothetical-deductive reasoning. Feelings of omnipotence, near-genius insightfulness, and absolute uniqueness are, in fact, hypotheses to be tested. As a consequence of the testing process, teenagers come to realize that the whole world is not focused on their thoughts or appearance but that others have their own preoccupations. Youthful idealism is modified as teenagers gain independence and begin functioning in a world full of the everyday problems of existence. They begin to see themselves in relation to other people, nations, and races and in relation to living and nonliving things. Gradually they become able to appraise their own situations more objectively.

Erikson's Concept of Identity Versus Role Confusion Erikson, like Freud, recognized the influence of sexual libido on the adolescent's self-concept. Unlike Freud, he did not believe that finding a heterosexual love partner would confer on one a mature self-concept. He proposed the nuclear conflict of **identity versus role confusion** as characteristic of adolescence. The search for a sense of identity, he felt, reaches crisis proportions at this time. Not only does the adolescent need to adjust to a sexually mature body, but he or she also needs to re-resolve all the previous nuclear conflicts (trust versus mistrust, autonomy versus doubt, initiative versus guilt, and industry versus inferiority) in light of the newly sensuous self.

The epigenetic nature of Erikson's theory is especially meaningful during this stage. Erikson proposed that many familial and societal, as well as sexual, forces exert an influence on identity achievement. Also, a sense of identity rests on the resolution of many subconflicts (Erikson, 1968). These include finding an appropriate sex role in keeping with family and societal expectations (beliefs about sex typing, behavioral adoptions, preferences and self-perceptions), finding one's own religious, political, economic, and social ideologies amid the confusing plethora of ideas and values espoused by others, finding a vocational or occupational direction that will please the self as well as others, and even finding the "best" way to walk or talk (with elongated legs and vocal cords) that fit one's changing self-concept.

Erikson (1980) felt that adolescents need time to experiment with many different roles before they settle into any particular niche. Adolescents use their same- and opposite-sex friends and their family members as sounding boards on which to test their changing identities. Erikson felt that young love is usually more conversa-

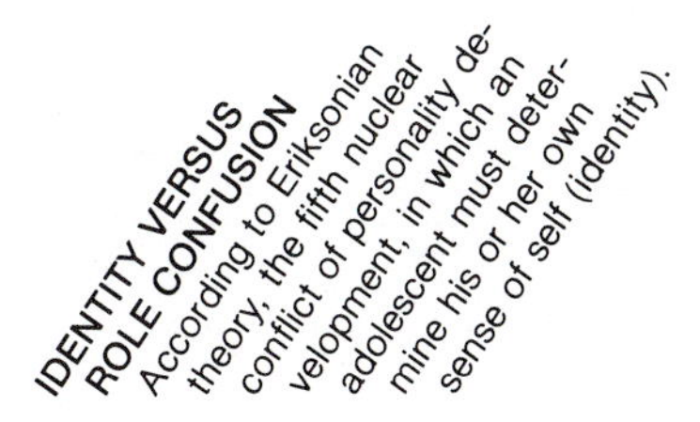

tional than sexual because young persons need to project their developing self-image confidentially to a second person and then have it reflected back to them. In love relationships teenagers feel more secure that their secrets can be confided without the danger of ridicule, abuse, or broadcasting. But many teenagers prefer to use their family members as sounding boards for the same reasons. Erikson (1963) wrote that the failure of so many teenage marriages may be linked to the fact that one cannot give oneself to another until one knows one's own identity.

Erikson (1968) has been criticized for being sexist in his discussion of the identity strivings of women and men. He wrote that a young woman should keep her identity "open" for the peculiarities of the man to be joined and of the children to be brought up. He did not discuss the converse, that something in the young man's identity must keep itself "open" for the peculiarities of the woman to be joined and of the children to be brought up.

Unless adolescents work at affirming and strengthening their identity in adolescence, Erikson feels they will suffer from role confusion. They may imitate others but feel confused about their own sense of self. This role confusion can lead to a progressive sense of identity dissolution. More and more the adolescent will look at others to find out which ones he or she is like, often choosing as most like the self the ones deemed least socially acceptable. When the identity disorder becomes stressful, some outside counseling or therapy is recommended, with parental and sibling participation as well. In some cases the parents project their unfulfilled wishes and desires onto their sons or daughters, asking them to become the person they wanted to, but could not, be. It is especially difficult for adolescents to discover their own uniqueness if parents and peers pressure them to conform and acquiesce to the will of others (see Table 7–2).

Marcia's Identity Statuses James Marcia (1980) built his identity theory on Erikson's beliefs about identity, youth, and crisis. He proposed four different ways in which adolescents resolve (or fail to resolve) their quest for identity: **foreclosure**, **moratorium**, **identity diffusion**, and **identity achievement**.

Foreclosed adolescents are similar to those Margaret Mead described in Samoa. They make a smooth transition from the status of child to that of adult by accepting and adapting the adult roles laid out for them by others—their parents, their friends, their culture. They do not seem to feel any need to question whether they should comply with the expectations of others; they simply oblige. They

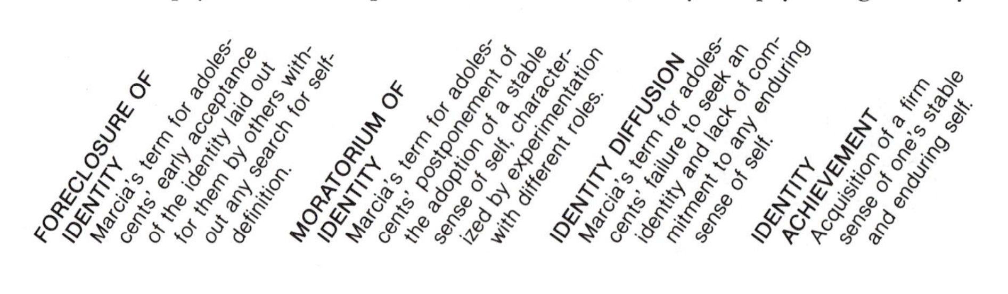

FORECLOSURE OF IDENTITY Marcia's term for adolescents' early acceptance of the identity laid out for them by others without any search for self-definition.

MORATORIUM OF IDENTITY Marcia's term for adolescents' postponement of the adoption of a stable sense of self, characterized by experimentation with different roles.

IDENTITY DIFFUSION Marcia's term for adolescents' failure to seek an identity and lack of commitment to any enduring sense of self.

IDENTITY ACHIEVEMENT Acquisition of a firm sense of one's stable and enduring self.

Table 7–2 Erikson's Fifth Nuclear Conflict: Identity Versus Role Confusion

Sense	Eriksonian descriptions	Fostering behaviors
Identity	Sense of sameness between self-concept and how one appears in the eyes of others	Others reflect back belief that youth has qualities that match self-image
	Felt continuity between identities prepared in the past, the meaning of one's identity for others, and the promise of career	Opportunity to pursue choice of activities in keeping with aptitudes and endowment
versus		
Role confusion	Doubt about sexual identity	Lack of interaction with opposite-sex peers
	Inability to find occupational identity	Overidentification with heroes of cliques or crowds
	Personality confusion as evidenced by delinquent, withdrawing, or suicidal behaviors	Lack of attention or feedback as being worthwhile

appear satisfied with the important life decisions that others have made for them, and they do not experience any crisis of identity.

A *moratorium* is a period of time in which one is permitted to delay meeting an obligation—a time out. Marcia believes that adolescents in an identity moratorium postpone the adoption of any set identity. They actively experiment with many different roles and explore a wide range of political, religious, and economic ideologies. They hesitate to make any commitments while searching for all the possibilities of an identity that might some day be theirs.

Identity-diffused adolescents also lack a commitment to any enduring sense of self. However, they are not actively searching for an identity either. They are more like reeds blowing in the wind. They are easily persuaded to follow others and assume whatever identity is required of them at the moment. They do not seem to experience distress at this status but are content with each new day-to-day self. They seem to be holding their identity open, as Erikson suggested women should do, for the peculiarities of the man (or woman), job, or activity to be joined. Marcia found the identity-diffused status to be more characteristic of the preadolescent, though it also describes some older adolescents.

Identity-achieved adolescents have firm commitments and an enduring self-concept. They are beyond the years of self-exploration and self-discovery but have experienced a conflict of identity versus role confusion (in the Eriksonian sense) prior to settling on a stable sense of self. They seem to feel comfortable with the choices they have made. They accept the weaknesses as well as the strengths of their personalities and no longer feel pressured to perfect themselves with changes.

Archer (1982), in an investigation of the frequency of Marcia's four identity statuses in adolescents, interviewed 160 teenage volunteers in New Jersey. The foreclosure and diffusion-identity statuses were most evident. Identity achievement occurred predominantly in the oldest subjects. She found that statuses also varied across content areas. Most adolescents had a foreclosed identity in sex-role preference, while in the political philosophy area more had a diffused identity. In religious beliefs, adolescents ranged from foreclosed to diffused to identity achieved; few were in a moratorium in their religious identity. Many adolescents were, however, in a moratorium in their vocational identity. Marcia (1980, 1983) wrote that ego identity requires a prolonged period for development. Obviously, it appears to develop more quickly in some areas than others and at different times in unique individuals.

The Progression of Ego Identity Savin-Williams and Demo (1984) identified three other areas of ego identity (or self-concept) in which development progresses unevenly: the experienced self, the presented self, and self-feelings. Adolescents are more apt to change the way they experience themselves, their self-images, and the way they present themselves than they are to change their self-feelings or self-esteem. The evaluation that adolescents make of themselves (self-esteem) is relatively stable from year to year. Higher self-esteem is associated with earlier authoritative-reciprocal parenting practices and use of inductive discipline rather than power assertion or deprivation of love (see Chapter 5). Adolescents who are socialized to be more androgynous (have attributes of both masculinity and femininity in their personalities) usually have higher self-esteem than do either males or females with many masculine, feminine, or undifferentiated traits (Lamke, 1982; Rust and McCraw, 1984).

In considering the progression and the impact of an identity search on any given adolescent, one must take into account the self-esteem with which he or she entered adolescence; the identity achieved in the past as it relates to Erikson's psychosocial stages; the level of cognitive functioning; the status of physiological functioning and physical maturation; the identities being urged on the adolescent by family members, school, friends, possible employers, the mass media, even the economic and political systems; and the expectations the adolescent may have for his or her future. Any generalization about how a "typical" identity conflict is resolved, or about the effect of the conflict on the adolescent, is, at best, speculative. Some identity searches continue into adulthood.

MORAL DEVELOPMENT

In Chapters 2 and 6 Kohlberg's cognitive-developmental approach to morality was presented. In his view moral thinking invariably develops according to the following sequence:

Stage 1: Obedience and punishment orientation
Stage 2: Instrumental relativist orientation
Stage 3: Interpersonal concordance orientation
Stage 4: Orientation toward authority, law, and duty
Stage 5: Social contract orientation
Stage 6: Universal ethical principles orientation

Kohlberg called stages 1 and 2 the premoral levels, stages 3 and 4 conventional morality, and stages 5 and 6 the levels of postconventional morality or **principled moral reasoning**. Refer back to Chapter 6 for a discussion of stages 1 through 4. In stage 5 reasoning, there is an emphasis on legalistic contracting of moral principles to guide behavior. Teenagers (or adults) at this level select for themselves the principles they will follow and show a concern that the rights and needs of others are not violated. In stage 6 reasoning, each moral dilemma is evaluated according to its own problems and the guiding principle for deciding what is "right" is the "conscience" and a concern for the universal dignity of humankind as well as one's own needs.

Although many adolescents (and also adults) give answers to the moral dilemmas posed by Kohlberg and others (see Chapter 6, p. 308 for an example) at the conventional moral reasoning stages, about 10 percent give answers based on the higher-level, principled reasoning stages (Kohlberg and Gilligan, 1971).

There are interrelationships between reaching the formal operations level of cognitive development (Piaget), being in the nuclear conflict of finding one's identity versus feeling role confusion (Erikson), and making moral judgments at the principled level (Kohlberg). Formal cognitive operating seems to be a necessary condition for reaching a principled level of morality. However, it certainly is not sufficient in and of itself to lead a person to high-level moral judgments. Such judgments are also based on environmental influences such as modeling, reinforcers, and discipline. A positive resolution of the identity conflict also facilitates higher-level moral judgments but is not essential to achieving principled morality (Cauble, 1976).

The form of discipline that correlates most highly with principled moral conduct is use of reasoning to induce changes in behavior (inductive discipline). Providing teenagers with explanations of the pros and cons of each alternative choice of action, exposing them to moral arguments, or even involving them in role-playing exercises to help them see the points of view of others will help them progress from conventional to postconventional moral reasoning (see Figure 7–9).

Any form of discipline is difficult during adolescence. Teenagers often feel that they are too old to be punished. They may rebel against the injustice of power-assertion and love-withdrawal techniques. They argue about almost every decision made that affects them. Therefore, some adults find it easier to allow teenagers to be responsible for their own behavior. They may hope that their past teaching plus example will be sufficient to keep their teenagers out

PRINCIPLED MORAL REASONING
Kohlberg's descriptive term for the last two stages of his sequence of moral development; also called postconventional morality.

Figure 7–9
While developing higher levels of moral reasoning, teens like to argue every rule laid down by adults, even when they approve of the rules. Familial and societal values that were accepted during childhood are now reexamined, debated, and sometimes revised in this process.

of trouble. Other adults may become dictatorial and insist that certain behavioral standards be met in exchange for allowance, clothes, privileges, meals, or shelter. Both permissive and authoritarian adults are usually guilty of maintaining a low level of actual involvement with the teenager. They communicate less, show less affection, and offer less companionship. Authoritative adults who can keep "reasoning" communication channels open and provide love, affection, and adequate explanations for their decisions, find that the discipline of teenagers is not impossible and that their behavioral standards are more often obeyed.

There have been many replications of Kohlberg's experiments suggesting stage sequences of moral development and the cognitive factors involved in progressing from one stage to the next. However, some studies have found that moral reasoning of a cognitive nature such as that required in the moral dilemmas that Kohlberg and others used in their research can be quite different from moral judgments made in actual day-to-day living. Leming (1978), for example, found that adolescents made higher-level moral judgments about classical moral dilemmas than they did about the actions of others in some practical moral situation. And Haan (1978) found that adolescent moral judgments are more frequently based on interpersonal formulations than on formal reasoning. Interpersonal moral solutions are achieved through group discussions that try to come to a consensus of all the participants on action(s) to be taken, rather than on any one individual's beliefs. Haan and associates (1985) proposed an alternative model of moral development with five levels of maturity. In their opinion, morality involves constructing agreements about respective rights. At the highest level a person is able to attend to the interests of all concerned parties.

Holstein (1976) reported that some adolescents regressed from a higher to a lower level of moral reasoning with age. Kitchener and her colleagues (1984) reported the opposite: that moral judgments typically increased over the late adolescent years, especially in subjects with higher education. Colby and her co-workers (1983), in longitudinal research, found downward moral stage change in only 4 percent of subjects. Social factors do play a role in moral reasoning. More mature moral judgments are possible when one has been taught about, and reinforced for, listening to and considering the obligations and rights of others, seeking the greater good for all humankind, and letting one's conscience be one's guide.

FAMILY INTERACTIONS

Adolescents typically fluctuate from wanting privileges in keeping with near-adult status (such as learning to drive and use of car, no curfews) to wanting the support and protection afforded them in childhood (no upkeep of room and clothes, meals prepared). Just as it is difficult for teenagers to achieve a sense of identity during adolescence, so too is it difficult for families to react to adolescents who are esthetes one day and slobs the next or adults one day and children the next. Parents who allow teenagers a great deal of freedom and pay little heed to their coming and going deprive them of some of the direction and support they both need and crave. Likewise, teenagers who have no freedom and too much surveillance feel thwarted in their efforts to be independent.

Independence training is hampered by authoritarian discipline (parents make the rules), by indulgent discipline (teenagers choose their own rules), and by parental indifference or neglect. The authoritative-reciprocal parent-child relationship more successfully allows teenagers to work through their identity crisis and gain a more secure independence—see Goethals and Klos's description of the ideally adequate parent in Table 7–3.

Table 7–3 Three Ways Parents Can Facilitate Adolescent Independence

1. Recognize the adolescent as a separate person in his or her own right, which implies that independent or competent behavior by the adolescent is gratifying rather than threatening.
2. Show genuine care or concern but not overinvolvement; give or offer but don't impose.
3. Terminate old ways of relating when they no longer are adequate or appropriate, which implies the openness to change that is necessary for any ongoing relationship.

SOURCE: G. W. Goethals & D. S. Klos, *Experiencing Youth: First-Person Accounts* (Boston: Little, Brown, 1970), p. 22.

Hauser and his colleagues (1984) found that family interactions emphasizing warmth, acceptance, and understanding supported identity development but found negative correlations between constraining behaviors (devaluing, withholding) and adolescent ego development. Garbarino, Sebes, and Schellenbach (1984) studied two-parent families in which an adolescent had been referred because of problematic adjustment. They found the parents in these families to be more punishing and less supportive of the adolescent, as well as more stressed by changes in their own lives. Adolescents want parental support (Coleman and Coleman, 1984). When they do not get it, they become more self-conscious and more self-focused (Riley, Adams, and Nielsen, 1984). They may also become pregnant, delinquent, substance abusers, or alienated. These possibilities will be discussed later in this chapter.

In the past males have been encouraged to assume more independent roles in adolescence than females. Women have been expected to transfer their dependency from their parents to a husband and to find their identity in his lifestyle. Increasingly, however, women are striving for more independent roles. Many now feel a need to define themselves as competent, self-respecting individuals before marriage. When and if they marry, they want to have a mutually interdependent relationship rather than one in which one partner becomes subject to the control of the other. Despite females' strides toward greater independence in careers and personality traits, however, most mothers still ask daughters to help with housework and do more traditionally sex-role stereotyped tasks, whatever the daughter's birth order (Sanik and Stafford, 1985).

Sibling order does influence achievement of independence. The oldest child is generally kept dependent longest. Younger children strive to have more privileges (and assume responsibilities) to be like the older siblings at earlier ages. Feelings of independence grow as teenagers earn, budget, and spend money, successfully achieve adult tasks and goals, and feel assurance that they will be accepted by the greater world of adults.

Teenagers spend less time with their parents when the mother works full time. Males also have longer arguments of greater intensity with their mothers if she works outside the home (Montemayor, 1984). On the other hand, maternal employment contributes to greater achievement motivation in daughters (Elder and MacInnis, 1983).

Teenagers from one-parent families are often asked to behave more independently than those from two-parent families (for example, to help support the family with a job, share a greater responsibility for housework and meals). However, this may result in greater social and emotional dependence on the custodial parent. How free an adolescent feels from the influence, control, or determination of a parent depends more on his or her self-concept and self-esteem than on quantity or quality of tasks performed. Garbarino and his colleagues (1984) found that adolescents with stepparents were often at high risk for dysfunction (identity confusion, emotional disturbance).

PEER INTERACTIONS

The importance of the peer group for testing new and different roles during adolescence was discussed in Erikson's and Marcia's theories of the identity conflict. The discussion of Piaget's theory of formal operations (combinatorial analysis, hypothetical-deductive reasoning) also stressed how important the peer group is for debating possible variations of decisions and situations. When families are minimally interactive, teenagers increasingly turn to their peers for information, advice, and companionship. **Peer pressures** in adolescence exert a strong force in shaping personalities.

Looking back through history, one finds that teenagers have been profoundly influenced through the ages by the convictions of peer groups (for example, Crusaders, Civil Warriors, Hitler's youth, hippies). There are many reasons for believing that teenagers today, and tomorrow, will be affected by the behaviors and beliefs of their peers.

In the 1960s and 1970s it was in vogue to talk about love children, the youth culture, or the counterculture, usually with some reference to problems created by urban unrest, race relations, drug use, or the Vietnam conflict. Prevailing social problems help shape the ideology of youth movements. However, social situations do not always affect all adolescents in the same way. There are always many variations, subcultures, and even countermovements to the mainstream norms.

Peer groups can evolve around shared interests such as athletics or the arts; they may grow between persons with common experiences such as doing well scholastically or becoming popular on the dating circuit. They may reflect neighborhood proximity or be organized around political or religious interests. Or they may reflect a fondness for certain modes of behavior like drinking or "turning on" with drugs. Any one teenager may belong to two or more peer groups simultaneously. Some groups identify themselves with a certain fashion of dress, grooming, or behavior. More often several groups prefer a certain form of dress that follows the fads and fashions of the day. The kinds and numbers of peer groups any one individual can join and the degree and type of influence the group exerts on the individual (or vice versa) depend on the individual. Personality factors like introversion, extraversion, self-esteem, and motivation, and situational factors like spending money, free time, sex, family rules, and school obligations all determine the strength of peer influence. Every teenager, whether leader or follower, active or passive, accepted or rejected, pays a great deal of attention to the behaviors and opinions of other young people with whom he or she comes in contact. Brown (1982) found that peer pressures for females appeared to be stronger than for males, especially in the areas of dating, sexual activity, and use of drugs and alcohol. "Doing one's own thing" is not typical until early adulthood, when a sense of identity and a degree of independence are established (see Figure 7–10).

PEER PRESSURES
Compelling influences to behave in certain ways brought to bear by one's age cohorts.

Friendships appear to become more stable, more intimate, and more mutually responsive in adolescence (Berndt, 1982). Adoles-

Figure 7–10
The adolescent peer group provides the individual teenager with a sense of security and acceptance and fosters a sense of group loyalty.

cents spend more time talking to their friends, and describe themselves as more happy doing so, than in any other activity (Csikszentmihalyi, Larson, and Prescott, 1977). Girls and boys appear to differ somewhat in their friendship patterns. Girls' friendships involve higher intimacy (Hunter and Youniss, 1982) and more conversations about themselves and their close relationships (Johnson and Aries, 1983). Boys' friendships, while more intimate than in childhood, appear to be less exclusive than are girls' friendships. Boys' conversations are more apt to center on activity-oriented topics than on personal emotions. Both girls and boys choose friends who are similar to them in age, sex, race, educational aspirations, attitudes toward school, actual academic achievement, and orientation toward contemporary teen culture (music, clothes, leisure-time activities). With increasing age, adolescents become more willing to share with and help their friends rather than compete with them. Adolescents with close and stable friends appear to have higher self-esteem than those with more fleeting, shallow friendships.

Popularity among peers has been investigated along dimensions of both personality characteristics and social status characteristics. Adolescents who are altruistic and able to see others' points of view are more popular than are self-centered teenagers. A high self-esteem also contributes to peer popularity. Social status correlates of popularity include being an athlete, being in the leading crowd, being a leader in activities, coming from the right family, having high grades, and having a nice car (Thirer and Wright, 1985). Athletic participation is more important for male than for female popularity (see Box 7–2).

BOX 7–2

Athletics in the Life of an Adolescent

Participation in an interschool or community-organized sports program has many advantages for a teenage male. Not only does the training exercise regimen increase physical fitness, but team rules help assure that the boy will eat well and not smoke, drink, use drugs, or even stay out late during the sports season. As a member of the team, the young man learns to compete through cooperation with others, rather than through self-aggrandizement. Such cooperation and team effort builds friendships and enhances self-esteem. To add to the benefits, research by Thirer and Wright (1985) suggests that the foremost criterion for male adolescent popularity with both males and females is having the status of athlete, or "letter-man."

Participation in interschool or community-organized athletic programs should have the same advantages for teenage females, but it does not. Thirer and Wright reported that being an athlete confers a fairly low social status on a girl. Butcher (1985), in a longitudinal study of girls from grades 6 through 10, found that their sports participation decreased or dropped out altogether by high school. In contrast, their secondary participation in sports increased—they vicariously followed sports by viewing live or televised sports events, listening to games on the radio, or reading about athletic contests. Girls who continued to actively participate, Butcher found, described themselves as assertive and independent. They preferred activity to a sedentary lifestyle and got personal satisfaction from their sports abilities. They also had parents and friends who encouraged their athletic participation and had sports equipment available to them. Many adolescent females do not have equipment or encouragement. The physical fitness advantage of sports is the same for female athletes as for males, with an added benefit—girls who engage in regular physical activity have fewer premenstrual and menstrual symptoms.

Are there disadvantages of athletic participation during adolescence as well? Some athletes neglect their schoolwork to have time for sports. However, many others learn to balance studying with playing and earn good grades. A second possible drawback to sports participation is the danger of sports injury. Accidents related to athletic participation rank second after auto accidents as the cause of serious adolescent physical impairment each year.

Psychologists worry about a further effect of some competitive athletic programs. Many coaches encourage a "killer instinct" in their players. Aggression and violence are legitimized on playing fields, whether ice, turf, wood, mat, or ring. Athletes who are highly aggressive toward opponents have been shown to have less mature moral reasoning than less aggressive athletes (Bredemeier and Shields, 1984).

The physical fitness and social status benefits of athletics in the life of an adolescent need to be weighed against the threats of injury and socially sanctioned violence. The latter disadvantage could be minimized with athletic educational programs that discourage intentional infliction of pain and that encourage codes of fair play.

SCHOOL INTERACTIONS

Organized school teams or groups provide ready-made (usually adult-supervised) peer groups for teenagers to join, if they can meet certain requirements (see Table 7–4). These groups give adolescents a chance to interact with young people who have interests similar to their own. If the organized unit works together for a common goal, its members usually develop a certain amount of cohesiveness and group loyalty. Such special interest groups, organized within a school system, help break down the racial, ethnic, and social class barriers that often characterize outside peer groups. Consequently, school groups can serve two very useful functions: (1) they allow not-so-popular teenagers a chance to experience a sense of belonging and acceptance; and (2) they allow teenagers from different backgrounds to learn more about one another as they work or play together.

Learning activities in school may not absorb as much of an adolescent's time and attention as social activities. By their own admission, high school students pay attention in class to the subject of study only 40 percent of the time (Csikszentmihalyi and Larson, 1984). They also daydream; feel sad, irritable, or bored; whisper; pass notes; or do homework for other classes. Many scholars have suggested that low teacher salaries, overcrowded classrooms, lack of supplies, and teacher burnout make school classes a negative experience. Others have suggested that adolescence is a poor time for requiring memorization of facts and attention to lectures given by teachers. The onset of formal operations is accompanied by a desire

Table 7–4 Possible School Groups with Which a Teenager May Become Affiliated

Honor society	Sororities
Student council	Fraternities
School newspaper	Audio-visual group
Yearbook staff	Future nurses
Debate team	Future farmers
Drama club	Home economics group
Stamp club	Voc-tech club
Coin club	Folk dancing
Rifle club	Modern dancing
Chorus	Exercise groups
Ensembles	Gymnastics
Band	Ski club
Orchestra	Swim club
Dance band	Outdoor clubs
School sports teams	International relations
Intramural sports teams	Chess club
Pep squads	Academic interest clubs
Cheerleaders	Ethnic identity groups

to discuss and debate possibilities in subjects as diverse as mathematics and history, art and gym. By reducing teacher load and encouraging student participation, classes could be made more interesting. However, budget, staff, time, and diploma considerations often make these simple goals unrealistic or impossible.

COMMUNITY INTERACTIONS

Organized community groups can enhance teenagers' sense of identity, sense of belonging, learning opportunities and career choices by putting them in close proximity with concerned adult leaders who are willing and able to share their time and experiences with the young. These extrafamilial activities (see Table 7–5) provide extra attention, guidance, and companionship from adult members of society. In the past, when the apprentice system functioned, towns were smaller, and neighborhoods were more cohesive, teenagers had many contacts with adults other than their parents. Today the trend is to segregate all social functions into age groupings. Neigh-

Table 7–5 Possible Community Groups Youths May Affiliate With

Boy and Girl Scouts	Various organized sports groups:
4–H program	Little League
Boys and Girls Clubs	Police Athletic League
Camp Fire program	Olympic training
Big Brothers/Big Sisters	Karate/judo clubs
American Field Service programs	Badminton leagues
Future Farmers	Roller skating
Future Homemakers	Ice (speed or figure) skating
Religious youth groups	Bowling leagues
Music groups	Boxing clubs
Drama groups	Wrestling clubs
Hospital volunteers ("Candy Stripers")	Track teams
Kennel clubs	Jogging groups
Demolay/Rainbow	Volleyball teams
Museum groups	Table tennis teams
Grange groups	Skiing clubs
Alateen	Lacrosse teams
Photography clubs	Polo teams
Stamp or coin clubs	Golf clubs
Puppeteers	Soccer teams
Red Cross volunteers	Tennis clubs
Chess clubs	Archery groups
Equestrian clubs	Bicycling groups
Rodeos	Trapshooting clubs
Neighborhood family centers	Fencing groups
Health spas	Hockey teams
Nature clubs	Basketball teams

bors mind their own business, and many adults (and children) avoid adolescents out of fear.

Many parents, because of their jobs, second jobs, stresses, commuting, and social or community obligations have little time left to interact with their teenagers. Bronfenbrenner (1972) feared this withering away of family and societal interaction with teenagers is contributing to a breakdown in the process of making them human. As he put it:

> By isolating our children from the rest of society, we abandon them to a world devoid of adults and ruled by the destructive impulses and compelling pressures both of the age-segregated peer group and the aggressive and exploitive television screen. . . . (p. 664)

By encouraging participation in age-integrated community activities, parents are helping their teenagers learn more about the lives, feelings, and values of other adults. When the parents participate themselves (even occasionally), the benefits accrue even more.

Many community groups consist of teenagers from the same neighborhood, social class, ethnic group, or religion. This kind of exclusive organization can give teenagers a feeling of identity within their own subculture. Although not furthering the American melting pot ideal, such group feeling can provide a sense of security and pride to many teenagers who are confused about their own sense of self. Community groups may also allow not-so-popular adolescents to belong to a group, especially if membership is determined by religion, social class, parents' group membership (such as Masons-Demolay), or ethnicity. Within age-segregated community groups with a heterosexual membership, teenagers can also find opposite-sex friends with whom they share common backgrounds or interests and of whom families may approve.

Age-segregated community groups, though good for some teenagers who meet membership requirements, can also have ego-deflating effects on adolescents who are denied membership. Social cliques can be extremely segregationist and cruel to the outsiders. Psychosocially mature teenagers can rebound from rejections by concentrating on their futures and by enjoying other friendships. Low-maturity teens, however, are much more dependent on peer acceptance for their own sense of self-worth (Josselson, Greenberger, and McConochie, 1977). Rejection, especially if it occurs over and over again, can leave a teenager feeling alienated and bereft of self-confidence. If he or she has also experienced pain and insecurity at home (such as strife with parents, separation or divorce of parents, substance abuse, mental illness, arrest of parents), a rejection by a peer group may precipitate a crisis (depression, physical or mental illness, suicide attempt). Teenage suicide will be discussed at the end of this chapter.

Some community groups (gangs) have the additional bad effect of accepting for membership status relatively meek adolescents (good followers) and putting them in positions of having to commit illegal acts or participate in violence in exchange for permanent membership in the gang. Within such groups many teenagers behave in very uncharacteristic ways or eventually begin to see getting into trouble as a desirable way of life because of the social acceptance and status it brings them. Delinquency is discussed later in this chapter.

SEXUAL INTERACTIONS

The one aspect of moral training that parents usually postpone until adolescence (or perhaps forever) is the area of sexuality. Sex education in schools, popular music with sexual themes, sex on television, sex in the cinema, street language, pornography, and the like are blamed for teenagers' sexual activity. The cultures of most of the Western industrialized nations exude sexuality. The explicit message is that one must have sex appeal. The implicit message is that postpubertal persons are sexually active. Peer pressures may encourage experimentation. "Use it or lose it" is a prevalent myth among some teenagers. Evrard (1985) reported that by age fifteen about 35 percent of American males and 20 percent of females have been sexually active. By age eighteen, 70 percent of males and 52 percent of females are sexually experienced. The converse is that by age eighteen, 30 percent of males and 48 percent of females are still virgins. Not every teenager is sexually active, despite cultural influences. Many feel good about saying "No" (see Figure 7–11).

Social scientists often point the finger of blame for sexual activity at the parents. Many teenagers do not get information at home about sex. They may be hesitant to ask questions of their embarrassed parents. They may also feel a need to break away from their parents advice, control, and protection. Erikson (1968) felt that an adolescent's sexual life is a self-seeking, identity-hungry activity in which each partner is really trying to understand himself or herself. McAnarney (1984) wrote that early coitus may be fulfilling a need to be touched, held, and cuddled. Parents become more restrained in physical contact with their children as they grow up and often terminate touching and hugging by adolescence.

Figure 7–11
After puberty one's basic maleness or femaleness leads to sexual urges. Coming to terms with emotional attitudes toward sex may be difficult for teenagers.

Contraceptive Use Very few sexually active teenagers take precautions to prevent conception. Guestimates are that only about one-third of teenagers having coitus protect themselves from becoming parents. Many reasons have been offered to explain this phenomenon. Teenagers themselves often plead ignorance. Parents may provide some sex education, but they often skirt around the issue of birth control, believing that knowledge about contraception will lead to its use. Only about 40 percent of American schools provide sex education courses, and 70 percent of students do not learn about contraception. Efforts to demonstrate that knowledge about birth control contributes to sexual activity have not been supported by research (Schwartz and Ford, 1983). Rather, there is evidence that there is a negative correlation between sex education and becoming pregnant (Zelnik and Kim, 1982). Many teenagers believe myths such as the following:

- Plastic wrap makes an effective condom.
- You can't get pregnant the first time.
- You can't get pregnant if you are standing.
- You can't get pregnant if you are still having your period.
- You can't get pregnant if the male withdraws in time.
- You can't get pregnant if you douche afterward.
- You can't get pregnant if you use foam afterward.
- You can't get pregnant if you take a birth control pill afterward.

Males often don't use condoms because they believe doing so takes all the pleasure out of intercourse. Many males also believe birth control decisions are female decisions (Cohen and Rose, 1984). Teenagers who do use contraceptives generally do not begin using them until several months after their first sexual experiences. Teenage couples with good communication patterns are most apt to eventually use effective birth control (Polit-O'Hara and Kahn, 1985).

When teenagers use contraception, they often choose the less-effective nonprescription methods, often without an understanding of how the methods work. Many adolescents are afraid to go to a public clinic, or to a physician, for advice. Many believe that the squeal rule is in effect: that their parents will be notified of the visit and its purpose. The most popular prescription contraceptive for teenage girls is still the Pill. However, Pill use has declined in recent years (Tyrer and Kornblatt, 1982). Many females are afraid of the Pill because of both myths (it makes your hair fall out) and realities (it may cause nausea; there is a correlation between extended Pill use and thromboembolism formation, visual disorders, gallbladder disease, and hypertension). Intrauterine devices are no longer available to American teenagers. Teenage girls frequently reject the diaphragm for other reasons. If they have a diaphragm, they are perceived to be announcing that they are sexually active and want sex. Diaphragms require privacy for consistent use. They are also considered messy and inconvenient. Girls with diaphragms may not heed instructions on how to insert them, or the need to use them with a spermicidal jelly or foam, or the need to leave them in place for six or more hours after intercourse. Consequently, they become pregnant. The same problems exist for contraceptive sponges. Some teenagers try to use the rhythm method, but due to insufficient information about a girl's fertile period, and the unpredictability of ovulation in adolescent females (especially those who are stressed), this method is often ineffective as well. Each year over 400,000 teenagers in the United States, having failed to prevent a pregnancy, resort to abortion (Alan Guttmacher Institute, 1981).

Teenage Pregnancy Each year over 1 million American teenagers have babies. Over one-quarter of these teenage mothers are under age fifteen. If current trends continue, about 40 percent of the teenage females in the United States will be pregnant before they reach age twenty (Alan Guttmacher Institute, 1981). While some persons believe this is predominantly a phenomenon among nonwhites, it is not. The rate of **teenage pregnancy** among white American teenagers is more than double the rate among white teenagers in Canada, England, France, Sweden, and the Netherlands. About 70 percent of teenage mothers are not married. In the approximate 30 percent of marital births, the marriage is often unstable. About 72 percent of teenage marriages end in divorce (McAnarney, 1983). Very few teenage mothers relinquish their babies to adoption agencies; about 93 percent of them choose to raise the child themselves (Zelnik and Kantner, 1978).

Why do so many American teenagers become pregnant? A lack of information about fertility and contraception is only one reason among many interacting factors. Having a baby may be viewed as a sign of maturity, a kind of status symbol. Many girls see motherhood as a way to achieve both an identity and a feeling of being loved and needed by someone else. Some girls may use pregnancy as an escape from an unhappy home situation. Many teenage mothers have a history of the indifferent-uninvolved pattern of parenting described in Chapter 5 or of being victims of child abuse or rape. Others may be reacting to the loss of a parent through divorce, death, hospitalization, or institutionalization. Teenage mothers can obtain federal and state aid to support themselves and their babies through programs such as food stamps, Medicaid, and aid to families with dependent children (AFDC). These programs consume over 8½ billion tax dollars for teenagers alone in the United States each year (Moore and Burt, 1982).

Babies carried by teenage mothers are at high risk for complications of pregnancy, birth, and infant development. Many mothers deny their pregnancies for several months, starving themselves and exercising rigorously to avoid appearing pregnant. Few seek prenatal care before the second trimester and then few keep regularly scheduled appointments. Mothers' diets often continue to be poor throughout pregnancy, overloaded with junk foods and deficient in protein, calcium, folic acid, and other vitamins. Teenage mothers are also often loathe to give up late-night dates in favor of sleep or to give up alcohol, cigarettes, or other drugs. As a consequence of these environmental factors, and of the mother's own physiological immaturity, infants are apt to be born early, with low birth weights,

or to be small for gestational age. They are twice as likely to have neurological difficulties as infants born to mothers in their twenties.

Teenage mothers are also at high risk for medical complications of pregnancy, birth, and postpartum recovery. They have a higher incidence of pregnancy-induced hypertension, preeclampsia or eclampsia, anemia, and childbirth mortality. The mortality rate is 60 percent greater in teenagers under age fifteen than it is in women over age twenty (Evrard, 1985).

Only about one-half of the teenagers who have babies return to finish high school (McAnarney, 1983). In some areas, high schools have established programs to provide prenatal care and nutritional counseling during pregnancy, and daycare facilities for infants after delivery, in order to allow girls to continue their education. Some programs have reduced the teen mother dropout rate to 10 percent with such services (Evrard, 1985).

Are teen mothers good mothers? Research suggests that the answer depends on the individual adolescent female. Those who have more knowledge about infants and child development and who have a positive attitude toward motherhood interact with their babies more (LeResche et al., 1983). Younger mothers tend to have less knowledge, to have more negative attitudes, and to be more controlling and less verbal with their babies (Fry, 1985). In general, teenage mothers have been shown to have a low level of verbal interaction with their infants (Landy et al., 1983). At five years of age, children raised by teenage mothers perform less well on tests of vocabulary (Wadsworth et al., 1984). They also tend to be shorter, have smaller head circumferences, and have more behavior difficulties. McAnarney (1983) found that the mother at highest risk of parental dysfunction is a fifteen- to sixteen-year-old, nonblack, lower-socioeconomic-status adolescent. She suggested that perhaps both younger and older adolescent mothers have more supports available to them than the 15- to 16-year-olds.

Research has all but ignored teen fathers. Many of them, even if they do not marry the mother, maintain contact with her and show a desire to name the baby and to meet some responsibilities toward the dyad (Barret and Robinson, 1982). However, their support usually drops off and they tend to disappear, usually after a year (Robinson and Barret, 1985).

Sexually Transmitted Diseases Ten to fifteen million new cases of **sexually transmitted diseases** (STDs) (formerly called *venereal diseases*) appear each year in the United States, with a preponderance of the diseases occurring in teenagers and young adults. The most common of the sexually transmitted diseases is not gonorrhea, not syphilis, not even genital herpes, but chlamydia. Many other STDs caused by viruses, bacteria, fungi, and even parasites occur with alarming frequency. Most of the STDs are easily diagnosed and treated. (All but herpes and AIDS can be cured.) However, fears and ignorance keep many teens away from physicians or clinics. They may try worthless home remedies or hope that the STD will cure itself. Instead, untreated STDs are spread to unsuspecting others and create problems such as sterility, cancer, brain damage, arthritis, ec-

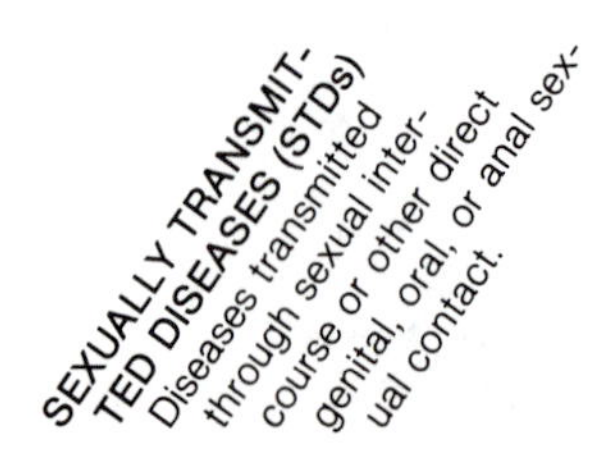

topic pregnancies, or miscarriages in adults and retardation, nervous system damage, blindness, or deafness on babies born to infected mothers. We will briefly review the symptoms, treatment, and sequelae of the more common STDs. AIDS is discussed in Chapter 8.

The first symptom of *chlamydia* is usually a purulent discharge from the penis or vagina. Pus means infection and should never be ignored. However, the tissue culture method for diagnosing chlamydia is expensive and not widely available, so often the diagnosis is made by looking for gonorrhea (a simple test) and assuming chlamydia if the gonococcal organism is absent. The symptoms of the two diseases mimic each other but treatment is not the same. Chlamydia can be eradicated only with erythromycin or a tetracycline-family antibiotic. If left untreated, it can cause inflammation of the oviducts in women (pelvic inflammatory disease) and tubal obstruction. This, in turn, can lead to an ectopic (outside of the uterus) pregnancy, or infertility. If a woman transmits chlamydia to a neonate, it can cause eye and ear infections, which may lead to blindness or deafness. In men, untreated chlamydia can lead to sterility.

Gonorrhea, also called "the clap," is not picked up from toilet seats or door knobs. The germs causing the disease die quickly unless they are in warm, moist areas such as those within the penis, vagina, anus, or mouth. Gonorrhea causes painful urination in men and women and a slight vaginal discharge or pelvic area pain in women. In men a discharge of thick yellow pus from the penis is usually noticed. Women more frequently notice no symptoms of disease. Cure can be achieved with penicillin or a penicillin substitute, but the disease may be contracted again and again. Physicians usually prefer to treat people who even suspect they may have been infected rather than wait for the dubious symptoms, especially in women. They also try to locate and treat the persons with whom the patient has had sexual contact. Untreated gonorrhea can sometimes lead to infection of the valves of the heart or joints (gonococcal arthritis), skin lesions, gallbladder disease, inflammation of the uterus and oviducts (in women), or sterility (men and women).

The first symptom of primary *syphilis* is usually a painless sore called a chancre. It heals fairly quickly and disappears. A man may notice such lesions on his penis, but women cannot detect chancres occurring within their vaginas. However, blood tests are available to detect syphilis infection. Long after the chancre has disappeared, new symptoms develop (secondary syphilis): a rash on the palms or soles, a complete body rash, and hair loss. If syphilis remains untreated, it goes into a latent stage. During this time it may be slowly and insidiously destroying parts of the body by its peculiar living habits in the blood vessel walls. Untreated syphilis eventually leads to a late tertiary stage characterized by inflammatory diseases of the skin, diseases of the aorta and cardiovascular system, and many forms of neurological disorders including blindness, paralysis, and an organic mental illness called general paresis. Treatment is penicillin or other antibiotic therapy. Physicians must trace sexual contacts and attempt to get them in for treatment when the disease is diagnosed. This is done confidentially, often with the help of specially trained personnel of local health departments.

Genital herpes, or type II herpes, is a close relative of type I herpes, commonly called the "cold sore." Like cold sores, genital herpes consists of painful lesions filled with a whitish fluid. They appear not on the mouth and nose area but around the genitals—on the penis or in the urethra in men; in the vagina or on the labia in women; occasionally on the thighs and buttocks. Type I herpes will not cause type II or vice versa. Nor will having cold sores give a person immunity from genital herpes. Genital herpes is highly contagious when the sores are wet but, like gonorrhea and syphilis, is spread by direct contact, usually through genital, oral, or anal sexual activity. The sores heal spontaneously in one to three weeks, just as do cold sores. The herpes II virus remains in the body in a latent stage. It can cause new bouts of painful genital sores periodically, at unpredictable times, such as when the carrier becomes stressed or physically ill. A new drug, Acyclovir, can lessen the severity and duration of bouts of herpes lesions at the time of infection and also prevent frequent recurrences. It cannot, however, cure the disease. The carrier should never have sex while herpes II sores are present to avoid spreading the disease. Genital herpes has been associated with an increased spontaneous abortion rate and with cancer of the cervix (Barclay, 1984). Women who have herpes type II should have Pap smears every six months rather than once a year. Babies carried to term by women with genital herpes are usually delivered by Caesarean section to prevent infection of the neonate through a birth canal with active lesions. The stresses of late pregnancy and labor may trigger reactivation of the virus. If a neonate is infected at birth, a systemic herpes virus infection may result in death or permanent central nervous system damage.

Genital warts (condylomata) are another rapidly spreading form of STD. Like other warts, they are benign, virus-induced epithelial tumors caused by the human papillomavirus (Cunningham, Hemsell, and Mickal, 1984). They may be flat, pointed, or cauliflower-like and can grow on the anus, in the perineal area, or in the vagina. They are uncomfortable as well as disfiguring. They tend to be especially exuberant during pregnancy. They can be removed by burning, freezing, chemicals, or surgery. Sometimes genital warts (like other body warts) disappear on their own. There is a positive correlation between genital warts and cervical cancer in women, so infected females should have semiannual Pap smears.

The first symptom of a vaginal infection caused by the *trichomonas* parasite is a greenish, frothy, malodorous discharge. In addition, the vulvar area may have burning, itching, and swelling. The parasite, a protozoan, can be transmitted to men during sexual contact. It can be eradicated with metronidazole given to all sexual partners.

Candida albicans, a yeast fungus, causes a thick white vaginal discharge and vulvar itching when it is present. It grows best premenstrually, or in women with a high dietary intake of simple sugars. It can be transmitted sexually. In women, it is treated with vaginal suppositories. It sometimes is also referred to as moniliasis.

The urethra (the canal leading from the bladder to the vulva or the penis) can become inflamed due to the presence of several different sexually transmitted organisms. Isolating the specific organism (if it is not the gonorrheal gonococcus) is difficult. In many cases of nongonococcal urethritis the causative factor is unknown. Urethritis causes painful urination with a thin white discharge present in the morning. It is more common in men than in women. It can usually sucessfully be treated with a week to ten-day course of tetracycline, but it often recurs. The dangers of nontreatment include pelvic inflammatory disease (possibly leading to sterility) for women or a form of disabling arthritis in both men and women. All sexual contacts of the infected person should also be treated.

All adolescents (or adults) who suspect that they may have a sexually transmitted disease can get information about where to go for confidential tests and treatment by calling the VD National Hotline's toll free number: 800–227–8922; 800–982–5883 in California. They should cease all sexual activity until they have been treated and advise all of their sexual contacts to seek treatment as well.

DELINQUENCY

Some teenagers are in trouble with the law before they are old enough to be arraigned in adult court. Juvenile court cases can roughly be classified into three types: status offenses, socialized conduct disorders, and unsocialized conduct disorders.

Status offenses involve conduct that would not be considered illegal if engaged in by an adult: school truancy, ungovernability, running away from home, sexual promiscuity. Depending on the state in which one lives, status offenders are labeled PINS (persons in need of supervision), CINS, JINS, or MINS (children/juveniles/minors in need of supervision). While both males and females commit status offenses, fourteen- to fifteen-year-old females are overrepresented in the caseloads of agencies that serve them. Most status offenders are returned to school or home when apprehended in spite of the fact that they usually flee again and again. Individual counseling may be offered if the adolescent shows signs of personal maladjustment. Family counseling is strongly advised to bring to light the family conflicts that the adolescent finds so intolerable. In some cases, when evidence of child neglect or abuse (physical, verbal, or sexual) is available, the child is assigned to a foster home temporarily while the parents receive counseling. Parents often have little in the way of empathy or positive regard for their runaway adolescents, or vice versa (Spillane-Grieco, 1984). This makes therapy challenging and difficult.

Socialized conduct disorders are illegal activities engaged in with the tacit or expressed approval of others, usually peers or parents. They may be minor infractions of the law (disorderly conduct, smoking marijuana, hunting without a license) or major crimes (drug dealing, car theft). The adolescent usually views the behavior as a way to win acceptance and approval from peers, parents, or significant others. Most adolescents report at least one infraction of the law while with a crowd or clique of similar lawbreakers because of pressures to conform.

Unsocialized conduct disorders are illegal activities committed by adolescents with an undercurrent of hostility and awareness of

STATUS OFFENSES
Conduct considered illegal if engaged in by a juvenile but not if engaged in by an adult.

wrongdoing. Such actions are usually referred to as **delinquent acts**. The extent of juvenile delinquency cannot be understated. The United States has the highest youth crime rate of all the industrialized nations in the world. More crimes are committed by adolescents than by all persons over twenty-five. They account for over one-half of the crimes in the United States. Juvenile males are three times more likely to perpetrate criminal activities than are underage females. Racial differences are not significant. A small percent of the teenage population (about 7 percent) commits about three-quarters of the crimes (Lipsitt, 1985). Juvenile delinquents are frequently out of school and unemployed. Many are alcoholics or drug abusers or have serious emotional problems. Many have no parents or have a history of being neglected or abused as children. Loeber (1982) found that most juvenile delinquents show an early onset of antisocial behaviors. They engage in more overt acts (fighting, disobedience) in childhood and escalate to more covert acts (theft, drug use) in adolescence. They not only have signs of maladjustment but also have negative self-images, especially in the areas of body image, moral and ethical self, and family self-concept (Jurich and Andrews, 1984).

A large percentage of **juvenile crimes** are serious: motor vehicle theft, burglary, grand larceny, drug pushing, aggravated assault, rape, and murder. The juvenile justice system is frequently lenient with youthful offenders. Rarely are they fingerprinted or photographed before age sixteen or eighteen (depending on the state). Frequently their cases are dismissed rather than tried. When tried, punishments are often merciful: stern lectures, paroles, short stays in juvenile rehabilitation camps or correctional facilities. Many social scientists see rehabilitation as multifaceted: increasing family responsibilities for wayward adolescents, increasing adolescents' feelings of responsibility for their own actions, and helping them see the enormity of their offenses and the effects on others. The availability of respected adult role models can be enormously beneficial to these teenagers, as in halfway houses with a few other delinquent adolescents and professionally trained surrogate parents. Nevertheless, because of the deep roots of the problems, rehabilitation is often unsuccessful (Gold and Petronio, 1980). Many juvenile delinquents eventually commit crimes as legal adults and are sentenced, in adult courts, to prison terms.

SUBSTANCE ABUSE

Although use of alcohol by minors and of most other psychoactive drugs without prescription is illegal, the concern of most adults to-

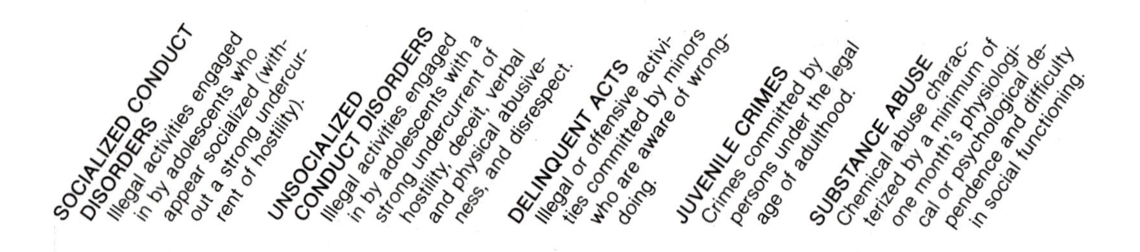

day is that **substance abuse** by minors is a sign of psychological disturbance. A normally curious teenager may want to experiment with mood-altering substances and may want to use them recreationally at parties. However, when an adolescent "needs" some substance on a regular basis and devotes his or her time to seeking and using the substance, an emotional problem is likely.

Substance abuse is more common among delinquent and alienated adolescents. It is also frequently associated with father absence (Stern, Northman, and Van Slyck, 1984). Jurich and associates (1985) also found that substance abusers more often come from families with either negligent or authoritarian discipline, hypocritical morality, gaps in communication, or mother-father conflicts. However, substance abuse may occur in "model" children from "model" families.

In general, substance abuse among teenagers has declined in recent years. Fewer teenagers smoke marijuana either regularly or occasionally. Fewer use sedatives, hypnotics, barbiturates, or narcotics, and fewer sniff intoxicants or experiment with hallucinogenic drugs. Even cigarette smoking has declined. It is no longer "cool." Many adolescents do not want to jeopardize their health with any drugs, including nicotine. Smokers in adolescence have been shown to be less stable, less intelligent, less self-confident, and more tense than nonsmokers (Tucker, 1984). The exception to the trend toward less substance use or abuse is that more adolescents are now seeking stimulant drugs to get them through the day; amphetamines or amphetamine substitutes, and cocaine. All of these substances of abuse will be discussed in greater detail in Chapter 8 ("Young Adulthood").

Despite its illegality, alcohol is widely available to many teenagers. It is frequently implicated in adolescent automobile fatalities. Some adolescents discover that they are addicted to it (are alcoholic) before they reach age twenty. Alcohol abuse may be responsible for a great deal of school truancy, diminished school performance, and accident proneness. Nevertheless, even alcohol use is decreasing among teenagers. The problems some young adults have with alcohol, including alcoholism, will be discussed in more detail in Chapter 8.

ALIENATION

While the majority of adolescents move toward others and search for their own identity during their teenage years, some act out against society (delinquency) or try to escape chemically (substance abuse). Still others simply withdraw from family, peer, and school interactions. These characteristic ways of behaving are not mutually exclusive. Many adolescents move toward people, away from people, and against people, depending on the situation and their moods. However, when a teenager's major mode of coping with life is to withdraw from it, he or she is said to be **alienated**.

ALIENATION
A feeling of estrangement from and hostility toward others.

Alienated adolescents have been described as more self-conscious, more anxious, and less apt to believe they can control the events in their lives (Moore and Schultz, 1983). They are often academic underachievers. Their tension and anxiety may be evidenced by many phobias, especially social phobias, or by obsessive or compulsive behaviors. They may be unpopular or even rejected by their peers (Faust, Forehand, and Baum, 1985). They are consequently more apt to spend time watching TV or listening to music. Loneliness in both boys and girls has also been associated with being stereotypically either very masculine or very feminine. Androgynous teenagers are less often alienated from others (Avery, 1982).

About 25 percent of teenagers feel both alienated and deeply depressed at some time during adolescence. **Depression** is more common in boys than in girls (Marcoen and Brumagne, 1985), perhaps because boys are less apt to have an intimate best friend with whom they can discuss their emotional concerns. Symptoms of depression include poor sleep with frequent wakings, decreased time spent in usual activities, impaired ability to concentrate, slower speech and gait, loss of interest in sex and friends, appetite changes, fatigue, boredom, hopelessness, ambivalence when faced with decisions, self-reproach, and an overwhelming feeling of sadness. In some adolescents depressed feelings lead to thoughts about, or attempts at, suicide.

Some social scientists refer to a current "plague" of **adolescent suicides**. The rate has quadrupled in the past thirty years without a corresponding rise in the suicide rate in the rest of the population (Lipsitt, 1985). Adolescent girls attempt suicide more frequently, but adolescent boys are three times more likely to complete it. Suicidal teenagers often have family problems. The loss of a parent or sibling through death, divorce, or institutionalization may trigger emotional disturbances within the family and within the teenager.

Older teenagers may also react with deep depression and suicidal behaviors to the loss of a significant person outside the family, such as a close friend (Triolo et al., 1984). Some sociologists believe that suicide may be contagious in adolescents (Phillips, 1985). In the late summer of 1985, for example, nine adolescents on the Wind River Indian Reservation in Wyoming followed each other into suicidal deaths. There are numerous other instances of one suicide seeming to trigger others among like-aged persons in the same schools or locale.

The "contagion effect" in suicide doesn't argue against the idea that many problems in suicidal teenagers stem from alienation. On the Wind River Indian Reservation, for example, mental health experts believed that the youths who killed themselves shared problems of low self-esteem, a lack of cultural identity, and a sense of hopelessness about their futures. Grob, Klein, and Eisen (1983) found that among predisposing factors to suicide those related to alienation from the family were most prominent, followed by low self-esteem, difficulty in peer relationships, and economic or ethnic differences.

DEPRESSION An emotional state characterized by intense and unrealistic sadness.

Some scientists feel that there may be physiological factors that predispose certain teenagers to depression and suicide. There is a modest but statistically significant relation between infants born at risk and eventual adolescent suicide (Lipsitt, 1985). It is pronounced in infants whose mothers failed to get prenatal care, whose mothers had a chronic disease during pregnancy, or who had respiratory distress for longer than one hour after birth. The first two of these problems may reflect an unhappy, disturbed, or sick mother who may have denied the child a sense of being wanted and loved throughout childhood and into adolescence. High-risk infants may also be more difficult to parent (see Chapter 4). This is not to deny that birth stress may create a biological vulnerability to later stress but to point out that interactive effects exist among life's great bundle of little events.

Most large cities have suicide prevention centers (SPC) and hotlines that troubled adolescents (and adults) can use when they feel they need help. Trained counselors will talk to them for as long as necessary and will help them consider alternatives to suicide. At high risk for suicide are persons who have contemplated their method, those who are depressed, and especially those who have attempted it before. It is a myth that people who talk about or make clumsy attempts at suicide rarely actually kill themselves. Attempts and suicidal talk may be cries for help. If help is not given, the adolescent, feeling hopeless and helpless, may take his or her life. Although SPCs and hotlines are helpful in the crises, most youth who have attempted suicide need individual professional counseling, and often family counseling as well, to help them overcome their wish for self-destruction and find an identity and purpose for living.

Summary

The teenage years include vast numbers of interacting changes. Consider, for example, the impact of sexual maturation, more logical reasoning abilities, and greater independence. The developing personality is particularly vulnerable to pressures from family members, peers, school, and community groups.

Puberty, the period of rapid growth preceding sexual maturity, may occur anywhere from ages eight to twenty and may last from a few months to a couple of years. Although both early and late maturers have special problems, late maturation can be especially traumatic because of its impact on self-concept and peer acceptance.

The focal points for health maintenance in adolescence are nutrition, the maintenance of good relations with others, stress management, prompt attention to infections, and safety precautions. Accidents are the leading cause of deaths and disabilities. Independence strivings and peer pressures make many teenagers particularly reckless. A plethora of health problems relate to nutrition: obesity, anorexia nervosa, bulimia, anemia, substance abuse, acne, and lowered resistance to infection (as in mononucleosis).

Piaget postulated a stage of formal operations to explain the cognitive changes of adolescence. It is characterized by interpropositional thinking, combinatorial analysis, and hypothetical-deductive reasoning. Persons using formal operations pay close attention to details and thoughtfully analyze new problems and situations.

Adolescent males frequently excel in cognitive tasks requiring spatial abilities. Females often show superior verbal skills. These sex differences in intellectual functioning may be genetically based,

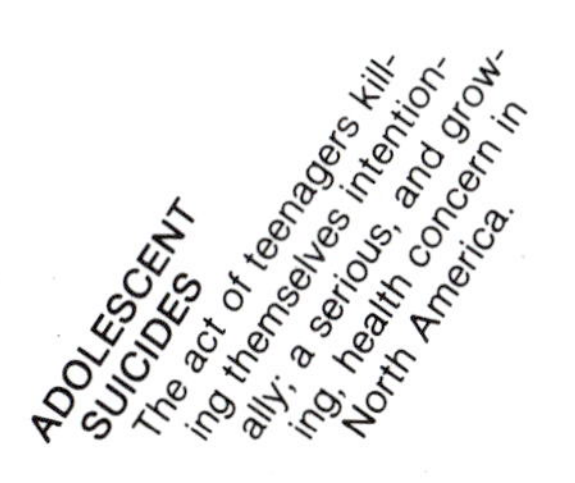

environmentally conditioned, or both. They often influence career choices.

Turbulence frequently mars the relationships between adults and adolescents in Western industrialized societies. New cognitive abilities contribute to a rise in youthful egocentrism, idealism, and argumentativeness.

The quest for identity, as described by Erikson and Marcia, is marked by a search for a sense of continuity and sameness between one's own and others' view of the self. Concerns about popularity, sexuality, future occupation, and independence are especially troublesome and time consuming. Family members and other significant adults may contribute to adolescents' role confusion by limiting or denigrating their self-directing behaviors.

Peer and school activities may contribute to an adolescent's positive self-concept. Sexual activity, unless protected, may result in pregnancy. Over one-half of all teenage mothers fail to finish high school. Sexual maturity may also bring in its wake careless sexual activity leading to one or more of the sexually transmitted diseases. Adolescents with negative self-concepts may turn against society (delinquency) or withdraw from society (substance abuse, alienation, suicide).

Questions for Review

1. Adolescents often feel ugly, ungainly, and not as physically attractive as their peers. Flattery by parents usually doesn't help. How can parents help their teenagers deal with these feelings without sounding superficial?
2. Compare the abilities children have in the concrete operations stage with the abilities adolescents have in the formal operations stage.
3. John, age fifteen, arrived home at the usual time after a school day. He had not been to school that day. His mother knew it. She asked if he had been at school and he answered yes. How should his mother handle this situation?
4. It might seem that the ability that is developed in adolescence to see all variations of a problem would enable an individual to feel more confident and secure. Is this what happens when adolescents develop this ability? Discuss.
5. Adolescents can be difficult to be around—argumentative at one moment, flattering at another, affectionate at another, and rejecting and angry at another—often all in the course of one day. Imagine you are an adolescent's parent. What sorts of things would you keep in mind to help *you* in dealing with all these mood changes and the feelings they stir up in you?
6. There is an increasing trend among pregnant teenagers to keep their babies, even if they decide not to marry. What are the implications of this decision for the girl, her family, the baby, and society?

7. You are a parent who has spent time discussing sex with your children. You have answered all their questions as openly and honestly as possible. Your daughter, age sixteen, comes to you, telling you she has decided to begin using birth control and wants your help in going to the doctor and choosing a method. She has no steady boyfriend. What would you say to her?
8. Young adolescents are sometimes removed from their homes because of neglect or abuse and placed in foster homes. Often they run away from these homes, skip school, and refuse to return to their parents' homes. They sometimes are then taken to court on a status offense, removed from the foster home, and placed in an institution. Describe a better method of intervention in these situations.

References

Alan Guttmacher Institute. *Teenage pregnancy: The problem that hasn't gone away.* New York: Alan Guttmacher Institute, 1981.

American Psychiatric Association. *Diagnostic and statistical manual of mental disorders* (DSM-III), 3rd ed. Washington, D.C.: American Psychiatric Association, 1980.

Andersen, A. E. Anorexia nervosa and bulimia: A spectrum of eating disorders. *Journal of Adolescent Health Care,* 1983, *4,* 15–21.

Andersen, A. E. *Practical comprehensive treatment of anorexia nervosa and bulimia.* Baltimore: Johns Hopkins University Press, 1985.

Archer, S. L. The lower boundaries of identity development. *Child Development,* 1982, *53,* 1551–1556.

Avery, A. W. Escaping loneliness in adolescence: The case for androgyny. *Journal of Youth and Adolescence,* 1982, *11,* 451–459.

Barclay, D. L. Disorders of the vulva and vagina. In R. C. Benson (Ed.) *Current obstetric and gynecologic diagnosis and treatment,* 5th ed. Los Altos, Calif.: Lange, 1984.

Barret, R. L., and Robinson, B. E. A descriptive study of teenage expectant fathers. *Family Relations,* 1982, *31,* 349–352.

Benbow, C. P., and Stanley, J. C. Sex differences in mathematical reasoning ability: More facts. *Science,* 1983, *222,* 1029–1031.

Berndt, T. J. The features and effects of friendship in early adolescence. *Child Development,* 1982, *53,* 1447–1460.

Bredemeier, B. J., and Shields, D. L. Divergence in moral reasoning about sport and life. *Sociology of Sport Journal,* 1984, *1,* 348–357.

Bronfenbrenner, U. The roots of alienation. In U. Bronfenbrenner (Ed.), *Influences on human development.* Hinsdale, Ill.: Dryden Press, 1972.

Brooks-Gunn, J., and Ruble, D. N. The development of menstrual-related beliefs and behaviors during early adolescence. *Child Development,* 1982, *53,* 1567–1577.

Brown, B. B. The extent and effects of peer pressure among high school students: A retrospective analysis. *Journal of Youth and Adolescence,* 1982, *11,* 121–133.

Bullough, V. L. Age at menarche: A misunderstanding. *Science,* 1981, *213*(17), 365–366.

Butcher, J. Longitudinal analysis of girls' participation in physical activity. *Sociology of Sport Journal,* 1985, *2,* 130–143.

Cauble, M. A. Formal operations, ego identity, and principled morality: Are they related? *Developmental Psychology,* 1976, *12*(4), 363–364.

Clarke, A. E., and Ruble, D. N. Young adolescents' beliefs concerning menstruation. *Child Development,* 1978, *49,* 231–234.

Clausen, J. A. The social meaning of differential physical and sexual maturation. In S. E. Dragastin and G. H. Elder, Jr. (Eds.), *Adolescence in the life cycle: Psychological change and social context.* New York: Wiley, 1975.

Cohen, D. D., and Rose, R. D. Male adolescent birth control behavior: The importance of developmental factors and sex differences. *Journal of Youth and Adolescence,* 1984, *13,* 239–252.

Colby, A.; Kohlberg, L.; Gibbs, J.; and Lieberman, M. A longitudinal study of moral judgment. *Monographs of the Society for Research in Child Development,* 1983, *48*(1–2, Serial No. 200).

Coleman, J., and Coleman, E. Adolescent attitudes to authority. *Journal of Adolescence,* 1984, *7,* 131–141.

Commons, M. L.; Richards, F. A.; and Kuhn, D. Systematic and metasystematic reasoning: A case for levels of reasoning beyond Piaget's stage of formal operations. *Child Development,* 1982, *53,* 1058–1069.

Csikszentmihalyi, M., and Larson, R. *Being adolescent: Conflict and growth in the teenage years.* New York: Basic Books, 1984.

Csikszentmihalyi, M.; Larson, R.; and Prescott, S. The ecology of adolescent activity and experience. *Journal of Youth and Adolescence,* 1977, *6,* 281–294.

Cunningham, F. G.; Hemsell, D. L.; and Mickal, A. Pelvic infections. In R. C. Benson (Ed.), *Current obstetric and gynecologic diagnosis and treatment,* 5th ed. Los Altos, Calif.: Lange, 1984.

Donelson, E. Development of sex-typed behavior and self-concept. In E. Donelson and J. Gullahorn (Eds.), *Women: A psychological perspective.* New York: Wiley, 1977.

Dusek, J. B., and Flaherty, J. F. The development of the self-concept during the adolescent years. *Monographs of the Society for Research in Child Development,* 1981, *46*(4, Serial No. 191).

Elder, G. H., Jr., and MacInnis, D. J. Achievement imagery in women's lives from adolescence to adulthood. *Journal of Personality and Social Psychology,* 1983, *45,* 394–404.

Elkind, D. *Children and adolescents: Interpretive essays on Jean Piaget.* New York: Oxford University Press, 1970.

Erb, T. O. Career preferences of early adolescents: Age and sex differences. *Journal of Early Adolescence,* 1983, *3,* 349–359.

Erikson, E. *Childhood and society,* 2nd ed. New York: Norton, 1963.

Erikson, E. *Identity, youth, and crisis.* New York: Norton, 1968.

Erikson, E. *Identity and the life cycle.* New York: Norton, 1980.

Evrard, J. R. Teen sexuality at root of rise in pregnancies. *Brown University Human Development Letter,* 1985, *1,* 8–9.

Faust, J.; Forehand, R.; and Baum, C. G. An examination of the association between social relationships and depression in early adolescence. *Journal of Applied Developmental Psychology,* 1985, *6,* 291–297.

Freeman, D. *Margaret Mead and Samoa: The making and unmaking of an anthropological myth.* Cambridge, Mass.: Harvard University Press, 1983.

Freud, A. *The ego and the mechanism of defense.* (C. Baines, trans.) New York: International Universities Press, 1946.

Fry, P. S. Relations between teenagers' age, knowledge, expectations and maternal behavior. *British Journal of Developmental Psychology,* 1985, *3,* 47–55.

Fulginiti, V. A. Infections: Viral and rickettsial. In C. H. Kempe, H. K. Silver, and D. O'Brien (Eds.), *Current pediatric diagnosis and treatment,* 8th ed. Los Altos, Calif.: Lange, 1984.

Garbarino, J.; Sebes, J.; and Schellenbach, C. Families at risk for destructive parent-child relations in adolescence. *Child Development,* 1984, *55,* 174–183.

Gillis, J. S. *Too tall, too small.* Champaign, Ill.: Institute for Personality and Ability Testing, 1982.

Goethals, G. W., and Klos, D. S. *Experiencing youth: First person accounts.* Boston: Little, Brown, 1970.

Gold, M., and Petronio, R. Delinquent behavior in adolescence. In J. Adelson (Ed.), *Handbook of adolescent psychology.* New York: Wiley, 1980.

Greif, E. B., and Ulman, K. J. The psychological impact of menarche on early adolescent females: A review of the literature. *Child Development,* 1982, *53,* 1413–1430.

Grob, M. C.; Klein, A. A.; and Eisen, S. V. The role of the high school professional in identifying and managing adolescent suicidal behavior. *Journal of Youth and Adolescence,* 1983, *12,* 163–173.

Haan, N. Two moralities in action contexts: Relationships to thought, ego regulation, and development. *Journal of Personality and Social Psychology,* 1978, *36,* 286–305.

Haan, N.; Aerts, E.; and Cooper, B. *On moral grounds: The search for practical morality.* New York: New York University Press, 1985.

Hall, G. S. *Adolescence* (2 vols.). New York: Appleton, 1904.

Halmi, K. A.; Falk, J. R.; and Schwartz, E. Binge-eating and vomiting: A survey of a college population. *Psychological Medicine,* 1981, *11,* 697–706.

Hauser, S. T.; Powers, S. I.; Noam, G. G.; Jacobson, A. M.; Weiss, B.; and Follansbee, D. J. Familial contexts of adolescent ego development. *Child Development,* 1984, *55,* 195–213.

Havighurst, R. J. *Developmental tasks and education,* 3rd ed. New York: McKay, 1972.

Hoffman, L. W. Early childhood experiences and women's achievement motives. *Journal of Social Issues,* 1972, *28,* 129–155.

Hoffman, L. W. Fear of success in males and females: 1965 and 1971. *Journal of Consulting and Clinical Psychology,* 1974, *42,* 353–358.

Holstein, C. B. Irreversible, stepwise sequence in the development of moral judgment: A longitudinal study of males and females. *Child Development,* 1976, *47,* 51–61.

Horner, M. Toward an understanding of achievement-related conflicts in women. *Journal of Social Issues,* 1972, *28,* 157–175.

Hunter, F. T., and Youniss, J. Changes in functions of three relations during adolescence. *Developmental Psychology,* 1982, *18,* 806–811.

Inhelder, B., and Piaget, J. *The growth of logical thinking from childhood to adolescence.* (A. Parsons and S. Milgram, trans.) New York: Basic Books, 1958.

Johnson, F. L., and Aries, E. J. Conversational patterns among same-sex pairs of late-adolescent close friends. *Journal of Genetic Psychology,* 1983, *142,* 225–238.

Johnson, S. *Facts about precocious puberty* (NICHD Fact Sheet 0-418-065). Washington, D.C.: U.S. Government Printing Office, 1983.

Josselson, R.; Greenberger, E.; and McConochie, D. Phenomenological aspects of psychosocial maturity in adolescence. *Journal of Youth and Adolescence,* 1977, *6,* 25–55; 145–167.

Jurich, A. P., and Andrews, D. Self-concepts of rural early adolescent juvenile delinquents. *Journal of Early Adolescence,* 1984, *4,* 41–46.

Jurich, A. P.; Polson, C. J.; Jurich, J. A.; and Bates, R. A. Family factors in the lives of drug users and abusers. *Adolescence,* 1985, *20,* 143–159.

Keating, D. P. Thinking processes in adolescence. In J. Adelson (Ed.), *Handbook of adolescent psychology.* New York: Wiley, 1980.

Kitchener, K. S.; King, P. M.; Davison, M. L.; Parker, C. A.; and Wood, P. K. A longitudinal study of moral and ego development in young adults. *Journal of Youth and Adolescence,* 1984, *13,* 197–211.

Koff, E.; Rierdan, J.; and Jacobson, S. The personal and interpersonal significance of menarche. *Journal of the American Academy of Child Psychiatry,* 1981, *20,* 148–158.

Kohlberg, L., and Gilligan, C. The adolescent as a philosopher: The discovery of the self in a postconventional world. *Daedalus,* 1971, *100,* 1051–1086.

Lamke, L. K. Adjustment and sex-role orientation in adolescence. *Journal of Youth and Adolescence,* 1982, *11,* 249–259.

Landy, S.; Clark, C.; Schubert, J.; and Jillings, C. Mother-infant interactions of teenage mothers as measured at six months in a natural setting. *Journal of Psychology,* 1983, *115,* 245–258.

Leming, J. S. Intrapersonal variations in stage of moral reasoning among adolescents as a function of situational context. *Journal of Youth and Adolescence,* 1978, *7,* 405–416.

LeResche, L.; Strobino, D.; Parks, P.; Fischer, P.; and Smeriglio, V. The relationship of observed maternal behavior to questionnaire measures of parenting knowledge, attitudes, and emotional state in adolescent mothers. *Journal of Youth and Adolescence,* 1983, *12,* 19–31.

Linn, M. C., and Petersen, A. C. Emergence and characterization of sex differences in spatial ability: A meta-analysis. *Child Development,* 1985, *56,* 1479–1498.

Lipsitt, L. P. Birth stress and adolescent suicide. *The Brown University Human Development Letter, Whole Special Report,* 1985.

Lipsitt, L. P. Who commits juvenile crime? *The Brown University Human Development Letter,* 1985, *1,* 9.

Loeber, R. The stability of antisocial and delinquent child behavior: A review. *Child Development,* 1982, *53,* 1431–1446.

Marcia, J. E. Identity in adolescence. In J. Adelson (Ed.), *Handbook of adolescent psychology.* New York: Wiley, 1980.

Marcia, J. E. Some directions for the investigation of ego development in early adolescence. *Journal of Early Adolescence,* 1983, *3,* 215–223.

Marcoen, A., and Brumagne, M. Loneliness among children and young adolescents. *Developmental Psychology,* 1985, *21,* 1025–1031.

McAnarney, E. R. The vulnerable dyad—adolescent mothers and their infants. In V. J. Sasserath (Ed.), *Minimizing high risk parenting.* Skillman, N. J.: Johnson & Johnson Baby Products Co., 1983.

McAnarney, E. R. Touching and adolescent sexuality. In C. C. Brown (Ed.), *The many facets of touch.* Skillman, N.J.: Johnson & Johnson Baby Products Co., 1984.

McAnarney, E. R., and Greydanus, D. E. Adolescence. In C. H. Kempe, H. K. Silver, and D. O'Brien (Eds.), *Current pediatric diagnosis and treatment.* Los Altos, Calif.: Lange, 1984.

Mead, M. *Coming of age in Samoa.* New York: Morrow, 1928.

Mead, M. *Sex and temperament in three primitive societies.* New York: Dell, 1949. (Originally published in 1935.)

Montemayor, R. Maternal employment and adolescents' relations with parents, siblings, and peers. *Journal of Youth and Adolescence,* 1984, *13,* 543–557.

Moore, D., and Schultz, N. R., Jr. Loneliness at adolescence: Correlates, attributions, and coping. *Journal of Youth and Adolescence,* 1983, *12,* 95–100.

Moore, K. A., and Burt, M. R. *Private crisis, public cost: Policy perspectives on teenage childbearing.* Washington: The Urban Institute Press, 1982.

Muller, J. Z. *Dysmenorrhea and premenstrual syndrome.* (NICHD Fact Sheet 461-338-814/25320). Washington, D.C.: U.S. Government Printing Office, 1985.

Newman, W. P., III, et al. Relation of serum lipoprotein levels and systolic blood pressure to early atherosclerosis. *The New England Journal of Medicine,* 1986, *314*(3), 138–144.

Overton, W. F., and Mechan, A. M. Individual differences in formal operational thought: Sex role and learned helplessness. *Child Development,* 1982, *53,* 1536–1543.

Overton, W. F., and Newman, J. Cognitive development: A competence-activation/utilization approach. In T. Field et al. (Eds.), *Review of human development.* New York: Wiley, 1982.

Phillips, D. P. The Werther effect: Suicide and other forms of violence are contagious. *The Sciences,* 1985, *25*(4), 32–39.

Physicians' Desk Reference. Oradell, N. J.: Medical Economics Co., 1986.

Piaget, J. *The growth of logical thinking from childhood to adolescence.* (A. Parsons and S. Seagrin, trans.) New York: Basic Books, 1958.

Piaget, J., and Inhelder, B. *The psychology of the child* (H. Weaver, trans.) New York: Basic Books, 1969.

Polit-O'Hara, D., and Kahn, J. R. Communication and contraceptive practices in adolescent couples. *Adolescence,* 1985, *20,* 33–43.

Rand, C. S. W. Obesity and human sexuality. *Medical Aspects of Human Sexuality,* 1979, *13*(1), 141–151.

Riley, T.; Adams, G. R.; and Nielsen, E. Adolescent egocentrism: The association among imaginary audience behavior, cognitive development, and parental support and rejection. *Journal of Youth and Adolescence,* 1984, *13,* 401–417.

Robinson, B. E., and Barret, R. L. Teenage fathers. *Psychology Today,* 1985, *19,* 66–70.

Rodin, J. The puzzle of obesity. *Human Nature,* 1978, *1*(2), 38–47.

Ruble, D. N., and Brooks-Gunn, J. The experience of menarche. *Child Development,* 1982, *53,* 1557–1566.

Rust, J. O., and McCraw, A. Influence of masculinity-femininity on adolescent self-esteem and peer acceptance. *Adolescence,* 1984, *19,* 359–366.

Sanik, M. M., and Stafford, K. Adolescents' contribution to household production: Male and female differences. *Adolescence,* 1985, *20,* 207–215.

Savin-Williams, R. C., and Demo, D. H. Developmental change and stability in adolescent self-concept. *Developmental Psychology,* 1984, *20,* 1100–1110.

Schwartz, M., and Ford, J. H. Family planning clinics: Cure or cause of teenage pregnancy? In U. S. House of Representatives, *Teen parents, and their children: Issues and programs.* Hearings before the Select Committee on Children, Youth, and Families, July 20, 1983. Washington: U.S. Government Printing Office, 176–197.

Selye, H. *Stress without distress.* Philadelphia: Lippincott, 1974.

Selye, H. Essay: They all looked sick to me. *Human Nature,* 1978, *1*(2), 58–63.

Silver, H. K. Growth and development. In C. H. Kempe, H. K. Silver, and D. O'Brien (Eds.), *Current pediatric diagnosis and treatment.* Los Altos, Calif.: Lange, 1984.

Spillane-Grieco, E. Characteristics of a helpful relationship: A study of empathic understanding and positive regard between runaways and their parents. *Adolescence,* 1984, *19,* 63–75.

Stern, M.; Northman, J. E.; and Van Slyck, M. R. Father absence and adolescent "problem behaviors": Alcohol consumption, drug use and sexual activity. *Adolescence,* 1984, *19,* 301–312

Stunkard, A. J., et al. An adoption study of human obesity. *New England Journal of Medicine,* 1986, *314*(4), 193–198.

Tanner, J. M. *Growth at adolescence,* 2nd ed. Oxford: Blackwell Scientific Publications, 1962.

Tanner, J. M. Sequence, tempo, and individual variation in the growth and development of boys and girls aged twelve to sixteen. In A. E. Winder (Ed.), *Adolescence: Contemporary studies,* 2nd ed. New York: Van Nostrand, 1974.

Thirer, J., and Wright, S. D. Sports and social status for adolescent males and females. *Sociology of Sport Journal,* 1985, *2,* 164–171.

Triolo, S. J.; McKenry, P. C.; Tishler, C. L.; and Blyth, D. A. Social and psychological discriminants of adolescent suicide: Age and sex differences. *Journal of Early Adolescence,* 1984, *4,* 239–251.

Tucker, L. A. Psychological differences between adolescent smoking intenders and nonintenders. *Journal of Psychology,* 1984, *118,* 37–43.

Tyrer, L. B., and Kornblatt, J. E. Teens' contraceptive needs. *Planned Parenthood Review,* 1982, *2,* 11–13.

Wadsworth, J.; Taylor, B.; Osborn, A.; and Butler, N. Teenage mothering: Child development at five years. *Journal of Child Psychology and Psychiatry,* 1984, *25,* 305–313.

Weideger, P. *Menstruation and menopause: The physiology and psychology, the myth and the reality.* New York: Knopf, 1976.

Weston, W. L. Skin. In C. H. Kempe, H. K. Silver, and D. O'Brien (Eds.), *Current pediatric diagnosis and treatment,* 6th ed. Los Altos, Calif.: Lange, 1980.

Whisnant, L., and Zegans, L. A. A study of attitudes toward menarche in white middle-class American adolescent girls. *American Journal of Psychiatry,* 1975, *132,* 809–814.

Zelnik, M., and Kantner, J. F. First pregnancies to women aged 15–19: 1976 and 1971. *Family Planning Perspectives,* 1978, *10,* 11–20.

Zelnik, M., and Kim, Y. J. Sex education and its association with teenage sexual activity, pregnancy and contraceptive use. *Family Planning Perspectives,* 1982, *14,* 3.

♥

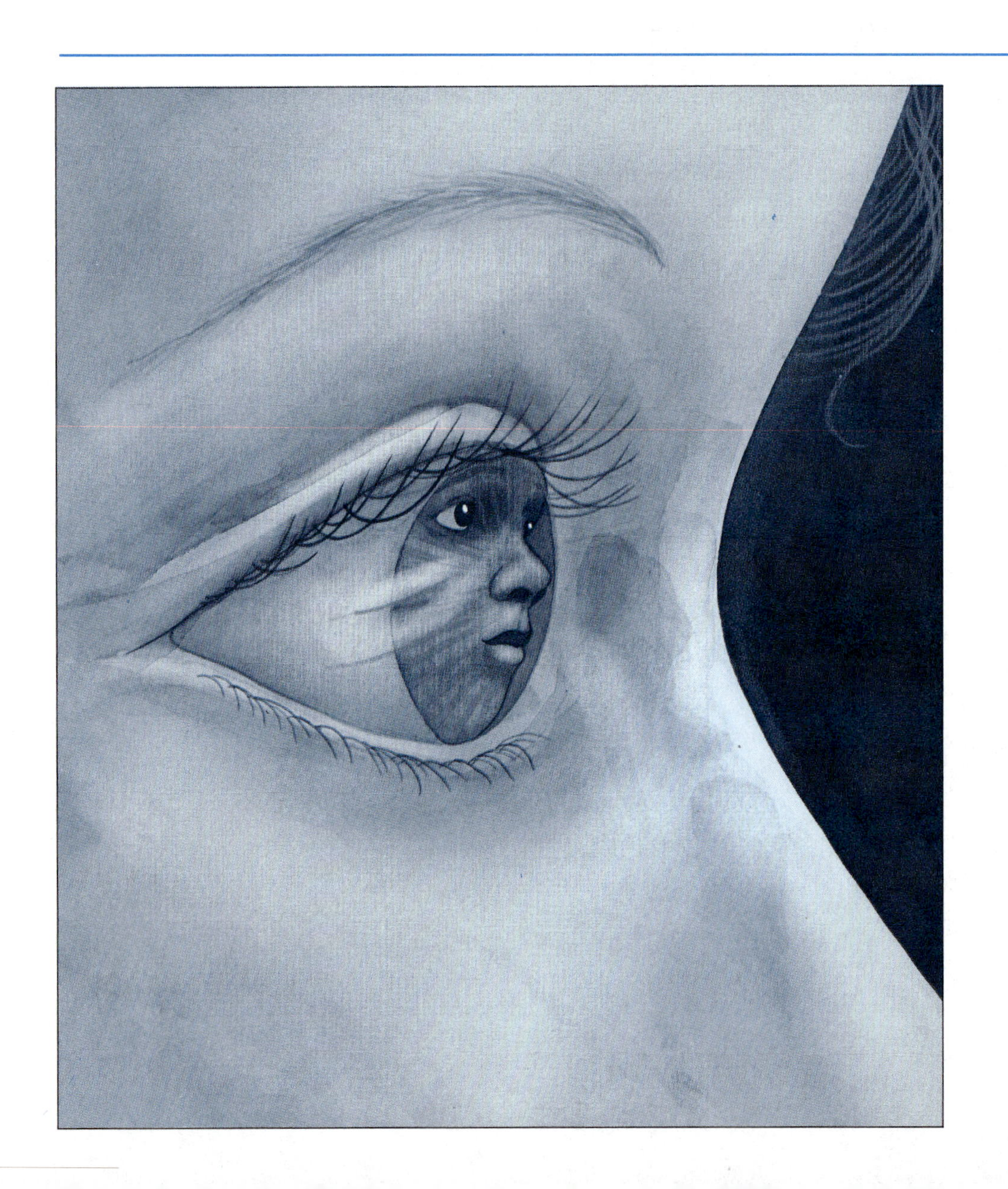

Young Adulthood

8

Chapter Outline

Physical Development
- *Physical Maturation*
- *The Effects of Stress*
- *Health Concerns*

Cognitive Development

Psychosocial Development
- *Independence from the Family of Origin*
- *Career Beginnings*
- *Maslow's Needs Hierarchy*
- *Erikson's Concept of Intimacy Versus Isolation*
- *Singlehood*
- *Marriage*
- *Parenthood*
- *Household Organization*
- *In-law Relations*
- *Infertility*
- *Substance Abuse*

Key Concepts

activists
adulthood moratorium
age thirty transition
alcohol abuse
alcoholic
alienated youth
apprenticeship learning
blackouts
circadian rhythms
cohabitation
conventionality
delirium tremens (DTs)
depressants
early adult transition
entering the adult world
family life cycle
family of origin
family of procreation
hallucinogens
hierarchy of needs
holistic medicine
honeymoon period
hormones
household organization
infertility
in-law relations
intimacy versus isolation
life change units
lunar cycles
marital morale
marriage
maturity
nuclear family
parenthood
peak experiences
physical dependence
posttraumatic stress disorder
psychological dependence
psychophysiologic disorder
self-actualization
singlehood
stimulants
superwoman syndrome
synergism (drug)
tolerance
youth stage

Case Study

Nell met her "ideal husband" on her second day of graduate school. He described himself as a "citizen of the world" because he had lived and worked in several countries prior to returning to school. He was seeking a doctorate in international studies. Nell had dreamed of becoming a foreign ambassador as an adolescent and had been told that as a girl from a politically unconnected family her chances were slim. She saw Caleb as a ticket into international politics. She fell in love. She was thrilled when he asked her out. She began to put his needs and desires before her own and spent as much time cleaning his apartment and making his meals as she did on her own studies. She offered to change her major to fit into his life's plans, but he discouraged this.

For two years, Nell tried to be as unconditionally caring, nurturant, compassionate, and loving toward Caleb as was humanly possible. He seldom took her out; but he welcomed her housecleaning and cooking assistance, as did his roommates. On several occasions, one or another of his friends would tell Nell that he was exploiting her while dating other women. She knew this was true but believed that ultimately he would be unable to resist her unselfish agapic style of love.

Caleb had proposed to Nell and had even asked her father's permission to marry her. He had never produced an engagement ring, however. On two occasions, they filled out the necessary papers for a marriage license, but Caleb found reasons to postpone the wedding until each license expired. At the end of the second year, Caleb left the university. He encouraged Nell to finish her degree and take a trip to Europe with her roommate before they married. She obeyed. While in Europe, she received a "Dear Jane" letter from Caleb calling off their engagement. She hurriedly flew to his new location to dissuade him. She still believed her agapic love could conquer all. Caleb coldly told her that he did not love her and did not want to marry her, and asked her to go away. She obeyed, very hurt but also hopeful of a reconciliation. It never occurred. What do you think of Nell's agapic type of love? Could it ever become the basis for an intimate, equitable, rewarding marriage?

In our society people in their twenties usually branch out from their families of origin and establish themselves in somewhat independent lifestyles. The activities and patterns of living pursued by various young people may be worlds apart. Consider, for example, the differences between struggling or unemployed ghetto youth and affluent young persons who can move into their families' businesses. Consider the gulf between persons secluded in convents or seminaries and those following rock bands. The examples can go on and on. The lives of young people who get married are vastly different from the lives of those who remain single. Those who pursue advanced education differ from those who never finish high school. Those who earn their own living differ from those whose parents support them. Children and adolescents, because of required school attendance, share similar experiences. People in their twenties often do not. This chapter will describe some of the concerns that face young adults regardless of the path they choose. The choice is profound. As Robert Frost (1949) put it:

> Two roads diverged in a wood, and I—
> I took the one less traveled by,
> And that has made all the difference.

Physical Development

Given proper diet, exercise, rest, and freedom from preexisting handicaps or disease, a person between ages twenty and thirty should enjoy a body performing in peak condition. The archetypal young adult can expect to have muscles and bones that are strong and resilient, freedom from serious infectious or degenerative disease, normally functioning endocrine glands (those that produce internal secretions such as hormones), and a digestive system that functions smoothly. Physical stamina should be sufficient to keep up with all the social, economic, and emotional tasks of this period (see Figure 8–1).

PHYSICAL MATURATION

Body shape and proportions finally reach their finished state, with the exception of weight and muscle mass. Fat accumulation and muscle mass are under more environmental influence (diet and exercise) and may fluctuate throughout a person's life. Skeletal development is slowly completed as the long bones of the upper legs and arms finish their ossification process (change from cartilage to bone with a concurrent increase in length). During childhood the head was larger than the trunk and the trunk more developed than the limbs. With the limbs, development proceeded from foot to calf to thigh and from hand to forearm to upper arm. For some adolescents this meant outsized feet or hands, which caused a degree of clumsiness until the near ends of their limbs caught up with peripheral growth. As bones ossify, the shaft of the bone develops first. In some bones ossification also takes place in scattered outlying areas called *epiphyses.* Between the ossified shaft and the epiphysis there is a nonossified area, the epiphyseal line, from which the bone continues to grow in length. In the twenties these epiphyses fuse with the main shafts of the bones. Once epiphyseal lines are calcified (or closed), the lengthwise growth of bones ceases. Attainment of final adult height coincides with this fusion. A simple radiograph can be used to analyze bone development and determine whether final adult skeletal growth has been achieved. A few millimeters may be added to the width of some bones later by surface deposition. Head length and breadth, facial diameters, and the width of bones in the legs and hands may increase slightly by this process throughout life.

Figure 8–1
The twenties are years usually characterized by peak physical performances with strong muscles and plenty of stamina for vigorous activity. Retention of these capacities is dependent on the continued use of appropriate exercises and diet and avoidance of health and safety hazards.

Muscles continue to gain strength throughout the twenties and reach peak strength at about age twenty-five, depending on exercise and genetic endowment. Men have larger muscles that can produce more force per gram than the muscle tissue of women. They also have a greater aerobic capacity for carrying oxygen in the blood to the muscles and for neutralizing the chemical products of exercising muscle.

Dental maturity is finally achieved in the twenties with the emergence of the last four molars, called wisdom teeth. For some young adults these last teeth are the cause of severe toothaches, as they sometimes become impacted (wedged between the jawbone and the next most forward molars) and need to be removed. The molars, which grind food, are not well utilized by persons who partake of soft, modern diets. The absence of one or more of the wisdom teeth causes no hardship.

The reproductive systems of both men and women are fully mature by the twenties. Kinsey and associates (1948) reported that male libido peaks between ages fifteen and twenty. For women Kinsey and colleagues (1953) reported a peak in the sex drive between ages twenty-six and forty. However, sex drive remains relatively high in most people for several decades, often into the sixties and seventies. Regardless of libido, the best years for reproducing children for both men and women are in the twenties. Testosterone secretion reaches its maximum daily output in men and ripening follicles in the ovaries produce more estrogen in women than they will later in life.

Evidence suggests that both men and women experience involuntary cyclic alterations in their sex hormone production. They also have involuntary fluctuations of certain adrenal hormones (for example, epinephrine). **Hormones** (internal chemicals secreted directly into the bloodstream) bring about functional changes within the body. They vary on a daily basis, called **circadian rhythms**, and on a monthly basis called **lunar cycles**. Changes in productions of hormones within each individual occur with clocklike regularity. Low production periods have been associated with apathy, indifference, and a tendency to magnify minor problems out of proportion. High periods correlate with increased energy, a decreased need for sleep, and a feeling of well-being. High estrogen levels have been associated with feelings of increased well-being and mood elevation in women. This phenomenon has been especially noted during pregnancy, when estrogen levels are much increased, and during the ovulatory phase of the menstrual cycle.

Although some women do experience increased anxiety, depression, or a tendency to magnify problems premenstrually, when estrogen levels are low, many do not. Persky (1974) studied twenty-nine healthy young women (average age twenty-two) and found remarkably little change in their moods related to their "periods." Their average values for the psychological variable assessed closely resembled the average values of a control group of male classmates. Some women and men show behavioral effects of high and low hormone levels; some evidence little or no outward signs of these daily and monthly fluctuations.

The effects of the sex hormones on the secretion of oils (sebum) from the sebaceous glands of the skin, which contributed to the acne of adolescence, become less pronounced in the twenties. Acne usually disappears, and young adults generally find they do not have to shampoo daily to keep their hair clean.

Brain cell development reaches its peak in the twenties. Actually, the final number of neurons and supportive cells that each individual possesses is determined by the end of the first postnatal year of life, but the cells themselves continue to grow and become more complex during childhood and adolescence. The cell fibers (axon and dendrites) and the myelin sheathing surrounding the fibers increase in size, number, and intricacy. Memory is thought to peak at the time when brain weight peaks. After this the brain cells slowly begin to degenerate. Myelin sheathing decreases, as do the size and number of cells. This gradual shrinking of the brain cells after about age thirty is not a cause for great concern. We hardly begin to use all of the brain cells we have at any point in our lives. The mental processes of many people in their sixties and seventies are still keen

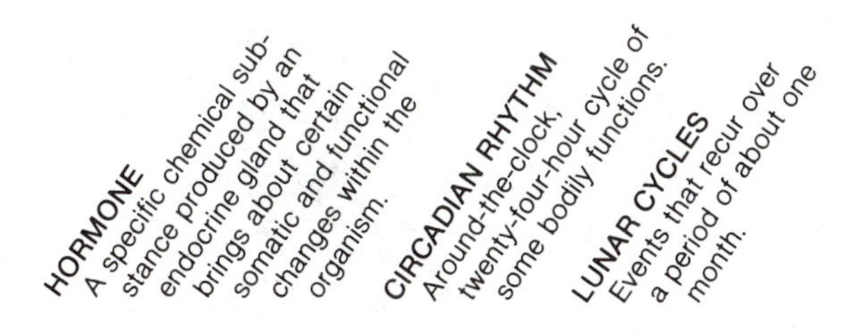

enough to control empires or make important contributions to the world of arts and science.

Just as a trend has been noticed toward increased height and earlier maturation in successive generations of well-nourished peoples over the past one hundred years, there also has been a trend toward increased head circumference and brain weight. Simultaneously, a distinct rise in IQ scores has occurred.

Many people with large heads would like to believe their increased brain size is evidence of superior intelligence. However, recent studies have found no correlation between brain size and IQ (Scott, 1983). Anatole France, the Nobel laureate French author of such novels as *Thaïs, Penguin Island,* and *The Revolt of the Angels,* and one of America's greatest poets, Walt Whitman, who inspired his readers to a fuller appreciation of the wonders of nature in *Leaves of Grass,* both had unusually small brains. France's brain, which weighed approximately 1000 grams, is one of the smallest on record for an adult human being (Dubos, 1982). The average adult male brain weighs approximately 1570 grams. The brains of both elephants and whales are considerably heavier, averaging 3980 grams and 2488 grams, respectively.

THE EFFECTS OF STRESS

People differ considerably in their abilities to handle stress, as described in Chapter 7 in the discussion of Selye's general adaptation syndrome. It has been suggested that such simple behaviors as exercising, talking, or crying can give vent to restrained emotions. Crying especially is credited with helping to restore mental and physical equilibrium during periods of tension and strain.

People who do not have ways to cope with their stresses, or who are placed in situations where the tension is seemingly too great to handle (loss of job, loss of family member, social disorganization, natural disasters) increase their probability of a wide range of illnesses, probably because their immune systems cease to function well (Miller, 1985). **Psychophysiologic disorders** are those that relate to the influence of the mind (emotions, fears, desires, frustrations, anxieties) on the malfunctioning of the body. Psychophysiologic diseases are very real, not imagined. However, successful treatment depends on the removal or alleviation of the sources of stress as well as on pharmaceutical and physical measures. Such comprehensive treatment is advocated by **holistic medicine**, which treats the whole person. Stress that is restrained and hidden from the outside world can affect individuals in any number of ways. Skin lesions may erupt, such as neurodermatitis (scales, itchy areas produced by ner-

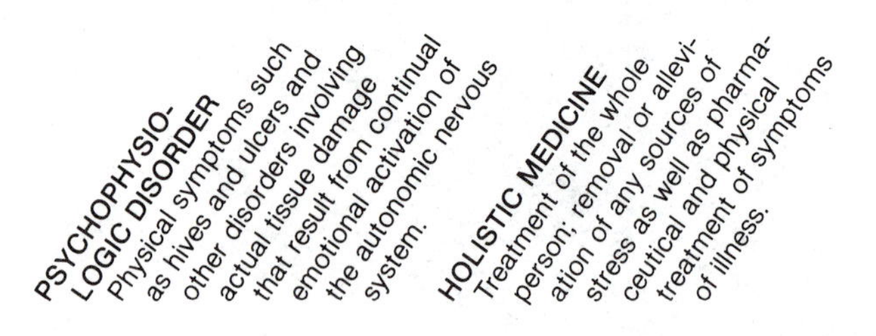

vous itching of the skin), herpes zoster (shingles), or urticaria (hives). Disorders of the gastrointestinal tract may appear, such as heartburn, chronic diarrhea, constipation, ulcers, bulimia, anorexia nervosa, or diabetes. The respiratory system may show its susceptibility to psychic stimuli, as in hyperventilation, influenza, pneumonia, or aggravation of asthma. The cardiovascular system may be affected by palpitations, fainting spells, hypertension, fibrillation, angina or large fluctuations in blood pressure, as may the nervous system (migraine headaches, tics, stuttering, tremors, or multiple sclerosis). Many other diseases have psychophysiologic bases. It is always difficult to separate the physical from the psychological causes of illness. However, we do recognize that stress predisposes a person to disease.

Our highly industrialized society with its rapid changes (in personal relationships, jobs, residences) requires that we make frequent decisions and alterations in our lives, leaving us in a state of distress. This state is similar to the "culture shock" a visitor to a foreign country experiences or the "shell shock" a war combatant feels. Too many changes at once drain our energy resources and strain our abilities to cope, adapt, and evolve. Holmes and Rahe (1967), both medical doctors, postulated that too much change in a relatively short period of time lowers one's resistance to infection or accidental injury and increases one's risk of a major health change. They listed the major changes an average adult may experience and asked hundreds of persons to rate them as to the amount of adaptation and readjustment they cause. The averaged mean values assigned to various life events are shown in Table 8–1. Many of these changes are positive. They still require an expenditure of energy, however, for coping with them.

Table 8–1 The Social Readjustment Rating Scale

Life event	**Mean value**
1. Death of spouse	100
2. Divorce	73
3. Marital separation	65
4. Jail term	63
5. Death of close family member	63
6. Personal injury or illness	53
7. Marriage	50
8. Fired at work	47
9. Marital reconciliation	45
10. Retirement	45
11. Change in health of family member	44
12. Pregnancy	40
13. Sex difficulties	39

Continued on page 404

Table 8–1 *Continued*

Life event	Mean value
14. Gain of new family member	39
15. Business readjustment	39
16. Change in financial state	38
17. Death of close friend	37
18. Change to different line of work	36
19. Change in number of arguments with spouse	35
20. Mortgage or loan for major purchase (home, etc.)	31
21. Forelosure of mortgage or loan	30
22. Change in responsibilities at work	29
23. Son or daughter leaving home	29
24. Trouble with in-laws	29
25. Outstanding personal achievement	28
26. Wife begins or stops work	26
27. Begin or end school	26
28. Change in living conditions	25
29. Revision of personal habits	24
30. Trouble with boss	23
31. Change in work hours or conditions	20
32. Change in residence	20
33. Change in schools	20
34. Change in recreation	19
35. Change in church activities	19
36. Change in social activities	18
37. Mortgage or loan for lesser purchase (car, TV, etc.)	17
38. Change in sleeping habits	16
39. Change in number of family get-togethers	15
40. Change in eating habits	15
41. Vacation	13
42. Christmas	12
43. Minor violations of the law	11

SOURCE: Reprinted with permission from *Journal of Psychosomatic Research,* 11 (2), 213–218, T. H. Holmes and R. H. Rahe, "The Social Readjustment Rating Scale," copyright 1967 Pergamon Press, Ltd.

Holmes and Rahe found that persons whose combined **life change units** (LCU) over a one-year period was 300 or more had about a 90 percent chance of having a major health change. If one's score was from 150 to 300 points in a year, the chance of a major health change was 50 percent. A score of 150 or less decreased one's chance of accident or illness to about 30 percent or less. Holmes and Rahe cautioned that persons who are about to make voluntary changes (such as a job change, a move, marriage) should pace them so they do not all occur at once. One can avoid the danger zone for a stress-related accident or illness by taking the time to readjust and adapt to one or two changes before subjecting oneself to more.

Many people turn to alcohol and drugs to alleviate stress. It is estimated that about 10 percent of the work force in the United States use alcohol on the job and another 2 percent struggle with drug addiction (Johnson, 1985). There are a number of other current ways in which people try to come to grips with too much stress at once. Some turn to professional counselors or talk over problems in specialized group therapy. There is a growing interest in holistic medicine to help individuals tolerate stress better; in the use of controlled diets; avoidance of drugs, alcohol, and nicotine; regular exercise; sufficient sleep; relaxation techniques; and developing a sense of self-control. Such practices can cut down or eliminate the need for tranquilizers in many young adults. In addition, many persons in their twenties are learning to choose their friends, work, and places of living to enhance their enjoyment of life. They seek out positive experiences and avoid the negative as much as possible.

Some persons, after intense stress, develop **posttraumatic stress disorder** (PTSD). This syndrome frequently characterizes soldiers who have returned from battle. However, it also is common in persons who have survived other stresses, such as unexpected deaths of family members or close friends, kidnapping, hostage situations, airplane, train, or automobile crashes, floods, fires, tornadoes, or earthquakes. Initially the survivor is disoriented and dazed. When this emotional confusion lifts, the victim goes through a period of denial. He or she may even seem happy, as if nothing happened. Eventually, however, the disaster becomes a reality. Several symptoms may develop and persist for extended periods of time: impaired memory, insomnia, reliving of the trauma in nightmares, somnambulism (sleepwalking), guilt for having survived, avoidance of close interpersonal relationships, digestive disturbances, accident proneness, frequent illness, work inhibition, aimlessness, displaced anger, increased use of drugs, depression. These aftereffects may not occur for a few days or even a few months after the trauma. They often persist several months or even over a year. Therapy for PTSD is aimed at helping the victim realize that these reactions are normal and that many others have also had extended posttraumatic stress symptoms. Group therapy is helpful if others with similar problems are available. Counselors try to help victims reestablish social networks and become involved in meaningful activities (McLeod, 1984). Without psychological help, some emotional distress can continue indefinitely.

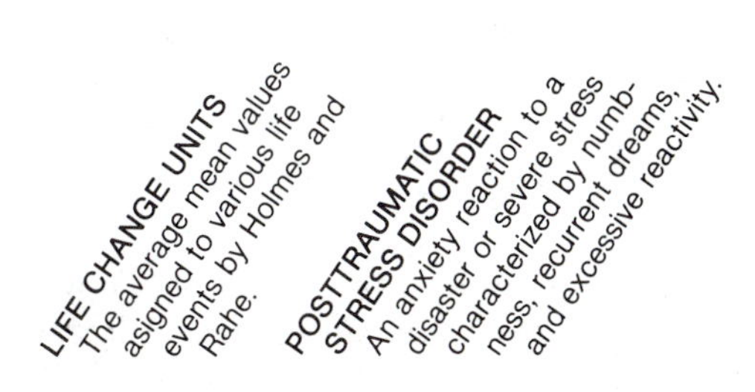

HEALTH CONCERNS

The twenties are generally years of excellent health and peak physical abilities. Most of the allergies of childhood disappear after the adolescent growth spurt (exceptions are hay fever and asthma). By the twenties young people have developed natural immunities to many of the infectious agents that troubled them in the past. When immune defense mechanisms are working well, infections that are contracted are not usually life threatening. An exception may be contact with the AIDS virus (see Box 8–1).

Persons who completed the primary series of immunizations against diphtheria, pertussis, tetanus, and polio (DPT-OPV) in the first year of life and against measles, mumps, and rubella (MMR) in the second year should have a booster immunization against diphtheria, and tetanus (DT) every ten years (American College of Physicians, 1985). Physicians now use the mid-decade birthday as the time for each DT booster (ages fifteen, twenty-five, thirty-five, forty-five, and so on). If the person has no record of immunization, no documented history of disease, or no laboratory evidence of immunity to measles, mumps, and rubella he or she should also have the MMR vaccination. This is especially important for young adults living together in close quarters (military barracks, college dormitories, and the like). The diseases are imported frequently by soldiers and students who have traveled abroad or by foreign students and other visitors. There have been many recent outbreaks, especially of measles, on military bases and college campuses. If a woman is pregnant, the MMR vaccination should be postponed until after the delivery of the baby. The live viruses in the vaccine are potentially teratogenic to the fetus.

Young adults living in close proximity to each other (students, military) and adults at high risk of exposure to influenza (health care professionals in training, policemen, firemen, school teachers, communications workers, ambulance workers, paramedics) are advised to receive influenza vaccine on an annual basis. The hepatitis B virus is also recommended for doctors, nurses, ambulance workers, and paramedics because of their frequent exposure to blood.

The threats to health that loom largest in the twenties are those related to poor control of emotional stress (accidents, alcohol and drug abuse, and stress-related illnesses). Preexisting chronic diseases (diseases of the heart, epilepsy, diabetes) continue to trouble some people in their twenties. Fertility problems may also be discovered in some otherwise healthy young adults. Accidents account for close to half of the total deaths of young adults. A large percentage of these fatalities involve motor vehicles. Many deaths are also due to self-inflicted (suicide) or other-inflicted injuries (homicide). Young adults are the only age group who have shown an increase in death rates during the last two decades, primarily due to accidents, homicides, and suicides.

Cognitive Development

In Chapter 7 it was noted that Piaget's highest level of intellectual functioning, formal operations, is acquired unevenly during adoles-

BOX 8–1

AIDS

Acquired immune deficiency syndrome (AIDS) is a sexually transmitted disease that reduces the body's ability to fight infection. AIDS patients invariably die. Many myths have sprung up in the few short years since AIDS first appeared. Let us look at the simple truths of the disease.

AIDS is caused by a virus with the unwieldy name HTLV-III/LAV (human T-lymphotropic virus type III/lymphadenopathy associated virus). In order to get AIDS, a person must receive the virus directly into the blood. The disease is, consequently, not a highly contagious one. Children who have it usually acquired it in utero or at birth from infected mothers, or from blood transfusions. As of March 1985, however, all blood transfusions have been free of HTLV-III/LAV. All blood has been tested with an ELISA (enzyme-linked immunosorbent assay) test, which can have false positives but *not* false negatives, and all blood that tests positive has been, and continues to be, destroyed.

Women who have AIDS usually acquired it by sharing needles with an affected person for intravenous injection of heroin, cocaine, or amphetamines. Women also spread it this way, or to infants in utero. No one has been able to isolate the virus from vaginal or cervical fluids (Finkbeiner, 1985).

The AIDS virus lives in T-helper cells, which are lymphocytes (white blood cells) that are concentrated most heavily in the blood and also in semen. Any practice that allows infected blood or semen to enter another person's blood (such as anal intercourse, sharing intravenous needles) will spread the virus. While saliva and tears may contain a few T-helper cells, there are no known cases of AIDS being spread through saliva or tears. The disease is *not* transmitted by casual contact with AIDS patients (for example, to family members, to doctors and nurses, to school children in play).

HTLV-III/LAV is a slow-growing virus at first. The incubation period for AIDS can be longer than five years (Food and Drug Administration, 1985). Infection by the virus has three preliminary forms before AIDS, and not all infected persons go on to develop AIDS. In the first form symptoms including fever, diarrhea, and weakness occur, which may be mistaken for a gastrointestinal flu. After this the patient will develop antibodies to HTLV-III/LAV and will test positive on the ELISA test. In the second form of infection, the person will be an asymptomatic carrier of the virus (the antibodies will not kill the virus). This second form may be the end of the infectious process. In a small percentage of infected persons, the disease continues to a third form called ARC (AIDS-related complex). The patient again has fever, diarrhea, and weakness, but this time it persists with weight loss and swollen lymph glands. Some people recover from ARC. Others go on to develop AIDS. They keep losing weight and their immune system fails. They usually die from opportunistic infections (pneumonia, tuberculosis, herpes, toxoplasmosis, Kaposi's sarcoma).

Efforts to cure AIDS are directed at killing the HTLV-III/LAV virus before it can progress to the fourth and final form of the disease. This must be accomplished without killing the T-helper cells and destroying the immune system. Prevention of AIDS requires that infected persons do not allow the entrance of their own blood or semen into the blood of noninfected persons.

cence. Acquisition of formal-operational thought is related to innate intelligence, education, and life experiences. Some people never attain this highest level of functioning. Some acquire it in midlife, given further opportunities that enhance its development. Others attain the ability to use interpropositional thinking, combinatorial analysis, and hypothetical-deductive reasoning and then lose their grasp of these skills in old age.

In the twenties young people make considerable use of their "gray cells," as brain cells are called, at whatever level of functioning is available to them (usually formal operations). Many American young people go to work (whether outside or inside the home) when they leave high school and must learn all the new skills, both technical and interpersonal, that relate to their jobs. They also have many roles to learn while achieving social and economic independence. Other American young people continue to pursue formal education in their twenties. They enroll in trade schools, colleges, and graduate schools and may temporarily postpone some of the other kinds of experiential learning.

Psychologists and educators call the need to learn well and do well *achievement need.* It is quite different from innate abilities and cognitive development but enhances the latter. Cognitive ability and achievement need contribute to successful learning far more than either contributes alone. Although caregivers can and do stimulate cognitive development in their children, they have a more potent role to play in stimulating the need to achieve. In a classic study by Rosen and D'Andrade (1959) the characteristics common to parents of boys with high achievement need were reported. Fathers were more competitive, took pleasure in problem-solving tasks, showed more father-son involvement, gave their sons more things to manipulate, and displayed more affection and emotion. Mothers were more dominant, stressed success and achievement over independence, and held high aspirations for their sons.

The achievement need of women, especially by the time they reach their twenties, is frequently below that of men. It is also below what could be expected of them considering their scores on IQ tests and their past performances in grade school and high school. A popular notion of women's intellect based on their achievement is that they start out ahead of boys, level off, and then retrogress with increasing age. There is no evidence to support this. Women and men can both reach the same levels of cognitive functioning, and women can maintain their high level functioning as well as men. Evidence does point to a falling off point of many women's need to achieve. Horner's (1972) classic study found that college women taking achievement tests could do well, but when asked to compete with men, they dissolved into bundles of nerves.

Horner's data suggesting a fear of success in women were presented in Chapter 7. Her research and replications of it have suggested that for some women, the desire to be liked, to be accepted, and to affiliate with others is stronger than the desire to succeed, especially if success brings in its wake jealousy or rejection. In a study of college women at Wellesley, Alper (1974) identified women who were high in feminine traits and who aspired to traditional wife/mother roles and others who were career oriented and more androgynous in their traits. The fear of success, which Horner (1972) and others found in women, was present in the traditional, highly sex-role socialized women. Such women told stories like the following when shown pictures to describe:

> The lady with the test tube is a movie star who has sunk to doing TV commercials. The other is an admiring nobody [who] pities her. . . . They both fade into oblivion. No one will like them or pay attention to them. . . . The star commits suicide and the other marries a florist and gets fat. (*Alper, 1974*)

However, the career-oriented women told "success" stories in response to the same pictures:

> The instructor is showing her student how to do a step in a complicated experimental procedure. The student has been working on the experiment for some time, and is nearly at the end of her work. She is observing carefully what the other woman does, and, when she returns to her own work, will repeat the procedure with care and precision and will be able to finish her project and achieve significant results. The instructor would like the student to become an able scientist, and is pleased with her work.

In addition, they were competitive and high achievers in college, indicating neither a fear of success nor a falling off in their levels of cognitive functioning.

Psychosocial Development

In Chapter 7 we defined adolescence as that period of time between the biological changes of puberty and independence from the family of origin. Using this definition, we find that some teenagers become young adults before reaching age twenty. Likewise, some people in their twenties remain adolescent. (Kenneth Keniston refers to some individuals in their twenties as youths rather than adolescents or young adults. His theory will be described shortly.) Although age is not as good a criterion for life stage as are the pursuits of an individual at a given time, certain pursuits tend to coincide with certain ages (see Figure 8–2).

Young adulthood suggests **maturity**. Opinions differ widely as to what constitutes maturity. Freud (1917/1966) wrote that maturity is the ability to discipline oneself in work and in heterosexual relationships. Neither the id (pleasure-seeking impulses) nor the superego (self-restricting impulses) dominate the balanced ego. He felt maturity is attained in the postadolescent period for most people. Carl Jung (1923), on the other hand, believed maturity should not be claimed by most people until middle adulthood. First, individuals must go through the trials of many life experiences.

MATURITY The quality of being fully developed, complete, full-grown, or perfect.

Erik Erikson (1963) wrote that maturity develops gradually as adults work their way through the last three nuclear conflicts of their lives: intimacy versus isolation, generativity versus stagnation, and

Figure 8–2
The young adult, unlike the adolescent, has a stable sense of self but is searching for vocational and social roles.

ego integrity versus despair. Lawrence Kohlberg (1973) described maturity as a cumulative process without a well-defined beginning or end. He saw the highest level of moral development as more mature than the conventional level, yet conceded that few people reach the higher stage. He warned that settling too quickly into a static lifestyle can inhibit the development of moral maturity.

Abraham Maslow (1970), whose theory is dealt with in detail later in this chapter, also saw maturity as a cumulative process. Individuals progressively meet their physiological needs, safety needs, love and belonging needs, self-esteem needs, and need for self-actualization. Maslow stated that although all people have the innate potential for becoming self-actualized, societal conditions are such that many people only advance as far as trying to meet their love and belonging or self-esteem needs.

The remainder of this social development section will examine the influence of the **family of origin** (one's biological parents or surrogate parents and any siblings), the commitment of oneself to a career, and the beginnings of a **family of procreation** (one's spouse and children). The information about twenty- to thirty-year-olds in these sections consists of generalizations, more applicable to describing the masses of young people in this age range than to describing any one particular person.

INDEPENDENCE FROM THE FAMILY OF ORIGIN

A great many young people continue to let their primary caregivers provide for part or all of their economic needs during their twenties.

Some look to their families to meet their social and emotional needs. It is becoming common practice for postadolescent young people to lay claim to a few years of an **adulthood moratorium** (a period of time in which a person is permitted to delay meeting an obligation). Such a moratorium lengthens the time until the youth declares himself or herself independent from the family of origin: able to satisfy his or her own needs for food, housing, health, safety, and the pursuit of happiness through some productive enterprise. The time may be filled with college, graduate school, exploratory travels, temporary jobs, and, in some cases, military service. The young person in an adulthood moratorium usually still considers "home base" to be the family of origin. However, even without making a declaration of independence, these young adults tend to function with more self-sufficiency than they did in their teenage years. Some live away from home, alone, or with friends.

Friends and various adult mentors can become important as role models and sources of social and emotional gratification. Moratoriums may also postpone the attainment of the developmental tasks of early adulthood, as set forth by Robert Havighurst (1972):

Selecting a mate
Learning to live with a marriage partner
Starting a family
Rearing children
Managing a home
Getting started in an occupation
Taking on civic responsibility
Finding a congenial social group

Arriving at a legal age (when one no longer has to have parental signatures for loans or licenses, when one can purchase alcoholic beverages, when one can marry and sign contracts) gives most young people a sense of increased independence from their families, even if they choose not to use it.

During the adolescent identity-seeking period the relationships between young persons and their families and friends tend to be somewhat shallow and one-sided. The young use other people as sounding boards for developing their own self-concepts. Once young people have established a stable and satisfying sense of self, they generally prefer to keep and develop more meaningful relationships with a few others. They usually choose friends who complement and enhance their own uniqueness. They become interested in who others are and where others are going. Whereas family contacts become less frequent as the young person moves out of the home, friendships with parents may go either way: dissolve or be strength-

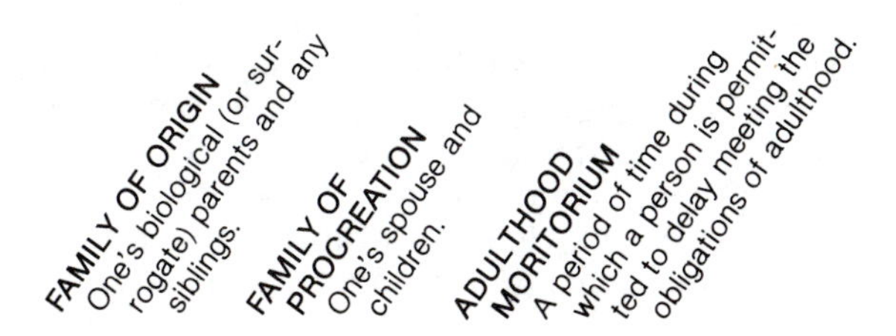

ened. Once the "self" is known, young adults can expend more time discovering who the parents are, in all their complex, three-dimensional (past-present-future) personalities. Gradually the hopes and dreams of the parents for themselves become meaningful to the young adult, as well as their hopes and dreams for the youth. In some cases, when the emerging adult rejects both what the parents want for themselves and what they want for him or her, an alienation occurs that may last until middle or late adulthood or forever.

But for other young people exploring the interests, feelings, values, problems, and welfare of the parents is exciting. They begin to give support and direction as well as take it. This turning of the tables ("You have cared for me, now I will care for you") indicates a real feeling of independence and maturity on the part of the young adult.

Roger Gould (1978), in cross-sectional studies of several hundred men and women between the ages of sixteen and sixty, found that the twenties were characterized by two different kinds of relationships with the family of origin. Young adults in their early twenties, Gould found, typically struggle with leaving their parents' world. However, they feel unprepared to contest the views of their family. They must challenge and resolve several false assumptions from their more childlike consciousness:

1. If I get any more independent, it will be a disaster.
2. I can see the world only through my parents' assumptions.
3. Only my parents can guarantee my safety.
4. My parents must be my only family.
5. I don't own my own body.

Gould found that these battles to change assumptions are replaced with new false assumptions to challenge by the middle to late twenties:

1. Rewards will come automatically if we do what we are supposed to do.
2. There is only one right way to do things.
3. Those in a special relationship with us can do for us what we haven't been able to do for ourselves.
4. Rationality, commitment, and effort will always prevail over all other forces.

The phrase/word that Gould used to characterize these subjects was "I'm nobody's baby now."

Daniel Levinson and associates (1978), after analyzing his data from a cross-sectional and retrospective study of forty men, labeled the years of separating from one's parents the **early adult transition** and the years of the middle to late twenties as **entering the adult world**. Levinson, like Gould, found questioning the nature of the family of origin and one's place in it a vital part of the novice phase of adulthood. In the early adult transition numerous separations, losses, and transformations are required to terminate old relationships. A process of initiation to the possibilities of the adult world with tentative steps toward choosing and participating before actually entering it also takes place. According to Levinson, entering the adult world involves exploration of self and world, making and testing provisional choices, searching for alternatives, increasing one's commitments, and constructing a more integrated life structure. Finally by age twenty-eight or twenty-nine the **age thirty transition** begins. The young adult feels uneasy, that something is missing or wrong. Levinson found men orienting toward the future to make life more worthwhile: finding a new life direction, making new choices, or strengthening commitments to choices already made.

George Vaillant (1977), in his longitudinal study of ninety-four men, found that from twenty to thirty men's major energy expenditures were directed at wooing and winning wives and learning how to get along with them. This happened, Vaillant found, as soon as the men won real autonomy from their parents. Once they found a sense of their own independent identity, they entrusted it to another (a wife). Once they recemented human relationships from the family of origin to a family of procreation they devoted themselves to the latter and began to consolidate careers. However, Vaillant found energy expenditures toward career commitments more characteristic of the decade of the thirties than of the twenties.

Duvall (1977) also saw the establishment of a family of procreation as paramount in the twenties. She described a **family life cycle**, divided into eight stages:

1. Married couples (without children).
2. Childbearing families (oldest child birth to thirty months).
3. Families with preschool children (oldest child thirty months to six years).
4. Families with school children (oldest child six to thirteen years).
5. Families with teenagers (oldest child thirteen to twenty years).
6. Families launching young adults (first child gone, last child leaving).
7. Middle-aged parents ("empty nest" to retirement).
8. Aging family members (retirement to death of both spouses).

Men and women in their twenties, Duvall felt, are primarily concerned with the first three stages of the family life cycle. We will

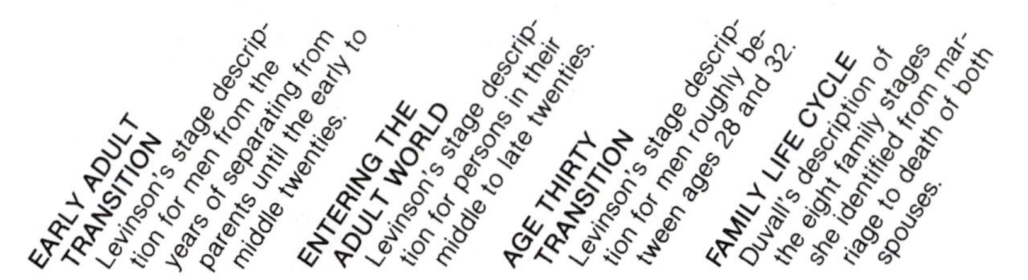

describe the concerns of these stages in the marriage and parenting sections of this chapter.

Intergenerational family relationships are easier if young adults in their twenties adopt lifestyles similar to, but independent of, their parents. Those who refuse to challenge the childish assumptions, as proposed by Gould (1978), and remain dependent on their families of origin worry their parents, both because they are not ready to settle into the reality of adulthood and because they seem wary of settling into the existing social order. The young adults may also feel chronically incompetent in their continued dependence while others around them are achieving adult independence and self-confidence. The greatest conflicts between parents and their twentyish offspring, however, often occur when the young adults not only declare their unconditional independence from the family of origin but also blatantly reject their parents' views and seek out diametrically opposed or antagonistic lifestyles.

Conventional Young Adults The majority of American young adults are considered **conventional** in their behavior. They work at becoming a part of the adult culture. Although they may deliberate the need for reform in such things as employment, taxation, welfare, political parties, the penal system, the military, the federal budget, and big business, they spend their time pursuing money or education or enjoying leisure activities. Some may experiment with the practices of less conventional youth, but they generally spend more time adhering to the customs of their parents and the mainstream culture. In terms of Marcia's identity statuses (see Chapter 7), conventional young adults are usually either identity foreclosed or identity achieved. They usually have high self-esteem (Prager, 1982). Identity-achieved females are not usually strongly stereotypically feminine sex-typed (Prager, 1983).

Keniston's Youth Stage Kenneth Keniston (1970) gave the classification **youth stage** to the young people in the Western industrialized world who are between adolescence and adulthood. To come under this rubric a young person must have acquired a stable sense of self (as opposed to being an adolescent seeking an identity). "Youth," however, are still searching—for a vocation, a social role they can comfortably play, the road to take, a place in the existing society. They feel tension and ambivalence about traditional practices vis-à-vis the developed sense of individuality. Some young people never experience this stage, some pass through it quickly, and others remain in it for prolonged periods.

The youth stage is not limited by race, sex, age, social class, or position in society. Not all college students go through this stage, for example. Some move from a prolonged adolescence into adulthood. As Keniston (1970) pointed out, poor and uneducated men and women, from Abraham Lincoln to Malcolm X, have experienced the youth stage. The classification is applied to those who have outgrown adolescent themes without yet having acquired adult themes. Instead, they have their own themes, which include:

- Tension between self and society.
- Pervasive ambivalence toward both self and society.
- Wary probe of the existing social order.
- Alternating estrangement and omnipotentiality.
- Refusal of socialization and acculturation.
- Youth-specific identities.
- Value of change and movement; abhorrence of stasis.
- Goals to be moved, to move others, to move through.
- Valuation of development.
- Fear of death equated with fear of stopping.
- View of adulthood as stasis (= death).
- Countercultures keep distance from existing social order.

Keniston found youths' sense of incongruency between the self and the world discontinuous. At times they feel isolated and estranged; at other times they feel capable of creating a new society for the world. They see multiple possibilities for change and feel free to achieve anything. The themes that characterize youth, Keniston pointed out, are rarely conscious. Also, different youths experience each of the themes with different intensity. What may be an enormous conflict for one may be relatively unimportant to another. The resolution of the youth stage and the acceptance of adulthood comes when the individual can acknowledge and cope with social reality without feeling that selfhood is static or compromised. It does not mean selling out, simply growing out. Thus the Jerry Rubin of the early 1970s, who in his youth stage presented himself as a Yippie, emerged in the 1980s as a sedate Wall Street analyst, recognizing that the transformation had a great deal of underlying continuity (Rubin, 1981).

Activist Youth Not all young adults are **activists**, just as not all twenty-year-olds are students (see Figure 8–3). Very few young adults make up activist groups. Activists are seldom found at small colleges, and at the large universities they are in fringe groups that rarely constitute more than 3 to 4 percent of the student population. Activists are generally intellectually above average. Like all youth as described by Keniston, they feel a tension between maintaining personal integrity and becoming effective and acceptable in society. However, activists either feel more passion or are more willing to do something about the estrangements they feel. They plan strikes, lectures, or marches; print and pass out statements; or organize cam-

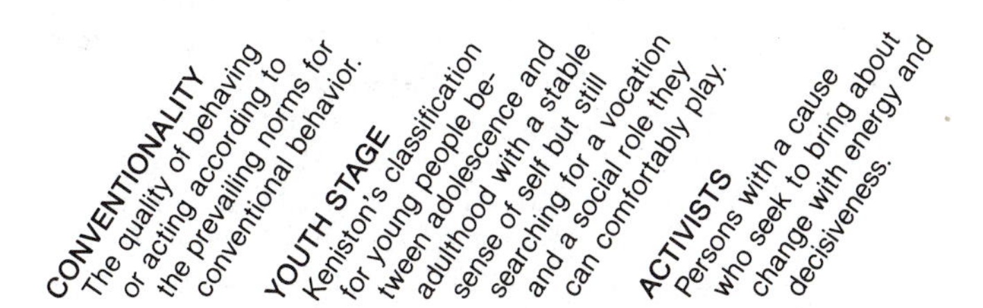

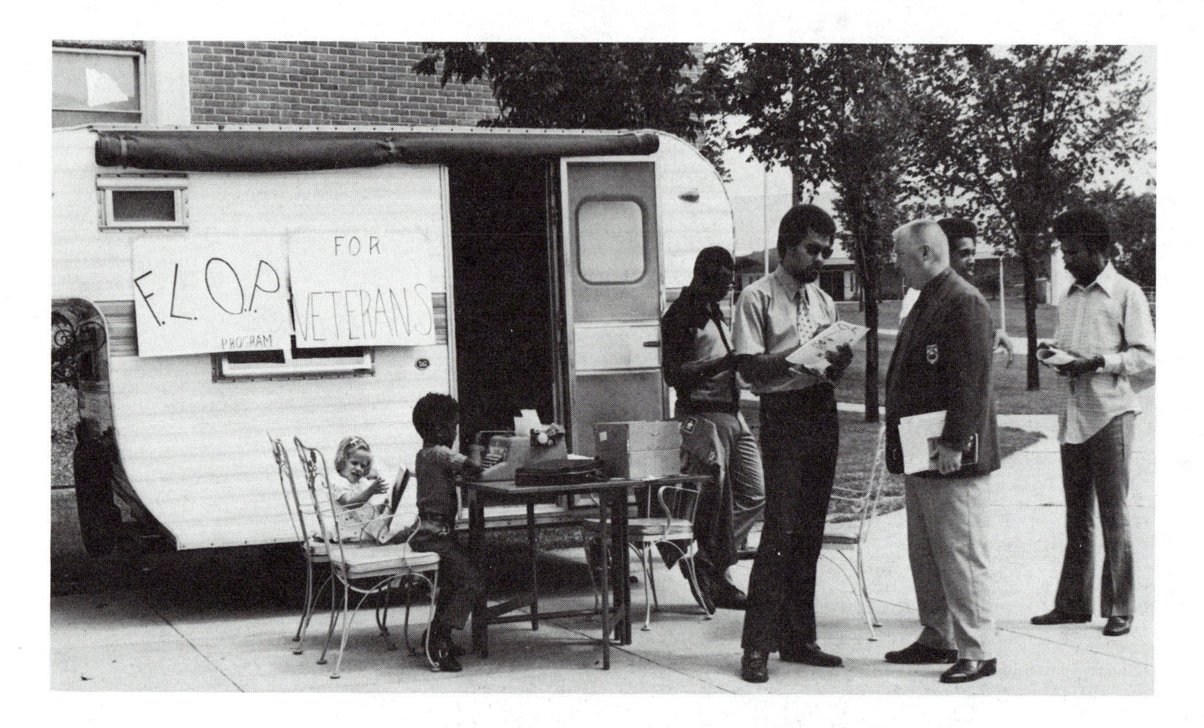

Figure 8–3
Young adults are as varied as their backgrounds and experiences. Some may prefer a stance of noninvolvement while others may be activists.

paigns for legislative change. If they find their jobs, government, religion, or education depersonalized or absurd they say so. They may roundly refuse to compromise their high ideals for status and success in a society that to them seems morally offensive. They embrace causes where they perceive they can benefit greater numbers of people. Activist campaigns are seldom undertaken simply for self-glorification. Keniston (1974) has commented that, given concomitant characteristics of love, compassion, and empathy, they may help make our society a happier place in which to live. However, he has also warned of the truth of the old adage: Compassion without morality is sentimental and effusive, while morality without compassion is cold and inhumane.

Alienated Youth Very different from activist young adults are those whose feelings of estrangement from society are manifested by a turning inward of frustrations. Rather than participate in organized groups, **alienated youth** try to escape from society through such means as drugs, alcohol, or living in seclusion. They are likely to have had preexisting personal, familial, or social problems before young adulthood descended with its tensions and ambiguities. They most often have had poor relations with their families of origin, and their alienation is both cause and effect of intrafamilial strife. For some, not being able to find a job adds to this alienation. The unemployment ranks are swollen with all kinds of young adults: those who have little preparation or experience for any kind of work; those who have been laid off; those who are overqualified for the available jobs because of their education; those who are handicapped. A high

ALIENATED YOUTH
Youth who feel estranged from society and try to escape through such means as drugs, alcohol, or adherence to socially unacceptable lifestyles and actions.

school, trade school, college, or even graduate school degree is no longer a guarantee of a ready career. Some (but not all) unemployed youth feel estranged and become alienated.

CAREER BEGINNINGS

A major element of Erikson's description of the nuclear conflict surrounding identity versus role confusion concerns career choice. During adolescence there may be a host of people to give advice about various careers: guidance counselors, primary caregivers, mentors, peers, siblings, and other relatives. Teenagers can feel tense about making such weighty decisions for their lives. Many steal time by simply saying they are college bound for a liberal arts education that will help them decide, or they declare themselves without choice, saying they must get a job wherever they can.

Parental expectations for their offspring are often in stark contrast to the expectations young adults have for themselves. The typical American parents would like to see their sons be successful and respected in their chosen field of work ("My son, the doctor"). Additionally sons should be intelligent, honest, responsible, independent, self-reliant, aggressive, and strong-willed. If they marry, the parents would like the chosen wife to be kind, unselfish, loving, attractive, well-mannered, and a good wife and mother. These same parents would like to see their daughters be kind, unselfish, loving, attractive, well-mannered, and good wives and mothers ("My daughter, Miss America") (Hoffman, 1975). Naturally, they would prefer their daughters to marry men who are successful and respected in their chosen fields. Even professional faculty women interviewed at the University of Michigan, in spite of professing to hold the same goals for offspring of either sex, were found to have higher occupational and academic goals in mind for their sons than for their daughters (Hoffman, 1977).

In today's changing social structure many more young women have higher academic and occupational goals for themselves than do their parents. Lavine (1982) found that if a daughter is raised perceiving her father to be dominant, she is more apt to select a stereotypically feminine career. However, daughters who perceive their mothers as having egalitarian power or as being dominant tend to select more stereotypically neutral or masculine careers. Parents are not always pleased with these choices. Likewise, many more young men have goals for themselves at odds with their parents' goals. Many of today's young men would prefer to have a wife with some career orientation both to have a second paycheck and to have a more compatible marriage with more shared interests. Many of these same men are more willing and interested than their fathers were in sharing homemaking and child-rearing tasks with their wives. Both young women and young men are expressing more desire than did their predecessors to postpone marriage, postpone parenthood, and have fewer children. This has become feasible because of safe and effective methods of birth control.

Advice on career choice often abates somewhat by the twenties. Would-be advisors believe that young people have already made up their minds about a career. Young people often give lip service to the vocational choice they are pursuing or are hoping to pursue. They hesitate to admit their uncertainty. They often seek less advice because they feel they should decide for themselves.

When looking for a career, young people consider what interests them and what they might be able to do with their given aptitudes. They also consider prospective salaries, the amount of education or preparation required, the prestige that goes with the job, and often the possibility of finding employment in their chosen field after spending considerable time preparing for it. Unfortunately, some still prepare for careers where job openings are diminishing (see Table 8–2).

Table 8–2 Projected Employment Opportunities for the Mid-1990s

Fewer job openings	More job openings
• College and university teachers	• Computer operators
• Postal clerks	• Computer service technicians
• Stenographers	• Computer systems analysts
• Butchers, meat cutters	• Business machine repairers
• Police officers	• Electrical engineers
• Industrial compositors	• Mechanical engineers
	• Physical therapists
	• Cashiers

SOURCE: Department of Labor, Bureau of Labor Statistics.

The young person who chooses to start a career after high school may try out several jobs before deciding on a preferred form of employment. Young persons who choose a career that entails advanced preparation at a trade school, college, or graduate school invest more of themselves (time, money, dreams) in a future job. They also may quit the job later but at a greater expense. Hopefully, their enthusiasm for their chosen career will grow as they progress through the educational process. Some will have had summer work or part-time jobs in related areas or will have seen and heard about the work from relatives and friends. They will feel more secure in the commitment they are making. Some will have practicum courses built into their education. These experiences will help them decide whether they have made the right vocational choice. If they do not like the work, they may be able to drop out of the particular career program and choose another. In some schools and for some curricula, practicum experience is missing. Some young persons may spend years preparing for a job without any feel for what it will be like.

APPRENTICESHIP LEARNING Education by legally agreeing to work a specified length of time for a master craftsman to learn the craft or trade.

One may look back wistfully to the days of **apprenticeship learning**. Children once worked alongside their parents and neighbors or at least had more access to places of employment. Children knew a great deal about the world of work and various occupations. When they decided what job they wanted to pursue as adults, they could, by an agreement between their caregivers and a skilled craftsman, go to learn the line of trade by working daily with the master craftsman (see Figure 8–4). Although some aspects of the apprenticeship system can be admired, contemporary culture offers so many possible kinds of work that deciding on any one direction to take is now appreciably more difficult. Specialization within jobs also makes apprenticeship learning more complicated. Increases in business size and rapid technological change also make it less practicable. Finally, education and child labor laws keep children out of much of the work world until they reach the age of sixteen. Most employment today involves on-the-job training no matter how little or how much education the employee has had ahead of time.

Working Women For women in our contemporary culture, a commitment to a career has concerns beyond those already described. In spite of the feminist movement, many people feel that women (especially women with young children) belong in the home (see Box 8–2). Employers hesitate to hire women of childbearing age for long-term jobs for fear that they will become pregnant and leave. If a woman asks for a maternity leave rather than quitting her job, she is often seen as uncommitted and may be refused leave (to test her loyalty) or laid off, usually for some "other" reason. Rosen, Jerdee, and Prestwich (1975) found that employers hesitate to give women jobs that require traveling. Women are also seldom considered for jobs that might entail transfers to other cities. A woman is expected

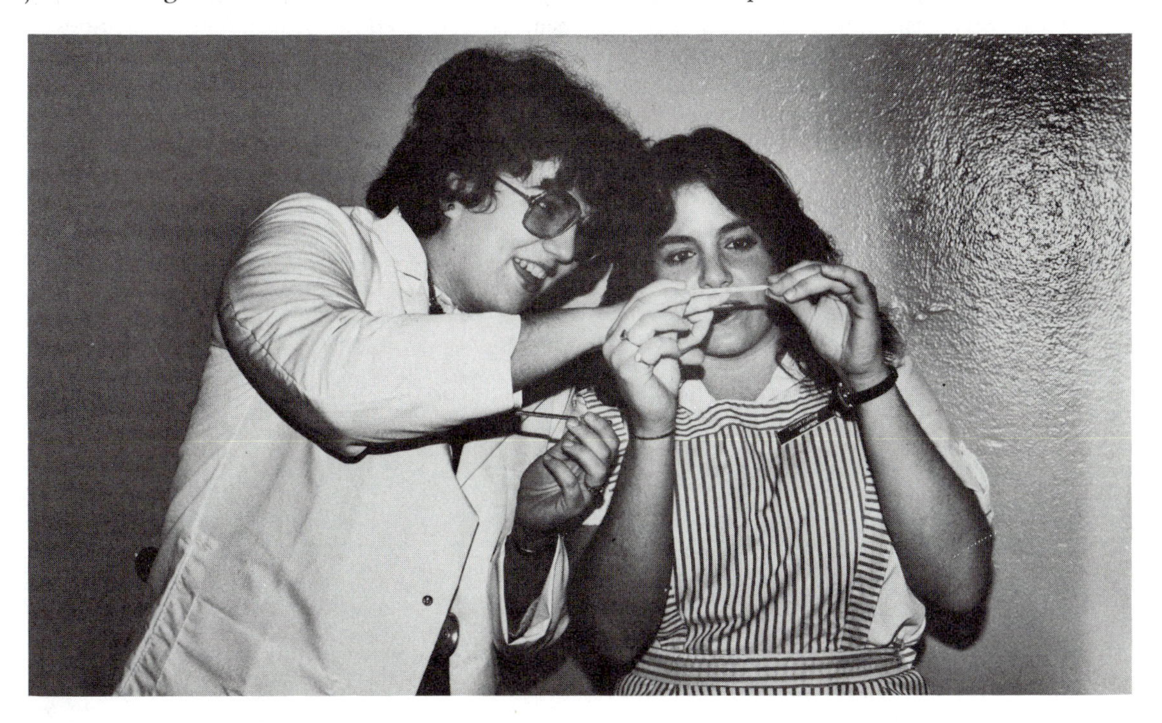

Figure 8–4
The apprenticeship system (wherein a person works for a specified length of time in a craft or trade in exchange for instruction) or a practicum experience (in which a student applies knowledge and becomes proficient at skills by working as part of the education) helps novice employees see what a job is like from the inside.

BOX 8–2

Women Versus Men in the Work Place

Are men being treated as full and equal members of the labor force? Should they be? Or are prejudices against men at work justified? How can a man give his best to an out-of-the-home job while he is also a husband and a father? Can a man ever be fully respected on the job unless we radically alter the basic structures of family life? As it stands now, would you want your staff to be composed entirely of men (unless, of course, the job for which you hired the man was something like child care, nursing, elementary school teaching, cleaning, secretarial work, and the like)?

By and large, men have been socialized to be husbands and fathers first, and workers second. Their attitudes toward a job will always reflect this bias toward home and family. Nonwhite men seem especially inclined to put their wives and children ahead of their jobs. Knowing this, employers do the logical thing. They pay a nonwhite male only about 79 percent of the white female's average entry hourly wage. If they hire a white male, they pay him 4 percent more, or about 83 percent of the white female's average entry hourly wage.*

To the credit of working men, they do not seem to develop as many stress-related illnesses (heart disease, hypertension, peptic ulcers) as women. They also have fewer accidents. Perhaps this is because they are so rarely found in managerial positions. Nevertheless, type A, "cardiac" personality women do make better managers, don't you agree?

Why do men want to be full and equal members of the labor force anyway? Why don't they stay home, raise children, and let the women handle the difficulties of the work place?

*Figures from 1980 census, in reverse (substitute female for male).

to leave her job and follow her husband if and when he is transferred. The reciprocal (that a man will leave his job and follow his wife if and when she is transferred) is not part of the attitude of most employers.

When equally qualified men and women apply for positions, placement officers not only favor the male applicants but also frequently ask more information on the females' personality, marital status, and children. They also ask what the husband and children will do if she is hired. Affirmative action committees have done a great deal to eliminate job discrimination because of sex (also race, age, and country of origin), but old prejudices against women persist. If a job requires aggressive interpersonal behavior and decisive managerial action, men are more apt to be hired.

Despite the increasing numbers of women in the labor force (over 50 percent of all adult women) and their increasingly higher educational attainments, women have not yet reached earnings

Table 8–3 Men's and Women's Average Wages for Full-Time Employment, 1972 to 1984

Year	Men	Women
1972	$10,202	$ 5,903
1974	11,889	6,970
1976	13,455	8,009
1978	15,730	9,350
1980	19,173	11,591
1982	21,655	13,663
1984	25,861	16,030

SOURCE: Department of Commerce, Bureau of the Census.

parity with men, nor has there been any appreciable progress in narrowing the earnings gap (Maymi, 1982). When women are hired, they are usually offered less money than comparable men (see Table 8–3).

Although being discriminated against at the time of job seeking is a threat, the more frequent concerns of women contemplating careers are role related. Do I want to be married? Can I combine marriage with a career? Do I want to have children? Can I combine child-rearing with a career? Will daycare centers or babysitters be available so that I can work while I have children? Are daycare centers or babysitters good for children? Are they reasonably priced? Will I be able to take a break from my career to raise children and then return to work?

While answers as to how physically and emotionally healthy it is for women to combine career, marriage, and child-rearing are not clear, more and more women are attempting to do all three. As pointed out earlier, many men are helping more with housework and child-rearing duties to enable their mates to continue in either part- or full-time jobs. Recently, social scientists have adopted the term **superwoman syndrome** to describe women who attempt to combine the roles of wife/mother, homemaker, and career woman. They perform all the traditional feminine role functions at home: entertain lavishly, keep an immaculate home, cook gourmet meals, and assume full responsibility for child care. Superwomen most frequently select career areas that are stereotypically feminine. Nearly 70 percent of all women are employed in traditionally female jobs. More than one-third of all women are employed in just ten feminine

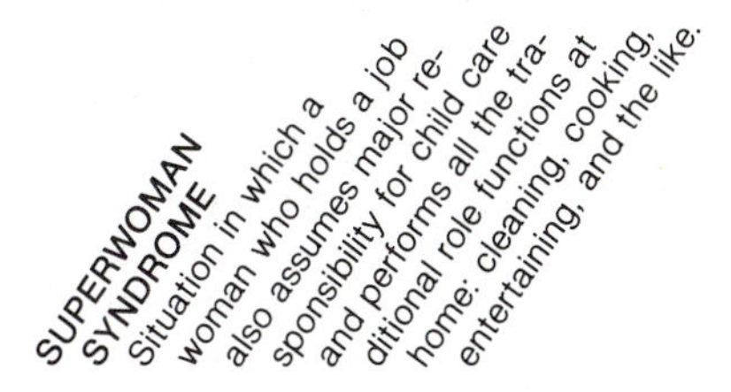

occupations. In order of prevalence, they are secretary, bookkeeper, sales clerk, cashier, waitress, nurse, elementary school teacher, private household worker, typist, and nursing aide (Maymi, 1982). In addition to trying to fill the stereotypic female role both at home and at work, superwomen also may conceal their achievements or temporarily reduce their efforts if their accomplishments put them in conflict over their social acceptability. The superwoman syndrome can be hard on a woman's physical and emotional health.

In 1985, a California group began a newsletter entitled *Superwomen Anonymous,* providing advice to working mothers on how to cope with the stresses of combining career, marriage, motherhood, hobbies, health, and the like. It stressed an overhaul of a woman's entire way of thinking (enough is enough!) and lowering expectations for performance and possessions rather than giving up (Orsborn, 1985).

Many women pursue career education beyond high school without really wanting a career for themselves. There are psychological risks inherent in this plan of action. A subtle indoctrination occurs as a woman studies for and interacts with other people studying for a particular career. She sees herself in the role. She develops self-confidence performing the role. It is then hard for her to give it up and become a full-time wife and mother. A woman who has had a career may miss her contacts with her working friends. She may long for the feelings of competence, self-esteem, respect, and importance she found in her job. Unfortunately, being exclusively a wife and mother is unpaid work and is not given due respect by many people in our culture. In her classic book *The Feminine Mystique,* Betty Friedan (1963) described the loneliness and lack of self-esteem of some housewives as "the problem that has no name." Although it is by no means a universal problem, it is not uncommon. The physical and emotional health of women who have no career other than wife and mother may be in more jeopardy today than that of the superwomen. Vast numbers of young housewives have clearcut symptoms of depression (to give one name to the problem Friedan described). Women have twice as much depression as men, even when matched for such influences as age, marital status, employment, and income (Radloff and Rae, 1979).

There are several reasons why depression may be so much higher in women than in men. Females are more restricted from birth and are subtly rewarded for being dependent. This can result in a learned helplessness. They are also more frequently taught to criticize themselves and to turn their anger inward rather than to be physically aggressive or blame others. They also learn to judge their self-worth by the opinions of others rather than by their own accomplishments. A wife/mother trained in dependency from girlhood typically expends her energy trying to meet the expectations of her husband and others (cooking, cleaning, caring for the children, staying sexually desirable). If the husband and the greater society withhold approval and praise instead the more independent, achieving, career-oriented superwoman, the resulting sense of low self-worth may be turned inward in self-blame, loneliness, insecurity, and helplessness, which can lead to depression. Encouraging women to be more self-sufficient and to seek activities that bring approval, acceptance, and a greater sense of self-worth are in the long run better antidotes to depression than medications.

Homemakers not only have more depression than working women, they also have more acute illnesses, more chronic conditions, more doctor visits, and more disabilities and activity limitations (Verbrugge, 1982). Part-time jobs with interesting assignments that bring a feeling of competence, respect, importance, and self-esteem may contribute to improved mental and physical health in women more than commitments to full-time careers or to full-time homemaking. Society has not yet, however, recognized the need for or provided many rewarding, paid, part-time jobs for women.

Disillusionment A worker does not usually perform well in a job until he or she has had a few months to learn the techniques of the work to be done. The settling-in process involves the establishment of interpersonal working relationships with the employer and other employees as well as learning the details of the job. Most young people need from a few months to a few years to enter a field of employment: Job changes, job training, and further education prolong the process. They then need a few years to stabilize in the job or career to which they finally make a commitment.

Disillusionment is not uncommon in the months or years spent entering a field of employment. Young persons may choose a job because of the stereotype it has, without a realistic understanding of the work entailed. Medicine, for example, is a high-prestige career with an aura of glamour. However, doctors and nurses must deal with death and disease, maintain patient rapport, work under the threat of malpractice suits, and handle many other less enchanting problems in their day-to-day lives. Modeling may look glamorous but requires hours of posing for takes and retakes under hot lights and a rigorous regimen of diet and exercise in off-duty hours. Stereotypes also differ outside and inside places of employment. The physician and the model may be envied their salaries and their limelight by the outside world but suffer from many forms of unpleasantness within their work-a-day worlds. Once a worker has settled in to a job, the disillusionment usually abates. The pros and cons of the job are eventually accepted as a fact of life.

In a study of what young adults preparing to enter the work force in the 1980s expected from their jobs, Bachman and Johnston (1979) found that the primary goals were not money, respect, and status but rather (1) interesting work that (2) uses skills and abilities with (3) good chances for advancement and (4) a predictable, secure future, where one can (5) see results of what one does, (6) make friends, and (7) be worthwhile to society. Both college and noncollege youth rated these seven attributes more important than money, respect, and status (which placed respectively 8, 9, and 10).

Many employers try to provide changes and challenges in assignments to keep their employees interested and motivated at their jobs, but this is not always possible. There are many highly technical jobs that must be done one way and one way only by well-trained, practiced hands. In some large businesses boredom is accompanied by a strong feeling of depersonalization. The worker feels simply like another cog in the machinery and cannot capture a sense of importance, uniqueness, or even self-direction at work.

Young persons in search of careers frequently change jobs. These shifts are more easily accomplished when people are young, do not require large salaries, and can travel. However, tight job markets, family pressures, too little or too much education, even fear of discrimination may make young people hesitate to look for better jobs. In spite of equal-opportunity employment laws, factors such as age, race, sex, marital status, health problems, and handicaps can jeopardize one's ability to find work.

Too many job changes can leave young people as dissatisfied with the world of work as can any one routine. Most human beings have a need for some sense of permanence, of belonging. There is often a certain comfort (as well as possible boredom) in having work that is well defined and predictable. Even in large, depersonalized organizations employees can meet in informal groups for coffee breaks or lunch and give one another personal attention and a sense of uniqueness. Some jobs are made more permanent by the promise of salary increments and pension benefits that accrue over time. Alvin Toffler (1970), in his book *Future Shock,* warned of the dangers of too many job changes, including those brought about by upward mobility, promotions within an organization, or promotions to other branches of the same company in faraway locations.

There is always a great deal of tension, sometimes more than a person can handle, accompanying a change of job. The work is unfamiliar. One must learn quickly and do well to assure job continuance. One's old friends are gone. Even a promotion within the same organization may put one's old friends at a distance because of sudden social inequality. If the job change involves a move to a new location, there are many extra-career tensions: finding housing, moving the family, starting children in new schools, locating new employment for a spouse. Increasingly young men and women are refusing promotions if they see them as threats to the more important goals of enjoying work and being able to use one's skills and abilities effectively.

In summary, career commitment is a problem sufficiently large to consume a considerable amount of the time and energy of people in their twenties.

MASLOW'S NEEDS HIERARCHY

The theories of development that stress the central role of one's self-concept have been given a great deal of attention in recent years. The theories of Abraham Maslow and Carl Rogers, discussed in Chapter 2, stress how a person's perceptions of himself or herself influence his or her choice of activities. Such terms as *perceptual* and *phenomenological* are also used to describe these theories. Each person's perception of his or her life space (past-present-future) and the phenomenological field of experiences that he or she denies, distorts, or incorporates into the self-concept are important factors that shape behavior. This self-concept is the agent that can have either a positive influence on a person in pursuit of an independent life or a negative effect, causing hesitancies and fears.

The late Abraham Maslow, defined his concepts as *epi-Freudian* and *epibehaviorist* (*epi* meaning "upon"). He stressed positive growth, normality, and excellence rather than abnormality, depression, and antisocial behaviors. Maslow (1970) postulated a **hierarchy of needs** common to all humans. The highest needs, those for self-actualization, self-understanding, and the understanding of others, can emerge only when lower needs are satisfied. The hierarchy of needs, from lowest to highest includes the following:

Physical needs—food, drink, sleep, activity; relief from pain and discomfort.
Safety needs—freedom from threats to supplies or life; secure, orderly, predictable environment.
Belonging and love needs—acceptance, affectionate relations with others.
Esteem needs—competence, confidence, task mastery, recognition, prestige; respect and approval from others.
Self-actualization—realized potential; feeling that one is what one is capable of being.

According to Maslow, gratification of a need makes a person feel good whereas deprivation breeds illness (mental or physical). A deprived person will prefer gratification of the missing need over everything else under conditions of free choice.

The emergence of **self-actualization** rests on prior satisfaction of the other four needs. It is a goal toward which healthy young adults continually strive. The characteristics Maslow (1968) attributed to self-actualizing persons include adequate perceptions of reality, comfortable relations with reality, a high degree of acceptance of themselves and others, a feeling of belonging to all humankind, close relationships with a few friends or loved ones, a need for privacy, an unhostile sense of humor, resistance to socialization pressures, autonomy, problem-centeredness, creativeness, spontaneity, a freshness of appreciation, a strong ethical sense, and frequent **peak experiences**. These experiences may occur when one has finished a long-term project, after a particularly good sexual experience, after one has done something special for someone else, even when one becomes aware that he or she is loved. While some young adults

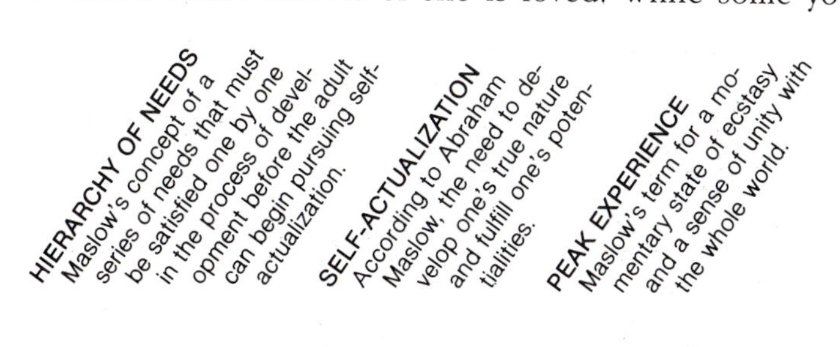

have "peaks" in the Maslowian sense, many more are prevented from approaching self-actualization because of their many unfulfilled needs for belonging and love or for self-esteem in both work and interpersonal relationships.

ERIKSON'S CONCEPT OF INTIMACY VERSUS ISOLATION

Erikson described the sixth nuclear conflict of life as that of finding **intimacy versus isolation** (see Table 8–4). Once young adults have acquired a sense of who they are and where they are going (identity), they begin to feel a need to reach out to others. Having severed their former close ties to caregivers, siblings, and schoolmates, they seek new friends with whom to share life's joys and sorrows. The persons with whom one relates do not necessarily have to be prospective marriage partners. A sense of intimacy (defined as closeness) can develop between people who work together, face various kinds of battles or stresses together, share leisure activities, or are roommates. Although intimacy usually develops between marriage partners, Erikson (1963) cautions that some marriages really amount to an *isolation à deux*—two people living in solitude together.

Table 8–4 Erikson's Sixth Nuclear Conflict: Intimacy Versus Isolation

Sense	Eriksonian descriptions	Fostering behaviors
Intimacy	A fusion of identity with that of others	Others confirm sense that mutuality of efforts is beneficial
	Commitment to concrete affiliations and partnerships	Others prove faithful to cooperative and intimate sharings and interactions
	Strength to abide by commitment to others	Experiences of making sacrifices and compromises to maintain relationships
versus		
Isolation	Avoidance of contacts that commit to intimacy	Experiences of being victimized, abused, or exploited by others
	Distantiation: readiness to isolate forces and people whose essence encroaches on one's own territory	Shaky sense of identity that causes a repudiation of things foreign to self
	Deep sense of self-absorption	Experiences showing need to compete, not cooperate

Erikson saw the development of the kind of mature intimacy described by Freud as being necessarily postponed until after youths have defined themselves in terms of their social, sexual, and career identities. They cannot fully relate to others until they have dealt with the problem of self. Remember from Chapter 7 that Erikson felt young love is usually more conversational than sexual. Young people use their friends as mirrors in which to reflect their developing self-images. Sexual relationships that are self-centered or exploitive or engaged in for erotic pleasure only are immature. Mature love includes self-giving, other-centeredness, or a fusion of one's self with another directed toward mutuality of pleasure. The utopia of intimacy as described by Erikson (1963) should include:

1. Mutuality of orgasm
2. with a loved partner
3. of the other sex
4. with whom one is able and willing to share a mutual trust
5. and with whom one is able and willing to regulate the cycles of
   a. work
   b. procreation
   c. recreation
6. so as to secure to the offspring, too, all the stages of a satisfactory development.

However, even as he defined the utopian, Erikson warned that in our complex society factors of tradition, opportunity, health, and temperament interfere with satisfactory mutuality. He also stated that although persons should be potentially able to accomplish mutuality, they should also be able to bear frustrations without undue regression.

Both Eriksonian conflicts of identity and intimacy seem to be more easily resolved by androgynous than by nonandrogynous young adults (Schiedel and Marcia, 1985). Females seem to resolve the conflict of intimacy versus isolation more readily than do males, especially when they are secure in their own ego identities beforehand. Males find it easier to resolve the conflict of intimacy versus isolation when they are secure in their occupational identity (Fitch and Adams, 1983).

Isolation as the alternative to intimacy is fairly common in our culture. Many young people have had little or no experience with making sacrifices or compromises, or practicing altruistic behaviors, prior to leaving their families of origin. Intimacy cannot develop without a sharing and giving of the self to others. Many young people have also had bad experiences trying to show regard for others. They have been exploited, abused, and victimized. They have learned that competition serves them better than cooperation. They fear any further openness with others.

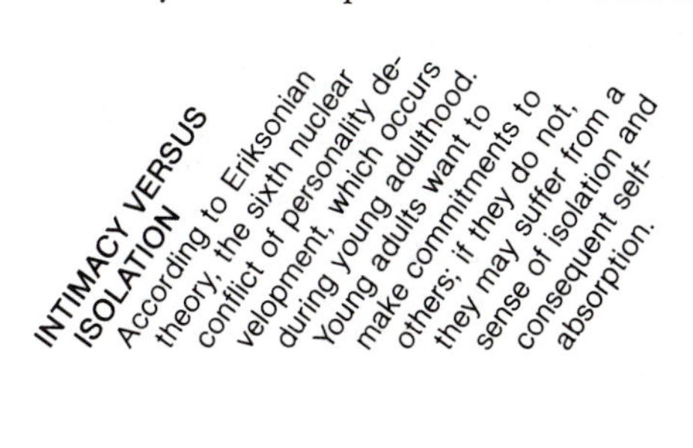

Many of our contemporary psychotherapies seek to help people find intimacy in contrast to isolation. Psychoanalysis typically goes back to an exploration of the parent-child relationship to help a person work through reasons for fearing intimacy. Behavior therapies seek to change the environments that stimulate or maintain isolation. Desensitization programs, assertiveness training, modeling, even biofeedback programs (where an individual becomes aware of his or her bodily processes by the use of monitoring instruments) can help people alter the external or internal environments that contribute to their isolation. Perceptual therapies help people assess themselves and improve their self-concepts, allowing them to move toward honest, open interactions with others. Nondirective counseling sessions can also help people understand and appreciate themselves so they can relate more fully to others.

SINGLEHOOD

To marry or not to marry is an important decision facing people in their twenties. There are a great many pressures, both internal and external, for marriage. Internal pressures include those described by Erikson—the seeking of an intimate companion to share life's joys and sorrows and a desire for a true intimacy and mutuality in sex—and those described by Maslow—desire to gratify the needs for belonging, love, self-esteem, and self-actualization. As the twenties pass, many young people feel the exigency to grab hold of life, master it, and establish themselves as self-reliant persons. For some, as Vaillant (1977) found, marriage appears to be one avenue toward accomplishing these goals. External pressures for marriage may include family prodding, watching one's friends get married, peer group prodding, dwindling social life as friends marry, religious convictions, external urgings to marry to improve financial or social status, even the pressure from an employer for marriage in order to enhance career opportunities.

At least 10 percent of young adults now resist the pressures for marriage in their twenties for various reasons. This rate has doubled in recent years (Stayton, 1984). Some will eventually marry, but many choose long-term **singlehood** willingly. They no longer fear the labels "spinster" and "bachelor," concerns about lack of libido or misplaced libido, or criticisms of their self-centeredness.

Attitudes toward singlehood have changed considerably over the past twenty years. Several hundred Americans were asked the question "Suppose all you knew about a [man/woman] was that [he/she] did not want to get married. What would you guess [he/she] was like?" Two-thirds of contemporary respondents gave either a positive or neutral description of such a person compared to over half of all respondents who gave a decidedly negative description of such a person in 1957 (Veroff, Douvan, and Kulka, 1981). Some young adults choose singlehood in order to pursue careers that involve frequent traveling or other activities that would be considerably more difficult with a spouse or children. Some may have the responsibilities of caring for older parents or younger siblings. Some may not meet a woman or man they want to marry or one who

SINGLEHOOD
The state of being alone; not united with another.

wants to marry them. Others may form intimate alliances with partners, possibly including cohabitation, but avoid the legal bonds of matrimony.

Singlehood has advantages: freedom to come and go at any hour, freedom to spend time and money as one chooses, autonomy concerning meals and housework, more career opportunities, more freedom to change, more freedom to move, more exciting lifestyle, sexual availability, self-sufficiency, plurality of roles, and psychological autonomy.

Schwartz (1976) found six groups that typified single men and women: professionals, socials, individuals, activists, passives, and supportives. Professionally oriented singles spend most of their time at work or in work-related activities. Social life for such persons may revolve around the neighborhood of the job: stores, bars, galleries. Friends are co-workers or people whom they see in the course of going to or from work. Social singles are those to whom one often applies the label "swinging singles." They spend a great deal of time on their social activities. Individuals are those who spend time getting to know themselves or in solitary hobbies. Activists, as described on page 415, devote a great deal of time to social causes that are particularly meaningful to them. Passive singles are apt to be the least satisfied being single and may be lonely or depressed. The supportive singles spend time doing for others and find meaning to their lives in these endeavors. With the exception of the passive singles, Schwartz found her subjects to be satisfied with their chosen lifestyles.

Most singles report some coital involvement by the age of twenty-five. Prior to 1970, twice as many young adult men as women reported that they were sexually active. The current proportions of men and women are nearly equal (Darling, Kallen, and Van Dusen, 1984). In a study of singles who are virgins compared to single nonvirgins, Jessor and Jessor (1975) found that the nonvirgins are more apt to be independent from their families, more reliant on peers, more unconventional, more tolerant of deviance from social norms by others, and higher in social criticism. The Jessors' data did not support the existence of a sexual revolution in young adults, however. Although they found a greater permissiveness toward premarital sex, they found many young adults expressing the belief that sex should be part of a loving relationship without exploitation.

Cohabitation (living together and sharing sex without being married) is becoming increasingly common. Cherlin and Furstenberg (1983) report that it has more than tripled since 1970. Macklin (1978) estimated that about 25 percent of college students cohabitate before marriage. Clayton and Voss (1977) found that cohabitants are also frequently nonstudents or high school dropouts especially in urban areas. However, cohabitation is generally short lived (two years) and ends with marriage or breaking up. Most cohabitants, Macklin found, rated their relationships as rewarding and maturing, with the benefits outweighing the problems. Watson (1983) found that only about 6 percent of cohabitants have regrets afterward.

COHABITATION
Living together and maintaining a sexual relationship without being legally married.

Watson studied eighty-four couples during their first year of marriage; fifty-four couples had lived together and thirty had not. The couples had a wide range of ages and educational and employment backgrounds. Comparisons revealed that the noncohabitant couples had significantly higher mean adjustment scores in their first year of marriage, due mainly to the responses of the wives. He suggested that this may be due to the noncohabitant woman's delight in establishing her own home and defining her new role, something her previously cohabitated counterpart did a year or two earlier.

MARRIAGE

Although singlehood or cohabitation may appeal to a portion of the young adults of the 1980s, a good marriage and family life is the goal of the majority (see Figure 8–5). Bachman and Johnston (1979) found that 79 percent of college students and 76 percent of noncollege youth ranked it as their number one goal, ahead of finding purpose and meaning to life, finding steady work, being successful at work, making a contribution to society, having lots of money, or being a leader in the community. The second-ranked goal of both college and noncollege youth was to have strong friendships. These goals indicate that the young adults of the 1980s are extremely interested in resolving the Eriksonian nuclear conflict of intimacy versus isolation on the positive side.

About 95 percent of all Americans eventually marry. The median age for first marriage is now twenty-five for men and twenty-two for women, older than it has been since the turn of the century. This reflects in part the practice of cohabiting before marriage and of postponing marriage until one has finished an education or become established in a career.

Figure 8–5
Marriage is one way in which young adults meet their needs for belonging and love, for esteem, respect, and approval from others. The average age for marrying is increasing, with growing numbers of young adults desiring singlehood for at least a portion of their twenties.

The joining of two persons into a legal partnership called marriage is easily accomplished by a justice of the peace or an authorized representative of a religious institution. The molding of the lifestyles of the two people into a workable team where joint interests are served involves a great deal of additional effort. Even couples who have lived together prior to marriage agree that the status of legality and the concomitant expectations of permanence of the union place new constraints on them. They must renew their attempts to live together harmoniously. **Marriage** means compromise and sacrifice. It means acceptance of each other's variations in moods, in needs, and in personality. It means emotional intimacy as well as sexual intimacy. A certain degree of romanticism accompanies mate selection. Wherever courtship takes place, the individuals involved put their best selves forward and try to hide their flaws. In marriage one sees the other's weaknesses as well as strengths. Our society offers very few courses in marriage preparation. To obtain a license one must simply prove legal age (or have consent of a guardian) and, in some states, freedom from sexually transmitted disease, as shown by a blood test. The best preparation most young adults have is a few premarital counseling sessions or a few lectures. Consequently, many people enter marriage with a very unrealistic idea of what is to follow.

The **honeymoon period** (mutual affection of newlyweds) of a marriage seldom lasts very long. Arguments soon break through the enchantment. Major sources of marital disagreements include money, use of leisure time, in-laws, chores, responsibilities, sex roles, and power. Quarrels, though viewed by many as destructive, can be constructive. They serve to clear the air and relieve tension. People tend to say what they really mean when they are angry. They boast. They gripe. They admit fears and uncertainties. Several other forms of communication, however, serve these ends as well as arguing. A willingness to engage in open, honest communication without game playing is one of the keys to a workable marriage. Another is to develop a relationship where partners are neither completely dependent on each other nor completely independent from each other but rather have a certain amount of interdependency.

Kieren, Henton, and Marotz (1975) described and defined **marital morale** as a more accurate indicator of the state of a relationship than happiness or stability. It is a measure of contentment with a marriage based on the number of personal and interpersonal goals that are being achieved. Happiness and stability may be marriage goals for some couples, but they are not necessarily goals in every

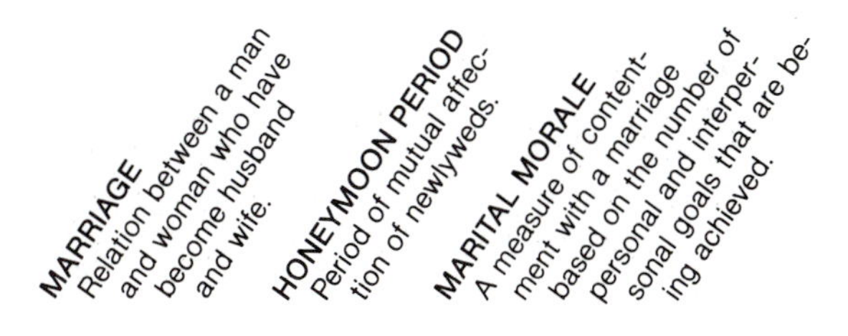

marriage. You can probably think of some people who maintain their unions in spite of what may appear to you to be overwhelming incompatibilities. Other couples are obviously still very much in love when they divorce. If the expectations of both partners in a marriage are being met, the marriage is likely to persist. If expectations fail, separation usually ensues.

Sternberg (1985) studied not only what leads to success or failure in romantic relationships but also components of love. He suggested ten factors that form a basic core of love:

1. Promoting the welfare of the loved one.
2. Experiencing happiness with the loved one.
3. High regard for the loved one.
4. Being able to count on the loved one in times of need.
5. Mutual understanding of the loved one.
6. Sharing oneself and one's things with the loved one.
7. Receiving emotional support from the loved one.
8. Giving emotional support to the loved one.
9. Intimate communication with the loved one.
10. Valuing the loved one in one's own life.

Sternberg kept sexual attraction separate from his list of central components of love because it is possible to have sexual attraction in the absence of love, and vice versa. He found that liking one's partner is a better predictor of satisfaction in a relationship than love. No matter how much a person loves a lover, the union is not apt to work out unless the loved one is liked as well.

Divorces (also annulments and separations) are frequent in our society. The number of new marriages ending in divorce is now approaching 50 percent. Young people seem to be more ready to divorce than older people. Divorce rates are highest for those marrying young (72 percent of teenage marriages) and for those with low incomes, poor education, or uncertain employment. Although grounds for divorce are now quite broad in some states (incompatibility, adultery, cruelty, desertion, nonsupport, alcohol or drug addiction, felony conviction, impotence, fraudulent contract, insanity, living separate and apart), the legal costs and the emotional trauma involved make most couples very stressed throughout the divorce process.

PARENTHOOD

Shortly after marriage most couples begin to feel pressure from parents, in-laws, peers, and friends to have children. The desire to oblige these people and conform to expectations is frequently given as a reason for procreation. Other reasons may include love of children, continuance of the family name, failure of birth control, proof of sexuality, improvement of the marriage, religious reasons, loneliness, security, status, tax exemption, and ability to afford them. Children can bring a great deal of love, warmth, and laughter

Figure 8–6
Many new parents experience an indescribable, intangible surge of wonder and joy as they first view the child they have created.

into a home. They also bring extra work and added financial responsibilities.

During a woman's pregnancy both fathers and mothers usually begin to anticipate and prepare for the changes that children will bring into their lives. Husbands may feel a surge in their masculinity as they watch their wife's abdomen expand and feel the unborn child move. They may begin helping and sheltering their wife more. In fact, pregnancy may usher in the first concept in the man that he is the guardian and defender of his wife and family. Women may feel either more or less feminine as they carry children. Some women decry the loss of their figures and the inconveniences of their pregnant stage. Others take great pride in their protruding figures and enjoy the legitimized reduction in their self-help activities.

Birth is usually a joyous occasion (see Figure 8–6). For a short while parent(s) enjoy receiving congratulations and hearing their very own infant's cry. But, sooner or later, the sound of the baby crying reminds parents of the exhausting work of **parenthood**. LeMasters (1970), after interviews with a wide spectrum of modern American parents, reported that 83 percent of them experienced a marital crisis after the birth of a baby. In contrast, a reader's poll conducted by *Parents Magazine* (Yarrow, 1982) found that 69 percent of the write-in respondents felt that having a baby strengthened

PARENTHOOD
Taking responsibility for protecting and raising one's offspring or adopted children.

their marriage. However, most of the women who wrote to the magazine (no husbands responded) also said they were more fatigued, went out less, and were more limited in their activities due to parenthood.

Suddenly the family relationships must all change to include the new member. There will be mother-father, mother-child, father-child, and, if there are other children, all the new child-child combinations. Schisms and jealousies ensue. The problems that existed in maintaining love and marital morale before the child's arrival often become aggravated. Sex is temporarily restricted while the mother heals. Freedom to come and go as one is accustomed to doing is inhibited by the need to provide baby care. The extra money from pay checks that once served as a basis for leisure activities is now diverted into baby necessities. In many families the infant's arrival marks a diminution of or end to the mother's wage-earning activities. Exhaustion from night feedings, extra housework, and tensions related to the baby are common. Feelings of inadequacy are also frequent.

Children seldom, if ever, really improve a marriage by themselves. What may happen is that couples on an upswing work even harder to communicate their problems and needs to each other or seek outside help for any difficulties they encounter. Couples on a downswing find their marital adjustment problems magnified. They also may finally turn to outside help to try to resolve their troubles. Once children are involved, divorce becomes more complicated. Adults can sever their relationships with each other but usually not with their offspring. Someone must provide for the well-being of a child for a full sixteen or more years after his or her birth.

There are reasons to be optimistic about parenthood as an influence on the development of young adults. Children can be immeasurably enriching and rewarding to adults who want them and are ready to love and care for them. (Erik Erikson's seventh nuclear conflict, that of *generativity versus stagnation,* will be discussed in this regard in the next chapter. Other effects of children on adult development will also be covered there.)

For some people in their twenties, parenthood is a foreign concept. For others, it is one of the most important influences on their own growth and development. Increasing numbers of young people are delaying having children until they have had a few years on their own to maximize their own development and to adjust to their marriages. The number of women deciding to postpone parenthood until they are between the ages of twenty-five and thirty-four has increased by about 8 percent in the past twenty years (Bloom, 1984). In 1980 the average age of a woman at the birth of her first child was 23.5 years, despite the increased numbers of teenage mothers. Women who are highly educated and pursuing careers are the most apt to delay the birth of a first child. Some couples are deciding that they do not want to have children at all.

Increasing numbers of young adults now approach parenthood with an earnest regard for the seriousness of the task. Courses in child-rearing are popular. Books on child-rearing are in demand in

Figure 8–7
Parenthood is a challenging task—one of the most important roles of adulthood. Fathers and mothers shoulder the responsibility for helping their offspring grow into trusting, autonomous, initiating, industrious, self-knowledgeable, and loving human beings in about eighteen short years.

libraries and bookstores. There is also evidence of a growing feeling among men that it is both admirable and desirable to help care for children (see Figure 8–7).

Parke (1982) in a review of the research on fathering, suggested that when men share with their wives the tasks of child care, the wife's self-esteem improves, the father's relationship with his children improves, and the children themselves may become both more independent and more internalized (have a belief that they can control the events in their lives).

HOUSEHOLD ORGANIZATION

Smooth, efficient running of a household is a major organizational task of persons in their twenties that can be accomplished in many ways. A home may be kept by a single person, by related or unrelated roommates, by spouses with or without children. In spite of the rhetoric on shared roles and equality, most individuals occasionally fall into patterns of boss and bossed. Only an individual living alone escapes this, and then only if her or his associates, parents, and friends resist the temptation to give directions.

In the traditional **nuclear family** setting (father-mother-children), the husband is usually perceived as having the managerial role. In fact, the wife, because of her more continuous proximity to the home, directs as many if not more of the tasks. Occasionally

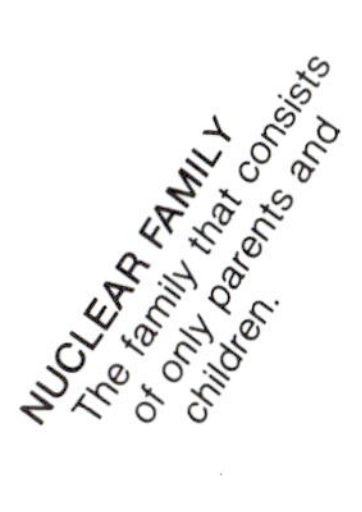

children are allowed to have input about what is to be done and who is to do it. In nontraditional families and in nonfamily settings, the management of the household has more chance of being egalitarian than in families trying to conform to conventional social customs. However, it is still possible for one person to assume an administrative role and for others to be dominated. Frequently, as in traditional families, one person will dictate some tasks and other(s) dictate the remaining ones.

Household organization is perceived by some to be simpler today due to modern appliances and convenience foods. Tasks are different from 100 years ago, but they are not necessarily simpler. Society is more complicated. Moves are frequent. Support systems in the larger community are disappearing. Managing a household, even a simple efficiency apartment, means paying bills, organizing time for work, meals, and leisure, socializing and entertaining, struggling with machines, repairing or replacing gadgets that seem engineered to break down, and dealing with meter readers, solicitors, repair people, telephone service, landlords, bill collectors, social service workers, neighbors, garbage collection, and pick-up or delivery of mail, newspapers, laundry, and groceries. People with children must organize and plan time for transportation to school and after-school events as well as cope with children's accidents, truancy, and disobedience. For single parents and for mothers with full-time jobs these may be gigantic chores. For some people household organization is untenable. Home means a crowded, chaotic, noisy place where tranquillity reigns only when inhabitants are asleep or absent.

IN-LAW RELATIONS

Married adults not only have to adjust to each other's idiosyncrasies, organize a household, and possibly rear children but also have to either design workable **in-law relations** or become estranged from their families of origin. In-law relations require a certain amount of diplomacy, compromise, and sacrifice, even with the best of all possible parents-in-law. The best parents-in-law shoulder many of the necessary sacrifices and compromises themselves for the sake of maintaining ties with their married children (see Figure 8–8).

When people marry, they should not expect complete freedom from their families of origin. What young marrieds can expect is that parents and in-laws will give them some freedom, tone down or hide some of their concerns, and refrain from too much advice giving.

Mothers-in-law come in for more accusations of interference than fathers-in-law, due partly to the fact that mothers usually have

Figure 8–8
Designing workable in-law relationships requires adjustment to pushes and pulls from many directions and some compromising.

more to give up in the way of parenting responsibilities. However, all relationships with in-laws are potentially problematic.

The wife's mother may feel that she has to help her daughter learn more about cooking, cleaning, and child care to the extent that the daughter has difficulty gaining a sense of self-esteem as a mature, capable wife/mother in her own right. She may resent her mother (even while being thankful for the help), and her husband may resent that his wife looks to her mother rather than to him.

The husband's mother often has difficulty seeing another woman take her place in her son's life. She may genuinely like her daughter-in-law yet find it hard not to mention all those things that her son liked when she was mothering him. These suggestions may be taken as veiled criticism or negative advice giving by a sensitive wife rather than as helpful suggestions. Some wives and mothers-in-law actually vie with each other for the attention, affection, and approval of the husband/son.

The wife's father may be problematic in the same way as the husband's mother. Fathers tend to persist in their beliefs that their daughters are naive, unaware, and immature. A father-in-law may be jealous of the husband who has taken his place in his daughter's life and may resent the husband for taking away the daughter's innocence. Daughters may aggravate the problem by comparing their husbands to "Daddy."

The husband's own father may give his son advice about finances, his role in marriage, fathering, or any number of other things to the point where the son has difficulty feeling self-confident about his role as husband/father. The wife may resent her husband's leaning on his father rather than managing his home and marriage alone.

Problems usually increase when families give financial aid to their married children. Aid may have certain spoken or unspoken strings attached. It also prolongs the young couples' feelings of dependence on their parents. Although financial aid may contribute to in-law friction, it is also very much needed and appreciated by some married persons. Living with in-laws may be easier than living close to them. When adults exist under the same roof, they work out their misunderstandings sooner and agree upon acceptable roles and responsibilities.

Designing workable in-law relations can be very rewarding. Some married couples really look forward to weekends and holidays with in-laws. They accept their parents' desire to spoil them, cook, clean, and give gifts and compliments. The older couple welcomes their role as grandparents. The younger couple showers their parents with gifts, compliments, love, and affection in turn. They may both give and accept help with finances, children, illness, and crises. Couples may be able to look on in-laws as role models and learn from them.

INFERTILITY

About 15 percent of couples fail to conceive a child after one year of regular coitus without contraception (Marshall, 1984). Reasons for **infertility** are about evenly divided between male and female problems (40 percent each), with both partners having problems in about 20 percent of marriages. Not long ago, more than half of all infertility problems were believed to be of psychological origin. Today, many more physical problems have been identified that contribute to the failure to conceive. Fortunately, many more therapeutic techniques have been developed as well. Now infertile couples have a very good chance of achieving a pregnancy after appropriate diagnosis and treatment (Speroff, Glass, and Kaseo, 1983).

INFERTILITY The state of being unable to produce offspring.

A common male physical factor in infertility is the presence of one or more varicose veins in the testicles, which decrease spermatogenesis. When the problem is treated, sperm counts improve in the majority of men. Another physical problem is autoimmunity of the man to his own sperm. In some cases, steroid therapy for two to three months will allow enough sperm production without antibodies to result in a pregnancy. Sometimes a woman develops an immunity to her husband's sperm. In these cases condom therapy is advised until the woman's antibody titer is significantly lowered. Thereafter, a condom is used except during the woman's fertile period to prevent the development of more of her antibodies.

Some men with very low sperm counts have a gonadotropin insufficiency and can be induced to produce more sperm with gonadotropin therapy. Another physical cause of infertility in men is a blocked vas deferens. It may be blocked because of an old inflammatory process from one of the sexually transmitted diseases or because of a congenital defect. Surgical techniques now allow for the reopening of many blocked tubes. Low sperm counts may also be caused by wearing tight briefs that keep the testicles too close to the body, and too warm, for adequate spermatogenesis. Excess exposure of the scrotum to heat from spas, whirlpools, hot baths, saunas, or sitting in hot environments can also adversely affect sperm production. So, too, can overexposure to radiation (from the sun or industrial or medical procedures). Some drugs, notably marijuana, also decrease fertility, as does overexposure to certain elements (lead, iron, zinc, copper). Men with low sperm counts are usually counseled to exercise regularly, lose weight if they are obese, avoid use of alcohol, marijuana, and tobacco, and refrain from excessive coitus. Intercourse every other day during the fertile period results in maximal fertility. Prolonged celibacy may depress sperm motility.

The most frequent physical reason for infertility in women is a blockage or abnormality of the oviducts (fallopian tubes). Blockages often occur as a sequelae to pelvic inflammatory disease (PID), which leaves scar tissue in the oviducts. Chlamydia and gonorrhea (see Chapter 7) are usually the infecting agents in PID. The oviducts may also be blocked as a result of endometriosis, tuberculosis, or polyps. Fortunately, many oviducts can now be cleared with microsurgery or laser surgery, restoring fertility to the woman. Another physical reason for failure to conceive in women is a failure to produce viable ova. If no primary oocytes exist, there is no possible treatment. However, available oocytes can be induced to undergo meiosis and produce mature ova with drug and hormone therapy when a woman is ready to conceive. If the reason for female infertility is insufficient cervical mucous, hormone therapy can also be used. If adhesions exist inside the uterus, a dilation and curettage (D&C) and hormone therapy should allow a re-created endometrial cavity to support a pregnancy after three to six months.

Women who exercise vigorously several hours a day (such as distance runners and dancers) may suffer temporary infertility as a result of altered endocrine functioning or of changes in the ratio of body fat to lean tissue that alters their menstrual cycles. Women who fast to diet, who are severely anemic, or who are anorexic may also suffer temporary infertility. These problems are treated by a return to normal eating, nutritional therapy, psychotherapy, or reducing the amount of daily exercise.

Many of the psychological reasons for infertility relate to stress. Fears, guilt, anxieties, hostilities, insomnia, a change in eating habits, an abuse of drugs, and so on will result in a change in endocrine functioning. The body reacts to stress by increasing the production of several "stress" hormones (ACTH, cortisol), which in turn reduce the production of the gonadotropins. Even the stress of wanting a pregnancy can interfere with fertility.

In women, stress can interfere with ova production and can also cause pain during intercourse, cause spasms in the vagina that prevent coitus, cause uterotubal spasms preventing fertilization, or cause uterine hypercontractility preventing implantation. In men, stress not only interferes with sperm production but also can lead to loss of libido, can cause functional impotence, or can lead to premature ejaculation.

Ignorance may be the primary reason for some failures to conceive. Not all couples understand the woman's monthly fertile period. Some use douches, exercises, or supposed aphrodisiacs to stimulate sex and fertility, yet these methods have the effects of decreasing libido, reducing fertility, or both.

At the end of a diagnostic workup about 2 percent of infertile couples will be told that one or both partners are sterile. In men, adult mumps complicated by orchitis (testicular inflammation) may prevent adequate spermatogenesis. This is an important reason why men should receive the mumps vaccine in childhood or as young adults. Sexually transmitted diseases, chromosomal disorders, generalized endocrine disorders, and chronic systemic diseases may also render some men and women permanently sterile.

Many couples who fail to conceive after a year or more of intercourse without contraception are fearful of seeking medical advice, for a variety of reasons. Is one partner sterile? Are they doing something wrong? Will an infertility study be embarrassing? Will it be expensive? Helpful? Because of the prevalent myth that infertility is usually a woman's problem, the wife is usually the first to seek an evaluation of her status. Physicians prefer to diagnose both partners together, seeing each separately but simultaneously, and reporting and discussing all procedures and findings to both. The uncertainty of the diagnosis creates problems of fear, denial, isolation, guilt, anger, depression, and lowered self-esteem in both men and women. Consequently, a physician must do a great deal of supportive counseling as well as physiological counseling during the evaluation period.

SUBSTANCE ABUSE

Substance abuse is not a phenomenon seen primarily among young people. It is practiced by all kinds of people in every age bracket. People use drugs for almost every possible effect today. They use them to go to sleep, to wake up, to get hungry, to lose their appetites, to stimulate activity, to relax, to deaden pain, to make sensations more acute—the list can go on and on. Drug abuse is not limited to illegally obtained substances. People abuse laxatives, vitamins, antacids, coffee and tea, cola drinks, aspirin, cigarettes, glue and other volatile (quickly evaporating) materials, and especially alcohol. Any drug is potentially harmful when misused. Some people have conscious or unconscious suicidal intents when they abuse drugs. The drug abuses of young adults often relate to peer pressures for use, to desires for a "high" (a feeling of good cheer) or to desires for a "down" (a feeling of relaxation).

The word *addiction* is being replaced by the term **physical** (or physiological) **dependence** to refer to the physical need for a chemical substance and the experiencing of withdrawal symptoms without it. Withdrawal symptoms differ for each substance on which a person can develop physical dependence but usually resemble a physical illness. Some substances create **psychological dependence** in addition to physical dependence. Withdrawal from a substance on which a person has become psychologically dependent will produce symptoms that resemble a mental disorder (anxiety, hyperactivity, depression). In addition to these two types of dependence, one can develop a **tolerance** to a drug. This means that one needs more and more of the substance to get the effect that one felt initially.

Alcohol The favorite mood-altering drug in the United States is alcohol (see Figure 8–9). Most people drink more in their twenties than they do during the rest of their life. This probably reflects a phase of learning about open social drinking (once they reach the legal drinking age) and a phase of wanting to appear mature. Social drinking leads to physical problems in only a small minority of persons in their twenties. It creates social problems for more, but certainly not all, young adults. The amount of **alcohol abuse** in our country far exceeds the amount of abuse of all other drugs. *Abuse* is defined as behavioral and interpersonal difficulties resulting from heavy use of alcohol (or any other chemical substance). About 25,000 fatal automobile accidents each year are directly related to alcohol abuse. In addition, diminished job performance, waning school performance, increased interpersonal aggressiveness, and physical trauma inflicted on self, spouse, children, friends, and others are common sequelae to alcohol abuse.

Although the general effects of alcohol ingestion are the same for all people, the individual effects of a glass of beer, wine, or spirits on each person are unique and change with time. In small quantities, and initially, alcohol has a stimulant effect. In larger quantities, and over time, it acts as a sedative. It is classified as a depressant

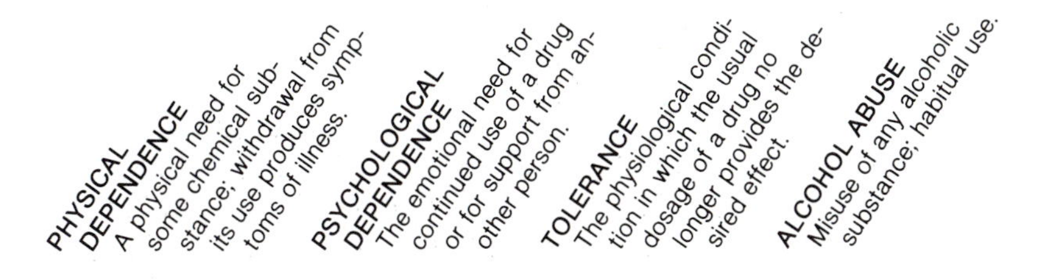

Figure 8–9
Social drinking is part and parcel of the life-style of many young adults. However, consumption of beer, wine, and spirits is not without adverse side effects—physical, social, and emotional.

drug. A large person can ingest more alcohol than a smaller person before becoming intoxicated. Alcohol taken on a full stomach will not reach the bloodstream and cause intoxication as quickly as a drink on an empty stomach. Eating while drinking also slows down the effects of alcohol. Many people believe they will get drunk more quickly on spirits than on beer or wine, or get drunk if they switch from one form of alcohol to another. In fact, it is the blood alcohol content that determines intoxication, not the form of alcohol consumed. A can of beer, a glass of wine, and a shot of spirits all contain about ½ ounce of pure alcohol. It takes about 1½ ounces of pure alcohol (three drinks) to make a smaller person (100 pounds) legally intoxicated (blood alcohol content of 0.10 percent). It takes about 3¼ ounces (six and a half drinks) to bring a very large person (240 pounds) to the same degree of intoxication. However, to impair each person's judgment (and driving ability), it would take only two drinks and four drinks, respectively. Switching drinks may make a person sick because of the mixture of nonalcoholic ingredients of beverages, not because various forms of alcohol are **synergistic** (enhance each other's effects).

Drunken behavior is no longer considered as socially acceptable as it once was. Rather than being "the life of the party" an intoxicated person is now more often viewed as an annoyance, especially if he or she loses muscular control and coordination, tries to drive, or gets sick.

The difference between a person who abuses alcohol and one who becomes alcoholic is physiological. An **alcoholic** is defined as a person who organizes his or her behavior around alcohol, continues to drink even though it causes serious personal problems, and suffers withdrawal symptoms whenever he or she sobers up. Withdrawal symptoms include perspiration, cramps, nausea, anxiety, and tremulousness ("the shakes" or "the jitters"). They can also include hallucinations (bad dreams, disordered perceptions), withdrawal seizures ("rum fits"), or **delirium tremens** ("DTs"). The delirious state called DTs is characterized by profound confusion, delusions, vivid hallucinations, tremor, agitation, sleeplessness, dilated pupils, fever, rapid heartbeat, and profuse perspiration. Each of these withdrawal symptoms can occur alone, but more generally they occur in various combinations. DTs can be fatal if the alcoholic experiences cardiac collapse after an extended period of excitement (rapid heartbeat and high blood pressure). Many other physical changes occur after chronic alcohol ingestion, particularly to the gastrointestinal system. Morning nausea and vomiting are common; they can be suppressed by drinking. Gastritis (inflammation of the stomach), peptic ulcers, pancreatitis (inflammation of the pancreas), and the liver disease cirrhosis are also common. Nutritional diseases (anemia, malnutrition, pellagra) frequently occur as well.

Most people can safely consume alcohol throughout their lives without becoming alcoholic (see myth 1, Table 8–5). Researchers are actively trying to discover the differences between alcoholics and nonalcoholic social drinkers or alcohol abusers. It is now accepted that alcoholism is a disease not unlike diabetes. It runs in families, suggesting some genetic vulnerability. However, not all offspring of alcoholic parents become alcoholics, so there are probably environmental factors that enhance its appearance. Most of the approximately 10 million alcoholics in the United States first suffered symptoms of the disease (loss of control over drinking, withdrawal sickness) by their middle to late twenties. Some even developed symptoms in adolescence after only minimal drinking. Obviously, their bodies handled alcohol differently than the bodies of nonalcoholics.

SYNERGISM
The interaction of drugs in which the total effect is much greater than the sum of their individual effects.

ALCOHOLIC
A person who has lost control over drinking behaviors and suffers withdrawal symptoms when blood alcohol levels approach or reach zero after alcohol use.

DELIRIUM TREMENS (DTs)
A severe form of alcohol withdrawal in which paranoia, disorientation, extreme agitation, hallucinations, anxiety, insomnia, anorexia, and tremors can occur.

Table 8–5 Common Myths About Alcoholism

1. Consumption of alcohol over a long period of time causes alcoholism.
2. Family members (wife, husband, parents, children) make a person turn to drink (become alcoholic).
3. Alcoholics are weak willed (lacking in will power).
4. Alcoholism is a self-inflicted disease.
5. Most alcoholics end up on the street.
6. Beer drinkers do not become alcoholics.
7. Alcohol is a stimulant drug.
8. Alcoholics must want help before they can be treated.
9. Tranquilizers are the best treatment for alcoholism.
10. Recovered alcoholics can safely return to social drinking.

A current guess about alcoholism causation is that neurotransmitters, which are altered by alcohol in all drinkers, become dependent on alcohol in order to function normally in alcoholics. Consequently, when alcoholics become sober, their neurotransmitters misfire, causing withdrawal symptoms (Noble, 1983). Another possibility is that alcoholics may be deficient in certain receptors for neurotransmitters. They may become dependent on alcohol because drinking helps correct the deficiency. Another theory about causation is that the breakdown products of alcohol differ in alcoholics and nonalcoholics, causing physical dependency in alcoholics (Schuckit, 1984).

Whatever the biological cause of alcoholism, certain reactions and behaviors have been identified that may forewarn that a person is at risk of developing the disease. Nonalcoholics more frequently feel dizzy, drowsy, or nauseated or turn red when their blood alcohol content (BAC) approaches 0.05 to 0.09 percent. They may not want to continue drinking. Future alcoholics more frequently reach legal intoxication (BAC over 0.1 percent) without unpleasant symptoms and continue drinking, often gulping drinks, way beyond intoxication. They may experience frequent **blackouts**. These are periods of amnesia (as opposed to passing out) from drinking too much. During a blackout, a person remains conscious and carries out activities of which he or she has no recollection when sober. The high blood alcohol content apparently interferes with the person's ability to store any memories. (This can happen to nonalcoholics who drink too much as well.)

BLACKOUT
A momentary lapse of consciousness. In alcoholism a period of temporary amnesia when the alcoholic functions but of which he or she has no memory when sober.

Prealcoholics, more often than nonalcoholics, drink alone, sneak drinks, avoid talking about their drinking, feel guilty about their drinking, and yet continue to drink more and more. Once they have the disease (are alcoholic), drinking is central to their lives. They often neglect job and family, lose interest in sex, neglect their health and nutrition, alibi and rationalize about their behavior, and become ill without alcohol.

Treatment of alcoholism requires permanent cessation of drinking. The initial detoxification process may be protracted with depression, irritability, and anxiety continuing long after the physiological withdrawal symptoms end. Physicians may prescribe short-term therapy with one of the minor tranquilizers (such as Valium). However, tranquilizers are useful only during the acute withdrawal period. Prolonged use will result in a physical dependence on the tranquilizers as a substitute for alcohol. ***Antabuse,*** a drug that causes nausea, cramping, and vomiting whenever alcohol is ingested, is frequently prescribed to aid in alcohol abstinence. Motivation to take the daily pill must be good for Antabuse therapy to be effective. Other therapies that seek to alleviate the psychological stresses that led to drinking are also effective for some patients. Alcoholics Anonymous (AA), an informal fellowship of reformed addicts that teaches a new way of life to combat the strong internal and external pressures to drink and gives continual support to alcoholics, has proved to be the single most effective force in alcoholic rehabilitation.

Very few alcoholics become street people. Estimates of alcoholics who are derelicts range from 3 to 8 percent. Most alcoholics, because of strong pressures from their family, friends, or employers (few are self-motivated) seek treatment of some sort. Most professionals feel it is extremely rare for an alcoholic to resume social drinking without reexperiencing cravings, loss of control, and renewed dependence on alcohol. Consequently, professionals do not use the past tense "recovered alcoholic" but rather the present tense "recovering alcoholic," indicating the lifelong nature of treatment.

In advanced alcoholism a form of irreversible mental illness, called Wernicke-Korsakoff syndrome, occurs. It involves atrophy (wasting away) of cerebral tissue. Although generally believed to be a disease of old men, it can, in fact, be found in young adult alcoholics of both sexes. Occasionally, Wernicke-Korsakoff symptoms are the first indication of the amount of drinking that has occurred over time. Some alcoholics, unable to abstain even one day without withdrawal, keep a moderate amount of alcohol in their bloodstream all day, every day, over several years while they continue to function as business executives, college professors, housewives, nurses, physicians, or other professionals or workers.

The disease alcoholism is not the only problem associated with alcohol use. Nonalcoholics who abuse alcohol may experience stomach and intestinal ulceration; nutritional deficiency diseases; breakdown of the liver (cirrhosis); lowered fertility in men; fetal alcohol syndrome in the fetuses of pregnant women (see Chapter 3); blackouts; fines, jail sentences, or loss of licenses for driving while intoxicated (DWI) charges; lost jobs; accidents; family problems; or social and economic problems from their excessive outlay of money for alcohol.

Depressants The second most commonly abused drugs after alcohol, and the most frequently prescribed substances in the United States, are minor tranquilizers, especially those of the benzodiazepine family (Valium, Librium, Dalmane, Tranxene). When taken as directed, usually a small dose three to four times a day for a short period of time, they are relatively safe (*Physician's Desk Reference,*

1986). However, both tolerance and dependence develop rapidly. Many young adults find that they need not only to have their prescriptions renewed but to have their dosages increased in order to cope with life's great bundle of little things. Withdrawal from one's tranquilizers, once psychological and physical dependence have developed, can result in nearly intolerable anxiety, insomnia, nightmares, tremors, delirium, hallucinations, convulsions, even death. An additional danger of minor tranquilizers is that they are synergistic with alcohol. As noted earlier, drug synergism refers to two drugs having a greater total effect than the sum of their individual effects. Mixing of alcohol and one of the benzodiazepines (or any other tranquilizer) can result in shallow respirations, cold clammy skin, dilated pupils, weak and rapid pulse, coma, and possible death.

Other **depressants** of abuse include antipsychotic drugs (Compazine, Mellaril, Stelazine, Thorazine), sedatives (Equanil, Miltown), and major tranquilizers, such as chloral hydrate, the barbiturates, and the barbiturate substitutes—glutethimide (Doriden), and methaqualone (Quaalude). Young adults more typically abuse depressants over the weekend or on days off to obtain sedation and relief from stress and anxieties. The barbiturates or barbiturate substitutes produce a state of intoxication quite similar to that of alcohol. They can also induce stupor, coma, and even death when taken in excess or with large amounts of alcohol.

Tolerance and addiction to all the depressants develops rapidly. Rapid withdrawal from a habitual large daily dose can lead to heart palpitations, muscle spasms, delirium, psychotic behaviors, convulsions, or death. Depressant overdose is the most common means of suicide among women. Studies indicate that the young American female population is being unnecessarily sedated (Schiefelbein, 1980). Approximately one-fourth of American women between ages eighteen and twenty-nine have been found to take some prescribed depressant medication in a given calendar year (Scarf, 1979).

Narcotics The strong tolerance and physical dependence properties of narcotics like opium, morphine, paregoric, codeine, and Demerol make them especially dangerous drugs. They relieve pain, and like the barbiturates they confer tranquillity. When first taken, they produce a feeling of exultation. This is followed by a euphoria, a buoyancy, and a feeling of well-being, which in turn is followed by apathy, drowsiness, slow respirations, constricted pupils, and blurred vision. Morphine, paregoric, codeine, and Demerol (a synthetic narcotic), when used medically, are injected into a muscle or taken orally. This slows down their entrance into the bloodstream and reduces the initial euphoric rush. Heroin (which has no accepted medical use) and the other narcotics can also be inhaled (snorted), mixed with other substances and smoked, injected under the skin (popped), or injected directly into blood vessels (mainlined).

The effects of narcotics depend on the method of administration, the dose, and the tolerance of the user. Repeated use leads to the need for larger and larger doses and eventually to physical dependence. Once a person is dependent, he or she must have a continual supply of the drug to avoid withdrawal symptoms. Early signs of withdrawal include dilated pupils, perspiration, watering eyes and nose, restlessness, and insomnia. Later symptoms include nausea, vomiting, diarrhea, violent abdominal cramps, leg spasms, tremors, and chills.

An overdose of narcotics produces unconsciousness and respiratory depression leading to death. The narcotics drug problem was once predominantly seen in urban ghettos. It is now also found increasingly in middle-class neighborhoods, especially among young adults. The sharing of needles to inject narcotics is one of the major ways in which AIDS and other infectious diseases are spread.

Stimulants Any drugs taken to make one more alert, including coffee, tea, cocaine, amphetamines, caffeinated soft drinks, and cigarettes, are **stimulants**. Although some are legal, others are strictly illegal or are controlled substances, available by prescription only.

Cocaine is derived from leaves of the South American coca plant. Some coca leaves are imported legally so that the cocaine can be extracted for medical purposes and the decocainized product used to flavor cola beverages. Currently more coca leaves or extracted cocaine are imported illegally, contributing to a vast drug use epidemic and law enforcement problem. Cocaine (coke) may be sold on the street as a crystal-like powder, "cut" with other white powders to varying degrees of purity. In its powder form it may be sniffed (snorted), ingested, or injected to produce a "flash" or "rush" of euphoria (mood elevation, heightened sense of well-being). The powder may also be freebased. To "free the base" is to alter the structure of the cocaine in order to smoke it. The cocaine is boiled down with substances such as ether, ammonia, or baking soda. Fires and explosions may result from this process. Crack, the product of heated street cocaine mixed with fillers, is nearly 90 percent pure. It is sold relatively inexpensively in rock form and is smoked through pipes. Smoking crack produces a shorter lived but more intense high than snorting coke. Because large amounts of cocaine reach the brain quickly after smoking crack, fatal overdose can occur. Death comes as a result of seizures, respiratory arrest, and coma, or sometimes by cardiac arrest.

Any form of cocaine use usually results in higher blood pressure, rapid breathing, elevated temperature, dilated pupils, de-

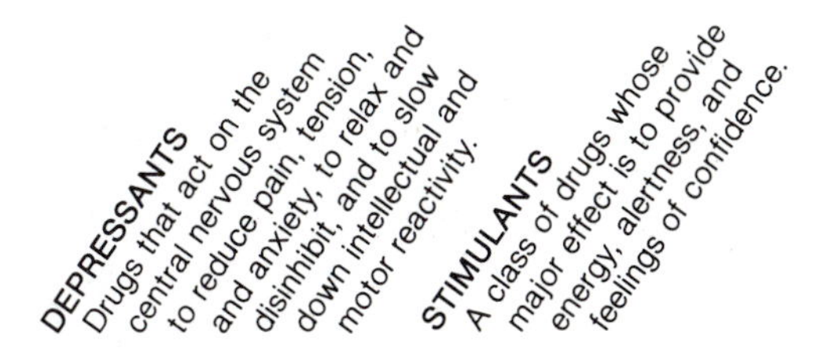

creased hunger, relief from pain, wakefulness, and a sense of mental and physical strength as well as euphoria. Paranoia or hallucinations may also occur with heavier use. When the pleasurable effects wear off, an unpleasant depression follows (crashing), which can be counteracted by using more cocaine. Many people quickly become psychologically dependent on cocaine for continued feelings of strength, self-confidence, and well-being, even after only short-term experimental use.

Amphetamines and methamphetamines decrease fatigue, reduce the appetite, generate excitement, and give an exaggerated sense of well-being. They may impair reflex activity and intellectual performance. In larger doses they can cause aggressive behavior, delusions of self-importance, hallucinations, or paranoia. In overdoses they cause tremors, rapid breathing, elevated blood pressure, confusion or panic, nausea, vomiting and diarrhea, convulsions, coma, and sometimes death (DEA, 1985).

Amphetamines can be obtained through prescriptions for "diet pills" or on the street. Abuse is common among students, housewives, athletes, truck drivers, and people who need to stay awake for long periods or want to keep their moods up and their weight down. Tolerance, physical dependence, and psychological dependence develop rapidly. Amphetamines are sometimes injected directly into the bloodstream, like cocaine or heroin, to enhance their effect. They are especially dangerous when taken this way.

Nonamphetamine drugs have been developed to suppress appetite or produce stimulation. Although they are generally less potent than the amphetamines, they have the same effects, create psychological dependence, and can lead to death from overdoses. Some are available without prescription. Withdrawal leaves the chronic user feeling irritable, anxious, apprehensive, and depressed.

The Hallucinogens **Hallucinogens** produce subjective perceptions of things that do not exist. Hallucinations may give the impression of things larger or smaller or more vividly colored than life and may simultaneously involve two or more senses. They may even give a person a sense of being outside the self, depersonalized. They may alter a person's sense of time and space and of objects and people in the environment (Schultes, 1980). Some people prefer to use the term *psychedelics* for the hallucinogenic drugs because the perceptual changes may be enticing, pleasure-giving insights into reality rather than hallucinations. In some cases perceptions also mimic psychosis, which has led to the labeling of the drugs as *psychotomimetic.*

The natural hallucinogens—peyote and mescaline, derived from a cactus, and psilocybin and psilocyn, derived from mushrooms—are sometimes used for religious services, mainly by Indian peoples in western Canada, the United States, and Mexico. The side effects of nausea and vomiting are considered purifying. The feeling of depersonalization is an aid to meditation and self-understanding. The altered perceptions that occur may be viewed as communion with supernatural powers.

HALLUCINOGENIC DRUGS Drugs that induce hallucinations; some are also called psychedelics.

Phencyclidine (PCP), once used legally in veterinary medicine, can lead to acute psychotic episodes. It is sometimes called "DOA" (dead on arrival) both because preoccupation with death is common during trips and because lethal trips sometimes result when dangerously potent or contaminated forms are produced in clandestine labs. Like cocaine, it has a high potential for creating psychological dependence.

Lysergic acid diethylamide (LSD) was once used with some success to produce schizophrenia-like states in studying model psychosis. It is now strictly illegal. Most of the preparations are potentially very dangerous. A poorly understood effect of LSD is its ability to cause *flashbacks.* A person may suddenly fall into another "trip" with hallucinations and disrupted perceptions months after taking the drug.

Methylene dioxymethamphetamine (MDMA, also called "ecstasy") is considered as potentially dangerous as LSD. Possession of even a trace of either of them can result in a $125,000 fine and fifteen years in prison (Drug Enforcement Administration, 1985). Many other illegal synthetic hallucinogens exist that produce similar hallucinogenic experiences. The hallucinogens with the exception of PCP do not produce physical dependence or withdrawal symptoms but are considered dangerous because they can unexpectedly cause toxicity and death and because persons having hallucinations may harm themselves or others.

Marijuana Marijuana is an unclassified drug, with properties of depressants, stimulants, narcotics, and hallucinogens. Individual states vary in their laws concerning its use. In some it is now legal to use it and to possess it in small quantities. In others it is criminal to have even traces of it found on one's person or belongings. A great deal of emotionalism surrounds the issue of decriminalization of marijuana. Many myths also exist about its effects.

Marijuana comes from the hemp plant (*Cannabis sativa*), a weed that grows freely in most parts of the world. Preparations are usually smoked but may also be consumed in drink or food. Users need to increase the dosage needed for a high over time, and they do experience withdrawal symptoms when they do not use it. A high from marijuana produces a euphoric release from inhibitions, tensions, and anxieties. Users also report feelings of bodily lightness, increased perceptual awareness, and heightened sensitivity to external stimuli. In very large doses it may cause vomiting, diarrhea, and loss of muscle coordination.

Withdrawal symptoms, after heavy long-term use, include irritability, restlessness, hyperactivity, decreased appetite, sweating, sudden weight loss, increased salivation, and increased pressure within the eye (National Academy of Sciences, 1982).

The active ingredient in cannabis leaves is thought to be delta-9-tetrahydrocannabinol (THC). For research purposes THC is synthesized in laboratories or extracted from marijuana grown by the government on special "pot" farms. This National Institute on Drug Abuse (NIDA) marijuana, which has a THC content of approximately

2 percent, has medical usefulness for treating glaucoma and reducing the nausea following chemotherapy in cancer patients. Street marijuana, especially that of South American origin, usually has a THC content of 4 to 8 percent. In spite of the popular notion that street pot is harmless, Turner and Waller (1979), after studying many thousands of scientific research reports on cannabis, did not find a single paper that gave it a clean bill of health. Its usual bad effects include lowered fertility, impaired pulmonary function, lowered resistance to infection, and a chronic cough. Long-term effects may include chromosomal mutations in the reproductive cells and impairment of intellectual functioning.

Marijuana has not been shown to incite people to violent or aggressive behavior. It causes less impairment of driving than alcohol. It does not unduly stimulate sexual desires, although some users report more enjoyment of sex, as well as more enjoyment of other sense-related stimuli (art, music) when high. For these reasons, and because it is so widely used, many people would like to see its use decriminalized.

One argument used to support the laws against marijuana in the United States is that its use leads to the use of the more dangerous narcotics. Supporters of tough laws point out that the pushers who handle marijuana also handle other illegal drugs and encourage buyers to try them as well. Proponents of laws to make marijuana legal argue that it is just as likely that alcohol, tobacco, or other drugs that help people relax will lead to narcotics use. The verdict on the relative safety or danger of marijuana and its future legality or illegality are issues yet to be resolved.

Summary

Young adults can be found in many settings: jobs, the military service, institutes of higher education, their families of origin, new families, singles' residences.

Given proper diet, exercise, rest, and freedom from preexisting handicaps or diseases, persons in their twenties should be in peak physical condition. The threat to health that looms largest in young adulthood is stress. Accidents, alcohol and drug abuse, and many psychophysiologic illnesses have their origin in persons' inabilities to cope successfully with the stresses of young adult living.

Cognitive functioning peaks in the twenties. Many young adults have a high need to achieve intellectually. Others subjugate academic achievement to focus on affiliation. Women especially may fear that too many intellectual achievements will stand in the way of their being accepted socially.

The transition from being dependent on one's family of origin to being an economically, socially, and emotionally independent adult is often difficult. Some young adults take a considerable length of time searching for their place in the existing adult social order. Others follow their parents' lead and adhere to conventional customs of the mainstream culture without any conflict. Still others become activists against some portion of the adult society, or become alienated from it entirely.

Entering a suitable field of employment often takes time. Young adults may change jobs frequently before making a commitment to one field. Some persons spend many years in educational centers preparing for a career before they take their first job. Women may choose to be wife/mother/homemaker, to pursue a career, to do one then the other, or to do both simultaneously. Men are increasingly taking more of a role as husband/father/homemaker when the wife also has a career.

Maslow proposed a needs hierarchy common to all humans. Physical needs, safety needs, belonging and love needs, esteem needs, and the need for self-actualization are progressively and cumulatively sought. Although some young adults may be struggling with physical and safety needs, most are trying to meet their needs for belonging and love and then for esteem. Self-actualization cannot occur unless the first four needs are being successfully met.

The nuclear conflict of young adulthood proposed by Erikson is that of finding intimacy versus feeling isolated. Intimacy involves making commitments and fusing one's identity with that of others and is usually seen in the context of marriage. However, it can also be achieved outside of marriage.

Many young adults enjoy singlehood, and others choose marriage. Marriages have the greatest potential for lasting when each partner's expectations are being met, when interdependency rather than a dependent or independent relationship exists, and when partners like as well as love each other.

For both men and women the twenties are the best years for reproducing offspring. However, childbearing is often postponed for social, economic, educational, career, or other considerations. Parenthood may bring persons in their twenties both joy, warmth, and laughter, and added work, stress, and financial burdens.

Household organization can be complicated, especially with frequent moves, limited space, and overextended budgets and with all adult members of the household holding part- or full-time jobs.

Prolonged postponement of childbearing may result in decreased fertility. As many as 15 percent of married couples may experience an infertility problem for a wide variety of reasons.

Substance abuse is greater in the twenties than at any other decade of the life span. Many young adults pose serious threats to their mental and physical well-being by abusing alcohol or other drugs.

Questions for Review

1. Consider the various definitions of maturity presented in Chapter 8. How would you define maturity? Do you agree that young adulthood suggests maturity? Discuss.
2. Horner found that women fear achievement and success. Some feminists suggest that women learn this fear from the time they are young. Do you agree or disagree? Discuss.
3. Some parents have difficulty accepting their young adults' independence, maturity, and sexuality. Imagine you are a counselor working with a couple who is having difficulty accepting their twenty-year-old's independence. How would you counsel this couple?
4. Newspapers often have headlines about young adults committing some violent act. Many of these individuals are characterized as being loners. Why do you think a loner would be more likely to commit a violent act than someone else?
5. Young adulthood is a time of establishing a feeling of self and independence, of choosing a career. Do you think men or women experience more stress and conflict in making these decisions and taking these steps? Or do you think both experience the same amount of stress? Explain your answer.
6. Our society has traditionally expected people in their twenties to find a mate, marry, and have children. Much has been written about freedom now to do otherwise. Do you believe individuals are truly any freer to be single, live together, or remain childless? Or do you believe these variations in lifestyle remain stigmatized by the majority of society? Why?
7. Why do you think individuals in their twenties, at a peak time of physical and mental ability, have a high incidence of suicide, drug abuse, and stress-related illnesses?
8. Describe how the resolution of the sixth nuclear conflict, intimacy versus isolation, is affected by the resolution of earlier nuclear conflicts described by Erikson.

References

Alper, T. G. Achievement motivation in college women: A now-you-see-it-now-you-don't phenomenon. *American Psychologist,* 1974, *29*(3), 194–203.

American College of Physicians. *Guide for adult immunizations.* Philadelphia: American College of Physicians, 1985.

Bachman, J. G., and Johnston, L. D. The freshmen, 1979. *Psychology Today,* 1979, *13*(4), 78–87.

Bloom, D. E. Putting off children. *American Demographics,* September 1984, 30–33, 45.

Cherlin, A., and Furstenberg, F. F. The American family in the year 2000. *The Futurist,* 1983, *17*(3), 7–14.

Clayton, R., and Voss, H. Shacking up: Cohabitation in the 1970s. *Journal of Marriage and the Family,* 1977, *39*(3), 273–283.

Darling, C. A.; Kallen, D. J.; and Van Dusen, J. E. Sex in transition, 1900–1980. *Journal of Youth and Adolescence,* 1984, *13,* 385–399.

Drug Enforcement Administration. *Drugs of abuse,* Washington, D.C.: U.S. Government Printing Office, 1985.

Dubos, R. Being human. *The Sciences,* 1982, *22*(1), 26–28.

Duvall, E. M. *Marriage and family development,* 5th ed. Philadelphia: Lippincott, 1977.

Erikson, E. *Childhood and society,* 2nd ed. New York: Norton, 1963.

Finkbeiner, A. AIDS: Just the facts. *Johns Hopkins Magazine,* 1985, *37*(6), 23–28.

Fitch, S. A., and Adams, G. R. Ego identity and intimacy status: Replication and extension. *Developmental Psychology,* 1983, *19,* 839–845.

Food and Drug Administration. Progress on AIDS. *FDA Drug Bulletin,* 1985, *15*(3), 27–32.

Freud, S. *Introductory lectures of psychoanalysis.* (J. Strachey, ed. and trans.) New York: Norton, 1966. (Originally published, 1917).

Friedan, B. *The feminine mystique.* New York: Norton, 1963.

Gould, R. *Transformations.* New York: Simon & Schuster, 1978.

Havighurst, R. J. *Developmental tasks and education,* 3rd ed. New York: McKay, 1972.

Hoffman, L. W. The value of children to parents and the decrease in family size. *Proceedings of the American Philosophical Society,* 1975, *119,* 430–438.

Hoffman, L. W. Changes in family roles, socialization, and sex differences. *American Psychologist,* 1977, *32*(8), 644–657.

Holmes, T. H., and Rahe, R. H. The social readjustment rating scale. *Journal of Psychosomatic Research,* 1967, *11*(2), 213–218.

Horner, M. Toward an understanding of achievement-related conflicts in women. *Journal of Social Issues,* 1972, *28*(2), 157–176.

Jessor, S. L., and Jessor, R. Transition from virginity to nonvirginity among youth: A social-psychological study over time. *Developmental Psychology,* 1975, *11*(4), 473–484.

Johnson, A. T. Municipal employee assistance programs: Managing troubled employees. *Public Administration Review,* 1985, *45,* 383–390.

Jung, C. G. *Psychological types or the psychology of individuation.* New York: Harcourt Brace, 1923.

Keniston, K. Youth: A "new" stage of life. *American Scholar,* Autumn 1970, *39*(4).

Keniston, K. Moral development, youthful activism, and modern society. In H. Kraemer (Ed.), *Youth and culture: A human development approach.* Monterey, Calif.: Brooks/Cole, 1974.

Kieren, D.; Henton, J.; and Marotz, R. *Hers and his: A problem-solving approach to marriage.* Hinsdale, Ill.: Dryden, 1975.

Kinsey, A. C.; Pomeroy, W. B.; and Martin, C. E. *Sexual behavior in the human male.* Philadelphia: Saunders, 1948.

Kinsey, A. C.; Pomeroy, W. B.; Martin, C. E.; and Gebhard, P. H. *Sexual behavior in the human female.* Philadelphia: Saunders, 1953.

Kohlberg, L. Continuities in childhood and moral development revisited. In B. P. Baltes and K. W. Schaie (Eds.), *Life-span developmental psychology: Personality and socialization.* New York: Academic Press, 1973.

Lavine, L. O. Parental power as a potential influence on girls' career choice. *Child Development,* 1982, *53,* 658–663.

LeMasters, E. E. *Parents in modern America.* Homewood, Ill.: Dorsey, 1970.

Levinson, D. J.; Darrow, C. N.; Klein, E. B.; Levinson, M. H.; and McKee, B. *The seasons of a man's life.* New York: Knopf, 1978.

Macklin, E. Review of research on nonmarital cohabitation in the U.S. In B. I. Murstein (Ed.), *Exploring intimate lifestyles.* New York: Springer, 1978.

Marshall, J. R. Infertility. In R. C. Benson (Ed.), *Current obstetric and gynecologic diagnosis and treatment,* 5th ed. Los Altos, Calif.: Lange, 1984.

Maslow, A. *Toward a psychology of being,* 2nd ed. New York: Van Nostrand, 1968.

Maslow, A. *Motivation and personality,* 2nd ed. New York: Harper & Row, 1970.

Maymi, C. R. Women in the labor force. In P. W. Berman and E. R. Ramey (Eds.), *Women: A developmental perspective.* Bethesda, Md.: NICHD/NIH (Publication No. 82-2298), 1982.

McLeod, B. In the wake of disaster. *Psychology Today,* 1984, *18,* 54–57.

Miller, N. Effects of emotional stress on the immune system. *Pavlovian Journal of Biological Sciences,* 1985, *20,* 47–52.

National Academy of Sciences, Institute of Medicine. *Marijuana and health.* Washington, D.C.: National Academy Press, 1982.

Noble, E. P. Social drinking and cognitive function: A review. *Substance and Alcohol Actions/Misuse,* 1983, *4*(2–3), 205–216.

Orsborn, C. Enough is enough. *Superwomen Anonymous,* 1985, *1,* 1–4.

Parke, R. D. The father-infant relationship: A family perspective. In P. W. Berman and E. R. Ramey (Eds.), *Women: A developmental perspective.* Bethesda, Md.: NICHD/ NIH (Publication No. 82-2298), 1982.

Persky, H. Reproductive hormones, moods, and the menstrual cycle. In R. C. Friedman, R. M. Richart, and R. L. Vande Wiele (Eds.), *Sex differences in behavior.* New York: Wiley, 1974.

Physician's Desk Reference. Oradell, N.J.: Medical Economics Co., 1986.

Prager, K. J. Identity development and self-esteem in young women. *Journal of Genetic Psychology,* 1982, *141,* 177–182.

Prager, K. J. Identity status, sex-role orientation, and self-esteem in late adolescent females. *Journal of Genetic Psychology,* 1983, *143,* 159–167.

Radloff, L. S., and Rae, D. S. Susceptibility and precipitating factors in depression: Sex differences and similarities. *Journal of Abnormal Psychology,* 1979, *88,* 174–181.

Rosen, B., and D'Andrade, R. The psychosocial origins of achievement motivation. *Sociometry,* 1959, *22,* 185–195, 215–217.

Rosen, B.; Jerdee, T. H.; and Prestwich, T. L. Dual-career marital adjustment: Potential effects of discriminating managerial attitudes. *Journal of Marriage and the Family,* 1975, *37,* 565–572.

Rubin, Z. Does personality really change after 20? *Psychology Today,* 1981, *15*(5), 18–27.

Scarf, M. The more sorrowful sex. *Psychology Today,* 1979, *12*(11), 44–52, 89–90.

Schiedel, D. G., and Marcia, J. E. Ego identity, intimacy, sex role orientation, and gender. *Developmental Psychology,* 1985, *21,* 149–160.

Schiefelbein, S. The female patient—Heeded? Hustled? Healed? *Saturday Review,* March 29, 1980, 12–16.

Schuckit, M. A. Relationship between the course of primary alcoholism in men and family history. *Journal of Studies on Alcohol,* 1984, *45*(4), 334–338.

Schultes, R. E. *The botany and chemistry of hallucinogens,* 2nd ed. Springfield, Ill.: Charles C Thomas, 1980.

Schwartz, M. A. Career strategies of the never married. Paper presented at the 71st Annual Meeting of the American Sociological Association, New York, September 3, 1976.

Scott, D. H. Brain size and "intelligence." *British Journal of Developmental Psychology,* 1983, *1,* 279–287.

Speroff, L.; Glass, R. H.; and Kase, N. G. *Clinical gynecologic endocrinology and infertility,* 3rd ed. Baltimore: Williams & Wilkins, 1983.

Stayton, W. R. Lifestyle spectrum 1984. *SIECUS Report,* 1984, *12,* 1–4.

Sternberg, R. J. The measure of love. *Science Digest,* April 1985, *60,* 78–79.

Toffler, A. *Future shock.* New York: Random House, 1970.

Turner, C., and Waller, C. *Marijuana: An annotated bibliography.* New York: Macmillan, 1979.

Vaillant, G. E. *Adaptation to life.* Boston: Little, Brown, 1977.

Verbrugge, L. M. Women's social roles and health. In P. W. Berman and E. R. Ramey (Eds.), *Women: A developmental perspective.* Bethesda, Md.: NICHD/NIH (Publication No. 82-2298), 1982.

Veroff, J.; Douvan, E.; and Kulka, R. A. *The inner American: A self-portrait from 1957 to 1976.* New York: Basic Books, 1981.

Watson, R. E. L. Premarital cohabitation vs. traditional courtship: Their effects on subsequent marital adjustment. *Family Relations,* 1983, *32*(1), 139–147.

Yarrow, L. What a baby does to your marriage. *Parents,* July 1982, *57,* 47–51.

♥

The Thirties and Forties

Chapter Outline

Physical Development
- *Physical Changes*
- *Nutrition*
- *Health Concerns*

Cognitive Changes

Psychosocial Development
- *Erikson's Concept of Generativity Versus Stagnation*
- *Career Concerns*
- *Family Concerns*
- *Community Concerns*
- *Friends*
- *The Midlife Transition*

Key Concepts

atherosclerosis
basal metabolic rate (BMR)
becoming one's own man (BOOM)
blood pressure
compression of the spinal column
coronary heart disease (CHD)
coronary prone behavior
cortisol
creativity
crystallized intelligence
divorce
fluid intelligence
generativity versus stagnation
hypertension
mental disorders
mentor
midlife transition
period of individuation
premenstrual syndrome
presbycusis
presbyopia
settling down
transactional analysis
ultraviolet light

Case Study

Urie and Phyllis met in a hospital. Urie's dad was the patient, Phyllis

the nurse. They began dating, fell in love, and married. They bought their first home close to Urie's parents, who were both older and in failing health. Although they would have liked to postpone parenthood until they were financially more secure, they had two daughters in the first three years of marriage so that Urie's parents could see and get to know their grandchildren. Urie was a salesman. Phyllis had to continue working as a nurse to help pay the bills. She worked nights so she could take care of her daughters and help Urie's parents during the day. Urie's dad died on their fourth wedding anniversary. His mother moved in with them soon afterward. Phyllis argued that the mother should go into a nursing home, as she was partially paralyzed and aphasic after two strokes. Urie could not do that to his mother. So they shared the care of their daughters, the mother, the housework, and the cooking for four years, while both held full-time jobs as well, until the mother died.

Urie and Phyllis then moved from the city into a dream cottage in the country so the girls could go to a small centralized school. Life seemed sweet. Phyllis took a job as a school nurse, which allowed her to work the same hours as her husband and be home when he and the girls were home. After six months, bad news came from Phyllis's mother. Her Dad had cancer. Her Mom did not drive and had no way to take him to and from his hospital treatments. Urie and Phyllis moved again, to be close to *her* parents. Phyllis once more worked nights so she could help her parents during the day. The daughters became teenagers who also shared in the care of "Gramps" and "Gram." Gramps died the year the oldest daughter left home for college. Gram had a series of small strokes shortly afterward and could no longer live alone, although the doctors said she might survive for another ten years. The youngest daughter took Gram in as her roommate and announced that she did not want to attend college but would stay home and help take care of Gram. Urie and Phyllis wondered whether they'd ever know what it was like to live like other married couples. Would they be selfish if they placed Gram in a nursing home and asked their daughter to become independent?

Have you ever heard it said, "Never trust a person over thirty"? This generalization about people in their thirties and forties suggests that they have surrendered to the establishment. They have given up youthful idealism and replaced it with a more realistic perception of what they must do to survive in their particular social network. They have relinquished the freedom of words and actions of their twenties and now behave according to the constraints of the social order. They have even developed a concern for their reputations. In effect, according to the popular notion, they have become more conventional.

It is always dangerous to generalize about members of any given group. There are apt to be as many differences between individuals within the group as between members and nonmembers. In fact, not all people in their thirties become conventional, just as not all individuals in their twenties wave banners proclaiming idealistic views. This chapter will describe some of the hallmarks of people in their thirties and forties. The reader should note that the characterizations that emerge will not be a perfect fit for any individual person (see Figure 9–1).

Physical Development

At some time between thirty and forty-nine, most people confront the fact that they have left youth behind them. The physical signs of aging gradually become more obvious. Most people begin to feel some signs of approaching middle age during these decades (for example, muscle or joint stiffness or decreased energy). Some welcome middle age as a sign of maturity, with its concomitant prestige and worldly wisdom. Others spend a good deal of time, effort, and money trying to hide the physical changes that indicate their age (such as gray hair or wrinkles).

PHYSICAL CHANGES

Physically, adults begin to gradually slow down in their thirties and forties. How the body functions is highly dependent on diet, exercise, rest, stress, genetic constitution, and freedom from disease or disabilities.

Figure 9–1
As this twentieth high school reunion shows, persons in their late thirties vary greatly in appearance. However, most confront the fact that their youth is behind them. Some may consider their accomplishments and their aspirations and renew the direction of their life during these years.

Muscle size, strength, and reflex speed can be maintained with regular exercise. Without it muscles begin a progressive decline. Size and strength that are lost from disuse, however, can be regained through rigorous exercise. Muscle strength of most men continues to be greater than that of most women throughout the life span.

Bone elongation ceases in the late teens. By the thirties and forties the bones have already lost some of their mass and density. As the cartilage between the vertebrae start to degenerate from normal wear, the vertebrae become compressed and the spinal column gradually begins to shorten. As adults age, they actually lose some of their height due to this **compression of the spinal column**. By age forty the average adult will stand one-eighth of an inch shorter than he or she did at age twenty (Batten, 1984). With increasing age the cartilage in all joints has a more limited ability to regenerate itself.

In addition to muscular and skeletal changes, there are age-related changes in the endocrine glands, the hormone-producing structures that control growth, metabolism, homeostasis, and reactions to stress. The secretion by the adrenal glands of **cortisol** (also called hydrocortisone), a major hormone that is essential for life, decreases about 30 percent over the entire adult life span. Plasma cortisol levels, however, remain virtually unchanged. This is because it is broken down and removed more slowly by the liver (Gregerman and Bierman, 1981). These continued adequate blood levels of cortisol allow aging persons to respond quickly to stress. Estrogens, progesterone, and androgens are produced in progressively decreasing amounts by the gonads. In women, the tapering off of estrogen may contribute to problems with premenstrual syndrome (see Box 9–1).

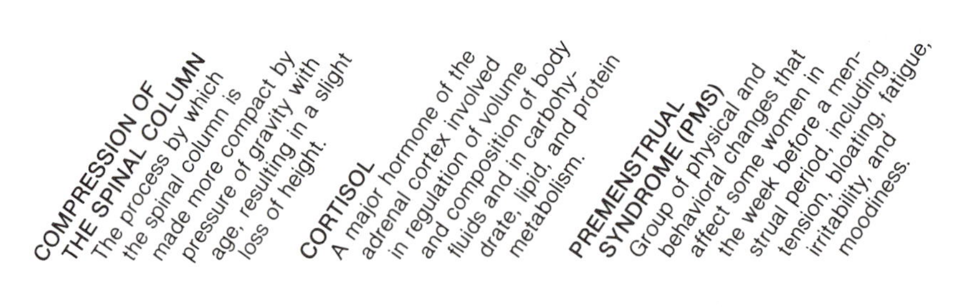

BOX 9–1

Moods, Behavior, and Premenstrual Syndrome

Should **premenstrual syndrome** (PMS) be allowed as a legal defense for a woman who commits crimes, abuses family members, or causes serious accidents two to ten days prior to the onset of her menses? Is she psychologically debilitated, physiologically impaired, or both? Is PMS real? Can it be prevented? Who gets it and why?

PMS is a group of symptoms that tend to occur together to some degree in about 40 percent of women in the week preceding their menses (Breen, 1981). It is more frequent in older women than in teens and young adults (Muller, 1985). It may affect any given woman in varying degrees, or not at all, in different months. Physiological symptoms may include abdominal bloating, breast enlargement, acne, cold sores, swelling of the hands and ankles, weight gain, food cravings, headache, fatigue, palpitations, backache, muscle pain, joint pain, and breast pain. Psychological symptoms may include depression, quickly shifting moods and emotions, tension, anxiety, panic attacks, aggressiveness, and irritability. Some women may behave in seemingly odd and erratic ways when suffering from PMS.

Before PMS can be prevented, researchers must understand its etiology. At present, its cause remains elusive. Efforts to demonstrate that PMS results from a progesterone deficiency have not been substantiated. A theory that neuropeptides from the pituitary play a role in PMS is currently being researched, but results may not be concluded for several years (Muller, 1985).

One of the difficulties PMS sufferers face is convincing others (or being convinced themselves) that they have a real physiological syndrome. There is currently no test available to diagnose PMS. Many women are told that PMS is all in their head. Many believe themselves to be going insane. Yet psychological intervention alone does not cure PMS.

Drugs touted as "cures" for PMS are as ineffective as psychotherapy. So far the only useful drugs are those prescribed for single debilitating symptoms: a diuretic for excess edema, bromocriptine for breast engorgement and pain, an analgesic for muscle or joint pain or headache, a sedative for emotional tension, and so on. However, these drugs are usually not given together. Tranquilizers should not be prescribed on a long-term basis to help patients cope with their symptoms.

The physiological problems of PMS can exacerbate psychological problems, and vice versa. Many PMS sufferers develop low self-concepts and high guilt levels because of their occasional inabilities to control their physical symptoms, moods, and behavior. Many have tried suicide as a last resort.

Psychotherapy, in conjunction with medical therapy, exercise, improved nutrition, and vitamins can reduce stress, enhance self-worth, reduce guilt, and identify and eliminate self-defeating behaviors. Support groups with other women who share the disorder can also be valuable in alleviating guilt, self-blame, and negative self-imagery.

While prevention may be years away, recognition of the reality of PMS, combined with symptomatic treatment, can reduce many of the physically and psychologically debilitating aspects of the disorder.

Both the **basal metabolic rate** (the body's consumption of oxygen) and secretions of thyroid hormones decrease with age. The decrease in the basal metabolic rate (BMR) is related to the decrease in the mass of muscle tissue, which is a large oxygen-consuming tissue. The decrease in thyroid hormones (which are involved in the regulation of the BMR) is an adjustment to the progressively slower rate at which it is broken down and removed from the blood (Ingbar and Woeber, 1981). Epinephrine and norepinephrine, the hormones of the sympathetic nervous system (a part of the nervous system that controls involuntary functions such as heart rate and blood pressure), are released more slowly in aging persons in response to stress. There is also evidence that the tissue response to these substances (such as contraction of arteries, release of glucose from its storage form, glycogen, in the liver) diminishes with age. Finally, there is a gradual decline of glucose tolerance and an increased prevalence of diabetes mellitus with age (Andres and Tobin, 1977). This is usually due to tissue resistance to insulin (the hormone secreted by the islets of Langerhans in the pancreas to help the body use sugar), which occurs with progressive age and often weight gain in susceptible individuals.

The respiratory system, heart, and circulatory system have parallel changes occurring with age. The lungs and bronchi become increasingly less elastic, causing a progressive decrease in maximum breathing capacity. Respiratory functioning may decrease as much as 25 percent between ages thirty and forty-nine (Smith, Bierman, and Robinson, 1978). It takes individuals longer to catch their breaths after exercise in their thirties and forties than when they were younger. The ability of the heart muscle to contract decreases, leading to a lower cardiac index (the cardiac output per minute per square meter of body surface). Cardiac function may decline by 15 to 20 percent between ages thirty and forty-nine (Smith, Bierman, and Robinson, 1978). Arteries, too, become less elastic; **blood pressure** (the force exerted by the heart in pumping blood and the pressure of the blood in the arteries) increases with every decade.

Weight increases are common in middle age due to a decreased energy expenditure without a concomitant decrease in caloric intake. The unneeded calories are stored as fat deposits, commonly in the wall of the abdomen ("spare tire") or the hips, thighs, and chest wall (the "middle-aged spread"). The entire digestive system may slow its process of digesting, absorbing, and eliminating foods. Constipation becomes an increasingly common complaint.

Skin begins to lose its resilience and elasticity. It can no longer stretch as tightly across the muscles and bones. Both women and men usually begin to notice wrinkles. Wrinkle creams and other patent or home remedies may temporarily shrink the skin, but they cannot effect a permanent cessation of these early signs of aging. Men usually worry less about wrinkles than women since society considers a lined masculine face to be more handsome or to have more character. The same lines in women are thought to signal a loss of beauty. Advertisers have long applauded the beauty of a tanned skin to help sell sunlamps or tanning lotions or creams. In fact, excessive exposure to **ultraviolet light** (light waves of extremely short wavelength) accelerates these skin changes: dryness, thinner skin, less rapidly growing skin, and skin that is more susceptible to skin cancer. Adults, especially light-skinned ones, should protect their skin with sun blockers when outdoors and should not use sunlamps.

Hair may grow more slowly, be lost, or occasionally lose its pigmentation (evidenced by gray hair) during the thirties and forties. Genetic predisposition toward baldness or early graying, nutritional factors, disease, drugs, or hormones may cause these changes. Androgens (male sex hormones) often affect hair growth. Masculine balding or receding hair line after puberty is a well-known and frequent occurrence.

The senses of hearing and vision may begin to show age changes as well. A hearing loss (**presbycusis**) limited first to high pitches may cause persons to stand or sit closer to the source of sound than previously. They may strain to hear or may talk in compensatory louder tones. A visual problem (**presbyopia**) primarily affecting near vision may necessitate reading glasses or bifocals for close work. The lens of the eyes gradually becomes less elastic with age, causing the eyes to lose their ability to bring near-point visual images into clear focus. Presbyopics can usually read far-off signs long before younger passengers in a car but have difficulty reading the numerals on the odometer or the car radio without glasses.

The discussion of the physical changes of the thirties and forties should not cause too much apprehension. Individual differences abound throughout life. Some people may be gray before forty, others not until their sixties or seventies. Some people may grow flaccid and flabby, whereas others remain lithe and limber (see Figure 9–2). Diet, exercise, sleep, adaptations to and control of stress, a cloudier climate, avoidance of sunlight, and artificial measures such as hair dyes, makeup, and corrective surgery can retard or hide the signs of aging. Or one may be proud of a physical appearance that shows signs of experience (and character) rather than looking young, naive, and "still wet behind the ears." Most persons in their thirties and forties still consider themselves young, approaching middle age rather than already part of it.

NUTRITION

Many adults slow down in their thirties and forties and consequently need fewer calories than they did in their teens and twenties. About 34 million adult Americans, or 25 percent of the population, are overweight (Van Itallie, 1985). This is defined as being 20 percent

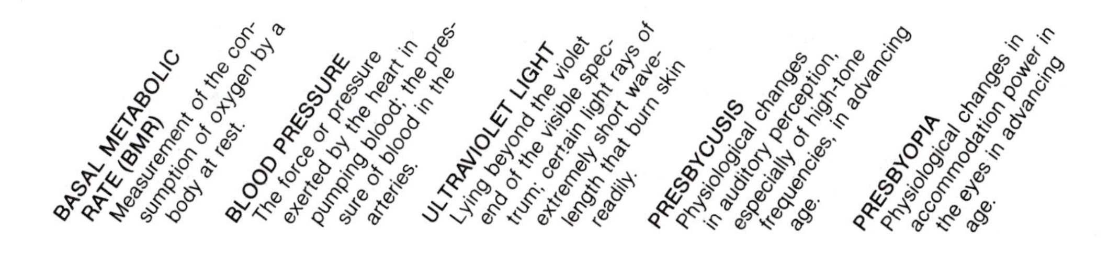

Figure 9–2
Diet and exercise often become greater concerns of adults as they pass through their thirties and forties. Exercise improves cardiovascular functioning and advances physical fitness and overall health.

over the desirable weight for one's sex, height, and body build (see Table 9–1). Many others are "pleasantly plump," or have "love handles" or a "spare tire" by these decades. Inadequate calorie intake, or undernutrition, is also becoming more common. In most cases, this is due to poverty. However, in some cases it is due to self-inflicted dieting.

Average adults without special dietary restrictions will be healthier if they choose plenty of dark green leafy vegetables, other fruits and vegetables, and whole grain breads, rice, pasta, and cereal products to make up a large proportion of their total daily caloric intake. As little as 5 grams per day of fat is sufficient (Walser et al., 1984); most of it should come from unsaturated fats such as soy, sunflower, corn, or safflower oil. The recommended daily protein intake for adults is 0.8 gram per kilogram of body weight, which is probably unnecessarily high (Walser et al., 1984). Symptoms of "meat intoxication" (headache, lethargy, nausea) may appear if one's protein intake is excessive. A high dietary intake of protein foods with saturated fats (organ meats, well-marbled steaks or roasts, pork, egg yolks, cream, shellfish) or of saturated fats (butter, lard, bacon fat, chicken fat) contributes significantly to the development of atherosclerosis, coronary heart disease, and some cancers.

A low dietary intake of fiber may also contribute to the development of atherosclerosis and to a greater incidence of some forms of cancer. Adults are now being encouraged to consume at least 25 grams of fiber per day (found especially in green leafy vegetables, fruits, berries, bran, whole grain bread and cereal products, and nuts).

Table 9–1 Desirable Weight According to Height*

Men

Height Feet	Inches	Small frame	Medium frame	Large frame
5	2	128–134	131–141	138–150
5	3	130–136	133–143	140–153
5	4	132–138	135–145	142–156
5	5	134–140	137–148	144–160
5	6	136–142	139–151	146–164
5	7	138–145	142–154	149–168
5	8	140–148	145–157	152–172
5	9	142–151	148–160	155–176
5	10	144–154	151–163	158–180
5	11	146–157	154–166	161–184
6	0	149–160	157–170	164–188
6	1	152–164	160–174	168–192
6	2	155–168	164–178	172–197
6	3	158–172	167–182	176–202
6	4	162–176	171–187	181–207

Women

Height Feet	Inches	Small frame	Medium frame	Large frame
4	10	102–111	109–121	118–131
4	11	103–113	111–123	120–134
5	0	104–115	113–126	122–137
5	1	106–118	115–129	125–140
5	2	108–121	118–132	128–143
5	3	111–124	121–135	131–147
5	4	114–127	124–138	134–151
5	5	117–130	127–141	137–155
5	6	120–133	130–144	140–159
5	7	123–136	133–147	143–163
5	8	126–139	136–150	146–167
5	9	129–142	139–153	149–170
5	10	132–145	142–156	152–173
5	11	135–148	145–159	155–176
6	0	138–151	148–162	158–179

*Weight in pounds according to frame (in indoor clothing weighing 5 lb for men and 3 lb for women; shoes with 1 in. heels).

SOURCE: Courtesy Metropolitan Life Insurance Co., 1983.

Some nutritionists argue that not every adult needs to radically alter his or her fat and fiber intake (Marshall, 1986). However, anyone with a high blood cholesterol level should reduce fat intake and increase intake of high-fiber foods (unless he or she has an acute colonic disease).

There is evidence suggesting that the average adult's salt intake (about 4 grams per day) is more than double the safe and adequate intake and may aggravate hypertension. Clearly, persons who have hypertension benefit from a reduction in salt (sodium) intake. Salting food is an acquired habit that can be broken. Many adults have learned to enjoy low-sodium foods in an effort to protect their hearts and their health.

The need for calcium, especially in women, is often not met by daily dietary intake of milk and milk products. Consequently, many adults are advised to supplement their diets with calcium carbonate. Other vitamin supplements are usually not necessary if one is eating a balanced diet. Excessive intake of vitamins A, D, and E can cause physiological problems, and megadoses of vitamin C have been associated with the formation of kidney stones.

While sugar does not cause diabetes, it does contribute to both dental caries and obesity. Many adults consume the bulk of their daily caloric intake from highly sweetened foods (candy, desserts, snack foods, crackers, sodas, alcoholic drinks). Obesity between the ages of twenty and fifty is considerably more dangerous than a similar degree of obesity in later life (Van Itallie, 1985). It puts one in a high-risk group for many illnesses, particularly hypertension, hypercholesterolemia, coronary heart disease, and diabetes. Body weights over 240 pounds in women and over 260 pounds in men constitute extreme health hazards (Kral, 1985). Obesity is considered the single most common preventable factor associated with chronic illness and death in adults under age fifty. Many professionals feel that public health information campaigns should be implemented to counteract this increasingly prevalent risk to life. Some have even suggested that legislation be enacted to control the marketing of "junk foods" (those with high calorie and low nutritive values).

HEALTH CONCERNS

The decades of the thirties and forties may be a period of excellent health or a time of increased disability due to illness. The state of one's health is strongly related to one's genetic predisposition to diseases, to lifestyle, to the body's immune responses, and to the availability and utilization of medical, surgical, and pharmacological therapeutic aids in the event of disease. As is true of the health of young adults in their twenties, physical well-being is threatened more by noninfectious agents (alcohol, drugs, smoking, pollution, accidents, obesity, stress, poor diet, and lack of exercise) than by pathogens or degenerative processes.

Accidents as a cause of death drop from about 44 percent among fifteen- to twenty-four-year-olds to about 16 percent among twenty-five- to forty-four-year-olds. Alcohol abuse, causing death through liver cirrhosis and gastrointestinal bleeding, climbs steadily. So, too, do heart diseases and mental disorders. Health maintenance in the thirties and forties often requires more than just avoiding or treating infectious diseases. Exercise, diet, and stress management are very important health considerations. Health spas, diet groups, and exercise programs are some of the ways people invest time and energy toward improving or maintaining their health.

The maximum benefits of exercise are derived from daily exercising, since physical stamina and fitness deteriorate with a lack of training. Many people schedule regular aerobic exercises (jogging, swimming, cycling) or isometrics (for example, weightlifting) into their daily lives.

Stress While high-saturated-fat/low-fiber diets are correlated with atherosclerosis and coronary heart disease, diet is not the only risk factor for these medical problems. There is now abundant evidence that when a person is under stress, the body produces more of the adrenal hormones cortisol and epinephrine. They in turn signal the liver to produce more cholesterol. They also signal fat cells to release stored fat and cholesterol into the bloodstream. Often a high blood cholesterol level is associated with daily stress more than with actual daily dietary intake of cholesterol.

People differ considerably in their abilities to handle stress (see Chapter 8). Friedman and Rosenman (1974) described two basic stable personality types. Type A's have a tendency to compete and challenge others; they have a sense of the urgency of time and strive to accomplish too much or do two things simultaneously; they are aggressive and impatient, disliking repetitious chores, waiting in line, or any inefficiency. Type B personalities are relaxed and easygoing. Rosenman (1978) stated that Type B's may also be ambitious and goal directed, but not aggressively so. They can tolerate delays without undue tension and they have learned how to react more calmly to the stresses of daily living. Friedman and Rosenman not only found stability in A's and B's over time; they also correlated the onset of stress-related diseases, especially coronary heart disease, with Type A personality. They did this both retrospectively (looking back over the lives of persons who had had heart attacks) and prospectively (successfully predicting what persons would eventually have heart attacks). Both men and women can have Type A personalities.

Many other researchers have supported Friedman and Rosenman's theory of the relationship between Type A behaviors and coronary heart disease. Type A behavior is now usually called the **coronary-prone behavior** pattern. In the approximate one-half of heart disease patients for whom no causes related to dietary, hereditary, obesity, or smoking factors can be traced, coronary-prone behavior pattern is usually present. People with this behavior pattern have such a sense of urgency for accomplishing tasks that they often ignore warnings of coronary heart disease. They also dislike repetitious exercises, relaxation techniques, or other procedures that

could help them reduce tensions (Mathews and Brunson, 1979). Often they make no real efforts to cope with or control their stresses until after they experience one or more nonfatal heart attacks.

Coronary Heart Disease The most prevalent chronic disease condition in men in their thirties and forties is **coronary heart disease** (CHD). It also affects women, though to a lesser extent. It is the single greatest killer of middle-aged adults (Friedman, 1978). Those factors related to CHD that a person can control are diet, cigarette smoking, stress, and exercise. Factors that a person can control with the help of modern medicine are hypertension, hypercholesterolemia (too much cholesterol in the blood), hyperlipidemia (too many lipid fats and oils in the blood), diabetes, and irregular heart rhythms. Factors that a person cannot control are heredity, race, sex, and age. Men and blacks are more susceptible, and the prevalence increases with age, especially in adults with a family history of CHD.

A myocardial infarction (MI, heart attack, coronary thrombosis, coronary occlusion) results from a sudden arrest or insufficiency of blood flow to the muscle of the heart because of a sudden obstruction or narrowing of the coronary artery supplying it. When it occurs, myocardial (heart muscle) cells become damaged, leading to their death (necrosis). The most common complaint of an MI is uncomfortable pressure, fullness, squeezing, or chest pain lasting 2 minutes or more. There is some radiation of pain into the upper arms, jaw, and neck. There may also be shortness of breath, sweating, weakness, and fainting. A few MIs may be painless. Victims of MIs should be kept warm, quiet, and horizontal and should be taken to a medical facility as quickly as possible.

The inability of the heart to pump effectively can lead to a backup of blood into various organs, a condition known as congestive heart failure. Congestive heart failure can occur as a result of an MI or can be unrelated to an MI. It can also occur as a result of rheumatic fever, a congenital heart defect, atherosclerosis, or hypertensive disease. The features of congestive heart failure are presence of edema (most commonly in the legs and ankles), shortness of breath, fatigue with exertion, an enlarged heart, a rapid and sometimes irregular pulse, and signs of congestion in various organs due to a backup of blood.

Angina pectoris is a condition characterized by a feeling of heaviness or pressure in the chest that falls short of an MI. Angina pectoris is almost always associated with severe atherosclerosis of the coronary arteries and results from transient relative inadequacy of blood to a portion of the heart, as in times of emotional excite-

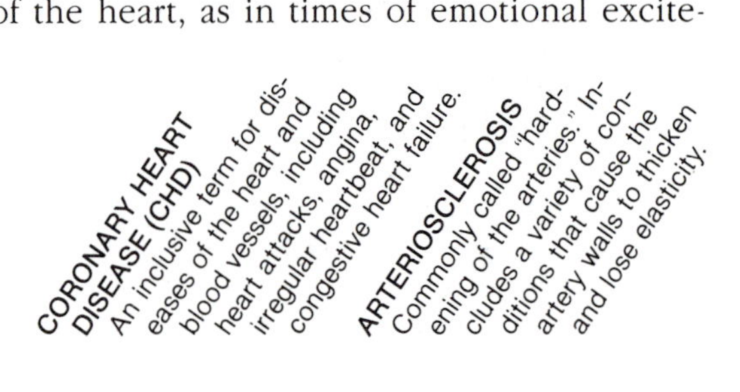

ment or unusual physical exertion. It is not severe enough to cause necrosis of the myocardium. The pain and pressure of the recurrent attacks can be controlled by medication that angina sufferers usually carry with them and use as needed.

Irregular heartbeats such as a slow beat or an irregular rhythm should be controlled by the use of drugs or an artificial pacemaker. A pacemaker is a small unit implanted on the chest wall with tiny wires to the heart that will produce the electrical impulses needed to make the heart pump normally.

Atherosclerosis The most common fatal factor underlying CHD is **atherosclerosis**. It is the most common form of arteriosclerosis ("hardening of the arteries"), a thickening and hardening of the walls of the arteries. Atherosclerosis involves mainly the intima (the innermost lining) of the arteries. (Sclerosis refers to any hardening of fibrous tissue.) Arteriosclerosis involves the media (muscle portion) of the arteries as well.

Although the exact mechanism of formation of atherosclerosis is unknown, studies suggest that diets high in saturated fats and cholesterol, obesity, physical inactivity, diabetes mellitus, hypertension, cigarette smoking, stress, and Type A behaviors are associated with a premature or accelerated course of the disease. One or more of these factors is generally present when persons in their thirties or forties are diagnosed as having atherosclerosis.

In addition, about 1 person in 500 has inherited familial hypercholesterolemia (Goldstein and Brown, 1985). Because it is an autosomal dominant disease (see Chapter 3), inheritance of only one gene for the condition (heterozygosity) will produce hypercholesterolemia. It is prevalent in all ethnic groups equally. The disease involves defects in cell receptors for low density lipoproteins (LDL) (particles that carry the greater percent of the cholesterol in the plasma). As a result of reduced cellular intake of LDL, the serum levels of LDL go up (hypercholesterolemia), and the patient is at high risk of developing atherosclerosis and CHD. The disease can be detected at birth because cord blood shows LDL concentrations two to three times normal. Most heterozygous hypercholesterolemia patients develop CHD in their thirties and forties unless adequately treated. Homozygous patients (those who have inherited the gene for the disease from both parents) may develop CHD in childhood unless they are treated. Current therapy may involve plasmapheresis (plasma is separated and removed from the blood cells and replaced with donor plasma) every two to three weeks for homozygotes and lifelong therapy with drugs that reduce cholesterol and stimulate LDL cell receptor synthesis in heterozygotes. In addition, all persons with hypercholesterolemia have stringent limitations of dietary cholesterol and saturated fat (Goldstein and Brown, 1985).

When serum LDL levels are high in any person, the excess cholesterol and other saturated fats are deposited on arterial walls in the form of patchy areas of fatty streaks and fibrous plaques (called atheromas). The aorta and the large arteries of the neck show fatty streaks in infancy. These reach a peak between twenty and thirty

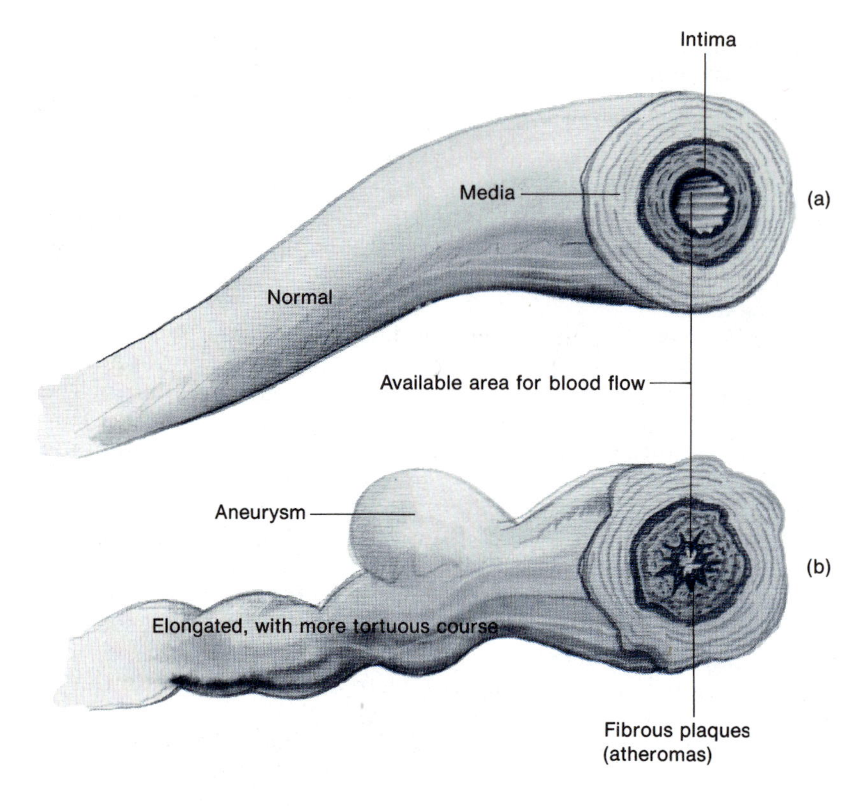

Figure 9–3 *Comparison of a healthy artery (a) and one with atherosclerotic changes (b).*

years. The coronary arteries and arteries of the brain may begin to contain fatty streaks and fibrous plaques during puberty. Subsequently, there is a progressive rigidity (loss of elasticity) of vessels with plaques. Other changes of arteries include dilation (widening), elongation, tortuosity (winding course instead of a straight course), calcification of the media, and aneurysm formation (ballooning out of vessels) (see Figure 9–3). Although a few fibrous plaques may be asymptomatic and produce no disease, the presence of too many of them may impede blood flow through the arteries and eventually cause coronary heart disease or strokes.

Hypertensive Disease High blood pressure, also known as **hypertension** or hypertensive disease, has many definitions. There is no one set line between normal and abnormal blood pressure. Rather, abnormal is the point at which any individual shows evidence of deleterious effects of his or her own high blood pressure. The values 160/95 are often used as the upper limits of normal for resting adults. However, Williams, Jagger, and Braunwald (1980) suggested that a more appropriate definition would take age and sex into account. Hypertension would then be:

HYPERTENSION (HIGH BLOOD PRESSURE) An unstable or persistent elevation of blood pressure above the normal range.

Women at any age > 160/95
Men below age 45 > 130/90
Men above age 45 > 140/95

Using these values, from 15 to 20 percent of the adult population of the United States have abnormal or high blood pressure. The upper value (systolic) refers to the peak pressure and the lower value (diastolic) refers to the bottommost pressure in the arteries during each heart contraction. Uncontrolled hypertension accounts for over 30,000 deaths each year in the United States and is a major contributing factor to atherosclerosis and CHD (American Heart Association, 1985).

High blood pressure appears to be caused by genetic factors or by an excess of substances that cause retention of salt by the body or narrowing of the arteries and arterioles. If hypertension is caused by recognized factors such as kidney disease, an adrenal tumor, a congenital defect of the aorta, or use of birth control pills, it is called *secondary hypertension.* If its cause cannot be determined, it is called primary or *essential hypertension.* As the years pass, the causes of hypertension are being increasingly recognized. The complex role of the central nervous system in the maintenance and regulation of blood pressure needs to be better understood.

The consequences of high blood pressure are higher incidences of CHD, atherosclerosis, stroke, and kidney failure. The results of untreated hypertension are worse for men than for women and for blacks than for whites. Young black males are most adversely affected by hypertension (Williams, Jagger, and Braunwald, 1980).

Accelerated hypertension (also called malignant hypertension) is not a separate disease but is a phase in the course of severe, untreated hypertension. It is manifested by any or all of the following: (1) rapidly worsening kidney failure, (2) heart failure, (3) retinal hemorrhages, and (4) dysfunction of thought processes and decreased level of consciousness. Failure to treat hypertension when it is at this stage will result in death.

The availability of a large number of antihypertensive medications facilitates the treatment of high blood pressure. Diuretics can get rid of excess fluid and sodium. Vasodilators widen narrow blood vessels. Other drugs can prevent vessels from narrowing. Each hypertensive individual may need his or her own unique drug or combination of drugs to keep the blood pressure within safe limits.

If the underlying cause of hypertension can be determined, its removal is preferred (surgically in the case of certain adrenal tumors or discontinuing the use of birth control pills). However, more than 90 percent of hypertensives require drug treatment and such measures as weight loss and reduced salt in the diet. In some cases, exercise programs and stress management programs (such as biofeedback) are also effective in lowering blood pressure to within a normal range (see Figure 9–4).

Other Chronic Illnesses Varicose veins are dilated, tortuous, superficial veins found most commonly on the calves and thighs. Overweight individuals, women who have been pregnant, and people with a family history of varicosities are most often affected. The causes are obscure, but in women they are aggravated by hormonal factors (Strandness, 1980). Normal veins have one-way valves that

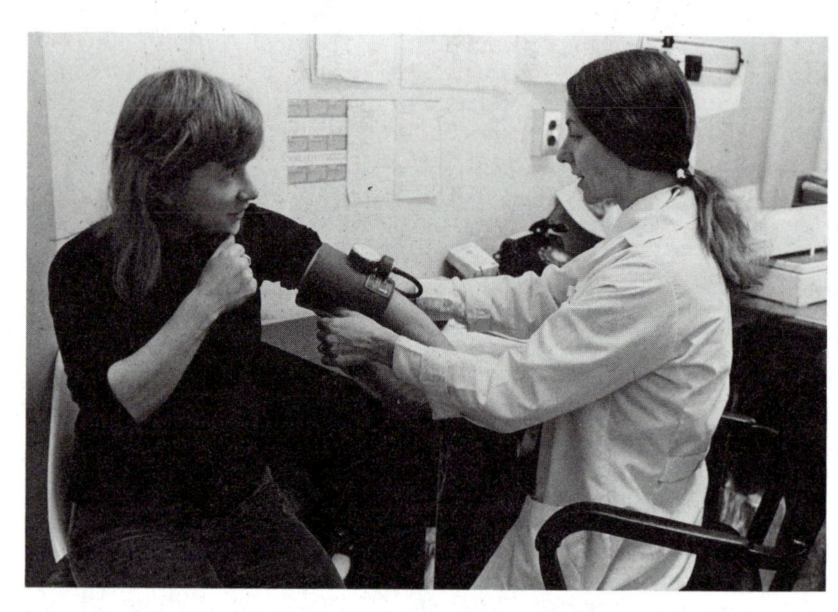

Figure 9–4
Persons in their thirties and forties should have annual physical examinations to detect and control potential problems such as high blood pressure at their onset.

prevent a backward flow of blood. This is especially important in humans, who are upright animals and consequently have high venous pressure in their legs. When these valves become incompetent and allow a backflow of blood, varicosities result. In mild cases they may produce little problem aside from cosmetic discoloration. However, they may also cause pain, a feeling of heaviness, fatigue, and swelling of the legs. Sometimes they require surgical removal. People with varicosities often wear support stockings to support the vessels and prevent swelling.

Other chronic diseases of the veins are thrombophlebitis and phlebothrombosis. In the former, veins become inflamed and become the site of blood clots. The features of thrombophlebitis include pain, tenderness, redness, and swelling of part of the leg. In phlebothrombosis there is an absence of inflammation associated with the presence of a clot (thrombus). Venous thrombosis (blood clot in the vein) has been associated with such things as varicose veins, injury to the legs, use of birth control pills, prolonged bedrest as after surgery or in the postpartum period, congestive heart failure, and certain malignancies. However, there may be no symptoms with phlebothrombosis, and a person may first know that a venous thrombosis is occurring when a clot dislodges and travels to the lung, producing symptoms of pulmonary embolism (obstruction of lung vessel by a clot). Venous thrombosis may also lead to destruction of venous valves, resulting in more varicose veins. Walking and leg muscle exercises are strongly recommended for pregnant women, obese persons, people with sedentary lifestyles, and postoperative and cardiac patients in an attempt to prevent venous thrombosis and its possible embolic consequences.

Rheumatoid arthritis (RA) is a chronic inflammatory disease with involvement of joints and other structures. The cause is unknown. Rheumatoid arthritis accounts for about one-third of all arthritic conditions, has an early onset, peaking in the forties, and affects women three times more often than men (Gilliland and Mannik, 1980). Flare-ups of RA may be precipitated by such things as emotional or physical stress. The initial symptoms include low-grade fever, muscle weakness, and pain on movement in the multiple joints of the hands. It progresses slowly and causes disabling stiffness, pain, swelling, and tenderness in other joints as well. The symptoms are frequently more pronounced in the morning and after other periods of physical inactivity. The therapy for rheumatoid arthritis is not a cure. There are frequent spontaneous remissions. Drug therapy may involve the use of aspirin or other analgesics, nonsteroidal anti-inflammatory drugs, gold salts, antimalarial drugs, penicillamine, or steroids (Harris, 1983). The use of medication allows relief of pain so function can be restored. In addition to medicines, treatment may include physiotherapy, exercises, and even surgery to correct joint deformities, repair ruptured tendons, and remove painful, inflamed tissue, thereby preserving joint and muscle function. Most patients with rheumatoid arthritis have some degree of disability all their lives.

Gout is a genetically determined disease resulting in a disorder in the metabolism of purine. The consequences of this disorder are gouty arthritis, which causes recurrent attacks of painful joint swelling, painful uric acid stones in the urinary tract, and chronic kidney failure. Attacks of gouty arthritis most often affect the joints of the large toes, ankles, feet, knees, fingers, wrists, and elbows. They appear without warning and may be precipitated by minor trauma, infections, or surgery. Gout cannot be corrected, but its manifestations can be controlled by medications. During acute attacks of painful joints, treatment consists of medication. Since gout is usually the result of increased production of uric acid, current drug treatment aims at decreasing this production. The advent of new drugs has changed the course and outcome of gout very favorably for many people.

Multiple sclerosis (MS) is frequently referred to as the great crippler of young adults. It is a chronic and slowly progressive disease of the central nervous system characterized by spotty and diffuse areas of loss of the myelin sheathing around axons of nerves in the brain and spinal cord and by a multitude of other symptoms. The age of onset for multiple sclerosis is typically between twenty and forty. The cause is unknown, though scientists suspect it may be related to a viral infection or an autoimmune process. The patient gets attacks of neurological disability followed by periods of remission of the disease but with subsequent recurrences and progressive worsening. As each successive attack occurs, more nerve tissue is damaged. The course of the disease varies greatly from patient to patient. Whereas some have a rapid downhill course, many experience a slow progression of symptoms with long periods of remission (decreases in severity) between exacerbations (increases in severity) of the disease. Poskanzer and Adams (1980) reported that

Figure 9–5
Many multiple sclerosis patients still walk, work, and remain actively involved in family and community activities for many years after the onset of their disease.

after twenty-five years with the disease one-third of MS patients can still work and two-thirds can still walk (usually with a cane) (see Figure 9–5).

Some common manifestations of multiple sclerosis include muscular weakness or paralysis, numbness, double vision, transient blindness, lack of coordination, and speech disorders. A great deal of research time and money is being spent to try to find the cause, an effective treatment, and a way to arrest the occurrence of multiple sclerosis. There are approximately 500,000 persons afflicted with MS in the United States.

Paraplegia refers to paralysis confined to the lower extremeties and, generally, the lower trunk. Quadriplegia refers to paralysis of all four extremities and of the trunk. The causes of paraplegia and quadriplegia include accidents affecting the spinal cord, fractures of the vertebral column causing pressure or severing of the spinal cord, a disease affecting the cord such as poliomyelitis, a tumor causing compression of the cord, or Pott's disease (tuberculosis of the spine). Quadriplegics and paraplegics not only lose their ability to move their lower trunk and extremities below the level of the spinal cord injury but also lose control of their bladder and bowel functions. They may develop urinary tract infections from retention of urine or from catheterization of the bladder. Chronic urinary infections and kidney failure may shorten their life spans. They also may develop bed sores at pressure points such as the buttocks.

Some handicapped persons using wheelchairs to locomote can lead moderately active lives, including participating in work, family, and community activities. Their intelligence and manual dexterity are unaffected by the paralysis. Campaigns such as "Hire the handicapped," public relations programs, supportive and self-help organizations, and simple architectural considerations such as the addition of ramps in buildings are helping to make it less difficult for people with these crippling conditions to adjust psychosocially and economically.

Mental Disorders Psychopathology is one of our nation's most important health problems, affecting 20 percent of the population and costing several billion dollars each year. Persons in their late thirties and early forties seek the most psychiatric care and also are the age group most frequently admitted to mental hospitals.

There is no single criterion for diagnosing **mental disorders**. Persons showing great personal distress, who disturb others, who become incapable of performing their usual jobs, or who have marked personality changes may be manifesting psychopathology.

Disorders that once were labeled neuroses are now called anxiety disorders, suggesting the role of stress in their onset. There are many varieties of anxiety disorders, including phobias, panic, obsessive-compulsive disorders, posttraumatic shock syndrome, and generalized anxiety. Other conditions once called neuroses are now called somatoform disorders (hypochondriasis, conversion reactions), dissociative disorders (amnesia, multiple personality), psychosexual disorders (transsexuality, paraphilias), or personality disorders (antisocial personality, narcissism). In general, these conditions are less severe than psychoses (in which a person loses touch with reality) and are more easily treated.

The psychotic conditions are characterized by more radical and damaging disruptions of consciousness, perception, thinking, and social behavior. The four major psychoses are schizophrenia, bipolar (manic-depressive) disorder, psychotic depression, and organic psychoses (those involving brain trauma or brain degeneration).

Schizophrenia may be of varying severity and have varying symptomatology. Its onset is frequently in early to middle adulthood. Its

MENTAL DISORDER
An unhealthy or abnormal condition of some mental function(s).

cause is unknown. Common features include misinterpretation or idiosyncratic distortions of reality, delusions, hallucinations, flat or inappropriate emotions, social withdrawal, ambivalence, and odd to bizarre behavior and appearance.

Hypotheses about the causes of schizophrenia are legion. Researchers have suggested genetic factors, metabolic abnormalities, endocrine dysfunctions, idiosyncratic responses to drugs or stress, and abnormality in one or more neurotransmitters. Despite long years of study, causes have not yet been successfully explained by any one theory. Psychodynamic theorists have suggested family pathologies and a failure of ego development. Behaviorists have suggested inappropriate social reinforcers. Family theorists have suggested the possibility of a schizophrenogenic (schizophrenia-causing) mother who is domineering and smothering, yet cold and rejecting.

Treatment of schizophrenia may include drug therapy, psychotherapy, and institutionalization. There are many antipsychotic drugs available today that effectively interrupt acute psychotic behaviors.

Psychotherapy can be beneficial to both a schizophrenic and his or her family. Although such intervention may not be a remedy, it does help those involved to learn to live with a condition characterized by phases of quiescent and active symptomatology. It may also contribute to more long-lasting remissions of symptoms. Institutionalization may be necessary for schizophrenia when psychotic behaviors threaten to disrupt the lives of others. However, physical restraints and locked doors are now less necessary due to the effectiveness of antipsychotic drugs.

Bipolar disorder (manic-depressive psychosis) brings periods of excitement (mania) and depression that may each extend for hours or days and have long or short periods of remission in between. In the manic, agitated state a person may show increased energy and actually accomplish great works. However, mania may also bring in its wake undesirable behaviors such as irrepressible, uninhibited, unconventional, assertive, or mischievous speech and actions that hurt others. In the depressed state the individual may be suicidal.

The causes of bipolar disorder, like the causes of schizophrenia, are not yet completely understood. However, an argument for biochemical disorder is supported by the great improvement or total recovery of close to 100 percent of patients who take daily maintenance dosages of lithium. Lithium is an alkaline metal element that has been called the "insulin of psychiatry" because it is so effective in treating and preventing the mood swings of mania and depression. It has few side effects, but the dosages must be carefully monitored to prevent lithium poisoning.

Depression is the most frequently treated psychiatric disorder. Unemployment heightens its risk and women are afflicted six times more often than men (O'Hara, 1984). Each year several million adults suffer one or more depressive episodes. Depression contributes heavily to problems of alcoholism, drug abuse, family breakups, CHD, cancer, and suicides. The usual features of depression are sadness, despondency, feelings of worthlessness, loss of appetite, insomnia, lack of concentration, indecision, and a waning interest in one's usual pursuits.

There are many types of depression. Depression that occurs as a consequence of the death of a loved one, loss of property, change in lifestyle, separation from a friend, or loss of job or status is called reactive or *exogenous* depression. Psychotic forms of depression are usually *endogenous* (arising from factors within the person). Internal problems creating depression are usually multifaceted. Stressors, drugs, sleep deprivation, or even diet may affect chemical activity in the brain, especially the neurotransmitters serotonin and norepinephrine, leading to mood disorders. However, persons with low self-esteem, unrealistically high standards for themselves, or a sense that they cannot control their own destiny are most prone to depression, be it exogenous or endogenous.

In addition to having the other symptoms of depression, persons who become psychotically depressed may develop anxious and agitated behaviors. They may have delusions of great sins and unworthiness and frequently discuss death. They may whine, pace, abuse themselves with scratching or head banging, and be verbally hostile and abusive to others. Suicide and physical deterioration from lack of food and rest loom large enough as threats in some persons that they are given protective hospitalization.

In some cases electroshock therapy brings marked improvement in the condition of depressed persons. Psychotherapy, behavioral therapy, and family counseling are also used to treat depression. Mood-elevating drugs are usually used in conjunction with these forms of therapy to enable the patient to participate. Depressions may be short or long lived but generally last less than six months with or without therapy. However, they also tend to recur in people who have experienced one psychotic depressive episode.

Some psychoses may be induced by physical illnesses that destroy brain tissue or by substance abuse (alcohol or drugs). Such organic brain syndromes are irreversible.

Acute Illnesses Even the most cautious thirty- and forty-year-olds who exercise regularly, eat and sleep well, and avoid excesses of alcohol, tobacco, and drugs may become ill from infectious disease. The following sections discuss respiratory infections, digestive disturbances, and urinary tract infections. Their severity depends on the number and virulence of the pathogenic organisms to which one is exposed and on the body's state of immune responsiveness and freedom from anatomical or physiological abnormalities.

Upper respiratory infections (URIs) are a common cause of lost time from work in this age span. Inflammations (denoted by the suffix "itis") of the upper respiratory tract include sinusitis, nasopharyngitis, pharyngitis, tonsillitis, and laryngitis (see Figure 9–6). Whereas the common cold is caused by a large number of infectious viruses, infections of the sinuses, pharynx, tonsils, and larynx are often caused by bacteria (for example, streptococcus and pneumococcus). Upper respiratory infections can range from mild and short lived to severe and life-threatening, depending on the organisms

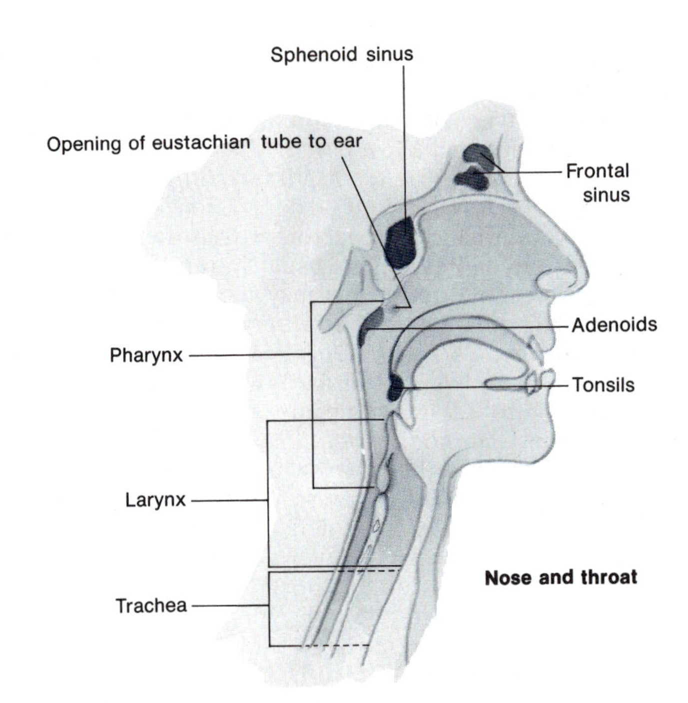

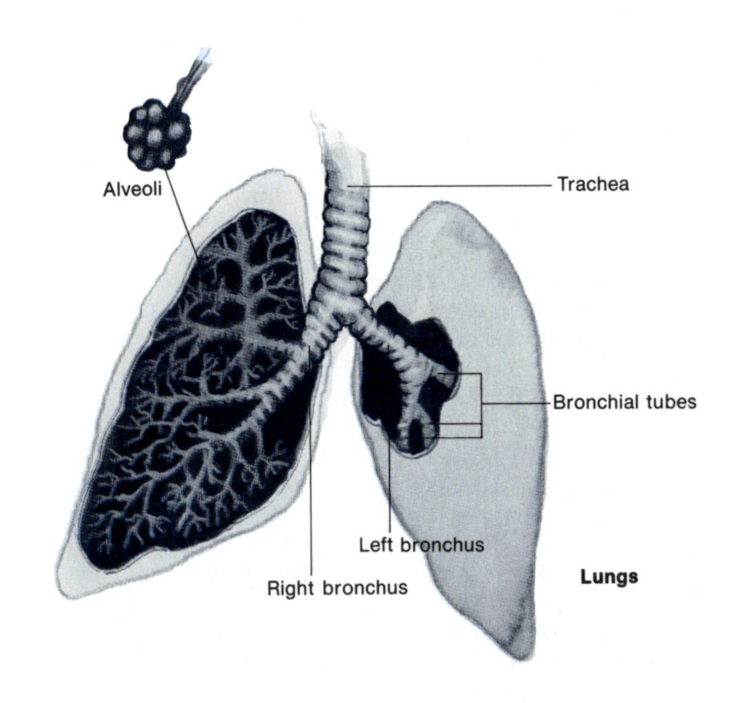

Figure 9–6
Organs of the respiratory tract that are often invaded by pathogenic microorganisms. The resultant respiratory tract infections are characterized by redness, heat, pain, swelling, and mucus production by the tissues invaded.

and the general health and immune responsiveness of the affected person. Infections caused by bacteria can be treated with antibiotics. Viral infections cannot be treated with antibiotics because the viruses themselves are not sensitive to these medications. Untreated streptococcal infections of the respiratory tract (or of the skin, as in impetigo) can be followed in a few weeks by such complications as rheumatic fever and glomerulonephritis. For this reason, "strep" infections are routinely treated for at least ten days with penicillin or other suitable antibiotics.

A true influenza (as opposed to a stomach upset called intestinal flu) is caused by an influenza virus of one of several subgroups. New viral strains are continually emerging. Influenza initially makes a person feverish and weak and gives muscle pain and headaches. Then, upper respiratory tract symptoms such as runny nose and sore throat and lower respiratory tract symptoms such as coughing and shortness of breath may appear. The illness may be relatively brief (three to seven days). However, it often predisposes the victims to secondary bacterial infections of the lungs (pneumonia) that may be serious, and occasionally fatal, even in young people.

Lower respiratory infections (LRIs) include bronchitis (infection of the large air passages in the lungs) and pneumonia (infection of the air sacs in the lungs). They can occur in anyone but are more common among smokers, persons with asthma, and persons continually exposed to irritating vapors or dust. Bronchitis and pneumonia may result from either viral or bacterial infections. Their symptoms include difficulty in breathing, production of sputum (mucus from air passages in the lungs), cough, chest pain, and fever. As with upper respiratory infections, responsiveness to antibiotics will depend on the causes.

Chronic bronchitis is quite different from acute bronchitis. It is more common among habitual smokers and inhabitants of smog-laden areas. The cough of a person with chronic bronchitis is persistent over a period of months and is usually associated with sputum production. Chronic bronchial irritations may predispose an individual to frequent secondary bacterial or viral bronchial infections. Chronic bronchitis is a forerunner of emphysema and often of lung cancer as well.

Adult asthma is characterized by wheezing, coughing, labored breathing, and mucus production. Asthmatics may have disabling attacks of wheezing and difficult breathing that last from hours to several days. Asthma may also lead to emphysema.

All the organs of the gastrointestinal (GI) tract (the esophagus, stomach, duodenum, jejunum, ileum, colon) and the organs that are intimately related to the GI tract (liver, biliary tract and gallbladder, pancreas) are subject to infection from pathogens, insult from irritants (alcohol, toxins in foods and drinks), and inherited diseases. A few of the more common problems that may occur among people in this age group are peptic ulcer disease, gastritis, gastroenteritis, appendicitis, gallstones, hernias, and ulcerative colitis.

Ulcers are eroded or scooped-out lesions on any mucous membrane or skin surface, with superficial loss of tissue and inflammation. Peptic ulcer disease (ulcers of the stomach and duodenum) is one of the most commonly reported illnesses in the United States, affecting from 10 to 15 percent of all people at some time in their lives. Duodenal ulcers make up 80 percent of all peptic ulcers (McGuigan, 1980). They are most common between the ages of twenty and fifty. They occur twice as often in men. Gastric ulcers are also more common in men. They tend to occur at later ages than duodenal ulcers. Duodenal ulcer disease is associated with high levels of gastric acid production, whereas gastric ulcers are found with generally low acid production, often with the use of aspirin and steroids. Peptic ulcers usually occur singly, although multiple ulcers of the duodenum or stomach, or both, are sometimes encountered. Emotional crises, stressful lifestyles, smoking, alcohol, excesses of spicy foods, steroids, aspirin, and certain other medications are among the agents suspected to cause or aggravate peptic ulcers.

The symptoms of an ulcer are feelings of an aching, gnawing, or burning in the upper abdomen. Often nausea also occurs. Antacids or eating may relieve the pain, but it will usually recur. Bleeding from ulcers may be manifested by skin pallor, general weakness, lightheadedness, tarry, black stools, or bloody vomitus.

Duodenal ulcer disease is usually a chronic problem. Although some persons may have a single bout of duodenal ulcer, the tendency is for periods of worsening to occur after periods of complete clearing (remission). A few gastric ulcers occur in cancer of the stomach, so physicians carefully follow the healing of gastric ulcers. Failure to heal a gastric ulcer indicates a need for exploratory surgery.

Ulcer patients must practice dietary regulation and moderation in their daily activities. They may take antacids to neutralize stomach acids, or prescription drugs that promote healing and shield the ulcer from gastric acids. Symptoms are most apt to recur in times of stress or after indiscretions in intake of foods, alcohol, or drugs that increase gastric acidity or cause gastric irritation.

Gastritis refers to an inflammation of the stomach caused by an excessive intake of alcohol, aspirin, or highly spiced food or by ingestion of foods containing certain bacterial toxins or poisons. Its symptoms include nausea, vomiting, upper abdominal pain, sometimes vomiting of blood, and weakness. Treatment is generally supportive, including antacids, medications to control nausea and vomiting, analgesics, and sometimes intravenous fluids if the symptoms persist.

Gastroenteritis is an acute inflammation of the lining of the stomach and intestine. The causes may be the same as for gastritis. Onset is generally rapid with symptoms such as weakness, loss of appetite, nausea, vomiting, abdominal cramping, and diarrhea. The treatment includes restraint from eating or drinking during the acute episode and medications for control of nausea, vomiting, intestinal spasms, and pain.

Appendicitis is the most common acute abdominal condition that usually requires surgery. The appendix, a small, nonfunctional structure attached to the first part of the colon, becomes inflamed and distended. Symptoms of appendicitis include abdominal pain, loss of appetite, nausea, and vomiting. Once a diagnosis of acute appendicitis is made, the appendix is surgically removed. The consequences of delayed treatment of appendicitis are peritonitis (inflammation of the membrane lining the abdominal cavity and its contents) and death from sepsis (overwhelming bacterial infection).

Diseases of the gallbladder and bile ducts also account for a large proportion of the abdominal surgery in this age group. The gallbladder stores bile, a digestive fluid produced in the liver and used in the gut for the digestion of fats. The biliary ducts transfer bile from the liver to the gallbladder and from the gallbladder to the duodenum. Inflammation of the gallbladder is called cholecystitis. An acute bacterial infection in the gallbladder produces an acute cholecystitis. Symptoms may include nausea, vomiting, chills, fever, and upper abdominal pain. Treatment first employs antibiotics, medication for pain, gastric suctioning, and intravenous fluids. Then, after the acute inflammation subsides, a cholecystectomy (surgical removal of the gallbladder) is done. About 20 percent of women and 8 percent of men harbor gallstones (Schoenfield, 1980). Obesity and pregnancy predispose a person to gallstone formation.

Another common reason for abdominal surgery in adults in their thirties and forties is the presence of a hernia (see Figure 9–7). A hernia is a protrusion of a part of an organ through a channel or "canal" in a structure that normally contains it. Hiatal hernias are protrusions of a part of the stomach up through the esophageal opening in the diaphragm. Inguinal hernias, femoral hernias, and

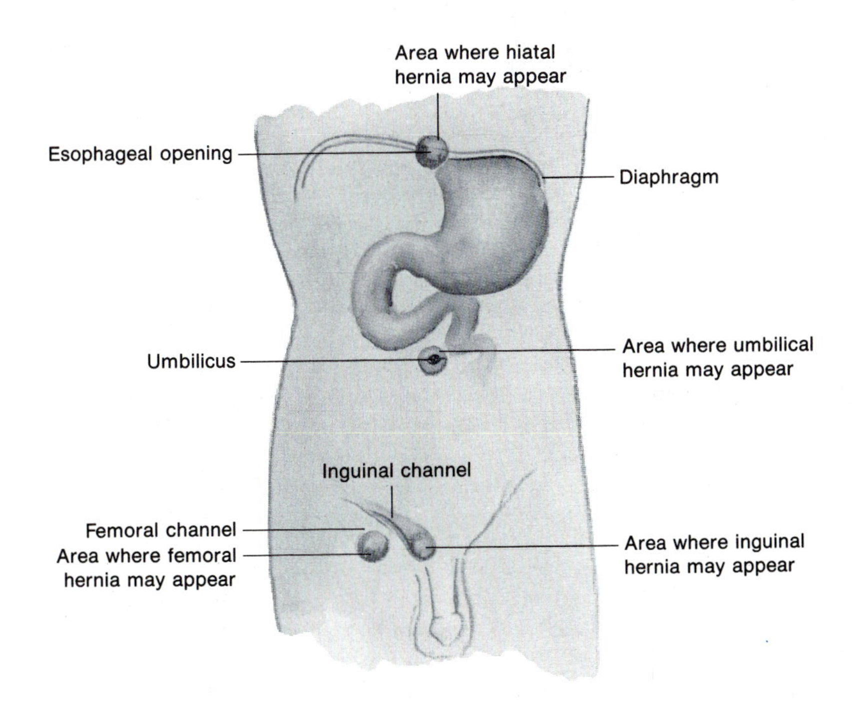

Figure 9–7
Areas of the body where four of the common varieties of hernias may occur.

umbilical hernias are protrusions of loops of the intestine through the inguinal and femoral channels and a defect in the anterior wall of the abdomen respectively. Frequently, herniated portions of the intestine retract (reduce) themselves spontaneously or can be returned to their normal location with pressure applied by skilled hands. Some people wear binders over areas prone to herniation to prevent the protrusion from recurring. A hernia usually causes some pain and tenderness at the site of protrusion. A herniated structure is in danger of becoming confined or imprisoned in its opening ("incarcerated"). As its veins become compressed, blood drainage is impaired. It swells up so that reducing it becomes even more difficult. Eventually its blood supply may be cut off, leading to tissue death. In cases of incarceration, emergency surgery must be performed to reduce the hernia and prevent tissue death and gangrene and to narrow the opening of the channel so that herniation will not occur again.

Ulcerative colitis is an inflammatory and ulcerative disease of the colon (large intestine). The disease usually has its onset in the early adult years. The cause is unknown. Etiologic suggestions of psychophysiologic, genetic, or autoimmune factors, of transmissible agents such as viruses, of reactions to allergens such as gluten or milk proteins, or of a combination of the above are all actively being researched. The disease may occur gradually with increased urgency for bowel movements, abdominal cramps, and bloody mucus, or very abruptly with lower abdominal pain, marked weakness, rapid weight loss, and stool containing pus, blood, and mucus. Complications include abscesses of the tissues surrounding the rectum, massive hemorrhaging, and colon perforation. Uncontrolled hemorrhaging or perforations are indications for an emergency colostomy (removal of a segment or all of the colon surgically). The risk of colon cancer is ten times greater in patients who have had ulcerative colitis for more than ten years (LaMont and Isselbacher, 1980) (colon cancer will be discussed in Chapter 10).

Urinary tract infections are more common in females than in males due to their shorter urethras that facilitate the retrograde entry of bacteria into the lower urinary tract. Bacteria may normally be present in the outermost part of the urethra. However, when they reach the urinary bladder, they may cause an infection known as cystitis. Symptoms of cystitis include lower abdominal pain, often radiating into the sacral region of the back; frequent, painful urination; bloody urine; fever; and chills. The urine normally flushes out bacteria from the urethra during voiding, thus preventing cystitis from occurring. Women are more prone to cystitis during pregnancy, after childbirth, and after frequent coitus. Treatment generally includes the use of antibiotics and liberal intake of fluids.

Pyelonephritis, an acute bacterial infection of the kidneys, may be a sequel to cystitis. Once the kidneys become infected, they may continually seed the bladder with infecting organisms and cause a prolonged bout of cystitis. Symptoms of pyelonephritis include flank pains; fever; chills; urgent, frequent, and burning urination; and nausea and vomiting. All kidney infections have the potential of damaging the kidney tissue, becoming chronic infections, and ultimately causing renal failure. They must be treated vigorously with appropriate antibiotics.

Cognitive Changes

Intellectually, the thirties and forties are very good years. Brain cell maturation and brain weight peaked in the late twenties, and a gradual but progressive shrinking is in progress. The weight of the brain decreases due to a loss in the number and size of brain cells and the loss of myelin sheathing. Impulses travel slightly more slowly across neurons and axons, causing a decrease in reaction time. However, mental acumen is still high (see Figure 9–8).

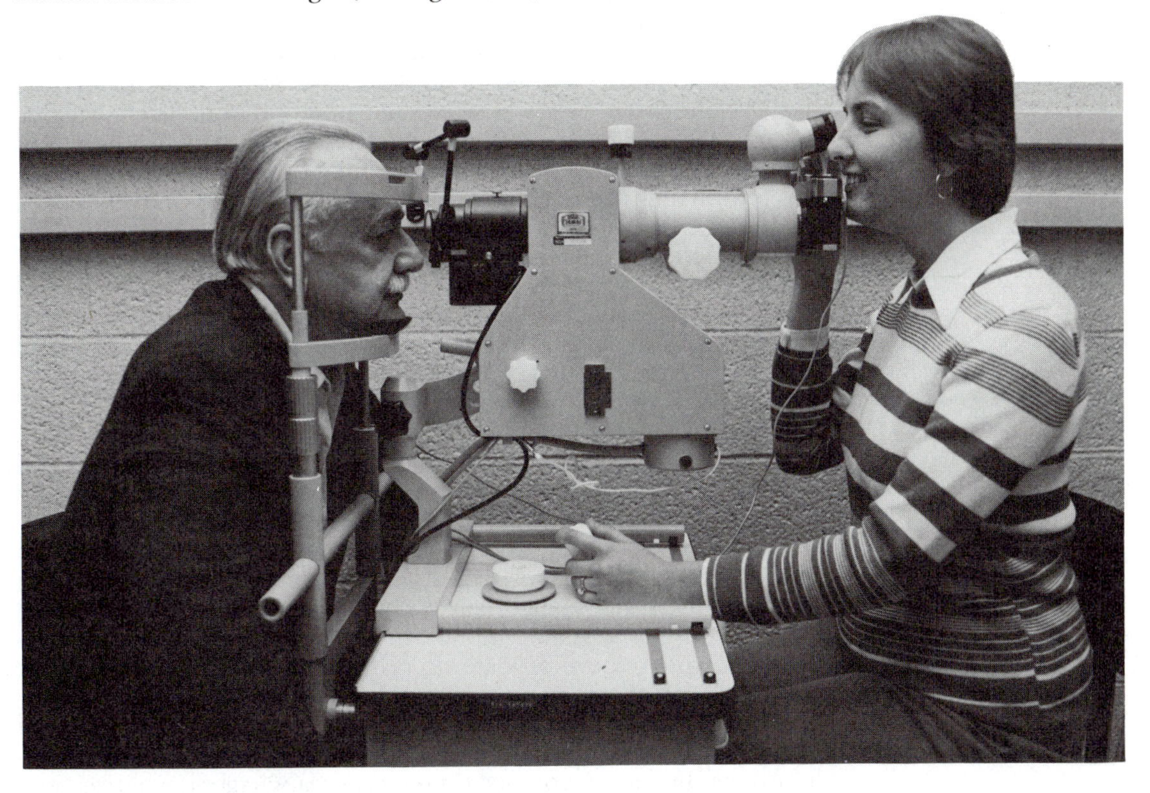

Figure 9–8
Research has indicated the possibility of both intellectual increments and declines in the adult years. Many persons take on tasks requiring greater use of their intellectual functions in midlife, whereas others use their abilities less and less.

Cattell (1963, 1971) proposed a model of the intellect that makes a distinction between two dimensions of mental abilities: fluid and crystallized intelligence. **Fluid intelligence** refers to capabilities such as associative memory, abstracting, inductive reasoning, dealing with figural relationships and problem solving. These capabilities are not as dependent on learning as they are on neurophysiological functioning and intactness of the central nervous system. **Crystallized intelligence** refers to skills such as verbal comprehension and the handling of word relationships, which are more dependent on learning and experience. Horn (1982) presented data suggesting that whereas fluid intelligence may diminish slightly following adolescence, crystallized intelligence may increase with advancing years.

It is difficult to detect any actual decline in intellectual performance in healthy adults before age sixty (Schaie and Hertzog, 1983). In fact, the experiences of life are often viewed as tools for improving intellectual performance over time. When work experiences require complex problem solving, adults do show more general intellectual flexibility in the middle decades (Kohn and Schooler, 1978).

Horner (1972) suggested that a fear of success may contribute to some decline in women's intellectual performance in adulthood. As cited in Chapter 8, Horner found that college women taking achievement tests could do well unless asked to compete with men, at which point they ceased to do their best. Some women's need to achieve drops far below that of men in adulthood. These women prefer the tasks of maintaining good relationships with their husbands, children, and others, and watching over the psychosocial aspects of living. Their affiliative needs may be more socially approved and rewarded. Young women who are more career oriented do not have a decreased achievement drive, nor do they fear success (Alper, 1974). Older women (in their forties and fifties) with high achievement needs have been shown to express an even greater independence and self-reliance than achievement-motivated women in their twenties (Erdwins, Tyer, and Mellinger, 1982). Erdwins and her colleagues found that achievement needs of traditional homemakers also increase with time, along with their affiliation needs. However, they want to achieve in areas characterized by conformity and cooperation rather than those characterized by self-reliance and independence.

Biologists feel that hormonal changes or the variation in the growth and development of the two cerebral hemispheres in the brains of men and women may account for some sex differences in adult intelligence. As mentioned in Chapter 7, the dominant cerebral hemisphere develops later in males than in females. The earlier development of the speech center, located in the dominant hemisphere, gives girls a head start in many verbal skills. Girls are more apt to enjoy studies related to the use of verbal skills (human relationships, literature, composition, foreign languages), whereas boys often prefer subjects that call for skills involving the handling of numerical or spatial relationships (mechanics, science, math). Hormones that may cause cyclic changes in moods of men and women at

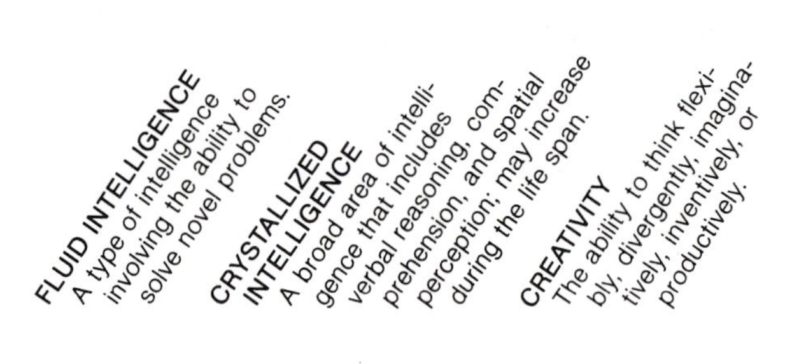

varying times of the month may also affect men's and women's intellectual abilities. Low levels of estrogens and androgens, known to contribute to feelings of depression, anxiety, and irritability in some people, may interfere with their concentration and the use of their other mental abilities.

Many women and some men return to school in their thirties and forties. Women find they have more time to study once their children enter the elementary grades. They may feel a need for more intellectual stimulation. Perhaps they did not finish high school or attend college earlier because of marriage. Occasionally women and men return to school not so much for intellectual stimulation as to enable themselves to obtain more interesting jobs. Whatever their reasons or combinations of reasons for returning to school, people in their thirties and forties often achieve higher grade point averages than younger students (Erdwins, Tyer, and Mellinger, 1982).

Creativity peaks in the thirties (see Figure 9–9). Lehman (1953), in a classic study of the lives of thousands of creative men and women, reported that the peak years for most creative ventures have been young. The mean peak years for symphony writing have been thirty to thirty-four. Great contributions to chemistry have been made most frequently between ages twenty-six and thirty. Mathematical revelations have come between ages thirty and thirty-four, and medical discoveries have been made by men and women most often between the ages of thirty-five and thirty-nine. Creativity and intelli-

Figure 9–9
During their thirties and forties many adults have a high output of fresh, creative ideas. Their productivity can enhance their careers and family lives and provide useful contributions to the community.

gence, as mentioned in Chapter 6, are not synonymous. Some highly creative individuals may have an average intellect. Creative persons usually continue to produce important works throughout their adult lives, even though they accomplish their more unique and original contributions early. In areas where creative endeavors require a larger amount of accumulated and systematized knowledge (such as medicine), the outstanding productions occur at slightly older ages.

Psychosocial Development

Havighurst (1972) suggested that the developmental tasks of middle adulthood are to:

1. Accept and adjust to physiological changes.
2. Attain and maintain a satisfactory occupational performance.
3. Assist children to become responsible and happy adults.
4. Relate to one's spouse as a person.
5. Adjust to aging parents.
6. Achieve adult social and civic responsibility.
7. Develop adult leisuretime activities.

Gould (1972) found the early thirties to be years when adults asked: "What is this life all about now that I am doing what I am supposed to do?" and "Is what I am the only way for me to be?" (p. 525). By the late thirties and early forties, the questions change to: "Have I done the right thing?" "Is there time to change?" (p. 526). By the late forties a "die is cast" feeling is present. By this time adults feel as if their personalities are pretty well set. Levinson and his colleagues (1978) characterized the thirties as a time for settling down and establishing a stable niche and the 40s as a time for reassessing one's life structure and then restabilizing into middle adulthood. He called the period of reassessment the *midlife transition*. We will discuss its implications later in this chapter.

It must be noted that although people may appear to be much the same in a given age range (or sex, or ethnic, or social group), they are in many ways quite different. Each individual has her or his own preoccupations, goals, tasks, hopes, dreams, problems, and ways of living.

ERIKSON'S CONCEPT OF GENERATIVITY VERSUS STAGNATION

Erik Erikson (1963) saw the nuclear conflict of the adult years as that of **generativity versus stagnation** (see Table 9–2). In Erikson's view, generativity involves more than just producing offspring. Mature adults who have achieved a sense of identity and intimacy begin to take a global view of their own lives. They ask "What is life all about?" Generativity involves answers that reflect some benefit to others. Adults can make a contribution to future generations through nurturing, teaching, and serving children or other adults. Producing offspring and guiding them through their nuclear conflicts of developing trust, autonomy, initiative, industry, and identity is an important aspect of generativity. However, Erikson wrote that one need not have one's own children. One can achieve generativity nurturing

Table 9–2 Erikson's Seventh Nuclear Conflict: Generativity Versus Stagnation

Sense	Eriksonian descriptions	Fostering behaviors
Generativity	Concern for establishing and guiding the next generation	Encouragement and devotion from progeny
	Concern for others including "belief in the species"	Others demonstrate their dependence on and need for the adult's guidance
	Concern for productivity and creativity as well as for progeny	Others reflect back the value of one's productions
versus		
Stagnation	Self-concern, self-indulgence	Early childhood impressions leading to excessive self-love
	Early physical or psychological invalidism, personal impoverishment	Feedback from others that own resources for generativity are doubtful
	Regression to obsessive need for pseudointimacy as opposed to real intimacy	

nieces, nephews, neighbor children, or school children. One can also accomplish a genuine sense of generativity through involvement with the welfare of future generations. Creative skills can be applied to improving the quality of life. Contributions can be made to civic and community causes. One can become a mentor to younger workers on the job. One can also nurture one's own aging parents. Any activity that has a positive influence on another can be viewed as a way to give back to the world as much as or more than one has taken from it.

Stagnation, in Erikson's view, involves a lack of productivity. Instead of leaving one's mark on the world in some contribution (children, work), one behaves parasitically. One takes without repaying. Stagnation is often associated with those persons of low self-esteem who believe any contribution they might make would be worthless anyway. Such persons almost vegetate, simply grinding through the minimal necessities of daily living. However, stagnation may also be applied to active, pompous, selfish individuals who work only to accumulate wealth and possessions for their own self-glorification. Such people exploit others and manage to shut out any thoughts of how they might provide for or give to others in return.

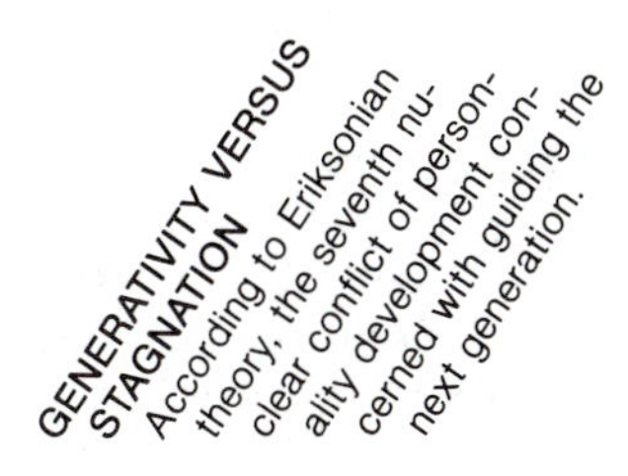

CAREER CONCERNS

Most men and over half of women pursue careers outside their homes in their thirties and forties. Their jobs become part of their identity. Vaillant, in an interview with one of his subjects, was told: "At twenty to thirty, I think I learned how to get along with my wife. From thirty to forty, I learned how to be a success at my job" (Vaillant and McArthur, 1972). Employees spend half of their waking hours at work and more time preparing for and getting to and from their places of employment. Careers certainly play a central role in social development of adults. Havighurst (1972) felt that attaining and maintaining a satisfactory occupational performance is one of the most important developmental tasks of these years. The work should not only be financially satisfying but also should be interesting to the worker.

People in many fields may devote considerable time in their thirties and forties to pursuing career advancement through night classes, journal reading, studying, experimenting, practicing, working overtime, or attending meetings. Professionals and nonprofessionals alike may concentrate on gamesmanship skills in hope of speeding up promotions: socializing, being seen at the right places, learning when to compete and when to cooperate, ascertaining when to comply with directives and when to ignore them, studying how to "win friends and influence people," and practicing authority roles. By and large the twenties are considered years for work preparation, job exploration, and settling in. The thirties and forties are considered years for work advancement (see Figure 9–10).

As reported in Chapter 8, young adults preparing to enter the work force want interesting work that uses their skills and abilities and has good advancement possibilities. By the thirties and forties many employees want other benefits. Work means different things to different people. Most people can pick and choose a few goals for their jobs. The hierarchy of order for their goals is apt to differ, however, even if people have the same goals. Here are some of many goals of employment:

Use of skills and abilities
Income
Security
Stability
Social affiliations
Prestige and status
Societal recognition
Enjoyable work
Serving others
Self-fulfillment
Challenging work
Worthwhile to society
Interesting work
Pension benefits
Insurance benefits
Convenient hours
Competitive work
Accomplishment at work
Responsibility at work
Self-direction at work
Independence
Creative outlet
Predictable future
Chances for advancement

Terkel (1974), in the book *Working: People Talk About What They Do All Day and How They Feel About What They Do,* pointed out that there are vast differences among humans in the ways they

Figure 9–10
Women in their thirties and forties are increasingly pursuing full-time careers. They work outside the home for a variety of personal, social, and economic reasons. Their jobs have a profound impact (sometimes positive, sometimes negative) on child-rearing, spouse relations, and household organization.

view their jobs. Job satisfaction is a feeling of contentment based on the number of personal and interpersonal goals that are being met by one's position and work. A quick cost accounting of job satisfaction can be done by jotting down goals and rating how well they are being met.

Not all people have the same attitude toward their jobs. Business is almost the only pleasure for some, while pleasure is almost the only business for others.

Satisfaction at work is influenced by expectations from others as well as by personal needs. One usually has a certain timetable in one's mind for work progression (for example, "I will probably be advanced to __________ level by the time I am __________"). One's spouse, parents, or friends may have radically different expectations. Associates may press for salary increments or work promotions when the individual would be quite content to stay put. It may be difficult for a person to feel job satisfaction without some visible signs of upward mobility if peers are being promoted. Campbell (1981) reported that the role expectations of significant others, and one's salary vis-à-vis one's peers and co-workers, are usually more significant factors in satisfaction than the nature of the role itself. Clausen (1981) suggested that blue-collar workers may be less affected by a lack of upward mobility because they have less personal investment of their identities in their jobs.

Promotions and advancements are usually greatest relatively early in a career. An employee usually settles in at an apprentice level in his or her first years of work, then gradually takes on more responsibilities. By the thirties and forties, some workers will take on a role of **mentor**; they will teach, advise, or sponsor a younger employee. When the younger employee no longer needs (or wants) the mentor, the middle-level employee may sponsor a new worker or may move further up the technical ladder and exert an influence on the functioning of the entire company (Graves, Dalton, and Thompson, 1980).

It is possible for career upward mobility to have a negative as well as (or instead of) a positive effect on social and emotional well-being. One may lose one's friends when promotions put a gap between social status standings. One may also lose one's sense of worth and achievement with a promotion. Peter and Hull (1969) in *The Peter Principle* stated the premise: In a hierarchy every employee tends to rise to his or her level of incompetence, and a corollary: In time, every post tends to be occupied by an employee who is incompetent to carry out his or her duties. Peter and Hull suggested that people who do a job well should stay at the job, not be rewarded upward to something they cannot handle. Although many management experts admit the occasional operation of the "Peter principle," they deny its pervasive influence and destructive effect.

Many women who seek outside jobs can find openings only in tedious, low-paying service or clerical work. As reported in Chapter 8, the gap between men's and women's earnings remains wide. The average male high school dropout often earns more than a female college graduate. Even though the government has been more progressive than the private sector in eliminating pay inequalities, equity has not been achieved. In the federal merit system, women make up 78 percent of the employees in government service (GS) levels 1–4 and 62 percent of the workers in GS levels 5–8. In contrast, men make up 96 percent of the employees in GS levels 14–15 and 97 percent of the workers in GS levels 16–18 (Friss, 1982). If a woman's salary and advancement are in step with her immediate co-workers of the same ability and seniority, her career satisfaction is usually high. Many companies have a tight-lipped policy about discussion of salaries to prevent dissatisfaction based on knowledge of unequal pay for comparable work.

Government and big businesses may strip individuals of a sense of personal accomplishment and self-worth. Persons often become numbers or simply insignificant cogs in a computerized machine. Good management strives to give employees self-esteem, pleasure, and a sense of achievement at work while also generating a feeling of loyalty to the company. Such things as shares of stock, rewards for completion of units of work, and individual praise in newsletters help bolster employee morale. Negotiation with and consensus of

MENTOR
An older adult who acts as a counselor or guide.

workers when developing directives rather than having them dictated by the power structure also enhance job satisfaction.

FAMILY CONCERNS

The establishment of a family of procreation (marriage and children) is the conventional behavior that society assumes adults in their thirties and forties will follow. The Puritans punished young people who failed to marry and procreate. Although society is no longer so condemning, it does exert pressures on people to fit the conventional molds. Society rewards adults who marry and procreate with celebrations, gifts, income tax breaks, and ease in obtaining bank loans and credit. People worry about the unmarrieds and persistently try to find eligible mates for them. They pressure married but childless adults to hurry up and have children before they pass their years of fertility or youthful energy for child rearing. Duvall (1977) saw the years of the thirties and forties as the time in the family life cycle when the majority of adults are in the intermediate stages that she called families with preschool children, families with school children, and families with teenagers.

Parenting Many of the concerns adults have about children were presented in Chapters 4 through 7 on child and adolescent development. Children can bring a great deal of warmth, love, and laughter into a home (see Figure 9–11). They continually demonstrate their dependency and make adults feel needed. They provide adults with both verbal and physical affection. They can also be very amusing. Children can help fulfill adults' belonging and love needs, as expressed by Maslow (1970). Children also bring extra work and added

Figure 9–11
Family life, while abounding with challenges and frustrations, can be an enormously satisfying experience. Laughing, a kiss, and "I love you" can wipe away unhappy memories. Even small successes by a member can generate a triumphant warmth in family relationships.

financial burdens. The day-to-day provision of food, clothing, shelter, and discipline can be very trying. Some parents find that they appreciate their children most when they are absent from them or when the children are asleep. Roger Gould (1972), in his description of adult time zones, described the early thirties as much more difficult than the twenties. Parents feel weary. They want to accept their children for what they are becoming and not impose roles on them. At the same time they want to be accepted for what they are as adults, not for what they are supposed to be.

Parenthood is one of the most important roles people can take upon themselves in life. Yet it is a role for which society offers little formal preparation. People assume it comes naturally. There is an abundance of books, articles, and columns on child-rearing. Few printed resources, however, can answer the specific questions of individual parents who have unique children, special problems, and extenuating circumstances. Parents often think the pediatrician, who might be a good resource, is too busy handling life-or-death illnesses to answer questions. Psychologists are thought to be too expensive. Free or inexpensive clinics where staff can answer questions about child-rearing generally receive little publicity. So parents call their own parents, their friends, or their neighbors for advice. Maybe the advice is good, and maybe it is not, but parents may take it because it's all they have. Then, almost invariably, parents hear or read somewhere that what they did was wrong or questionable. Some parents can maintain their self-esteem and confidence that they are doing their best, given their own particular circumstances. Others may find that their years of child-rearing are fraught with feelings of guilt, anger, and self-doubt. Supportive counseling would help some parents deal with their feelings and be more comfortable in their roles as parents.

As babies grow into toddlers, they begin to test limits. Their pushes for autonomy frequently occur at the worst possible times of day. Their developmental milestones may be too early or too late according to the time that experts report they should occur. Young children interrupt or postpone all kinds of adult behaviors, from sexual intercourse to quarreling. One child seems to double the problems of organizing time and money. A second child seems to magnify the problems at least three- or four-fold. With one child, the intrafamilial problems and jealousies revolve around mother-father, mother-child, and father-child. With a second child, relationships double to child one–child two, child one–mother, child one–father, child two–mother, child two–father, and mother-father. Each subsequent child further multiplies and complicates the interrelationships (see Figure 9–12).

As children enter school, they begin to make demands on parents for money, unsupervised excursions with friends, and parental participation in school activities. They want trips to all the places their friends have been; they request fad clothes, junk foods, and all the things that they see so enticingly advertised on television. They also bring home to their parents new worries about grades, adequacy of the school system, disputes with teachers, and behavioral prob-

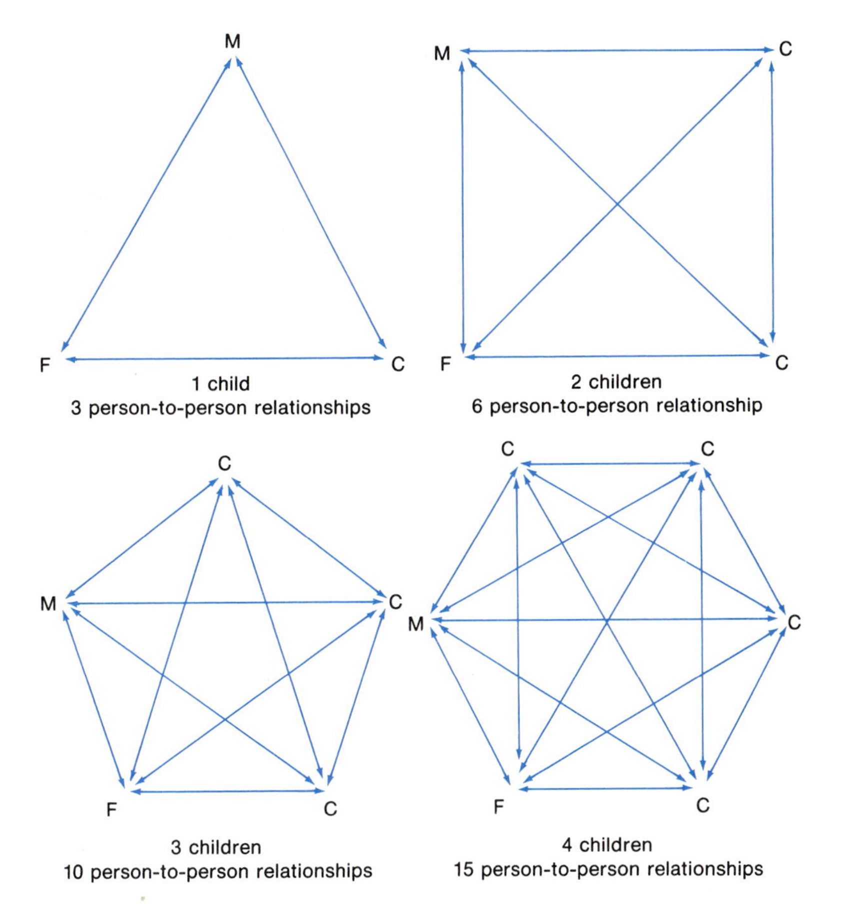

Figure 9–12
The number of different person-to-person relationships in a family increases dramatically with the addition of each new person.

lems that involve parents with neighbors, other parents, school personnel, community workers, or possibly even the law. Discipline—how much and what kind—becomes a primary concern. As with other child-rearing concerns, one can find abundant advice on discipline. Parents may try, at various times, reasoning, behavior modification, love deprivation, power assertion, indulgence, or neglect. The disciplinary techniques that become habitual depend on what the parents have read or heard, how they were disciplined as children, what they perceive as working best, and what their spouse, relatives, neighbors, and friends reward them for using. In order for the more effective inductive techniques of discipline to predominate over love withdrawal, power assertion, indulgence, or neglect, the parents must have positive feedback from the significant people in their environment. They must also be concerned about their children's internalization of moral rules and must be convinced that reasoning is the best way to help children develop conscience.

Mothering, when it means the exclusive care of young children without the help of the father, grandparents, or a babysitter, is extremely hard work. Our culture predominantly stresses the mother's responsibility for all child care. Consider the job: night feedings, diapers, meals, snacks, shopping, laundry, mending, cleaning, entertaining, comforting, assisting, directing, advising, chauffering, cheering, encouraging, motivating, and protecting from danger. The entertainments preferred by children are seldom really stimulating or amusing to "Mom." She may miss the hours she once had for work, rest and relaxation, sleep, beauty routines, social activities with companionable adults, and exclusive attention from her husband. The "happy homemaker" image of a wife and mother, so glorified by advertising in the mass media, imposes an unrealistic ideal. Mothers who remember that the actresses or models in the media are being paid to play a role can accept themselves better than mothers who measure themselves against the yardstick of the glorified Hausfrau. Likewise, mothers whose husbands, families, and friends value their efforts and achievements are usually happier than mothers whose associates criticize and devaluate their daily efforts and exertions.

For years, our society has placed greater emphasis and value on paid labor. Generally, this means that men's status is greater than women's, since men are the ones who earn more money. If a woman earns no money, she is afforded the least prestige. Research by Tavris (1972) indicates that the more children a woman has, the less satisfied she becomes with her homemaking role. Ragozin and her colleagues (1982) found that the older a woman is when she has her children, the more satisfied she is with mothering and the more time she commits to her role. However, as reported in Chapter 3, older mothers are at an increased risk of producing infants with chromosomal abnormalities or other congenital problems. Bernard (1972) reported that anxiety disorders are very high among housewives. They are nine times as likely to attempt suicide as women who are not housewives.

Women need to have the options of being able to go to work or stay at home with equal valuing and satisfaction. The same should hold true for men. In order to effect this kind of freedom for individuals, society must grapple with how it determines the value and labors of an individual—by the amount of money he or she earns or by other criteria.

Fathering is increasingly being recognized as an important role in the day-to-day socialization of children (see Figure 9–13). The social changes emphasizing equality of all peoples, regardless of sex, race, age, religion, or country of origin and the feminist movement have helped to break down some of our old masculine/feminine stereotypes about work. Hoffman (1977) found that while fathers rarely assume complete childcare responsibilities if a mother is present in the family, they do take a more active role in caregiving if the mother works outside the home. They can even admit doing so, in some social settings, without damage to their masculine egos.

Although "househusband" is not a label all men welcome, and to be "macho" (masculine, male, vigorous, with connotations of not doing women's work) is still very important to some men, times are changing. Chapter 3 cited the increasing numbers of American parents who attend prepared childbirth classes before their babies are

born. These classes strongly recommend the father's participation in the birth event. This is a good start for a father-child relationship. Rooming-in and relaxed hospital visiting arrangements for fathers further allow them to become familiar with their infants. The more a man interacts with his newborn, the more comfortable he feels providing food, clothing, and contact comfort later. Mothers learn in the same way, through observation and experience. If the father participates regularly in caregiving activities, the baby will form a strong attachment bond to him. The infant will smile and coo at him and later show signs of missing him (normal separation anxiety) when he departs.

Most studies of parenting are really studies of mothering. Studies of fathers tend to deal with fathering in lower animals, fathering in the past, fathering in other cultures, theories about father roles, father-mother relationships, and absent fathers rather than with contemporary father-child interactions and interaction effects. Research needs to take a longer, harder look at fathers. More good-quality research in this area is necessary, especially since fathers are increasingly asking, "What should I do to help rear my children besides earn money and mete out discipline?"

Single parenting (as discussed in Chapter 6) is difficult but can be done successfully and can be very rewarding. Boys raised without fathers can benefit greatly from having some interactions with a supportive male role model (an uncle, grandfather, teacher, neighbor, "big brother"), especially as they approach adolescence. Girls raised without mothers likewise benefit from close relationships with female role models.

Good parenting (fathering and mothering) becomes particularly important as children approach adolescence. Bronfenbrenner (1977) wrote that the junior high school years are the most critical in terms of a young person's development. Parents must give directions, advice, or assistance on such things as dating and sexuality, drug or alcohol use, independence strivings, and school problems. They may remember things from their own adolescence that they promised they would never inflict on their own children. Yet, due to the changing times, such promises now seem difficult to fulfill. The lifestyle of today's fortyish parents was quite different twenty years ago. Life was simpler, slower, and more predictable. The mass media were less pervasive. Fewer families had television, and it was usually black and white, with predominantly family programming, which was often obstructed with static and "snow." Public transportation was slower and less readily available. Cultural norms, mores and folkways, standards of conduct, etiquette, and moral codes—all were more clearcut. Norms today are in a state of flux and are often ambiguous. Parents can see or hear a full range of questionable language, sex, and aggressive behaviors exhibited by their associates, the media, their leaders, and their friends. Are children to do as grownups say, or as grownups do?

Parents often see their teenage children as marks of their own success. They may have nurtured dreams about their offspring's future for many years. It is not easy for them to realize that their hopes and dreams may not come to fruition. If they must settle for less, or something else, they try to assure themselves that the something else will still reflect well on them. Many parents cannot help but see their children as end products of their own parenting and reflections of their own adult worth. The pressures that such attitudes place on both parents and children are sometimes unbearable. Depression in middle-aged women is frequently related to overinvolvement and overidentification with children. Mothers may become bewildered by their sons' and daughters' behavior: refusing to obey, refusing to study, exhibiting temper tantrums, running away, having scuffles with the law, using drugs. They may feel helpless, despondent, and ultimately worthless as mothers.

Some parents in our society have a different problem with their teenage children: they see them as standing in the way of their own success. Family for them is simply an aggregate of persons living under the same roof. Parents may not want to be bothered with the social, emotional, or financial problems of their offspring when they are struggling so hard to make sense of their own lives. Mothers and fathers may resent their children receiving the complimentary glances that were once thrown their way. They may feel angered and threatened by teenage idealism that criticizes their own lifestyle. They also may intensely dislike having their authority questioned.

Parents who are overinvolved or overidentified, parents with a degree of resentment, and parents who reject their demanding teenage children all find it difficult to assist adolescents. Communicating with teenagers is best achieved by democratic, authoritative parents (see Chapter 6) who have kept an interchange of thoughts and opinions alive since early childhood (see Figure 9–14). Adolescents (as discussed in Chapter 7) need freedom to exercise their own judgment, within limits. They also need listening persons to turn to as

Figure 9–14
Many competent, able, compassionate parents find their adolescent children difficult to manage. Communication channels must be kept open, in spite of the grievances that seem to be ever present during these years.

sounding boards. Whatever the difficulties of parenting teenage children and assisting them to assume their own independent lifestyles, it is sometimes comforting to realize that, in time, most children do appreciate the problems their parents had in raising them.

Marital Relations Infidelity reaches a peak in the forties. Levinson and associates (1978) found that more than one quarter of the fortyish men they studied were actually involved in affairs, and most of them had fantasies of affairs and wondered about their commitments to their wives. The years when children are adolescents (usually the decade of the forties) are the years in which most married couples are least satisfied with each other. Approximately one quarter of all divorces involve persons over age forty (Jacobson, 1983).

Consider the factors that may contribute to rough going in marriage between a person's thirtieth and fiftieth year: signs of loss of youth, reproductive changes, health, career, and/or personality changes, coping with spouse's uncertainties, maintaining an economic standard of living, managing a home, and guiding children through all the conflicts of childhood and adolescence. Such increased marital stress leads to a decreased self-esteem that in turn leads to more marital stress.

Husbands often use their wives as scapegoats for their frustrations and lack of satisfactions with personal accomplishments. A husband might, for example, blame his wife for a vocational rut: "She held me back." "She pushed me too much." "She didn't keep pace with me intellectually." "She refused to be cooperative or gracious to my associates." He may somehow feel it is his wife's fault that he is losing his youthful appearance: "She feeds me all the wrong foods." "She interferes with my exercise program." "She causes my gray hairs." Or he might blame his wife for his children's shortcomings: "She was too permissive." "She was too busy with her own concerns to be a good mother." "She did not teach them to value money or property." Or a husband might criticize his wife for spending too much time with the children and not enough time with him. He may suspect her of having affairs or be having one of his own. There are many ways in which husbands blame their wives for their own problems. Some husbands take their frustrations out on their wives physically (see Box 9–2).

Meanwhile, wives, confused about their own life goals and accomplishments, may find their husbands at fault for their shortcomings, too: "He ignored me for his job." "He held me back and wouldn't let me develop my potentials." "He never felt my job was as important as his." "He never gave me enough respect." "He never helped me with the children or the house." They may suspect their husbands of having an extramarital affair or may be having an affair of their own. They may feel that their husbands are totally unsympathetic to their problems of household organization, fatigue, depression, child-rearing, loneliness, or approaching menopause. This is an especially rough period for women who have defined their role predominantly in terms of mothering. As children prepare to leave the home, some of these traditional homemakers are left facing a gap in their lives, a feeling of worthlessness.

Women who have continued their careers, combining them with marriage, also have many faults to find with their husbands at this time in their lives: "He always put his needs ahead of mine." "He was jealous or resentful of my successes." "He certainly did not contribute his fair share to household and child-rearing tasks." The happiness of dual-career marriages is strongly related to the husband's attitude about his wife's job. If he accepts her employed status, compromises with some adaptations to the demands of her job, helps at home, and gives her emotional support, the arrangement can add to rather than detract from marital happiness (Birnbaum, 1975). In gen-

BOX 9–2

Men Who Abuse Women

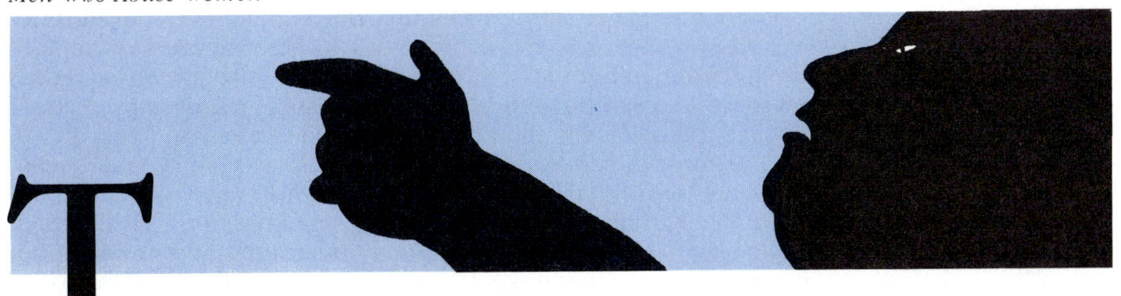

The doors are closed. Inside the home the male reigns supreme. Home is his sovereignty, his sanctuary. Here his women (wife, daughters, sisters, mother, grandmother) should defer to him and protect, nourish, and cherish him. If they don't, they must be disciplined and taught (again) intrafamilial male versus female roles. This bit of chauvinism is firmly entrenched in the minds of the majority of North American men, women, and children.

Guesstimates are that about one-half of all men resort to some physical form of discipline of at least one adult woman at some time in their lives. About 15 percent commit some violent act against their wives more frequently. Researchers have studied the differences between men who do not insist on being the supreme rulers behind the family walls of privacy, men who protect their home authority roles verbally rather than physically, and men who resort to abuse of women.

Abusive men exist in every race and in every social and economic walk of life. The wives of prominent doctors, lawyers, politicians, performers, businessmen, and the like are least apt to expose their victimization to public scrutiny. They remain silent becaue of terror, demoralization, embarrassment, shame or guilt, or confide only in private therapy. The wives of poor or unemployed husbands are somewhat more likely to report their assaults to the police or some other protective social service agency. However, many of them also remain silent. The tendency to downplay family violence, or lock trouble in, gives a false impression that few abusive men exist and that violent males are all down and out losers.

Males who perpetrate violence (throwing things, shoving, knocking down, kicking, punching, slapping, choking, and so on) against female family members (wives, grown daughters, sisters, mothers, even grandmothers) usually have several of the following characteristics: few close friends, social isolation, low self-esteem, an inner core of rage about not being given their proper share of something, a lack of trust in others, suspicion of others, anxiety, tension, frustration, insecurity, a belief that violence against women is justified, and a history of having witnessed intrafamilial violence (Queijo, 1984).

Efforts to demonstrate that abused women are masochists who have a secret desire for punishment have not been successful. Abused women do not fit any neat categories. They may be young or old, rich or poor, beautiful or homely, shy or sassy. Some fight back; these women are usually accused of being domineering, hostile, power hungry, and deserving of an occasional battering. Some seek immediate physical protection and take steps to terminate living arrangements with the abuser. Often they are counseled by lawyers, physicians, therapists, family members, religious leaders, or friends to try to work things out. Some women blame themselves, feel ashamed or guilty, and resolve to be still better wives (or daughters, sisters, mothers) to avoid the repetitions of violence. Often they are accused of being "doormats."

After violent acts most men feel guilty about what they have done and become very loving toward the woman they have hurt (Straus and Gelles, 1980). They usually promise never to repeat the behavior. The majority of women want to believe these reassurances. However, without a change in attitudes, social situations, self-esteem, security base, and the like, most men cannot keep such promises. Prevention requires personal counseling, family therapy, societal attitude changes, and enforcement of laws against beating anyone—even someone inside the family.

eral, evidence suggests that marriage is a happier state for men than for women in the decades of the thirties and forties (Pearlin, 1975). Even though men may view their marriage or family responsibilities as a hassle, once they are divorced or widowed they tend to replace the lost wife quite quickly (Reiss, 1980). The adverse consequences on social and emotional health from the loss of a spouse by death or divorce are greater in men than in women (Brown and Fox, 1979).

Divorce in persons approaching middle age is made more painful by the sense of failure after having spent so many years trying to make the marriage work and by fears of loneliness without a partner. Kelly (1982) found that self-doubt, mood swings, and problems adjusting to single living were common sequelae to midlife divorce. Postdivorce adjustment may take several years (Kitson, 1982). Bohannan (1971) wrote that a divorce really involves several facets: the legal divorce, the psychic divorce, the emotional divorce, the economic divorce, the community divorce, and the co-parental divorce (if the couple has children). While some people may perceive it as a growth experience, others may be left bereft and quite devastated by a midlife divorce.

If children are involved, the problems of divorce are often compounded. Current projections are that 50 percent of today's children will experience a separation or divorce of their parents. Children ask for and need to receive reasons for the change in their family. It is important that the reasons given for the divorce do not portray either parent in a bad light. It is also essential to reassure children, even adolescent children, that they are not at fault (see Chapter 6). Decisions about custody or financial responsibilities are often problematic. Once custody and visiting privileges have been settled, the absent parent should keep promises to visit as faithfully as possible. The custodial parent should help the children understand if visits are postponed and not use such occasions to try to turn the children against the absent parent.

Dating is problematic for divorced parents with children, especially teenage children. Although some children have a strong desire for a stepparent, they may also fear having one. They may fear losing their parent's love to the stepparent. Or they may feel jealous of the new person in their family (Einstein, 1979). Older children may make a parent feel embarrassed or guilty as they compare and pass judgments on the person who is replacing their absent parent (Gardner, 1977). Approximately 75 percent of women and 80 percent of men who divorce eventually remarry, nearly one-half of them within three years (Furstenberg, Spanier, and Rothschild, 1982).

Remarriage can also bring jealousies and competition between parent, stepparent, and children. Each new couple resolves in its own way the question of who is to discipline, how, and for what. Children may refuse to obey the stepparent. Bids for attention are frequent. White and Booth (1985) found that remarriages are usually as satisfactory as first marriages. However, stepchildren do reduce the quality of family life and decrease the quality of parent-child relationships. Remarried couples with children more frequently say they would not remarry if they had to do it over again. They also

DIVORCE Legal dissolution of a marriage.

move their teenage stepchildren out of the home faster than they would natural teenagers. With or without children, remarriage involves the same kinds of issues as first marriages—designing relations with in-laws, organizing the household around belongings, and making decisions about housework, leisuretime, and sex.

For those who survive the stresses of marriage during middle adulthood—and many do—marital satisfaction frequently climbs to a new high once the children are gone. Then husbands and wives may redefine their relationship in terms of friends, lovers, or companions rather than as persons trapped together with legal and psychosocial responsibilities toward their offspring.

Relations with Aging Parents In their thirties and forties most adults come to sympathize with their own parents and other aging people in a new way. Neugarten (1968) quoted the following statement from an interviewee: "My parents, even though they are much older, can understand what we are going through; just as I now understand what they went through" (p. 95). This change in the way some persons in their thirties and forties view their parents may lead to more confiding in them and more advice seeking from them. The younger adults can appreciate that the older adults understand their problems (see Figure 9–15). Cohler and Grunebaum (1981) suggested that close kinship ties between generations can be an important source of ego strength, identity, and feelings of personal congruence.

At the same time that the younger adults may acquire a new sympathy for their parents, their parents may turn to them for aid—social, financial, emotional, physical, or the like. This may present new problems for the younger generation. It is difficult to criticize

Figure 9–15
Individuals in their thirties and forties often find themselves giving advice and assistance to aging parents as well as seeking guidance.

one's own parents, realizing that time is running out for them. Yet Gould (1972) found that on occasion the adults he studied were quite vocal in their criticism of their parents. They blamed the older generation for many of their own life problems. Levinson's group (1978) found that having to assist one's own aging parents caused a great deal of emotional turmoil. The men he studied had to let go of their dependency on their own parents at this point. They had to redefine their relationships with their parents in terms of changed roles. The father, for example, could no longer be viewed as the man whom the interviewee had known as a child.

Cicirelli (1981) reported that younger generation adults are willing to provide health care assistance to their parents, but older generation adults most frequently request assistance with day-to-day services. A call to assist one's aging parents is a test of maturity that may arrive for some adults very early in life, for others very late in life, and for others not at all. How a person responds to the request for assistance depends not only on the person's maturity and moral values but also on the extent of the request, the previous relationship with the parents, other responsibilities, and other resources for the aging parents.

COMMUNITY CONCERNS

People in their thirties and forties define themselves not only in terms of their family successes or failures but also in terms of their position in the community. Common yardsticks against which individuals measure themselves are associates at work, neighbors, brothers, sisters, same-age cousins, former schoolmates, and friends. Although "keeping up with the Joneses" is a common theme of one's community relations, it is not the only concern. Many adults are kept so busy with projects for the good of their community or society that they have no time to worry about their status vis-à-vis others. Civic responsibilities can be viewed as all the ordinary affairs of social living: obeying laws, paying taxes, working, and maintaining a courteous demeanor toward other people. They also may include more specific roles such as voting, supporting causes for the common good, joining civic-minded groups, and running for public office.

Our culture tends to confer positions of leadership to persons of both advanced age and experience. By the time individuals reach their thirties or forties they may receive credit for their accomplishments by election or appointment to status or authority positions such as school board member, church board member, officer of an organization, member of a board of directors of a professional or service organization, or political leader. Many others accrue notable records of accomplishments in community activities of their choice without any desire for external recognition (see Figure 9–16).

Neugarten and Moore (1968) found that persons approaching middle adulthood become more strongly identified with their political party. They are also more apt to run for and get elected to political office. In general, the more responsibility the office carries, the older the person is who gets elected to fill it. At the state levels,

Figure 9–16
As adults become more efficient at home- and job-related duties, they often seek new challenges or opportunities for service. Many assume civic responsibilities or become involved in politics, education, the arts, religion, or volunteer activities.

Neugarten and Moore found that representatives were younger than senators, who in turn were younger than governors.

Women's roles in the community have changed a great deal in the past thirty years. In the 1950s women generally sought volunteer roles in the community after their children were grown. They worked in church groups, PTAs, charitable organizations, political parties, hospital auxiliaries, garden clubs, sewing circles, craft classes, and children's groups such as scouts or 4-H. Many women still do these things. Increasingly, however, they seek careers, elected offices, or paid part-time positions.

The decline of volunteerism among women may be due in part to the societal feeling that only paid work is valuable work. Women, like men, want to feel that their efforts are valuable. Much volunteer work is busy work—address typing, envelope stuffing, or similarly routine, monotonous tasks. When a task is not fulfilling in terms of meeting the needs to feel wanted, needed, goal-oriented, intellectually stimulated, or financially rewarded, a person is not as inclined to persist with the task. Many government and social agencies that once depended heavily on volunteers now have hired employees to do the work to avoid the confusion, tensions, mix-ups, carelessness, and absenteeism that characterized some volunteers.

The volunteer work that survives is the work that allows the performer to meet his or her needs for affiliation and usefulness in well-appreciated, serious jobs. There are now about 93 million volunteer jobs, about evenly filled by men and women, according to a 1983 Gallup poll. However, a schism exists in work assigned to volunteers. Men tend to get jobs at the policy level, while women work at the operations level, especially in political volunteerism (Towle, 1985). Women volunteers have made an enormous impact, especially in their historic preservation efforts, saving many houses and museums for posterity.

FRIENDS

Gould (1972) found that an interest in an active social life climbs in adults between the ages of thirty and fifty (see Figure 9–17). Both men and women show a greater interest in church and church-related activities. They join more clubs and participate in more family gatherings. They also tend to socialize more in career-related functions. This may be due to a desire to hang on to or improve a job or to bid for a promotion, or it may simply be because they want time away from home.

Many of the social, leisuretime pursuits of people in their thirties and forties show a decrease in energy expenditure over their activities of earlier years. They prefer less rigorous sports—golfing, bowling, hunting, fishing. They play more cards or tabletop games. Although some people seek new concerns and hobbies to get them "out of a rut," their interests and attitudes usually reflect their preferred undertakings in the past.

Friends can have a strong impact on each other's emotional adjustment. They do a great deal of data processing for one another. A person tells a friend of joys and sorrows, problems, and worries. A friend may respond in a way that is rewarding and has a favorable influence on a personality or in a way that is upsetting. The same friend may be helpful one day and unresponsive the next. Friends

Figure 9–17
Friends and social gatherings can be very important to persons in their thirties and forties.

tend to react to each other not only on information imparted and received but also with aspects of their own needs, desires, and motivations coming into play. Lynch (1985) wrote that heart-to-heart talking, listening, and otherwise responding to others not only enhances emotional health but also protects one's heart and blood vessels. He argued that the entire cardiovascular system benefits from human dialogue.

Harry Stack Sullivan (1963) based his theory of personality primarily on interpersonal relationships. He felt social approval is as great an influence on one's behavior as shelter, food, and sleep. One's self-concept is formed by the nature of one's relations with others. Rewarding interactions contribute to mental health, whereas upsetting relations lead to a disturbed personality and physical and psychophysiologic disorders. Social living necessitates contacts with many different people. A mature person with a stable self-concept can use communication skills to protect the self from conflict with others. An insecure person with a less well-consolidated self may not be able to escape so readily from the hostility of others.

Sullivan's theory provided the basis for much of the **transactional analyses** of social relationships currently practiced by psychotherapists. These are methods of examining transactions between individuals to determine how they perceive themselves and others. Many adults are insecure and face their friends with the attitude "You're O.K., but I'm not O.K." They continually seek advice and help from their associates. Therapists try to help them develop an "I'm O.K., you're O.K." attitude instead. Two other unhealthy perceptions underlying the behavior of some adults toward their friends are "I'm not O.K., you're not O.K." or "I'm O.K., you're not O.K." Harris (1967) identified the former as a give-up position and the latter as an antisocial position.

Ellis and Harper (1966) stated that rational living necessitates a consideration of interpersonal relationships to determine whether altercations are due to one's own behavior or to another person's problems. If a person's own behavior is immature, she or he should try to change. However, if the source of a conflict is the jealousy, vindictiveness, or greed of another, the mature rational adult should accept it as such and maintain his or her own feelings of adequacy and self-esteem.

Parlee (1979) surveyed 40,000 adults to determine what qualities are sought in friends. The characteristics desired most were confidentiality, loyalty, and warmth. Much less important were similarities in education, politics, income, and occupation (see Figure

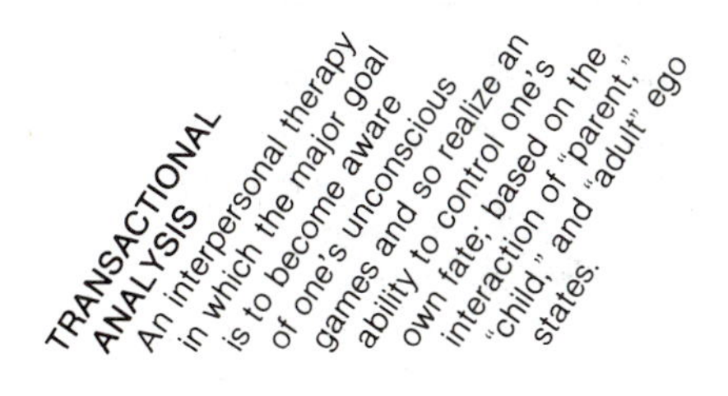

INGREDIENTS OF FRIENDSHIP

"HOW IMPORTANT TO YOU IS EACH OF THESE QUALITIES IN A FRIEND?"

Numbers represent percentage of respondents who said a quality was "important" or "very important."

Quality	Percentage
Keeps confidences	89%
Loyalty	88%
Warmth, affection	82%
Supportiveness	76%
Frankness	75%
Sense of humor	74%
Willingness to make time for me	62%
Independence	61%
Good conversationalist	59%
Intelligence	57%
Social conscience	49%
Shares leisure (noncultural) interests	48%
Shares cultural interests	30%
Similar educational background	17%
About my age	10%
Physical attractiveness	9%
Similar political views	8%
Professional accomplishment	8%
Abilities and background different from mine	8%
Ability to help me professionally	7%
Similar income	4%
Similar occupation	3%

Figure 9–18 *Ingredients of friendship.* SOURCE: From "The friendship bond," by M. B. Parlee. Reprinted from *Psychology Today Magazine,* copyright © 1979 Ziff-Davis Publishing Company.

9–18). Friends certainly play a central role in the psychosocial development of adults. They help determine self-concept, self-esteem, perception of and trust in others, and, to some extent, overall social demeanor in the community.

THE MIDLIFE TRANSITION

Most students of the middle adult years agree that a **midlife transition** occurs somewhere between thirty-five and fifty years of age. At some point the adult realizes that she or he is no longer young. This realization sets in motion a process of exploring the past, the present, and the future. One's self-concept in terms of one's current occupation and responsibilities are at the core of this review process. "Am I doing the right thing?" "Is there time to change?" The transition is often called the *midlife crisis,* suggesting the pain, the stress, the crucial decisions, and the alterations that accompany the transition. It is not easy to dispose of dreams that may never come to pass or to face the reality of what is now and what may be left of one's life.

Carl Jung (1923) was one of the earlier social scientists to describe a midlife transition. He believed that it is a crucial period of development. Here decisions for a lifetime must be made to assure a well-adjusted psyche. Jungian theory proposed that maturity does not evolve until the forties, after the trials of a number of life experi-

Figure 9–19
Jungian theory proposes that it is not until the forties that men and women come to terms with the outer world and begin to experience inner growth. This maturing brings a unification of ideas and greater self-realization, which Jung referred to as "individuation."

ences. Before this, adults work at divesting themselves of their childish ways. They face outer reality and come to terms with it. Then, after their forties (transition), they begin to discover their inner selves. They try to unify their former fantasies, hopes, and dreams with the actuality of their lives. Jung described the period of life after the transition as the **period of individuation**. It is only during these years, in his opinion, that adults finally move toward self-realization (see Figure 9–19).

Erik Erikson (1963) wrote that the nuclear conflict of trying to achieve generativity (bringing up the next generation—fathering or mothering children or younger adults) versus the opposite extreme, stagnation (lacking growth or personal contribution), is still the

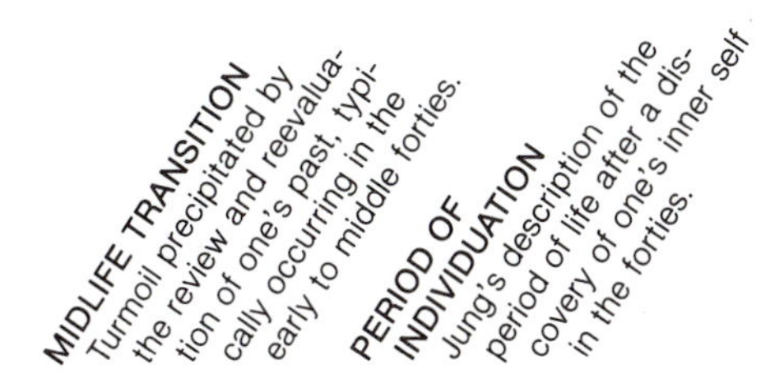

dominant life theme of most people in their midlife. Although Erikson does not spell out the details of a midlife transition, he implies that such a change occurs before and during the final nuclear conflict of life, "integrity versus despair." In this last stage the adult goes through a thorough soul-searching of "one's one and only life cycle."

Bernice Neugarten (1968), in her many studies of middle-aged adults, found that the transition may begin with a painful awareness of being viewed as old by youth. For example, in her investigations one respondent said, "When I see a pretty girl on the stage or in the movies . . . and when I realize 'My God, she's about the age of my son, it's a real shock. It makes me realize that I'm middle-aged" (p. 94). Another said "Mentally I still feel young, but suddenly one day my son beat me at tennis . . . " (p. 96).

Neugarten found that the midlife transition is clocked more by life contexts—body, career, family—than by chronological age. For example, the family cycle runs its course differently in various social classes. In one study Neugarten and Moore (1968) calculated that the wife of an average unskilled worker finishes school by fifteen, marries by eighteen, and has her last child by twenty-three. An average professional worker's wife finishes school by twenty, marries by twenty-three, and has her last child by thirty. For unskilled workers middle age may arrive in the early to middle thirties. For skilled workers there is usually some realization of whether one will become a supervisor or remain at a lower-level position by the middle to late thirties. A business or professional worker may have a clear picture of whether he or she can reach a desired salary or professional goal (or both) sometime in the forties.

Cultural pressures can also influence the age at which this self-assessment and reposturing occurs. Professional athletes must face a restructuring of their lives relatively early, usually by their mid-thirties. Lerch (1984) wrote that most of them view this transition as a form of social death. For example, Bradley (1977) wrote:

> For the athlete who reaches thirty-five, something in him dies; not a peripheral activity but a fundamental passion. It necessarily dies. The athlete rarely recuperates. He approaches the end of his playing days the way old people approach death. (p. 204)

However, athletes do continue to live. After the crisis of retirement, they restructure their lives and, unless they've earned enough to avoid subsequent employment, they move on to new projects, often in some way related to their sport.

Men and women share the problems of a midlife transition. However, the bulk of research on the stresses that occur emphasizes the male's transitions. The cross-sectional and retrospective studies of men done by Levinson and his colleagues (1978) led them to describe the stages of **settling down** in the early thirties, **becoming one's own man** (BOOM) in the late thirties, and **midlife transition** in the forties.

During the settling down stage younger employees often have

a mentor. The mentoring relationship usually evolves in the work setting. The mentor supports and facilitates his or her protégé's dream. The relationship lasts two to three years on the average, eight to ten years at most. It may end with moves or job changes on a cooled-off but still friendly basis. However, more often it ends with conflict. The younger employee enters the stage of BOOM. He or she may begin to see a mentor as tyrannical, destructively critical, and demanding or as standing in the way of his or her individuality and independence. The mentor, in turn, may see her or his protégé as inexplicably touchy, unreceptive, rebellious, and ungrateful.

Levinson and colleagues (1978) saw the BOOM stage extending from about thirty-six to forty-one. Employees attempt career advancements, a measure of authority, and less dependence on others. Their desires for affirmation and advancement make them especially vulnerable to social pressures during this time. They want desperately to be understood and appreciated. The culmination of the BOOM stage can be a great success, a failure, or, more frequently, a flawed success—good enough for others but blemished in the eyes of the dreamer. For about 80 percent of Levinson's subjects this period ended, and the midlife transition began, with a moderate to severe crisis. The adults felt a need to modify their lives, choose new paths, and redirect their futures.

Levinson found adults in this midlife transition anxious and fearful. They became irritated with others whom they perceived as trying to lead them, or push them, or hold them back. Their assessment was personal—something they had to do alone. Many suffered from depression during their transitions. Some tried to bury their troubles in various ways: overeating, alcohol, psychophysiologic disease. Some made radical changes in their lifestyles: leaving a job, leaving the family, having extramarital affairs. In fact, Levinson suggested that most transitional men tend to have fantasies about young, erotic girls and also about older nurturing women. He believed that this may still be related to an attempt to cut free from mother. No matter how the assessment process is manifest in a person's life, Levinson stated that it is normal. From his viewpoint, a person cannot go through this stage of life unchanged. Life will have a different meaning, and he or she will emerge on a new personality level.

Gould (1978), in his cross-sectional studies of adults, found that the early thirties involve processes of opening up. The major false assumption challenged is that "Life is simple and controllable. There are no significant coexisting contradictory forces within me." Any changes made are tentative and reflect uncertainties. However,

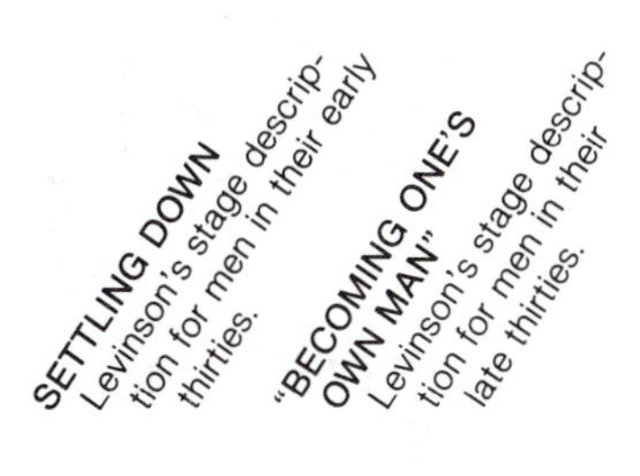

by ages thirty-five to forty-five, Gould found, adults see that life is neither simple nor controllable, and this new vision frightens them. They feel an inner demand for action and the pressures of passing time. They see the falsehood of five assumptions left over from the naiveté of younger adulthood:

1. The illusion of safety can last forever.
2. Death can't happen to me or my loved ones.
3. It is impossible to live without a protector (women).
4. There is no life beyond this family.
5. I am an innocent.

Gould found that this reappraisal (an acceptance of death, an acceptance of the "hoax" of life, a recognition of evil within and around) caused periods of passivity, rage, depression, and despair (midlife crisis). However, letting go of childish desires for absolute safety and innocence brought about a new understanding of the meaning of life—uncontaminated by the need for magical solutions and protective devices. Gould found that his subjects emerged from this midlife transformation with more access to their innermost selves and with an adult sense of freedom. By the late forties a "die is cast" feeling was present and was seen as a relief from the internal tearing apart of the immediately preceding years.

Vaillant (1977), in his longitudinal study of men, found behaviors similar to those described by Levinson and Gould in the decades of the thirties and forties. He found the thirties to be years for career consolidation. In order to achieve success in the work place, the younger adult is aggressive and assertive. He or she uses the defenses of rationalization and projection to escape inner feelings about career thrusting and maneuvering. By age forty (give or take a few years), workers leave the compulsive busywork of their occupational pursuits and once more become explorers of the world within. Vaillant, however, felt a "crisis" was the exception, not the rule. He saw the growth and change of the forties as a normal progression in the life cycle. Dissatisfactions with careers often reflect desires to be of more service to others. Behaviors that are discontinued are often those that were ill fitting to begin with. By the late forties many of the men Vaillant studied seemed to have experienced a sort of rebirth from bland, colorless characters to vibrant and exciting men. They no longer blamed others for their career ruts and felt in control of their own destinies. They typically were more committed both to their careers and to their personal relationships after their rebirth.

Women may experience midlife transitions that are more closely related to their roles and responsibilities in younger adulthood. Professional career women, like professional career men, usually ascertain by some time in their forties whether or not they will attain their desired goals and status and then alter their directions or alter their self-perceptions accordingly (see Box 9–3). The single woman may have to adjust to the knowledge that her chance of

BOX 9–3

Samples of Feelings Accompanying or Following the Midlife Transition

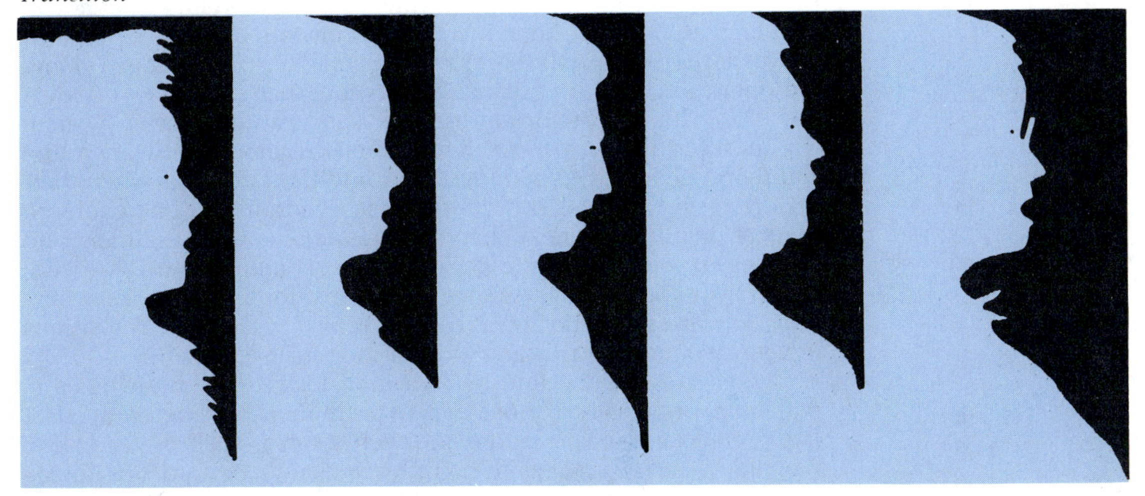

1. "There is a difference between wanting to *feel* young and wanting to *be* young. Of course it would be pleasant to maintain the vigour and appearance of youth; but I would not trade those things for the authority or the autonomy I feel—no, nor the ease of interpersonal relationships nor the self-confidence that comes from experience."

2. "I discovered these last few years that I was old enough to admit to myself the things I could do well and to start doing them. I didn't think like this before. . . . It's a great feeling . . ."

3. "You feel you have lived long enough to have learned a few things that nobody can learn earlier. That's the reward . . . and also, the excitement. I now see things in books, in people, in music that I couldn't see when I was younger. . . . It's a form of ripening that I attribute largely to my present age . . ."

4. "I know what will work in most situations, and what will not. I am well beyond the trial and error stage of youth. I now have a set of guidelines. . . . And I am practised . . ."

5. "It is as if there are two mirrors before me, each held at a partial angle. I see part of myself in my mother who is growing old, and part of her in me. In the other mirror, I see part of myself in my daughter. I have had some dramatic insights, just from looking in those mirrors. . . . It is a set of revelations that I suppose can only come when you are in the middle of three generations."

6. "I sympathize with old people, now, in a way that is new. I watch my parents, for instance, and I wonder if I will age in the same way."

7. "My parents, even though they are much older, can understand what we are going through; just as I now understand what they went through . . ."

8. "When I think back over the errors I made when I was 28 or 30 or 35, I am amazed at the young men today who think they can take over their companies at 28. They can't possibly have the maturity required. . . . True maturity doesn't come until around 45. . . . And while some of the young men are excellently educated, we middle-aged men are no longer learning from a book. We've learned from past experience . . ."

9. "I moved at age 45 from a large corporation to a law firm. I got out at the last possible moment, because after 45 it is too difficult to find the job you want. If you haven't made it by then, you had better make it fast, or you are stuck."

SOURCE: From B. L. Neugarten, The awareness of middle age, in B. L. Neugarten (Ed.), *Middle Age and Aging* (Chicago: University of Chicago Press, 1968), pp. 93–98. Reprinted by permission.

marrying is now low, that she will probably not have the status of wife and most probably not that of mother. The married professional career woman may experience some doubts as to whether she has done or is doing well in the dual roles of wife and professional or, if she has children, in the triple role of wife/mother/professional. Women who work full-time in skilled, semiskilled, or unskilled jobs will experience the same reevaluations of their lives ("Am I doing the right thing?") but perhaps somewhat earlier since they discover earlier the limits of their advancement and earning power. Women who devoted themselves to the homemaker/mother role in young adulthood generally experience their midlife transitions when their children leave home. (The "empty nest" syndrome will be explored in more detail in the next chapter.) The pain, confusion, stress, and changes that women undergo are often explained away as menopausal symptoms. However, women's transitions, like men's transitions, are also rooted in social and emotional aggregates. Transitions in many women occur ahead of their menopause.

The stress of the midlife transition leads some individuals to real crises or decisions to make changes in their lives before it is too late. Some midlife alterations are well known, such as Gauguin's moving to the South Seas to paint or England's Edward VIII giving up his throne to marry a commoner. Executives may give up positions of power to start their own small corporations, priests and nuns may leave their orders to marry, and the affluent may leave their riches and become involved in work for charitable causes. Change is not an uncommon phenomenon in midlife.

Behavior in midlife is often closely aligned to behavior in the earlier years. Woodruff and Birren (1972) followed a group of adults longitudinally for twenty-five years and found that people who were more neurotic in midlife tended to have been neurotic in college also. Those who were well adjusted in college had better mental health in midlife. Costa and McCrae (1980) also studied a group of adults longitudinally and found that very little change occurred in basic personalities over the years. The cross-sectional studies that reveal the phases of the adult life span should always be considered with an appreciation of the uniqueness and mental health of every individual approaching or going through the phase.

Summary

The typical ecological settings of the thirties and forties are the home, places of outside employment, and places for leisure pursuits and social activities. During these decades people struggle for their own place in the social order. At some point they reassess their aspirations and accomplishments, their status, and their life satisfactions. After this midlife survey some make changes. Others work harder at whatever it is they are doing.

Physically, signs of aging appear. However, these years can be years of near peak physical prowess. Many health problems are related to inability to handle the stresses of everyday living or to dietary factors or to both. Chronic illnesses affecting adults in their thirties and forties include coronary heart disease, atherosclerosis, hypertensive disease, and mental disorders. Acute respiratory, digestive, or urinary disturbances can turn into chronic problems.

Intellectually, the thirties and forties are very good years. Many people return to school or continue studying. The output of unique, original materials by creative persons peaks in the decades between thirty and fifty.

Erikson saw the nuclear conflict of the adult years as that of achieving a sense of generativity versus stagnation. Generativity involves nurturing, teaching, and serving children or other adults. One can achieve generativity without having one's own children through creative skills, productivity, and contributions to the quality of life of others and of future generations. Stagnation involves a lack of productivity or concern for others.

Careers are usually very important to persons in their thirties and forties. While the twenties were years for choosing a vocation, the thirties and forties represent the decades for becoming a success at one's job.

Parenthood can bring extra work, financial burdens, and feelings of guilt, anger, and self-doubt. Many parents need the support and reassurances of persons around them that they are competent in child-rearing. Children can also add warmth, love, and laughter to adults' lives.

Marital relationships are often troublesome. Spouses may be anxious about loss of youthfulness and about their reproductive slowdowns. Spouses' midlife reassessments may take them in different directions, creating problems of readjustment. Often parents have to deal with their children's identity crises as well as their own psychological changes. Research indicates that these are years when married couples are least satisfied with each other. Marriages are not as stable today as in past years. Divorce is made more complicated by the presence of children. A majority of divorced persons do remarry.

Aging parents often turn to their adult children for support—financial, emotional, and physical—as they reach their retirement years. Assisting aging parents can cause emotional turmoil. It is hard to realize that the persons on whom one depended for many years are now dependent instead.

Contributions to civic and community causes often peak in the thirties and forties. Persons may become more strongly identified with their political party, their religion, their social clubs. They may run for, or be appointed to, positions of leadership. The nature of social activities engaged in changes from the more active pursuits of youth to less rigorous sports—golfing, bowling, games—or social or family gatherings.

Friends are often chosen on the basis of aspirations for upward mobility, status, and job advancement. They also are apt to be people with similar concerns (religion, neighborhood, children). Closest friends are usually those with whom one dares to be more honest. Friends can create situations of conflict and stress or give support and reassurance to each other.

Most persons explore their inner lives sometime during their thirties and forties and restructure their outer lives accordingly. The midlife transition may involve a crisis or may be a relatively smooth transition from the dreams of younger adulthood to the realities of older adulthood.

Questions for Review

1. For a number of years the emphasis in American society has been on the maintenance of youth and the needs of the young. What sorts of implications do you think this has for people in their thirties and forties who are beginning to see signs of their own aging?
2. Describe some of the ways you think environment contributes to the high incidence of coronary heart disease in this country. Consider in your answer the social, emotional, and physical components of environment.
3. Individuals who seek help from psychiatrists often receive various kinds of labels for their condition. They might be called schizophrenic, depressive, or anxiety disordered. Describe the pros and cons of the use of such labels. Consider especially the effects on the individual.
4. Erikson stated that stagnation is often associated with persons of low self-esteem. Why do you think this is so? Relate this to the resolution of nuclear conflicts described in earlier chapters.
5. Imagine you have been married for 15 years, are 40 years old, and are going through a divorce. Would you turn to your friends or family for emotional support? Why?
6. Increasing numbers of married women are entering the labor force. What issues does this raise for men in terms of their role within a family? Include suggestions on how men may deal with these issues.
7. Describe some of the ways husbands and wives can avoid using each other as scapegoats for their failures and insecurities as they go through their lives together, especially during these middle decades of unrest.
8. Describe a midlife transition you have witnessed occurring in a friend or a relative. If possible, illustrate your example with conversations you have had with this person about or during this transition.

References

Alper, T. G. Achievement motivation in college women: A now-you-see-it-now-you-don't phenomenon. *American Psychologist,* 1974, *29*(3), 194–203.

American Heart Association. *Heart facts: 1985.* Dallas: American Heart Association's Office of Communications, 1985.

Andres, R., and Tobin, J. D. Endocrine systems. In C. E. Finch and L. Hayflick (Eds.), *Handbook of the biology of aging.* New York: Van Nostrand Reinhold, 1977.

Bardwick, J. M. *The psychology of women.* New York: Harper & Row, 1971.

Batten, M. Life spans. *Science Digest,* February 1984, 46–51, 98.

Bernard, J. *The future of marriage.* New York: World, 1972.

Birnbaum, J. Life patterns and self-esteem in gifted, family-oriented, and career-committed women. In T. S. Mednick, S. Tangri, and L. W. Hoffmann (Eds.), *Women and achievement.* Washington, D.C.: Hemisphere, 1975.

Bohannan, P. *Divorce and after.* New York: Anchor Books, 1971.

Bradley, W. *Life on the run.* New York: Bantam Books, 1977.

Breen, J. L. Premenstrual tension. *Medical Aspects of Human Sexuality,* 1981, *15*(6), 52.

Bronfenbrenner, U. Nobody home: The erosion of the American family. *Psychology Today,* 1977, *10*(12), 41–47.

Brown, P., and Fox, H. Sex differences in divorce. In E. S. Gomberg and V. Franks (Eds.), *Gender and disordered behavior: Sex differences in psychopathology.* New York: Brunner/Mazel, 1979.

Campbell, A. *The sense of well-being in America.* New York: McGraw-Hill, 1981.

Cattell, R. B. Theory of fluid and a crystallized intelligence: A critical experiment. *Journal of Educational Psychology,* 1963, *36,* 1–22.

Cattell, R. B. *Abilities: Their structure, growth and action.* Boston: Houghton Mifflin, 1971.

Cicirelli, V. G. *Helping elderly parents: The role of adult children.* Boston: Auburn House, 1981.

Clausen, J. A. Men's occupational careers in the middle years. In D. E. Eichorn et al. (Eds.), *Present and past in middle life.* New York: Academic Press, 1981.

Cohler, B. J., and Grunebaum, H. U. *Mothers, grandmothers, and daughters: Personality and child care in three-generation families.* New York: Wiley, 1981.

Costa, P. T., Jr., and McCrae, R. R. Still stable after all these years: Personality as a key to some issues in adulthood and old age. In P. B. Baltes and O. G. Brim, Jr. (Eds.), *Life span development and behavior,* vol. 3. New York: Academic Press, 1980.

Duvall, E. M. *Marriage and family development,* 5th ed. Philadelphia: Lippincott, 1977.

Einstein, E. Stepfamily lives. *Human Behavior,* 1979, *4,* 63–68.

Ellis, A., and Harper, R. *A guide to rational living.* Hollywood, Calif.: Wilshire, 1966.

Erdwins, C. J.; Tyer, Z. E.; and Mellinger, J. C. Achievement and affiliation needs of young adult and middle-aged women. *Journal of Genetic Psychology,* 1982, *141,* 219–224.

Erikson, E. *Childhood and society,* 2nd ed. New York: Norton, 1963.

Erikson, E. *Identity and the life cycle.* New York: Norton, 1980.

Friedman, M. Type A behavior: Its possible relationship to pathogenic processes responsible for coronary heart disease. In T. M. Dembroski (Ed.), *Coronary-prone behavior.* New York: Springer-Verlag, 1978.

Friedman, M., and Rosenman, R. H. *Type A behavior and your heart.* New York: Knopf, 1974.

Friss, L. Equal pay for comparable work: Stimulus for future civil service reform. *Review of Public Personnel Administration,* 1982, *2,* 39.

Furstenberg, F. F.; Spanier, G.; and Rothschild, N. Patterns of parenting in the transition from divorce to remarriage. In P. W. Berman and E. R. Ramey (Eds.), *Women: A developmental perspective.* Bethesda, Md.: NICHD/NIH (Publication No. 82–2298), 1982.

Gardner, R. *The parents' book about divorce.* New York: Doubleday, 1977.

Gilliland, B. C., and Mannik, M. Rheumatoid arthritis. In K. J. Isselbacher et al. (Eds.), *Harrison's principles of internal medicine,* 9th ed. New York: McGraw-Hill, 1980.

Goldstein, J. L., and Brown, M. S. Familial hypercholesterolemia: A genetic receptor disease. *Hospital Practice,* 1985, *20*(11), 35–46.

Gould, R. L. The phases of adult life: A study in developmental psychology. *American Journal of Psychiatry,* 1972, *129*(5), 33–43.

Gould, R. L. *Transformations.* New York: Simon & Schuster, 1978.

Graves, J. P.; Dalton, G. W.; and Thompson, P. H. Career stages in organizations. In C. B. Derr (Ed.), *Work, family and career.* New York: Praeger, 1980.

Gregerman, R. I., and Bierman, E. L. Aging and hormones. In R. H. Williams (Ed.), *Textbook of endocrinology,* 6th ed. Philadelphia: Saunders, 1981.

Harris, E. D., Jr. Evaluation of pathophysiology and drug effects on rheumatoid arthritis. *The American Journal of Medicine,* 1983, *75*(4B), 56–61.

Harris, T. *I'm O.K.—You're O.K.* New York: Harper & Row, 1967.

Havighurst, R. J. *Developmental tasks and education,* 3rd ed. New York: McKay, 1972.

Havighurst, R. J. The world of work. In B. Wolman (Ed.), *Handbook of developmental psychology.* Englewood Cliffs, N.J.: Prentice-Hall, 1982.

Hoffman, L. W. Changes in family roles, socialization, and sex differences. *American Psychologist,* 1977, *32,* 644–657.

Horn, J. L. The aging of human abilities. In B. Wolman (Ed.), *Handbook of developmental psychology.* Englewood Cliffs, N.J.: Prentice-Hall, 1982.

Horner, M. Toward an understanding of achievement-related conflicts in women. *Journal of Social Issues,* 1972, *28*(2), 157–176.

Ingbar, S. H., and Woeber, K. A. The thyroid gland. In R. H. Williams (Ed.), *Textbook of endocrinology,* 6th ed. Philadelphia: Saunders, 1981.

Jacobson, G. F. *The multiple crises of marital separation and divorce.* New York: Grune & Stratton, 1983.

Jung, C. G. *Psychological types, or the psychology of individuation.* New York: Harcourt Brace, 1923.

Kelly, J. B. Divorce: The adult perspective. In B. Wolman (Ed.), *Handbook of developmental psychology.* Englewood Cliffs, N.J.: Prentice-Hall, 1982.

Kitson, G. C. Attachment to the spouse in divorce: A scale and its application. *Journal of Marriage and the Family,* 1982, *44,* 379–393.

Kohn, M., and Schooler, C. The reciprocal effects of the substantive complexity of work and intellectual flexibility: A longitudinal assessment. *American Journal of Sociology,* 1978, *84,* 24–52.

Kral, J. G. Morbid obesity and related health risks. *Annals of Internal Medicine,* 1985, *103*(6), 1043–1047.

LaMont, J. T., and Isselbacher, K. Diseases of the colon and rectum. In K. J. Isselbacher et al. (Eds.), *Harrison's principles of internal medicine,* 9th ed. New York: McGraw-Hill, 1980.

Lehman, H. *Age and achievement* (Vol. 33, Memoirs Series). Princeton, N.J.: Princeton University Press, 1953.

Lerch, S. Athletic retirement as social death: An overview. In N. Theberge and P. Donnelly (Eds.), *Sport and the sociological imagination.* Ft. Worth: T.C.U. Press, 1984.

Levinson, D. J.; Darrow, C. M.; Klein, E. B.; Levinson, M. H.; and McKee, B. *The seasons of a man's life.* New York: Knopf, 1978.

Lynch, J. J. *The body's response to human dialogue.* New York: Basic Books, 1985.

Marshall, E. Diet advice, with a grain of salt and a large helping of pepper. *Science,* 1986, *231,* 537–539.

Maslow, A. *Motivation and personality,* 2nd ed. New York: Harper & Row, 1970.

Mathews, K. A., and Brunson, B. I. Allocation of attention and the type A coronary-prone behavior pattern. *Journal of Personality and Social Psychology,* 1979, *37,* 2081–2090.

McGuigan, J. E. Peptic ulcer. In K. J. Isselbacher et al. (Eds.), *Harrison's principles of internal medicine,* 9th ed. New York: McGraw-Hill, 1980.

Muller, J. Z. *Dysmenorrhea and premenstrual syndrome.* Washington, D.C.: U.S. Government Printing Office, 1985.

Neugarten, B. The awareness of middle age. In B. Neugarten (Ed.), *Middle age and aging.* Chicago: University of Chicago Press, 1968.

Neugarten, B., and Moore, J. The changing age-status system. In B. Neugarten (Ed.), *Middle age and aging.* Chicago: University of Chicago Press, 1968.

O'Hara, J. The depression mystery. *World Press Review,* June 1984, 29–30.

Parlee, M. B. The friendship bond. *Psychology Today,* 1979, *13*(4), 43–54, 113.

Pearlin, L. I. Sex role and depression. In N. Datan and L. H. Ginsberg (Eds.), *Lifespan developmental psychology: Normative life events.* New York: Academic Press, 1975.

Peter, L., and Hull, R. *The Peter principle.* New York: Morrow, 1969.

Poskanzer, D. C., and Adams, R. D. Multiple sclerosis and other demyelinating diseases. In K. J. Isselbacher et al. (Eds.), *Harrison's principles of internal medicine,* 9th ed. New York: McGraw-Hill, 1980.

Queijo, J. The paradox of intimacy. *Bostonia Magazine,* 1984, 21–25.

Ragozin, A. S.; Basham, R. B.; Crnic, K. A.; Greenberg, M. T.; and Robinson, N. M. Effects of maternal age on parenting role. *Developmental Psychology,* 1982, *18,* 627–634.

Reiss, I. L. *Family systems in America,* 3rd ed. New York: Holt, Rinehart & Winston, 1980.

Rosenman, R. H. The interview method of assessment of the coronary-prone behavior pattern. In T. M. Dembroski (Ed.), *Coronary-prone behavior.* New York: Springer-Verlag, 1978.

Schaie, K. W., and Hertzog, C. Fourteen-year cohort-sequential analyses of adult intellectual development. *Developmental Psychology,* 1983, *19,* 531–543.

Schoenfield, L. J. Diseases of the gallbladder and bile ducts. In K. J. Isselbacher et al. (Eds.), *Harrison's principles of internal medicine,* 9th ed. New York: McGraw-Hill, 1980.

Smith, D. W.; Bierman, E. L.; and Robinson, N. M. *The biologic ages of man,* 2nd ed. Philadelphia: Saunders, 1978.

Strandness, D. E. Vascular diseases of the extremities. In K. J. Isselbacher et al. (Eds.), *Harrison's principles of internal medicine,* 9th ed. New York: McGraw-Hill, 1980.

Straus, M., and Gelles, R. J. *Behind closed doors: Violence in the American family.* New York: Doubleday, 1980.

Sullivan, H. S. *The interpersonal theory of psychiatry.* New York: Norton, 1963.

Tavris, C. Woman and man. *Psychology Today,* 1972, *5*(10).

Terkel, S. *Working: People talk about what they do all day and how they feel about what they do.* New York: Pantheon, 1974.

Towle, L. H. The dilemma of the female volunteer. *Cornell Human Ecology Forum,* 1985, *15*(2), 24–25.

Vaillant, G. E., and McArthur, C. C. Natural history of male psychologic health: I. The adult life cycle from 18–50. *Seminars in Psychiatry,* 1972, (4), 415–427.

Vaillant, G. E. *Adaptation to life.* Boston: Little, Brown, 1977.

Van Itallie, T. B. Health implications of overweight and obesity in the United States. *Annals of Internal Medicine,* 1985, *103*(6), 983–988.

Walser, M.; Imbembo, A. L.; Margolis, S.; and Elfert, G. A. *Nutritional management: The Johns Hopkins Handbook.* Philadelphia: Saunders, 1984.

White, L. K., and Booth, A. The quality and stability of remarriages: The role of stepchildren. *American Sociological Review,* 1985, *50,* 689–698.

Williams, G. H.; Jagger, P. I.; and Braunwald, E. Hypersensitive vascular disease. In K. J. Isselbacher et al. (Eds.), *Harrison's principles of internal medicine,* 9th ed. New York: McGraw-Hill, 1980.

Woodruff, D., and Birren, J. Age changes and cohort differences in personality. *Developmental Psychology,* 1972, *6*(2), 252–259.

♥

The Fifties and Sixties

10

Chapter Outline

Physical Development
- *Physical Changes*
- *Reproductive Changes*
- *Health Considerations*

Psychosocial Development
- *Launching Children*
- *Stability of Personality*
- *Marital Relationships*
- *Relations with Adult Children*
- *Grandparenting*
- *Relations with Aging Parents*
- *Career Concerns*
- *Leisure and Social Activities*
- *Erikson's Concept of Ego Integrity*

Key Concepts

alloplastic mastery
autoplastic mastery
benign
cancer
carcinogen
cardinal traits
cataract
central traits
cirrhosis
climacteric
colostomy
common traits
complementary leisure
coordinated leisure
diabetes mellitus
distant figure grandparents
ego integrity versus despair
estrogen replacement therapy (ERT)
extraversion
formal grandparents
fun-seeking grandparents
glaucoma
hot flash
hysterectomy
individuation
interiority of personality
launching
leukemias
lung cancer epidemic
lymphomas
malignant
menopause
metastasis
neoplasm
neuroticism
omniplastic mastery
oncogenes
oncology
oophorectomy
openness
parent-surrogate grandparents
periodontal disease
personality stability
preparation and recuperation
preretirement planning
reservoir-of-wisdom grandparents
retinal hemorrhage
secondary traits
second honeymoon phenomenon
tumor
unconditional leisure

♥

Case Study

Marvin and Coretta had their last child when they were ages forty-two and forty, respectively. Consequently, their nest did not empty until they were in their sixties. Marvin had been in business for himself for many years. When his other friends had midlife transitions in their late forties and made subtle or radical changes in their lifestyles, Marvin kept doing the same thing each day.

When all of his children finally became financially independent, Marvin decided it was time for him to do what he had wanted to do for years. He went back to school. His wife was supportive of his decision, although many of his relatives and friends rebuked him. They told him that he was too old to start a new career or that he was crazy to give up a comfortable business.

For three years, Marvin and Coretta struggled while he both studied and continued his business on a part-time basis. When he completed his studies, he sent out résumés and interviewed for jobs everywhere. His friends laughed. Nobody wanted to hire a sixty-five-year-old man who had just finished school and had no experience in his new field. Nobody, that is, except a firm in Alaska. Marvin leased his business and went to Alaska. Coretta followed as soon as she could pack.

The new job lasted only one year. Again Marvin looked for work. At age sixty-six, he could not find another position. He eventually resumed his old work because he had only leased, not sold, the business. His four-year absence had hurt but not destroyed his clientele. Both Marvin and Coretta were glad they had made their little adventure, even if it hadn't worked out as well as they had hoped. Had they been foolish, or was there wisdom in their mid-sixties undertaking?

At some point in the decade of the fifties, most people think of themselves as middle-aged. Those in excellent health may consider themselves young. However, such hallmarks as children leaving home and grandchildren being born serve to remind people where they are chronologically, regardless of how they feel physically and emotionally.

The fifties and sixties are often years of peak status and power (see Figure 10–1). Past accomplishments may be recognized and rewarded with a measure of respect. Financial earnings are probably greater than they have been in the past, while expenses may be lower because of departure of children from the home. Persons in their fifties and sixties are often regarded as reservoirs of wisdom and good judgment. A vast number of them really are in the prime of their lives due to an emphasis on diet, exercise, physical fitness, stress management, and health maintenance.

Physical Development

The physical changes that normally occur in the fifties were once difficult to ascertain because of the high incidence of debilitating diseases that masked normal development. Now, however, because of improved medical techniques and an increased concern for protection from disease and environmental hazards (including unhealthful diets, stress, and lack of exercise), a larger number of normal people are surviving, and the changes that are due to aging rather than to illness are becoming more prevalent.

PHYSICAL CHANGES

Whereas neurons stop dividing in the prenatal days, other cells (like those of the liver, pancreas, bowel, and skin) continue to actively divide for a much longer time. There are conflicting theories about why some body cells eventually die without replacing themselves and why some lose their ability to function properly over time. There are probably several different causes for cell deterioration and cell malfunctions. Many biologists believe that the aging of cells, including structure and function, may be programmed in the cells' genetic material. Changes seen in the aging of cells include changes in the

Figure 10–1
The fifties and sixties are often years of peak earnings and great productivity. Men and women in this decade may experience a sense of achievement at work and pleasure and pride in their family and community relationships.

rate of protein synthesis and breakdown, changes in endocrine organ functions, and changes in cellular responses to many hormones.

The changes of aging are gradual and continue in the same directions described for the thirties and forties, except to a greater and greater extent. Muscles have diminished strength, size, and reflex speed; bones lose mass and density, break more easily, and heal more slowly; skin becomes less elastic, drier, and more wrinkled; nails grow more slowly; hair thins out and grays; vision and hearing become less keen; the sense of taste becomes less acute; digestive disturbances occur more frequently; fat accumulation is a threat unless eating habits are curbed; there are changes in the patterns and amounts of production of many hormones (see Table 10–1); voice changes; energy is diminished; and with each passing year it takes slightly longer to catch one's breath after exertion.

Considerable retarding of physical decline can be accomplished through regular exercise, elimination of excess calories from the diet, diminished cholesterol and saturated fat intake, adequate calcium and protein intake, cessation of cigarette smoking, curbed use of alcohol and drugs, and avoidance, whenever possible, of stress and environmental pollutants such as external radiation, emis-

Table 10–1 Changes in Hormones with Age

Hormone	Change*	Function
Antidiuretic hormone	↔	Regulates water reabsorption in the kidney
Growth hormone	↔	Stimulates body growth
Gonadotropins	↑	Influences growth and activity of ovaries and testes
Testosterone	↔ ↓ (?)	Masculinization
Adrenal androgen	↓	Masculinization
Estrogens (female)	↓	Feminization
Estrogens (male)	↔ ↑ (?)	Feminization
Thyrotropin (TSH)	↑	Stimulates growth and function of thyroid gland
Triiodothyronine (T_3)	↔	Involved in regulation of metabolic rate
Thyroxin (T_4)	↔ ↓	Involved in regulation of metabolic rate
Parathyroid hormone	↓	Regulates calcium, phosphate, and bone metabolism
Cortisol	↔	Involved in regulation of volume and composition of body fluids and in carbohydrate, lipid, and protein metabolism
Aldosterone	↓	Regulates body sodium and potassium levels (mainly in kidneys)
Insulin	↔	Regulates carbohydrate, lipid, and protein metabolism
Glucagon	↔	Counters action of insulin (for example, causes elevation of blood sugar)

SOURCE: Modified from R. Gregerman and E. Bierman, Aging and hormones. In R. H. Williams (Ed.), *Textbook of endocrinology,* 6th ed. (Philadelphia: Saunders, 1981). Reprinted by permission.

*Symbols: ↑ = increase; ↓ = decrease; ↔ = no change.

sions from internal combustion engines, exhaust fumes, smoke, and contaminated waters.

Intellectually, people in their fifties and sixties continue to add words to their vocabulary and organize and process new information they take in from the external environment. Baltes and Schaie (1974) reported that a reduction in motor skills may begin to interfere with certain aspects of functioning. Eye-hand coordination, for example, is not as acute as it was in earlier years, and reaction time slows due to a decline in the velocity of nerve impulses. A person's intellectual functions in middle age are still more dependent on motivation,

Figure 10–2
By the fifties and sixties many adults show signs of aging; for example, wrinkles and thin or graying hair. However, for most of them physical activities, intellectual activities, and life satisfaction remain high.

education, experience, intellectual activity, and intellectual stimulation than on age (see Figure 10–2).

REPRODUCTIVE CHANGES

The reproductive organs of both men and women begin to atrophy with advancing age, eventually rendering them sterile. Although records indicate that some men father children into their ninth decade and some women in their sixties bear children, the norm is for humans to cease bearing children in their thirties or forties. The end of the female reproductive cycle is relatively clearly marked by the menopause. The male climacteric is gradual and is less obviously concluded. Neither menopause nor the male climacteric need appreciably affect the sex drive. It may remain powerful throughout the life span.

Menopause The term **menopause** comes from two Greek words meaning month and cessation and refers broadly to all of the physiological changes that occur when a woman experiences a discontinuation of her monthly menstrual function. At the menopause the levels of two pituitary gland gonadotropins—luteinizing hormone (LH) and follicle-stimulating hormone (FSH)—are high. However, the ovary, which has been gradually shrinking since the late twenties, responds less to the pituitary hormones and produces less estrogen and progesterone of its own. There are fewer graafian follicles left to ripen and fewer ova left to be released. Menopause is thus a consequence of ovarian "failure."

For at least ten years prior to the cessation of the menses, a woman receives signals that her reproductive organs are changing. In most women, the time between menses gradually shortens. Thus, the menstrual cycle may take 30 days at age thirty, 25 days at age forty, and 23 days by the late forties (Whitbourne, 1985). The mean age for menopause is fifty years, but the age of onset varies widely. It may occur as early as forty-two or as late as sixty (Gregerman and Bierman, 1981). In some women, the time between menses may lengthen. Most women experience more irregular and more anovulatory cycles prior to menopause. They also more frequently develop mild to severe premenstrual distress. When the amount of estrogen/progesterone produced by the few remaining graafian follicles/corpora lutea is so diminished that the endometrium of the uterus is no longer stimulated to prepare it to receive a fertilized ovum, menstruation ceases.

Notman (1980) suggested that a more appropriate term for the menopause would be "the perimenopausal years" since the changes that occur are gradual, not abrupt. An early menopause may occur as the result of an **oophorectomy** (removal of the ovaries), a **hysterectomy** (removal of the uterus), prolonged nursing, poor health, living in an extremely cold climate, or excessive exposure to radiation. The age of onset of menopause also has a link to genetic inheritance.

The decreased levels of estrogen accompanying menopause (or oophorectomy) cause some vasomotor and other physical changes in some, but not all, women. One of the more common vasomotor symptoms of menopause is the **hot flash**, also frequently referred to as the hot flush. The body becomes warm and flushed, usually from the breasts up. Perspiration may be followed by chills. The flush may last from a few seconds through several minutes and may recur several times a day. Hot flashes are more common during the months of missed periods and in the evening hours or at night. More rarely, women may experience nausea and vomiting, constipation or diarrhea, gas, frequent or painful urination, low backache, an appetite change, headaches, dizziness, heart palpitations, numbness

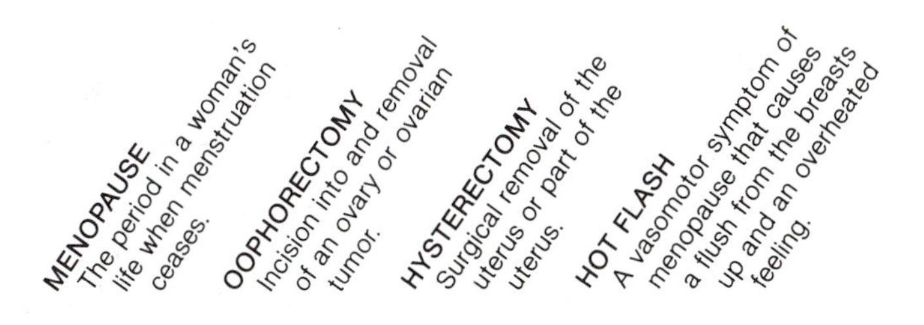

in the fingers or toes, breast tenderness, depression, anxiety, and insomnia during the perimenopausal period.

While the majority of women experience symptoms of menopause for only a year or two, some may continue to have vasomotor instability (hot flashes or night sweats) for up to eight years following the cessation of their menses. **Estrogen replacement therapy** (ERT) can ameliorate these episodes. Medical opinion differs as to the desirability of giving estrogen, however. Some doctors do not want to interfere with nature. Some give estrogen only for major problems with vasomotor changes. Others give estrogen replacement daily as soon as unpleasant menopausal symptoms begin. Some continue it for several years. Women continue to produce some estrogen from the adrenal glands or by conversion of adrenal androgen in the liver even after menopause. ERT supplements the naturally produced hormone, bringing it up to a level that provides relief from symptoms related to insufficient hormones.

ERT may be useful in reducing the incidence of osteoporosis (a disease characterized by a porous condition of the bones). There has been much controversy over whether estrogen therapy also increases the risk of gallstone formation, elevated blood pressure, or cancer of the endometrium (the tissue that lines the uterus). Recent evidence strongly suggests that there is a risk that depends on the amount and duration of estrogen use (Gregerman and Bierman, 1981). Shapiro and his colleagues (1985) reported that use of ERT not only increases the risk of endometrial cancer during replacement therapy but also increases the risk for up to ten years after estrogen use is discontinued. Wilson, Garrison, and Castelli (1985) reported that long-term ERT also puts women at a 50 percent elevated risk of death from cardiovascular disease and at a twofold risk of death from cerebrovascular disease. Increased rates for myocardial infarctions were also observed, particularly among estrogen users who smoked cigarettes. A study of nurses done by Harvard Medical School did not find that ERT increased the risk of fatal heart disease (Stampfer et al., 1985). However, the cutoff age for the nurses studied was fifty-nine. Perhaps further longitudinal studies will reveal increased risk of cardiovascular disease in the estrogen-using nurses as they reach their sixties and beyond.

Currently between 2 and 3 million perimenopausal women in the United States rely on ERT to control hot flashes. They have been assured that it will reduce their risk of developing osteoporosis. They are sure to notice that supplemental estrogens also help preserve more youthful skin, hair, and breast tissue. The benefits of ERT have to be weighed against the risks of its use. These risks keep many physicians from prescribing estrogen.

The psychological changes of menopause are often grossly exaggerated. The menopausal woman was once expected to be tearful, unpredictable, forgetful, unattentive, frigid, and haggard. Yet the majority of postmenopausal women feel that menopause did not change them in any important way. Only about 50 percent of women are troubled by hot flashes or night sweats. The remaining 50 percent are not troubled. Young women are more concerned about it

Figure 10–3
A second honeymoon is sensed by many spouses following the years of midlife transition, menopause, and climactic changes. This is characterized by a renewed interest in loving, caring, and sharing activities.

than women experiencing it, perhaps due to fear of the unknown. Many women find an end to their menses a happy relief. Their sex life remains good or even improves after menopause (see Figure 10–3).

Many of the psychological symptoms once associated with menopause (anxiety, depression, and so on) are probably more related to other concurrent events in a woman's life. Woods (1982) found that women who are satisfied with their lives are less apt to report psychological reactions to menopause than are unhappy women.

Many women experience a renewed interest in sex (and in making their physical appearance more attractive to men) after the menopause. Masters and Johnson (1966) described this unleashed sexual drive as a **second honeymoon phenomenon** and attributed it in part to women's newly found freedom from fears of pregnancy. Social factors such as no children in the home, fewer tiring household tasks, and a desire for more companionship may also contribute to this new burst of sexuality.

Bachman and Leiblum (1981) found that the direction sexual interest takes is greatly dependent on the woman's premenopausal sexual activity and satisfaction. Some women may feel that menopause suggests an end to their sex appeal or sexual usefulness (reproduction). These psychologic factors may influence their sexual expression.

Male Climacteric The term **climacteric** comes from the Greek term for a critical time. There is a great deal of confusion and debate

about whether there really is a male climacteric and, if so, when it occurs. Many men go through a transitional period in their forties. They reexamine their lives, and some may make radical changes in the way they do things. The transition period may erupt in a crisis (such as leaving wife, leaving job, turning to drugs or alcohol) for some men. These emotional repercussions were discussed in more detail in Chapter 9. The question of whether these changes are brought about or influenced by the normal male reproductive changes that accompany aging is as moot as the question of whether psychological changes accompany menopause.

Androgen levels decline slowly with age. As the production of testosterone decreases, men may experience a delayed erection time, a reduced ejaculatory volume, fewer viable sperm, a small decrease in the size of the testes, and a gradual loss of facial, pubic, and underarm hair. These changes occur so slowly that they may go unnoticed or may be detected only in the seventh or eighth decades. A few men may have an abrupt onset of physical symptoms, suddenly becoming impotent. The latter occurrence is felt to be a reaction to aging triggered more by financial, social, familial, or marital problems than by hormones. Such impotence may last for a few weeks, a few years, or forever, depending on emotional health, physical health, and therapy. For most men, however, the ability to engage in and enjoy sexual activity continues into old age.

Some men do experience a gradual waning of sexual interests with age, uninterrupted by dramatic changes in their levels of sexual libido. This waning of interest varies greatly from man to man and reflects, in part, sexual stimulation received and past history of sexual responsiveness. Masters and Johnson (1966) reported that men in their studies with the most evidence of maintained sexuality in their later years were those who had had more sexual activity in their early years. The men with problems of sexual inadequacy and impotence usually had one or more of the following problems:

1. Monotony of a repetitious sexual relationship (usually translated into boredom with his partner).
2. Preoccupation with career or economic pursuits.
3. Mental or physical fatigue.
4. Overindulgence in food or drink.
5. Physical and mental infirmities of either the man or his spouse.
6. Fear of performance associated with or resulting from any of the former categories.

CIRRHOSIS
A disease of the liver characterized by degeneration, fatty infiltration, atrophy, and inflammation.

The cure rate for impotence in men over age fifty is high. Therapy successes indicate that a willingness to return to active sexual practice and an interested partner are more important than any hormonal, medicinal, or other physiological routines in restoring potency.

The physical changes that occur due to decreasing testosterone levels are accompanied by many other normal physical changes of the fifties and sixties (such as decreased muscle strength, wrinkles, graying hair, possible digestive disturbances, or fat accumulation). Potentially traumatic psychological events may occur (seeing a son or daughter become sexually mature, having new financial burdens appear). These events may cause depression, unpredictability, forgetfulness, loss of attention, irritability, headaches, insomnia, fatigue, loss of sex drive, appetite changes, flushes, or, in short, any of the symptoms once associated with female menopause. The male climacteric, when and if it occurs, may have physical and psychological causes compounding the hormonal effects just as the female menopause does.

HEALTH CONSIDERATIONS

The four leading causes of death of persons in their fifties and sixties are coronary heart disease, cancer, vascular disease, and accidents. The fifth leading cause of death is liver cirrhosis, usually related to a long-time abuse of alcohol (see Box 10–1). During the last few years

BOX 10–1

Alcohol and Cirrhosis

If one is not an alcoholic (does not have a current or past physical dependence on alcohol), one has nothing to fear from regular drinking. Right? What harm can there be in drinking as long as one does not become drunk and disorderly? The fact is that without technically being an alcoholic, a steady drinker can do severe damage to his or her liver.

Cirrhosis of the liver is a disease characterized by cell degeneration, scarring, tissue loss, and deformity of this vital gland. Cirrhosis interferes with the liver's many versatile and vital functions (protein production, sugar storage, bile formation, detoxification). Cirrhosis is often fatal. It is also often the result of chronic alcohol abuse.

The quantity and duration of drinking necessary to cause liver cirrhosis is unknown but usually involves daily imbibing over at least a ten-year period (LaMont, Koff, and Isselbacher, 1980). However alcoholic cirrhosis has been diagnosed in some patients by their early twenties. Some individuals have a low liver tolerance for alcohol and a genetic predisposition for alcohol-induced cirrhotic changes. Annual deaths due to cirrhosis of the liver are highest for black males (about 30 per 100,000), are similar for white males and black females (about 14 per 100,000), and are lowest for white females (about 7 per 100,000) (Jackson, 1982).

Early symptoms of cirrhosis include a firm and enlarged liver, jaundice, intermittent ankle edema, increasing weakness, and fatigability. The treatment is simple: a nutritious diet, added vitamins, and absolutely no further use of any alcoholic beverages for life. Early diagnosis and meticulous attention to diet will prolong life and prevent further liver degeneration. However, the prognosis is poor for persons who are diagnosed after the appearance of complications of liver disease or who refuse to stop drinking.

the general health of Americans has improved a great deal. The number of deaths from heart disease has declined but the number of cancer deaths has increased. There has also been a significant rise in the number of deaths from respiratory ailments. In spite of these increases, life expectancy continues to climb slowly. In a study of longevity among nearly 17,000 Harvard alumni, Paffenbarger and his colleagues (1986) found that death rates from all causes were about one-third lower in those who expended 2,000 or more kilocalories per week in exercise (walking, climbing stairs, sports) than among less active men, even without consideration of the presence of hypertensive disease, smoking, obesity, or early parental death. (These other factors all increase the risk of earlier death.) By the age of eighty, the amount of additional life attributable to exercise alone was more than two years.

In a similar study, Frisch, Wyshak, and their colleagues (1985) surveyed over 5,000 women who had graduated from ten American colleges between 1925 and 1981. Those who began athletic training early and remained lean and physically active had two and a half times less cancer of the uterus, ovary, cervix, and vagina, and two times less cancer of the breast, than their less active classmates. Many North Americans take these findings seriously and are now protecting their health with regular sports or exercise programs. They have also learned that such things as a healthful diet (with more lean proteins and green, leafy vegetables and less saturated fat) and a cutback on smoking and drinking can help them avoid costly medical bills.

All adults should remember that they need periodic immunizations against preventable disease. The American College of Physicians (1985) recommends that all healthy adults receive mid-decade birthday (forty-five, fifty-five, sixty-five) booster doses of tetanus and diphtheria toxoids. In addition, health care personnel should receive an annual immunization against influenza, and persons who have frequent direct contact with blood or infected tissues should be immunized against hepatitis B.

Chapter 9 discussed the causes and ways to avoid coronary heart disease and vascular disorders. It is reemphasized that exercise, dietary discretion, and avoidance of smoking are foremost as preventive measures. The other main causes of death and disability in the younger old (persons in their fifties and sixties) are cancer and accidents. We will discuss cancer causation and common forms of cancer in the following pages. Diabetes, visual disorders, and dental problems, which affect many persons in these decades, will also be presented.

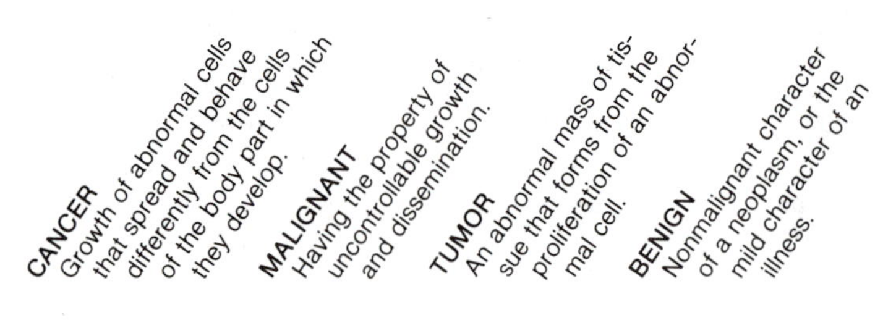

Cancer **Cancer**, from the Latin "crab," is a general term used to describe many diseases that are characterized by an abnormal transformation of a cell and an unlimited proliferation of the same abnormal cell. Because of exponential growth, one abnormal cell advances from 2 to 4 to 16 to 256 to 65,536 abnormal cells, and so on, rather rapidly. The word **malignant** refers to the immortality of the abnormal cells—their ability to continue to grow and invade normal tissue without ceasing; from 65,536 cells to 4,294,967,296 cells to the next exponential leap, and so on. The word **tumor** refers to an abnormal mass of tissue that may form from the proliferation of an abnormal cell. Tumors are not inflammatory. The suffix *-oma* also refers to a tumor (as in lymphoma). If the tumor ceases to grow, or grows very slowly and does not invade normal tissue, it is called a **benign** (as opposed to malignant) tumor, and it is not considered cancerous. Some forms of cancer (such as leukemia) do not form tumors (masses). The term **neoplasm**, from the Greek *neo* for "new" and *plasma* for "thing formed," refers to any new growth of cells or tissues, or any cancer. However, it is generally used to refer to a malignant tumor. **Metastasis** means the movement of the abnormal proliferating cells from the part of the body in which they originated to another. Metastasis has occurred, for example, when a brain tumor results from a primary lung neoplasm, or when a liver neoplasm results from a primary ovarian tumor. **Oncology**, from the Greek "onkos" for mass, is the study of the causes, properties, and treatment of neoplasms.

A revolution is currently taking place in oncology. For years, the search for the cause of cancer (or different causes for different neoplasms) concentrated on factors external to the human host of the disease. Many **carcinogens** (cancer-producing substances, from the Greek *karkinos* for "cancer" plus *genesis* for "origin") were identified. Tobacco, radiation, some food additives, some industrial products, and some synthetic hormones, for example, are known to trigger cell transformations and then initiate proliferation of the abnormal cells. However, we all live, breathe, and eat in a world full of carcinogens. Why do carcinogenic agents cause cancer in some, but not all, people? Today, the search for the cause of cancer has shifted to internal factors. Scientists trained to study such diverse disciplines as ion transport and lipid biochemistry are joining in the search (Macara, 1985).

Oncogenes, cancer-causing genes, have been discovered in the cells of all humans. By late 1985, forty distinct oncogenes had been identified (Weinberg, 1985). As you read this, more will undoubt-

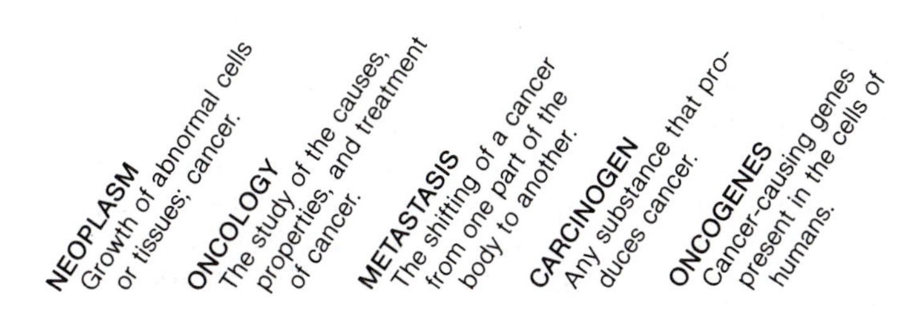

edly have been discovered. However, the properties of specific oncogenes and a full understanding of how they mediate abnormal cell change and growth may be years away. To say that oncogenes cause cancer is only a part of the story. Oncogenes must be triggered; they normally lie dormant within all cells. While we all have oncogenes as part of our genetic make-up, some of our normal genes may also be converted to oncogenes. How specific carcinogenic agents, or diet, or stress, or lack of exercise interact with or trigger specific oncogenes, or turn normal genes into oncogenes, is not currently well understood.

One avenue of exploration in discovering how oncogenes initiate cancer is directed at known fragile sites on human chromosomes (Yunis and Hoffman, 1985). All persons are born with a predisposition for their DNA to break under certain conditions (a role for stress, diet, or carcinogens?) during cell division, at certain weak points. A segment from a chromosome may break off at a fragile site and get lost, leaving a deletion or gap on the chromosome. The missing piece may have been essential to control the activity of the oncogene and keep it dormant. Alternately, segments from two broken chromosomes may trade places. This rearrangement may also allow previously dormant oncogenes to begin malignant transformations within the cell.

Another area of oncogene research explores how oncogenes give normally mortal cells immortality after inducing an abnormal cell transformation. For some cancers, two oncogenes seem to work in complementary fashion to create abnormal cells and to induce other genes to produce growth factors for proliferation, or they trick the cell into believing there are growth factors available for proliferation when there are none present. Alternately, in some cancers, a single oncogene may be able to make a cell tumorigenic and free it from growth factor dependence, giving it immortality.

Many other theories about tumorigenesis are also being actively investigated. Once the mechanisms of oncogenes are better understood, it may be possible to develop more effective treatment of cancer, or even to prevent it from occurring. This knowledge may also lead to insights into many other disease processes. Such knowledge, however, is probably years away.

At present, close to 400,000 Americans die from some form of cancer each year (see Figure 10–4). We will briefly mention the symptoms of and treatments for the most common forms.

The alarming incidence of lung cancer in smoking men and women has been referred to as the **lung cancer epidemic** (Tisi, 1980). Lung cancer is the most prevalent and among the most lethal forms of neoplasm. The good news is that its incidence in white males is decreasing. The bad news is that in females and nonwhite males, lung cancer rates are continuing at high levels. Smoking habits are credited with both the decline of lung cancer in white males and the increasing incidence in nonwhite males and all females.

LUNG CANCER EPIDEMIC The alarming, rapid increase in lung cancer in recent years.

Early symptoms of a lung carcinoma—a persistent cough, coughing up blood, difficult breathing—may be mistaken for other problems. These symptoms especially mimic the symptoms of

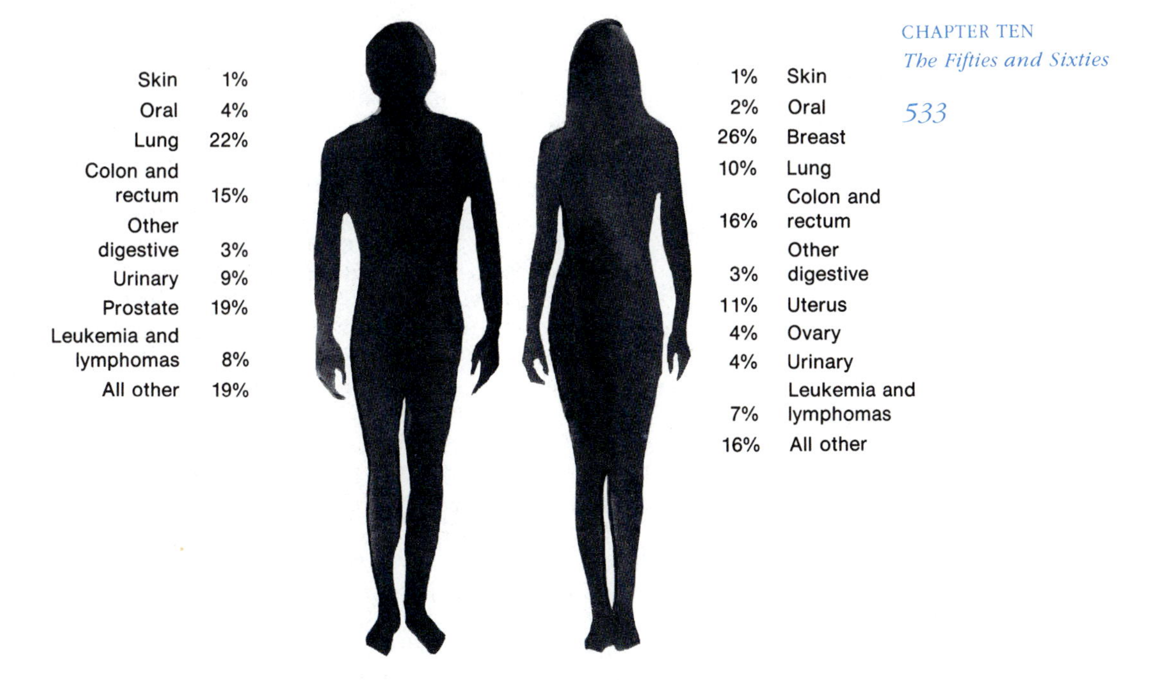

Figure 10–4
The American Cancer Society's estimates of cancer incidence by site and sex.

chronic obstructive pulmonary disease (emphysema, chronic bronchitis), which are quite common in smokers. Early diagnosis of lung cancer may be difficult, even with annual or semiannual chest x-rays. Researchers have sought new means of early detection. Experienced pathologists can detect early cases by examining the sputum of smokers for malignant cells. Late signs and symptoms of lung cancer include weight loss, hoarseness, and chest pain. Often a malignancy is diagnosed after discovering a metastatic lesion resulting from the spread of the lung cancer to another, usually distant, site. The great distance of metastasis is possible because cancer cells are carried by the bloodstream or lymphatic system.

Lung cancers are sometimes resected through a lobectomy (removal of one of the lobes of the lung). Inoperable pulmonary cancers may be treated with radiation and chemotherapy. The earlier the tumors are detected and treated, the better. However, the overall five-year survival rate for lung cancer is very low, only 13 percent. Most patients die within two years.

The American Cancer Society has produced a list of seven caution signals to alert persons to any possible cancer:

1. hange in bowel or bladder habits.
2. ny sore that does not heal.
3. nusual bleeding or discharge.
4. hickening or lump in breast or elsewhere.
5. ndigestion or difficulty in swallowing.
6. bvious change in a wart or mole.
7. agging hoarseness or cough.

Breast self-examinations and breast screening by mammography have helped to reduce the death rate from breast cancer. When detected early, malignant breast tumors can be completely removed. However, undetected breast cancer frequently metastasizes to the nearby lymph nodes, lowering the cure rate. Breast cancer is more common in certain groups of women than others. Seven high-risk groups are: (1) whites, (2) those who have had uterine cancer, (3) those with a family history of breast cancer, (4) those with previous breast disease, (5) those who have delayed child-bearing, (6) nonparous (nonchildbearing) women, and (7) those with early menarche (before age twelve) (Giuliano, 1984). Women who have normally lumpy breasts (fibrocystic breasts) can easily spot cancerous tumors if they become familiar with the pattern of their lumps. When one lump stands out as separate from the general pattern, a woman should have it medically evaluated.

Cancer of the cervix (opening to the uterus) has shown a marked decline in the last twenty years because of the widespread use of Pap smears. This examination detects abnormalities in cervical cells that can lead to cancerous growths. The cure rate is almost 100 percent with early detection. All adult women should have a Pap smear annually. Abnormal uterine bleeding and vaginal discharge may be early symptoms of cancer of the cervix.

Cervical cancer is more common in women who have had breast cancer, who have used ERT for over a year, who have chronic vaginal infections, who have borne many children, who have early, frequent, or numerous sexual partners, and who have herpes simplex Type II (see Chapter 7). It is less common in Jewish women and celibate women (Hill, 1984). If it is detected very early while it is still localized in the external cervix, it can be treated with laser beam therapy, cyrosurgery, or electrosurgical cauterization. If it has penetrated deeper, a hysterectomy (removal of the uterus) or radiotherapy is necessary. Physicians are increasingly suggesting that women over age forty who undergo a hysterectomy should also have an oophorectomy (removal of the ovaries). The two operations together are called total hysterectomy. Ovarian cancer is the fifth leading cause of female death (Woodruff, 1984). Only a small percentage of malignant tumors actually begin in the ovaries. Many cancers spread to the ovaries from the uterus, the breast, or the intestinal tract. Ovarian cancers are "silent," producing no symptoms until they are large enough to palpate. They are extremely difficult to diagnose before they metastasize to other organs, and the cure rate is low.

Cancer of the male prostate (a gland located below the bladder, surrounding the urethra, which secretes a lubricating fluid during sexual intercourse) may be confined to the prostate, may spread through the capsule but not metastasize, or may metastasize to the lymph nodes, bones, lungs, liver, or adrenal glands (Griffin and Wilson, 1980). Any extension beyond the capsule is associated with a poor prognosis (outcome). This cancer is rarely manifest in early adulthood but becomes increasingly common from the fifties onward. An early symptom may be some difficulty associated with urination caused by pressure on the urethra from the growing tumor. However, back pain due to metastasis to the vertebral column may also be the first symptom noticed. Surgical removal of the prostate gland is used to treat cancers that are small or still confined within the prostatic capsule. With demonstrated bone or extrapelvic involvement, radiation or chemotherapy may be used in addition to or instead of surgery. The cure rate is best if the cancer is detected and removed early. Hormone therapy is an important mode of treatment in this disease. These cancers depend on testosterone from the testes for their growth. Treatment may therefore consist of removal of the testicles (orchidectomy) and/or the use of estrogens, the feminizing hormones that induce a chemical castration.

Cancer of the large intestine (colon) accounts for about 20 percent of cancer deaths in the United States (LaMont and Isselbacher, 1980). Symptoms usually occur a considerable time after the cancer develops. They include changes in bowel habits, constipation, or blood in the stool. They are frequently ignored or considered normal bowel function disturbances. Many persons do not seek medical advice until they also experience weakness, loss of appetite, weight loss, lower abdominal pain, or obstruction of the bowel. If the tumor can be removed before it spreads to the regional lymph nodes, liver, or lungs, the five-year survival rate is high (75 to 80 percent). Cancer of the colon shows a markedly high intrafamilial occurrence rate. With a known family history of colonic carcinoma, adults should request periodic sigmoidoscopic examinations from middle age onward.

If the colon cancer involves the lower rectum, surgery usually involves removing the end of the colon (rectum), closing the anal opening, and establishing an artificial opening (**colostomy**) for the colon through the lower abdominal wall. A person with a colostomy must wear a small bag attached to this opening (stoma) to collect fecal materials that may be discharged at any time. Colostomy patients regularly irrigate their colons to evacuate the feces. Many establish once-a-day elimination patterns that make any appreciable amount of discharge into their bags a rare occurrence. If a colon cancer is higher up the colon, surgery may require removal of the portion of the colon around the tumor and only a temporary colostomy. After a few months such a colostomy can be reversed. The two portions of intestine from which the cancer was removed can be reunited so normal bowel function can resume.

Ostomy societies have sprung up all over the country to enable persons who have had artificial openings made into their gastrointestinal canals (colostomy, ileostomy) to share problems, solutions, and self-care techniques with one another. The comradeship of these groups often helps to alleviate the problems of withdrawal and depression that may plague a person after an ostomy operation (see Figure 10–5). If others can accept that an ostomy is not calamitous or even too unusual, the patient can usually return to his or her normal activities and lifestyle quite successfully.

COLOSTOMY The surgical operation of forming an artificial anal opening in the colon.

Cancer of the bladder or kidneys seldom occurs prior to middle

Figure 10–5 *Persons who have had an ostomy operation often find that others like themselves provide the most help and encouragement in readjusting to the normal activities of daily living.*

age. It is much more common in men than in women, and its occurrence has been linked especially to exposure to certain carcinogenic chemicals and dyes (Brenner and Humes, 1980). There are suggestions that heavy cigarette smoking may also contribute to an increased incidence of bladder cancer due to irritation from tobacco tars that may be excreted in the urine. Additionally, it has been suggested that coffee, artificial food colorings, synthetic sweeteners, or chronic irritation from certain bladder parasites may somehow trigger oncogenes to start tumor formation in the urinary collecting system.

Symptoms of bladder cancer may include blood in the urine, frequent urination, or painful urination. As with other cancers, the cure rate depends on the rapidity of the treatment after detection and the extent of any metastasis. Treatment may be radiation, fulguration (destruction of tissue by means of heat from an electric current), surgical resection of the tumor, or instillation of chemotherapeutic agents into the bladder at intervals. Many bladder cancers have a tendency to recur following excision, making the prognosis for a cure somewhat guarded.

Leukemias are malignant changes in blood cells originating in any of the blood-forming organs (bone marrow, liver, spleen, lymph nodes, thymus). An abnormally large number of immature and abnormal leukocytes proliferate in the circulating blood and elsewhere. Eventually, fewer normal blood cells are produced. As a consequence of the large numbers of cancerous blood cells, there is often enlargement of the liver, spleen, and lymph glands. The crowding out of normal cells by the cancerous cells produces a loss of red blood cell production leading to anemia and its attendant fatigue. A loss of the normal white cell response causes increased susceptibility to infection, and a loss of the platelet-producing cells leads to easy bruising and abnormal bleeding.

Many different forms of leukemia exist, involving any of the different types of blood cells. Researchers have recognized several probable contributing factors, including viruses, ionizing radiation, chemical agents, and genetic factors (Clarkson, 1980). The various leukemias are usually classifed as acute or chronic and are subclassified according to the predominant, abnormally growing cells. In the acute form symptoms develop rapidly; in the chronic form their onset is very gradual and may go unrecognized. Prior to currently available chemotherapy and bone marrow transplant, the categorizations into acute and chronic forms had some value in terms of predicting life expectancy. Now, patients with some acute leukemias may have their diseases brought into remission and may live longer than patients with some chronic forms of leukemia.

The different subclassifications of leukemia require different therapy programs, all aimed at reducing the number and controlling the rate of growth of abnormal cells. Adjunctive therapy may include antibiotics, blood transfusions, and radiation, aimed at controlling the complications of leukemia.

Lymphomas are malignancies arising in the lymph-cell-producing areas (lymph nodes, spleen, liver, and areas in the intestine). For unexplained reasons, the incidence of lymphomas appears to be increasing each year (Ultman and DeVita, 1980). They are the seventh commonest cause of death from cancer in the United States. The lymphomas include the lymphosarcomas and the forms of Hodgkin's disease. In lymphosarcoma, complaints and physical findings vary with the location of the tumors, which usually appear as large, firm lymph nodes matted together. Hodgkin's disease can take several forms. It is a chronic, progressive illness with the same organs involved as in lymphosarcomas but with a different course. Pruritus (itching), fever, anemia, and weight loss are all typical features. Treatment of both lymphosarcomas and the forms of Hodgkin's disease may include chemotherapy, radiation therapy, and sometimes surgical excision of isolated growths. The inherent nature of the tumor itself seems to be the most important factor in determining

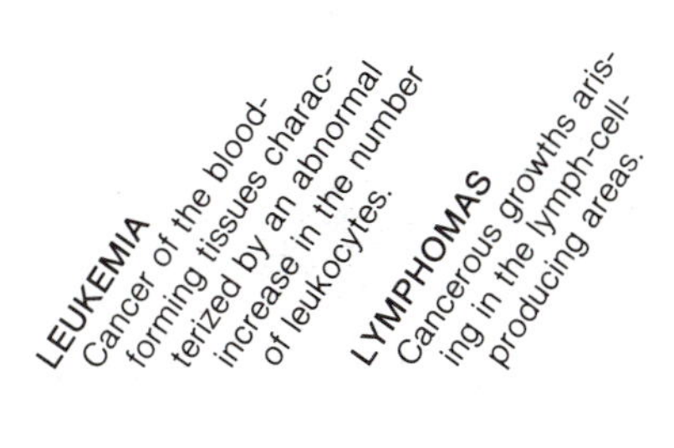

the course of these disorders. Some cases run chronic courses. Others run rapid, downhill courses despite treatment. In general, cases of localized disease with no fever or weight loss seem to have the best chance for long survival. For Hodgkin's disease, a combination program will now achieve a complete remission of symptoms for up to 80 percent of patients (Ultman and DeVita, 1980).

Current beliefs about multifactorial cancer causation have led experts to emphasize healthful living to prevent any neoplastic growths. Since tobacco is implicated in as many as 30 percent of all cancers, a major thrust of public health campaigns is to persuade people to give up smoking or chewing. Since obesity is implicated in cancers of the breast, cervix, ovary, prostate, colon, and gallbladder (Garfinkel, 1985), persons are advised to maintain weights within the desirable range for their sex, height, and body build. Diets low in saturated fats and high in fiber, whole grains, and green leafy vegetables are recommended. In addition, regular exercise should be a part of each person's daily activities.

Diabetes **Diabetes mellitus**, or simply diabetes, is one of the most common diseases of metabolism, affecting between 2 and 6 percent of the population (Foster, 1980). Diabetes increases in frequency with every decade of life.

The main feature of diabetes is a high blood glucose level (hyperglycemia) often associated with glucose in the urine. This hyperglycemia results from underutilization of glucose by the tissues and overproduction of glucose by conversion from other substances (mainly from amino acid subunits of protein and from liver glycogen). The central hormones associated with diabetes are insulin and glucagon, both products of the pancreas. Insulin has many actions, but its essential role is in promoting entry, storage, and utilization of glucose, amino acids, and fats in tissues. In its absence, in situations of tissue resistance to the effects of insulin, or in the excessive presence of its antagonist glucagon, glucose, amino acids, and fats cannot enter tissues. On the contrary, the muscle and fat tissues break down and release amino acids and fats into the blood. Many of these products then spill into the urine in large amounts to the detriment of the body. The condition is really one of starvation of the body's cells.

In Chapter 6 insulin-dependent diabetes was briefly discussed. It should be pointed out now, for clarification, that in the insulin-dependent form of diabetes there is a total or almost total lack of

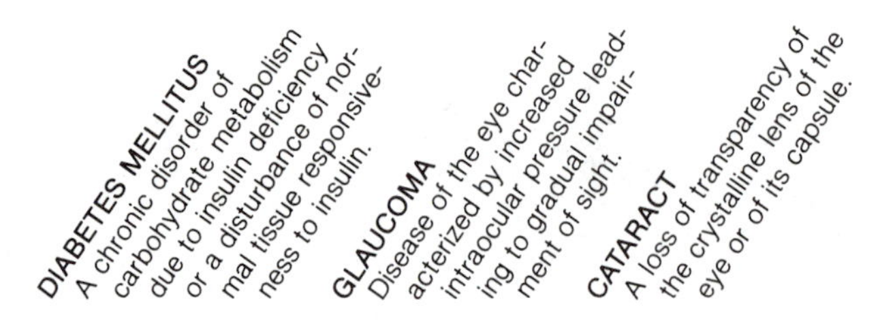

insulin production. In order to survive, the diabetic must have insulin administered. Noninsulin-dependent diabetes is very often completely asymptomatic. It may be discovered in routine diabetes screening, or during a physical examination. It has become recognized that tissues become less sensitive to insulin with aging. In order for normal blood sugar levels to be maintained, more insulin must be secreted. Obesity, diet, physical inactivity, various forms of stress, hormonal imbalance, drugs, and toxins may also contribute to the development of noninsulin-dependent diabetes, but only if a person is already genetically predisposed to the disease (Horton, 1983).

The relative lack of insulin seen in noninsulin-dependent diabetes causes the hyperglycemia that is thought to lead to the complications of diabetes. Recently, attention has also been given to the production of a substance called sorbitol from glucose present in high amounts in the blood. Sorbitol has also been implicated in some of the complications of diabetes. It accumulates in some tissues that do not require insulin to get the glucose into their cells and causes swelling and dysfunction. It is currently accepted as the biochemical explanation for the high prevalence of cataracts and peripheral neuropathies in diabetics. Peripheral neuropathies are manifested by feelings of tingling, numbness, or burning, generally of the feet. Diabetics with peripheral neuropathies often develop painless breakdown of the skin of their feet, leading to infected ulcers or gangrenous toes. Elevated blood glucose levels also lead to abnormalities in the walls of some blood vessels, resulting in renal (kidney) and retinal (eye) disease.

Often the adult with mild noninsulin-dependent diabetes (mildly elevated blood sugar, no complications) can reduce blood sugar by weight loss, by decreased intake of refined sugars (glucose) in the diet, and by exercise to promote glucose utilization. Drug treatment of noninsulin-dependent diabetes includes agents that promote the release of insulin from the pancreas.

Visual Impairment The causes of impairment of vision may vary with age. By the fifties and sixties cataracts, glaucoma, and disorders of the retina such as retinal hemorrhage are the most frequent causes of impaired vision. This is in contrast to the first five decades, when the most frequent problems were refractive errors that can be corrected easily with glasses.

Glaucoma is characterized by increased intraocular pressure due to some impediment of the outflow of the aqueous humor (see Figure 10–6). In 90 percent of cases the cause is unknown (Victor and Adams, 1980). It may be asymptomatic and progress slowly in producing loss of vision. Glaucoma can and should be easily recognized by physicians and is easily treated with medication. Because it so often is not detected early in its course, it is a major cause of visual loss in the later adult years.

A **cataract** is a density of the lens of the eye, most often due to degenerative changes in the lens. It also may follow trauma to the eye or lens and is often associated with diabetes. A cataract first

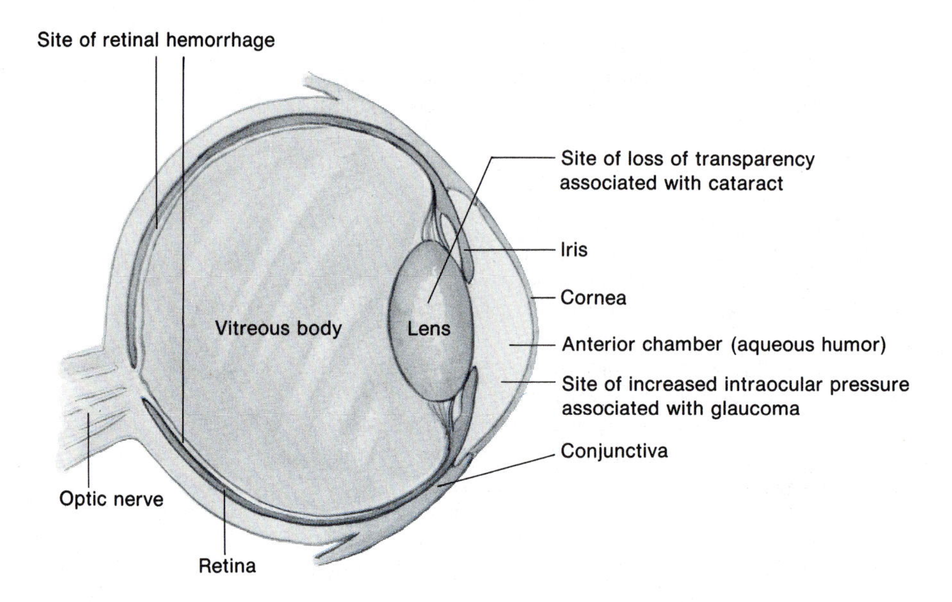

Figure 10–6
Schematic drawing of a section through the eye to illustrate sites of visual disorders common in adulthood.

occurs as a progressive, painless loss of vision. The amount of visual loss depends on the location and the degree of density in the lens. Surgical replacement of the lens is indicated when there is significant density producing loss of vision.

Retinal hemorrhage is bleeding within the retina. It is more frequent in persons with kidney disease and diabetes. It can be treated with laser coagulation of the peripheral blood vessels or by evacuation of clots from the vitreous body.

Teeth Problems During the first forty years of life, the main cause of tooth loss is dental caries, the bacterial decay of the tooth's outer enamel. During early adulthood the worst problems most people have with their dentists are paying for the bills on their children's teeth or submitting to cleaning, fillings, repairs on broken teeth, or replacement of lost fillings. After the fourth decade, however, gum diseases may become problematic. A new type of dental caries, the caries of maturity, occurs. It attacks the cervix (neck) of the teeth near the gumline. It is caused by a type of oral pathogen, *Odontalmyces viscosis,* which is not prevalent in the mouths of younger adults (Massler, 1975). An enzyme, collagenase, produced by bacteria that accumulate in plaques on the surface of the teeth, breaks down the attachments of the teeth to the surrounding bone. As a result, middle-aged people may begin to develop **periodontal disease**, sometimes referred to as pyorrhea. It is usually manifested by loosening of the teeth and sometimes with pus from the gingiva (gum) around the teeth (see Figure 10–7). Treatment is largely preventive and is aimed at removing the plaque (tenacious film of food and bacteria at the neck of the tooth) with frequent, careful brushing and careful flossing. Although proper tooth and mouth cleansing and avoidance of excess sugars can eliminate the need for

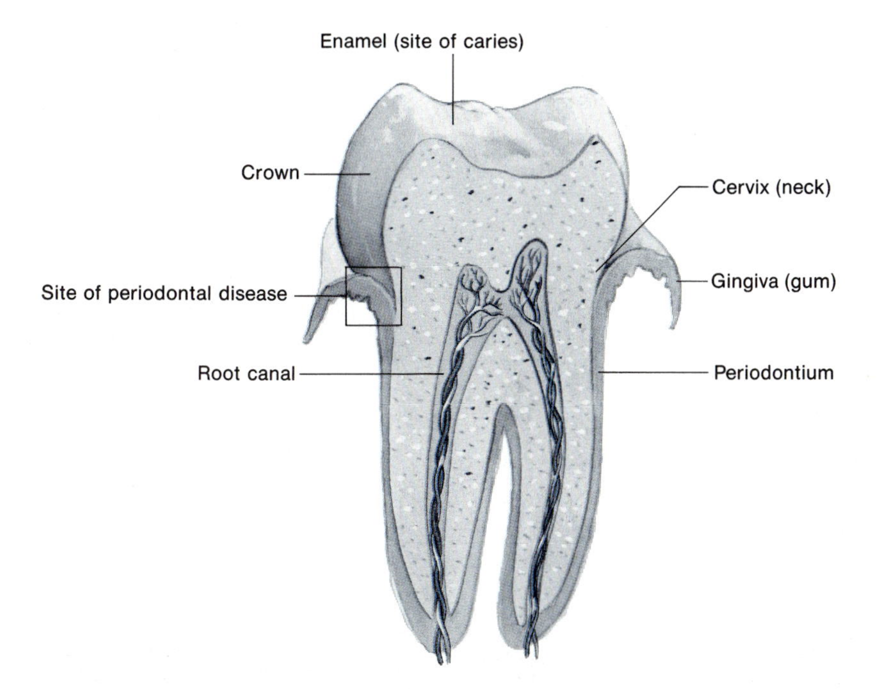

Figure 10–7
Schematic drawing of a section through a tooth to illustrate sites of tooth and gum diseases common in adulthood.

extractions of teeth in the later years, many Americans fail to pay attention to their teeth until it is too late. Consequently, it is common for people to begin wearing partial or complete dental plates by middle age.

Improperly fitted dentures can cause chronic mouth irritations and possibly contribute to the development of malnutrition in some people. Denture plates may need periodic adjustments to assure continued good fit with advancing age because the bony structure of the mouth often changes as bone mass and amount of mineralized tissues decrease. In spite of preparations that are advertised to hold dentures firmly in place "no matter what one eats," people with false teeth may find it impossible or extremely difficult to chew some foods, such as charcoal-broiled steaks and corn on the cob.

Psychosocial Development

People in their fifties and sixties typically have fewer family responsibilities than in earlier decades and more time for leisure and social activities. They also tend to view themselves more favorably than

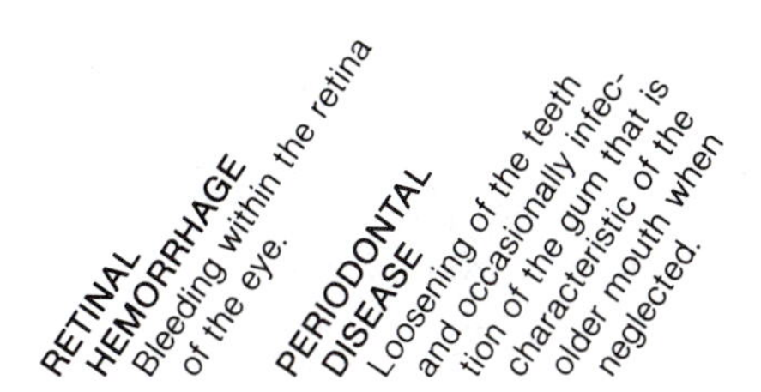

they have done in the past. They begin to reap the rewards of their lifetime of stresses and joys, pains and pleasures. Just as there is a feeling among youth that people over thirty cannot be trusted, there is a feeling among adults that many people over fifty have accumulated the wisdom of the ages through their experiences. Many people are glad to have arrived at fifty and have no desire to relive their thirties and forties.

LAUNCHING CHILDREN

Evelyn Duvall (1977) conceptualized the family life cycle as having eight major divisions (see Chapter 8, p. 413). Although fifty- and sixty-year-olds may fit into any of these divisions, they are more apt to fall into the classifications of the family launching its children and adjusting to the emptying household. Some adults are single during these years (see Box 10–2).

BOX 10–2

Middle-Aged Singlehood

Most studies of psychosocial development through the adult years are studies of married persons. Changes in their lives are assessed in relation to possible family events: marriage, birth of children, divorce, remarriage, parenting of children, parenting of teens, launching of children, grandparenting. Yet about 5 percent of adults never marry. Another 20 percent remain single for the remainder of their lives after the death of a spouse or a divorce. What are single childless persons like in their fifties and sixties? Are they swinging singles or lonely and forgotten bachelors and spinsters? How do they maintain a positive self-concept in a marriage-minded society?

With some exceptions, middle-aged single adults behave very much like middle-aged married adults. They invest about the same amount of their time and energy in their jobs. They entertain and are entertained by their friends. They spend time pursuing hobbies and leisure interests. They may be active in political, religious, community, or social groups. They have household responsibilities (meals, cleaning, laundry, bill paying, repairs, shopping). They may be involved in surrogate parenting of nieces and nephews or of their own aging parents.

Unmarried adults at any age are no longer a breed apart. It is less necessary to have a husband or a wife today than at any other time in our history (Blake, 1982). There is legitimate social status for a fully adult, happily unmarried, independent human being. Single women can often achieve and maintain a more clear-cut occupational status without the confounding factor of spouse's status (is she a college professor or an insurance salesman's wife?) Many older single women have purposely chosen never to marry. Likewise, many older single men are single by choice, not because they are unattracted by, unattractive to, or rejected by women. More research should be focused on all the psychosocial ramifications of being older and single. Meanwhile, it is a mistake to believe that only the married "live happily ever after" or that all unmarrieds share some stereotypic lifestyle.

Today's fifty-year-old married couples can often look forward to spending as much time alone together after launching their children as they spent raising their children. The time when the first child (usually the oldest) leaves home is a momentous occasion in most families. For some parents (especially the mother) the child's going away may be an occasion for grieving. For years the parents may have hovered over and protected the child. To then release the child into the world and to cut off the daily ministrations may bring an acute sense of bereavement. **Launching** (sending off) the first child is usually the hardest. However, families may feel acute pain as each offspring moves out of the family home. Although circumstances in every individual household make the timing and paths of launching children slightly different, many married couples or single parents approach this phase of their lives during middle age. In general, women are more affected by this change in the family than are men. Whether or not they work outside the home, women usually assume the primary role as nurturant caregivers for their families. These nurturing tasks take up a considerable amount of time, both in contemplation and in actual deeds of sustenance and support. Many mothers find surrogate children to nurture (their husbands, infirm friends, their aging parents, pets). Some try to hold on to their own children and may meet with varying degrees of success. Mothers who refuse to let go are often considered meddlesome and intrusive by their newly independent children. Some mothers feel more relief than grief after launching the last child (Harkins, 1978). They now have more time to devote to their careers, husbands, or other interests. Fathers may also be pained by the empty nest, although the majority of men are less involved with the nurturant aspects of parenting and more heavily involved in their careers. Consequently, they may be less upset by the absence of their offspring in the home. Some fathers also welcome launching children enthusiastically, especially when they have had a stormy relationship with their offspring (Osherson, 1980). The launchings may be felt most acutely by men as a proof of acquired middle age.

STABILITY OF PERSONALITY

Carl Jung (1923) described the early years of adulthood as years in which expansion into the outer world is the prevailing manner of operating or working. He felt that middle age brings noticeable personality changes. The maturing adult's manner of dealing with the outer world becomes more constricted or contracted. Older individuals begin to integrate outer reality with inner fantasy and move toward more self-realization, which Jung called **individuation**.

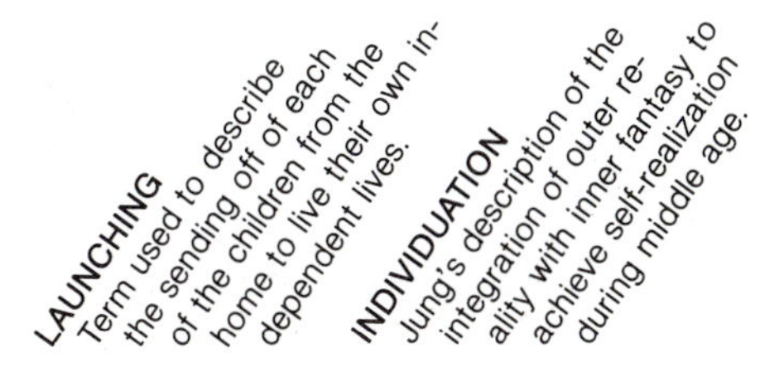

Bernice Neugarten (1968) after her extensive interviews with middle-aged and aging persons in the 1950s and 1960s, came to a similar conclusion. She described the move inward toward more preoccupation with satisfying personal needs as an increased **interiority of the personality**. This movement generally occurs around the fifth decade. "It is in this period of the life line that introspection seems to increase noticeably and contemplation and reflection and self-evaluation become characteristic forms of mental life" (p. 140). Neugarten held that the reflections of the middle years of life differ from the reminiscences of old age. Middle-aged adults restructure their personalities after their reflecting. They develop new concepts of self, time, and death with a consideration for how they will spend the rest of their lives. Very often they emerge with a feeling of being more fully in charge of their destinies than ever before. This satisfaction does not lend itself to many desires to relive their thirties and forties.

Roger Gould's (1972, 1978) descriptions of the principal characteristics of personality over the life span also support the conclusions of Jung and Neugarten that persons in their fifties and beyond turn inward. He stated that these patterns of reflection and contemplation result in a mellowing and warming up, with much more self-acceptance and self-approval. They "look within themselves at their own feelings and emotions, although not with the critical 'time pressure' eye of the late thirties or with the infinite omnipotentiality of the early thirties but with a more self-accepting attitude of continued learning from a position of general stability" (Gould, 1972, p. 526).

David Gutmann (1976) found that the increased interiority of personality that occurs during the middle years of life has important repercussions for men's social behaviors. In early adulthood men tend to be bold and active in their dealings with the external environment. They tend to act on their impulses and maintain a position of control over their own lives to the greatest extent possible. However, after midlife men become less bold. They begin to see the external world as more complex and dangerous. They conform and accommodate to it more. Gutmann (1977) studied men in a preliterate Mexican farming culture, Navajo Indians, and urban American men and determined that this personality shift occurs with age in all three cultures, regardless of the male sex-role expectations. Although the degree of shift varies from male to male and men without any obvious personality changes can be found, men generally become less self-assertive and more conforming with age. Gutmann

Figure 10–8
Some research suggests that personality characteristics remain stable over time. Thus the gentle, sensitive, reserved twenty-year-old may remain gentle, sensitive, and reserved through the next several decades.

describes this as a shift from **alloplastic** (active) mastery to **autoplastic** (passive) and **omniplastic** (magical) mastery of the outer world.

Feldman, Biringen, and Nash (1981) found that as men become less self-assertive and more conforming, affiliative, and compassionate in middle age, women become more instrumentally competent and more autonomous. This finding has led some people to conclude that middle-aged women become more "masculine." Perhaps with society's decreasing emphasis on sex-role stereotyped behaviors, this perception of postmenopausal women will disappear. Or perhaps society's emphasis on androgynous behaviors will make obsolete the attributions of competence, compassion, nurturance, and other such behaviors as more appropriate for one sex than for the other.

In contrast to the view of some change (expressed by Jung, Neugarten, Gould, and Gutmann) is the view that basic personality characteristics remain stable over the adult life span (see Figure 10–8). As the French put it: *Plus ça change, plus c'est la même chose* (the more it changes, the more it remains the same).

Allport (1937) postulated the existence of stable personality across the life span (see Chapter 2). In his view, every individual personality is a unique cluster of traits. Traits, he felt, are autonomous, consistent, self-sustaining forms of readiness for response. He felt that individuals may have all-pervasive, "ruling passion" qualities called **cardinal traits**; characteristics that are consistent and rest on upbringing called **central traits**; behaviors typical of all members of the culture called **common traits**; and minor ways of behaving known only to a few, called **secondary traits**. Allport's theory was

once highly controversial, but today more and more psychologists are coming around to this view.

Costa and McCrae (1980a) have argued that personality traits should be viewed as independent variables that function jointly with age and stage to influence some of the outcomes of life, rather than as variables dependent on age or stage. They have presented an 18-facet model of personality, with three domains encompassing six dimensions each, as follows:

Neuroticism	**Extraversion**	**Openness**
Anxiety	Attachment	Ideas
Depression	Assertiveness	Feelings
Self-consciousness	Gregariousness	Fantasy
Vulnerability	Excitement-seeking	Esthetics
Impulsiveness	Positive emotions	Actions
Hostility	Activity	Values

A person's score (high to low) in each of the three domains (**neuroticism**, **extraversion**, and **openness**) is seen as a pervasive part of personality and can be used to predict behaviors. Costa and McCrae (1980b) have demonstrated the endurance over time of these dispositions. They developed their model of **personality stability** after they studied the responses of men from their twenties to their seventies on self-report personality inventories filled out at intervals of from six to ten years apart (Costa and McCrae, 1977, 1978). They found a high degree of stability in how the men responded to the questions over time. Assertiveness in the early years, for example, remained assertiveness in midlife and in old age. Block (1981) reported a similar pattern of stability in his longitudinal study of subjects from their early teens through their mid-forties. Vaillant (1977) also followed men from adolescence through their mid-forties. He suggested that individual adaptability styles remain relatively constant and influence how one copes with each new stress.

Haan (1981) reported more personality stability in women than in men. She looked at longitudinal data collected on both sexes from adolescence through middle age. Personality characteristics of self-confidence, emotional under–over-control, open-closed self, and cognitive investment were more stable, while characteristics related to a nurturant-hostile dimension and under–over-controlled heterosexuality changed slowly. All the characteristics retained a relative continuity with past patterns of behavior, however.

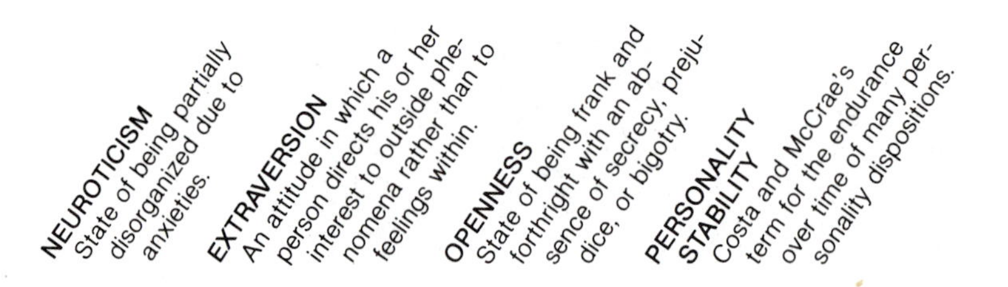

In an ongoing Minnesota study of identical twins reared apart and reunited in adulthood, preliminary evidence suggests that some personality characteristics may have a genetic basis (Bouchard et al., 1981). The scores of some sets of twins on personality measures have been closer than might be expected from the same person taking the personality assessment on two different days. Many reunited twins are similar in sociability, energy level, interests, and abilities. The consistency of each twin's behavior over time has not been assessed, since they were only recently brought into the research study. However, their concordance with each other suggests a lifelong stability of personality.

Brim and Kagan (1980) presented a collection of several papers suggesting constancy and several papers suggesting change in human personality. Changes do seem to characterize movement through the life span but with some more stable personality measures influencing the quality and quantity of the change. Scarr (1981) suggested that students of human development should view continuity and change as complementary themes, each making the other more complete and understandable.

MARITAL RELATIONSHIPS

Campbell (1981) reported an overall greater satisfaction with life in middle age. Many researchers have found that part of this satisfaction can be attributed to an upsurge in marital happiness after the nest empties. Whether or not personalities change, husbands and wives usually become more companionable. They use each other as confidants, relating joys and sorrows, interests, worries, satisfactions, annoyances, and dissatisfactions. They are more apt to really listen to each other and to rely on each other's advice. This promotes a sense of mutuality and interdependence in the marriage, which increases marital satisfaction (Garrett, 1982). External situations may also contribute to an improved marriage—especially more adequate income with fewer expenses when children are gone, and an increased level of job satisfaction (Skolnick, 1981). Sexual relations frequently improve once children are out of the home and the wife has ceased menstruating. Participation in sexual relationships reinforces feelings of mutuality, sharing, and concern over each other's well-being, and it enhances a positive sexual self-image and feeling of continued sexual competence (Weg, 1983). Wives are usually less tired in their fifties than they were in their forties because of fewer childrearing and household responsibilities, which they often balanced with other jobs. This reduction in role strain also contributes to greater marital satisfaction (Spanier and Lewis, 1980).

Gould's (1972) clinical work revealed that increased marital happiness and contentment is often associated with a change in attitude toward the spouse. The postparental individual is less likely to view the spouse as a parent or a source of supplies.

Although marital happiness increases for the majority of postparental couples, others seek divorce soon after the children leave home. In many cases the spouses involved have lost touch with each other long before the empty nest. They have only been biding their time, waiting for the children to depart before separating and divorcing. In other cases the empty nest proves distressing. The husband may experience a new identity crisis. He may have stayed at work more in the preceding years to meet the expense of college educations or weddings. Suddenly he has more time on his hands and no interests to fill the empty hours and days. Hobbies that just occupy time do not relieve boredom and restless feelings. Or the husband may be afraid to quit working so hard for fear of demotion, early retirement, or loss of job. These frustrations may be displaced from their real cause and blamed on the wife.

Identity crises after the empty nest also may occur in women. As mentioned earlier in this chapter, a woman whose predominant role has been mothering is left unemployed when children depart. She may find it difficult to redirect her child-rearing time into new efforts and to find new outlets for her nurturing skills. She may begin to make more and more demands on her husband. She may try to recapture her youth through extramarital affairs. Although many women find new interests in jobs, school, politics, or religion after the children leave, others wait for their husbands to solve the crises for them and may divorce them if they fail. Marital dissatisfaction in middle age may also be related to heavy drinking, absence of intimate associates, physical disabilities, mental disorders, chronic illnesses, or job dissatisfactions.

RELATIONS WITH ADULT CHILDREN

The family both contracts and expands with the launching of children. Although the number of persons living in the family home dwindles, the extended family may grow through marriage to include the new spouses and their offspring. Some parents are pleased with their children's choices of mates. They welcome them into the family with open arms. They may also seek to establish closer ties and frequent contacts with the parents of their children's new spouses. Not infrequently, however, parents have a degree of difficulty in accepting the persons their children choose as mates. This is more common when the chosen mate is from a different social class, ethnic group, religion, or region of the country or demonstrates beliefs, values, or lifestyles that are distasteful to the parents. As reported in Chapter 8, there are often special problems that arise between mother and daughter-in-law and father and son-in-law. A mother may resent her son's preference for his new bride's attention over her own. A father may be especially bitter toward the groom who has taken away his "little girl." All interactions may be problematic. A mother and father may resent the fact that their daughter now seeks her husband's over her parents' advice. Or they may want their son to take social and economic favors and advice from them, not from his wife or his wife's family.

Designing good in-law relations takes a great deal of tact and skill. Parents must remember that their children need to be independent to be mature human beings. It is difficult, however, not to be intrusive or overgenerous with advice or financial or social assistance after years of practice. On the other hand, parents who sever all ties with their offspring at the launching phase may be contributing to their own future conflicts and stresses with their sons- and daughters-in-law and alienation from their possible grandchildren. Some interactions and interdependencies are conducive to the establishment of good relationships that remain workable over time.

Parents are more apt to intrude in their offsprings' lives if children marry while still quite young rather than waiting until they have finished their education and begun a career. Mothers are more apt to intrude in their offsprings' lives than are fathers, probably due to their greater investment in the nurturing role. This is in keeping with the previously reported idea that mothers feel the absence of children more acutely than fathers.

It is not always the parents who intrude into or sever ties with their offsprings' mates. Many problematic relations with adult children relate to the fact that the departed child and spouse cut off ties with the parents. Many mothers and fathers would dearly love to give housekeeping or childcare aid, gifts, and the like to their children but find their offers of assistance rejected. This causes some parents to suffer great pangs of confusion and disappointment. Invariably they question their own performance as parents. Many initial problems of parent-in-law and children-in-law discord are worked out in time with honest communication and a bit of pride swallowing on both sides. However, some problems may persist for many years.

GRANDPARENTING

In spite of an American vision of grandparents as old, heavy-set, rosy-cheeked, wrinkled, benevolent persons enjoying rocking chairs and home-baked foods, most grandparents are quite different from this stereotype. Today's grandparents are more apt to be actively working, involved, middle-aged persons. Troll (1983) reported that the average age for a first-time grandmother is about forty-nine to fifty-one, and the average age for first becoming a grandfather is only a couple of years older, fifty-one to fifty-three. Great-grandparents, or great-great-grandparents (and there are many of them alive today) are much more apt to approach the popular notion of how a grandparent should appear.

Approaching grandparenthood may improve relationships with adult children. Many middle-aged parents see a grandchild as a wonderful gift—a way of assuring their own immortality (Kivnick, 1982). Grandmothers, in particular, are often anxious to have some input into both the care of the new grandchild and the nurturing of the daughter (or daughter-in-law) who has produced progeny. Daughters, in turn, may not want as much independence from the mother (or mother-in-law) when they realize the weighty responsibility of sheltering, protecting, feeding, changing, and loving a baby. Contact between new mothers and their own mothers generally increases after the birth of a grandchild (Fischer, 1981). Many young mothers depend on their own mothers to share infant care—especially teen mothers, single-parent mothers, and mothers who return to full-time employment shortly after their infant's birth. Grandmothers (and grandfathers) can be important sources of support and socialization for their children and grandchildren (Tinsley and Parke, 1983). Fathers and sons also frequently grow closer when a grandchild is born. A son may turn to his father for financial advice (life insurance, budgeting) or personal-social advice (fathering, dealing with the changed marital relationship).

Approaching grandparenthood speeds on the realization of aging for many people. It may also contribute to the process of personality restructuring that leads to a more conforming, accommodating, passive mastery of the external world. Grandparents are seldom asked to assume the primary, active, assertive responsibility for shaping their grandchildren's behaviors. Rather, they are expected to conform to, supplement, and support the decisions and rules about their grandchildren handed to them by their children. The former pattern of active mastery and control is usurped by the younger generation.

Figure 10–9
The majority of today's grandparents enjoy their roles with grandchildren. They can often spend more pleasant, relaxed moments with them than they could with their own children.

In a study of American grandparents, Neugarten and Weinstein (1968) found that about 60 percent of the persons they interviewed felt comfortable in the role of grandparenthood (see Figure 10–9). About 30 percent of them were uncomfortable, however. They had difficulty viewing themselves as grandparents, or they had conflicts with the parents about the rearing of the grandchildren, or they found that their grandparenting responsibilities were void of positive rewards and full of disappointments.

Neugarten and Weinstein (1968) identified five major classifications into which grandparents fall in terms of style of interacting with grandchildren: (1) formal, (2) fun seeking, (3) parent surrogate, (4) reservoir of family wisdom, and (5) distant figure. **Formal grandparents** do no child-rearing beyond occasional babysitting.

They maintain a clearly demarcated line between parents' responsibilities and grandparents' roles. **Fun-seeking grandparents** see their role primarily as playmates. Authority lines are irrelevant. The principal goal is that all parties involved should enjoy the interaction. **Parent surrogates** take on all the caregiving responsibilities for the grandchildren and are more parents than grandparents. The **reservoir of family wisdom** is a grandparent who is really an authority figure for parents as well as grandchildren. Such a grandparent uses interaction time to pass on special skills or resources. The **distant figure** stays remote from grandchildren except possibly to observe rituals such as birthdays or religious holidays. The fun seekers, surrogates, and distant figures are now the most common styles adopted by grandparents in their fifties and sixties. Older grandparents and great-grandparents are more apt to be formal or reservoirs of family wisdom.

Grandmothers tend to have warmer relationships with their grandchildren than do grandfathers (Troll, 1980). On topics where the generations perceive some similarity of attitudes, there is an exchange of views. However, on topics where grandparents and grandchildren perceive some disagreement, discussions are avoided. General religious or political attitudes are more apt to be discussed than specific practices. Grandmothers' influences cover such topics as lifestyle, values, interpersonal relationships, work, and education. Grandfathers more often limit their areas of advice to work and education (Troll and Bengtson, 1979).

RELATIONS WITH AGING PARENTS

Bernice Neugarten and Roger Gould both report that during the fifties people tend to mellow and develop warmer, more sympathetic feelings toward their own parents. One of Neugarten's (1968) interviewees said:

> I was shopping with my mother. She had left something behind on the counter and the clerk called out to tell me that the "old lady" had forgotten her package. I was amazed. Of course, the clerk was a young man and she must have seemed old to him. But the interesting thing is that I myself don't think of her as old. . . . She doesn't seem old to me. . . . (p. 95)

Gould (1972) reported that middle-aged adults less frequently see their parents as the source of their problems. They begin to call them "Mom" and "Dad" with more warmth and affection.

Although feelings mellow, responsibilities for aging parents often grow. The older generation may gradually require more assistance in managing their own lives and households. Housekeeping chores, shopping trips, transportation, financial arrangements, holiday preparations, and the like may be more and more troublesome for aging parents. They are more likely to turn to their children than to strangers for help and are more likely to request assistance from their daughters than from their sons. However, not all older parents turn to their middle-aged children for help. Many prefer to remain independent as much as possible.

Some aging parents, particularly lonely widows or widowers, move in with their daughters or sons. The tensions of trying to share a household are usually not as great as they are among younger adults with live-in parents. The middle-aged adult tends to be more sympathetic toward the older adult. Many over-fifty persons also tend to be more conforming and accommodating toward one another. If the nest is empty, the frictions commonly associated with differences of opinion about child-rearing do not exist. The older parent can often be a help with light housework and may even be able to contribute some financial assistance for equipping and maintaining the household. In addition, aging parents can be a source of companionship to the postparental middle-aged adults with whom they live.

CAREER CONCERNS

Men and women who have worked throughout their adult lives generally reach the zenith of their careers in their fifties and sixties. Not only are their salaries as high as they will probably go, but their prestige and power are at their peak. Persons at all other stages of life—children, adolescents, young adults, older adults—look to middle-aged experts for advice and direction and expect middle-aged adults to help them solve their problems and make appreciable changes in the society. Many middle-aged adults become mentors (see Chapter 9) to younger co-workers. Middle-aged persons are expected to be stable. They seldom make the radical shifts or impulsive moves that may have characterized their younger years. They are expected to remain in their present jobs until retirement. By this time the threat of being fired or laid off is usually low due to seniority status. For many people the years of work between approximately ages fifty and seventy are the most comfortable, satisfying years of their careers. They are doing what they know how to do best with fewer threats of moves or job changes. A few people find this lack of upward mobility frustrating. They may become bored with their work and simply mark time until they can retire. Some people are now opting to retire early, at fifty, or sixty, or whenever they have enough money to live at a level they deem comfortable. Work for them is so boring that they give it up gladly to do what they prefer to do for the rest of their lives.

For women who have been predominantly homemakers until the launching of children, a first retirement occurs in their forties. Many women then choose to embark on a second career (see Figure 10–10). For this group, jobs are approached with expectations of

Figure 10–10
The view of women in their fifties and sixties as cozy, dependent homebodies is being transformed. Many creative, energetic "grandmothers" are demonstrating that they can both bake cookies and contribute in meaningful ways in paying careers, volunteer work, or leisure-time activities.

upward mobility during their fifties. They do not view themselves as winding down their lives but rather as renewing themselves—starting all over again. Some go back to school before embarking on a job. Career boredom is less apt to occur in women who begin their out-of-the-home jobs later in life.

Some men may be jealous of their wives' new careers. A history professor told Neugarten (1968): "I'm afraid I'm a bit envious of my wife. She went to work a few years ago, when our children no longer needed her attention, and a whole new world has opened to her. But myself? I just look forward to writing another volume, and then another volume . . ." (p. 97).

LEISURE AND SOCIAL ACTIVITIES

Neugarten (1974) noted an interesting facet of our changing American society: "A hundred years ago, the higher one's education and income, the more leisure one had. Now . . . the best educated and the most skilled professionals . . . put in sixty- and eighty-hour weeks. As you go down the occupational scale, people are working fewer hours. . . . It is the blue-collar worker who has gained leisure over the past 100 years" (p. 36).

The postparental phase often brings in its wake increased pocket money and greater opportunities to get up and go out for many adults. Although some couples may strike out anew and attempt to go places and do things foreign to their previous lifestyles, it is more common for leisure pursuits to follow tried and trusted paths of the past: watching television, eating out, socializing with friends, attending cultural or sports events, or participating in activities centered around religion or politics or involving their extended families.

Many people in their fifties have a limited amount of actual time for pursuing enjoyable leisure activities. The professional workers may be tied up with eighty-hour work weeks. They may spend considerable amounts of their free time getting to and from work or doing work-related reading, entertaining, or preparations. Blue-collar workers, who supposedly have the most leisure time, often take on a second job to supplement the income from the first job. Most people have work to do at home (such as do-it-yourself projects, home repairs, meals, housework) that also cuts into their planning of leisure and social activities (see Figure 10–11). Kelly (1972) proposed that people's use of their nonworking time can be classified according to whether they choose an activity for the free time or have it determined for them and whether the free-time activity is independent of their career or dependent on it. A great deal of a person's nonworking time is actually used in career-related activities. Kelly labeled the work-related activities that are freely chosen by individuals **coordinated leisure**. Examples include reading in one's professional field or improving one's technical skills at home. Work-related activities that are determined by others are called **preparation and recuperation** by Kelly. Examples may include entertaining clients or preparing teaching aids at home. Free time that is not used in career-related activities is classified by Kelly as either **complementary leisure** or **unconditional leisure**. Complementary leisure is determined directly or indirectly by others. It includes those activities that a person pursues because he or she is expected to do so (religious activities, voluntary services, community activities). Unconditional leisure includes the things freely chosen because one enjoys doing them. Many people have little free time left over in their lives for unconditional leisure. When one does have this ideal leisure, he or she often chooses to do enjoyable work (hobbies, handicrafts).

Kimmel, Price, and Walker (1978) suggested that as much as a decade or two before retirement adults should develop a few interests that may deepen into satisfying leisure pursuits during retirement. Kimmel called this projecting **preretirement planning**, or anticipatory socialization. This pursuit of new enjoyable leisure activities should include making a new circle of friends to help replace those one will be leaving at the job. Kimmel and his col-

Figure 10–11
Each person decides how he or she will spend free time. Most people have plenty of work to do at home or have social or job-related obligations that take up many of their leisure hours.

leagues felt preretirement planning should also include considering source of income after retirement, maintaining health after retirement, and planning, in general, to accept the changed role that an eventual retirement will bring. The actual event of retirement with its economic and psychosocial implications will be discussed further in Chapter 11.

ERIKSON'S CONCEPT OF EGO INTEGRITY

Erik Erikson (1963) saw the nuclear conflict of the later adult years as that of achieving **ego integrity versus despair** (see Table 10–2). In

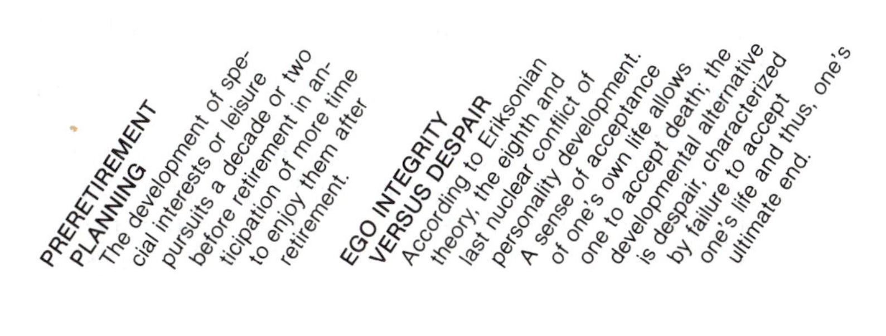

Table 10–2 Erikson's Eighth Nuclear Conflict: Ego Integrity Versus Despair

Sense	Eriksonian descriptions	Fostering behaviors
Ego integrity	Ego's accrued assurance of its proclivity for order and meaning	Associates recognition of ego's lifestyle as meaningful
	Love of the human ego	Demonstrations of human integrity and human dignity by others
	Acceptance of one's life cycle as something that had to be and that permitted no substitutions	Recognition of style of integrity of culture and civilization
versus		
Despair	Fear of death	Lack of ego integration
	Nonacceptance of one's life cycle	Feedback from others that own resources have not been sufficient during life cycle

Erikson's view, ego integrity involves an acceptance of the fact that one's life and work and leisure have been one's own responsibility. Attaining a sense of ego integrity necessitates a great deal of self-acceptance as well as acceptance of other family and community members. Although aspects of this last stage of Erikson's "eight ages of man" theory are suggestive of the midlife transition described in Chapter 9, this final Eriksonian stage goes far beyond the self-assessment of a midlife transition. An individual in the process of achieving ego integrity looks back over his or her own life experiences with a degree of acceptance. Mistakes, faults, failures, and disappointments are not denied or overlooked. Accomplishments, assets, successes, and satisfactions are appreciated. The individual feels content with the outcome of his or her life. Family and community members are accepted in a new way. The individual no longer wishes that others had been different but accepts responsibility for the course of many life experiences based on the way he or she acted and reacted to the deeds of others. When a person can look back on life with the satisfaction of ego integrity, death becomes much more acceptable. The whole process of achieving ego integrity may take many years. Some middle-aged adults have the nuclear conflict of generativity versus stagnation still very much at the forefront of their lives during their fifties and sixties. Others are more concerned with developing ego integrity.

The opposite of ego integrity in Erikson's theory is despair. If an individual cannot accept what his or her life has been by the later adult years, a feeling of hopelessness develops. Despair, in Erikson's view, involves low self-esteem, feelings of incompetence, speculation on what one would like to change about one's past, a lack of acceptance of one's age and status and lifestyle, and a fear of death. Failures are emphasized more than assets, accomplishments, and successes. Despairing persons would like to start life again but realize that it is too late and consequently become discouraged, frustrated, and resentful rather than accepting. Just as different people work at the eight nuclear conflicts of life at varying chronological ages, different persons resolve their conflicts in contrasting ways. It is possible for a person to go through a period of despair before achieving a sense of ego integrity.

Summary

The fifties and early sixties can be years of peak status and power: in the family, in the community, in the world of work. Health may continue to be excellent or begin to decline. Physical strength and stamina are diminished, and signs of aging are evident. Yet vast numbers of persons in this age span still feel in the prime of their lives.

The end of the female reproductive cycle is marked by menopause. Menopausal symptoms vary from woman to woman and range from negligible to psychologically and physically difficult. The end of the male reproductive cycle is gradual and less obviously concluded.

Health maintenance requires attention to diet, exercise, safety precautions, rest and relaxation, and prompt treatment of disease symptoms. The more common health problems of this age span are heart disease and cancer. Dental problems, visual disorders, and noninsulin-dependent diabetes are also more likely to occur.

Family responsibilities can be problematic—launching children (to jobs, to marriage, to higher education), adjusting to the empty nest, reworking marital relationships, relating to adult children, helping aging parents. Typically, however, family responsibilities will decrease, and time for work, leisure, and social activities will increase during these years.

Women are generally more affected by the launching of children than men. Many seek new postparental roles. This embarkation on a second career may be seen as a revving up rather than as a winding down of one's life.

Introspection, a looking into one's own mind, feelings, and reactions, is a common tendency in the fifties. Individuals often restructure their concepts of self and time and plan for the rest of their lives. Most tend to be more self-accepting and less self-assertive than in previous decades. Neugarten called this an interiority of personality. Whether or not basic personality changes occur (such as decreased assertiveness) is currently an area of active research with data arguing both for and against the stability of traits.

Marriages tend to become more stable and satisfying after the launching of offspring. Spouses may look to each other more for advice, approval, and companionship.

Specific circumstances and points of view may make the experiences of maintaining son- and daughter-in-law relationships problematic, gratifying, or both for persons in their fifties and sixties.

Likewise, grandparenting and caring for one's own aging parents can be hard work, rewarding, or a little of each.

Women and men who have been pursuing careers for many years often find that in their fifties and early sixties they are regarded as experts by their co-workers. Many become mentors (loyal advisors and friends) to younger adults who are just starting careers.

Erik Erikson saw the conflicts associated with these years as leading either to a state of ego integrity or one of despair. When adults can acknowledge that their work and leisure are their own responsibility and accept the accomplishments, failures, satisfactions, and disappointments of their lives, they move toward ego integrity. Despair results if they overemphasize their failures, blame others for them, and refuse to accept their life patterns.

Questions for Review

1. Menopause is often more feared by younger women than by women experiencing it. Why do you think society has perpetrated so many myths about the horrors of menopause?
2. Individuals who have cancer often speak of rejection, avoidance by others, and inability to discuss their feelings with others. Why do you think these reactions prevail? What sorts of information would help others react more openly to the cancer patient?
3. Differentiate between insulin-dependent diabetes and non-insulin-dependent diabetes. What are the variations in treatment for these two diseases?
4. Recent articles indicate a revitalized interest in physical activity and exercise. What do you think are some of the causes of this renewed interest?
5. What sorts of things might mothers do to prepare themselves for coping with the launching of children? What might fathers do?
6. Why do you think it is true that parents tend to intrude more in their children's married lives when their children marry young?
7. Erikson describes ego integrity as self-acceptance and labels the opposite as a sense of despair. How do you think family and friends affect the achievement of ego integrity? Discuss.

References

Allport, G. *Personality: A psychological interpretation.* New York: Holt, Rinehart & Winston, 1937.

American College of Physicians. *Guide for adult immunizations.* Philadelphia: American College of Physicians, 1985.

Bachman, G. A., and Leiblum, S. R. Sexual expression in menopausal women. *Medical Aspects of Human Sexuality,* 1981, *15*(10), 96B–96H.

Baltes, P. B., and Schaie, K. W. Aging and IQ: The myth of the twilight years. *Psychology Today,* 1974, *7*(10), 35–38, 40.

Blake, J. Demographic revolution and family evolution: Some implications for American women. In P. W. Berman and E. R. Ramey (Eds.), *Women: A developmental perspective.* Bethesda, Md.: NICHD/NIH (Publication No. 82-2298), 1982.

Block, J. Some enduring and consequential structures of personality. In A. I. Rabin et al. (Eds.), *Further explorations in personality.* New York: Wiley, 1981.

Bouchard, T. J., et al. *The Minnesota study of twins reared apart: Project description and sample results in the developmental domain. Twin Research 3: Intelligence, personality and development.* New York: Alan Liss, 1981.

Brenner, B. M., and Humes, H. D. Tumors of the urinary tract. In K. J. Isselbacher et al. (Eds.), *Harrison's principles of internal medicine,* 9th ed. New York: McGraw-Hill, 1980.

Brim, O. G., Jr., and Kagan, J. (Eds.), *Constancy and change in human development: A volume of review essays.* Cambridge, Mass.: Harvard University Press, 1980.

Campbell, A. *The sense of well-being in America.* New York: McGraw-Hill, 1981.

Clarkson, B. The acute leukemias. In K. J. Isselbacher et al. (Eds.), *Harrison's principles of internal medicine,* 9th ed. New York: McGraw-Hill, 1980.

Costa, P. T., Jr., and McCrae, R. R. Age differences in personality structure revisited: Studies in validity, stability, and change. *Aging and Human Development,* 1977, *8,* 261–275.

Costa, P. T., Jr., and McCrae, R. R. Objective personality assessment. In M. Storandt, I. C. Siegler, and M. F. Elias (Eds.), *The clinical psychology of aging.* New York: Plenum, 1978.

Costa, P. T., Jr., and McCrae, R. R. Still stable after all these years: Personality as a key to some issues in adulthood and old age. In P. B. Baltes and O. G. Brim, Jr. (Eds.), *Life-span development and behavior,* vol. 3. New York: Academic, 1980. (a)

Costa, P. T., Jr., and McCrae, R. R. The influence of extroversion and neuroticism on subjective well-being: Happy and unhappy people. *Journal of Personality and Social Psychology,* 1980, *38*(4), 668–678. (b)

Duvall, E. *Marriage and family development,* 5th ed. Philadelphia: Lippincott, 1977.

Erikson, E. *Childhood and society,* 2nd ed. New York: Norton, 1963.

Feldman, S. S.; Biringen, Z. C.; and Nash, S. C. Fluctuations of sex-related self-attributions as a function of stage of the family life cycle. *Developmental Psychology,* 1981, *17,* 24–35.

Fischer, L. R. Transitions in the mother-daughter relationship. *Journal of Marriage and the Family,* 1981, *43,* 613–622.

Foster, D. W. Diabetes mellitus. In K. J. Isselbacher et al. (Eds.), *Harrison's principles of internal medicine,* 9th ed. New York: McGraw-Hill, 1980.

Frisch, R. E.; Wyshak, G.; et al. Lower prevalence of breast cancer and cancer of the reproductive system among former college athletes compared to nonathletes. *British Journal of Cancer,* 1985, *52*(6), 885–891.

Garfinkel, L. Overweight and cancer. *Annals of Internal Medicine,* 1985, *103,* 1034–1036.

Garrett, W. R. *Seasons of marriage and family life.* New York: Holt, Rinehart & Winston, 1982.

Giuliano, A. E. Diseases of the breast. In R. C. Benson (Ed.), *Current obstetric and gynecologic diagnosis and treatment,* 5th ed. Los Altos, Calif.: Lange, 1984.

Gould, R. L. The phases of adult life: A study in developmental psychology. *American Journal of Psychiatry,* 1972, *129*(5), 33–43.

Gould, R. L. *Transformations.* New York: Simon & Schuster, 1978.

Gregerman, R., and Bierman, E. Aging and hormones. In R. H. Williams (Ed.), *Textbook of endocrinology,* 6th ed. Philadelphia: Saunders, 1981.

Griffin, J. E., and Wilson, J. D. Diseases of the testes. In K. J. Isselbacher et al. (Eds.), *Harrison's principles of internal medicine,* 9th ed. New York: McGraw-Hill, 1980.

Gutmann, D. Individual adaptation in the middle years: Developmental issues in the masculine mid-life crisis. *Journal of Geriatric Psychiatry,* 1976, *9,* 41–59.

Gutmann, D. The crosscultural perspective: Notes toward a comparative psychology of aging. In J. E. Birren and K. W. Schaie (Eds.), *Handbook of the psychology of aging.* New York: Van Nostrand Reinhold, 1977.

Haan, N. Common dimensions of personality development: Early adolescence to middle life. In D. Eichorn et al. (Eds.), *Present and past in middle life.* New York: Academic Press, 1981.

Harkins, E. B. Effects of empty-nest transition on self-report of psychological and physical well-being. *Journal of Marriage and the Family,* 1978, *40,* 549–558.

Hill, E. C. Disorders of the uterine cervix. In R. C. Benson (Ed.), *Current obstetric and gynecologic diagnosis and treatment,* 5th ed. Los Altos, Calif.: Lange, 1984.

Horton, E. S. Role of environmental factors in the development of noninsulin-dependent diabetes mellitus. *The American Journal of Medicine,* 1983, *75*(5B), 32–40.

Jackson, J. J. Death rate trends of black females, United States, 1964–1978. In P. W. Berman and E. R. Ramey (Eds.), *Women: A developmental perspective.* Bethesda, Md.: NICHD/NIH (Publication No. 82-2298), 1982.

Jung, C. G. *Psychological types or the psychology of individuation.* New York: Harcourt Brace, 1923.

Kelly, J. Work and leisure: A simplified paradigm. *Journal of Leisure Research,* 1972, *4*(1), 50–62.

Kimmel, D. C.; Price, K. F.; and Walker, J. W. Retirement choice and retirement satisfaction. *Journal of Gerontology,* 1978, *33*(4), 575–585.

Kivnick, H. O. *The meaning of grandparenthood.* Ann Arbor: University of Michigan Press, 1982.

LaMont, J. T., and Isselbacher, K. J. Diseases of the colon and rectum. In K. J. Isselbacher et al. (Eds.), *Harrison's principles of internal medicine,* 9th ed. New York: McGraw-Hill, 1980.

LaMont, J. T.; Koff, R. S.; and Isselbacher, K. J. Cirrhosis. In K. J. Isselbacher et al. (Eds.), *Harrison's principles of internal medicine,* 9th ed. New York: McGraw-Hill, 1980.

Macara, I. G. Oncogenes, ions, and phospholipids. *American Journal of Physiology,* 1985, *248,* C3–C11.

Massler, M. Dental considerations in the later years. In E. Brown and E. Ellis (Eds.), *Quality of life: The later years.* Acton, Mass.: Publishing Sciences Group, 1975.

Masters, W. H., and Johnson, V. E. *Human sexual response.* Boston: Little, Brown, 1966.

Neugarten, B. Adult personality: Toward a psychology of the life cycle. In B. Neugarten (Ed.), *Middle age and aging.* Chicago: University of Chicago Press, 1968.

Neugarten, B. The awareness of middle age. In B. Neugarten (Ed.), *Middle age and aging.* Chicago: University of Chicago Press, 1968.

Neugarten, B. The roles we play. In E. Brown and E. Ellis (Eds.), *Quality of life: The middle years.* Acton, Mass.: Publishing Sciences Group, 1974.

Neugarten, B., and Weinstein, K. The changing American grandparent. In B. Neugarten (Ed.), *Middle age and aging.* Chicago: University of Chicago Press, 1968.

Notman, M. T. Adult life cycles: Changing roles and changing hormones. In J. E. Parsons (Ed.), *The psychobiology of sex differences and sex roles.* Washington: Hemisphere, 1980.

Osherson, S. D. *Holding on or letting go: Men and career change at midlife.* New York: The Free Press, 1980.

Paffenbarger, R. S., Jr., et al. Physical activity, all-cause mortality, and longevity of college alumni. *The New England Journal of Medicine,* 1986, *314*(10), 605–613.

Scarr, S. Steps in the stream. *The Sciences,* 1981, *21*(3), 24–26.

Shapiro, S., et al. Risk of localized and widespread endometrial cancer in relation to recent and discontinued use of conjugated estrogens. *The New England Journal of Medicine,* 1985, *313*(16), 969–972.

Skolnick, A. Married lives: Longitudinal perspectives on marriage. In D. E. Eichorn et al. (Eds.), *Past and present in middle life.* New York: Academic Press, 1981.

Spanier, G. B., and Lewis, R. A. Marital quality: A review of the seventies. *Journal of Marriage and the Family,* 1980, *42,* 825–839.

Stampfer, M. J., et al. A prospective study of postmenopausal estrogen therapy and coronary heart disease. *The New England Journal of Medicine,* 1985, *313*(17), 1044–1049.

Tinsley, B. R., and Parke, R. D. Grandparents as support and socialization agents. In M. Lewis (Ed.), *Beyond the dyad.* New York: Plenum, 1983.

Tisi, G. M. Neoplasms of the lung. In K. J. Isselbacher et al. (Eds.), *Harrison's principles of internal medicine,* 9th ed. New York: McGraw-Hill, 1980.

Troll, L. E. Grandparenting. In L. W. Poon (Ed.), *Aging in the 1980s.* Washington, D.C.: American Psychological Association, 1980.

Troll, L. E. Grandparents: The family watchdog. In T. Brubaker (Ed.), *Family relationships in later life.* Beverly Hills, Calif.: Sage, 1983.

Troll, L. E., and Bengtson, V. Generations in the family. In W. Burr, R. Hill, F. I. Nye, and I. Reiss (Eds.), *Contemporary theories about the family.* New York: Free Press, 1979.

Ultman, J. E., and DeVita, V. T. Hodgkin's disease and other lymphomas. In K. J. Isselbacher et al. (Eds.), *Harrison's principles of internal medicine,* 9th ed. New York: McGraw-Hill, 1980.

Vaillant, G. *Adaptation to life.* Boston: Little, Brown, 1977.

Victor, M., and Adams, R. D. Common disturbances of vision, ocular movement, and hearing. In K. J. Isselbacher et al. (Eds.), *Harrison's principles of internal medicine,* 9th ed. New York: McGraw-Hill, 1980.

Weg, R. B. The physiological perspective. In R. B. Weg (Ed.), *Sexuality in the later years.* New York: Academic Press, 1983.

Weinberg, R. A. The action of oncogenes in the cytoplasm and nucleus. *Science,* 1985, *230,* 770–776.

Whitbourne, S. K. *The aging body: Physiological changes and psychological consequences.* New York: Springer-Verlag, 1985.

Wilson, P. W. F.; Garrison, R. J.; and Castelli, W. P. Postmenopausal estrogen use, cigarette smoking, and cardiovascular morbidity in women over 50. *The New England Journal of Medicine,* 1985, *313*(17), 1038–1043.

Woodruff, J. D. Diseases of the ovaries. In R. C. Benson (Ed.), *Current obstetric and gynecologic diagnosis and treatment,* 5th ed. Los Altos, Calif.: Lange, 1984.

Woods, N. F. Menopausal distress: A model for epidemiologic investigation. In A. M. Voda, M. Dinnerstein, and S. R. O'Donnell (Eds.), *Changing perspectives on menopause.* Austin: University of Texas Press, 1982.

Yunis, J. J., and Hoffman, W. R. Birth of an errant cell: A new theory about the cause of cancer. *The Sciences,* 1985, *25*(6), 28–33.

♥

The Later Years

11

Chapter Outline

Physical Development
- *Benign Senescence*
- *Cognitive Changes*
- *Health Considerations*

Psychosocial Development
- *Retirement*
- *Leisure Activities*
- *Personality in Old Age*
- *Marriage*
- *Widows and Widowers*
- *Extended Family Relationships*
- *Living Facilities*

Key Concepts

accidental drug overdosing
ageism
Alzheimer's disease
aneurysm
armored-defended personalities
benign senescence
cerebrovascular accident
disengagement
elder abuse
embolus
Foster Grandparents Program
geriatrics
gerontology
hemorrhage
hypothermia
integrated personalities
later maturity
life expectancy
logorrhea
longevity
multiple infarcts
myth of a sexless old age
myth of serenity
nursing homes
osteoarthritis
osteoporosis
passive-dependent personalities
presenile dementia
retirement
senescence
senile dementia
senility
terminal decline
thrombus
transient ischemic attacks (TIAs)
unintegrated personalities

♥

Case Study

Bertram began working as a bookkeeper for a small, new roofing company during World War II. He was a forty-one-year-old immigrant with only a grade school education. Nevertheless, he quickly learned the ins and outs of the roofing business. When the war ended, the young aggressive owner of the company could have hired a real bookkeeper/accountant. He chose to keep Bert instead.

The working relationship became more interdependent each passing year. Bertram eventually became indispensable to the business as it expanded into areas other than roofing. When Bert turned sixty-five, nothing was said about retirement. Bert was healthy (he had never taken a sick day) and he wanted to keep working. The boss realized that it would take two or more employees to replace him as well as a considerable period of time for Bertram to teach any new employees the unique bookkeeping system he had devised.

When Bert was seventy-one, his boss of thirty years had a serious heart attack. The boss quickly retired himself and sold the business. The new owners wanted their own books and bookkeepers. Bert was without a job. He had had little advance time to prepare for retirement. He had never really considered life without his work because his boss had been so much younger than he was. Seventeen days after his last day on the job, Bertram was admitted to the local hospital for emergency surgery of a strangulated hernia. His family doctor suggested that the reason for the sudden change in his health was the shock of retirement. The surgeon said retirement shock syndrome is a myth and that the hernia would have protruded and become incarcerated sooner had he still been working. What do you think?

Consider the meaning of old age in septuagenarian political leaders such as Ronald Reagan, Josip Tito, David Ben-Gurion, Golda Meir, Charles DeGaulle, Leonid Brezhnev, Winston Churchill, Mohandas Gandhi, or Mao Tse-tung. Or consider old age in octogenarian entertainers such as Bob Hope, Claudette Colbert, Cary Grant, Sir John Gielgud, Sir Lawrence Olivier, or Henny Youngman. George Burns started a new television series at age ninety. Classical guitarist Andres Segovia embarked on a U.S. concert tour at age ninety-three. Pablo Picasso painted masterpieces until age ninety-two and Grandma Moses painted her landscapes until age one hundred-one (she did not start painting until age seventy-four). Many individuals maintain their vim, zest, vitality, and a real *joie de vivre* (keen enjoyment of the pleasures of life) well into and beyond their seventies.

Old age is a difficult topic for many to discuss. Our youth-loving culture generally fears growing old. Sometimes this fear translates into a fear of older persons. Society creates euphemisms for them: senior citizens, retirees, golden-agers, the sunshine crowd. Some older individuals are healthy, some are ill. Some are crippled, some rich, some poor. Some are venerable and revered, some feisty, some mellow, some ageless. In every imaginable category they account for over 30 million persons in the United States. Some people refer to the increased numbers of old people in the population as the graying of America. One in five adult Americans are now over age sixty-five. By the year 2030, it is projected that half of all Americans will be over age sixty-five. Half of all the people on earth who ever lived past the age of sixty-five are alive today. **Life expectancy**, the average number of years that a person may be expected to live, has climbed dramatically in the past two millennia. During the lifetime of Christ, life expectancy averaged in the early twenties. Today it averages in the mid-seventies (see Figure 11–1). Life expectancy for women exceeds that of men by seven years. Life expectancy for white women is about 78.8 years compared to 71.5 years for white men. Blacks' average life expectancy is about 73.5 years for women compared to 64.9 years for men.

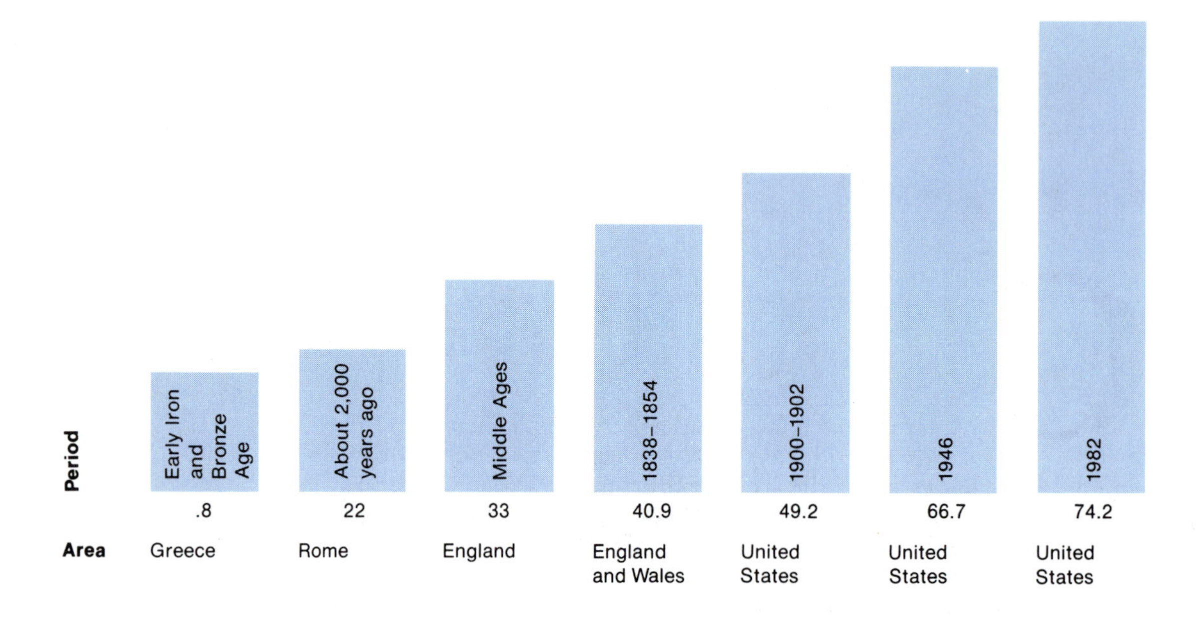

Figure 11–1
Average length of life from ancient to modern times.
SOURCE: *Adapted from L. Hayflick, The Aging of humans and their cultured cells,* Resident and Staff Physician, *1984, 30(8), 37. Reprinted with permission.*

Physical Development

The physical appearance of individuals over seventy varies as much as does the appearance of individuals of earlier decades. Call forth mental images of two or more persons you know who are beyond seventy. Consider various aspects of these persons: hair color, skin tone, posture, mobility, voice, eating behavior. These persons may have few if any similarities. The physical changes that accompany aging appear at different times and in different ways in all individuals. Persons who have escaped chronic or debilitating diseases into their later years and who exercise regularly and watch their diets carefully stand a much better chance of retaining a vigorous appearance than persons who have experienced debilitating illness, chronic disease, obesity, or muscular atrophy.

Chronological age cannot be used as a predictor of physical decline. In some cultures (Hunza, Pakistan; Georgia, Russia; Vilcabamba, Ecuador) and even in some American families **longevity** (long life) is the rule, not the exception. Three U.S. White House conferences on aging, held in 1961, 1971, and 1981, called for more research on aging and helped spark the growth of **gerontology**, the study of the normal aging process.

BENIGN SENESCENCE

The term **senescence** refers to the process of growing old, with the accompanying decrease in functional abilities. The term **benign** often precedes senescence to prevent confusing the term with senility. *Benign* means manifesting kindness and gentleness, or favorable, from the old French *bene* for "well," plus the root of *genus* for "kind." Medically, benign is used to suggest mild character, as opposed to a malignant character. Benign senescence is sometimes confused with senility. **Senility** refers to a degenerative condition of the brain with problems of confusion and disorientation. It does not occur in all old persons, as senescence does. We will discuss senility further in the health section of this chapter.

Explanations of why senescence occurs differ among biologists studying the phenomenon. Some feel that human cells have a genetically preprogrammed, fixed life span. Even if we could eliminate accidental, homicidal, and disease-related causes of death, humans might still die from normal physiological decrements in the vicinity of one hundred years of age (Hayflick, 1984). Some feel that the endocrine system, and one or more of the hormones it secretes, directs the gradual decline of a variety of human physiological functions. Research by Denckla (1975) suggested that aging occurs partially because the pituitary gland controlled by the hypothalamus of the brain stimulates the production of smaller quantities of thyroid hormones by the thyroid gland. Denckla's research also pointed to the possibility that the pituitary gland secretes another hormone, called DECO (decreased consumption of oxygen), that may have a blocking effect on cells, preventing them from using the thyroid hormones normally circulating in the blood.

Zatz and Goldstein (1985) suggested that hormones secreted by the thymus gland, which controls the immune system, might wane with age. The thymus gland is known to shrink to about one-tenth of its original size by old age. A slowdown in the functional abilities of the immune system to produce antibodies increases one's vulnerability to disease, and, in turn, diseases speed decrements in organ, tissue, and cellular functioning. Some biologists feel that some mechanism in the brain may be responsible for both the production or lack of production of certain hormones or for the slowing of the immune response. Some scientists believe we are aged by an increased production of, or altered sensitivity to, prostaglandins, especially of the E series (Licastro and Walford, 1986). Prostaglandins are potent biologically active compounds, derived from fatty acids, that can function as "local" hormones.

Many researchers are looking at dietary elements that may speed up, or prolong, the aging process. Walford suggested that marked food restriction with some vitamin supplementation might increase the human life span (Batten, 1984). Others have suggested that only certain foods or condiments (saturated fats, refined sugars, salt) be restricted while others (green leafy vegetables, whole grains) be increased. Most agree that tobacco use and alcohol abuse

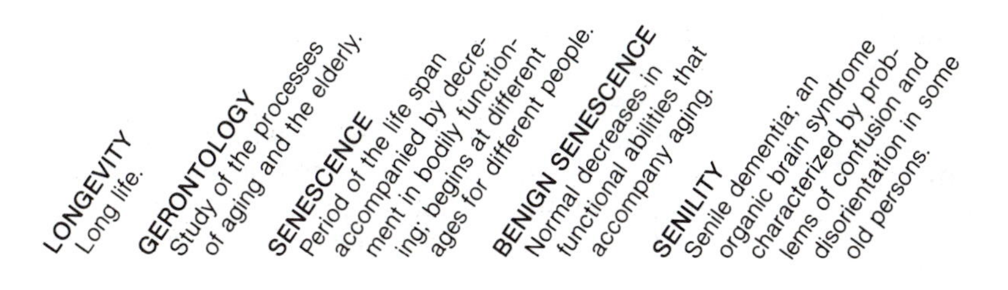

shorten life expectancy. Regular exercise, on the other hand, seems to slow down the rate at which tissues become debilitated, especially those of the cardiovascular, respiratory, muscular, and psychomotor systems (Whitbourne, 1985).

As knowledge accumulates, it seems likely that many interactions will be found between the brain, the endocrine system, the immune system, genes, environmental stresses, exercise, and diet, all of which affect the aging process. Let us now look at some of the specific functional changes that are a part of benign senescence.

As the body ages, the heart reduces its stroke volume due to diminished output by the left ventricle. Muscles consequently receive less aerobic support, and the lungs are relatively less efficient in their ventilatory function (Whitbourne, 1985). A loss of bone density and mass causes a compression of the bones, especially in the vertebral area, creating up to an inch decrease in height. The loss of mass in the mandible (jawbone) produces a change in the physical appearance of the chin. All bones break more easily and recuperate more slowly. A loss of cartilage makes painful joint complaints more common and contributes to a decreased range of movement and mobility. Connective tissue throughout the body is slowly debilitated by alterations in collagen and elastin. A curved posture becomes increasingly more common with age: the head down, the back and knees bent, and a forward pitch in walking. Older persons usually do most tasks slowly. James McCracken (1976) offered this remembrance of how his elderly father got out of a chair. "He'd sit in his chair just thinking about getting up. He'd run his hands up the arms of the chair a little way, brace, and push. And he'd stand up. Well, he was up. He'd stand for a moment, put his hands on the back of his hips. He'd still be bent over a little bit. But then he'd straighten up and be off about his business."

Because of the loss of lean body mass, there is a relative increase in the body's fat content. Although some old persons have problems with obesity, appetites are usually less keen. Less than half of the taste buds are still functionally active, making all foods less interesting. A reduced amount of saliva and digestive juices of the stomach makes indigestion more common (Stare, 1977). Reduced absorption from the intestines, lowered esophageal and intestinal peristalsis, and decreased secretions of the intestinal mucosa add to constipation problems, as do diets low in bulk and fiber. A loss of teeth may make chewing of many foods difficult. Diets are often unbalanced for reasons of economics (lean protein foods and fresh fruits and vegetables may be expensive) and preparation procedures (cooking may be difficult due to arthritis and visual problems).

Stamina and strength are greatly diminished regardless of caloric intake. The basal metabolic rate (BMR), a measure of the body's uptake of oxygen for metabolism, decreases. Sweat glands become less active. Most older persons find it increasingly difficult to adapt to changing temperatures. They are especially sensitive to cold temperatures and may require higher room temperatures, or more clothing, than would be comfortable for a younger person. Some older persons occasionally develop **hypothermia**, a subnormal body

HYPOTHERMIA Subnormal body temperature.

temperature. This can be life-threatening. It is more common in persons with preexisting chronic conditions such as diabetes, cardiovascular disorders, coronary heart disease, cirrhosis, or alcoholism. (External rewarming is contraindicated because it diverts blood from the vital organs: hypothermic people should be taken to a hospital or a physician for intravenous therapy and internal rewarming.)

Vision deteriorates steadily beyond the sixties, and problems of cataracts, glaucoma, retinal disease, and presbyopia (change in accommodation power in the eyes) continue (see Figure 11–2). The iris constricts or dilates more slowly, affecting the rapidity with which older persons can accommodate to variations between dark and light. Vision may be adequate when the darkness or light is constant, but sudden changes to more or less light may cause temporary blindness. Night driving is more hazardous for older persons for this reason. Color perception and depth perception also become less accurate.

The muscles supporting the eardrum lose fibers, contributing to hearing losses. They are so common among old people that persons should habitually raise their voices to all older persons. Higher tones are more difficult to hear than low ones. Many old people, either unaware of their hearing losses or hesitant to admit such a loss, may pretend to understand messages that they actually did not hear completely. Or they may try to piece together what they did hear and fill in the blanks. As a result, they may receive the wrong

Figure 11–2
Most persons in their later years need to wear corrective lenses to improve their vision.

message. Noise pollution (background noises) may further distort or mask messages for hard-of-hearing persons.

A loss of vestibular function in the inner ears may give old people difficulty with balance. They may sway slightly when standing or experience dizziness when rising, when viewing heights, or when trying to climb hills. They may lose their balance and fall quite frequently.

Hair may turn white in old age due to loss of pigmentation at the roots. Pigment deposits in skin cells may produce "age spots" on the hands, face, and other body surfaces. The skin gradually loses its elasticity and may develop folds or jowls. If an old person loses weight, skin is apt to hang loosely for a long time in areas once filled by deposits of fat. Skin surfaces that are allowed to dry out may take on a withered look. Decubitus ulcers (bedsores, pressure sores) may develop in places where circulation is impaired by prolonged sitting or lying.

All of the above descriptions may make growing old seem painful or cruel. Remember that the types and degrees of change occur gradually over a long period of time and vary greatly from individual to individual. A great many of the 30 million older people in the United States enjoy their lives. They would not all agree that being old is intolerable. They might argue, quite persuasively, that being sixteen or twenty-six or thirty-six is more difficult.

COGNITIVE CHANGES

Just as not all people become senile, not all people experience cognitive disabilities in old age. Some abilities (such as vocabulary) may decline little or not at all in a majority of older people, while other abilities (such as inductive reasoning) may gradually decline in a large segment of the aging population (Horn and Donaldson, 1980). Baltes and Schaie (1976) pointed out that there is a great deal of plasticity as well as vast individual differences related to adult cognitive behaviors. Current knowledge is not sufficient to make broad generalizations about which cognitive abilities increase or decrease with age, or about all the factors related to the stability or decline of intellectual functioning.

The idea that intellectual abilities do fall off with age has been suggested and demonstrated in numerous cross-sectional research studies. For example, using the Wechsler Adult Intelligence Scale (WAIS) to appraise the IQs of large numbers of aging persons, Wechsler (1958), Eisdorfer (1963), and Botwinick (1970) all reported that the oldest people scored less well than the younger subjects. However, the latter two research reports expressed concern about interpreting the results as conclusive of an age-related IQ decline. In cross-sectional research the subjects cannot be matched for their initial level of intelligence, since they are tested only once. Even if chance allowed for an equal distribution of persons of initially high, average, and low IQ to be tested in each age group, other factors make the results inconclusive. Even high IQ older persons may not have had any formal schooling beyond early adolescence. The availability of high school and college educations to anyone other than the privileged classes is a relatively new phenomenon. In addition, the 1930s Depression interfered with the education of many of today's old people. The WAIS questions are geared to various academic subjects. They are not particularly relevant to the concerns or types of knowledge of old people. Many elderly become bored or tired or lack the motivation to do well on the examinations. Younger adults are apt to be both better educated and more "test-wise" than their elders. (Test wisdom refers to one's knowledge of how to choose right answers on an examination. Some students may know less of the material covered in a course than others but score better on tests because of this "test wisdom.") There are also sex, culture, and language-bound factors that influence how well or how poorly one performs on standardized intelligence tests (see Chapter 6). These factors may also influence the IQ test scores of older immigrant or first-generation adults.

Some cross-sectional researchers have used nonverbal and less culture-bound tests of intelligence with old people. For example, Heron and Chown (1967) used the Raven Progressive Matrices Test, and Schaie (1958) used Thurstone's Primary Mental Abilities Test. These studies also reported results indicating that older subjects perform below the level of younger adults.

Longitudinal studies are felt to be more accurate indicators of intellectual changes, since they test and retest the same persons over a period of years. Schaie and Strother's (1968) seven-year longitudinal study and Eisdorfer and Wilkie's (1973) ten-year study are representative of this kind of research. They found some IQ decline with age, but results were highly individualistic. Although some persons showed a decline, others demonstrated quite stable ability levels over time.

Intellectual loss in old age may reflect many interrelated physiological decrements due to disease. For example, blood vessels that are narrowed by atherosclerosis can cause diminished blood supply to the brain as well as cause coronary heart disease with its associated sequelae. Malignancies may metastasize to the brain as well as to other body parts or may interfere with circulation. Cardiovascular disease, hypertension, emphysema, acute infections, poor nutrition, lack of exercise, injuries, or surgery may all temporarily or permanently diminish, to some degree, the blood supply of oxygen to the brain. Hearing or visual losses may interfere with comprehension of incoming information. Senescent changes in the functioning of the brain cells may diminish the speed at which they code and store new memories. Even the decreased speed of transmission of information along the neurons may slow down learning. Horn (1982) suggested that these types of biological processes cause a slow but gradual decline in fluid intelligence throughout the adult years (see Chapter 9). Crystallized intelligence, which is more culture bound (dependent on learning and experience) may show less decline or may even increase in some persons. Although biological processes may set limits on the intellectual functioning of some elderly persons, learning, problem solving, reasoning, and judgment are still possible (see Figure 11–3). The worldwide reliance on the wisdom of many

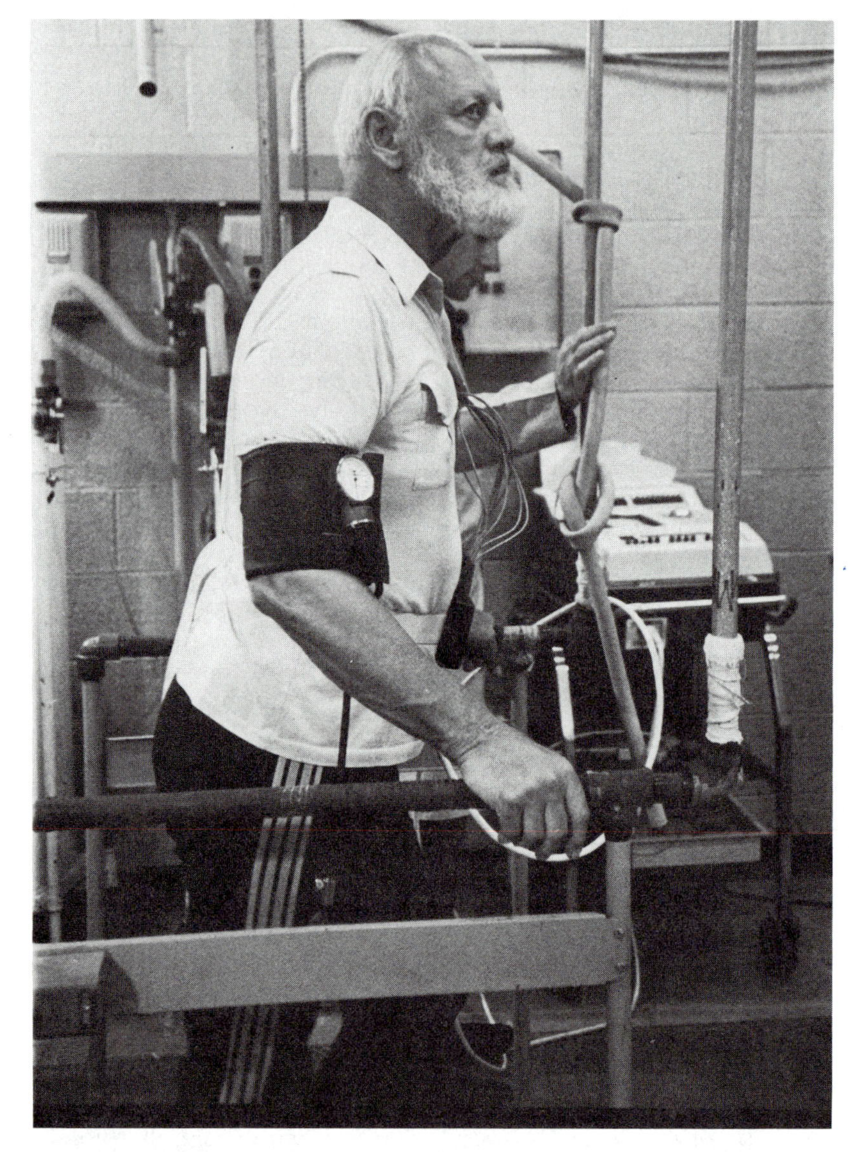

Figure 11–3
Physical fitness, especially cardiovascular fitness, helps maintain intellectual functioning with age. A great many healthy aged persons resemble healthy young persons in their cognitive and physical behaviors. Their wisdom about various topics is often widely respected.

persons who are well into their seventies or even eighties attests to many people's beliefs that this is so.

Wilkie and Eisdorfer (1974), Riegel and Riegel (1972), and others have suggested that there is probably a **terminal decline** in intelligence. A significant drop in IQ test scores often precedes death. Researchers have repeatedly observed that old persons who score poorly have died before the next year's retest period.

TERMINAL DECLINE
A significant decline in intellectual abilities that often precedes death.

HEALTH CONSIDERATIONS

Over 75 percent of the elderly have at least one chronic health problem. Many of the problems that are chronic in the elderly had their origins in childhood or earlier adulthood. Most of these long-term or recurrent disease conditions have been discussed in preceding chapters. The following are common problems of the elderly (the chapter in which they were discussed is indicated in parentheses): obesity (7); general malnutrition (7); anemia (7); digestive disturbances (9); diabetes (6 and 10); gallbladder disease (9); cirrhosis (10); urinary tract disorders (7 and 9); psychophysiologic illnesses (6 and 8); excessive self-medication or drug abuse (8); depression (9); mental disorders (9); rheumatoid arthritis (9); lung and respiratory diseases (9); cancers (10); hypertension (9); coronary heart diseases (9); and atherosclerosis and other vascular diseases (9). The most common chronic health problems of the elderly are CHD and vascular conditions, hypertension, cancer, degenerative diseases of bones and joints, disabilities from accidents, and mental defects. Whereas the number of deaths from heart disease declined in the past decade (due both to improved medical care and to greater individual concern for disease-prevention measures), the incidence of cancer, arthritis, respiratory diseases, cirrhosis, and diabetes climbed.

Many of the respiratory diseases of the elderly can be made less severe by prompt medical attention to symptoms (cough, congestion, sore throat). The immune response is much slower than it was in previous decades, and the common cold can become complicated by bronchitis or pneumonia before it runs its course. Elderly persons are especially susceptible to infections of the lower respiratory tract. Influenza, which is the fifth leading cause of death in persons over sixty-five, and pneumococcal pneumonia can be prevented by immunization. Physicians are now advised to immunize all persons age sixty-five or older with pneumococcal vaccine once (no boosters are necessary) and with influenza vaccine each year (American College of Physicians, 1985). In addition, they should receive booster doses of tetanus and diphtheria toxoids during mid-decade years (sixty-five, seventy-five, eighty-five).

Two important concerns for health maintenance in the elderly are the rising costs of health care and the decreasing numbers of physicians willing to provide such care. Young doctors today tend to continue their education in one of the specialized branches of medicine or surgery after completion of their internships. Few choose either family practice or **geriatrics** (the branch of medicine that treats old people). Geriatrics is both a difficult subspecialty and an unprofitable one. Many elderly persons are inaccurately diagnosed as being senile rather than given the expensive and time-consuming tests required to completely evaluate the causes of their presenting symptoms. Many elderly people do not take the medicines prescribed for long-term daily use because they cannot afford them. Many old people also neglect dental care or neglect vision and hearing problems, because dentures, glasses, or hearing aids are too expensive.

GERIATRICS Branch of medicine concerned with the elderly.

The remainder of this section will examine problems not previously discussed: osteoporosis, osteoarthritis, the special accident hazards of the elderly, cerebrovascular accidents (strokes), and Alzheimer's disease.

Osteoporosis **Osteoporosis** is characterized by a progressive reduction in the mass of bone and greater inner bone porosity. It occurs in both men and women. However, since women have less bone mass than men, their loss is more apt to lead to pain and discomfort and to fractures, especially of the wrist, hip, or vertebrae. The shorter, the thinner, and the smaller the bone structure of the person, the more apt she or he is to develop osteoporosis. Simple, painless photon absorptiometry procedures can measure bone mass and mineral density and provide valuable information about bone loss at any time of life (see Figure 11–4).

Whites and Orientals are born with less bone mass than are blacks, so their incidence of osteoporosis is higher. Most blacks have

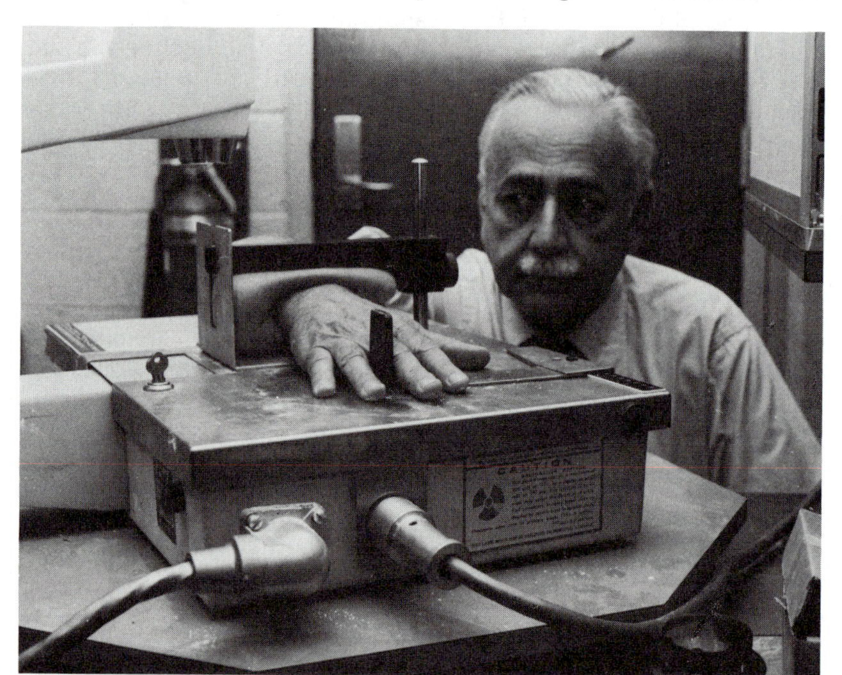

(A)

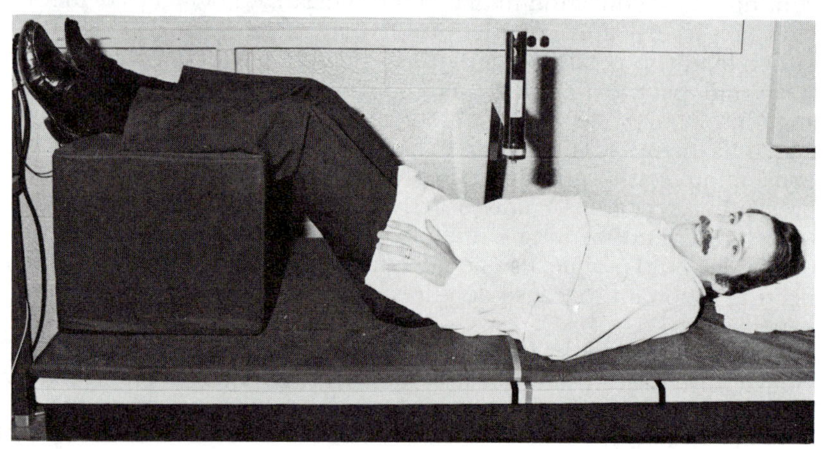

(B)

Figure 11–4 *Photon absorptiometry procedures. (A) In single photon absorptiometry a painless narrow beam scans the forearm and an attached computer reads out an index of bone width and mineral content. (B) In dual photon absorptiometry, bone mass and mineral content are measured in the lower spine, a susceptible part of the body in people with osteoporosis.*

a greater blood concentration of the hormone calcitonin, which provides a relatively greater protection to their skeletons from bone loss (Avioli, 1986). Osteoporosis seems to be especially common in whites originating from northern Europe. In Britain, for example, 40 percent of women over age seventy have suffered vertebral fractures and another 10 percent have had fractures of the wrist (Gregerman and Bierman, 1981). In the United States, 25 percent of women over age seventy have had at least one vertebral fracture, and an estimated 200,000 break their hips annually.

The demineralization that leads to osteoporosis begins early, between the ages of twenty-five and thirty. Current emphasis is being placed on preventing osteoporosis, as little can be done to treat it once the bone has become porous and fragile. To prevent bone loss in younger adults and retard further demineralization in osteoporotic patients, calcium supplementation is recommended. Women in their twenties, thirties, and forties should consume 1 gram of calcium every day. This is the amount found in 1 quart of milk. Women in their fifties, sixties, seventies, and eighties should consume at least 1.5 grams of calcium daily (Avioli, 1986). Since few women care to drink 1½ quarts of milk each day, they are advised to supplement their diets with other calcium-rich foods (hard cheese, salmon, sardines, nuts, broccoli) or to take calcium tablets. They must also be sure to take at least 400 units of vitamin D each day to help the body absorb the calcium (milk is already fortified with vitamin D).

Other factors that may play a role in bone loss are chronic hyperthyroidism, a lack of exercise, alcohol abuse, the use of certain drugs (thyroid hormone, cortisone, tetracycline, diuretics, and antiepileptics), drinking of phosphorated carbonated beverages, heredity, cigarette smoking, and low estrogen levels. Cigarette smoking lowers blood estrogen levels (Jensen, Christiansen, and Rodbro, 1985). Although estrogen is greatly diminished in all postmenopausal women, the use of ERT to prevent osteoporosis is a debated issue (see Chapter 10). While ERT can reduce osteoporotic fractures, it may also contribute to cancer or cardiovascular disease. Many osteoporotic persons wear special corsets or back braces to help support their curving spines and prevent compound fractures.

Osteoarthritis At least 31 million people suffer from one or more of the over one hundred kinds of arthritis, and 4.4 million adults are severely disabled by these conditions (Pardini, 1984). Some are stricken with rheumatoid arthritis (see Chapter 9). After age seventy it has been estimated that 85 percent of persons have some **osteoarthritis** (Mannik and Gilliland, 1980). This condition, also called

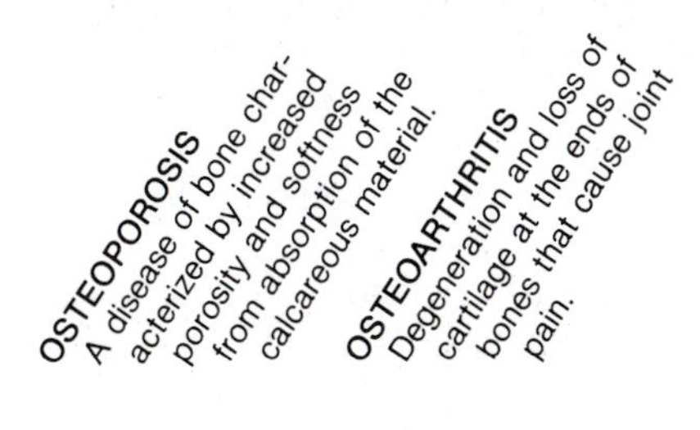

osteoarthrosis, hypertrophic arthritis, or degenerative joint disease, is characterized by degeneration and loss of the cartilage at the ends of the bones and by sharp "spur" formations. Pain is confined to the joints, especially on motion and weight bearing. It most typically affects the large joints, particularly the hips and knees, but can also affect the vertebrae, ankles, and fingers. Obesity, or the bearing of too much weight on weakened joints, is thought to contribute to osteoarthritis, as do the normal jolts of everyday living. Treatment is usually aimed at relieving pain and preventing further joint trauma. Medication used for pain may involve the use of aspirin or other analgesic drugs. Aspirin causes gastrointestinal irritation in some people. Newer nonsteroidal anti-inflammatory drugs that have fewer gastric side effects may be prescribed to protect the gastric mucosa (Salzman, 1983). Diet regimens for weight loss may be prescribed for the obese person. Regular range-of-motion exercises for the affected joints may also be recommended. Surgical replacement of a painful, crippled joint with a prosthesis is occasionally used for persons with a great deal of pain and joint destruction. Many persons with osteoarthritis use walking aids to help them get around (see Figure 11–5).

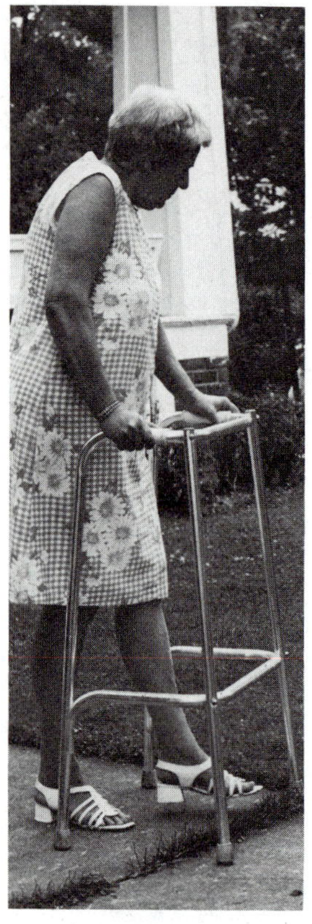

Figure 11–5
Physical disabilities can restrict movement of persons of all ages but are more common in aged populations. Age-related changes in bone and joint structure contribute to diseases such as osteoarthritis and rheumatism. Many persons continue to get around in their communities by using canes or walkers.

Accidents Accidents are the fourth leading cause of death in Americans, with an annual death rate of about 39 persons per 100,000 population. They also cause disabling injuries in approximately 8.7 million persons each year, costing about $97 billion in medical expenses, lost wages, and insurance administration. While motor vehicle accidents account for about half of all accidental deaths, they cause only about 20 percent of the disabling injuries. Home accidents cause more disabilities, especially in the elderly. Falls, fires, burns, poisonings, and firearms are the causes of many accidental home deaths or permanent impairments. Choking, suffocation, and heat stroke also cause many accidental deaths in the home.

Several million of the drivers on our roads are elderly. Most of them drive cautiously. They fear revocation of their licenses if they are arrested for some traffic violation. Driving for them signifies freedom, independence, and a relatively safe, accessible, inexpensive form of transportation. Most elderly people prefer back roads to super highways and often avoid night driving. Many are also duly cautious not to drive when tired or under the influence of alcohol. Nevertheless, the traffic accident rate per miles driven is approximately the same for old people as it is for teenagers. Whereas teenagers generally err in the direction of high speed, elderly drivers have more mishaps while failing to yield, signaling incorrectly, changing lanes, making turns, missing stop signs and stop lights, or parking.

Waxed floors, loose carpets, slippery bathtubs, high beds, misplaced furniture, and stairs cause many old people to trip and fall. Living quarters of the elderly should have elevators or well-lighted, well-railed, wide, nonslippery steps or should be on one level to help avoid some of these disasters.

A frequent cause of poisoning in older persons is **accidental drug overdosing**. Over 95 percent of the elderly population take some daily medications. The most frequent prescription drugs are those for cardiovascular or degenerative joint diseases, tranquilizers, and sedatives. In addition, most older people buy over-the-counter drugs and practice self-medication for respiratory or gastrointestinal disorders, for general aches and pains, or to help them sleep. Sleeping pills are more often a cause of insomnia, than a cure for it (see Box 11–1).

While taking both prescription and nonprescription medicines, many elderly people also drink alcoholic beverages that increase central nervous system depression. The effects of combining alcohol with tranquilizers, sleeping pills, some allergy medications, and some other central nervous system depressants are often multiplied rather than just combined. This result is known as drug synergism (see Chapter 8). Drug synergism may result in stupor, or even coma or death.

Many drugs that were safely used in one dosage in early or middle adulthood need to be prescribed in lesser (or greater) dosages in old age. This is because the relative decrease in lean body mass and the increase in the body's fat content make drugs stay in the body longer, make them act differently, or cause them to be absorbed by body tissues differently. It is never safe to save prescription drugs to use later. This can result in fatalities in older persons. An especially dangerous practice is to share prescriptions (as when a middle-aged person gives an elderly parent some leftover antibiotics).

Many elderly persons forget whether they have taken medicines. For this reason, they should keep a daily record of all the drugs they are taking and their schedule for taking them. Physicians should question their patients about the drugs they take regularly before prescribing new medications. They should also educate their patients as to the interactive or synergistic effects of various medicines, giving them clear instructions as to what may and may not be taken together. Some misdiagnosed "senility" is simply a result of overmedication, improperly used medicines, or adverse reactions to drugs as a result of synergistic effects.

Many pharmacists now keep computer records of the prescription drugs their customers use. They can call a physician if he or she prescribes a drug that will have a synergistic or toxic reaction with a drug prescribed by another physician. This is especially important when persons go to many different specialists (cardiologist, oncologist, ophthalmologist, gynecologist, endocrinologist) and fail to re-

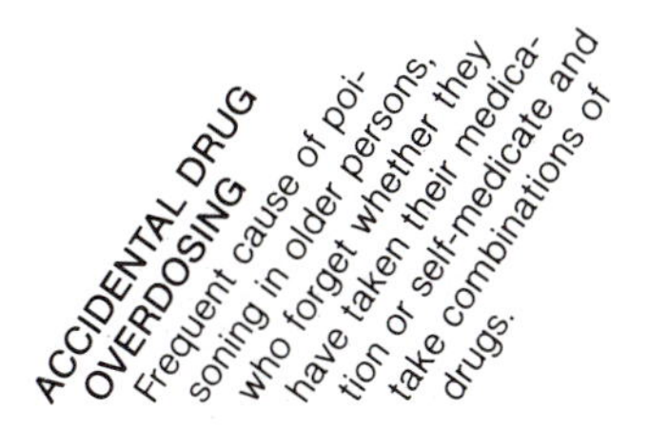

BOX 11–1

Sleep Disturbances in Old Age

Compared to younger adults, persons over age seventy need more time to fall asleep, spend much less time in deep sleep, have a reduced amount of rapid-eye-movement sleep (that in which dreams take place), and awaken more readily; they may spend up to 20 percent of the night in periods of wakefulness (Whitbourne, 1985). They also frequently have sleep disorders, especially sleep apnea and nocturnal myoclonus (Dement, Miles, and Bliwise, 1982).

Apnea means without breath or respiration. There are two kinds of apnea that occur during sleep. In *obstructive apnea,* the muscles that control the airway relax to the point that the airway is blocked. This usually occurs in persons born with narrow airways that become further narrowed by obesity or in those who sleep on their backs through the night. In *central sleep apnea,* which for unknown reasons becomes increasingly common in old age, the brain simply misses occasional signals to the lungs and diaphragm to breathe during sleep. Persons suffering from either form will not be aware of the apnea; even though they may stop breathing for from a few seconds to up to two minutes, on up to a few hundred occasions each night, they realize only that they have had a restless sleep, have experienced periods of wakefulness, or are fatigued throughout the day.

Any prescription for sleeping pills will aggravate both of these sleep disorders. Sleeping pills may even cause death in central sleep apnea by sedating the central nervous sytem to the point where it forgets for longer than a few minutes to send out signals for breathing. Both types of sleep apnea can be treated with a nasal mask for sleep, which provides positive air pressure. Obstructive sleep apnea may also be treated surgically by widening the narrow airway (Mendelson, 1984).

Nocturnal myoclonus is characterized by leg spasms, jerks, and kicking, up to a hundred times a night. As with sleep apnea, the victim is unaware of his or her behavior. The cause is unknown. The myoclonus (muscle contractions and relaxations) may awaken the sleeper, who does not know the reason for so much wakefulness. Victims of nocturnal myoclonus may complain of some tingling in their legs, or of leg pressure, while awake. Sleeping pills cannot cure nocturnal myoclonus. If given over an extended period of time, they leave the patient physically and psychologically dependent on them for sleep. They can also cause patients to be disoriented during the day, to have impaired memories, and in some cases, to have slurred speech. Regular bedtime and wake-up schedules to stabilize body rhythms, daily exercise, weight loss, and a change of sleeping position may help reduce nocturnal myoclonus. Sometimes warm milk before bed promotes a more peaceful sleep. In some cases, physicians may prescribe a mild anticonvulsive drug for nocturnal myoclonus.

Any persons, young or old, who have difficulty sleeping should be evaluated at a sleep disorder clinic for the possible physiologic causes of their insomnia. These specialized centers now exist in most of the major cities in the United States and Canada.

member or fail to tell each doctor what the others have prescribed. However, a pharmacist can supply this invaluable service only if the person has all prescriptions filled at the same pharmacy. Persons should ask their pharmacist about the drugs they are taking. They should tell him or her about the over-the-counter preparations they use as well. Many of these so-called "safe" medicines may be potentially dangerous when combined with other drugs. While the pharmacist can be a safety officer for preventing accidental drug overdosing or drug synergism in the elderly, he or she must be given accurate and complete information and must be asked for advice. Drugstore clerks cannot substitute for pharmacists. And the pharmacist must know about all other medicines the person is taking, even those supplied by doctors (as in samples), by friends, or by other family members or those purchased from other pharmacies or grocery stores.

Cerebrovascular Accidents The term *stroke* is often used to describe a **cerebrovascular accident** (CVA), a sudden, crippling, sometimes fatal occurrence. Although young people, and even children, occasionally suffer strokes, they most typically affect older people. When a stroke occurs, the blood flow through one or more blood vessels of the brain is disrupted, either by obstruction or by rupture. If the disruption is severe and prolonged enough to deprive adjacent brain tissue of blood and oxygen, the involved tissue will cease to function and subsequently die. The death of the tissue that results from the arrest of circulation in the artery supplying the part is called an *infarct.*

The warning signals of a CVA include dizziness, unsteadiness, sudden falls, temporary dimness or loss of vision, particularly in one eye, temporary loss of speech or trouble in speaking or understanding speech, or sudden weakness or numbness of the face, arm, and leg on one side of the body (American Heart Association, 1985). Many major CVAs are preceded by mild forms of any of these warning signals that may occur days, weeks, or even months before the stroke. Early warning signals are called **transient ischemic attacks** (TIAs) or "little strokes." Many little strokes are referred to as **multiple infarcts**. They may be a cause of dementia (see page 581).

Impaired blood flow can occur in any part of the brain and can be brief, prolonged, or recurrent. The symptoms and degree of disability from a stroke depend on where in the brain the disruption occurs, how long it lasts, and how extensive an area of tissue is damaged. Although about 500,000 Americans suffer some form of

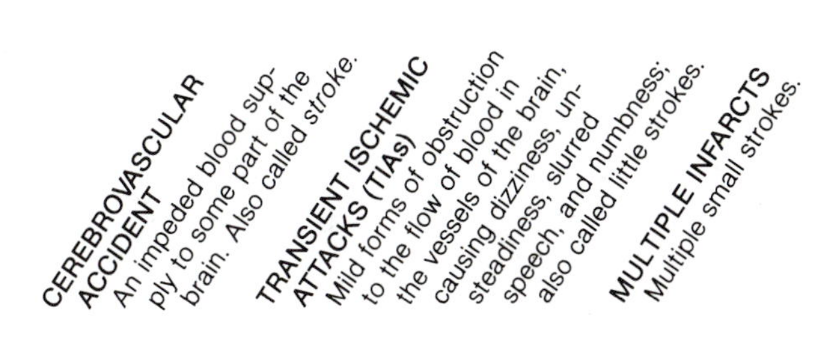

stroke each year, only about 32 percent of strokes are fatal (American Heart Association, 1985). Approximately 1.9 million stroke victims are still living with some form of disability. In many patients stroke sequelae demand lifelong extensive care.

There are several factors that may cause the disruption of blood flow in the brain. The majority of strokes, especially in the elderly, are caused by thrombosis, embolism, hemorrhage, or a ruptured aneurysm. Cerebral thrombotic strokes occur because a **thrombus** (a blood clot) forms in one of the blood vessels of the brain. Thrombotic strokes may have either a rapid or a gradual onset but more generally are slow to develop, over a few hours. They may be referred to as thrombosis-in-evolution or stroke-in-evolution. TIAs may precede the actual stroke (confusion, weakness, numbness, difficulty with speech, trouble understanding speech, dizziness, or diminished vision).

An **embolus** is a blood clot that forms in the blood vessels in one part of the body and travels to another. An embolus may also be a plaque from atherosclerotic deposits in the arteries of the neck leading to the brain, or, rarely, some other foreign material such as air, fat, or tumor cells (Mohr, Fisher, and Adams, 1980). Embolic strokes are usually abrupt in onset, like a bolt out of the blue. They are often associated with a recent heart attack, a diseased or recently surgically replaced heart valve, or the abnormal heart rhythm called atrial fibrillation.

A **hemorrhage** is a flow of blood, often profuse. Hemorrhagic strokes may result when a cerebral blood vessel ruptures and bleeds into an area of the brain. Such ruptures are frequently the consequence of an **aneurysm** (an outpouching of an artery) produced in the cerebral blood vessels in severe and prolonged hypertension. The onset of hemorrhagic strokes may be rapid or gradual, depending on the speed and extent of the bleeding. Disabilities will vary according to the region and extent of the damage.

Treatment of strokes may be aimed at the primary cause, at preventing complications and recurrences, and at restoring as much function as possible. If blockage occurs in the carotid artery of the neck, carotid endarterectomy may be used to remove the blocking substances. Damaged heart valves may also require surgery if they are the suspected origin of emboli. Anticoagulant drugs may be given to prevent a thrombus from enlarging and to prevent new thrombi from forming. Antihypertensive drugs may be given to control high blood pressure, which is a leading cause of hemorrhagic strokes and aneurysm formation.

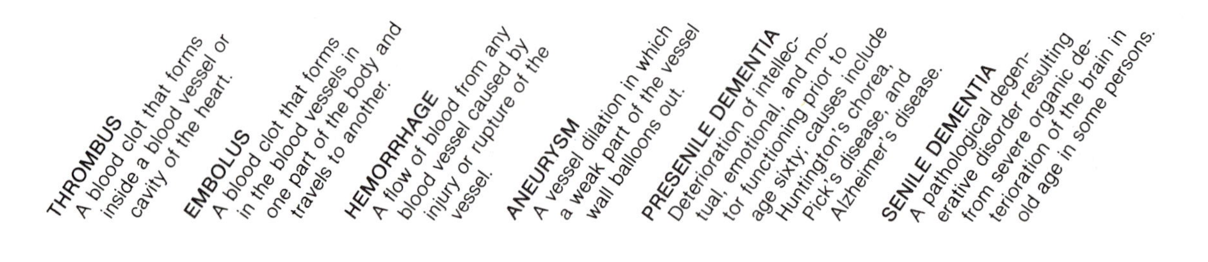

Aspirin has been shown to be an effective drug for preventing the formation of emboli and thrombi (Fields, 1983). It can be used both in persons who have had TIAs and in those who have had CVAs. Only a small amount (about half of a 325 mg aspirin tablet per day) is needed to achieve an antithrombotic effect. Larger doses may cause gastrointestinal irritation in some people.

Restoration of function after a stroke depends on the extent of the CVA and the parts of the body affected. The kinds of therapy given the patient may include physical rehabilitation therapy, speech therapy, and psychotherapy. It is important that family members be involved so that they will know what to expect, both physically and mentally, when the stroke victim returns home. The current goal is to make survivors as independent and productive as possible, considering their residual deficits, and to prevent additional CVAs in the future.

Senile Dementia The term *senility* is from the Latin root *senilis,* meaning old. It is used most often to describe old persons with organic brain syndromes (damage to, or degeneration of, brain tissue). It is not, however, a true diagnosis of any specific disease condition. Unfortunately, senility is often used as a label or diagnosis for older persons who have acute, reversible problems that cause them to become temporarily restless, forgetful, irritable, or disoriented. It may also be used to describe old persons who are depressed or anxious. Medical personnel are sometimes guilty of gross neglect for diagnosing old people as being "senile" without searching for reversible disease conditions and providing them with proper and adequate medical care (see Table 11–1).

Dementia refers to a general mental deterioration usually characterized by confusion, memory loss, disorientation, and disordered thinking. If it occurs in younger adults it is called **presenile dementia**. Some degenerative diseases of the nervous system that affect younger adults (presenile dementias) include Alzheimer's disease, Pick's disease, and Huntington's chorea. These have a genetic basis. Schizophrenia can also cause some dementia in young adults. Lacerations into brain tissue (as from a bullet or accidental head injury) or an overdose of some drugs (PCP, LSD) can also leave permanent organic brain damage and dementia. There is no known cure for degenerative diseases of the brain or for extensive damage to brain tissue as from lacerations or drug overdoses.

If dementia occurs in older persons, it is usually called **senile dementia**. Senile dementia can be caused by any of the things that cause presenile dementia. *Multiple infarct dementia* is caused by multiple small areas of infarction after TIAs. The blood supply is blocked by clots in the blood vessels with resultant tissue death, but the damage is slow and cumulative and, therefore, not recognized as a cerebrovascular accident. This cause may not be discovered until autopsy. Today researchers believe that as much as 80 percent of senile dementia may be due to Alzheimer's disease, another 10 percent to multiple infarcts, and another 10 percent to loss of neural tissue for other reasons (many still unknown).

Table 11–1 Reversible, Treatable Conditions That May Be Mistaken for Senile Dementia

1. Nutritional disorders
   - Malnutrition
   - Pernicious anemia (B_{12} deficiency)
   - Iron deficiency anemia
   - Pellagra (niacin deficiency)
   - Beriberi (thiamine deficiency)
2. Endocrine disorders
   - Myxedema (hypothyroidism)
   - Grave's disease (hyperthyroidism)
   - Addison's disease (insufficient adrenal hormone)
   - Cushing's syndrome (excessive adrenal hormone)
3. Toxicities
   - Bromide intoxication
   - Lead encephalopathy
   - Mercury poisoning
   - Manganese poisoning
   - Carbon monoxide poisoning
   - Overmedication
   - Adverse side effects of medicine
4. Depression
5. Sensory deprivation ("cabin fever")
6. Alcohol or drug abuse
7. Head injury; concussion, contusion
8. Brain tumor
9. Other untreated or poorly controlled diseases
   - Diabetes
   - Chronic obstructive pulmonary disease
   - Epilepsy
   - Hypertension
   - Atherosclerosis
   - Congestive heart failure
   - Angina pectoris
   - Abnormal heart rhythms
   - Nephrosis
10. Cerebrovascular accident
11. Transcient ischemic attacks
12. Normal-pressure hydrocephalus (obstruction to cerebrospinal fluid)
13. Subdural hematoma
14. Brain infection
15. High fever
16. Hypothermia
17. Migraine headaches
18. Anxiety disorders
19. Situational reaction (bereavement, move to new residence)

Persons with senile dementia show a progressive loss of intellectual abilities and increasingly abnormal behaviors over time. They may have swings from relatively good hours to bad hours, or from good days with near normal functioning to bad days marked by memory loss and antisocial behaviors, before the brain changes become extensive. They become increasingly neglectful of their personal hygiene. They may show marked personality changes—for example, becoming obsessed with some unusual interest or activity or becoming exceedingly aggressive or impulsive. They may have auditory or visual hallucinations. Their memory loss is often limited to recent events. It is not uncommon for a patient with senile dementia to have a near perfect recall of some childhood event yet be unable to remember what was just seen on television. Communications often become jumbled, with unusual verbal patterns being repeated over and over. This disorder is called **logorrhea**. Emotions become labile (unstable, liable to change quickly). Old structures (clothes, furnishings, foods, surroundings, persons) are preferred to new ones.

In the later stages of senile dementia, patients may become incontinent (lose the ability to prevent discharge of urine or feces) and incoherent and often fail to recognize even their closest friends and relatives. Even memory for the distant past fails. They may exist in a vegetative state, requiring total custodial care.

Alzheimer's Disease **Alzheimer's disease** (AD) can affect adults of any age but is more common in old age, where it is called senile dementia of the Alzheimer's type (SDAT). It probably affects between 5 and 10 percent of persons over age sixty-five and is guessed to be the cause of dementia in about 1.5 million elderly Americans. There is currently no way to absolutely diagnose the disease prior to death. On autopsy, however, the diagnosis can be confirmed by a loss of neurons in the basal forebrain and the presence of neurofibrillary tangles (Whitehouse et al., 1982). The diagnosis of AD/SDAT is usually made after eliminating the presence of other diseases that could cause dementia; screening techniques include medical and psychological tests, computerized tomography (CT scan), and positron emission tomography (PET scan).

The cause of Alzheimer's disease remains elusive. There is probably a genetic predisposition to the degenerative process, as it tends to affect persons with a family history of the disease. However, it can also strike persons with no known family involvement. Many possible causes are now being investigated, including an enzyme defi-

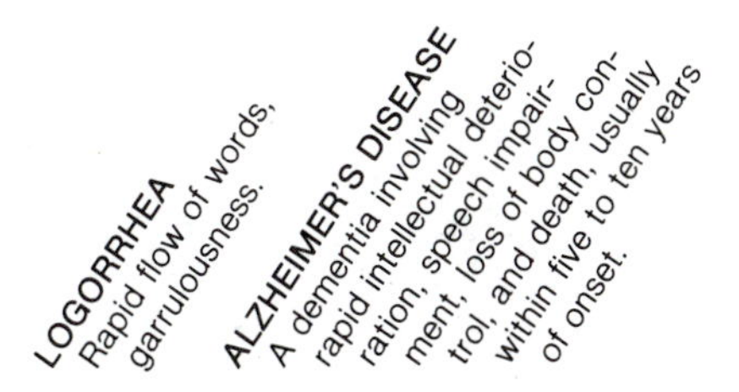

ciency, a neurotransmitter deficiency, a mineral deficiency (magnesium or calcium), a heavy metal toxicity (manganese or aluminum), a slow-growing virus, a serious head injury, an immunologic defect, or a chromosomal defect. Since Down syndrome individuals who live into adulthood develop AD, the defect might involve the twenty-first chromosome pair (Sinex and Myers, 1982).

There is no cure for AD/SDAT. Most drug therapies are still in the experimental stages of development. Patients may live up to fifteen years after a probable diagnosis is made. They gradually become more irascible and uncooperative until the late stage of the disease. Then they are apathetic and lose all abilities for self-care. They usually die from some other infectious process. The care of an AD/SDAT patient can be grueling physically, mentally, and financially on family members (see Box 11–2). Many Alzheimer's victims end their lives in nursing homes because their care becomes more than family members can handle.

Psychosocial Development

Robert Havighurst (1972) called old age **later maturity**. He saw the following as developmental tasks to be accomplished:

- Adjusting to decreasing physical strength and health.
- Adjusting to retirement and reduced income.
- Adjusting to death of spouse.
- Establishing an explicit affiliation with one's age group.
- Adjusting and adapting social roles in a flexible way.
- Establishing satisfactory physical living arrangements.

RETIREMENT

The move from an agrarian society to a predominantly industrialized society brought in its wake the phenomenon of **retirement**. Not too long ago people worked until they became physically incapacitated or until they died. They worked less as they got older or did more advising and less manual labor, but they still felt a part of the job. When Congress established the Social Security Act in 1935, it arbitrarily set sixty-five as the age when men could first collect government retirement benefits (sixty-two for women). Many industries in turn set sixty-five as a mandatory retirement age for men and sixty-two for women. In the late 1970s Congress banned rules setting sixty-five as mandatory retirement age for many employees at the request of senior citizens' groups. Many companies have readjusted

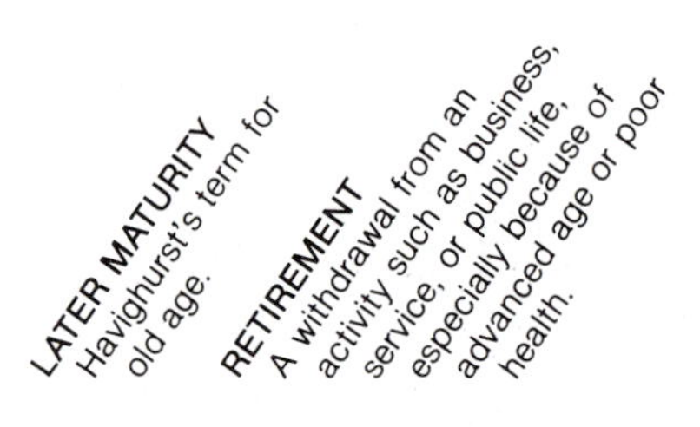

BOX 11–2

The Home Caregiver of the Demented or Disabled Elderly

Increasingly attention is being focused on the impact on the spouse, or other family members, of caring for victims of stroke, Alzheimer's disease, or any of the other dementing or debilitating diseases of the elderly.

When the diagnosis of an incurable, irreversible disease with a steady downhill course is first given, most people react with disbelief and denial. However, when the patient is sent home, the spouse (or other family caregiver) gradually becomes aware that the diagnosis is real. The disease begins to permeate the caregiver's entire quality of life, for the worse. Despite love for the affected patient, resentment and hostility are typical. The caregiver often feels a lack of support from others, confusion about the disease and what to expect, a lack of information from the physician, a trapped feeling, isolation, depression, and a loss of self-identity (Barnes et al., 1981). Feelings of resentment are almost invariably followed by feelings of guilt. One of the most difficult aspects of caring for a beloved but severely disabled person is the constant view of the disintegrated condition of one who was once so healthy. The emotions of love, pity, resentment, and guilt become inextricably intertwined.

If a patient is mentally but not physically disabled (as in Alzheimer's disease), the caregiver may be both physically abused by the belligerent, irascible victim of disease and physically exhausted trying to prevent the confused, disoriented person from wandering outside the home. When the disabled patient is less mobile, the caregiver may be physically exhausted from the work of transferring him or her from bed to wheelchair, from changing wet clothing or sheets, from feedings, and from all other aspects of daily care. If the spouse or caregiver has a preexisting chronic illness (such as coronary heart disease, osteoporosis, arthritis, diabetes, hypertension), this care is especially difficult. It is not unusual for a caregiver to die before the disabled person for whom he or she has been providing care.

The financial price of caring for a disabled person, even at home, is steep. One must hire or impose on another person to watch the patient while buying medicines, groceries, and supplies or leaving the home for any reason. Alternately, one can have others do the shopping, banking, and other out-of-the-home duties. The caregiver can no longer be employed at any job other than caregiving, and the expenses must be covered by pension or social security checks. Medicaid will provide funding for home health care only for persons who qualify as indigent (very low income and few salable assets). Often long-term care of the disabled leaves little income for the unaffected spouse when the patient dies.

Some support groups now exist in which caregivers of the disabled give each other mutual emotional encouragement and practical problem-solving advice. Some telephone support lines have been established for the same purpose. However, most caregivers need the respite of an occasional outside meeting in addition to a telephone support network. In a book entitled *The 36 Hour Day,* Mace and Rabins (1984) have outlined many ways to more successfully manage the care of an elderly disabled family member. Each home caregiver needs all the support and encouragement he or she can get to manage each "36 hour" day. Changes in social and funding policies for long-term care of the disabled at home are needed to lighten caregivers' burdens in the future.

their policies and now require retirement at age seventy. Some corporations with their own pension plans ask employees to retire earlier, or allow them to work beyond age seventy. Each state has the right to set its own rules concerning mandatory retirement. Many workers in occupations where the lives of others can be affected by their efficiency (fire fighters, police, pilots) are forced to retire as early as age fifty-five. Census bureau figures show that about 86 percent of Americans retire or are already retired when they first become eligible for social security benefits, and about 12 percent continue to work past age seventy (see Figure 11–6).

There is a prevalent belief in our society that the transition from working to retirement brings in its wake a shock syndrome, a decline into illness, depression, loneliness, anxiety, self-doubt, feelings of uselessness, or financial worries. Although it is true that the transition requires substantial changes in one's lifestyle, it is not true that shock and decline are necessary results. For some people retirement is a welcomed relief from the toils of daily labor. For others it means a change from a disliked job to a more enjoyable, self-directed form of employment. For still others it does generate feel-

Figure 11–6
Many of the two million-plus Americans who do not retire from their jobs after age seventy are professionals, are self-employed, or own their own businesses. For them work is a source of personal and social identity, pleasure, creativity, and profit, and they choose not to give it up.

ings of uselessness and impending death. There are a great many differences in how retirees view their changed status, just as there are a wide variety of ways in which people handle every other developmental milestone throughout life.

Feelings of job deprivation and lack of a role are often related to the kind of work performed prior to retirement. Upper-status workers (executives and professionals) object more to forced retirement and often fear the consequences of being unemployed. They enjoyed the wages, recognition, and power of their jobs. However, once retired, they are more often able to organize and plan for their leisure hours in such a way as to minimize some of their feelings of job deprivation. Unskilled workers may have less to lose with retirement. They had low autonomy, low pay, and often little or no satisfaction in their jobs. However, they are often less able to adapt to loss of work roles.

Many factors help retirees make better retirement adjustments. Persons who retire early tend to adjust better than employees who continue at their jobs beyond the expected age of retirement. Persons who voluntarily retire fare better than persons who are compelled to retire. Preretirement planning and a prior involvement with hobbies and leisure activities makes the transition easier, as does experience in allocating time. Cox and Bhak (1979) found that the single most critical determinant of retirement adjustment may be the attitudes of the retiree's significant others. When one's family members, close personal "confidants," cliques at work, and friends in social clubs or religious organizations view the retirement as a positive move, the retiree's adjustment is significantly better than it is when significant others feel that the end of employment is a mistake or a disaster.

Retirement is a financial blow for many persons. On the average, social security benefits and other pension plans total less than 50 percent of preretirement income. It is often easier for persons with low preretirement wages to learn to live on their social security checks than it is for persons who were once more affluent.

Many of today's retirees fought in World War II or worked in war-related industries at home. They participated in the booming postwar economy and contributed a portion of each paycheck to social security. Many believed that this government-sponsored program would enable them to retire at sixty-five and live for several vital years in a sunny southern climate. For many retirees, this bright dream has not materialized. Inflation has made the cost of retirement much steeper than anticipated. The average social security payment provides for income just slightly above the poverty level. The majority of today's retirees do not have corporate pension plans to supplement their social security payments. In general, corporate pension funds go to the people who need them the least. Few of today's retirees established their own individual retirement accounts (IRAs), even though social security was never designed to be one's only source of retirement income. Retirement hits widows especially hard. Close to two-thirds of them have no income other than a social security check. About 70 percent of today's elderly poor are women.

A majority of retired persons who are struggling to pay their bills would rather have part-time, flexible-hour jobs with diminished physical demands than more government handouts. Many of them are in good health, have retained their job skills, and are bored with nothing "important" to do all day. The 1981 White House Conference on Aging recommended that both the public and private sectors in the United States make available more part-time, flexible-time jobs for older workers.

Returned elderly employees are usually good employees. They have fewer accidents and less absenteeism than their (supposedly) healthier younger colleagues. However, the jobs most often offered them (handyman, cleaning lady, gardener) are low paying and often too physically taxing. The Older Americans Act, under Title V: Senior Community Service Employment Program (SCSEP) has authorized funds to all fifty states, and also to several national contractors, to provide part-time employment for older low-income persons and to encourage a transition from subsidized to unsubsidized jobs after training, support, and counseling (Coombs-Ficke and Lordeman, 1984). The SCSEP has spurred the development of many innovative programs to ease the poverty-stricken senior employment problem. However, much remains to be done for other relatively healthy, capable older persons who are struggling below, at, or barely above the poverty level.

A less mentioned problem of retirement is the adverse effect it has on some marriages. After years of one or both partners hurrying to dress, eat breakfast, and depart for a job, the change to both persons staying home can be stupefying. Many adjustments are needed. Conversation may be difficult. Partners may get in each other's way. Disputes often arise about how to do even the most trivial of household tasks. One woman in a study by Maas and Kuypers (1974) summarized her feelings this way: "I think it's harder on a wife when her husband retires. I mean, he was around underfoot all the time—and it meant three meals a day—you know—to get and prepare and clean up after and so on" (p. 140). Peterson and Payne (1975) found that marital satisfaction prior to retirement is the best predictor of the effects of more leisuretime on a marriage. Couples who were accustomed to sharing pleasurable leisure hours together before retirement had less trouble adjusting to more time together than couples who had a history of marital conflict.

LEISURE ACTIVITIES

Humans are social beings. Retirement often means the loss of contact with friends at work. With each passing year older people are also apt to experience through death the loss of other relatives, friends, and acquaintances. The diminution of social contacts can mean low morale, low self-esteem, and depression for many persons. Gerontologists (people who study aging and the personalities of the aged) now believe that old people who remain socially active usually have a higher level of satisfaction with their lives. One must remember that factors such as family relationships, health, and finances influence how much activity a given old person can have.

Some elderly persons resent the fact that their health, their spouse's health, or their lack of finances prevent them from getting about and socializing more. Other elderly persons accept their limitations and their diminished social contacts and are still satisfied.

At-home activities pursued by many old persons include socializing with friends, gardening or raising plants, reading, watching television, caring for a spouse or younger family member (grandchild or great-grandchild), household tasks, repairs, cooking, laundering, hobby or handicraft projects, reading, and card or game playing. Television programming is not, in general, aimed at older viewers. Many elderly persons prefer news programs to entertainments such as game shows, detective stories, or situation comedies (Kubey, 1980). Often reruns of older shows bring more enjoyment to them than do the newer programs. Out-of-the-home activities are usually less rigorous sports (bowling, shuffleboard, golfing), walking, cycling, doing volunteer work, touring, visiting, shopping, and church, political, or club-sponsored events.

Many retired persons become active as volunteers in hospitals, churches, and other community facilities. One of the more popular volunteer projects for old people today is the **Foster Grandparents Program**, in which old people become involved in teaching, supporting, and helping children with special needs (see Figure 11–7). Eisdorfer (1975) recalls having lunch with a group of Foster Grandparents working at a home for profoundly retarded children. He

Figure 11–7
Many old persons, with or without prior training, are excellent teachers of children. They are patient, persevering, and compassionate and can elicit a desire to learn on the part of the child, unequaled by a teacher with many students to motivate at once.

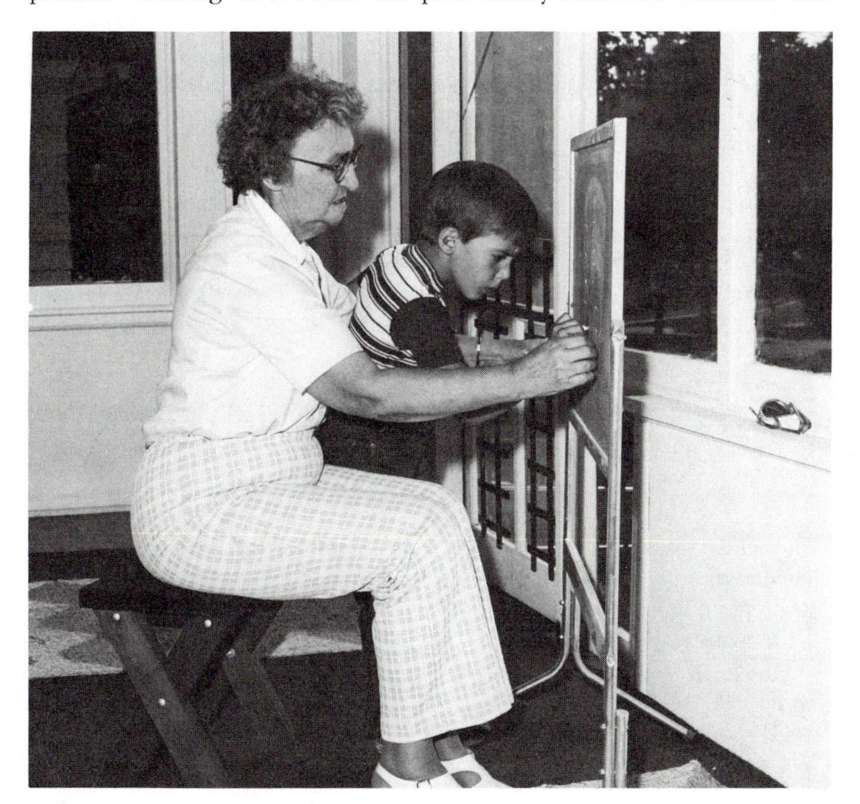

FOSTER GRANDPARENTS
Program that allows older persons to become involved in teaching, supporting, and helping children with special needs.

found them a lively, aggressive, involved group of human beings. A child psychiatrist then told him that when the elderly first arrived for in-service training sessions, they were a pitiful, depressed group of old people. They even refused to talk. However, after giving love and affection to the children, they changed as much as, if not more than, the children. Although some volunteer projects provide a tremendous feeling of self-worth to the elderly, others are fraught with difficulties. Volunteer work does not necessarily enhance well-being. The work must be meaningful and must be rewarded with honest praise and genuine appreciation if the worker is to feel good about it.

The recreational activities available to retired persons who want them and are able to enjoy them vary from community to community and from state to state. Because of the great numbers of retirees who move to warmer regions of the country, areas such as southern California, southern Arizona, and Florida have a considerable selection of recreations planned, produced, and maintained for older adults. Northern and midwestern cities and larger towns are also becoming increasingly aware of the large numbers of retired persons who compose their population and are planning, building, and supporting centers for senior citizens.

The types of leisure and recreational activities chosen by old people are dependent on their lifestyles and health. Maas and Kuypers (1974) identified ten different lifestyles into which the elderly can be clustered (see Table 11–2). The lifestyles into which women can be clustered are (1) husband-centered wives, (2) uncentered mothers, (3) visiting mothers, (4) work-centered mothers, (5) disabled-disengaging mothers, and (6) group-centered mothers. Husband-centered women are more interested in their spouses than in their children, grandchildren, and siblings. Uncentered women are most apt to live alone, be in poor health, and have meager financial resources. Visiting women have frequent social interactions in churches, clubs, and their own or others' homes. Work-centered women are more apt to be widows or divorcees, to live alone, to be in good health, and to enjoy full- or part-time jobs. The

Table 11–2 Lifestyles of the Later Years

Women	Men
Husband-centered wives	Family-centered fathers
Uncentered mothers	Hobbyists
Visiting mothers	Remotely sociable fathers
Work-centered mothers	Unwell-disengaged fathers
Disabled-disengaging mothers	
Group-centered mothers	

SOURCE: Adapted from Henry S. Maas and Joseph A. Kuypers, *From Thirty to Seventy* (San Francisco: Jossey-Bass, 1974).

disabled-disengaging mothers are frequently in poor health but live with others from whom they withdraw. The group-centered mothers differ from the visiting mothers in that they prefer formal social functions. They are usually from well-to-do families and have high education levels.

Lifestyles of the men may develop independently from those of women. A husband-centered wife is not, for example, necessarily married to a family-centered man. The four groupings into which elderly men can be placed are (1) family-centered fathers, (2) hobbyists, (3) remotely sociable fathers, and (4) unwell-disengaged fathers. Family-centered men have children living nearby whom they visit often. The hobbyists live farther from their children. The core of their lives is their leisuretime interests and activities. Remotely sociable fathers are more involved in formal activities (politics, social organizations) than in interpersonal relationships. Unwell-disengaged fathers are most apt to be in poor health and have few friends.

The lifestyles of the men Maas and Kuypers studied showed more continuity over the forty-year span into old age than did the living modes of women. This may be due to the fact that women are more affected by loss of children, loss or gain of outside employment, and changing marital and family situations than men. They make more adaptations in their lifestyles to cope with these altered circumstances.

PERSONALITY IN OLD AGE

Maas and Kuypers (1974) took a position contrary to that of Costa, McCrae, Block, and others (see Chapter 10) who argued that personality remains relatively stable over the adult life span. Maas and Kuypers reported that personality changes very definitely occurred over a forty-year time span in the persons they studied longitudinally (N = 142). The Maas and Kuypers data argued against the stereotype of all old people being negative, depressed, and disabled. Everyone was very individual, even down to his or her physical manifestations of age.

Maas and Kuypers saw **disengagement** (withdrawal and detachment from obligations, occupations, relationships) as a phenomenon mainly affecting people in ill health. Cumming and Henry (1961) had theorized that disengagement is a natural part of aging and that old people and society mutually separate from each other. Kuhlen (1964), after looking at the data on disengagement, felt that the separation is due more to the attitudes of society—old people withdraw because they feel unloved and unwanted, and they often resent being neglected. Havighurst, Neugarten, and Tobin (1968) found that in older people feelings of satisfaction with life were correlated more with participation than with disengagement. Robert Butler (1975a) stated in his Pulitzer Prize-winning book, *Why Survive? Being Old in America,* that disengagement of all old people from society is a myth. It is merely one of many patterns of reaction to old age.

In some North American families, the old reside with the young and are an integral, honored part of the household, whether they

help with the work load, add to it, or do a little of both. More often in our culture the old live apart from the young. Families differ in the ways they regard their elders: as beloved grandparents or great-grandparents; as dons or dowagers to be courted with favors in hopes of an inheritance; as burdens; as persons already three-quarters dead. Just as younger adults vary in the ways they regard their elders (sentimentally, solicitously, disdainfully, neglectfully), old persons vary in the ways in which they treat the younger generations (lovingly, angrily, disparagingly). Research has not shown that any particular racial, religious, or cultural group categorically experiences a better old age or increases the likelihood that a particular personality orientation will emerge by old age. Factors such as health, wealth, longevity, past experiences, surviving kin, and intrafamilial communication patterns make the personalities and lifestyles of persons in the United States over seventy highly diverse (see Figure 11–8).

Charlotte Bühler (1968) reported that most of the older European people she studied in the 1920s and 1930s could be classified into one of four personality types: (1) those who were satisfied with their past lives and were content to relax; (2) those who would not sit back but felt they must strive to the end; (3) those who were dissatisfied with their past lives and sat back with an unhappy air of

Figure 11–8
Persons over seventy do not fit into any one stereotype. They are just as different as the persons included in any other age bracket. They vary in activities, abilities, person orientations, health, and life satisfaction.

Table 11–3 Comparison of Personality Orientations of Aging

Bühler (1968)	Neugarten, Havighurst, and Tobin (1968)	Maas and Kuypers (1974)	
		Women	Men
Satisfied and content to relax	Integrated	Person-oriented mothers	Person-oriented fathers
Striving to the end	Armored-defended	Autonomous mothers	Active-competent fathers
Dissatisfied, unhappy, resigned	Passive-dependent	Fearful-ordering mothers	Conservative-ordering fathers
Dissatisfied, frustrated, guilt-ridden	Unintegrated	Anxious-assertive mothers	

NOTE: Categories across studies are not synonymous.

resignation; and (4) those who were dissatisfied with their lives and continued to experience regrets, frustrations, and feelings of guilt until the end (see Table 11–3).

Neugarten, Havighurst, and Tobin (1968) described four similar personality patterns of older Americans: (1) **integrated personalities**, (2) **armored-defended personalities**, (3) **passive-dependent personalities**, and (4) **unintegrated personalities**. Although integrated persons all faced up to their emotions and experienced rich inner lives, their outward behaviors varied from those who reorganized their activities, substituting new ones for old ones, to those who focused on one or two activities, to those who just chose a rocking-chair approach to old age. Armored-defended personalities held their impulses and emotions in tight harness. They were achievement driven and either wanted to work until the end or take care of themselves without outside assistance. Passive-dependent personalities were less satisfied with life. Although some remained apathetic, others gleaned a measure of contentment from leaning on and receiving succor from other persons. Unintegrated personalities were characterized by dissatisfactions, disorganization, and low activity levels.

Maas and Kuypers (1974) described seven typical personality patterns of the elderly. Four of the patterns picture women and three depict men. (The 142 subjects they studied ranged in age from sixty

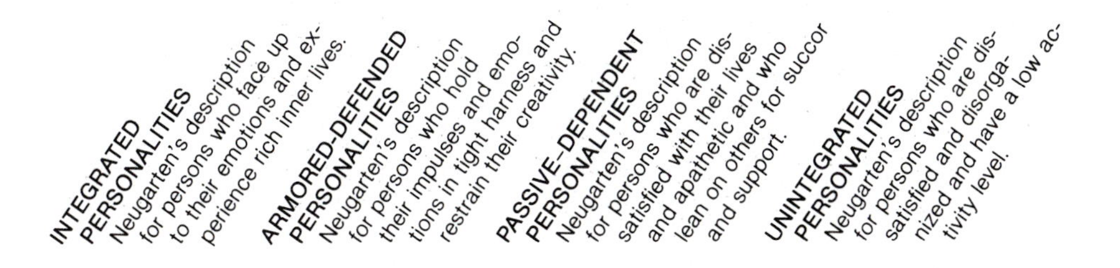

to eighty-two, with the average woman's age being sixty-nine and the men's seventy-one. All of the persons studied had been married and had had children.) Women were grouped as (1) person-oriented mothers, (2) fearful-ordering mothers, (3) autonomous mothers, and (4) anxious-assertive mothers. Person-oriented mothers were giving, sympathetic, warm, and well liked. Fearful-ordering mothers were withdrawn, reassurance-seeking, anxious individuals who found little personal meaning to their worlds. Autonomous mothers were involved more with formal organizations than with their families. They kept others at a distance and were independent, productive, and self-defensive. Anxious-assertive mothers were histrionic and self-dramatizing and were apt to be moody and hostile as well as talkative. Men were grouped as (1) person-oriented fathers, (2) active-competent fathers, and (3) conservative-ordering fathers. The person-oriented men were warm, sympathetic, and giving. They sought more reassurance and were less poised than the person-oriented women but were still popular. Active-competent men were power-conscious, aloof, verbally fluent, charming persons who were also rebellious and nonconforming to social expectations. Conservative-ordering men were similar to the fearful-ordering women; however, they were not as anxious and afraid. They were self-satisfied but also overcontrolled, repressive, and conventional.

While some personality changes may seem pronounced after retirement, there is probably a great deal of stability in the underlying behavior patterns of each person. These patterns influence the quality and quantity of each person's changes. Attempts to place old people into personality groupings such as those just described may successfully characterize the majority of the elderly, but there will always be unique old people. You may be able to think of at least one old person you know who does not fit comfortably into any of the descriptions just presented.

MARRIAGE

A myth about personality in the aged attacked by Butler (1975b) but still prevalent in our society is a belief that emotions become dulled as people age. The **myth of serenity** portrays the elderly as peaceful, carefree, relaxed persons, beyond the storms and stresses of their earlier years. In fact, emotional reactions of the elderly remain powerful enough to trigger divorces, marriages, outbursts of rage, psychophysiologic illnesses, deep depressions, or any of the other manifestations of tumultuous emotions that affect younger adults. As reported in Chapter 10, marriages do tend to be happier and more companionable in older persons. Wives often become more assertive with or more dominant over their husbands, especially if the husband's health is failing (Troll, Miller, and Atchley, 1979). Some aging husbands and wives perceive that they have lost their power to accomplish change by giving vent to their emotions. Consequently, they may practice silence. However, this reaction does not indicate that they are immune to insults, injustices, rejections, feelings of despair, and the like.

Many people also believe sexual desire is dulled with age. However, as reported in Chapter 10, sexual interests and performance need not cease after menopause or after a male experiences slowed erection time and diminished ejaculatory volume. When partners are interested and willing to help each other find ways to stimulate libido, sexual activity can often be maintained throughout the life span. Corbett (1981) reported that because of the **myth of a sexless old age**, some nursing homes do not even allow married couples to share the same bedroom. Sexuality in the elderly is not an expression of senility, nor of deviance, but of continued good health: sexual, physical, social, and emotional.

Butler (1969) described a sentiment in our society similar to racism and sexism, but aimed at old persons, that he termed **ageism**. It is a process of systematic stereotyping of old people in which they are seen as old-fashioned in morality, rigid in thought and manner, and senile. Ageism allows younger adults to see the elderly as something very different from themselves and consequently permits them to stop identifying with old people. Ageism takes a cruel turn where sexuality is concerned. Old people often become the butt of sexual jokes; they are portrayed as exhibitionists, dirty old men (or women), or as never-say-quit impotents who will try every quack nostrum or gadget offered them to improve libido. In fact, young adults buy more of the sexual gadgets and potions. The average age of a "dirty old man" is about twenty-seven, not seventy.

Some elderly persons become increasingly bored by or hostile toward each other as they spend more and more of their waking hours together under the same roof after retirement. Some file for divorce. Some separate. Extramarital affairs are not unknown among the elderly. Nor are late-life marriages or remarriages (see Figure 11–9). About 60 percent of late-life remarriages are of older men to younger women. Occasionally, an older woman will marry a younger man. However, many persons in this culture still frown on this breech of custom. Many older women prefer not to remarry after the death of a spouse. They are tired from their nursing/caregiving duties to their late husbands and want respite. Many cannot imagine replacing a loved and lost companion with anyone new, believing it would show disrespect. Despite some widow's hesitations, however, many older women do remarry. Butler (1975b) found that a poignant sense of tenderness—a feeling that each encounter is precious because it may be the last—often exists and characterizes the relationships between persons who have met and married late in life.

MYTH OF SERENITY False belief that all old people are peaceful, relaxed, and carefree.

MYTH OF A SEXLESS OLD AGE False belief that all old people have lost their sexual desires.

AGEISM A widely prevalent social attitude that overvalues youth and discriminates against the elderly.

Figure 11–9
Marital relationships in the later years are often devoted and blissful, as exemplified by this couple's remarriage on their fiftieth wedding anniversary.

WIDOWS AND WIDOWERS

There are many more widows than widowers in the world, due to the longer life expectancy of women. Since the custom for women to marry men older than themselves is so prevalent, many women outlive their husbands for a decade or more. There are many widows and widowers under, as well as over, age seventy. Our discussion, however, will be concerned with older persons who have lost their spouses.

The last chapter of this book will focus on death and bereavement and examine the ways in which people handle their emotions at the time of death. The death of a beloved person, especially one with whom another has had an amiable, daily relationship over a number of years, leaves the survivor feeling lost and threatened and as if some part of the self had died. A numbness and disbelief follow immediately after death. The more difficult period of bereavement occurs a month or more after the loss (see Chapter 12).

A person's reactions to losing a spouse vary according to how companionable the marriage was, how much warning the person had of the approaching death, how independent the person is, how supportive family and friends are, and how many financial burdens are left to the widow or widower. In general, women have a less difficult time adjusting to the death of a spouse than men. A widow is more apt to have observed the reactions of other women to their husbands' deaths for several years. She is more apt than her husband to put herself in the widow's place and imagine what life will be like without a spouse. This is referred to as rehearsal for widowhood. Men seldom do such rehearsing because it is expected that a man will predecease his wife. A woman usually knows other widows to whom she can talk confidentially, pouring out her grief, fears, loneliness, or bitterness. Women are more apt to have harmonious ties to many persons, both within and outside the family from whom they can seek sympathy, nurturance, and support (see Figure 11–10). Many acquire pets, plants, or new time-absorbing hobbies.

Having adequate financial resources makes adjustment to widowhood much easier. Greater difficulties arise for those who are saddled with debts from their marriage or from the funeral or who must cope with diminished income because of the spouse's death.

The widower is more apt to suffer acute loneliness after the death of his wife. Often she was his only true confidante. A man may feel that he has to be courageous and unemotional in his bereavement. He may refuse to discuss death with others, hiding his intense emotions rather than defusing them through conversation. Widowers

Figure 11–10
Women tend to have more interactions with other women for purposes of questioning, consoling, sharing, advising, or "visiting."

are also apt to be less independent in terms of living alone than are widows. Only about 8 percent of older widowers live alone compared to about 32 percent of older widows. Although they may have been fiercely independent in their work-a-day worlds, many know little about cooking, laundry, and the like.

The mortality rate for older widowed men is about 60 percent higher than it is for married men of the same age (Helsing and Szklo, 1981). It is not unusual for a man to die of cardiovascular disease, coronary heart disease, a cerebrovascular accident, cancer, or some other disorder within a year of his wife's death. Suicide rates are higher in widowed persons than in the married; they are especially high in older widowed men. The incidence of alcohol abuse is also high in widowers over age seventy. In addition, elderly widowers have a high rate of depression, anxiety disorders, and lowered self-esteem.

Many cities have organized self-help groups for widows and widowers to encourage them to help each other through the bereavement and readjustment process after the loss of a spouse. Although widowers may be more in need of such support groups, they are less apt to attend meetings or establish social networks with other grieving men or women. They are more apt to try to hide their emotions or to seek out a single woman for support and remarry.

EXTENDED FAMILY RELATIONSHIPS

The fact that about three-quarters of the 30 million old people in the nation today live apart from their families is not entirely a matter of ageism or the young rejecting the old. Sussman (1976) found that a majority (60 percent) of younger adults indicated they would be willing to let their elderly parents move in with them. Many old people prefer to live alone, maintaining their privacy and their sense of competence and self-determination for as long as possible. They do not want to impose on their children or become burdens to them. It is difficult for parents to move in and become dependent on children, a reversal from the years and years of having their children dependent on them.

A preference for self-help as long as possible generally does not mean that old people prefer to be disengaged from their families and friends. Parents still appreciate phone calls, letters, and visits from children, grandchildren, and great-grandchildren if the reasons for the communications are friendly, not demanding or hostile.

A majority of elderly people live within a half hour's drive of at least one of their family. Families differ greatly in the amount and quality of time spent with their old people. In general, contact with children and grandchildren is frequent and may increase as the second generation move into their teens, become young adults, or have children of their own. Close to half of the grandparents over age seventy are also great-grandparents. It is not uncommon for the elderly to enjoy even great-great-grandchildren as the years pass. There have been at least eleven cases of great-great-great-grandparents reported in the past twenty years (see Figure 11–11).

Due to some physical disabilities, elderly persons frequently

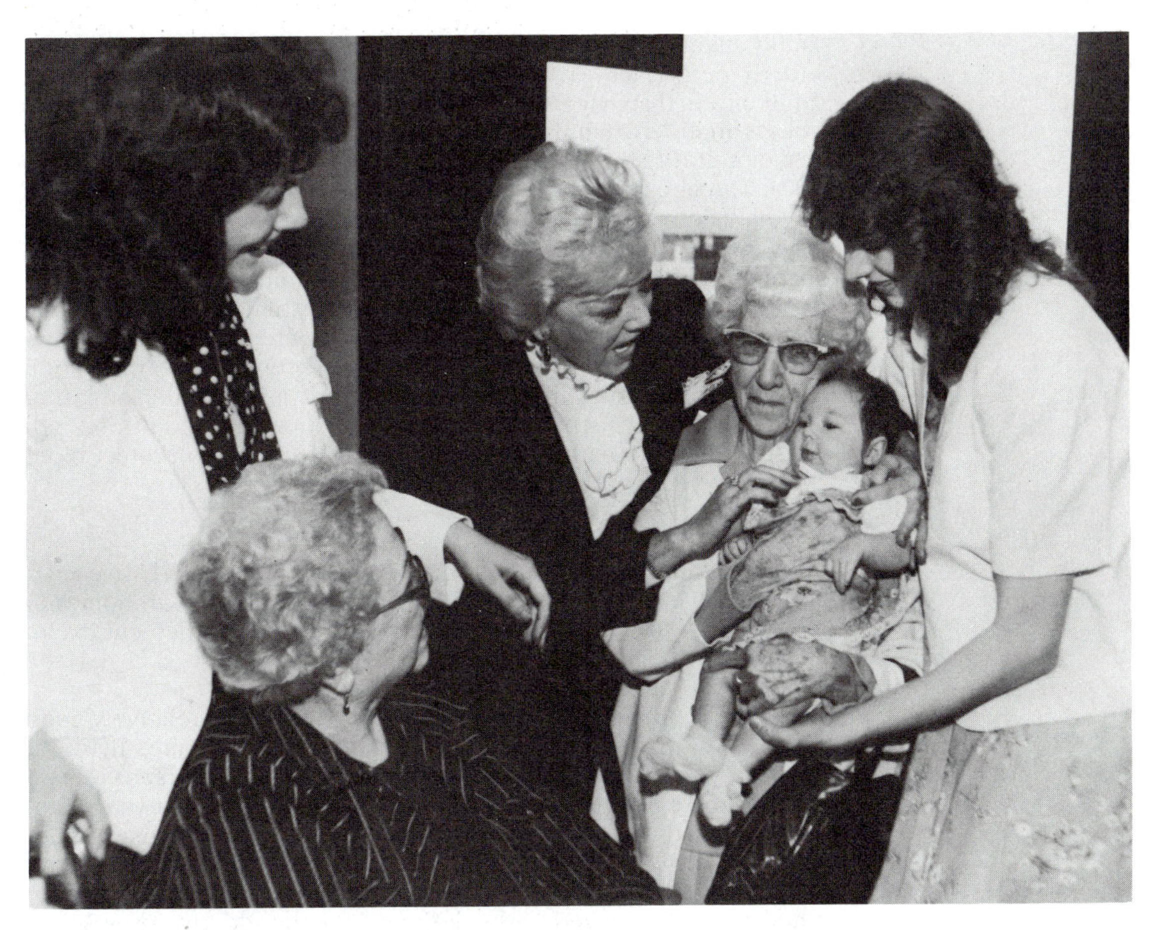

Figure 11–11
This jubilant grandmother, great-great-grandmother, great-grandmother, great-great-great-grandmother, and mother welcome a sixth living generation of women to their family. The ninety-year-old great-great-great-grandmother was congratulated on her status by the President calling from the White House after the great-great-great-granddaugher was born.

prefer to have family members come to visit them rather than making trips to see family. Here are three typical comments about children and grandchildren made by people in the Maas and Kuypers (1974) study: (1) "The children always want me to come more than I can. . . . " (2) "I kind of wish they did live closer. . . . When you get older, you miss them." (3) "It's a pleasure for a little while [to be with grandchildren]. I wouldn't say that I could be with them all day long" (pp. 137, 143, 146). Cicirelli (1981) pointed out that the extended family, particularly adult children, help elderly parents with physical chores (home repairs, housework, shopping). Increasingly, elderly persons help each other, revoking the need for visits from children for these reasons.

Much has been written recently about the abuse of elderly persons perpetrated by younger extended family members. The extent of **elder abuse** is unknown but is guessed to affect about 4 percent of the aged population. It is more apt to happen when an older person lives with a younger abusive adult. Very few cases come to the attention of appropriate authorities. This may be because the older, frailer person is intimidated by the younger, stronger family member or because the elderly person still loves the son, daughter, or other family member too much to report him or her. In some cases, the elderly person may actually be taking care of the younger abusive adult more than vice versa. The characteristics of a person who abuses an elderly family member are often similar to the characteristics of a child abuser (see page 255) or a spouse abuser (see page 499). The abusive person may feel powerless; may have a history of having been abused or of witnessing abuse; may have low self-esteem; may have few friends or confidants; may have his or her own marital problems, employment problems, mental or physical health problems; or may believe that the abuse is justified because of something the older person said or did, or neglected to say or do. Laws such as those that now exist for child and spouse abuse protection are needed to assure that neighbors, friends, medical personnel, or even other family members will report elder abuse. In addition, shelters or other alternative living facilities are needed to protect older persons while younger abusers are prosecuted.

LIVING FACILITIES

As has been mentioned, about three-quarters of the elderly live apart from their families. The large majority of them still live in homes of their own. Many live in rented facilities. Only about 5 percent live in hospitals or homes for the aged.

One's Own Home For many persons who still enjoy relatively good health, few physical disabilities, and adequate income, living at home may bring great happiness. While many retired persons remain in the homes in which they have lived during early adult years, others opt to sell and buy something smaller, or something in a sunnier climate. They must compete for available housing with younger families. Increasingly, retired persons are choosing mobile homes located in well-managed parks designed for the elderly. They select them for the low demands on their energy, the low maintenance costs, and the independence associated with owning one's own home rather than renting from someone else. Most older mobile home dwellers show a high level of satisfaction with their living arrangements (Mackin, 1985). Another alternative that allows older persons to own their own home, to maintain privacy and independence, and to cut maintenance chores and upkeep costs, is an ECHO house (Hare and Haske, 1984). This is a small, temporary living unit placed in the yard of another single-family home. An older person can rent out his or her larger home and live in the ECHO. Or the elderly can choose to live in an ECHO close to, but separate from, extended family members. Some middle-aged adults build accessory apartments within their homes for their aging parents. This allows older persons to maintain some privacy and independence but may make them feel more like renters, or boarders, than homeowners.

Condominiums, cooperative homes, or small homes in specially designed retirement communities are popular places for people to move in their later years. Carp (1975) studied many old persons who moved from their own housing to a federally financed community for the elderly. He reported that after one year 99 percent of them rated the project as a very good place to live. More importantly, after eight years 90 percent of them still rated the place as very good, with another 9 percent rating it as okay. They were able to adjust to the modern setting and make new friends in the adjacent Senior Center. He suggested that morale is generally good in planned societies where the inhabitants are of a comparable age range and activities are designed to be of interest and within the ability levels of the participants.

Renting Some elderly people rent dwellings for their later years. Satisfaction or dissatisfaction is very much dependent on what one can afford to rent and the services provided by the landlord or landlady. Although all repairs and services should be the responsibility of the owner of the rental property, rentals are often neglected. Tenants may be afraid of being evicted if they complain. Most retirees have a fixed income. Regular rent increases are a hardship to them. Some rental properties available to the elderly are in poor condition (walk-up flats, dingy rooms in hotels or boarding houses, poorly heated trailers), and they reflect the meager amount spent to maintain them.

The government has tried to alleviate some of the problems of inadequate housing for the aged by making public housing units available to them. Tenants generally are not required to pay more than 25 percent of their income to rent such federally supported dwellings. However, annual income eligibility requirements generally are fixed at levels that exclude both the elderly who are too poor or slightly too rich. There is a shortage of federally subsidized senior rental housing (and also of units subsidized by state or local governments). Even when the elderly meet income eligibility requirements, they may be placed on a waiting list. It may then be well over a year before a unit is available. In the meantime, they must live where they are, despite conditions, or try to find some other temporary alternative living facility. When an older person knows that a subsidized rental unit may become available on short notice, he or she usually tries to find an alternative place to live with a short-term lease that can be terminated without undue penalty. Owners of rentals usually prefer long-term leases. This can prove very frustrating for seniors with limited fixed incomes.

Living Alone Over 10 million elderly Americans live alone. Approximately 32 percent of older women and 8 percent of older men are adapting, more or less successfully, to this situation due to singlehood, divorce, or death of their spouse. For those who have neighbors, friends, and family around or some outside employment, the arrangement may be satisfactory. Some older Americans, in fact, jealously guard their freedom to live alone. They prefer to come and go as they please. They can often escape their feelings of solitude by doing volunteer work or visiting geriatric activity centers. Some programs such as meals-on-wheels, dial-a-bus, and homemaker services bring help to people who are too disabled or ill to get about easily by themselves when living alone.

Poverty is prevalent in the aged population. As reported earlier, about 70 percent of today's elderly poor are women. Moen (1983) reported that nearly a quarter of the elderly women she surveyed were living in poverty. Characteristics that often correlate with female poverty are widowhood, low level of education, chronic health problems, being nonwhite, service occupation employment, and less time in the labor force than other women. Many exist in substandard dwellings. It may be difficult for them to keep up their homes if they are chronically ill or crippled. Without an adequate income, household repairs and upkeep costs can be staggering. Plumbing, electrical wiring, furnaces, pipes, or drains may require attention. Termites or rodents may create problems; heavy winds or rains can do considerable damage, and one must face continual bills for utilities, heating, and taxes. Many women live in dwellings without inside flush toilets, or hot water showers or baths, and with minimal heat in winter and general conditions of deterioration. Many of the severely impoverished elderly homeowners are trapped in their poverty. They cannot find buyers for their homes or afford to move elsewhere (see Figure 11–12).

Many of the elderly poor in the United States live in rural areas, especially in the central states. Federally subsidized programs for the aged are concentrated in urban areas, especially on the Atlantic, Pacific, and Gulf coasts. Few serve the needs of persons living, for example, in the mountains or on small farms scattered across the prairies. Many old people live without any means of transportation (public or private), far from the nearest towns or sources of supplies and medical care. Some do not even have telephones to link them to others. Similar situations exist for poor rural old people in Canada. Often, mail delivery is the closest link to society.

Nursing Homes When health problems become too disabling, the best solution, especially for poverty-stricken old people with no significant family, is to move into a nursing home. Three times more females than males choose this option, reflecting both the greater number of widows compared to widowers in the aged population and the widows' relative greater poverty.

Nursing homes are generally regarded as a last resort as a living facility for the aged. They are not, however, the place where most old people go to die. Only about 20 percent of the aged population ever live in a nursing home. At any given moment in time, only about 5 percent of the aged population reside in such facilities. They have pros and cons. The pros generally concern problems of family, economics, and health. About 50 percent of all older persons in nursing homes have only a distant cousin, niece, or nephew somewhere but no relation who could reasonably be expected to take care of them. The poverty that affects about one-third of the elderly also makes nursing homes desirable. Widows, in particular, often cannot financially cope with self-care. Finally, the majority of nursing home residents have several chronic health problems. These leave them in need of around-the-clock medical supervision beyond that which can be provided by family members or nurses in private homes. Thus,

NURSING HOMES Residences providing needed care for the infirm, disabled, and chronically ill.

Figure 11–12
Many older Americans live in homes that are falling into disrepair. They are unable to pay for renovations, unable to sell, and unable to move elsewhere.

for many persons nursing homes or extended-care hospital settings are the best living facilities available for the later years.

Nursing homes are frequently envisioned as foul-smelling, barrackslike dumping grounds for old persons who are demented and "vegetating" until death. This is not an accurate picture. While facilities differ from those that are genuinely homelike to those that are more barrackslike, and patients differ from those who are mentally keen to those who are demented and from those who are mobile to those who are paralyzed, most nursing homes attempt to provide the best possible physical and mental care for each resident.

Just as nursing homes differ and patients differ, so too do families who place an older and infirm relative in a nursing home. Some love the old person very much but simply cannot provide adequate care at home any longer. Others do, unfortunately, abandon the elderly relative for whom they have little compassion or tolerance. For some middle-aged children, placing a parent in a nursing home can have positive effects. Smith and Bengtson (1979) found that in about 15 percent of the families they studied, the middle-aged child found a new sense of love and affection to replace an old sense of duty and obligation to the aged parent after nursing home placement. Rather than having to cope with the strenuous physical task of caregiving, the child visited the parent to really communicate. In another approximate 30 percent of families, Smith and Bengtson found that nursing home placement of the aged and infirm parent strengthened the already existing bond of love and affection, and in about 25 percent of families, it kept the bond at a continued level of closeness. In the remaining 30 percent of families, nursing home placement continued the pattern of separateness, gave parent and child feelings of anger and guilt, respectively, or was seen as a virtual "dumping" of parent by child, where the child abdicated all his or her responsibilities to the nursing home.

If a man or woman has no living family or few friends left, he or she may find friends among the residents of the home. If a person cannot afford nursing home care, Medicaid will pay the costs. Medicaid rules vary from state to state. In some states a person must be rid of all financial assets (home, stocks and bonds, savings for a funeral) before he or she is eligible. In others they may hold on to a few possessions and valuables and still receive Medicaid. Medicaid, Medicare, veteran's benefits, and social security checks go right to the nursing home once a person resides there. However, the law provides that each resident of a nursing home should receive an allowance for his or her own personal use. While medical attention varies from home to home, nursing homes are theoretically better equipped to handle medical emergencies than most private residences could be. At present, our society is struggling to answer the question of how much and for how long Medicare or Medicaid should be provided to sustain the lives of terminally ill patients in both hospitals and nursing homes. This question has both moral and legal implications and will not be easily resolved.

Nursing homes are big business in North America. They receive over $30 billion annually for the care of the aged and infirm. Most have over one hundred beds. Many are chain-operated—run by the same owner as several other nursing homes in the region. Chains large enough to be listed on either the New York or the American Stock Exchange control about 10 percent of all the old-age homes in the country. With such far-reaching interests, it is sometimes difficult for the administrators of the large homes to be concerned about the problems of individual patients. In the best homes the staff members show concern for each person, using respectful titles ("Mr. Hine") rather than nicknames ("Pops," "Gramps") or diagnoses ("the coronary in room 402"). However, personnel in some homes are poorly trained, disrespectful, or even cruel to the patients. In the best homes the residents are helped to get out of bed, dress, and participate in exercises (walks, physical therapy) and social activities (outings, games, parties, recreational therapy). However, in some homes the highlights of the day are meals and bed changes. Some patients, for medical reasons, are never dressed and taken out of bed. All nursing homes are subject to annual inspections to insure that they meet state and federal standards for patient care, fire, safety, and sanitation.

There are questions to ask and observations to make in assessing a potential nursing home for a friend or relative. Do a number of patients have bedsores? Are most patients up and dressed? What's the rapport with the staff? Do patients receive regular baths? Are their teeth well cared for? Do doctors make regular visits? Is there enough staff? Is the place clean and well maintained? Often, local social service agencies or offices for the aging can provide information about nearby facilities and answers to these questions above and beyond what one sees on a visit through a facility. Some facilities are not certified to care for patients covered by medicaid. If it is likely that private resources will be exhausted and medicaid payments will be needed eventually, it is important to select a facility that is certified.

Every nursing home must guarantee its residents and their families certain rights. Among these are 30 days' advance notice before a transfer or discharge, the right to receive up-to-date information on diagnosis, treatment, and prognosis, the right to refuse treatment, a written explanation of services provided, the right to privacy in medical examinations and treatment, the right to privacy with visitors, the right to receive and send unopened mail and untapped telephone calls, and the right to present grievances about the facility to a state agency that protects the rights of nursing home residents.

Summary

The later years may be spent in active pursuit of goals or at a leisurely pace, by healthy individuals or by those debilitated by disease. The ecological settings of persons over seventy are as varied as those of other humans at any time in the life span.

Benign senescence, the process of growing old, is marked by body changes in structure and function (loss of bone density and mass, loss of some visual and auditory acuity, loss of pigmentation in skin and hair). The changes are gradual and often begin quite early in adulthood.

Research has not sufficiently demonstrated that cognitive disabilities occur in all older persons. Many persons demonstrate stable cognitive abilities until very late in life.

The majority of the elderly have at least one chronic health problem. The most common include degenerative diseases of bones and joints, coronary heart disease, and vascular disorders. Exercise, proper diet (adequate fluid and nutrients, fewer excess calories, fewer fats, and increased fiber), and prompt treatment of disease symptoms can often extend one's life span. While cerebrovascular accidents (strokes), transient ischemic attacks (TIAs), Alzheimer's disease, or unknown factors may cause senile dementia in some old people, "senility" is not synonymous with old age. Strokes and TIAs are not always associated with senile dementia. Alzheimer's disease is.

Mandatory retirement at age seventy can be a blow to the finances, social life, and self-esteem of men and women after years of steady, paid employment outside the home. Conversely, it may be welcomed. Some workers opt to retire even before age seventy. Many retired persons become active as volunteers in hospitals, churches, and community projects. Many also pursue leisure activities planned for older adults in their communities.

Personality orientations and lifestyles in the later years are diverse. Personalities may range from integrated to armored-defended, to passive-dependent, to unintegrated, with many variations of personalities in each category. Lifestyles may range from spouse-centered to family-centered, visiting-centered, work-centered, group-centered, uncentered, hobby-centered, remotely sociable, disabled, unwell, and disengaged.

Marriages in old age, like marriages in younger adult years, range from very happy to very unhappy. Divorces and remarriages are not foreign to the aged population.

Life expectancy for women exceeds that of men by about seven years, making many more widows than widowers in our country. Many more widowers than widows remarry. Becoming single after years of marriage alters one's lifestyle in many ways, most of which require difficult readjustments.

About three-quarters of the nation's older people live apart from their children, but a majority live close enough to visit or be visited by one or more children or grandchildren at frequent intervals. Most older persons maintain their own homes or apartments for as long as possible. Only a small minority live in hospitals or homes for the aged.

Questions for Review

1. Describe the various physical changes that occur during benign senescence.
2. Various studies have concluded that intelligence declines over time. Critique these studies. Are they valid? In your critique use the information in Chapter 1 on evaluating research.
3. Describe the various causes of strokes and the various forms of treatment.
4. Your neighbor's father has been diagnosed as probably having Alzheimer's disease. She asks you what to expect. What will your answer be?

5. Mandatory retirement has recently come under attack. Do you believe people should be forced to retire at a specific age? Discuss the pros and cons of mandatory retirement.
6. Disengagement mainly affects people of ill health. Why do you think this is so?
7. Describe life experiences that might lead to the four personality patterns outlined by Neugarten, Havighurst, and Tobin.
8. Describe some examples of ageism that exist in the mass media—newspapers, magazines, and television.
9. Why do you think it is true that widowers tend to remarry at a much greater rate than widows?
10. Communities and housing projects have been established for older persons. Some people feel that these living situations tend to isolate older individuals, whereas others feel they are beneficial. What is your opinion? Discuss.
11. Would you feel guilty about putting one of your parents in a nursing home? Explain your answer.

References

American College of Physicians. *Guide for adult immunizations.* Philadelphia: American College of Physicians, 1985.

American Heart Association. *Heart facts.* Dallas: American Heart Association, 1985.

Avioli, L. V. Osteoporosis: A guide to detection. *Modern Medicine,* February, 1986, 28–42.

Baltes, P. B., and Schaie, K. W. On the plasticity of intelligence in adulthood and old age: Where Horn and Donaldson fail. *American Psychologist,* 1976, *31*(10), 720–725.

Barnes, R. F.; Raskind, M. A.; Scott, M.; and Murphy, C. Problems of families caring for Alzheimer patients: Use of a support group. *Journal of the American Geriatric Society,* 1981, *29*(2), 80–85.

Batten, M. Life spans. *Science Digest,* 1984, 46–51, 98.

Botwinick, J. Geropsychology. *Annual Review of Psychology,* 1970, *21,* 239–272.

Bühler, C. The developmental structure of goal setting in group and individual studies. In C. Bühler and F. Massarik, (Eds.), *The course of human life.* New York: Springer, 1968.

Butler, R. Ageism: Another form of bigotry. *Gerontologist,* 1969, *9,* 243–245.

Butler, R. *Why survive? Being old in America.* New York: Harper & Row, 1975. (a)

Butler, R. Sex after sixty-five. In L. Brown and E. Ellis (Eds.), *Quality of life: The later years.* Acton, Mass.: Publishing Sciences Group, 1975. (b)

Carp, F. Housing and living arrangements. In L. Brown and E. Ellis (Eds.), *Quality of life: The later years.* Acton, Mass.: Publishing Sciences Group, 1975.

Cicirelli, V. G. *Helping elderly parents: The role of adult children.* Boston: Auburn House, 1981.

Coombs-Ficke, S., and Lordeman, A. State units launch employment initiatives. *Aging Magazine,* February/March 1984, 18–22.

Corbett, L. The last sexual taboo: Sex in old age. *Medical Aspects of Human Sexuality,* 1981, *15*(4), 117–131.

Cox, H., and Bhak, A. Symbolic interaction and retirement adjustment: An empirical assessment. *International Journal of Aging and Human Development,* 1979, *9*(3), 279–286.

Cumming, E., and Henry, W. *Growing old: The process of disengagement.* New York: Basic Books, 1961.

Dement, W. C.; Miles, L. E.; and Bliwise, D. L. Physiological markers of aging: Human sleep pattern changes. In M. E. Reff and E. L. Schneider (Eds.), *Biological markers of aging.* Bethesda, Md.: NIH (Publication No. 82-2221), 1982.

Denckla, W. D. Pituitary inhibitor of thyroxine. *Federation Proceedings: Federation of American Societies for Experimental Biology,* 1975, *34*(1), 96.

Eisdorfer, C. The WAIS performance of the aged: A retest evaluation. *Journal of Gerontology,* 1963, *18,* 169–172.

Eisdorfer, C. Making life worth living. In L. Brown and E. Ellis (Eds.), *Quality of life: The later years.* Acton, Mass.: Publishing Sciences Group, 1975.

Eisdorfer, C., and Wilkie, F. Intellectual changes. In L. Jarvik, C. Eisdorfer, and J. Blum (Eds.), *Intellectual functioning in adults.* New York: Springer, 1973.

Fields, W. S. Aspirin for prevention of stroke: A review. *American Journal of Medicine,* 1983, *74*(6A), 61–65.

Gregerman, R. I., and Bierman, E. L. Aging and hormones. In R. H. Williams (Ed.), *Textbook of endocrinology,* 6th ed. Philadelphia: Saunders, 1981.

Hare, P. H., and Haske, M. Innovative living arrangements: A source of long-term care. *Aging Magazine,* January 1984, 3–8.

Havighurst, R. *Developmental tasks and education,* 3rd ed. New York: McKay, 1972.

Havighurst, R.; Neugarten, B.; and Tobin, S. Disengagement and patterns of aging. In B. Neugarten (Ed.), *Middle age and aging.* Chicago: University of Chicago Press, 1968.

Hayflick, L. The aging of humans and their cultured cells. *Resident and Staff Physician,* 1984, *30*(8), 36–44.

Helsing, K. J., and Szklo, M. Mortality after bereavement. *American Journal of Epidemiology,* 1981, *114,* 41–52.

Heron, A., and Chown, S. *Age and function.* Boston: Little, Brown, 1967.

Horn, J. L. The aging of human abilities. In B. Wolman (Ed.), *Handbook of developmental psychology.* Englewood Cliffs, N.J.: Prentice-Hall, 1982.

Horn, J. L., and Donaldson, G. Cognitive development II: Adulthood development of human abilities. In O. G. Brim, Jr. and J. Kagan (Eds.), *Constancy and change in human development: A volume of review essays.* Cambridge, Mass.: Harvard University Press, 1980.

Jensen, J.; Christiansen, C.; and Rodbro, P. Cigarette smoking, serum estrogens, and bone loss during hormone-replacement therapy after menopause. *The New England Journal of Medicine,* 1985, *313*(16), 973–975.

Kubey, R. W. Television and aging: Past, present and future, *The Gerontologist,* 1980, *20,* 16–35.

Kuhlen, R. Developmental changes in motivation during the adult years. In J. Birren (Ed.), *Relations of development and aging.* Springfield, Ill.: Thomas, 1964.

Licastro, F., and Walford, R. L. Effects exerted by prostaglandins and indomethacin on the immune response during aging. *Gerontology,* 1986, *32,* 1–9.

Maas, H., and Kuypers, J. *From thirty to seventy.* San Francisco: Jossey-Bass, 1974.

Mace, N., and Rabins, P. *The 36 hour day: Family guide to caring for persons with Alzheimer's disease, related dementing illnesses and memory loss in later life.* Baltimore: Johns Hopkins University Press, 1984.

Mackin, J. Mobile homes popular with elderly. *Human Ecology Forum,* 1985, *15*(2), 21–22.

Mannik, M., and Gilliland, B. C. Degenerative joint disease. In K. J. Isselbacher et al. (Eds.), *Harrison's principles of internal medicine,* 9th ed. New York: McGraw-Hill, 1980.

McCracken, J. The company tells me I'm old. *Saturday Review,* August 7, 1976.

Mendelson, W. B. Sleep after forty. *American Family Physician,* 1984, *29*(1), 135–139.

Moen, P. Is employment the solution to the poverty of female headed households? Paper presented at the meeting of the National Council on Family Relations, Minneapolis, October 1983.

Mohr, J. P.; Fisher, C. M.; and Adams, R. D. Cerebrovascular diseases. In K. J. Isselbacher et al. (Eds.), *Harrison's principles of internal medicine,* 9th ed. New York: McGraw-Hill, 1980.

Neugarten, B.; Havighurst, R.; and Tobin, S. Personality and patterns of aging. In B. Neugarten (Ed.), *Middle age and aging.* Chicago: University of Chicago Press, 1968.

Pardini, A. Exercise, vitality, and aging. *Aging,* 1984, 19–29.

Peterson, J. A., and Payne, B. *Love in the later years.* New York: Associated Press, 1975.

Riegel, K. F., and Riegel, R. M. Development, drop, and death. *Developmental Psychology,* 1972, *6,* 306–319.

Salzman, R. T. Management of rheumatoid arthritis and osteoarthritis. *American Journal of Medicine,* 1983, *75*(48), 91.

Schaie, K. W. Rigidity-flexibility and intelligence: A cross-sectional study of the adult life span from twenty to seventy years. *Psychological Monographs,* 1958, *72* (Whole No. 462).

Schaie, K. W., and Strother, C. R. The effects of time and cohort differences on the interpretation of age changes in cognitive behavior. *Multivariate Behavior Research,* 1968, *3,* 259–294.

Sinex, F. M., and Myers, R. H. Alzheimer's disease, Down's syndrome, and aging: The genetic approach. In F. M. Sinex and C. R. Merril (Eds.), *Alzheimer's disease, Down's syndrome and aging.* New York: New York Academy of Sciences, 1982.

Smith, K. F., and Bengtson, V. L. Positive consequences of institutionalization: Solidarity between elderly parents and their middle-aged children. *The Gerontologist,* 1979, *19*(5), 438–447.

Stare, F. Three score and ten plus more. *Journal of the American Geriatric Society,* 1977, *25,* 529–533.

Sussman, M. B. Family life of old people. In R. H. Binstock and E. Shanas (Eds.), *Handbook of aging and the social sciences.* New York: Van Nostrand Reinhold, 1976.

Troll, L. E.; Miller, S. J.; and Atchley, R. C. *Families in later life.* Belmont, Calif.: Wadsworth, 1979.

Wechsler, D. *The measurement and appraisal of adult intelligence,* 4th ed. Baltimore: Williams & Wilkins, 1958.

Whitbourne, S. K. *The aging body: Physiological changes and psychological consequences.* New York: Springer-Verlag, 1985.

Whitehouse, P. J., et al. Alzheimer's disease and senile dementia: Loss of neurons in the basal forebrain. *Science,* 1982, *215,* 1237–1239.

Wilkie, F., and Eisdorfer, C. Terminal changes in intelligence. In E. Palmore (Ed.), *Normal aging II: Reports from the Duke longitudinal studies, 1970–1973.* Durham, N.C.: Duke University Press, 1974.

Zatz, M. M., and Goldstein, A. L. Thymosins, lymphokines, and the immunology of aging. *Gerontology,* 1985, *31,* 263–277.

Death and Bereavement Throughout the Life Span

12

Chapter Outline

Key Concepts

acceptance stage
acute grief
anger stage
anticipatory grief
anticipatory guidance
bargaining stage
bereavement
bereavement process
bereavement reactions
broken heart
denial stage
depression stage
euthanasia
grief
grief work
grieving process
hospice
intermediate grief phase
Kübler-Ross stages of dying
living will
morbid grief reactions
mourning
personal fable
personification of death
recovery phase of bereavement
thanatology

♥

Case Study

For a year Margaret had been providing around-the-clock care for her husband, who was disabled and demented after several small strokes. Her children, all of whom lived far away, had offered to help her with Papa's care. They had also hired a nurse to help her. But she resisted assistance and asked the nurse to leave. She wanted to live as independently as possible and care for her husband herself.

Margaret was exhausted. She had had a cold for several weeks that never got better. She asked someone to watch her husband while she went to a physician. The doctor wanted to hospitalize her for some tests. An adult son and his wife came home to care for Papa while she was in the hospital. Her diagnosis was an acute leukemia, a rare type for which there is still no known effective therapy. Her physician wanted to transfer her to a major research facility in a distant city for experimental chemotherapy. The drugs could possibly extend her life for a few additional months. Margaret said no. She preferred to stay close to her husband, home, and friends.

An adult daughter arrived from out of town to spend time with Margaret in the hospital. Margaret did not want to be in the hospital. After a week, she persuaded her physician to discontinue the blood transfusions and let her go home to be cared for by her daughter.

Although she was very weak, Margaret had her daughter help her prepare to die. They made phone calls and wrote good-bye letters to her friends. They went through closets and drawers and she asked that her daughter give certain things away to special people or organizations. When relatives and friends visited her, she spoke reassuringly. She was not afraid to die. She viewed death only as a transition.

She often lay with her head in her daughter's lap while her daughter read to her. They held hands as much as possible. She liked physical contact. Even though her husband was barely aware of what was happening, he also participated in the hand holding.

Margaret died in her daughter's arms two weeks after her discharge from the hospital. Her husband was close by, watching. Although Papa was usually confused, he was able to reassure his daughter at that time with "Don't cry, we knew she was going to die."

Margaret probably could have lived a little longer if she had agreed to a transfer to a cancer center for chemotherapy, or even if she had remained in her local hospital for blood transfusions. Should her family have allowed her to make the choices she made?

Thanatology, the study of death, is growing as a field of scientific inquiry. Mortality and immortality have always been concerns of theologians and philosophers. Meanwhile, most humans remain afraid of death. According to Shakespeare, Julius Caesar reasoned thus with his wife:

> Cowards die many times before their deaths;
> The valiant never taste of death but once.
> Of all the wonders that I yet have heard,
> It seems to me most strange that men should fear;
> Seeing that death, a necessary end,
> Will come when it will come.

Freud (1920/1966) felt that fear of death is related to a destructive or death instinct (Thanatos), which has parity with a self-preservation instinct (Eros). Humans hate to die, they see no reason for it, and it becomes their agony (Burton, 1974).

Today social scientists probing death and dying are attempting to break down the taboos against discussing the end of life. By discussing death they hope to help us all learn to live more comfortably with a knowledge of our finite state.

Perspectives on Death

The emotions a loved one's death evokes are disquieting: fear, anger, anxiety, jealousy, recriminations, weeping, and anguish. And yet, Kübler-Ross (1969) tells us that a proximity to death can be an enriching, growth-promoting experience. If a dying person can honestly express a host of negative and confusing feelings, he or she can move to a final stage of peaceful acceptance of death. If a mourner can express a host of chaotic, anguished feelings, he or she can move to a stage of rebuilding a rewarding new life. And if doctors, nurses, social workers, and religious and other counselors can learn to accept their own fears and concerns about death, they can be enormously helpful to dying patients. They can help terminally ill persons grow and find harmony. In addition, they can be inspired rather than depressed by such experiences.

THANATOLOGY
The study of death.

In this chapter death will be examined first from the point of view of the dying individual: child, young adult, and older adult. Then the ways in which death affects bereaved persons will be discussed. Finally, death will be viewed as a catalyst to enjoying life more fully. The text will end by exploring some perspectives on living.

APPROACHES OF CHILDREN TO DEATH

Children are minimally aware of death when very young. In fact, the finality of death may not be fully realized until children approach the end of their elementary school years. Each child's understanding is very much dependent on his or her own experiences with death, teachings about death, and cognitive maturity.

Preschool children view death as changed circumstances. They believe that powerful adults should be able to make things that have disappeared reappear. They do not recognize death as a final process. Rather, they seem to believe that the deceased will someday return or come to life again at the will of a parent, or perhaps, God. For many preschool children death is experienced first through the loss of a pet. The fact that pets are often rather quickly replaced with new pets may help perpetuate a young child's belief that death is temporary or causes only a slight change in ongoing situations. Children also have ample opportunities to witness reincarnations of heroes, heroines, villains, cartoon characters, and the like on television programs. It is difficult for preoperational children to separate fantasy from reality.

School-age children develop a greater understanding of death with each passing year, with each new experience of learning about death of another animal or person, and with each new lesson about death taught to them by caregivers, teachers, religious leaders, or friends. Lansdown and Benjamin (1985) found that approximately 60 percent of the children they studied had a fairly well developed notion of the reality and finality of death by age five, and all of them understood it by age nine. Between the ages of five and nine, many children still believe that death does not have to be permanent. They believe that a dead person may come back to life for a while to finish some tasks or see loved ones. A school-age child whose grandfather and mother had both died within a few months of each other wrote a letter to God asking to have his mother returned. He explained, "You have my Grandpa and all the angels to make you laugh. I need my mother."

Very often school-age children engage in the **personification of death**. They see it as some invisible force that carries people off. The personification of death is often individualistic and ranges from night, to invisible animal, to monster, to some deceased person, to skeleton, to angel of death, to God. Another school-age child exemplified the personification of death as God. "You know, God is going to come down from Heaven and take me back with Him" (Morrissey, 1965). Death is generally believed to occur at night, in the dark, when nobody can see. Many nightmares and night terrors of children may stem from their anxieties about the link between death and

PERSONIFICATION OF DEATH Embodiment of death as a human figure.

night. Many bedtime prayers reinforce this link; for example, "If I die before I wake, I pray Thee Lord my soul to take." Adults often refer to death as eternal rest or eternal sleep. In fact, more deaths occur while people are pursuing waking activities than while they sleep, most often in the hours between six in the evening and midnight (Lamberg, 1984). It is important for adults to avoid equating death with sleep while talking with children, lest they become anxious about taking naps or falling asleep at night.

Some children equate death with risk taking or with traveling far from home and caregivers. The media coverage of distant wars, coups, police actions, assassinations, or violence in general and reports of other distant disasters with many fatalities, such as volcanic eruptions, earthquakes, tornados, hurricanes, and airplane crashes, make many children fearful of the safety of leaving home. After the 1986 tragedy of the space shuttle *Challenger,* many children felt that Christa McAuliffe would not have died if she had only stayed home (Lipsitt, 1986).

With increasing cognitive maturity, teenagers begin to work on personal philosophies of both life and death. They often take an intense interest in the concepts of death and afterlife offered by their own and other people's religions. The age at which a child is allowed to attend a funeral varies from family to family, and from situation to situation, but most adolescents are considered old enough to attend a funeral if they so desire. They may be profoundly affected by the experience. Elkind (1981) described a **personal fable** that develops in most adolescents. This is a belief in one's own immortality and invulnerability. Adolescents may view a casket, or the deceased's body, and decide that death happens only to others. Freud (1920/1966), as well as Elkind, noted adolescents' increased risk taking as evidence of the great distance they feel from death. Freud believed that id impulses of the Thanatos (death force) type are discharged against others (aggression) or against the self (risk taking) in increased intensity during the adolescent years. While some adolescents may believe in their own immortality, others may develop a serious intention to experience death, as evidenced by the high attempted and completed suicide rates among teenagers (see Chapter 7).

When preschool children are hospitalized with terminal illnesses, their reactions of fright are more related to fear of separation and pain than to any awareness or fear of death. Preschoolers are still very much dependent on their caregivers for physical help with their activities and for emotional support and nurturance. When caregivers are replaced by doctors or nurses, as often occurs in hospital settings, the preschooler may become anxious. Not only are caregivers absent, but all of the other familiar sights and sounds (home, siblings, pets, friends) that add to a feeling of security are lost. The terminal illness itself is seldom a threat beyond the fact that the child may experience pain and discomfort from it.

School-age children who are hospitalized with terminal illnesses often have more anxiety about medical procedures (such as

PERSONAL FABLE
A belief in one's own immortality and invulnerability.

intravenous feedings, blood transfusions, x-rays, bone marrow aspirations, blood tests, or injections) than about approaching death.

If a school-age child still views death as a temporary phenomenon, he or she will have hopes of everything eventually returning to normal, despite death. In some children, especially younger ones, parents and medical personnel may have tried to shield the patient from knowledge of the seriousness of the illness. Even when a prognosis has not been given, the child often perceives the seriousness of the situation. The grief of the parents becomes obvious to the child. Eventually the child may ask "Am I going to die?" Opinions differ as to how to answer. Marlow (1973) stated her belief that honesty is best. For example, an answer may be worded carefully: "Yes, but we are not certain when this will be." Many professionals feel that a physician or nurse should not answer without the parents' permission, since they may prefer to do it themselves or with the help of a religious counselor. If parents or professionals hedge on such questions, the child may become extremely anxious. The fear of death is worse when one cannot discuss it. It is especially traumatic for a child to be left alone when family members are feeling that the future is too terrible to mention. Many children want the reassurance that someone will be with them when they die, and that death will not be too painful.

Binger, Mikkelsen, and Waechter (1970) believed that in some cases children may realize that they are dying and try to shield their parents from the fact. This lack of open, honest communication can prevent acceptance of death and the growth that accompanies acceptance in both the dying child and the parents. The dying child is still alive. He or she needs comforting, loving, and help to overcome some of the overwhelming terrors being experienced. If parents and professionals listen, children will usually give an indication of how much they know or suspect and what particular concerns they wish to discuss. Morrissey (1965) found that once a child is aware of impending death, one of three patterns of reaction may occur:

1. Anxiety expressed symbolically and physiologically.
2. Anger with acting-out behaviors.
3. Depression with withdrawal.

The reactions may be sequential or may even occur almost simultaneously. There is a danger that the dying child who misbehaves and makes demands on the parents may be overindulged. Such overindulgence can be more upsetting than helpful. Children feel more secure when they know their limits. When limits are stripped away, the child may make preposterous requests to test the parents. Parents will invariably come to resent this. The dying child usually senses this resentment and may accuse them of not being loving any longer. If demands are then met to prove love, a vicious cycle is begun. The child will make more and more requests. The parents will become more and more frustrated and resentful. The child will sense the hostility, and, ultimately, the child may die feeling unloved or rejected. The parents may feel relieved of their demanding child as well as grieved. The guilt feelings for feeling relieved can be enormous (Marlow, 1973).

APPROACHES OF YOUNGER ADULTS TO DEATH

Keller, Sherry, and Piotrowski (1984) studied adults' evaluation of death in general, beliefs in the hereafter, and death anxiety related to the self in six age groups from young adulthood through old age. Many middle-age and late middle-age persons show little anxiety about death. Becker (1973) suggested that they may deny the threat of death by sublimating themselves in activities related to family, career, and achievements. Younger and older adults show more anxiety about death in general, but older adults show less anxiety about their own death. Older adults evidence more beliefs in a hereafter than younger adults, and women have more beliefs in an afterlife than do men at all ages.

Adults with fatal illnesses usually know or guess their prognosis quite soon after physicians and family members learn the facts of the disease. The physician is responsible for imparting the information to the patient. However, this is often fraught with difficulties. Many adults resist having the prognosis spoken aloud. They may not be ready to deal with this information about their own death. In addition, many physicians, nurses, aides, orderlies, clerks, and volunteers avoid talking to dying patients, especially about death. If they do so, they soften the blow with euphemisms such as hyperplastic tissue, lesion, or neoplastic proliferation (for cancer) or cerebral ischemia (for strokes). Most doctors admit that they learned not to inform patients through experience, not in school. Kastenbaum and Aisenberg (1972) questioned 200 nurses and attendants about how they responded to patients' talk about death. Five patterns emerged:

1. Reassurance: "You're doing so well now. You don't have to feel this way."
2. Denial: "You don't really mean that . . . you're not going to die. Oh, you're going to live to be a hundred."
3. Changing the subject: "Let's think of something more cheerful. You shouldn't say things like that; there are better things to talk about."
4. Fatalism: "We are all going to die sometime, and it's a good thing we don't know when. When God wants you, He will take you."
5. Discussion: "What makes you feel that way today? Is it something that happened, something somebody said?"

Changing the subject and avoiding any discussion of death was most common. Such neglect, denial, fatalism, or "turning off" the patient makes it especially difficult for an adult to approach the fact that he or she has but a short time left to live.

Kübler-Ross (1969) identified five stages through which persons pass when they realize death is imminent. (Teenagers and cognitively mature children may also progress through the same stages in trying to accept their deaths.) The **Kübler-Ross stages of dying** are:

1. Denial.
2. Anger.
3. Bargaining.
4. Depression.
5. Acceptance.

Some persons may move back and forth across these stages or experience two of them simultaneously. On occasion they occur out of the sequence presented. Some persons may never reach the acceptance stage. It is important to remember that Kübler-Ross's research was conducted with patients who knew that they were dying. Kübler-Ross, a Swiss physician and psychiatrist who immigrated to the United States in 1958, spent many years working with dying patients at the University of Chicago. She voluntarily spent many hours with each person, rather than providing the more typical brief physician-to-patient visit. She listened and gave each dying person her own unconditional love and acceptance. The same stages may not be descriptive of persons who have not been told, or have not learned or even guessed, the terminal nature of their illness. These stages may also not be descriptive of patients who, because of extreme pain, are kept in a heavily drugged state of impaired consciousness.

In the first stage, that of **denial**, the patient may vigorously oppose the notion that the end is near. As Thomas Bell (1961) wrote:

> This can't be happening to me. Not to me. Me with a malignant tumor? Me with only a few months to live? Nonsense. . . . Such things happen, should happen, only to other people. . . . People who are strangers . . . born solely to fill such quotas.

The patient may hop from doctor to doctor, or from clinic to health spa to faith healer to miracle worker seeking a different diagnosis or a cure for the illness. He or she may become isolated, insulated from friends or acquaintances who know the truth. A reaction of shock, in which the person seems dazed, temporarily helpless, and confused, may ensue. Eventually, when the patient can no longer say, "This can't be happening to me," he or she will move to the second stage, that of anger. The question becomes, "Why me?"

At the **anger** stage the patient may strike out especially hard at loved ones. The terminal illness may be blamed on the spouse, parents, siblings, children, best friend, doctor, employer, or God. Fre-

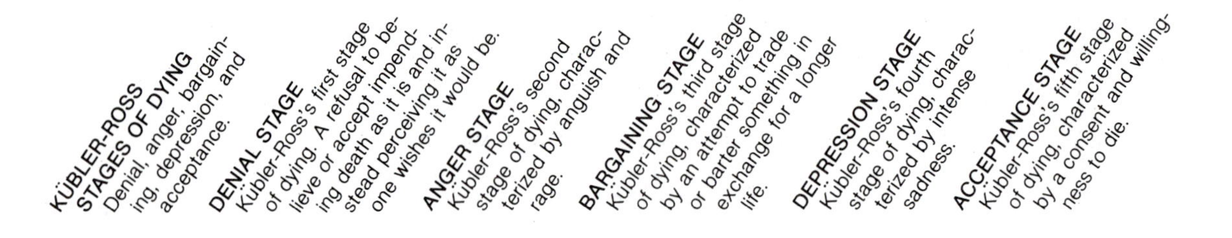

quently, a vehement jealousy arises against people in good health. It is not unusual at this time for the patient to behave in agonized ways: cursing, accusing, condemning, screaming, being aggressive, being destructive, demonstrating hate and bitterness. The venting of such strong emotions may help a patient pass through this stage more readily than trying to choke back the mental and physical anguish being experienced.

The next stage, that of **bargaining**, usually begins when the anger is somewhat dissipated. The patient may agree to alter his or her behavior, become devoted to a religion, give away goods, buy some special treatments, or purchase wares for doctors or persons responsible for his or her health; in short, he or she may attempt to strike almost any kind of bargain in exchange for a longer life. Bargaining with God through prayers or exchanges with a religious representative are common. Variations of the Faust legend, in which people try to sell their souls to the devil to prolong their lives, have also been written, based on the tumultuous experiences of this stage.

Eventually, when the patient gives up hope that bargaining will effect a cure, he or she usually becomes **depressed**. The patient may refuse visitors, show little interest in external events, be silent even with loved family members and friends, and simply cry softly or stare into space. It is as if the patient is preparing for the time when he or she will no longer be able to see or hear others. This mourning need not be the final stage before death.

Kübler-Ross identified a fifth stage that many persons reach: **acceptance**. The patient leaves depression behind and goes about saying and doing all the unfinished business of his or her life. Rather than acting defeated or bitter, the patient seems to accept impending death with a peaceful, quiet sense of expectation. Often the acceptance is expressed by a desire to have just one or two close friends near, to hold hands, to listen to reading or quiet music, or simply to share silence. The patient seems to have reached the ego integrity described by Erikson (1963): "acceptance of one's one and only life cycle as something that had to be and . . . permitted of no substitutions." The fear of death is removed.

Kübler-Ross (1975) not only described the stages through which dying persons pass but also identified what others can do to be of most help to terminally ill patients. She taught many physicians, nurses, social workers, religious counselors, and others at seminars at the University of Chicago, the University of Colorado, and eventually around the world to overcome their own anxieties about death. In doing so, they became better able to communicate with dying persons. By the late 1970s, she also began "Death, Dying, and Transition" workshops for persons from a wider variety of backgrounds, often including terminally ill patients and their relatives. Workshops lasted five days and allowed participants to share their own personal despairs and come to a deeper acceptance of themselves and others. Kübler-Ross extended her work with the dying to include work with living persons who have experienced dying close at hand (Vietnam veterans, parents of murdered children, families of suicide victims). She has shared her insight through many articles and books as well as through lectures, seminars, and workshops. In *Death: The Final Stage of Growth,* she presented five rules that summed up one minister's notes from a seminar:

1. . . . I must try to be myself. If the dying patient repulses me, for whatever reason, I must face up to that repulsion. I also must let the other person be himself. . . .
2. . . . when we talk to each other about ourselves, we will find something in common.
3. . . . let the patient "tell him" (the counselor) how he feels . . . "let the patient be." This simple rule does not imply granting all of the patient's demands and jumping whenever the patient wants the counselor to jump. . . . The belief that "I know what's best for the patient" is not true. The patient knows best.
4. I must continually ask myself "What kind of a promise am I making to this patient and to myself?" . . . [Is it to] save this person's life or to make him happy in an unendurable situation(?) . . . stop trying to attempt both. If I can learn to understand my own feelings of frustration, rage, and disappointment, then I believe I have the capacity to handle these feelings in a constructive manner.
5. My fifth and last rule . . . is expressed in the Alcoholics Anonymous prayer: "God grant me the serenity to accept the things I cannot change, the courage to change the things I can, and the wisdom to know the difference." (p. xviii)

The persons who are most helpful to a dying person are those who allow communication to proceed honestly and those who have come to terms with their own frustrations and anxieties about death. However, family members may find it extremely difficult to talk about death or to understand and control their own emotions. Family members may go through stages similar to those of the patient—denial, anger, bargaining, and depression—before they can accept the prognosis. Their own grief sometimes leaves them as much in need of support as the patient approaching death. Occasionally, the patient may try to help family members or may simply prefer not to see them.

In order to meet the needs of dying patients, medical practitioners, social service personnel, family, and friends should be aware of their major requirements: (1) the need to control pain, (2) the need to retain dignity or feelings of self-worth, and (3) the need for love and affection (Schulz, 1978). Dignity can be enhanced by allowing the dying adult as much control over decisions affecting his or her death as over the rest of the life span. Love and affection can be shown by listening, imparting desired information, perceiving the patient's approach to death (allowing for self-respect and dignity), and reassuring him or her that others will be there to provide love until the end. It can also be shown through the tender touch: holding hands, stroking, touching.

APPROACHES OF OLDER ADULTS TO DEATH

Old people are generally less afraid of their own death than are young adults who have not yet had a chance to pursue careers or raise families. However, the approaches of individual old people to death vary tremendously. Whereas some welcome death, others try to postpone it in every way possible. Some will discuss it; to others it is a taboo topic. The variety of views old people have about death were expressed in interviews conducted by Maas and Kuypers (1974):

> "Life and death go together—that's the natural way. . . . I don't like to—just—think about death. . . . We're ready to die before we're ready to know how to live." (p. 137)
> "Talk about death? No, we never have thought to talk about it. We're always too busy doing something." (p. 140)
> "I think death would be wonderful. . . . There's a time for everybody and when it comes, it comes. . . . Death to me is a way out of this troubled world that we're in. And I think sometimes that death is going to be peaceful." (p. 141)

Old persons have all had experience witnessing the deaths of others around them. This practice in mourning and coming to terms with grief and anxieties concerning the death of others gives them some degree of preparation for facing their own finiteness. Butler (1975) pointed out that many older persons show a perseveration in reminiscing about their lives. Did they make sense? Were they worthwhile? By so doing, they seek to find some way to assure themselves that their lives were meaningful and thus prepare themselves to face death more peacefully.

When an elderly person is terminally ill, he or she is still apt to go through the stages described by Kübler-Ross: denial, anger, bargaining, depression, and finally acceptance. Physicians or family members may try to shelter the elderly from knowledge that his or her illness will be fatal. However, dying persons usually know their status, whether or not the prognosis is put in words. They may have financial or personal matters that they want to put in order. Older persons should be allowed to take care of such matters when they desire. The terminal decline in intelligence that precedes death (see Chapter 11) may render some individuals incapable of accomplishing logical, ordered thinking when death is near.

Fanslow (1984) wrote about the calming effect of touch on elderly dying patients. She has taught nurses and family members to hold hands, to place one's hand near the patient's heart, or to stroke the patient rhythmically to decrease predeath anxiety and provide the sense of peace needed for finishing business with surviving relatives and friends. Touching is therapeutic for family and friends as well. It helps them adjust to the reality of the imminent death and to cope with the difficulties of holding on and letting go.

BOX 12–1

Should Life Be Sustained in a Permanent Vegetative State?

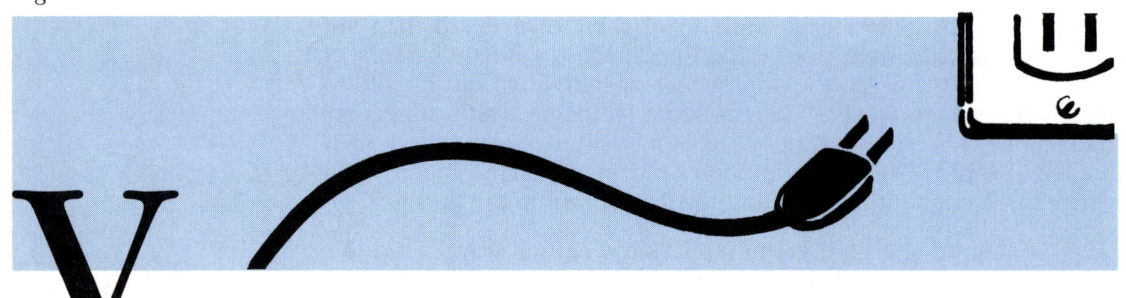

Varying versions of the Hippocratic Oath have been taken for over 2,000 years by physicians entering the practice of medicine. A current version, approved by the American Medical Association, includes the phrase "That you will exercise your art solely for the cure of your patients." Physicians are also advised by the American Medical Association's Council on Ethical and Judicial Affairs that they may discontinue all life supports (respirators, dialysis, even food and water) for patients who are in irreversible comas from which there is no hope of cure. This directive does not mandate the discontinuance of life supports; it simply allows that such action is not unethical in cases where it is the wish of the family and perhaps also the previously stated wish of the patient (as in a living will signed prior to the comatose state). Each physician must examine his or her own conscience to decide for how long to attempt to cure a patient, or alternatively, whether to allow the comatose incurable patient to die with dignity minus a roomful of life-sustaining apparatuses.

The problem goes beyond a moral-ethical dilemma. Should life be maintained at costs in excess of $250 per day ($8,000 a month, $100,000 a year) in patients who are in a permanent vegetative state? Few families can afford such long-term care. The fees, instead, are usually paid by private insurance companies or government medicare/medicaid. Some patients can survive for many years in irreversible comas. The price of insurance for each insuree goes up in proportion to these costs of comprehensive medical care.

In some cases, families must go to court to obtain an injunction requiring that physicians, hospital staff, or nursing home staff cease to care for persons in permanent vegetative states. The life-or-death decision is then in the hands of neither the medical community nor the close family members but rather an impartial judge or jury.

Who should determine how and for how long to treat a person in an irreversible coma? The question will not be an easy one to decide, and the answers may be years in coming.

Some persons sign what is called a **living will** earlier in their lives or when a terminal illness is diagnosed. This legal document, which must be witnessed and notarized, states that the individual does not desire to be kept alive by artificial means but rather would like to be allowed to die with dignity (see Figure 12–1). A standardized "living will" is available from euthanasia societies throughout the United States. The concept of **euthanasia** (literally meaning good death) has become a controversial one, generally associated with mercy killing. Various religious and philosophical groups differ in their opinions of where one draws a line between not making heroic efforts to sustain life and actually allowing or even hastening death by withholding life-supporting equipment or medicines (see Box 12–1).

LIVING WILL
A legal document stating that a person does not wish to be kept alive by artificial means; signed while a person is healthy.

To Make Best Use of Your Living Will
You may wish to add specific statements to the Living Will in the space provided for that purpose above your signature. Possible additional provisions are:

1. "Measures of artificial life-support in the face of impending death that I specifically refuse are:
   a) Electrical or mechanical resuscitation of my heart when it has stopped beating.
   b) Nasogastric tube feeding when I am paralyzed or unable to take nourishment by mouth.
   c) Mechanical respiration when I am no longer able to sustain my own breathing.
   d) ____."
2. "I would like to live out my last days at home rather than in a hospital if it does not jeopardize the chance of my recovery to a meaningful and sentient life or does not impose an undue burden on my family."
3. "If any of my tissues are sound and would be of value as transplants to other people, I freely give my permission for such donation."

The optional Durable Power of Attorney feature allows you to name someone else to serve as your proxy in case you are unable to communicate your wishes. Should you choose to fill in this portion of the document, you must have your signature notarized.

If you choose more than one proxy for decision-making on your behalf, please give order of priority (1, 2, 3, etc.).

Space is provided at the bottom of the Living Will for notarization should you choose to have your Living Will witnessed by a Notary Public.

Remember . . .
- Sign and date your Living Will. Your two witnesses, who should not be blood relatives or beneficiaries of your property will, should also sign in the spaces provided.
- Discuss your Living Will with your doctors; if they agree with you, give them copies of your signed Living Will document for them to add to your medical file.

My Living Will

To My Family, My Physician, My Lawyer and All Others Whom It May Concern

Death is as much a reality as birth, growth, maturity, and old age – it is the one certainty of life. If the time comes when I can no longer take part in decisions for my own future, let this statement stand as an expression of my wishes and directions, while I am still of sound mind.

If at such a time the situation should arise in which there is no reasonable expectation of my recovery from extreme physical or mental disability, I direct that I be allowed to die and not be kept alive by medications, artificial means, or "heroic measures." I do, however, ask that medication be mercifully administered to me to alleviate suffering even though this may shorten my remaining life.

This statement is made after careful consideration and is in accordance with my strong convictions and beliefs. I want the wishes and directions here expressed carried out to the extent permitted by law. Insofar as they are not legally enforceable, I hope that those to whom this Will is addressed will regard themselves as morally bound by these provisions.

(Optional specific provisions to be made in this space – see other side)

DURABLE POWER OF ATTORNEY (optional)

I hereby designate ______________________ to serve as my attorney-in-fact for the purpose of making medical treatment decisions. This power of attorney shall remain effective in the event that I become incompetent or otherwise unable to make such decisions for myself.

Optional Notarization:

"Sworn and subscribed to

before me this _______ day

of _____________, 19_____."

____
Notary Public
(seal)

Signed ______________________

Date ______________________

Witness ______________________

____
Address

Witness ______________________

____
Address

Copies of this request have been given to ______________________

____

(Optional) My Living Will is registered with Concern for Dying (No. _______)

Figure 12–1
The living will.

The **hospice** movement, which is growing in popularity throughout Europe and North America, is not as controversial as the euthanasia movement. A hospice attempts to provide the most supportive climate possible, both physically and psychologically, for dying patients. Although a hospice is usually a specialized physical facility, hospice services may also be provided in one's own home (Munley, 1983). Since the amendment of the Social Security Act in 1982, medicare funds have been allowed to pay for hospice services for up to three years. Specially trained doctors, social workers, nurses, chaplains, physical therapists, and volunteers associated with the hospice system carefully regulate activities and pain-relieving substances (morphine, cocaine, gin) in such a way as to alleviate fears, pain, and discomfort without creating an agitated, depressed, or comatose state. They also attend carefully to the patient's emotional, social, and spiritual status. The environment is kept as loving and caring and peaceful as possible. Each dying person is respected as a unique individual with his or her own needs and desires. Each patient has autonomy regarding decisions about his or her own care. Constant attention is given by staff, volunteers, relatives, or friends: listening, holding hands, touching. By including family members in the hospice services, the hospice staff can also simultaneously help them work through their relationship to the dying person, help them work through the difficulties of holding on and letting go, and give some anticipatory guidance for what will occur after the death.

Anticipatory guidance involves preparing survivors for their probable actions and reactions at the time of death and bereavement. Families can do some **anticipatory grief** work as well: recognizing and coping with their feelings of denial, anger, and depression. Counselors can make them aware of their needs for supportive relationships and good communication within their support networks after the death. Although hospice staff are trained to help relatives and friends as well as dying patients cope with death, their anticipatory guidance procedures can and should be used by all medical, paramedical, and social services personnel.

Most people would prefer to die in their own homes (Kalish and Reynolds, 1976). Unfortunately, few families allow this to happen. Many people are afraid of continued close proximity to a dying person. Family members may not want to feel in any way responsible for the death finally occurring. In addition, many persons feel that the spirit of the dead person will continue to dwell in the room or in the home where death took place. They prefer that death occur in a distant place (hospital, nursing home, hospice) where the spirit can then reside with unknown others.

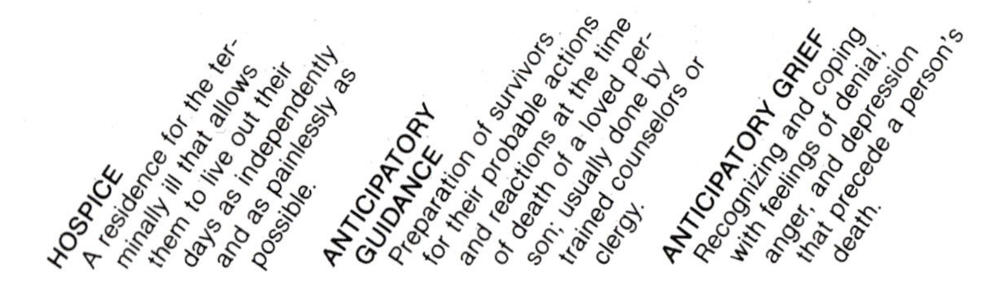

Bereavement and Grief

Thus far this chapter has focused on the reactions of persons to their own approaching deaths. The reactions of others to the death of a loved one are also painful. It does not matter that death may relieve the suffering of the loved person or, depending on religious convictions, that the departed is believed to have made a transition into a better afterlife. Surviving individuals will still miss a loved person who dies and will need to work through an assortment of disturbing emotions concerning the meaning of that death to their lives.

Each year an estimated 8 million Americans experience the death of an immediate family member. Each type of death (death of a child, death of a spouse, death of a parent, suicide) carries with it a special kind of acute pain. Most researchers concerned with stress agree that it is the most potent kind of stress in life. It can produce major changes in the functioning of the immune, respiratory, cardiovascular, endocrine, and autonomic nervous systems, with resulting changes in physical and mental health. The Committee for the Study of Health Consequences of the Stress of Bereavement (1984) provided the following definitions for reactions to a death of a loved person:

> **Grief**: The feeling (affect) and certain associated behaviors such as crying.
> **Grieving process**: The changing affective state over time.
> **Bereavement**: The fact of loss through death.
> **Bereavement reactions**: Any psychological, physiological, or behavioral response to bereavement.
> **Bereavement process**: The emergence of bereavement reactions over time.
> **Mourning**: The social expressions of grief, including mourning rituals and associated behaviors. (pp. 9–10)

Mourning rituals are vastly different across cultures, ethnic groups, social classes, and even rural and urban areas. What many people do not realize is that bereavement reactions and grief vary greatly as well, not only across social settings but also from individual to individual. The grieving process may involve rapid mood swings and uncontrollable emotions (from sadness, to anger, to fear, to love, to happiness, even to hatred). These feelings are unexpected and intense. They may leave the bereaved confused or ashamed. They may also leave other bereaved family members angry at what they perceive to be inappropriate affective reactions.

Researchers have found that the grieving process may begin before death (in cases of terminal illnesses) and may continue over a year beyond death. Books, magazines, articles, and radio and television shows are just beginning to educate the public about the different reactions to bereavement. Physicians, nurses, psychiatrists, psychologists, social workers, chaplains, and the like are becoming better prepared to support bereaved families as more and more information about reactions becomes known. In the remainder of this chapter we present an overview of some of the current knowledge about the grieving and bereavement processes.

Lindemann (1944), in a classic paper, described seven emotions that characterize **acute grief**:

1. Somatic distress occurring repeatedly and lasting from 20 minutes to an hour at a time.
2. A feeling of tightness in the throat.
3. Choking with shortness of breath.
4. A need for sighing.
5. An empty feeling in the abdomen.
6. Loss of muscular power.
7. An intense subjective distress described as tension or mental pain.

The duration of these symptoms may vary depending on the relationship between the deceased and the survivors. They are common during the first week after death when the funeral or memorial service is being held (see Box 12–2).

Intermediate grief reactions soon replace acute grief. They differ depending on age, sex, ethnicity, relationship to the dead person, prior experiences with grief, support networks, and preexisting mental or physical health. Let us look at different grief reactions depending on the age of the deceased.

REACTIONS TO THE DEATH OF A CHILD

There is no one way all parents react when one of their offspring dies. Some parents go into a state of shock. They may be confused and unable to act or react. Some parents may refuse to believe that the death has occurred. They behave as if the child will return soon. Some parents have emotional breakdowns. Some become agitated and acutely anxious about trivia and innumerable smaller events. Some parents are overcome with unreasonable fears. Others may be-

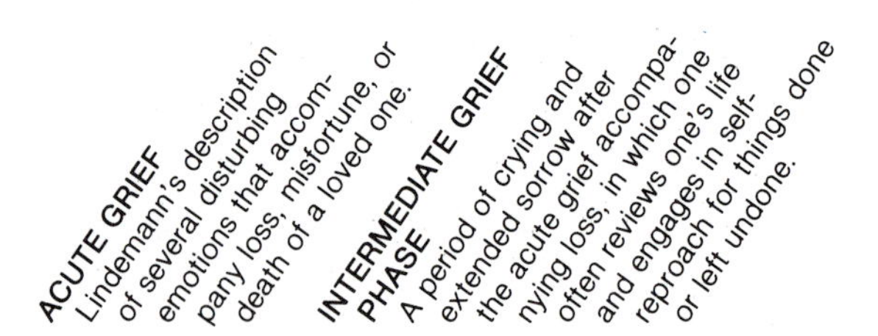

BOX 12–2
Funerals

Much attention has been directed toward the high price of burial in recent years. The average traditional funeral costs about $3,500, but it is not unusual for families to pay two to three times this amount for a funeral to honor their dead. Why so much? Consider the services: a casket, a cemetery plot, opening the grave, a vault to line the grave, embalming of the body, use of the funeral home, flowers, transportation of the body to the cemetery, additional limousines for the family, the graveside service, closing of the grave, a grave marker, and fees for services for funeral director, staff, and clergy.

Many persons are opting to bequeath their bodies to science. In such cases, the funeral is usually replaced by a memorial service. Such a service allows mourning rites and early grief work, as well as a memorialization of the life and contributions of the deceased.

The number of persons who opt for cremation has doubled in recent years, to close to 10 percent of bodies. Many funeral directors handle cremations as well as body burials. In fact, they may sell the family a casket for transportation to the crematorium, provide use of the funeral home, sell a cemetery plot and marker for the cremated remains, and direct the graveside service as well. Most states do not require a casket for a body destined for a crematorium. If the family elects to bypass a funeral home and deal directly with the crematorium, fees may be considerably lowered. Most persons elect to hold a memorial service for the dead even if they do not have a funeral to bury the ashes. Laws governing the disposal of ashes vary from state to state. Many crematoriums will provide a variety of urns or vaults for cremated remains in a wide range of prices. Cremation can be as expensive as a funeral.

Many states require neither embalming nor vaults to line the grave for body burials. A simple pine box in a simply lined grave will meet most legal requirements. It is often difficult, however, to find a place to purchase an inexpensive casket or to find a cemetery willing to bury the dead in an inexpensively lined grave.

Funeral expenses are occasionally covered by union pension funds, fraternal orders, insurance policies, bank trusts, or church organizations. In some cases, families can also qualify for financial assistance from social security or the Veterans Administration.

Many persons opt to arrange their own prepaid funerals. Some mortuaries are owned by giant corporations that also own cemeteries, monument works, flower stores, and other related services. For one fee, all services will be provided at the time of death. The law requires that all the money prepaid on such contracts be placed in escrow. Preneed contracts can be cancelled and all money refunded on written demand at any time prior to death. Preneed contracts allow people to elect a lower-priced (or even higher-priced) service than they believe their families might select in the emotionally taut, vulnerable days following death. Unless a will specifies funeral, cremation, or donation-to-science plans, the next of kin will be asked to make all decisions about burial soon after death occurs.

come deeply depressed. Some parents show their grief in emotional outpourings. Others give no visible signs of their feelings although the grieving process is very much a part of their inner lives. The sorrow parents feel manifests itself in different ways for different people.

The death of a child generally leaves everyone feeling that a great injustice has been done. The child had no chance to live out his or her life. A sense of bitterness compounds the grief. The work of mourning, called **grief work**, is to come to terms with the fact of death and to work out some of the bitterness, sense of loss, hurt, disappointment, and frustrations that accompany it. Parents who can express their feelings and allow others to comfort them are able to emerge more rapidly from their bereavement process.

The initial acute grief reaction to a death of a child usually gives way to a period of crying and extended sorrow. Parents should be encouraged to weep openly. After a few days to weeks the rate of crying will decrease and an intermediate grief phase will commence (Glick, Weiss, and Parkes, 1974). In this phase the parents will review the child's life. It is common for parents to search for their dead child in places where he or she spent a lot of time (bedroom, playroom, parks). It is also common for them to experience a sense of personal guilt for their child's death. Johnson-Soderberg (1983) found that guilt is more common in parents who had only a short time (under two weeks) to prepare for their child's death and is more common in mothers than in fathers. Guilt often complicates the intermediate grief phase and may contribute to the high divorce rate of parents who have lost a child by death. Johnson-Soderberg found that parents often believe that they committed an actual wrongdoing, either by commission (action) or omission (inaction) to oneself or to another, that caused harm and somehow related to the child's death. In many cases, the guilt is based on some deviation from a social norm (getting pregnant before marriage, mother working outside the home). Many parents experience obsessive, repetitive "movie scenes" about the death of their child during the intermediate grief phase. If such torturing movies are kept secret, the parent may feel that he or she is going crazy. Sharing them with others, and learning that other bereaved parents have also experienced such recurrent scenes, both alleviates guilt and provides relief from the movie.

Eventually the survivors will understand the futility of self-reproach or searches and will work at accepting the fact of death. This is the **recovery phase of bereavement**. They may make an appointment with the dead child's doctor and repeat the same questions they asked earlier. They may seek support from other bereaved parents, social workers, religious counselors, community counseling centers, psychologists, psychiatrists, relatives, or friends. They come to terms with their guilt, disappointment, and loss and accept the finality of the death. They decide to live again and resume social activities. They feel more self-assured. They turn their attention back to their living children (if they have others), to each other, and to other relatives and friends.

A danger inherent in parents' recovery from the death of a child is that they will overprotect or smother surviving persons. Parents need to be helped and encouraged to allow each other, other children, and all other surviving family members to live out their own unique lives, in their own way, without feeling pressured to take on the roles of the lost child. Some parents may try to have another child as soon as possible. They realize that they cannot replace the dead child, but they hope that a new baby will fill their void and further ameliorate their feelings of loss.

Parents who refuse to discuss or deal with their child's death and who refuse sympathy, compassion, and support from others may delay recovery interminably. Grief that is hidden may burst forth later in psychophysiologic illness or in mental illness. Lindemann (1944), in his classic study of bereavement, called delayed emotional responses abnormal and termed them **morbid grief reactions**. Most of the morbid-grief patients he studied had serious disorders that led to permanent physiological damage. Parkes (1972) reported morbid-grief patients who, in addition to serious psychophysiologic illnesses, suffered from alcoholism, deep depressions, severe anxiety disorders, or psychosis. Lynch (1977) found that the medical consequences of unresolved (morbid) grief, or the **broken heart**, can be death. Lynch pointed out that years ago "grief" was openly recognized as a cause of death. Today our sophisticated, medical-wise society does not tolerate such an ill-defined diagnosis. However, Lynch's research suggested that in many cases heart attacks and other fatal accidents or illnesses are correlated with bereavement.

Siblings and close playmates of a dead child also experience a number of painful emotions associated with death. They often feel guilty, believing that some action or angry thought of theirs caused their brother, sister, or playmate to die. Or they may believe that the dead child was killed as a punishment for some action or thought that he or she had. This perception leaves many surviving children fearful of dying whenever they are angry or misbehave. Or they develop an overwhelming fear of death in general: "If it could happen to my friend, it can happen to me." In addition, due to their limited understanding of death, they may be angry at the dead child for leaving them. Children need to be given a chance to express all of their emotions openly and freely after experiencing the death of a sibling or friend. They will cue adults on what they most want to know: Will it happen to me? Was I responsible? Will I ever see him (or her) again? Bereaved children should be encouraged to cry. They should be given extra emotional support and a sense of secu-

rity in their environments. A sudden change in rules, discipline, schedules, or living arrangements can be especially upsetting to them when they are trying to cope with their anxieties and fears about death.

Some children who cannot cope with their guilt, fears, or anger at the death of another child turn their emotions inward and become depressed and withdrawn or develop physical symptoms (Sargent, 1979). They may need professional help if these behaviors persist over several months.

REACTIONS TO THE DEATH OF A YOUNGER ADULT

In general, the death of a young adult is met with feelings of injustice in much the same way as is the death of a child. It is especially hard to understand or accept the death if the adult had many persons dependent on him or her (children, employees), if he or she were doing some great humanitarian work, or if the person was a gifted athlete or performer.

The death of young adults not only brings grief to loved ones but also causes a sense of discomfort in most associates. An untimely death makes us all remember how unpredictable our own futures are, how finite we all are. However, the process of grieving, although painful, can be a growth-promoting experience. If people can come to terms with their pain and loss and their concerns and fears of death, they may gain a sense of competence from mastering what seemed insurmountable grief. With this new self-confidence they may begin to live each day in a more meaningful way.

The ways in which adults express their grief over the loss of an adult friend or relative vary from tight control of emotions to hysterical outpourings of anguish. Gorer (1965) described a certain underlying sequence of events that is typical. Initially, persons are disbelieving. When the realization of death finally sets in, protests (silent or otherwise) are common. Confusion follows. Survivors may feel hopeless, without direction, desperate. They may move slowly, think slowly, and detach themselves from others and from the course of everyday living. Finally, grief work must occur. Bereaved persons must break their emotional ties to the past and rid themselves of the hopes and dreams they shared with the dead person. Grief work can take months to accomplish.

Sometimes bereaved persons fear they are losing their minds. They find it difficult to be aware of the things that are happening around them or to react to events in their customary fashion. Personalities may change (for example, short tempers replacing patience, or reserve replacing gregariousness). During the acute and intermediate periods of grieving, persons may be unusually sensitive to criticism. One of the more difficult aspects of grief is a feeling of regret at things left undone or unsaid to the departed. Guilt feelings are a common ingredient in grief work and must be resolved. So, too, are feelings of anger. Most people can accept their feelings of guilt or regret more readily than they can their hostility toward the deceased. They feel they have no right to be angry at someone who paid the ultimate price, death. Yet, anger is a common affective response to a loved person's departure. Survivors may feel betrayed, stranded, rejected; as if, by dying, the deceased has pulled the rug from under their own feet.

Fear is yet another emotion common to grieving. On occasion real reasons for fear exist (economic instability, deprivation of primary social-emotional support). On other occasions the dread is nameless and often more frightening because its cause cannot be ascertained.

Another difficult aspect of grieving may be coping with the attitudes of certain friends and neighbors. Some persons do not recognize the necessity of grief work. They feel it would be better if the bereaved would forget, quit talking about the dead, and stop crying. In a study of attitudes about grieving in persons in England and the United States, Gorer (1965) found that the grieving process is most often viewed as a form of self-indulgence. On the other hand, having supportive medical, social, or religious counselors or family and friends helps a grieving person to work through his or her emotions, to reach the stage of accepting the death. Gorer suggested three things that must be done:

1. The bereaved must separate himself or herself from the deceased by breaking the bond that holds them together.
2. He or she must readjust to an environment from which the deceased is missing.
3. He or she must form new relationships.

Reactions during the grieving process may include a host of behavioral, physiological, and psychological responses. There is evidence that many accidents are linked to bereavement. In general, grief tends to exacerbate preexisting accident proneness. Thus, poor drivers may make even more errors in judgment on the road, or persons with poor motor coordination may have more falls. In a similar vein, grief may aggravate preexisting physical illnesses. Thus, the person with asthma or emphysema may have greater respiratory distress, the person with coronary heart disease may have more angina attacks, or the person with chronic ulcers may have flareups or bleeding of their lesions. Healthy individuals may also find themselves more susceptible to infections during the grieving process. An increase in use of alcohol, tobacco, tranquilizers, or sleeping pills by persons who previously used them is common. Finally , while grief is normal and is distinct from depression, a small percentage of bereaved individuals may develop symptoms of real clinical depression four months or more after their loss. This is more common in persons who have a prior history of mental illness, especially depression. Clinical depression puts a person at high risk for suicidal behaviors and should be treated professionally.

When a child loses a parent, grief and bereavement are especially agonizing. Behaviors that have been identified as possible childhood consequences of the death of a parent include changes in personality, changes in sex-role identity and sexual behaviors, moral changes, antisocial behaviors, emotional disturbances, and cognitive and achievement-related changes (Berlinsky and Biller, 1982). Each child's reaction to the loss of a parent will be individual and will depend on the child's age, sex, personality, developmental concept of death, and previous experience with death; on the sex of the lost parent; on the reason for the parent's death; and on the support provided by surviving caregivers.

Some children react with loud protesting and increased aggressiveness. Others withdraw and become detached. Recurrent attacks of anxiety are common whenever a reminder of the dead parent appears. Guilt is often a part of a child's grief. The child may believe that he or she was in some way responsible for the parent's disappearance. Children must be told the truth about the parent's death. They must also be encouraged to cry, to ask questions, and to do grief work to come to terms with all their feelings of anger, hurt, loneliness, guilt, fear, and anxiety. Grief work, though painful, can prevent more serious consequences.

Children's grief work is a threefold task: to cope with the immediate impact of the loss of a mother or father, to mourn, and to resume and continue their emotional development in harmony with their level of maturity (Furman, 1974). Some alternate caregivers should become available to meet the kinds of bodily and psychological needs previously met by the deceased parent. In addition, the surviving caregiver should create a milieu in which the child can mourn. This can be especially difficult with a very young child who may not know how to communicate and who cannot differentiate or tolerate emotions such as anger or sadness. Some thanatologists feel that children cannot mourn until about age three (Furman, 1973). Others place it much later, even until adulthood is reached (Wolfenstein, 1969; Miller, 1971). Many children have difficulties going forward with their emotional development. They often regress in emotional behaviors instead, becoming more fearful, more insecure, more withdrawn and detached, more aggressive, more hostile, and less able to control their shifting moods.

Like adults, children often become more susceptible to illness and become more accident prone while going through the grieving process. It is not unusual for them to have appetite and sleep disturbances and complain of pain, especially abdominal cramps. School-age children often have difficulty remembering or concentrating, and their academic performance may decline. With adequate support and encouragement, children can eventually recover from the loss of a parent, usually within a year, without long-term negative consequences. How much support, for how long, and whether professional assistance will be required depend on each child's unique circumstances. Some long-term consequences of the death of a parent on a child that have been reported include mental illness (especially depression), an increased risk of suicide, impaired sexual identity, impaired capacity for intimacy, and lack of autonomy (Committee for the Study of Health Consequences of the Stress of Bereavement, 1984). The adequacy of the caregiving to the bereaved child will have a major effect on any ultimate outcome.

REACTIONS TO THE DEATH OF AN OLDER ADULT

Mourning for an aged person who dies is generally less painful than mourning the death of a child or younger adult. One senses that the death is more a part of a natural process. The aged person can be viewed as having lived a long and full life. Death in the elderly is also generally anticipated, consciously or unconsciously, for a period of time before it occurs and some anticipatory grief work may be done. Family and friends have a chance to prepare for what life will be like when the older person dies. Grief is most severe in the persons who may still have been dependent on the departed (for example, the widow, or widower). Clayton and her colleagues (1972) and Bornstein, Clayton, and others (1973) studied depression in older widows and widowers. They found that almost half of the older surviving spouses were depressed at some point during the first bereavement year and nearly 15 percent were depressed for the entire year. Many of the older persons who were not clinically depressed had many individual depressive symptoms of varying durations. Lynch's (1977) research suggested that the increased incidence of older married persons dying within one or two years of each other may reflect the difficulty surviving spouses have in working through the intense emotional stresses of grieving. Grief and bereavement reactions may contribute to what Lynch called the "broken heart." Death occurs due to coronary heart disease, a stroke, an accident, cancer, some other illness, or even suicide, but the physical or mental illness was exacerbated by the bereavement.

The sequence of feelings in accepting the death of an elderly person is the same as the sequence following the death of a younger person: acute grief with disbelief, possibly anger and confusion, and finally the painful process of the intermediate grief phase and disassociating oneself from the departed. Survivors must find a meaning and purpose to their own lives that no longer center around or include the dead person in order to recover and return to normal functioning.

BEREAVEMENT INTERVENTIONS

Most friends are not sure how to help a recently bereaved person. What can one say or do? Thanatologists recommend that the process of mourning should not be cut short. It is a healing process. Good friends should be there for the bereaved and should listen to whatever feelings are expressed in an accepting, nonjudgmental way. It is not helpful to say, "I know how you feel." One can never know the most intimate thoughts of another. "How do you feel?" is a more helpful query, but one must phrase this question gently, with a desire to learn and be supportive, not with a desire to instruct. It is wrong to try to tell a bereaved person how he or she should feel. Rebukes or advice will complicate bereavement, not ease the pain. In most cases, the less said, the better. "I'm sorry," and an embrace or tender touch shows caring. One may also inquire whether the bereaved needs any assistance (such as with meals, shopping, transportation). In time, the bereaved may express his or her desire to have help resolving feelings of guilt, anger, and despair. It is helpful then to put the bereaved in contact with trained counselors or support groups. Hospices, physicians, specially trained nurses, community support groups, clergy, social workers, psychologists, and psychiatrists may be able to provide assistance, insight, and understanding with the grieving process.

The hospice movement includes services to family members as well as to terminally ill patients. Although medicare will not reimburse hospice personnel for the counseling they provide to family members of a dying patient, the hospice cannot qualify for medicare collection unless it includes such counseling services. While participating in a hospice program, anticipatory guidance is provided and some anticipatory grief work can be done. During the early bereavement period, family members can continue to talk to and receive support from hospice workers. This support can be invaluable to each person involved and can help prevent many of the complications of bereavement as well as morbid grief reactions. Hospices do not exist in all cities and towns, however, and their bereavement services may be limited to the families of patients who have used the hospice.

Some individuals seek medical care for their behavioral, physiological, and psychological reactions to bereavement. Many physicians order extensive tests to diagnose specific disorders responsible for whatever symptoms are reported. While this is often necessary (bereavement exacerbates physical illnesses), physicians should also assist the individuals to accomplish their grief work. They can provide counseling themselves or they can help the bereaved find either a support group or an individual counselor such as a qualified nurse, social worker, religious counselor, psychologist, or psychiatrist. Physicians should exercise caution in prescribing medications for bereaved individuals. There is some evidence that antidepressant drugs, sometimes used to suppress intense, distressing, or disabling grief reactions, may have later adverse consequences. Many physicians prescribe minor tranquilizers (Valium, Librium) to help grieving persons. There have been virtually no controlled trials on the efficacy of these drugs with the bereaved (Committee for the Study of Health Consequences of the Stress of Bereavement, 1984). Sedatives or hypnotics prescribed for insomnia may, over time, become the cause of insomnia. The grieving and bereavement processes should not be completely suppressed but should be worked through for healing and growth to occur.

Support groups consisting of other bereaved persons, usually headed by a specially trained nurse, social worker, psychologist, psychiatrist, or member of the clergy, can enhance each individual's personal coping strategies. It can be beneficial to learn that other persons are experiencing frightening emotions, rapid mood swings, forgetfulness, or physical and psychological symptoms. Attentive listening and empathy for others are often as helpful as sharing one's own turmoil. The person-to-person exchange can reveal coping techniques and solutions to problems and enhance the bereaved person's sense of personal worth. Many support groups are focused on specific circumstances (suicide, homicide, the death of a child, SIDS death, widow-to-widow support). Others serve the needs of any bereaved person.

When a person has a prolonged grieving process, experiences exaggerated bereavement reactions, cannot find a compatible support group, does not have supportive friends, or does not want to share a personal grief with nonprofessionals, psychotherapy can be an effective form of bereavement intervention. Therapy may be brief (crisis intervention) or long term and open ended. It may follow any number of theoretical frameworks (psychodynamic, behaviorist, cognitive, humanist) or may be eclectic. It may be offered to a single bereaved individual or to a whole family. In many cases, grief work can be accomplished more rapidly and with a more favorable outcome with such professional assistance.

Perspectives on Living

An understanding and acceptance of the totality of the human life span, including death, can make us all feel more akin to each other, whatever one's age or life stage.

Experiencing death can be a catalyst to living more fully. Most people have good intentions. Most people would like to live lives that benefit others, but, when one lives as if life will go on forever, it is too easy to postpone until tomorrow all those things that ought to be done today. Eventually one looks back and sees one's life as being self-absorbed, selfish, and not very beneficial to others. This is embarrassing and unpleasant. Such negative feedback from the internal valuing system prevents one from becoming all one is capable of being. It is difficult, if not impossible, to like others if one does not first like oneself.

One of the secrets of building up a positive self-image and of learning to like oneself is to spend a portion of each day reaching out to others. This must be done lovingly and unselfishly, not because one wants something in return but simply because it is in us to care for someone else. When concern and affection are given another, the reward is not only the appreciation of the one served but, more important, a feeling of self-satisfaction, self-esteem, and personal growth. By repeated giving, one is able to learn to like and understand other human beings more fully. There will still be times when one experiences frustrations, exploitation, insults, and inhumanity. However, loving, caring relationships will also develop.

The concept of death, the idea that our future is uncertain, is one of the major keys to continued personal growth and giving. As Kübler-Ross (1975) put it, "When you fully understand that each day you awaken could be the last you have, you take the time, that day, to grow, to become more of who you really are, to reach out to other human beings."

Take a moment to consider who you really are. How and where do you fit into the fellowship of humankind? Consider the wise command spoken by the sage Chilon (but generally attributed to Socrates), "Know thyself."

Humans have existed on earth for about 100,000 years. Most of this time they have been hunters and gatherers. Farming in a primitive form only started about 10,000 years ago. The industrial revolution occurred only 200 years ago. Now every single generation sees fantastic changes in the way society lives. Society must take risks to survive. The future is uncharted, unpredictable. There are difficult problems in trying to adapt and cope with the ever-changing world around us. Life exists with conditions of stress and fear. Alvin Toffler (1970) called this "future shock" and suggested that our fear may at times nearly paralyze us. It can prevent us from behaving rationally with respect to future planning. How does one plan for a world that may in a short time include such things as cloning, computerized robots capable of holding important societal offices or positions, a cancer cure, a prevention of aging, an accidental nuclear explosion, the spread of anarchism over large parts of the world, a diminution of fresh air and fresh water?

Try to view living in the world today and tomorrow from a more positive perspective (see Figure 12–2). The American Medical Association (1974) suggested several special kinds of environments that should be reasonably available to all people to help shape a more peaceful, humane world.

The first is an environment of health. All members of society should be able to breathe clean air, drink clean water, eat uncontaminated, nutritious food, and afford medical attention to prevent and

Figure 12–2
We all should consider our responsibilities for providing the next generation with a quality of life that matches (or surpasses) our own: clean air and water, nutritious food, energy, freedom, love, quality education, employment, and equal treatment under the law.

treat disease processes. A vital part of this environment of health is physically fit parents of our future generations. Our world should make it possible for parents, especially the mother, to confer optimum health benefits to offspring. The mother should not have to conceive at too young an age or in a malnourished, drug-addicted, or chronically diseased state, all of which could adversely affect the baby-to-be.

A second environment that all people should reasonably expect to enjoy is a quality home life. One cannot become all one is capable of being without love and freedom from physical dangers at home. People also need family environments that are free from racial discrimination and abject poverty.

A third specialized kind of environment that should be available to all humans is that of quality education. One of our priorities must be to provide educational opportunities flexible enough to allow all people to learn to read and reason about the various areas of human knowledge.

A fourth environment needed by everyone for a better world is that of constructive outlets for aggression, which Lorenz (1966) and others have postulated to be an instinctual behavior in the human species. It is not in the best interests of humankind for persons to turn their aggression and hostilities against each other. The world has become close-knit by rapid transportation and instant communications. Persons of all races, ethnicities, and religions are now economically and politically interdependent, whether or not they desire to be so. Humans should work together to solve the food, fuel, pollution, and population problems of the future rather than against each other for self-centered profits. Society needs to provide opportunities for persons to direct their aggressive urges at righting environmental obstacles to a quality life, at improving health and physical fitness, at enriching the lives of others, or at defusing the tense emotions triggering aggressive urges and possibly world destruction.

Another environment generally recognized as both a necessary and reasonable right is that of employment. Jobs should be structured to allow for more cooperative efforts among employees rather than more competition for employment in a tight job market or competition for every salary increment. The dual hardships of unemployment plus inflation can quickly destroy the quality of one's life. Feelings of impotence among the unemployed may contribute to destructive aggressive outbursts such as child abuse, spouse abuse, elder abuse, and crimes against society.

Still another environment that should be available to all persons is equal treatment under the law. We must not have a double standard of justice, one divided between the rich or powerful and the poor or powerless of the world. Every human being deserves a feeling of her or his own dignity as part of the fellowship of humankind.

Where do you see yourself in the scheme of the total fellowship of humankind? What will you give to the population of the world in order to assure a better quality of life and a peaceful, humane future for all peoples?

Overview

This text has discussed development throughout the life span, considering physical, cognitive, and psychosocial components. The notion of development throughout adulthood is rather a recent one. For years developmental texts focused primarily on growth during childhood, ending with adolescence. Adulthood was generally viewed as encompassing years of gradual decline rather than of change and progress. The trend now is to view development as occurring throughout life and even into a terminal illness.

Along with this change is a trend that encourages persons to discuss and consider the end of the life span—death—something that for years was considered morbid or taboo. The processes of grieving and bereavement are the focus of much new research. Social supports through these processes are important to prevent negative physical or psychosocial consequences. This healthy opening up of discussions about death and bereavement has enabled individuals to consider their own mortality, their own finiteness. Rather than being gruesome, such awareness can be liberating. An awareness of one's finiteness can help a person to look at each day as one filled with opportunity and potential for accomplishment. Rather than living one's life in the future, this awareness can help people to focus on what can be achieved here and now.

Although development occurs as a process throughout life, it is generally divided into definable stages. Development through infancy and childhood proceeds from sensory and motor responses to verbal communication, thinking, conceptualizing, and relating to others (modeling, responding, asserting the self). These basics normally continue in the individual throughout her or his life.

In adolescence the individual begins to test out a separate life. Values are questioned and identity is engraved.

Early adulthood usually establishes the individual as an independent person. Employment, further education, the beginning of one's own family are all aspects of setting up a distinct life, with both its own characteristics and the characteristics and customs of previous generations.

During middle adulthood persons have new situations to face, new transitions with which to cope. Children grow up and leave home. Signs of aging become apparent. Relationships change, sometimes becoming stronger, sometimes ending. Roles shift. New abilities may be found, and opportunities sought.

Finally, during late adulthood, people assess what they've accomplished. Some are pleased. Some feel they could have done more or lived differently. In the best of instances, individuals accept who they are and are comfortable with themselves.

The human developmental process, therefore, is always changing. It is exciting and scary, joyful and disappointing. In each person's developmental process, he or she brings unique characteristics and contributions to surrounding settings. The environment that is established and continues for future generations hinges upon what each person contributes and values. Step by step, the process continues—with potential for a high quality of life for everyone.

Questions for Review

1. Some researchers say that proximity to death can be a growth-producing experience. Do you agree or disagree with this? Explain your answer.
2. Describe some ways in which parents, other relatives, friends, and medical personnel can help alleviate the fears of a seven-year-old who is terminally ill and is hospitalized.
3. Kübler-Ross defined five stages that terminally ill persons experience in facing their own deaths. Describe the characteristics of each of these stages.
4. Are you in favor of "living wills," or do you believe it is best left up to a person's loved ones to determine whether to maintain life-sustaining equipment? Explain your answer.
5. American society tends to be more critical of men who cry and show emotion than are some other societies. What sorts of implications does this have for a man who has recently lost someone he loved?
6. Describe the special kinds of environments that should be reasonably available to all people for a more peaceful, humane, and healthy life span.

References

American Medical Association. *Quality of life: The early years.* Acton, Mass.: Publishing Sciences Group, 1974.

Becker, E. *The denial of death.* New York: Free Press, 1973.

Bell, T. *In the midst of life.* New York: Atheneum, 1961.

Berlinsky, E. B., and Biller, H. B. *Parental death and psychological development.* Lexington, Mass.: D. C. Heath, 1982.

Binger, C.; Mikkelsen, C.; and Waechter, E. Terminal illness: Implications for patient, family, and staff. In H. Thorpe (Ed.), *The hospitalized child, his family, and his community.* San Francisco: American Association for Child Care in Hospitals, 1970.

Bornstein, P. E.; Clayton, P. J.; Halikas, J. A.; Maurice, W. L.; and Robins, E. The depression of widowhood after 13 months. *British Journal of Psychiatry,* 1973, *122,* 561–566.

Burton, A. *Operational theories of personality.* New York: Brunner/Mazel, 1974.

Butler, R. *Why survive? Being old in America.* New York: Harper & Row, 1975.

Clayton, P. J.; Halikas, J. A.; and Maurice, W. L. The depression of widowhood. *British Journal of Psychiatry,* 1972, *120,* 71–78.

Committee for the Study of Health Consequences of the Stress of Bereavement. *Bereavement: Reactions, consequences and care.* Washington, D.C.: National Academy Press, 1984.

Elkind, D. *The hurried child.* Reading, Mass.: Addison-Wesley, 1981.

Erikson, E. *Childhood and society,* 2nd ed. New York: Norton, 1963.

Fanslow, C. A. Touch and the elderly. In C. C. Brown (Ed.), *The many facets of touch.* Skilman, N.J.: The Johnson & Johnson Baby Products Co., 1984.

Freud, S. *Introductory lectures on psychoanalysis.* (J. Strachey, Ed. and Trans.) New York: Norton, 1966. (Originally published, 1920.)

Furman, E. *A child's parent dies: Studies in childhood bereavement.* New Haven, Conn.: Yale University Press, 1974.

Furman, R. A. A child's capacity for mourning. In E. J. Anthony and C. Koupernik (Eds.), *The child in his family: The impact of disease and death. Yearbook of the International Association for Child Psychiatry and Allied Professions,* vol. 2. New York: Wiley, 1973.

Glick, I. O.; Weiss, R. S.; and Parkes, C. M. *The first year of bereavement.* New York: Wiley, 1974.

Gorer, G. *Death, grief and mourning.* Garden City, N.Y.: Doubleday, 1965.

Johnson-Soderberg, S. Parents who have lost a child by death. In V. J. Sasseroth (Ed.), *Minimizing high-risk parenting.* Skilman, N.J.: The Johnson & Johnson Baby Products Co., 1983.

Kalish, R. A., and Reynolds, D. K. *An overview of death and ethnicity.* Farmingdale, N.Y.: Baywood, 1976.

Kastenbaum, R., and Aisenberg, R. *The psychology of death.* New York: Springer, 1972.

Keller, J. W.; Sherry, D.; and Piotrowski, C. Perspectives on death: A developmental study. *Journal of Psychology,* 1984, *116,* 137–142.

Kübler-Ross, E. *On death and dying.* New York: Macmillan, 1969.

Kübler-Ross, E. *Death: The final stage of growth.* Englewood Cliffs, N.J.: Prentice-Hall, 1975.

Lamberg, L. *The American Medical Association's straight-talk, no-nonsense guide to better sleep.* New York: Random House, 1984.

Lansdown, R., and Benjamin, G. The development of the concept of death in children aged 5–9 years. *Child: Care, health and development,* 1985, *11,* 13–20.

Lindemann, E. Symptomatology and management of acute grief. *American Journal of Psychiatry,* 1944, *101,* 141–148.

Lipsitt, L. P. The Challenger Tragedy. *The Brown University Human Development Letter,* 1986, *2*(2), 2.

Lorenz, K. *On aggression.* (M. Wilson, trans.) New York: Harcourt, Brace & World, 1966.

Lynch, J. J. *The broken heart: The medical consequences of loneliness.* New York: Basic Books, 1977.

Maas, H. S., and Kuypers, J. A. *From thirty to seventy.* San Francisco: Jossey-Bass, 1974.

Marlow, D. *Pediatric nursing,* 4th ed. Philadelphia: Saunders, 1973.

Miller, J. B. M. Reactions to the death of a parent: A review of the psychoanalytic literature. *Journal of the American Psychoanalytic Association,* 1971, *19*(4), 697–719.

Morrissey, J. Death anxiety in children with a fatal illness. In H. Parad (Ed.), *Crisis intervention.* New York: Family Service Association, 1965.

Munley, A. *The hospice alternative: A new context for death and dying.* New York: Basic Books, 1983.

Parkes, C. M. *Bereavement: Studies of grief in adult life.* New York: International University Press, 1972.

Sargent, M. *Caring about kids: Talking to children about death.* (DHEW Publication No. ADM 79-838). Washington, D.C.: U.S. Government Printing Office, 1979.

Schulz, R. *The psychology of death, dying and bereavement.* Reading, Mass.: Addison-Wesley, 1978.

Toffler, A. *Future shock.* New York: Random House, 1970.

Wolfenstein, M. Loss, rage and repetition. *Psychoanalytic Study of the Child,* 1969, *24,* 432–462.

Glossary

ABO incompatibility: Undesirable physiological effects of giving A-type blood to a person with anti-A antibodies or B-type blood to a person with anti-B antibodies. Antibodies against type A or type B blood can also cross the placental barrier and produce undesirable effects in the fetus.
Acceptance stage: Kübler-Ross's fifth stage of dying, characterized by a consent and willingness to die.
Accidental drug overdosing: Frequent cause of poisoning in older persons, who forget whether they have taken their medication or self-medicate and take combinations of drugs.
Accommodation: Piagetian term referring to changes in existing schemas to include new experiences.
Achievement needs: Needs to learn well and perform well.
Acne: A common skin disease characterized by chronic inflammation of the sebaceous glands, usually causing pimples on the face, back, and chest.
Activists: Persons with a cause who seek to bring about change with energy and decisiveness.
Acute grief: Lindemann's description of several disturbing emotions that accompany loss, misfortune, or death of a loved one.
Adaptation: Piagetian term for the complementary processes of assimilation and accommodation.
Adolescent egocentrism: Belief held by the adolescent that other people are preoccupied with his or her appearance and behavior.
Adolescent suicides: The act of teenagers killing themselves intentionally; a serious, and growing, health concern in North America.
Adulthood moritorium: A period of time during which a person is permitted to delay meeting the obligations of adulthood.
Affiliation needs: Needs to unite or associate oneself with other persons.
Age thirty transition: Levinson's stage description for men roughly between ages 28 and 32.
Ageism: A widely prevalent social attitude that overvalues youth and discriminates against the elderly.
Alcohol abuse: Misuse of any alcoholic substance; habitual use.
Alcoholic: A person who has lost control over drinking behaviors and suffers withdrawal symptoms when blood alcohol levels approach or reach zero after alcohol use.
Alienated youth: Youth who feel estranged from society and try to escape through such means as drugs, alcohol, or adherence to socially unacceptable lifestyles and actions.
Alienation: A feeling of estrangement from and hostility toward others.
Alleles: Pair of genes affecting a trait. When alleles are identical, an individual is homozygous for a trait; when alleles are dissimilar, the individual is heterozygous.
Alloplastic mastery: Gutmann's description for active world mastery.
Alzheimer's disease: A dementia involving rapid intellectual deterioration, speech impairment, loss of body control, and death, usually within five to ten years of onset.
Amniocentesis: Analysis of a sample of fluid drawn from the sac surrounding the fetus during the fourth month of pregnancy to detect aberrations.

Analysis of variance (ANOVA): A statistical examination of normally distributed, independently obtained observations, having the same variance, where the mean of each observation can be represented as a linear combination of certain unknown parameters.
Androgens: Male sex hormones, produced primarily by the testes.
Androgyny: A sex-role identity that incorporates some positive aspects of both traditional male and traditional female behaviors.
Androsperm: Y-carrying sperm.
Aneurysm: A vessel dilation in which a weak part of the vessel wall balloons out.
Anger stage: Kübler-Ross's second stage of dying, characterized by anguish and rage.
Animism: Piagetian term for the attribution of life to inanimate objects.
Anorexia: Loss of appetite.
Anorexia nervosa: Chronic failure to eat for fear of gaining weight; characterized by an extreme loss of appetite that results in severe malnutrition, semistarvation, and sometimes death.
Anoxia: A severe deficiency in the supply of oxygen to the tissues, especially the brain.
Antibodies: Modified proteins in the body fluids formed in response to an antigen that subsequently will react with and destroy or inhibit the functioning of that antigen.
Anticipatory grief: Recognizing and coping with feelings of denial, anger, and depression that precede a person's death.
Anticipatory guidance: Preparation of survivors for their probable actions and reactions at the time of death of a loved person; usually done by trained counselors or clergy.
Antigen: Any foreign material (such as a microorganism, virus, toxin, or protein) that stimulates the immune system to produce antibodies when it comes in contact with appropriate body tissues.
Apgar scale: Medical technique to measure adjustment of neonate at birth; measures appearance, pulse, grimace, activity, and respiration.
Apnea: Temporary stopping of breathing.
Applied research: Research based on the desire to know for the sake of being able to do something better or more efficiently.
Apprenticeship learning: Education by legally agreeing to work a specified length of time for a master craftsman to learn the craft or trade.
Archetypes: Jungian term for prototypes of great figures in human literature, religion, mythology, and art believed to emerge from the collective unconscious in dreams and fantasies.
Armored-defended personalities: Neugarten's description for persons who hold their impulses and emotions in tight harness and restrain their creativity.
Arteriosclerosis: Commonly called "hardening of the arteries." Includes a variety of conditions that cause the artery walls to thicken and lose elasticity.
Artificialism: Piagetian term for the belief that the universe is made by humans, for humans.
Assimilation: Piagetian term referring to the process of making new information part of one's existing schemas.
Associative play: Play with some shared activity and communication.
Asthma: A chronic disorder characterized by wheezing, coughing, difficulty in breathing, and a suffocating feeling.
Astigmatism: Unequal curvature of one or more of the refractive surfaces of the eye that interferes with visual focus.

Atherosclerosis: A form of arteriosclerosis. The intima (inner layer of artery walls) is made thick and irregular by deposits of a fatty substance. The internal channel of arteries becomes narrowed, and blood supply is reduced.
Attachment: An active, affectionate, reciprocal relationship specifically between two individuals; their interaction reinforces and strengthens the bond.
Attentional deficit disorder: A childhood disorder characterized by an inability to focus attention.
Authoritarian-autocratic parents: Parents who are directive and who firmly control their children, yet are somewhat emotionally distant and cold.
Authoritative-reciprocal parents: Parents who combine high controls with warmth, receptivity, and encouragement.
Autonomy versus shame and doubt: According to Eriksonian theory, the second nuclear conflict of personality development. The child develops either a sense of autonomy (independence, self-assertion) or the feeling of doubt and shame.
Autoplastic mastery: Gutmann's description for passive world mastery.
Autosomes: The chromosomes of a cell, excluding those that determine sex.

Bargaining stage: Kübler-Ross's third stage of dying, characterized by an attempt to trade or barter something in exchange for a longer life.
Basal metabolic rate (BMR): Measurement of the consumption of oxygen by a body at rest.
Basic anxiety: Term used by Horney to describe the apprehensions and misgivings that originate at birth and expand through childhood if love and affection are insufficient.
Basic four food groups: Milk; vegetables and fruit; meat or protein foods; and breads and cereals.
Basic research: Research conducted for the satisfaction of knowing or understanding.
Battered child syndrome: Physical abuse of infants and children.
"Becoming one's own man": Levinson's stage description for men in their late thirties.
Behaviorism: School of psychology based on the study of observable behavior and the patterns of stimulus and response that govern it.
Behavior modification: An intervention approach using procedures based on principles of learning.
Benign: Nonmalignant character of a neoplasm, or the mild character of an illness.
Benign senescence: Normal decreases in functional abilities that accompany aging.
Bereavement: The fact of loss through death.
Bereavement process: The emergence of bereavement reactions over time.
Bereavement reactions: Any psychological, physiological, or behavioral response to bereavement.
Birth order effects: Any behaviors believed to be brought about by birth position in the family relative to siblings (firstborn, middle child, and so on). These effects are seen in children of the same birth order across a majority of families and are therefore felt to be a result of birth position.
Blackout: A momentary lapse of consciousness. In alcoholism a period of temporary amnesia when the alcoholic functions but of which he or she has no memory when sober.
Blastocyst: The cluster of cells resulting from cell division of a zygote in the first week after conception.
Blood pressure: The force or pressure exerted by the heart in pumping blood; the pressure of blood in the arteries.

Bonding: An attachment between neonate and parent that occurs in the first few hours after birth under conditions of close physical contact.
Bone age: A measure of the maturity of the skeleton determined by the degree of ossification of bone structures.
Brain stem: The pons, medulla, midbrain, and diencephalon.
Braxton-Hicks contractions: Irregular painless contractions of the uterus during the last trimester of pregnancy before the onset of labor. Also called false labor.
Brazelton scale: Scale used to assess the neurological integrity and behaviors of neonates.
Broken heart: Lynch's term for the severe, unresolved, morbid grief that can result in the death of the bereaved.
Bulimia: Excessive overeating or uncontrolled binge eating followed by purging.

Caesarean section: Surgical delivery of a baby through an incision made through the abdomen and uterus of the mother.
Cancer: Growth of abnormal cells that spread and behave differently from the cells of the body part in which they develop.
Carcinogen: Any substance that produces cancer.
Cardinal traits: Allport's term for all-pervasive, "ruling passion" qualities.
Case study: An observational study in which one person is studied intensively.
Cataract: A loss of transparency of the crystalline lens of the eye or of its capsule.
Catch-up growth: A period of rapid growth, following a period of illness or malnutrition, that continues until the child has attained the height of his or her previous normal growth curve.
Central nervous system (CNS): The brain and spinal cord.
Central traits: Allport's term for consistent characteristics that rest on upbringing.
Centration: Tendency to focus on one aspect of a situation and to neglect the importance of other aspects; characteristic of preoperational thought in Piaget's theory.
Centromere: The spindle arrangement with the chromatids at the equator of chromosomes as produced in the metaphase of mitosis and meiosis.
Cephalocaudal development: Development that proceeds in a head-to-toe direction; upper parts of the body develop before lower parts do.
Cerebellum: A brain mass lying in the back of the head underneath the posterior cerebrum and above the pons and medulla in the brainstem. Like the cerebrum, it has two hemispheres.
Cerebral cortex: In humans and higher mammals the large outer layer of the two cerebral hemispheres, in major part responsible for our characteristically human behavior.
Cerebrovascular accident: An impeded blood supply to some part of the brain. Also called *stroke.*
Cerebrum: The largest, uppermost part of the brain divided into two large lobes called the cerebral hemispheres.
Character education: Term used in the Soviet Union to describe socialization practices.
Child abuse: The intentional, nonaccidental physical or sexual abuse of a child by a parent or other adult entrusted with the child.
Chi-square test: Statistical test used to evaluate how variables are distributed in a population and to evaluate the independence or dependence among various qualitative features within the elements of a population.
Chromosomal aberration: Atypical development caused by extra chromosomal material or insufficient chromosomal material on one of the chromosome pairs.

Chromosomes: The long strands of hereditary material containing the genes and composed exclusively of nucleic acids; found in the nucleus of the cell.

Circadian rhythm: Around-the-clock, twenty-four-hour cycle of some bodily functions.

Cirrhosis: A disease of the liver characterized by degeneration, fatty infiltration, atrophy, and inflammation.

Classical conditioning: A basic learning process in which a previously neutral stimulus (conditioned stimulus) is paired with one that elicits a known response (unconditioned stimulus) until the neutral stimulus comes to elicit a similar response (conditioned response).

Client-centered therapy: A nonjudgmental, nondirective, humanistic psychotherapy developed by Carl Rogers.

Clinical investigations: In-depth studies of the course of an individual's life experiences and personal history.

Cognition: Knowing the world through the use of one's perceptual and conceptual abilities.

Cohabitation: Living together and maintaining a sexual relationship without being legally married.

Cohort: A person born about the same time and into the same social environment as another person.

Colic: Unexplained, explosive, inconsolable crying in infants from two weeks to three months of age.

Collective unconscious: According to Jung, the unconscious life of all human beings, which is composed of many common elements and not just sexual strivings as Freud contended.

Colostomy: The surgical operation of forming an artificial anal opening in the colon.

Colostrum: The fluid secreted by the breasts for several days after childbirth preceding the secretion of milk.

Combinatorial analysis: A Piagetian term referring to the ability to see all the possible variations of a problem and to test each one systematically.

Common traits: Allport's term for behaviors typical of all members of a culture.

Compensation: A defense mechanism in which one attempts to win respect or prestige in an activity as a substitute for inability to achieve in some other realm.

Complementary leisure: Leisure that is determined directly or indirectly by others.

Compression of the spinal column: The process by which the spinal column is made more compact by pressure of gravity with age, resulting in a slight loss of height.

Concrete operational stage: Third stage of Piagetian cognitive development, during which children develop logical, but not abstract, thinking.

Confidence intervals: Mathematical expressions that define a range into which a certain proportion of the values of a variable or feature of a population will fall.

Conservation: Piagetian term for the awareness that two substances of equal amount remain equal in the face of perceptual alteration, so long as nothing has been added to or taken away from either substance.

Contact comfort: Cuddling and other forms of warm physical caressing that bring consolation, relaxation, or ease.

Control group: Those subjects in an experiment who are not exposed to whatever conditions are being studied.

Conventionality: The quality of behaving or acting according to the prevailing norms for conventional behavior.

Conventional morality (or stage of moral development): According to Kohlberg, the second level of moral development; characterized by observation of standards of others because one wants to please other people.
Cooperative play: Play in which there are rules or goals.
Coordinated leisure: Work-related activities that are freely chosen by individuals.
Coronary heart disease (CHD): An inclusive term for diseases of the heart and blood vessels, including heart attacks, angina, irregular heartbeat, and congestive heart failure.
Coronary prone behavior: Highly competitive, tense, hyperalert, achievement-oriented, aggressive, hostile, and impatient habitual modes of action.
Corpus luteum: Body formed in the ovary by a ruptured Graafian follicle that has discharged its ovum.
Correlational research: Studies of the relationship (positive or negative) of variables to each other rather than of cause and effect.
Correlation coefficient: A statistical index for measuring correspondence in changes occurring in two variables. Perfect correspondence is $+1.00$; no correspondence is 0.00; perfect correspondence in opposite directions is -1.00.
Cortisol: A major hormone of the adrenal cortex involved in regulation of volume and composition of body fluids and in carbohydrate, lipid, and protein metabolism.
Creativity: The ability to think flexibly, divergently, imaginatively, inventively, or productively.
Critical period: Specific time during development when an event will have its greatest impact.
Crossing over: A process during meiosis in which individual genes on a chromosome cross over to the opposite chromosome. This process increases the random assortment of genes in offspring.
Cross-sectional study: A research method using different subjects studied at the same time.
Croup: An inflammation of the respiratory passages, with labored breathing, hoarse coughing, and laryngeal spasm.
Crystallized intelligence: A broad area of intelligence that includes verbal reasoning, comprehension, and spatial perception; may increase during the life span.
Cultural universals: Something that applies to every culture in the world.
Cystic fibrosis: Genetically inherited disease of children characterized by increased fibrous connective tissue, malfunctioning of the pancreas, and frequent respiratory infections.

Deductive reasoning: Thinking about experiences from generalities to particulars, or from the universal to the individual.
Defense mechanisms: In psychodynamic theory, self-protective, usually unconscious psychological devices that help block the ego's awareness of anxiety-provoking memory and instinct.
Delinquent acts: Illegal or offensive activities committed by minors who are aware of wrong-doing.
Delirium tremens (DTs): A severe form of alcohol withdrawal in which paranoia, disorientation, extreme agitation, hallucinations, anxiety, insomnia, anorexia, and tremors can occur.
Denial stage: Kübler-Ross's first stage of dying. A refusal to believe or accept impending death as it is and instead perceiving it as one wishes it would be.
Dental caries: Decay of teeth.
Deoxyribonucleic acid (DNA): Molecules coding for hereditary information that determines makeup of cells.

Dependent variable: In an experiment, the behavior that is measured and is expected to change with manipulation of an independent variable.
Depressants: Drugs that act on the central nervous system to reduce pain, tension, and anxiety, to relax and disinhibit, and to slow down intellectual and motor reactivity.
Depression: An emotional state characterized by intense and unrealistic sadness.
Depression stage: Kübler-Ross's fourth stage of dying, characterized by intense sadness.
Deprivation discipline: Method of discipline that predominantly uses a loss of material objects or privileges as a consequence of misbehaving.
Descriptive psychology: The study of human and animal behavior that describes patterns of behavior.
Descriptive statistics: Numerical data assembled, classified, and tabulated to describe and summarize the characteristics of the data.
Diabetes mellitus: A chronic disorder of carbohydrate metabolism due to insulin deficiency or a disturbance of normal tissue responsiveness to insulin.
Dick-Read method of delivery: Approach to childbirth in which expectant parents attend classes to learn to eliminate the fear of labor through education and exercise.
Dilation: Enlargement of an opening, blood vessel, canal, or cavity.
Diploid: Relating to the pairs of chromosomes in the nucleus of somatic cells.
Discipline: Training that develops self-control, character, orderliness, and efficiency.
Disengagement: The false belief that all old people withdraw and detach themselves from obligations, occupations, and relationships.
Disequilibrium: Piagetian term referring to the lack of balance between assimilation and accommodation. Most learning takes place in states of disequilibrium.
Displacement: A defense mechanism that involves the transfer of emotion from an unsafe or unacceptable object to a safer, but inappropriate one.
Distant grandparents: Those who stay remote from their grandchildren except for birthday, religious, or other special observances.
Divorce: Legal dissolution of a marriage.
Dominant-recessive relationship: Denoting the ability of one gene (dominant) to manifest itself and its characteristics to the exclusion of its paired gene (recessive).
Down syndrome: Moderate to severe retardation caused by extra 21st chromosome material; marked by distinct physical characteristics.
Dreikurs's attention-seeking stages: Stages—from being annoying, to rebelliousness, to revenge, to noninteraction—that Dreikurs believed a child goes through unless he or she feels accepted as a valued and equally loved family member.
Drug synergism: The interaction of drugs in which the total effect is much greater than the sum of their individual effects.

Early adult transition: Levinson's stage description for men from the years of separating from parents until the early to middle twenties.
Eclecticism: Method or system of thought in which one chooses various sources and selects materials from many doctrines.
Eczema: A stress-related skin disorder that encompasses conditions ranging from an itching rash to a cluster of open wounds.
Effacement: A thinning out; becoming less conspicuous.

Ego: In psychodynamic theory, the rational self-concept, which mediates between the instinctual demands of the id and the externally oriented pressures of the superego.
Egocentrism: Inability to consider another's point of view.
Ego integrity versus despair: According to Eriksonian theory, the eighth and last nuclear conflict of personality development, characterizing old age. Successful resolutions, a sense of acceptance of one's own life, allowing one to accept death; the developmental alternative is despair, characterized by failure to accept one's life and thus, one's ultimate end.
Eidetic memory: The ability to remember minute details of a situation.
Eight ages of man: Erikson's theory that there are eight nuclear conflicts to be resolved during a human's life span.
Elder abuse: The intentional, nonaccidental physical or sexual abuse of an aged person by adult children or other adults entrusted with elder care.
Electra complex: In Freud's theory of psychosexual development, the female child's conflict between desire for the father and fear of punishment by the mother.
Embolus: A blood clot that forms in the blood vessels in one part of the body and travels to another.
Embryonic period: The second prenatal period, which lasts from the beginning of the second week to the end of the seventh week after conception. All the major structures and organs of the individual are formed during this time.
Empiricism: Search for knowledge through experience (observation or experiment) rather than through reasoning processes.
Encoding: Categorizing incoming information.
Encopresis: Lack of bowel control.
Endometrium: The mucous membrane lining the uterus.
Entering the adult world: Levinson's stage description for persons in their middle to late twenties.
Enuresis: Lack of bladder control.
Epilepsy: A brain disorder in which abnormalities in the electrical activity of the brain produce loss of consciousness and convulsions.
Epinephrine: The chief adrenal hormone that stimulates the sympathetic nervous system.
Episiotomy: Surgical incision of the vulva for obstetrical purposes to prevent uneven laceration during delivery.
Equilibrium: Piagetian term referring to a relative state of balance between assimilation and accommodation. This state seldom lasts long.
Eros: According to Freud, the constructive life instinct of the id, which deals with survival, self-propagation, and creativity.
Estrogen replacement therapy (ERT): Estrogen given to bring the level of hormone up to that once produced naturally, to provide relief from symptoms associated with insufficient estrogen.
Estrogens: Female sex hormones, produced primarily in the ovaries.
Euthanasia: Literally, "good death"; the practice of hastening the death of hopelessly ill individuals by withholding life-sustaining procedures so that death will occur naturally.
Experiment: A study in which the investigator manipulates one or more variables to determine the effect on another variable.
Experimental psychology: The study of human and animal behavior that stresses knowledge based on countable or measurable actions and reactions.
Expressive language: The ability to produce meaningful utterances.

Expressive orientation: Stereotypic "feminine" behaviors aimed at serving others' physical, social, and emotional needs through nurturance and emotional expressions.
Extinction: Dying out of a response as the result of withdrawal of positive reinforcement.
Extraversion: An attitude in which a person directs his or her interest to outside phenomena rather than to feelings within.
Extravert: A personality whose thoughts, feelings, and interests are directed toward other persons, social affairs, and external phenomena.

Failure-to-thrive syndrome: Decline of growth, with height and weight below the norm, in infancy.
Family life cycle: Duvall's description of the eight family stages she identified from marriage to death of both spouses.
Family of origin: One's biological (or surrogate) parents and any siblings.
Family of procreation: One's spouse and children.
Fantasy: Defense mechanism in which one imagines what was said or done or what will be said or done.
Fear of success: Horner's term for the avoidance of too much success found in some men and women.
Fetal period: Final stage of prenatal development (eight weeks to birth).
Field study: Observational study in which only specified activities and events are cataloged.
Fixation: The inability to progress normally from one stage of psychosexual development to the next due to overgratification or excessive deprivation during that stage.
Flagellum: Long hairlike process attached to sperm, protozoon, or bacterium that whips and thus creates movement.
Fluid intelligence: A type of intelligence involving the ability to solve novel problems.
Forceps: A pair of tongs used to seize anything. Obstetrical forceps are used for grasping and rotating the fetal head without compressing it during delivery.
Foreclosure of identity: Marcia's term for adolescents' early acceptance of the identity laid out for them by others without any search for self-definition.
Formal grandparents: Those who maintain a clear line between parents' responsibilities and grandparents' roles and who do little or no child-rearing.
Formal operations: According to Piaget, the stage of cognition characterized by the ability to think abstractly.
Foster Grandparents: Program that allows older persons to become involved in teaching, supporting, and helping children with special needs.
Free association: A technique in which therapy patients verbalize whatever thoughts come to mind without editing, structuring, or censoring.
Fun-seeking grandparents: Those who serve as playmates to their grandchildren.

Galactosemia: An inborn error in the metabolism of galactose, which accumulates in abnormally large amounts in the blood. It can lead to nutritional defects and mental and physical retardation.
Gaucher's disease: A storage disease characterized by derangement of lipid metabolism and an enlarged spleen.
Gene abnormalities: Atypical development or growth caused by the inheritance of one or more aberrant genes.
General adaptation syndrome: Selye's description of the three stages one goes through when faced with intolerable stress: alarm, resistance, and exhaustion.
General to specific development: Development from the main, overall use of body parts to use of limited, distinct body parts.

Generation gap: A distance or difference in ideas between parents and offspring or between people spaced a generation (about thirty years) apart.
Generativity versus stagnation: According to Eriksonian theory, the seventh nuclear conflict of personality development concerned with guiding the next generation.
Genes: The basic units of heredity located on the chromosomes.
Genetic epistemology: The study of the origin, nature, method, and limits of knowledge.
Genital stage: Freudian term for the adolescent stage of mature sexuality.
Genotype: Actual genetic composition of an individual.
Geriatrics: Branch of medicine concerned with the elderly.
Gerontology: Study of the processes of aging and the elderly.
Giftedness: Characterized by one or more of the following attributes: above-average intellectual ability; specific academic aptitude; creative or productive thinking; leadership ability; skill in visual or performing arts.
Glaucoma: Disease of the eye characterized by increased intraocular pressure leading to gradual impairment of sight.
Graafian follicle: One of the small round sacs in the ovaries, each of which contains an ovum.
Grief: Affective feeling of sadness and certain associated behaviors, such as crying.
Grief work: The work of mourning in which one comes to terms with the fact of death and deals with the bitterness, hurt, disappointment, and sense of loss.
Grieving process: The changing affective state over time after a loss.
Growth trajectory: The curved path that a normal child follows in terms of growth in height and weight.
Gynosperm: X-carrying sperm.

Hallucination: A sensory perception that occurs in the absence of any appropriate external stimulus.
Hallucinogenic drugs: Drugs that induce hallucinations; some are also called *psychedelics.*
Haploid: Having a number of chromosomes equal to half the number in an ordinary somatic cell.
Hay fever: Seasonal nasal obstruction, itching, sneezing, mucous discharge, and eye irritations as a result of exposure to specific wind-borne pollens.
Hearing defect: Lack of something essential to normal hearing ability.
Heightism: Prejudice directed against shorter people.
Heimlich maneuver: A method of dislodging objects blocking the airway by applying sudden pressure just under the rib cage.
Hemorrhage: A flow of blood from any blood vessel caused by injury or rupture of the vessel.
Heterozygous: Having allelic genes carrying messages for different manifestations of a trait.
Hierarchy of needs: Maslow's concept of a series of needs that must be satisfied one by one in the process of development before the adult can begin pursuing self-actualization.
Holistic medicine: Treatment of the whole person; removal or alleviation of any sources of stress as well as pharmaceutical and physical treatment of symptoms of illness.
Holophrases: One-word sentences that occur early in language acquisition.
Homozygous: Having allelic genes carrying the same message for the manifestation of a trait.
Honeymoon period: Period of mutual affection of newlyweds.
Hormone: A specific chemical substance produced by an endocrine gland that brings about certain somatic and functional changes within the organism.

Hospice: A residence for the terminally ill that allows them to live out their days as independently and as painlessly as possible.
Hospitalism: The progressive stages of protest, despair, and denial that characterize young children's behavior when they are separated from their primary caregivers for an extended time.
Hot flash: A vasomotor symptom of menopause that causes a flush from the breasts up and an overheated feeling.
Household organization: The task of arranging and maintaining a home.
Humanistic psychology: The view that humans have free will and both the right and capacity for self-determination based on purpose and values (in contrast to Freud's emphasis on unconscious forces).
Hyperopia: Farsightedness.
Hypertension (high blood pressure): An unstable or persistent elevation of blood pressure above the normal range.
Hypothermia: Subnormal body temperature.
Hypothetical-deductive reasoning: Piagetian term referring to the ability to test each possible variation of a problem in order to discover the correct solution.
Hysterectomy: Surgical removal of the uterus or part of the uterus.
Id: According to Freud, the mass of biological drives with which the individual is born.
Identification: Process by which an individual acquires the characteristics of a model.
Identity achievement: Acquisition of a firm sense of one's stable and enduring self.
Identity diffusion: Marcia's term for adolescents' failure to seek an identity and lack of commitment to any enduring sense of self.
Identity versus role confusion: According to Eriksonian theory, the fifth nuclear conflict of personality development, in which an adolescent must determine his or her own sense of self (identity).
Implantation: The embedding of the prenatal organism in the uterine wall.
Independence training: Education to become self-sufficient.
Independent variable: The variable that is controlled by the experimenter to determine its effect on the dependent variable.
Individual differences: Denotes the belief that each human being is born with and has through his or her life a unique physical and psychological makeup, unlike that of any other human being.
Individuation: Jung's description of the integration of outer reality with inner fantasy to achieve self-realization during middle age.
Inductive discipline: A technique employed by parents who try to control their children's behavior by carefully explaining how they want their children to behave and by use of reason.
Inductive reasoning: Reasoning from particular facts to a general conclusion or from the individual to the universal.
Indulgent-permissive parents: Parents who combine little or no control with high levels of responsiveness, receptivity, and encouragement.
Industry versus inferiority: According to Eriksonian theory, the fourth nuclear conflict of personality development, which occurs during middle childhood. Children must learn the skills of their culture or face feelings of inferiority.
Infant daycare: A situation in which several infants are cared for all day by a few adults in a setting specifically designed for infant care.
Infectious mononucleosis: A disease characterized by the presence in the blood of an excessive number of cells having a single nucleus, usually causing fever and enlargement of the lymph nodes.

Inferential statistics: The assembling, classifying, and tabulating of numerical data in such a way that the characteristics of the population from which the data were drawn can be inferred.
Inferiority: Term used to describe the feelings of inadequacy that characterize some people.
Infertility: The state of being unable to produce offspring.
Initiative versus guilt: According to Eriksonian theory, the third nuclear conflict of personality development, which characterizes children from three to six years. Children develop initiative when they try out new things and are not overwhelmed by failure.
In-law relations: The reciprocal dealings of persons related by marriage, not blood.
Inner speech: Term denoting the language of the mind, the private speech we use only for our own thinking and reasoning.
Instrumental orientation: Stereotypic "masculine" behaviors characterized by task-orientation, self-serving actions, independence, and tight rein on emotional expression.
Insulin: A hormone from the pancreatic islets of Langerhans involved in the regulation of carbohydrate metabolism.
Integrated personalities: Neugarten's description for persons who face up to their emotions and experience rich inner lives.
Intellectualization: A defense mechanism characterized by separation of events or ideas from the emotions that impinge on them.
Intelligence quotient (IQ): Mathematical score computed by dividing an individual's mental age by chronological age and multiplying by 100: $IQ = MA/CA \times 100$.
Interiority of personality: Neugarten's description of the move inward toward more preoccupation with satisfying personal needs that occurs in middle age.
Intermediate gene effects: Denoting the ability of both homologous genes to exert an influence on some characteristic or function rather than working in a dominant-recessive relationship.
Intermediate grief phase: A period of crying and extended sorrow after the acute grief accompanying loss, in which one often reviews one's life and engages in self-reproach for things done or left undone.
Interpersonal relationships: Close interactions with other persons in one's social network.
Interpropositional thinking: According to Piaget, the ability to handle relationships among several different properties at the same time.
Intimacy versus isolation: According to Eriksonian theory, the sixth nuclear conflict of personality development, which occurs during young adulthood. Young adults want to make commitments to others; if they do not, they may suffer from a sense of isolation and consequent self-absorption.
Introjection: Defense mechanism whereby the characteristics of another's personality are adopted as one's own.
Introvert: One whose orientation is toward the self rather than toward association with others.
Intuitive phase: Piaget's second substage of the preoperational stage of development.
Iron deficiency anemia: A reduction in the amount of hemoglobin in the blood due to a dietary deficiency of iron.

Jensenism: Belief that genetic factors play the major role in the average black/white intelligence difference.
Juvenile crimes: Crimes committed by persons under the legal age of adulthood.

Klinefelter syndrome: A condition that affects males in which the nondisjunction of the sex chromosomes results in an extra X chromosome, causing underdeveloped testes, sterility, and sometimes mild mental retardation.
Kübler-Ross's stages of dying: Denial, anger, bargaining, depression, and acceptance.

Labor: The process of expulsion of a fetus from the uterus at the normal termination of pregnancy.
Lactation: Secretion of milk by the mammary glands; breastfeeding.
Lamaze method of delivery: Approach to childbirth in which expectant parents attend classes prior to labor and delivery to learn fundamentals such as anatomy of childbirth, breathing techniques to use during labor contractions, and physical, psychological, and emotional preparation for childbirth.
Language Acquisition device (LAD): Inborn mental structure that enables children to build a system of language rules.
Lanugo: Fuzzy prenatal and neonatal body hair.
Late maturation: Puberty that occurs significantly later than the average.
Later maturity: Havighurst's term for old age.
Launching: Term used to describe the sending off of each of the children from the home to live their own independent lives.
Learning disability: Extreme difficulty in learning with no detectable physiological abnormality.
LeBoyer method of delivery: Delivery in a warm, darkened, soothing environment reducing as much trauma to the neonate as possible.
Leukemia: Cancer of the blood-forming tissues characterized by an abnormal increase in the number of leukocytes.
Libido: In psychodynamic theory, the energy of the life instinct, which Freud saw as the driving force of personality.
Life change units: The average mean values asigned to various life events by Holmes and Rahe.
Life expectancy: The average number of years that a person of a given age may expect to live.
Limbic system: The part of the brainstem forming a border around the lower forebrain that is involved in emotional reactivity.
Living will: A legal document stating that a person does not wish to be kept alive by artificial means; signed while a person is healthy.
Logorrhea: Rapid flow of words, garrulousness.
Longevity: Long life.
Longitudinal study: A research method that involves examining the same subjects over a long period.
Long-term memory: Type of memory that involves long-term storage of material; appears to hold up well with advanced age.
Love-withdrawal discipline: Method of discipline that predominantly uses a loss of affection and approval as a consequence of misbehaving.
Lunar cycles: Events that recur over a period of about one month.
Lung cancer epidemic: The alarming, rapid increase in lung cancer in recent years.
Lymphomas: Cancerous growths arising in the lymph-cell-producing areas.

Male climacteric: Term used to suggest a critical time or transitional period in men's lives influenced by male reproductive changes.
Malignant: Having the property of uncontrollable growth and dissemination.
Marital morale: A measure of contentment with a marriage based on the number of personal and interpersonal goals that are being achieved.
Marital relationships: The interactions between spouses.
Marriage: Relation between a man and woman who have become husband and wife.

Maturity: The quality of being fully developed, complete, full-grown, or perfect.
Mean: An arithmetic average.
Median: The measure of central tendency that has half of the cases above it and half below it; the 50th percentile.
Mediation: Using skills already developed to help acquire new skills or greater dexterity at old skills.
Meiosis: A type of cell division in which each reproductive cell receives one chromosome from each of twenty-three chromosome pairs.
Menarche: The time of the first menstrual period.
Mendelian laws: The principles of hereditary phenomena discovered and formulated by Gregor Mendel.
Menopause: The period in a woman's life when menstruation ceases.
Menstruation: The periodic discharge of the bloody endometrial lining from the uterus.
Mental disorder: An unhealthy or abnormal condition of some mental function(s).
Mental images: Symbolic representations in the mind for objects and events that may be verbal (words) or nonverbal (mental pictures).
Mental retardation: A condition of mental deficiency, defined as being two standard deviations below the norm in IQ with deficits in adaptive behavior, manifest in the developmental period.
Mentor: An older adult who acts as a counselor or guide.
Meritocracy: Term used by Herrnstein to describe an educational system that would provide teaching in keeping with inherited abilities to learn.
Metamemory: Conscious or intuitive knowledge about memory.
Metastasis: The shifting of a cancer from one part of the body to another.
Midlife transition: Turmoil precipitated by the review and reevaluation of one's past, typically occurring in the early to middle forties.
Mitosis: The process of ordinary cell division that results in two cells identical to the parent cell.
Mode: The numerical term that occurs most frequently in a set of data.
Modeling: In behavioral theory, the learning of a new behavior by imitating another person performing that behavior.
Molding: The temporary misshaping or elongation of a newborn's head caused by pressure on the soft bones during the birth process.
Montessori preschools: Specialized preschools modeled after those Maria Montessori developed in Europe 100 years ago, emphasizing individual experiences in cognitively stimulating activities.
Moral realism: According to Piaget, the earliest stage of moral development in which rules are seen as sacred and unalterable.
Moral relativism: According to Piaget, the belief that rules are modifiable to bring the greatest good to the greatest number.
Morality of constraint: Piagetian term for the earliest of his two moral levels of childhood.
Morality of cooperation: Piagetian term for the second of his two moral levels of childhood.
Morality training: Education that directs behaviors toward the good and away from the bad.
Moratorium of identity: Marcia's term for adolescents' postponement of the adoption of a stable sense of self, characterized by experimentation with different roles.
Morbid grief reactions: Delayed emotional responses to loss, misfortune, or death of a loved one that may be expressed in a serious psychosomatic disorder, drug abuse, depression, or serious anxiety disorder.

Mores: Fixed customs imbued with an ethical significance, often having the force of law.
Morula: The mass of cells resulting from early mitotic cell division of the zygote.
Mosaicism: Chromosomal abnormality resulting from faulty mitotic cell division early in the embryonic period that as a consequence leaves some but not all body cells with an abnormal number of chromosomes.
Motor milestones: The major developmental tasks of a period (such as infancy) that depend on muscular movements.
Motor norms: Standards for achievement of activities related to muscle use set for various ages of childhood.
Motor tics: Involuntary, repetitive, nonpurposeful movements of a part of the body.
Mourning: The social expressions of grief, including mourning rituals and associated behaviors.
Multiple infarct dementia: The cumulative effect of multiple small strokes, eventually closing down many of the brain's faculties.
Myelin: A white fatty covering on many neural fibers that serves to channel impulses along fibers and to reduce the random spread of impulses across neurons.
Myopia: Nearsightedness.
Myth of serenity: False belief that all old people are peaceful, relaxed, and carefree.
Myth of a sexless old age: False belief that all old people have lost their sexual desires.

Narcotics: A class of drugs that induce relaxation and reverie and provide relief from anxiety and physical pain.
Natural family planning: Contraception that utilizes knowledge of ovulation for planning abstinence from intercourse.
Nearness training: Education to remain dependent on others.
Neglect: Failure to attend to important aspects of child care such as provision of physiological needs, safety, a sense of love and belonging, and discipline.
Neglecting parents: Parents who show little interest in controlling their children's behaviors or in responding to them.
Neonate: Newborn infant up to two to four weeks.
Neoplasm: Growth of abnormal cells or tissues; cancer.
Neurons: Nerve cells; the basic unit of the nervous system and the fundamental building blocks of the brain.
Neuroses: Nonpsychotic emotional disorders characterized by anxiety and associated avoidance behaviors; called *anxiety disorders* in newer diagnostic terminology.
Neuroticism: State of being partially disorganized due to anxieties.
Neurotransmitters: A group of chemicals that facilitate the transmission of electrical impulses between nerve endings in the brain.
Norepinephrine: A hormone secreted by the adrenal medulla that stimulates the nervous system and exerts a metabolic influence on liver and muscles. It also acts as a vasoconstrictor.
Nuclear conflict: Eriksonian term applied to major conflicts characterizing each of his eight stages.
Nuclear family: The family that consists of only parents and children.
Numbering: Designating the place of objects in a series, taking into account classes and orders within classes.
Nursery schools: Programs for young children emphasizing socialization attended before enrolling in grammer school; usually before age five.
Nursing homes: Residences providing needed care for the infirm, disabled, and chronically ill.

Obesity: The quality or state of being more than 20 percent above one's ideal body weight for size, age, and sex.
Objectivity: State or quality of observing phenomena without bias, in a detached, impersonal way.
Observation: An act of noting and recording facts and events as they occur in the real world; also the data so noted and recorded.
Oedipal complex: In Freud's theory of psychosexual development, the male child's conflict between his desire for his mother and his fear of punishment by his father.
Omniplastic mastery: Gutmann's description for magical world mastery.
Oncogenes: Cancer-causing genes present in the cells of humans.
Oncology: The study of the causes, properties, and treatment of cancer.
Ontogeny: Origin and development of an individual organism and its functions from conception to death.
Oophorectomy: Incision into and removal of an ovary or ovarian tumor.
Openness: State of being frank and forthright with an absence of secrecy, prejudice, or bigotry.
Onlooker play: Watching others play but not interacting.
Operant conditioning: Basic learning process in which behaviors are repeated or reduced as a function of environmental consequences (reinforcements and punishments).
Organization: Piagetian term for one of two basic processes (the other being adaptation), which denotes the arranging, systematizing, and structuring of one's knowledge and thoughts.
Osteoarthritis: Degeneration and loss of cartilage at the ends of bones that cause joint pain.
Osteoporosis: A disease of bone characterized by increased porosity and softness from absorption of the calcareous material.
Otitis media: Inflammation of the middle ear.
Ovaries: The female reproductive organs in which egg cells, or ova, are produced.
Oviduct: The tube leading from the ovary to the uterus.
Ovulation: Expulsion of an ovum from the ovary, which occurs once about every twenty-eight to thirty-two days from puberty to menopause.
Ovum: Egg; female sex cell.

Pap smear: A method of studying cells from the cervix to detect the presence of a malignant process.
Parallel play: The side-by-side play of two or more children with some independence of action yet heightened interest because of each other's presence.
Parenthood: Taking responsibility for protecting and raising one's offspring or adopted children.
Parent-surrogate grandparents: Those who take on all the caregiving responsibilities for the grandchildren.
Participant observation: A method of research in which the researcher attempts to become immersed in the way of life to be studied to the point of temporarily becoming part of the social context.
Passive–dependent personalities: Neugarten's description for persons who are dissatisfied with their lives and apathetic and who lean on others for succor and support.
Peak experience: Maslow's term for a momentary state of ecstasy and a sense of unity with the whole world.
Peer: Any individual of about one's same level of development.
Peer group: The group of persons who constitute one's associates, usually of the same age and social status.
Peer pressures: Compelling influences to behave in certain ways brought to bear by one's age cohorts.

Period of individuation: Jung's description of the period of life after a discovery of one's inner self in the forties.
Periodontal disease: Loosening of the teeth and occasionally infection of the gum that is characteristic of the older mouth when neglected.
Permissive parenting: Parenting with few demands about rules and regulations and lax control.
Personal fable: A belief in one's own immortality and invulnerability.
Personality stability: Costa and McCrae's term for the endurance over time of many personality dispositions.
Personification of death: Embodiment of death as a human figure.
Phenomenological approach: A therapeutic procedure in which the therapist attempts to see the patient's world from the vantage point of the patient's own internal frame of reference.
Phenotype: Observable characteristics of a person; may vary from genotype (genetic characteristics).
Phenylketonuria (PKU): A genetic metabolic disorder that leads to central nervous system damage and mental retardation unless phenylalanine in the diet is carefully restricted in infancy and early childhood.
Phonetic drift: The tendency of an infant learning language to predominantly produce the phonemes that he or she hears spoken.
Phylogeny: Evolution of traits and features common to a species or race.
Physical causality: Piagetian term denoting belief that causal agency of an event was physical.
Physical dependence: A physical need for some chemical substance; withdrawal from its use produces symptoms of illness.
Pica: Hunger for nonfood substances.
Pincer grasp: Hold between thumb and forefinger.
Placenta: Organ that conveys food and oxygen to the prenatal organism and carries away its body wastes.
Placental barrier: Barrier to substances created by the fact that only molecules small enough to diffuse through the villi of the placenta reach the fetus.
Polar bodies: The nonfunctioning daughter cells that result from meiotic division of female gametes.
Postpartum: Immediately after birth; the first week of life.
Posttraumatic stress disorder: An anxiety reaction to a disaster or severe stress characterized by numbness, recurrent dreams, and excessive reactivity.
Power-assertive discipline: Disciplinary techniques used by parents who rely on punishment, fear, and material rewards.
Precipitous delivery: Sudden delivery of a fetus after less than three hours of labor.
Precocious puberty: Puberty beginning before age nine in girls or before age ten in boys.
Preconceptual phase: Piaget's first substage of the preoperational stage of development.
Preconventional morality (or stage of moral development): According to Kohlberg, the first level of moral development; involves observation of standards of others to avoid punishments or gain rewards.
Premature, or preterm, baby: Baby born before the thirty-seventh week of pregnancy, dated from the mother's last menstrual period.
Premenstrual syndrome (PMS): Group of physical and behavioral changes that affect some women in the week before a menstrual period, including tension, bloating, fatigue, irritability, and moodiness.
Premorality: According to Kohlberg, the first level of moral development, characterized by hedonistic, self-serving urges.
Prenatal: Before birth; the stage of human development lasting from conception to birth.

Preoperational thought: Thought that characterizes children from about two to seven, in Piaget's second stage of cognitive development; characterized by use of mental images but with no concrete, systematic, orderly means for arranging thoughts.

Preparation and recuperation: Kelly's term for work-related activities that are determined by others.

Prepared childbirth: Method of childbirth in which the woman is prepared for delivery by knowledge about the physiological processes involved and by learning a series of exercises that make the delivery easier.

Preretirement planning: The development of special interests or leisure pursuits a decade or two before retirement in anticipation of more time to enjoy them after retirement.

Presbycusis: Physiological changes in auditory perception, especially of high-tone frequencies, in advancing age.

Presbyopia: Physiological changes in accommodation power in the eyes in advancing age.

Presenile dementia: Deterioration of intellectual, emotional, and motor functioning prior to age sixty; causes include Huntington's chorea, Pick's disease, and Alzheimer's disease.

Primary sexual characteristics: Changes in breasts, ovaries, uterus, and vagina, or in the penis and testes, that occur at puberty and lead to reproductive maturity.

Principled moral reasoning: Kohlberg's descriptive term for the last two stages of his sequence of moral development; also called postconventional morality.

Probability: Chance stronger than possibility but falling short of certainty.

Progesterone: A principal female sex hormone, secreted by the corpus luteum.

Progestins: All hormones that help prepare the lining of the uterus for implantation of a fertilized ovum.

Projection: A defense mechanism characterized by attributing one's own unacceptable thoughts and motives to another.

Prolonged labor: Dilation and effacement of the cervix of the uterus that progresses very slowly, lasting over 24 hours.

Prosocial behaviors: Altruistic behaviors; giving, caring, sharing.

Proximal-distal development: Development that proceeds in a near-to-far direction; parts of the body near the center develop before the extremities do.

Psychodynamic: Referring to the assumption that human behavior is a function of events occurring inside the mind and is explainable only in terms of those mental events.

Psychological causality: Piagetian term denoting belief that the causal agency of an event was psychological.

Psychological dependence: The emotional need for continued use of a drug or for support from another person.

Psychophysiologic disorder: Physical symptoms such as hives and ulcers and other disorders involving actual tissue damage that result from continual emotional activation of the autonomic nervous system.

Psychoses: Severe emotional disturbances characterized by loss of contact with reality.

Puberty: The period of life during which an individual's reproductive organs become functional.

Pubescence: Time of life span characterized by rapid physiological growth, maturation of reproductive functioning, and appearance of primary and secondary sex characteristics.

Pyloric stenosis: A narrowing of the end of the stomach in infants that leads to projectile vomiting and constipation.

Quickening: The first fetal movements that a mother can readily perceive.

Random sample: A method of selecting research subjects that allows all members of the population under study an equal chance of being selected, with the expected result that the selection will be representative of the population.

Range: In statistics, the highest and lowest values of the data.

Rational-emotive theory: Albert Ellis's theory that self-defeating thoughts and feelings underlie disordered behavior.

Rationalism: Search for knowledge through the intellect and reasoning processes rather than through experience.

Rationalization: A defense mechanism in which socially acceptable reasons are offered for something that is done for unconscious or unacceptable motives.

Reaction formation: A defense mechanism in which the individual distorts a drive to its opposite; for example, acting kindly toward a person whom one dislikes.

Realism: Piagetian term for the confusion of psychological events with objective reality.

Receptive language: The ability to understand the spoken word.

Reciprocity: Corresponding, complementary, inverse relationships.

Recovery phase of bereavement: A period during which survivors work at accepting the fact of death.

Reflex: Involuntary reaction to stimulation.

Reflex abilities: Those actions performed involuntarily by infants in response to some external excitation.

Regression: A defense mechanism that involves the return to an earlier, less threatening developmental stage that one has already passed through.

Regression analysis: Statistical method used to test for significance of relationships among two or more variables.

Reinforcement: In operant conditioning the experimental procedure of immediately following a response with a reinforcer.

Replication: Reproduction of the original; important in experimental research to give more confidence that the research results did not occur by chance.

Repression: A defense mechanism in which unacceptable impulses are pushed down into the subconscious or unconscious levels of thought.

Reservoir-of-wisdom grandparents: Those who serve as authority figures for parents and grandchildren, passing on special skills, resources, and family history.

Respiratory distress syndrome: Breathing disorder of the newborn caused by underexpansion of the lungs and reduced lung volume, resulting in hypoventilation and hypoxia; also called hyaline membrane disease.

Retinal hemorrhage: Bleeding within the retina of the eye.

Retirement: A withdrawal from an activity such as business, service, or public life, especially because of advanced age or poor health.

Retrospective study: A research method of investigating subjects' development, examining their histories by means of interviews and records.

Reversibility: Piagetian term for the awareness that an operation can be reversed to bring back the original situation.

Rh factor: An agglutinizing factor present in the blood of most humans; when it is introduced into blood lacking the factor, antibodies form in the blood.

RHogam: Specially prepared gamma globulin that prevents the RH-negative mother from producing anti-RH antibodies after the miscarriage or birth of an RH-positive child.

Ribonucleic acid (RNA): A substance formed from, and similar to, DNA. It acts as a messenger in a cell and serves as a catalyst for the formation of new tissue.
Rooming-in: Hospital policy that permits the neonate to remain with the mother in her room during the post-delivery days.
Rosenthal effect: Treatment of a person according to some preconceived notion of how he or she will behave intellectually, which has the effect of eliciting the expected academic performance.
Rudimentary behavior: According to Sears, behaviors used by infants to interact with others in order to achieve gratification and reduce displeasure.

Schema: Piagetian term for a basic cognitive unit.
School phobia: Fear of going to school or separating from parents, accompanied by somatic symptoms of illness.
Scientific method: Practice of acquiring knowledge by systematically stating problem, forming hypotheses, experimenting, observing, and drawing conclusions.
Second honeymoon phenomena: Renewed interest in one's spouse and in sexual activities after the menopause.
Secondary behavioral systems: According to Sears, dyadic interactions between child and caregiver.
Secondary motivational systems: According to Sears, methods for dealing with others acquired from interactions outside the family.
Secondary sexual characteristics: Physical characteristics that appear in humans around the age of puberty and that are sex differentiated but not necessary for sexual reproduction.
Secondary traits: Allport's term for minor personality characteristics or ways of behaving known only by a few close associates.
Secular growth trend: A trend toward increased height in each new generation of children seen in well-nourished children over the past 100 years.
Secure attachment: Affectionate reciprocal relationship between parent and infant in which infant resists being separated from caregiver.
Self-actualization: According to Abraham Maslow, the need to develop one's true nature and fulfill one's potentialities.
Senescence: Period of the life span accompanied by decrement in bodily functioning; begins at different ages for different people.
Senile dementia: A pathological degenerative disorder resulting from severe organic deterioration of the brain in old age in some persons.
Senility: Senile dementia; an organic brain syndrome characterized by problems of confusion and disorientation in some old persons.
Serous otitis media: Chronic inflammation of the middle ear characterized by fluid effusion and a certain degree of conductive hearing loss.
Settling down: Levinson's stage description for men in their early thirties.
Sex chromosomes: The one pair of chromosomes that determines the sex of the organism (XX is female; XY is male).
Sex education: Education about reproduction and all the things that distinguish a male from a female.
Sex-linked diseases: Any of the diseases carried by the X chromosome as a genetic recessive that occur predominantly in men because the smaller Y chromosome has no homologous gene to cancel out the deleterious one.
Sex-role models: Any persons who demonstrate sex-appropriate behaviors and are imitated in the acquisition of sex identity.

Sex typing: Process by which children acquire attitudes and behaviors deemed by their culture to be appropriate for members of their sex.
Sexual abuse: The intentional sexual molestation of a minor child.
Sexually transmitted diseases (STDs): Diseases transmitted through sexual intercourse or other direct genital, oral, or anal sexual contact.
Short-term memory: Memory retained for brief periods of time—for example, less than a day.
Sibling rivalry: Competition between brothers and sisters.
Sickle-cell anemia: A form of anemia in which abnormal red blood corpuscles of crescent shape are present.
SIDS (sudden infant death syndrome): A major cause of death for American infants between one week and six months of age. The cause is unknown. Also called "crib death."
Singlehood: The state of being alone; not united with another.
Single parenting: Taking on the responsibilities for protecting and raising one's offspring or adopted children alone.
Socialized conduct disorders: Illegal activities engaged in by adolescents who appear socialized (without a strong undercurrent of hostility).
Social-learning theory: School of thought based on the belief that new behaviors are acquired through observational learning and imitation and are maintained through either direct or vicarious reinforcements.
Soft spots: The spaces (fontanelles) between the bones of the neonate's skull that have not yet changed from cartilage to bone.
Solitary play: Play engaged in by oneself or alone.
Somatic: Pertaining to the body.
Specific reading disorder: Difficulty with basic reading skills or reading comprehension.
Sperm: Male reproductive sex cell.
Spermatogenesis: The formation of mature sperm.
Spina bifida: A limited defect in the spinal column through which the spinal membranes, with or without spinal cord tissue, protrude.
Stages of labor: Dilation and effacement of the cervix, birth of the neonate, delivery of the placenta, and the immediate postpartum period.
Standard deviation: A statistical technique for expressing the extent of variation of a group of scores from the mean.
Statistical significance: Characteristic of a variable if it is unusual and if its chance of occurring is less than a value arbitrarily set by a researcher, usually less than 1 percent or 5 percent of the time.
Status offenses: Conduct considered illegal if engaged in by a juvenile but not if engaged in by an adult.
Stepparenting: Taking on the responsibilities for protecting and raising the children of one's spouse.
Stimulants: A class of drugs whose major effect is to provide energy, alertness, and feelings of confidence.
Stimulus-response learning: Learning that takes place when a reaction is repeatedly paired with a stimulus to obtain some reward or avoid some undesirable consequence.
Strabismus: Crossed eyes; walleyes.
Stranger anxiety: Phenomenon that often occurs during the second half of the first year, when infants express fear of strange people and places and protest separation from parents.
Strep infection: An infection, usually in the throat, resulting from the presence of streptococci bacteria.
Stuttering: The interruption of speech fluency through blocked, prolonged, or repeated words, syllables, or sounds.

Subconscious: Partial unconsciousness. The state in which mental processes take place without the conscious perception of the individual.
Subjectivity: State or quality of imposing personal prejudices, thoughts, and feelings into one's work.
Sublimation: A defense mechanism in which impulses are channeled away from forbidden outlets and toward socially acceptable ones.
Substance abuse: Chemical abuse characterized by a minimum of one month's physiological or psychological dependence and difficulty in social functioning.
Superego: In Freudian theory, that part of the mind in which the individual has incorporated the moral standards of the society.
Superiority: Adlerian term used to describe the exaggerated feeling of high value and worth that characterizes some people.
Superwoman syndrome: Situation in which a woman who holds a job also assumes major responsibility for child care and performs all the traditional role functions at home: cleaning, cooking, entertaining, and the like.
Symbolic play: Piagetian concept to describe play in which child makes something stand for something else.
Symbolic thought: Thought dependent on mental images rather than sensory or motor input.
Synapse: The junction of two neurons; the locale where an electrical impulse is transmitted from one neuron to another.
Syncretism: Belief that co-occurring events belong together.
Syntax: The rules for combining morphemes to form words and sentences.

Tay-Sachs disease: A genetic disorder of lipid metabolism.
Teenage pregnancy: Conception and gestation by an adolescent female.
Telegraphic speech: Early speech that uses salient words (nouns and verbs) and omits auxiliary parts of speech.
Teratogenic drugs: Any drugs that cause malformations of the embryo/fetus within the uterus.
Teratology: The study of any agent that disturbs prenatal development and produces abnormalities; literally, "the study of monsters."
Terminal decline: A significant decline in intellectual abilities that often precedes death.
Testes: The male reproductive organs in which spermatozoa are manufactured; testicles.
Testosterone: An androgen steroid that affects accessory organs of sex and secondary sex characteristics; produced mainly by the testes.
Thanatology: The study of death.
Thanatos: In Freudian theory, the death instinct of the id.
Thrombus: A blood clot that forms inside a blood vessel or cavity of the heart.
Toilet training: Teaching a child to defecate and urinate in the proper place.
Tolerance: The physiological condition in which the usual dosage of a drug no longer provides the desired effect.
Traits: Distinctive and enduring characteristics of a person or his or her behavior.
Transactional analysis: An interpersonal therapy in which the major goal is to become aware of one's unconscious games and so realize an ability to control one's own fate; based on the interaction of "parent," "child," and "adult" ego states.
Transient ischemic attacks (TIAs): Mild forms of obstruction to the flow of blood in the vessels of the brain, causing dizziness, unsteadiness, slurred speech, and numbness; also called little strokes.

Transductive reasoning: Piaget's term for a preoperational child's tendency to use associative reasoning instead of induction or deduction.
Trust versus mistrust: Erikson's first nuclear conflict of personality development in which an infant develops a sense of whether the world and its inhabitants are safe and can be trusted.
T-test: Statistical test that is used to evaluate the difference in mean outcomes of two treatments.
Tumor: An abnormal mass of tissue that forms from the proliferation of an abnormal cell.
Turner syndrome: A congenital physical condition resulting from the individual's having only one X chromosome and no Y chromosome. The afflicted individual looks like an immature female and is characterized by a lack of reproductive organs, abnormal shortness, and mental retardation.
Ultraviolet light: Lying beyond the violet end of the visible spectrum; certain light rays of extremely short wavelength that burn skin readily.
Unconditional leisure: Leisure that is freely chosen by an individual.
Unconscious: In Freudian theory, the largest level of consciousness, containing all memories not readily available to the perceptual conscious because they have been either forgotten or repressed.
Unintegrated personalities: Neugarten's description for persons who are dissatisfied and disorganized and have a low activity level.
Unsocialized conduct disorders: Illegal activities engaged in by adolescents with a strong undercurrent of hostility, deceit, verbal and physical abusiveness, and disrespect.
Uterus: Pear-shaped muscular organ in which the fertilized egg develops until birth; also known as the *womb*.
Vas deferens: The duct that carries sperm from the testicle to the ejaculatory duct of the penis.
Vernix caseosa: Waxy covering of the fetus in utero; also found on the neonate.
Villi: Capillaries of the placenta that link the developing umbilical veins and arteries of an embryo with the uterine wall.

Warts: Small tumors on the skin or mucous membranes.
Wilms' tumor: A malignant kidney tumor of young children.

Youth stage: Keniston's classification for young people between adolescence and adulthood with a stable sense of self but still searching for a vocation and a social role they can comfortably play.

Zygote: A new individual formed by the union of ovum and sperm at fertilization.

Author Index

Subject Index